Palaeoproterozoic Supercontinents and Global Evolution

The Geological Society of London
Books Editorial Committee

Chief Editor
BOB PANKHURST (UK)

Society Books Editors
JOHN GREGORY (UK)
JIM GRIFFITHS (UK)
JOHN HOWE (UK)
PHIL LEAT (UK)
NICK ROBINS (UK)
JONATHAN TURNER (UK)

Society Books Advisors
MIKE BROWN (USA)
ERIC BUFFETAUT (FRANCE)
JONATHAN CRAIG (ITALY)
RETO GIERÉ (GERMANY)
TOM MCCANN (GERMANY)
DOUG STEAD (CANADA)
RANDELL STEPHENSON (UK)

IUGS/GSL publishing agreement

This volume is published under an agreement between the International Union of Geological Sciences and the Geological Society of London and arises from IGCP 509.

GSL is the publisher of choice for books related to IUGS activities, and the IUGS receives a royalty for all books published under this agreement.

Books published under this agreement are subject to the Society's standard rigorous proposal and manuscript review procedures.

It is recommended that reference to all or part of this book should be made in one of the following ways:

REDDY, S. M., MAZUMDER, R., EVANS, D. A. D. & COLLINS, A. S. (eds) 2009. *Palaeoproterozoic Supercontinents and Global Evolution*. Geological Society, London, Special Publications, **323**.

TRAP, P., FAURE, M., LIN, W. & MEFFRE, S. 2009. The Lüliang Massif: a key area for the understanding of the Palaeoproterozoic Trans-North China Belt, North China Craton. *In*: REDDY, S. M., MAZUMDER, R., EVANS, D. A. D. & COLLINS, A. S. (eds) *Palaeoproterozoic Supercontinents and Global Evolution*. Geological Society, London, Special Publications, **323**, 99–125.

GEOLOGICAL SOCIETY SPECIAL PUBLICATION NO. 323

Palaeoproterozoic Supercontinents and Global Evolution

EDITED BY

S. M. REDDY
Curtin University of Technology, Australia

R. MAZUMDER
Indian Statistical Institute, Kolkata, India

D. A. D. EVANS
Yale University, USA

and

A. S. COLLINS
Adelaide University, Australia

2009
Published by
The Geological Society
London

THE GEOLOGICAL SOCIETY

The Geological Society of London (GSL) was founded in 1807. It is the oldest national geological society in the world and the largest in Europe. It was incorporated under Royal Charter in 1825 and is Registered Charity 210161.

The Society is the UK national learned and professional society for geology with a worldwide Fellowship (FGS) of over 9000. The Society has the power to confer Chartered status on suitably qualified Fellows, and about 2000 of the Fellowship carry the title (CGeol). Chartered Geologists may also obtain the equivalent European title, European Geologist (EurGeol). One fifth of the Society's fellowship resides outside the UK. To find out more about the Society, log on to www.geolsoc.org.uk.

The Geological Society Publishing House (Bath, UK) produces the Society's international journals and books, and acts as European distributor for selected publications of the American Association of Petroleum Geologists (AAPG), the Indonesian Petroleum Association (IPA), the Geological Society of America (GSA), the Society for Sedimentary Geology (SEPM) and the Geologists' Association (GA). Joint marketing agreements ensure that GSL Fellows may purchase these societies' publications at a discount. The Society's online bookshop (accessible from www.geolsoc.org.uk) offers secure book purchasing with your credit or debit card.

To find out about joining the Society and benefiting from substantial discounts on publications of GSL and other societies worldwide, consult www.geolsoc.org.uk, or contact the Fellowship Department at: The Geological Society, Burlington House, Piccadilly, London W1J 0BG: Tel. +44 (0)20 7434 9944; Fax +44 (0)20 7439 8975; E-mail: enquiries@geolsoc.org.uk.

For information about the Society's meetings, consult *Events* on www.geolsoc.org.uk. To find out more about the Society's Corporate Affiliates Scheme, write to enquiries@geolsoc.org.uk.

Published by The Geological Society from:
The Geological Society Publishing House, Unit 7, Brassmill Enterprise Centre, Brassmill Lane, Bath BA1 3JN, UK

(*Orders*): Tel. +44 (0)1225 445046, Fax +44 (0)1225 442836)
Online bookshop: www.geolsoc.org.uk/bookshop

The publishers make no representation, express or implied, with regard to the accuracy of the information contained in this book and cannot accept any legal responsibility for any errors or omissions that may be made.

© The Geological Society of London 2009. All rights reserved. No reproduction, copy or transmission of this publication may be made without written permission. No paragraph of this publication may be reproduced, copied or transmitted save with the provisions of the Copyright Licensing Agency, 90 Tottenham Court Road, London W1P 9HE. Users registered with the Copyright Clearance Center, 27 Congress Street, Salem, MA 01970, USA: the item-fee code for this publication is 0305-8719/09/$15.00.

British Library Cataloguing in Publication Data

A catalogue record for this book is available from the British Library.
ISBN 978-1-86239-283-0

Typeset by Techset Composition Ltd., Salisbury, UK
Printed by CPI Antony Rowe, Chippenham, UK

Distributors

North America
For trade and institutional orders:
The Geological Society, c/o AIDC, 82 Winter Sport Lane, Williston, VT 05495, USA
Orders: Tel. +1 800-972-9892
             Fax +1 802-864-7626
             E-mail: gsl.orders@aidcvt.com

For individual and corporate orders:
AAPG Bookstore, PO Box 979, Tulsa, OK 74101-0979, USA
Orders: Tel. +1 918-584-2555
             Fax +1 918-560-2652
             E-mail: bookstore@aapg.org
             Website: http://bookstore.aapg.org

India
Affiliated East-West Press Private Ltd, Marketing Division, G-1/16 Ansari Road, Darya Ganj, New Delhi 110 002, India
Orders: Tel. +91 11 2327-9113/2326-4180
             Fax +91 11 2326-0538
             E-mail: affiliat@vsnl.com

Contents

REDDY, S. M. & EVANS, D. A. D. Palaeoproterozoic supercontinents and global evolution: correlations from core to atmosphere — 1

EGLINTON, B. M., REDDY, S. M. & EVANS, D. A. D. The IGCP 509 database system: design and application of a tool to capture and illustrate litho- and chrono-stratigraphic information for Palaeoproterozoic tectonic domains, large igneous provinces and ore deposits; with examples from southern Africa — 27

KUSKY, T. M. & SANTOSH, M. The Columbia connection in North China — 49

WAN, Y., LIU, D., DONG, C., XU, Z., WANG, Z., WILDE, S. A., YANG, Y., LIU, Z. & ZHOU, H. The Precambrian Khondalite Belt in the Daqingshan area, North China Craton: evidence for multiple metamorphic events in the Palaeoproterozoic era — 73

TRAP, P., FAURE, M., LIN, W. & MEFFRE, S. The Lüliang Massif: a key area for the understanding of the Palaeoproterozoic Trans-North China Belt, North China Craton — 99

GLADKOCHUB, D. P., DONSKAYA, T. V., REDDY, S. M., POLLER, U., BAYANOVA, T. B., MAZUKABZOV, A. M., DRIL, S., TODT, W. & PISAREVSKY, S. A. Palaeoproterozoic to Eoarchaean crustal growth in southern Siberia: a Nd-isotope synthesis — 127

DIDENKO, A. N., VODOVOZOV, V. Y., PISAREVSKY, S. A., GLADKOCHUB, D. P., DONSKAYA, T. V., MAZUKABZOV, A. M., STANEVICH, A. M., BIBIKOVA, E. V. & KIRNOZOVA, T. I. Palaeomagnetism and U–Pb dates of the Palaeoproterozoic Akitkan Group (South Siberia) and implications for pre-Neoproterozoic tectonics — 145

BAYANOVA, T., LUDDEN, J. & MITROFANOV, F. Timing and duration of Palaeoproterozoic events producing ore-bearing layered intrusions of the Baltic Shield: metallogenic, petrological and geodynamic implications — 165

SALMINEN, J., PESONEN, L. J., MERTANEN, S., VUOLLO, J. & AIRO, M.-L. Palaeomagnetism of the Salla Diabase Dyke, northeastern Finland, and its implication for the Baltica–Laurentia entity during the Mesoproterozoic — 199

PETTERSSON, Å., CORNELL, D. H., YUHARA, M. & HIRAHARA, Y. Sm–Nd data for granitoids across the Namaqua sector of the Namaqua–Natal Province, South Africa — 219

LOMPO, M. Geodynamic evolution of the 2.25–2.0 Ga Palaeoproterozoic magmatic rocks in the Man-Leo Shield of the West African Craton. A model of subsidence of an oceanic plateau — 231

GIUSTINA, M. E. S. D., DE OLIVEIRA, C. G., PIMENTEL, M. M., DE MELO, L. V., FUCK, R. A., DANTAS, E. L. & BUHN, B. U–Pb and Sm–Nd constraints on the nature of the Campinorte sequence and related Palaeoproterozoic juvenile orthogneisses, Tocantins Province, central Brazil — 255

DOS SANTOS, T. J. S., FETTER, A. H., RANDALL VAN SCHMUS, W. & HACKSPACHER, P. C. Evidence for 2.35 to 2.30 Ga juvenile crustal growth in the northwest Borborema Province, NE Brazil — 271

SINGH, S., JAIN, A. K. & BARLEY, M. E. SHRIMP U–Pb c. 1860 Ma anorogenic magmatic signatures from the NW Himalaya: implications for Palaeoproterozoic assembly of the Columbia Supercontinent — 283

MAZUMDER, R., RODRÍGUEZ-LÓPEZ, J. P., ARIMA, M. & VAN LOON, A. J. Palaeoproterozoic seismites (fine-grained facies of the Chaibasa Formation, east India) and their soft-sediment deformation structures — 301

PAYNE, J. L., HAND, M., BAROVICH, K. M., REID, A. & EVANS, D. A. D. Correlations and reconstruction models for the 2500–1500 Ma evolution of the Mawson Continent — 319

Index — 357

Palaeoproterozoic supercontinents and global evolution: correlations from core to atmosphere

S. M. REDDY[1]* & D. A. D. EVANS[2]

[1]*The Institute for Geoscience Research, Department of Applied Geology, Curtin University of Technology, GPO Box U 1987, Perth, WA 6845, Australia*

[2]*Department of Geology and Geophysics, Yale University, New Haven, CT 06520-8109, USA*

*Corresponding author (e-mail: S.Reddy@curtin.edu.au)

Abstract: The Palaeoproterozoic era was a time of profound change in Earth evolution and represented perhaps the first supercontinent cycle, from the amalgamation and dispersal of a possible Neoarchaean supercontinent to the formation of the 1.9–1.8 Ga supercontinent Nuna. This supercontinent cycle, although currently lacking in palaeogeographic detail, can in principle provide a contextual framework to investigate the relationships between deep-Earth and surface processes. In this article, we graphically summarize secular evolution from the Earth's core to its atmosphere, from the Neoarchaean to the Mesoproterozoic eras (specifically 3.0–1.2 Ga), to reveal intriguing temporal relationships across the various 'spheres' of the Earth system. At the broadest level our compilation confirms an important deep-Earth event at *c.* 2.7 Ga that is manifested in an abrupt increase in geodynamo palaeointensity, a peak in the global record of large igneous provinces, and a broad maximum in several mantle-depletion proxies. Temporal coincidence with juvenile continental crust production and orogenic gold, massive-sulphide and porphyry copper deposits, indicate enhanced mantle convection linked to a series of mantle plumes and/or slab avalanches. The subsequent stabilization of cratonic lithosphere, the possible development of Earth's first supercontinent and the emergence of the continents led to a changing surface environment in which voluminous banded iron-formations could accumulate on the continental margins and photosynthetic life could flourish. This in turn led to irreversible atmospheric oxidation at 2.4–2.3 Ga, extreme events in global carbon cycling, and the possible dissipation of a former methane greenhouse atmosphere that resulted in extensive Palaeoproterozoic ice ages. Following the great oxidation event, shallow marine sulphate levels rose, sediment-hosted and iron-oxide-rich metal deposits became abundant, and the transition to sulphide-stratified oceans provided the environment for early eukaryotic evolution. Recent advances in the geochronology of the global stratigraphic record have made these inferences possible. Frontiers for future research include more refined modelling of Earth's thermal and geodynamic evolution, palaeomagnetic studies of geodynamo intensity and continental motions, further geochronology and tectonic syntheses at regional levels, development of new isotopic systems to constrain geochemical cycles, and continued innovation in the search for records of early life in relation to changing palaeoenvironments.

Supercontinents occupy a central position in the long term processes of the Earth system. Their amalgamations result from lateral plate-tectonic motions manifesting mantle convection, and their disaggregation, although thought by some to be externally forced by plume or hotspot activity (e.g. Storey *et al.* 1999), is widely considered to result from their own thermal or geometric influences upon such convection (Gurnis 1988; Anderson 1994; Lowman & Jarvis 1999; Vaughan & Storey 2007; O'Neill *et al.* 2009). Continental collisions compress land area and thus are expected to lower sea level; fragmentations involve crustal thinning and generate young seafloor, and therefore should raise sea level (Fischer 1984; Worsley *et al.* 1984). These eustatic effects can influence global climate by changing global albedo and rates of silicate weathering (a sink for atmospheric carbon dioxide) on exposed versus drowned continental shelves (Nance *et al.* 1986; Marshall *et al.* 1988). Chemical weathering of pyrite and organic carbon in mountain belts is a major influence on the atmospheric oxygen budget as is their burial in marine sediments. These processes are in turn dependent on global tectonics (Berner 2006). Climatic changes, plus geographic patterns of ocean circulation, will exert a profound influence on the biosphere (Valentine & Moores 1970). Microbes play an important role in concentrating low-temperature mineral deposits through near-surface redox reactions (e.g. Dexter-Dyer et al. 1984; Labrenz et al. 2000) and many mineral deposits are associated with specific tectonic environments related to broader supercontinent cycles (Barley & Groves 1992; Groves et al. 2005).

The connections illustrated above are based on concepts or a few well-constrained examples. The actual record of supercontinents on Earth is not yet well enough known to verify or modify the models in deep time. Prior to Pangaea (approximately 0.25–0.15 Ga) and Gondwana-Land (0.52–0.18 Ga), the possible configurations and even existence of Neoproterozoic Rodinia (c. 1.0–0.8 Ga) are intensely debated (Meert & Torsvik 2003; Li et al. 2008; Evans 2009). Prior to Rodinia, an earlier supercontinental assemblage at 1.9–1.8 Ga has been suggested. Although only preliminary and palaeomagnetically untested models of this supercontinent have been published (Rogers & Santosh 2002; Zhao et al. 2002, 2004), its assembly appears to have followed tectonic processes that are remarkably similar to those of the present day (Hoffman 1988, 1989). This supercontinent is referred to by various names (e.g. Columbia, Nuna, Capricornia) but here, and below, we refer to it as Nuna (Hoffman 1997).

It is unclear whether Nuna's predecessor was a large supercontinent, or whether it was one of several large, but distinct coeval landmasses (Aspler & Chiarenzelli 1998; Bleeker 2003). Nonetheless, numerous large igneous provinces, with ages between 2.45 and 2.2 Ga, perforate the world's 35 or so Archaean cratons and could represent an episode of globally widespread continental rifting at that time (Heaman 1997; Buchan et al. 1998; Ernst & Buchan 2001).

Given that the Palaeoproterozoic era is defined chronometrically at 2.5–1.6 Ga (Plumb 1991), it thus encompasses one or more episodes, perhaps cycles, of global tectonics. As enumerated by the following examples, these tectonic events coincide with fundamental changes to the Earth as an integrated system of core, mantle, lithosphere, hydrosphere, atmosphere, and biosphere. Understanding these changes requires the integration of seemingly disparate geoscience disciplines. One pioneering review of this sort (Nance et al. 1986) has been followed by an incredible wealth of precise geochronological data constraining ages within the Palaeoproterozoic geological record. In this chapter we have compiled the currently available data from core to atmosphere, from late Archaean to late Mesoproterozoic time (3.0–1.2 Ga), to provide an overview of Earth-system evolution and an up-to-date temporal and spatial framework for hypotheses concerning the global transition through Earth's 'middle ages'.

Evolution of the core

Today, the metallic core of the Earth consists of a solid inner part, with radius of about 1200 km, surrounded by a liquid outer part with outer radius of about 3500 km. As such, it occupies about one-sixth of the Earth's volume, and dominates the planetary concentration of the siderophile elements Fe and Ni. It is generally accepted that the core was originally entirely fluid, and that through time the solid inner core nucleated and grew as the Earth cooled gradually (Jacobs 1953; Stevenson 1981; Jeanloz 1990). However, the rate of this inner-core growth is highly uncertain and depends upon various estimates of radioactive element (mainly K) concentrations in the core versus mantle and crust (Nimmo et al. 2004; Davies 2007).

Recent estimates of the age of inner core nucleation, on a theoretical basis constrained by the geochemical data, are typically about 1 Ga (Labrosse et al. 2001; Nimmo et al. 2004; Butler et al. 2005; Gubbins et al. 2008). Much discussion on this topic is confounded with discussions of the intensity of Earth's ancient geomagnetic field. This is because two of the primary energy drivers of the geodynamo are thought to be thermal and compositional convection in the outer core due to inner core crystallization (Stevenson et al. 1983). Early attempts to determine the palaeointensity of Earth's magnetic field, which is among the most laborious and controversial measurements in geophysics, suggested an abrupt increase in moment near the Archaean–Proterozoic boundary (Hale 1987). Subsequent refinements to techniques in palaeointensity (e.g. the single-crystal technique applied by Smirnov et al. 2003) have generated mixed results, but several of the measurements indicate a strong field in the earliest Proterozoic (Fig. 1a). More traditional palaeointensity techniques complete the later Proterozoic time interval with results of generally low palaeointensity (Macouin et al. 2003). There is currently no systematic test among the various palaeointensity techniques, so absolute Precambrian palaeointensity values remain ambiguous. In addition, the large apparent increase in palaeointensity at c. 1.2 Ga (Fig. 1a) is underpinned by sparse data that lack some standard reliability checks (Macouin et al. 2004). However, we tentatively explore the possibility that all the available records provide reliable estimates of ancient geomagnetic field strength.

Given that there is growing consensus that the inner core began to crystallize relatively late in Earth history, we speculate that the increase in palaeointensity at 1.2–1.1 Ga could be due to the inner-core growth-related energy inputs toward geodynamo generation. In that case, the Archaean–Proterozoic interval of strong intensity, if sampled adequately, could be due to an anomalous episode of geodynamo-driving energy from the top of the core: perhaps, for example, a result of enhanced thermal gradients following a global peak in Neoarchaean subduction (Condie 1998). A more complete dataset of Archaean–Palaeoproterozoic palaeo-intensity measurements, with better indications of reliability for the obtained values, will be necessary to further our understanding of such ancient core processes.

Mantle evolution

The early Earth was undoubtedly hotter than at present, due to the relative abundance of radioactive isotopes that have since decayed, and the secular cooling of primordial heat from planetary accretion. Because it forms $c.$ 80% of Earth's volume, the nature of secular cooling of the Earth's mantle carries fundamental implications for core formation, core cooling and the development of the geodynamo, the development of a modern style of plate tectonics and crustal evolution. Critical to the evolution of the mantle are the total concentrations of radioactive elements in the Earth, their spatial and temporal distribution and residence times in the core, mantle, and crust, and the degree to which layered versus whole-mantle convection can redistribute them. The precise quantification of the mantle's secular cooling is therefore intimately linked to its structure and geochemistry. These remain a subject of significant debate and have been arguably 'the most controversial subject in solid Earth sciences for the last few decades' (Korenaga 2008c). Without doubt, mantle convection is a requirement of a secularly cooling Earth (see Jaupart et al. 2007; Ogawa 2008 for recent reviews) but establishing the nature of modern convection in the modern Earth has proved difficult and is more so in its extrapolation to the geological past (Schubert et al. 2001).

Numerical modelling is one approach that has received considerable attention in attempting to constrain Earth cooling models. Such modelling commonly utilizes the simple relationship between radiogenic heat production, secular cooling and surface heat flux. Critical to these models is the way in which surface heat flux is calculated and how this heat flow is scaled to mantle convection over time. 'Conventional' models commonly calculate surface heat flux using scaling laws that assume a strong temperature dependency on viscosity such that hotter mantle convects more vigorously, thereby increasing surface heat flux. As clearly enunciated by Korenaga (2006), applying conventional heat flux scaling from current conditions back through geological time predict unrealistically hot mantle temperatures before 1 Ga and lead to the so called 'thermal catastrophe' (Davies 1980). Several numerical models have therefore addressed different ways of modifying the scaling laws to avoid these unrealistic mantle temperatures. One way of doing this is by assuming a much higher convective Urey ratio (i.e. the measure of internal heat production to mantle heat flux) in the geological past (e.g. Schubert et al. 1980; Geoffrey 1993) than modern day estimates (Korenaga 2008c). Although this assumption overcomes problems of the thermal catastrophe in the Mesoproterozoic, the solution leads to high Archaean mantle temperatures that appear inconsistent with empirical petrological data for mantle temperatures (Abbott et al. 1994; Grove & Parman 2004; Komiya 2004; Berry et al. 2008) (Fig. 1b).

An alternative way of alleviating the problem of a Mesoproterozoic 'thermal catastrophe' is to assume layered mantle convection rather than whole-mantle convection (e.g. Richter 1985). A layered mantle has been inferred as a means of maintaining distinct geochemical reservoirs, particularly noble gas compositions measured between mid-ocean ridge basalts and plume-related ocean island basalts (e.g. Allegre et al. 1983; O'Nions & Oxburgh 1983) and large-ion lithophile elements budgets of the crust and mantle (e.g. Jacobsen & Wasserburg 1979; O'Nions et al. 1979). The nature of this layering differs between different models (c.f. Allegre et al. 1983; Kellogg et al. 1999; Gonnermann et al. 2002). However, critical to this argument is that the modelling of layered versus whole-mantle convection and its affect on present day topography supports the latter (Davies 1988), and seismic tomographic data provide evidence of subducting slabs that penetrate the lower mantle (e.g. van der Hilst et al. 1997). These observations are difficult to reconcile with the classic model of layered convection (see van Keken et al. 2002 for review).

With increasing geochemical data from a range of different sources, geochemical constraints on mantle evolution are becoming more refined, and the two-layer mantle reservoir models have necessarily become more complex (see reviews of Graham 2002; Porcelli & Ballentine 2002; Hilton & Porcelli 2003; Hofmann 2003; Harrison & Ballentine 2005). These data have led to the formulation of models in which some of the chemical heterogeneity is stored in the core or deep mantle (Porcelli & Elliott 2008), or is associated with a unmixed lower mantle 'magma ocean' (Labrosse et al. 2007), or is associated with lateral compositional variations (Trampert & van der Hilst 2005), or is explained by filtering of incompatible trace elements associated with water release and melting associated with magnesium silicate phase changes in the mantle transition zone (Bercovici & Karato 2003). However, a recent development in layered mantle geodynamics and the understanding of core–mantle interaction is the significant discovery of the post-perovskite phase transition (Murakami *et al.* 2004) and the realization that such a transition seems to account for many of the characteristics of the seismically recognized D″ of the lower mantle (see Hirose & Lay 2008 for a recent summary). The large positive Clapeyron slope of the perovskite–post-perovskite phase transition means that cold subducting slabs that reach the core–mantle boundary are likely to lead to enhanced post-perovskite formation and form lateral topography on the D″ layer (Hernlund *et al.* 2005). With the addition of slab material to the D″ layer, and its potential for enhanced melting (Hirose *et al.* 1999), it is likely that there will be significant geochemical implications for lower mantle enrichment and melting (see for example Kellogg *et al.* 1999; Labrosse *et al.* 2007), in addition to the potential for mantle plume generation (Hirose & Lay 2008), both of which are yet to be resolved. The post-perovskite stability field intersects only the lowermost depths of the present mantle geotherm (Hernlund *et al.* 2005) and the secular cooling rate of the lowermost mantle is not well enough known to estimate the age at which this post-perovskite phase first appeared in Earth history. However, this is likely to be of importance in the secular evolution of the deep Earth.

To summarize, current geodynamical and geochemical models for the modern mantle are therefore necessarily complex and it seems that the answer may reside in a mixture of whole-mantle and episodic layered convection that is closely linked to large-scale plate tectonic processes and affects chemical heterogeneities preserved at a range of scales (Tackley 2008).

Despite the advances in understanding the geodynamics of the present-day mantle, extrapolation to the Palaeoproterozoic mantle remains elusive. In the last few years, developments in the modelling of secular cooling of the mantle have utilized surface heat flux scaled to a 'plate tectonic' model involving mantle melting at mid-ocean ridges (Korenaga 2006) or intermittent plate tectonic models in which subduction flux, indicated by geochemical proxies (e.g. the Nb/Th ratios of Collerson & Kamber 1999 Fig. 1e), varies over time (Silver & Behn 2008*a*). These models overcome the problems of Mesoproterozoic thermal runaway, calculated back from present day conditions via conventional scaling laws, in the first case by taking account of depth-dependent mantle viscosity variations as a function of melting (Hirth & Kohlstedt 1996) and in the second by reducing the amount of heat loss at times of plate tectonic quiescence (Silver & Behn 2008*a*). Importantly, such models predict significantly reduced plate velocities during the Neoarchaean and Palaeoproterozoic eras than conventional models do (Fig. 1c) because of either the increased difficulty of subducting thicker dehydrated lithosphere (Korenaga 2008*b*) or the episodic nature of the supercontinent cycle (Silver & Behn 2008*a*). However, they remain controversial (Korenaga 2008*a*; Silver & Behn 2008*b*), particularly in light of recent evidence for a temperature control on lower crustal thermal conductivity (Whittington *et al.* 2009) and empirical estimates of Archaean mantle temperatures (Berry *et al.* 2008) that are higher than the model predictions (Fig. 1b). Even so, the development of models that are intimately linked to plate tectonic processes and that involve the formation and subduction of strong, plate-like lithosphere that controls mantle convection (e.g. Tackley 2000; Bercovici 2003), provide predictions from the core to the atmosphere that can be empirically constrained by the extant geological record, for example secular variations in crust and mantle geochemistry (Fig. 1d), the amalgamation and dispersal of supercontinents (Fig. 1i) and chemistry of the oceans and atmosphere (Fig. 1k, q).

Another feature of mantle evolution that has received considerable scientific attention is the formation and secular development of mantle plumes (e.g. Ernst & Buchan 2003 and references therein). One of the expressions of the impingement of mantle plumes on Earth's lithosphere is the development of large igneous provinces (LIPS) (Ernst & Buchan 2001, 2003) manifest as continental or oceanic flood basalts and oceanic plateaus. The geological record of LIPS (summarized by Ernst & Buchan 2001; Abbott & Isley 2002) recognizes a general decrease of size in younger LIP events, and an episodic distribution in time (Fig. 1d). Time-series analysis of LIPS recognizes a *c.* 330 Ma cycle in the period from 3.0–1.0 Ga upon which weaker, shorter duration cycles are superimposed (Prokoph *et al.* 2004). As pointed out by Prokoph *et al.* (2004), no simple correlation exists between the identified LIP cycles and possible forcing functions, though current global initiatives to date mafic volcanism precisely (e.g. Bleeker & Ernst 2006) should improve the likelihood of establishing such correlations. Currently, the correlations reported in Figure 1 show that LIP activity at *c.* 2.7–2.8 Ga coincides with global banded iron formation (BIF) and orogenic gold formation, while a second event at 2.45 Ga corresponds with a the major formation of Superior-type BIF. In addition, the major 2.7 Ga peak in LIP activity lies temporally close to the peak in juvenile crust formation (Fig. 1g), a correlation that has led to the suggestion that these features may be geodynamically related (Condie 2004) and which may explain observed gold enrichment (Brimhall 1987; Pirajno 2004).

Crustal growth and emergence

Continental crust represents only a small proportion of Earth's volume, yet it contains significant proportions of Earth's incompatible elements recorded in long-lived igneous, sedimentary and metamorphic rocks (e.g. Turekian & Wedepohl 1961; Taylor 1964). Establishing the secular evolution of compositional changes in continental rocks therefore provides fundamental constraints on the growth of the continental crust, the evolution of the mantle from which it was extracted, and the chemical evolution of the oceans and atmosphere. However, establishing the composition of continental crust and how this changes over time is not straightforward (e.g. McLennan 1989) and has been debated intensely (see reviews of Hawkesworth & Kemp 2006a; Rollinson 2006 for recent summaries).

One widely used approach has been the analysis of fine-grained sedimentary rocks, thought to provide a statistical representation of the upper crust, to establish the increase in volume of the continental crust over time (e.g. Allegre & Rousseau 1984; Taylor & McLennan 1985, 1995). Other approaches include the direct analysis of zircon age populations from juvenile continental crust (Condie 1998), or indirect analysis by assessing the chemical evolution of the depleted upper mantle from which continental crust is believed to be ultimately sourced (e.g. Bennett 2003 and references therein) (Fig. 1f). More recent developments include the integration of *in situ* stable and radiogenic isotope analysis (Valley *et al.* 2005; Hawkesworth & Kemp 2006b).

Despite the difficulties in constraining the composition of continental crust (Rudnick *et al.* 2003) and the likely requirement of two-stage differentiation (e.g. Arndt & Goldstein 1989; Rudnick 1995; Kemp *et al.* 2007), several fundamentally different growth models dominate the literature (for recent summary see Rino *et al.* 2004): rapid differentiation of the crust early in Earth's history and subsequent recycling such that there has been little subsequent increase in volume over time (e.g. Armstrong 1981, 1991); growth mainly in the Proterozoic (Hurley & Rand 1969; Veizer & Jansen 1979); high growth rates in the Archaean followed by slower growth in the Proterozoic (Dewey & Windley 1981; Reymer & Schubert 1984; Taylor & McLennan 1995) and growth accommodated by discrete episodes of juvenile crust formation (McCulloch & Bennett 1994; Condie 1998, 2000).

The first of these models (Armstrong 1981, 1991) seems unlikely in the light of relatively constant $\delta^{18}O$ from Archaean rocks that show little evidence of extensive crustal recycling prior to 2.5 Ga (Valley *et al.* 2005). These data also point towards significant crustal growth in the Archaean and so are also inconsistent with the models of dominantly Proterozoic growth (Hurley & Rand 1969; Veizer & Jansen 1979). In the case of progressive or episodic crustal growth, mantle depletion events should also mimic continental growth. Despite possible complexities associated with crustal recycling and questions regarding the nature of mantle convection, Re–Os data from peridotites and platinum group alloys indicate mantle depletion events that cluster at 1.2, 1.9 and 2.7 Ga (Pearson *et al.* 2007) (Fig. 1e). A temporally similar peak at 2.7–2.5 Ga is recorded in Nb/Th data (Collerson & Kamber 1999) following polynomial fitting of the data (Silver & Behn 2008a) and in the $^4He/^3He$ ocean island basalt data inferred at 2.7 Ga and 1.9 Ga (Parman 2007) (Fig. 1e). Although the age constraints on the mantle depletion events recorded by the He data are poor, the pattern of Os and Nb/Th data are similar to that documented by the temporal distribution of juvenile continental crust (Fig. 1g) and considered to reflect the formation of supercontinents (Fig. 1i) (Condie 1998; Campbell & Allen 2008), possibly linked to large-scale mantle overturn events (Condie 2000; Rino *et al.* 2004), and the cessation of, or decrease in, subduction flux (Silver & Behn 2008a). Recent models integrating chemical differentiation with mantle convection also predict the episodicity of juvenile crust formation (Walzer & Hendel 2008), as do large-scale mantle overturn events that are thought to take place on a timescale of several hundred million years (Davies 1995).

Despite the above correlations, the pattern of juvenile crust ages has been argued to be a consequence of the preservation potential within the supercontinent cycle, in particular the ability to preserve material inboard of arcs (Hawkesworth *et al.* 2009). Although this seems a reasonable interpretation based on the crustal record, the temporal link between mantle depletion events (Collerson & Kamber 1999; Parman 2007; Pearson *et al.* 2007) and juvenile crust (Condie 1998; Campbell & Allen 2008) is less easy to explain by preservational biases and remains a compelling observation for linking these processes. Our current preference is therefore for continental growth models that involve continental crust formation in the Archaean with subsequent reworking, recycling and the addition of juvenile material via episodic processes through the Proterozoic.

The growth of continental crust, its evolving volume and its thickness are intimately related to the evolution of the mantle (e.g. Hynes 2001). These characteristics also play a critical role in continental freeboard, the mean elevation of continental crust above sea level, and the emergence of the continents (e.g. Wise 1974; Eriksson *et al.* 2006). The emergence of the continental crust above sea level in turn influences the nature of sedimentation (Eriksson et al. 2005b) and enables weathering that affects ocean and atmospheric compositions. Several lines of evidence point to continental emergence around the Archaean–Proterozoic boundary, primarily in the form of distinctive geochemical trends that require low-temperature alteration and crustal recycling. One such line of evidence comes from a recent compilation of $\delta^{18}O$ data from the zircons of juvenile rocks, which show a clear trend of relatively constant Archaean values with increasing values at c. 2.5 Ga (Valley et al. 2005) (Fig. 1f). This trend mimics those recorded in Hf and Nd isotopic data from juvenile granitic rocks (Fig. 1f), which is thought to represent depletion of the mantle associated with extraction of the continental crust (Bennett 2003). The pattern of oxygen isotopic variation in zircon is explained by complex contributions of various processes (Valley et al. 2005). However, a requirement is for a component of low-temperature fractionation commonly interpreted to be associated with continental weathering, the recycling of supracrustal rocks and subsequent melting.

The most recent data to place temporal constraints on continental emergence comes from an analysis of submarine versus subaerial LIP (Kump & Barley 2007) (Fig 1f), which shows an abrupt increase in the secular variation of subaerial LIPs at c. 2.5 Ga. This is thought to have had significant global repercussions with respect to the increase in atmospheric oxygen levels (Kump & Barley 2007) and is discussed further below. Recent modelling of continental emergence that links continental freeboard with different models for cooling of the Earth indicates that the emergent continental crust was only 2–3% of the Earth's surface area during the Archaean, a stark contrast to present day values of c. 27% (Flament et al. 2008).

Sedimentary rocks document the physical, chemical, and biological interface between the Earth's crust and its changing Precambrian surface environments (see Eriksson et al. 2005b for an overview). Palaeoproterozoic sedimentary basins share many of the same features as their modern counterparts, whether in rift settings (Sengor & Natal'in 2001), passive margins (Bradley 2008), strike-slip basins (e.g. Ritts & Grotzinger 1994), or foreland basins (e.g. Grotzinger et al. 1988). Ironically, Precambrian sedimentary structures can be much easier to decipher than their Phanerozoic counterparts due to the lack of bioturbation in pre-metazoan depositional environments. However, there are some notable differences between Precambrian and modern clastic sedimentation systems. Most notably, sandstones of pre-Devonian river systems are commonly characterized by sheet-braided geometry with greater channel widths ascribed to the lack of vegetative slope stabilization (e.g. Long 2006). The deposits of specific palaeoenvironments such as aeolianites have a temporal distribution modulated by long-term preservational potential and possible relationships to phases of supercontinental cyclicity (Eriksson & Simpson 1998), as is also the case for glaciogenic deposits, which are summarized below. Also discussed below are factors determining the temporal variation in redox-sensitive mineral clasts such as detrital pyrite and uraninite (Fig. 1p), and the abundances and chemical compositions of banded iron-formations (Fig. 1k), evaporites (Fig. 1n), and carbonate chemistry and structure (Grotzinger 1989; Grotzinger & James 2000).

Supercontinents

All supercontinents older than Pangaea are conjectural in both existence and palaeogeography. The concept of Precambrian episodicity, and hence supercontinental cycles, arises from global peaks in isotopic age determinations, in which the basic result has not changed throughout a half-century of compilation (Gastil 1960; Worsley et al. 1984; Nance et al. 1986, 1988; Condie 1995, 1998, 2000; Campbell & Allen 2008). The most prominent age peaks are at 2.7–2.6 and 1.9–1.8 Ga, with some studies also indicating a peak at 1.2–1.1 Ga. The age peaks are global in distribution, although many regions contain regional signatures, and they are best correlated to collisional accretion of cratons in North America (Hoffman 1988, 1989). This led to the naming of successively older pre-Pangaean supercontinents Grenvilleland, Hudsonland and Kenorland (Williams et al. 1991). The younger two of these entities are most commonly given alternative names Rodinia (for recent reviews, see Li et al. 2008; Evans 2009), and Nuna or Columbia (Hoffman 1997; Rogers & Santosh 2002); 'Grenvilleland' has been forgotten, and 'Hudsonland' almost so (cf. Pesonen et al. 2003). Here we favour 'Nuna' for the supercontinent that assembled at 1.9–1.8 Ga (see Zhao et al. 2002 for a global review) because it represents a preferred renaming of 'Hudsonland' by one of the co-authors of the original study (Hoffman 1997) and has priority over 'Columbia' (Rogers & Santosh 2002), the latter of which was also defined by a specified palaeogeography that has been modified in subsequent references (e.g. Meert 2002; Zhao et al. 2002, 2004).

There are additional names for putative Palaeoproterozoic continental assemblages. Capricornia (Krapez 1999) refers to a model for the early amalgamation of Australia and hypothesized adjacent cratons Laurentia, India, Antarctica, and Kalahari. Arctica is defined as the postulated assemblage of Siberia with northern Laurentia, and Atlantica comprises the proposed long-lived amalgamation of West (and parts of northern) Africa, Amazon, São Francisco–Congo, and Rio de la Plata (Rogers 1996). Of these, a Siberia–northern Laurentia connection, if not directly adjacent then slightly separated, is allowed by numerous independent palaeomagnetic comparisons from 1.5 to 1.0 Ga (Pisarevsky & Natapov 2003; Pisarevsky et al. 2008; Wingate et al. 2009), and lack of tectonic activity after 1860 Ma in southern Siberia would suggest that this position had been established by that time (Poller et al. 2005; Pisarevsky et al. 2008). A long-lived Atlantica continent is difficult to test palaeomagnetically, due to a dearth of reliable results from its constituent cratons (Meert 2002; Pesonen et al. 2003). Minor transcurrent motions between West Africa and Amazon, sometime between original craton assembly at 2.1 Ga (Ledru et al. 1994) and their Gondwanan amalgamation in the Cambrian (e.g. Trindade et al. 2006), are proposed based on limited palaeomagnetic data from those two blocks (Onstott & Hargraves 1981; Onstott et al. 1984; Nomade et al. 2003). Such minor amounts of relative displacement would preserve the proposed tectonic correlations among the Atlantica cratons, but further palaeomagnetic testing is needed.

Nuna, commonly under the guise of 'Columbia', is commonly reconstructed with many cratonic juxtapositions taken from inferred models of Rodinia (Rogers & Santosh 2002; Zhao et al. 2002, 2004; Hou et al. 2008a, b). These models lack robust palaeomagnetic constraints, and the reconstructions are commonly distorted by unscaled cut-outs from a Mercator projection of Pangaea. When palaeomagnetic data are incorporated into geometrically accurate reconstructions, the sparsity of reliable results has led authors to conflicting conclusions of either non-existence of a 1.8 Ga supercontinent altogether (Meert 2002), or one that accommodated substantial internal shears (Laurentia–Baltica motions illustrated in Pesonen et al. 2003), or the consideration of at most a few fragments with applicable data from discrete time intervals (Li 2000; Salminen & Pesonen 2007; Bispo-Santos et al. 2008).

The most robust long-lived juxtaposition of Palaeo–Mesoproterozoic cratons is that of Laurentia and Baltica throughout the interval 1.8–1.1 Ga. First proposed on geological grounds and named 'NENA' (northern Europe–North America, Gower et al. 1990), this juxtaposition finds palaeomagnetic support from numerous results throughout that interval, defining a common apparent polar wander path when poles are rotated according to the reconstruction (Evans & Pisarevsky 2008; Salminen et al.). The NENA reconstruction is distinct in detail from the commonly depicted juxtaposition of Hoffman (1988) that has been reproduced in Zhao et al. (2002, 2004), which is not supported palaeomagnetically. NENA is a specific reconstruction between two cratons and is distinct from Nuna, the supercontinent proposed to have assembled at 1.9–1.8 Ga without particular palaeogeographic specifications. By coincidence, the two names are similar, and NENA appears to be a robustly constrained component of Nuna. Payne et al. extend the Palaeo–Mesoproterozoic apparent polar wander comparisons to include Siberia, North and West Australia, and the Mawson Continent (Gawler craton with original extensions into Antarctica). The Australian proto-continent is restored in that analysis by the same sense of relative rotations as proposed by Betts & Giles (2006). As more data from other cratons accumulate through the 1.8–1.5 Ga interval, this approach should lead to a successful first-order solution of Nuna's palaeogeography.

Kenorland is the name given to the palaeogeographically unspecified supercontinent that might have formed in the Neoarchaean era (Williams et al. 1991). Its etymology derives from the Kenoran orogeny (Stockwell 1982) that represents cratonization of the Superior craton at about 2.72–2.68 Ga (Card & Poulsen 1998; Percival et al. 2006) and of that age or younger in the Neoarchaean on other cratons (Bleeker 2003). Breakup of Kenorland would be represented by numerous large igneous provinces starting with a global pulse at 2.45 Ga (Heaman 1997). Subsequently, the meaning of 'Kenorland' has varied. Aspler & Chiaranzelli (1998) referred to Kenorland as the specified palaeogeography of ancestral North America, with an interpretation of the Trans-Hudson and related orogens that accommodated at most accordion-like oceanic opening and reclosing, but not extensive reshuffling of cratons. Baltica and Siberia were included in unspecified palaeogeographic configurations. A second proposed supercontinent ('Zimvaalbara' comprising Zimbabwe, Kaapvaal, Pilbara and perhaps São Francisco and cratons in India) is proposed to have assembled and begun to break up somewhat earlier than Kenorland, at 2.9 and 2.65 Ga, respectively; yet its final fragmentation was conjectured at 2.45–2.1 Ga, simultaneous with that of Kenorland (Aspler & Chiarenzelli 1998).

Bleeker (2003) chose a different nomenclature that referred to 'Kenorland' as a possible solution to Neoarchaean–Palaeoproterozoic palaeogeography whereby all (or most) cratons were joined together; he also proposed an alternative solution of palaeogeographically independent 'supercratons' that would each include a cluster of presently preserved cratons. Three supercratons were exemplified and distinguished by age of cratonization: Vaalbara (2.9 Ga), Superia (2.7 Ga), and Sclavia

(2.6 Ga). Vaalbara has the longest history of recognition and palaeomagnetic testing (Cheney 1996; Wingate 1998; Zegers et al. 1998) with the most recent tests allowing a direct juxtaposition (Strik et al. 2003; de Kock et al. 2009). Superia has become more completely specified by the geometric constraints of precisely dated, intersecting dyke swarms across Superior, Kola–Karelia, Hearne, and Wyoming cratons through the interval 2.5–2.1 Ga (Bleeker & Ernst 2006). Aside from Slave craton, the additional elements of Sclavia remain unknown. Lastly, a completely different view of Kenorland (Barley et al. 2005) entails inclusion of all (or most) of the world's cratons, following assembly as late as 2.45 Ga, and breakup at 2.25–2.1 Ga. This definition would be consistent with a global nadir in isotopic ages during the 2.45–2.25 Ga interval, proposed earlier as a time of supercontinent assembly (Condie 1995), or a more recently and radically, a cessation of plate tectonics altogether (Condie et al. 2009).

Palaeogeographic constraints on these proposed Neoarchaean–Palaeoproterozoic supercontinents and supercratons, other than the aforementioned Vaalbara, are limited by a small number of reliable palaeomagnetic data (Evans & Pisarevsky 2008). The best progress has been made in the Superior craton, where a series of precise U–Pb ages on mafic dyke swarms has been closely integrated with palaeomagnetic studies at the same localities, and with particular attention to baked-contact tests to demonstrate primary ages of magnetic remanence (e.g. Buchan et al. 2007; Halls et al. 2008). If a similar strategy is applied to other cratons then the solution of Kenorland or supercraton configurations will be much closer to realization.

Finally, Piper (2003) incorporated the extant Archaean–Palaeoproterozoic cratons into a long-lived supercontinent (duration 2.9–2.2 Ga) named 'Protopangaea'. This assemblage mirrors Piper's proposed Meso-Neoproterozoic supercontinent 'Palaeopangaea' (Piper 2007). Both are based on broad-brush compilations of the entire database containing a complex mixture of primary and secondary magnetizations rather than on precise palaeomagnetic comparisons of the most reliable data. Meert & Torsvik (2004) point out some of the problems with this approach, and Li et al. (2009) demonstrate specific quantitative refutations of the reconstructions.

Linked to the supercontinent debate is the controversy regarding the timing of initiation of a modern-style of plate tectonics (cf. Stern 2005; Cawood et al. 2006). A range of geological, geodynamic and geochemical constraints (recently summarized by Condie & Kroner 2008; Shirey et al. 2008; van Hunen et al. 2008), suggest the strong likelihood of plate tectonic behaviour in the Palaeoproterozoic and possibly back into the Archaean; an inference also made by Brown (2007) based on the presence of characteristic high pressure–lower temperature and high temperature–lower pressure metamorphic mineral assemblages associated with subduction and arcs respectively (Fig. 1h). Recent numerical models are consistent with this interpretation (e.g. Labrosse & Jaupart 2007). For a recent summary of the different aspects of the ongoing debate the reader is referred to Condie & Pease (2008) and references therein.

In summary, although Rodinia's configuration remains highly uncertain, a working model of its predecessor Nuna is being assembled rapidly with the acquisition of new geochronological and palaeomagnetic data considered in the context of global tectonic constraints. It remains uncertain whether there was a supercontinent at all during the Archaean–Proterozoic transition, and if so, when it existed and how its internal configuration of cratons was arranged. Figure 1i depicts these uncertainties using the template illustration of Bleeker (2003). Creation of a global stratigraphic database (Eglington et al. 2009) not only illustrates the growing wealth of global age constraints on the Palaeoproterozoic rock record, but also shows which stratigraphic units are in greatest need of dating, and which units are most promising for successful preservation of primary palaeomagnetic remanence directions that will allow further progress in supercontinent reconstruction.

Minerals and mineral deposits

Understanding the secular development of mineral abundances and concentrations has importance for the exploration and exploitation of economic ore deposits. A recent review of the evolution of Earth's minerals (Hazen et al. 2008) recognizes 10 stages of increasing mineral diversity associated with major global changes. In the timeline of interest here, the largest increase in mineral types has been linked to the development of the oxygenated atmosphere (Hazen et al. 2008). This event, the timing of which is best displayed by changes in the non-mass-dependent fractionation of sulphur isotopes (see below), entailed a significant increase in the number of mineral species (Fig. 1j) and correlates with the timing of the major deposits of Superior-type BIFs (Fig. 1k) and systematic changes in mineral deposits through time (Fig. 1j) (Barley & Groves 1992; Groves et al. 2005). The main peaks of BIF formation and deposition of unconformity-related, sediment-hosted uranium deposits at <1800 Ma are likely to represent the changing environmental conditions (see below). In contrast, the spike of orogenic gold deposits at c. 2.7 Ga and volcanic-hosted massive sulphide deposits at c. 2.6 Ga and c. 1.7 Ga, which temporally correspond to peaks in juvenile continental crust production (Fig. 1g) and mantle depletion events (Fig. 1e) may represent fundamental differences in the nature of tectonic processes around the Archaean–Proterozoic boundary (Groves et al. 2005). These are speculated to be associated with the development of Earth's first supercontinent (Barley & Groves 1992) and an increased preservation potential due to thicker subcontinental mantle lithosphere in the geological past (Groves et al. 2005).

The evolving ocean and atmosphere

Proxies for the secular evolution of the Neoarchaean–Proterozoic surface palaeoenvironment are numerous (Fig. 1k–q), commonly contentious, and often deeply interrelated in the development of conceptual models. Nevertheless, profound change is apparent, reflecting a combination of secular and cyclic trends. Here the development of this paper will deviate from its template of discussing the Earth system from its interior to its exterior, because one proxy in particular, the non-mass-dependent sulphur isotope system, is robust enough to provide a conceptual and temporal context for all the others.

Anomalous variation among the four stable sulphur isotope ratios, deviating from purely mass-dependent effects, are common in Archaean and earliest Proterozoic sedimentary pyrites (Farquhar et al. 2000). The deviations are thought to arise from gas-phase sulphur reactions in the upper atmosphere with concomitant mixing into seawater, with further constraints on sulphur speciation provided by increasingly refined datasets (e.g. Ono et al. 2003, 2009a, b). The termination of this non-mass-dependent fractionation signal is now estimated at c. 2.4–2.32 Ga (Bekker et al. 2004). Atmospheric modelling suggests that it was caused by either the rise of O_2 above 10^{-5} of present atmospheric levels (Pavlov & Kasting 2002) or loss of CH_4 below a critical threshold of c. 10 ppmv due to a shrinking ecological role of methanogenic producers (Zahnle et al. 2006). The collapse of methane from formerly high levels in the Archaean atmosphere (reviewed by Kasting 2005) probably plays a large role, not only the oxygenation history, but also the occurrence of Palaeoproterozoic ice ages, which will be discussed below. The combined data point to the 2.4–2.3 Ga interval as host to a 'Great Oxidation Event' (GOE: Cloud 1968; Holland 2002) during which Earth's surface environment changed profoundly and irreversibly.

Possible changes in atmospheric oxidation state can be estimated by age variations in the non-mass-dependent signal (Fig. 1o). Many data now exist for the crucial interval of 3.0–2.4 Ga (Ohmoto et al. 2006; Farquhar et al. 2007; Kaufman et al. 2007; Ono et al. 2009a, b) and perhaps the most compelling evidence for a 'whiff' of oxygen prior to 2.4 Ga is furnished by Mo and Re concentration peaks in black shales at 2.5 Ga (Anbar et al. 2007; Wille et al. 2007). Much of the inferred global nature of these peaks depends on the correlation of signals between the Pilbara and Kaapvaal cratons, whereas a subsequent palaeomagnetic reconstruction of the blocks places them in direct juxtaposition (de Kock et al. 2009). More records of similar nature are needed from other cratons through this interval, to substantiate the global nature of any hypothesized oxygenation events. Localized, photosynthetically produced 'oxygen oases' could attain ppm-level O_2 concentrations within plumes dissipating into the methane-rich Neoarchaean troposphere in a timescale of hours to days (Pavlov et al. 2001).

The causes of the GOE remain unclear. Traditionally attributed to development of photosystem II in cyanobacteria (Cloud 1968; Kirschvink & Kopp 2008) or enhanced burial rates of organic carbon produced by that process (Karhu & Holland 1996), other potential long-term sources of oxidizing agents involve dissociation of atmospheric methane coupled to H_2 escape (Catling et al. 2001; Catling & Claire 2005), or the changing oxidative state of the upper mantle coupled to hydrothermal alteration of seafloor basalts (Kasting et al. 1993; Kump et al. 2001; Holland 2002). Finally, a qualitative empirical relationship between continental collisions and global oxygen increases over the past three billion years (Campbell & Allen 2008), linked to carbon and sulphur burial through enhanced physical weathering and sediment transport in mountain belts, is rendered particularly speculative for the GOE due to the poorly understood history of Palaeoproterozoic supercontinents or supercratons (see above). Widespread consideration of the non-biological alternatives has been motivated primarily by the apparent antiquity of molecular biomarkers for photosynthesizing organisms as old as 2.7 Ga, about 300–400 million years prior to the GOE (Brocks et al. 1999); however, the reliability of those records is currently debated (Eigenbrode et al. 2008; Rasmussen et al. 2008; Waldbauer et al. 2009).

Returning to the ocean, banded iron formations are a *prima facie* example of non-uniformitarianism in the Earth's palaeoenvironment. Deposited widely prior to 1.85 Ga (Fig. 1k), they are nearly absent in the subsequent geological record, apart from a few small occurrences or unusual associations with

Neoproterozoic low-latitude glacial deposits of putative Snowball Earth events (Kirschvink 1992; Klein & Beukes 1992). The marked concentration of iron formation in the Palaeoproterozoic era has traditionally been construed as an indication of atmospheric oxidation at about that time (e.g. Cloud 1968) and that element is no doubt part of the story. However, recent research on iron formations using geochemical and isotopic tracers is painting a more refined picture of oceanic evolution.

Iron formations are broadly categorized into two classes, one with close association to volcanic successions (Algoma-type), and one without (Superior-type) (Gross 1965; Gross 1983). The Algoma-type iron formations are distributed throughout the Archaean sedimentary record, whereas Superior-type deposits only become significant at 2.6 Ga and later (Huston & Logan 2004) (Fig. 1k). By 2.5 Ga, Superior-type iron formations predominate, to the extent that 'Siderian' was informally proposed as the first period of the Palaeoproterozoic era (Plumb 1991). With refined global geochronology, the global peak in Palaeoproterozoic iron formation (e.g. Klein 2005 and references therein) appears to split into two modes, at c. 2.45 Ga and c. 1.9 Ga (Isley & Abbott 1999; Huston & Logan 2004), although some large deposits remain imprecisely dated (e.g. Krivoy Rog, Ukraine; Simandou, West Africa). The temporal distribution of iron formation is closely matched by the occurrence of thick, massive seafloor calcium carbonate cements (Grotzinger & Kasting 1993), commonly with crystalline microtextures linked to high Fe concentrations in seawater (Sumner & Grotzinger 1996).

Detailed petrological and facies analysis of iron formations has led to models with varying degrees of ocean stratification (e.g. Klein 2005; Beukes & Gutzmer 2008), but of concern to the present compilation are the broader trends in Earth's palaeoenvironmental evolution. Do the time-varying abundances of iron formations through the Archaean–Proterozoic transition represent fundamental changes in oceanography? At the older end of that age range, the answer is likely to be 'no'. Trendall (2002) discussed the requirement of tectonic stability to allow accumulation of the giant iron formations, and with the exception of the c. 1.9 Ga deposits, many of the large iron formations appear at the first submergence of each craton following initial amalgamation. If so, the initial peak in iron formations between 2.7 and 2.4 Ga (Fig. 1k) has causative correlation with other records such as continental emergence (Fig. 1f). The apparently abrupt end of the first pulse of iron formations at c. 2.4 Ga remains to be tested by further geochronology (Krivoy Rog, Simandou), but its close temporal association with the rise of atmospheric oxygen strongly suggests a causal connection. Possible feedbacks involving phosphorus limitation on primary production (Bjerrum & Canfield 2002) have been disputed on quantitative grounds (Konhauser et al. 2007). However, fractionations of Fe isotopes in sedimentary pyrite appear to support the close coincidence between the timing of oxygenation in the atmosphere and the ocean (Rouxel et al. 2005).

Renewal of iron-formation deposition at 2.0–1.8 Ga could also indicate profound changes in the hydrosphere. The so-called 'Canfield ocean' model (Canfield 1998, 2005; Anbar & Knoll 2002) is the suggestion that after atmospheric oxidation, the increased weathering of continental sulphide minerals brought reactive sulphate ions into the marine realm, where sulphate-reducing bacteria responded with enhanced pyrite formation, thus stripping seawater of its hydrothermally generated Fe. In this model, the disappearance of iron formations at c. 1.8 Ga reflects the change to more reducing, sulphidic conditions rather than the deep ocean turning oxic. Meanwhile, the upper layer of seawater would remain oxygenated in contact with the post-GOE atmosphere; this first-order stratification is proposed to have persisted until the end of Precambrian time (Canfield et al. 2008). Despite some criticism (e.g. Holland 2006), the model has passed initial tests using depth-dependent iron speciation and sulphur stable isotopic variations in Mesoproterozoic marine sediments (Shen et al. 2002, 2003), patterns of trace metal concentrations (Anbar & Knoll 2002), isotopes (Arnold et al. 2004) and molecular biomarkers (Brocks et al. 2005).

Amid these novel geochemical proxies for hydrospheric and biospheric evolution, more traditionally studied isotopic systems show equally dramatic variations. One of the largest seawater carbon-isotopic anomalies in Earth history is known as the 'Lomagundi' or 'Jatuli' event respectively after the c. 2.1 Ga strata in Zimbabwe or Karelia where it was first documented (cf. Schidlowski et al. 1975) with typical values as high as +10‰ (Fig. 1m). An elegantly simple model attributed the peak to organic-carbon burial associated with evolution of oxygenic photosynthesis and, consequently, the rise of atmospheric oxygen (Karhu & Holland 1996). As discussed below, however, there is (contested) evidence that the advent of oxygenic photosynthesis could have preceded the isotopic peak by at least several hundred million years. Although the most enriched ^{13}C values were probably generated in restricted, evaporative basins, such enrichments may well have been mere additions to a global peak, as indicated by the global extent of the signal (Melezhik et al. 1999, 2005a, b).

Rather than a single, long-lived Lomagundi-Jatuli event spanning the entire interval of 2.2–2.06 Ga (Karhu & Holland 1996), evidence has now accumulated for multiple positive excursions with intervening non-enriched values through that period (Buick et al. 1998; Bekker et al. 2001, 2006). Recent U–Pb dating on interstratified volcaniclastic rocks in Fennoscandia has constrained the end of the event to c. 2.06 Ga (Melezhik et al. 2007), but the earliest enrichments have been more difficult to date. The oldest strongly ^{13}C-enriched carbonate units, in the Duitschland Formation of South Africa (Bekker et al. 2001), are older than 2316 ± 7 Ma based on Re–Os of diagenetic pyrite from the stratigraphically overlying Timeball Hill shales (Bekker et al. 2004). Those same pyritic shales lack a non-mass-dependent sulphur-isotope fractionation signal (Bekker et al. 2004) and thus the onset of the Lomagundi-Jatuli event(s) coincides exactly, to the best knowledge of available ages, with atmospheric oxidation above 10^{-5} of present atmospheric levels. These events directly follow the only Palaeoproterozoic ice age with a ^{13}C-depleted cap carbonate, which left a record on two palaeocontinents if the correlations of Bekker et al. (2006) are correct. The relationship between Palaeoproterozoic ice ages and Earth system evolution will be explored further, below.

The strontium isotopic composition of seawater, as recorded in carbonate rocks, involves more difficult analytical methods, and although well established for the Phanerozoic Eon, is relatively poorly constrained for Precambrian time (Shields & Veizer 2002). Data from the Palaeoproterozoic and surrounding intervals are sparse, and subject to uncertainties in age as well as the minimum-altered values. Nonetheless, a noticeable pattern of increasing ^{87}Sr/^{86}Sr, away from the inferred (slowly increasing) mantle value, characterizes the general Palaeoproterozoic trend ('seawater' curve in Fig. 1l). As Shields & Veizer (2002) noted, this probably indicates greater continental emergence and riverine runoff of radiogenic strontium through the Archaean–Palaeoproterozoic interval. Shields (2007) provided a more sophisticated model of these processes, including rough estimates of carbonate/evaporite dissolution as a contributor to the riverine runoff component, to produce a percentage estimate of the riverine contribution to seawater ^{87}Sr/^{86}Sr ratios through time (dashed curve in Fig. 1l). Even with this more detailed model, the first-order interpretation remains valid. However, the critical period for testing the connection with continental emergence (Fig. 1f), that is 2.7–1.9 Ga, is represented by a scant number of data (Shields & Veizer 2002).

Long-term secular evolution of oceanic chemistry can also be measured by compositional and sedimentological trends in carbonates and evaporites. Grotzinger (1989) and Grotzinger & James (2000) noted the abundance of carbonate platforms mirroring the growth of large sedimentary basins due to stable cratonization, much like the pattern observed for iron formations. The latter study also illustrated the successive peaks in ages of: aragonite crystal fans/herringbone calcite, tidal flat tufas, molar tooth structures, giant ooids, and various biogenic features of Archaean–Proterozoic sedimentary history. Not all of these records are well understood, but as noted above, disappearance of aragonite crystal fans and herringbone calcite is best correlated to the end of iron formation deposition and thus broadly to the rise of atmospheric oxygen.

Ocean palaeochemistry can also be inferred from the record of evaporite deposits, which in the Palaeoproterozoic era consist almost entirely of pseudomorphs after the original minerals. Evans (2006) compiled volume estimates for the largest evaporite basins through Earth history, summarized here in Figure 1n. A previous compilation (Grotzinger & Kasting 1993) described sulphate evaporites as old as c. 1.7 Ga and no older, but there is more recent recognition of common sulphate pseudomorphs in sedimentary successions at c. 2.2–2.1 Ga (reviewed by Pope & Grotzinger 2003; Evans 2006). Those gypsum- or anhydrite-bearing strata are usually associated with redbeds and ^{13}C-enriched carbonates of the Lomagundi-Jatuli event (Melezhik et al. 2005b). Evaporite deposits of both younger (Pope & Grotzinger 2003) and older age (Buick 1992; Eriksson et al. 2005a) contain an evaporative sequence from carbonate directly to halite, excluding sulphate. Temporal correlation of halite-dominated evaporites with the peaks in iron-formation (Fig. 1k, n) conforms to the model of Fe-oxide seawater with low sulphate content (pre-2.4 Ga, plus 2.0–1.8 Ga), alternating with Fe-sulphidic deepwater driven by mildly oxidized surface water with higher sulphate content (Anbar & Knoll 2002). A rising oceanic sulphate reservoir through the Proterozoic Eon is also inferred by modelling rates of sulphur isotope excursions through sedimentary sections (Kah et al. 2004; Fig. 1n).

The evolving palaeoclimate and biosphere

Palaeoproterozoic climate changes were as extreme as any in Earth history, with low-latitude ice ages interrupting an otherwise dominantly ice-free record (Evans 2003). Indications of a background state of hot (55–85 °C) Archaean oceans, from ^{18}O records of chert and phosphate (Knauth & Lowe 2003; Knauth 2005), are controversial (cf. contrasting views of Lowe & Tice 2007; Shields

& Kasting 2007; van den Boorn et al. 2007). Regardless, basal clades of both eubacteria and archaea were likely thermophilic (Boussau et al. 2008). The early fossil record of life on Earth, especially from ages older than 3.0 Ga, is fragmentary and ardently debated (e.g. Brasier et al. 2002; Rose et al. 2006; Westall 2009). Microbial activity of some sort is evident from the presence of wrinkle mat textures in sedimentary rocks as old as 2.9 and even 3.2 Ga (Noffke et al. 2003, 2006a, b, 2008). Stromatolites of the Tumbiana Formation, Western Australia, show an impressive diversity of forms (Buick 1992) that occupied a varied array of littoral marine environments (Sakurai et al. 2005); they also contain putative nano/microfossils (Lepot et al. 2008). Molecular biomarkers also support the existence of extant late Archaean prokaryotes: although Rasmussen et al. (2008) have reinterpreted the methylhopane (and sterane, see below) record from 2.7 Ga shales in Western Australia (Brocks et al. 1999) as a post-metamorphic feature, additional records from 2.7–2.5 Ga in the same succession show facies-dependent and thus possible palaeo-ecological distributions of 2-alpha and 3-beta methylhopanes (Eigenbrode et al. 2008) that are less likely due to post-metamorphic infiltration. A similar test of varying biomarker proportions among sedimentary and volcanic facies in the Abitibi greenstone belt of southern Canada, indicates possible archaeal and bacterial activity coincident with hydrothermal gold precipitation at 2.67 Ga (Ventura et al. 2007).

The eukaryotic fossil record, prior to 1.2 Ga, is equally controversial. Examples are described here in order of increasing age. At the younger end of the interval covered by this review (Fig. 1t), the record of bangiophyte red algae, in northern Canada, presents the oldest phylogenetically pinpointed eukaryotic body fossils (Butterfield 2000) and a robust starting point for considering older examples. The taxonomic affiliations of such older 'eukaryotic' fossils is inferred from their sizes and complexities (Knoll et al. 2006), including many simple and ornamented acritarchs, and various filamentous forms. Within the macroscale, the next older putative eukaryotic fossil is *Horodyskia*, found in c. 1.5–1.1 Ga strata in Western Australia and North America (reviewed by Fedonkin & Yochelson 2002; Grey et al. 2002; Martin 2004). Informally known as 'strings of beads', *Horodyskia* is difficult to place taxonomically; Knoll et al. (2006) consider it to be 'a problematic macrofossil whose eukaryotic affinities are probable, but not beyond debate.'

The next two older macrofossils have been purported to preserve the trails of motile, multicellular organisms and are more contentious than the younger taxa outlined above. The first, in the Chorhat sandstone of the lowermost Vindhyan basin in India (Seilacher et al. 1998) attains Palaeoproterozoic antiquity on the merits of two concurrent and independent high-precision U–Pb studies (Rasmussen et al. 2002b; Ray et al. 2002). However, Seilacher (2007) has subsequently introduced a viable alternative hypothesis that the Chorhat traces could represent (biogenic?) gas structures trapped beneath a microbial mat. The second, consisting of discoidal and furrowed impressions in sandstone of the Stirling Ranges in Western Australia (Cruse & Harris 1994; Rasmussen et al. 2002a) has recently been dated to the interval 1960–1800 Ma (Rasmussen et al. 2004). An extensive discussion on these putative trace fossils retains the original interpretation of their being produced by 'motile, mucus-producing, probably multicellular organisms', which on the basis of size alone were probably eukaryotic (Bengtson et al. 2007). However, recent discovery of furrowed trails produced by extant *Gromia* amoebas may provide an adequate explanation for the Stirling biota (Matz et al. 2008); such an explanation needs further testing.

The oldest likely eukaryotic body fossil, *Grypania spiralis*, is found in the c. 1.88 Ga Negaunee iron-formation (Han & Runnegar 1992; Schneider et al. 2002), coeval with the spectacular palaeontological record of the nearby Gunflint Chert (Tyler & Barghoorn 1954; Fralick et al. 2002) and only slightly younger than the equally impressive Belcher Islands microflora (Hofmann 1976) at c. 2.0 Ga (Chandler & Parrish 1989). Classification of *Grypania* is based on its morphological similarity to Mesoproterozoic occurrences from North China and North America (Walter et al. 1990). A recent report describing a spinose acritarch in amphibolite-grade Archaean metasedimentary rocks of South Australia (Zang 2007) seems less convincing.

Apart from body fossils, evidence for eukaryotic life in the Palaeoproterozoic also includes the molecular biomarker record of steranes. Sterol biosynthesis is largely, although not entirely, limited to the eukaryotic realm (see discussion in Kirschvink & Kopp 2008; Waldbauer et al. 2009). Sensationally old steranes were identified in the 2.7 Ga Jeerinah Formation of Western Australia (Brocks et al. 1999, 2003a, b). However, Rasmussen et al. (2008) attributed these signals to a secondary fluid migration into the boreholes, at some unknown time after c. 2.16 Ga regional metamorphism. There are other Palaeoproterozoic sterane biomarker records. Dutkiewicz et al. (2006) and George et al. (2008) found them in sediments of the basal Huronian Supergroup (c. 2.4 Ga), with a signal that pre-dates c. 1.9 Ga Penokean metamorphism, and Dutkiewicz et al. (2007) discovered them in the

c. 2.1 Ga Francevilian series of Gabon, with a signal that pre-dates supercritcality of the Oklo natural nuclear reactor at 1.95 ± 0.04 Ga (Gauthier-Lafaye & Weber 2003).

Waldbauer et al. (2009) conducted a benchmark study in attempts to demonstrate the syngeneity of their observed sterane biomarker signal, obtained from c. 2.65–2.45 Ga strata in South Africa. The molecular fossils are described as pre-metamorphic, and vary according to sedimentary facies in correlative sections from adjacent drillcores. Nonetheless, the carbonate formations in those drillcores have been pervasively remagnetized at about 2.2–2.1 Ga (de Kock et al. 2009), indicating basinwide low-grade hydrothermal fluid infiltration-at the same age within error as, and possibly in direct palaeogeographic proximity to, the regional metamorphic event on the adjacent Pilbara craton as described by Rasmussen et al. (2008).

We return to the Palaeoproterozoic glacial deposits, which are classically used to infer the palaeoenvironmental conditions in which these biological innovations occurred. Following an almost entirely ice-free Archaean history, the Palaeoproterozoic world was exposed to at least three ice ages, which appear to have penetrated deep into the tropics (Evans et al. 1997, global constraints reviewed by Evans 2003). These ice ages, lesser known but seemingly of equal severity to their more widely publicized Neoproterozoic 'snowball Earth' counterparts (Hoffman & Schrag 2002), are generally rather poorly constrained in age to within the interval 2.45–2.22 Ga. As with the Neoproterozoic ice ages, estimating the number of Palaeoproterozoic glaciations is complicated by the fact that the diamictites themselves are commonly the principal items of correlation among cratons through this interval (e.g. Aspler & Chiarenzelli 1998; Bekker et al. 2006).

The end of the second among three ice ages, recorded in the Huronian succession and correlative strata in Wyoming, is marked by a ^{13}C-depleted 'cap carbonate' unit that may be broadly comparable to those better developed after Neoproterozoic ice ages (Bekker et al. 2005). The South African sections contain two distinct sequences of diamictite and overlying ^{13}C-depleted carbonate (Bekker et al. 2001), unconformably overlain by a third diamictite, the Makganyene Formation, in turn overlain by flood basalt and variably Mn-rich carbonate and ironstone units (Kirschvink et al. 2000) with near-zero δ^{13}C values (Bau et al. 1999). Palaeomagnetic data from the flood basalt indicates deep tropical palaeolatitudes, constituting the best evidence of its kind for a Palaeoproterozoic snowball Earth event (Evans et al. 1997; Kirschvink et al. 2000; Evans 2003). Within the limits of existing age constraints, the Makganyene ice age could be correlative with the uppermost Huronian glaciation at 2.23 Ga (Bekker et al. 2006), or, all three Huronian glacial levels could be distinctly older (Kopp et al. 2005). Given that the Lomagundi-Jatuli positive carbon-isotope excursion(s) began as early as 2.32 Ga, the near-zero δ^{13}C values in the post-Makganyene carbonate units are anomalously negative and warrant comparison with other Proterozoic postglacial cap carbonate sequences. Considering the high greenhouse forcing required to offset the Palaeoproterozoic 'faint young Sun' (Sagan & Mullen 1972), escape from any 'snowball' climate regime of that age would have required tens of millions of years of volcanic outgassing uncompensated by silicate weathering (Tajika 2003). If all of the Palaeoproterozoic glacial deposits represent so-called 'hard' snowball ice ages, then Earth's panglacial climate mode would have occupied a substantial fraction of time in the 2.45–2.22 Ga interval.

As discussed above, the disappearance of non-mass-dependent sulphur isotope fractionation and the onset of highly ^{13}C-enriched carbonates of the Lomagundi isotopic event are both located stratigraphically within the broad age range of these glaciations. More precisely, if the Bekker et al. (2006) correlations between North America and South Africa are correct, then the oldest carbonate-capped glacial deposits (Bruce and Rooihoogte Formations) constitute the stratigraphical boundary between two fundamentally distinct states of Earth's palaeoenvironment. Closely below this level are the final vestiges of detrital pyrite/uraninite deposition (Roscoe 1973) and non-mass-dependent sulphur isotope fractionation (Papineau et al. 2007). Closely above the level are the entirely mass-dependent-fractionated pyrites of the Timeball Hill Formation, dated at 2.32 Ga and representing the rise of atmospheric oxygen (Bekker et al. 2004; Hannah et al. 2004) and the oldest carbonates with strongly enriched ^{13}C values indicating the onset of the Lomagundi-Jatuli isotopic excursions (Bekker et al. 2001). Glacial deposits with cap carbonates thus appear to be closely related to the rise of atmospheric oxygen. Collapse of the methane-rich, pre-2.4-Ga greenhouse due to atmospheric oxygenation (see above and Kasting 2005) could well be a trigger for the low-latitude ice ages, perhaps in addition to the silicate weathering removal of carbon dioxide due to the widespread and largely subaerial large igneous provinces at 2.45 Ga (Melezhik 2006; Kump & Barley 2007). But the ice ages themselves could also have contributed further to rapid pulses of oxygen production: Liang and co-workers (Liang et al. 2006) postulated the mechanism of hydrogen peroxide trapping in ice throughout the duration of a 'hard' snowball stage, which would be released suddenly to the oceans upon deglaciation.

This glaciation-oxygenation scenario has been developed further, as reviewed by Kirschvink & Kopp (2008). In that model, the hydrogen peroxide plume into the oceans upon panglacial melting would constitute the evolutionary driver of intracellular oxygen-mediating enzymes, which are seen as a necessary precursor to oxygenic photosynthesis. The post-Makganyene sequence, extraordinarily rich in Mn, would represent the final and irreversible oxidation of the deep oceans (Kirschvink et al. 2000). Two outstanding problems with the timing of the model are as follows: (1) the Lomagundi-Jatuli positive isotopic excursion, apparently requiring burial of photosynthetically produced organic carbon, is found in pre-Makganyene strata (Bekker et al. 2001); and (2) increasingly more rigorous biomarker studies provide compelling evidence for photosynthetic organisms as old as 2.7 Ga (e.g. Eigenbrode et al. 2008; Waldbauer et al. 2009).

Following the well known 2.45–2.2 Ga ice ages, the next nearly 1.5 billion years has traditionally been noted as entirely ice-free (Evans 2003). Recent documentation of periglacial features at $c.$ 1.8 Ga (Williams 2005) contests this conclusion. Nonetheless, the dominant climate state was nonglacial throughout most of Palaeoproterozoic–Mesoproterozoic time. Even in a mildly oxygenated, post-GOE atmosphere, methane is increasingly favoured as a minor but powerful greenhouse gas to combat the low luminosity of the Mesoproterozoic Sun (Pavlov et al. 2003; Kasting 2005; Kah & Riding 2007).

Extra-terrestrial influences: bolides and Earth-Moon orbital dynamics

The two largest known bolide impact craters on Earth are Palaeoproterozoic in age, their sizes eclipsing that of the end-Cretaceous Chicxulub structure (Earth-Impact-database 2009). The largest, Vredefort in South Africa, is dated at 2.02 Ga; the second-largest, Sudbury, has an age of 1.85 Ga (both impacts reviewed by Grieve & Therriault 2000). No ejecta blanket has been discovered from Vredefort, but three regions of ejecta localities within 500–800 km from Sudbury are now reported (Addison et al. 2005; Pufahl et al. 2007). An impact spherule-bearing locality on the Nain or North Atlantic craton (Chadwick et al. 2001) has age constraints of 1.88–1.85 Ga (Garde et al. 2002), barely within range of the Sudbury event. The palaeogeographic proximity of Nain craton to the Sudbury impact site on Superior craton is unknown at 1.85 Ga (Wardle et al. 2002). In addition to these two largest craters, Figure 1s shows three smaller ($\geq$30 km diameter) craters in the 3.0–1.2 Ga time interval: Yarrabubba at an unknown age younger than its $c.$ 2.65 Ga target rocks (Macdonald et al. 2003), Keurusselkä at an unknown age younger than its $c.$ 1.88 Ga target rocks (Hietala et al. 2004) and Shoemaker with a maximum age of 1.63 Ga (Pirajno et al. 2003, 2009).

As there are no older preserved impact craters than $c.$ 2.4 Ga (Earth impact database 2009), all knowledge of prior impact history must be inferred from the lunar record, or determined from ejecta beds containing either spherules (Simonson & Glass 2004) or anomalous concentrations of siderophile elements (Glikson 2005). For the time interval investigated here, the most prominent impact record is found in the sedimentary cover of the Vaalbara supercraton. At least three distinct spherule beds can be recognized; some readily correlated between Australia and South Africa, within the interval 2.63–2.49 Ga (Simonson et al. 2009). It is unknown what effects these impacts, undoubtedly a small subset of the total Archaean–Palaeoproterozoic bolide flux to the Earth, had on the ancient surface environment. Completion of the IGCP509 global stratigraphic database (Eglington et al. 2009) will help identify suitable sedimentary basins for finding ejecta blankets from the large craters described above.

The orbital parameters of Earth and the Moon can be gleaned from the sedimentary record of tidal rhythmites, which are usually in fine-grained mudstones and siltstones (e.g. Williams 2000). They can also be found in sandstone crossbed foresets (Mazumder 2004; Mazumder & Arima 2005), including the oldest rhythmites in the geological record at $c.$ 3.2 Ga (Eriksson & Simpson 2000). The most complete calculation of orbital parameters from tidal rhythmites can be found in Williams (2000), who listed two alternative calculations for the 2.45 Ga Weeli Wolli banded iron formation in the Hamersley Ranges of Western Australia: one assuming that the lamina couplets (microbands) represented annual increments, the other assuming that they represented fortnightly cycles. Trendall (2002) has discussed how the annual microband model is consistent with U–Pb age constraints through the Hamersley succession, and by implication, that the fortnightly alternative is not. The annual microband model predicts 17.1 ± 1.1 hours in the solar (Earth) day, 514 ± 33 solar days per year, and an Earth–Moon distance of 51.9 ± 3.3 Earth radii (compared to 60.27 at present) for the earliest Palaeoproterozoic (Williams 2000).

Discussion

The data summarized above and in Figure 1 provide an overview of secular changes in the Earth system from the Neoarchaean to Mesoproterozoic eras. Many of the relationships illustrated by two or three strands of data have been noted previously. In this overview, we have attempted to compile the first comprehensive summary of the secular changes from Earth's core to atmosphere over the Palaeoproterozoic supercontinent cycle, and the data reveal intriguing temporal relationships between changes in deep Earth and its surface environment.

Many of the physical models used to extrapolate modern planetary dynamics back into deep time suffer from the necessary simplifications of tractability, and many of the historical data proxies are incomplete or contentious. However, an emphasis on interdisciplinary Earth-system science has led to multiple working hypotheses for the interrelationships among the various proxy records, and we are particularly inspired by the following 10 recent developments.

1) Analytical and numerical models of Earth's thermal history and mantle convection are approaching the ability to generate plate tectonics self-consistently and to account for distinct geochemical reservoirs. These have potential to solve several long-troublesome paradoxes of geophysics and geochemistry.
2) Improved laboratory calibration of the perovskite to post-perovskite transition has refined temperature estimates at the core–mantle boundary, which will provide a better 'initial' boundary condition for such thermal evolution models (and also constrain core heat loss and thus geodynamo history). The geodynamic and geochemical consequences of the associated D'' layer are likely to play critical roles in future models of Earth's thermal history and core–mantle interaction.
3) Resolution of the historical plume flux by a focused and global campaign to date mafic large igneous provinces (LIPs) will provide valuable constraints on mantle evolution, and will facilitate accurate pre-Pangaean continental reconstructions and a record of supercontinent amalgamation and dispersal that will in turn provide templates for mineral deposit belts and the development of new tectonic models.
4) An abrupt increase in the proportion of subaerial versus subaqueous LIP deposits at the Archaean–Proterozoic transition adds another indication of widespread continental emergence at that time; the temporal distribution of various types of iron formations can be understood better in the context of that transition.
5) The termination of non-mass-dependent fractionation of sulphur isotopes is now constrained to between 2.4 and 2.32 Ga, providing a robust stratigraphical marker by which all other proxies can be compared. Prior to this marker event, atmospheric oxygen levels were below 10^{-5} of present atmospheric levels, a largely non-glacial world was kept warm by a substantial methane greenhouse effect, marine sulphate levels were low and halite evaporites followed directly after carbonate precipitation. Hydrothermal iron from the deep oceans upwelled onto the recently stabilized continental shelves to produce the vast Hamersley-type banded iron formations. After the marker event, oxygen levels rose to an unspecified level that generated Earth's oldest lateritic palaeosols, at least two ice ages occurred (one with cap carbonates and negative ^{13}C excursions, the other demonstrably extending to tropical palaeolatitudes), sulphate evaporites became abundant, and the background state of marine carbon isotopes became highly enriched as the unparalleled Lomagundi-Jatuli positive excursion.
6) The stratified, sulphidic-ocean model for 1.8–0.8 Ga has passed several geochemical and palaeoecological tests.
7) Further recognition of impact spherule beds provides a new strategy for evaluating Earth's impact history prior to the age of its largest and oldest (well dated) preserved crater (Vredefort, 2.02 Ga).
8) Despite warranted caution concerning both modern and ancient possible contamination of molecular fossils extracted from drillcores, some recent studies provide impressive benchmarks for testing the syngeneity of ancient biomarker records.
9) The two alleged Palaeoproterozoic animal fossil occurrences, uncomfortably more than a billion years older than the most reliable molecular clock studies would indicate, are now both explained by non-metazoan microbial processes.
10) Development of a global stratigraphic database for the Palaeoproterozoic era (Eglington et al. 2009), as a final product of the IGCP Project 509, will allow ready correlation of rock units and tectonic settings across the world's cratons, which will be useful to researchers across all these proxy records of planetary evolution through Earth's post-Archaean transition.

Among many of the solid-Earth proxies shown in Figure 1, an important event occurred at c. 2.7 Ga. These include an abrupt increase in geodynamo palaeointensity (Fig. 1a), an unrivalled peak in the global LIP record (Fig. 1c), especially when scaled for preserved continental area by age, a broad maximum in several mantle depletion proxies (Fig. 1d), a strong peak in juvenile continental crust production (Fig. 1g), and a 'bonanza' of orogenic gold, massive-sulphide and porphyry copper deposits (Fig. 1j). It is tempting to link these records together in a model of enhanced mantle convection, perhaps due to a series of mantle plumes (Barley et al. 1998), slab avalanches (Condie 1998), or both. Stabilization of cratonic lithosphere was widespread from 2.7 to 2.5 Ga (Bleeker 2003), and the consequent emergence of continents (Fig. 1f, l) and expansion of sedimentary basins at their margins created accommodation space for the accumulation of voluminous banded iron-formations (Fig. 1k) and flourishing of photosynthetic life (Fig. 1t). The latter development, at about 2.3 Ga, led to irreversible atmospheric oxidation (Fig. 1o, p), possibly dissipating a former methane greenhouse atmosphere and ushering in the extensive Palaeoproterozoic ice ages (Fig. 1r). Extreme events in global carbon cycling appear at this time (Fig. 1m). After the great oxidation event, marine sulphate levels rose (Fig. 1n), sediment-hosted and iron-oxide-rich metal deposits became abundant (Fig. 1j), and the transition to sulphide-stratified oceans (Fig. 1p) cradled early eukaryotic macrofossils (Fig. 1t).

Supercontinents may indeed be the centrepiece of the long-term Earth system, but their history is one of the least constrained elements in Figure 1. Nonetheless, there is hope for eventual understanding. New isotopic methods for precise geochronological calibration of deep time have made these comparisons possible. The stratigraphies of most of the world's Precambrian cratons are now increasingly constrained by acquisition of such precise rock ages, and there is no sign of slowing. Dedicated global working groups such as IGCP projects have fostered frequent, direct communication among researchers around the world. An emphasis on interdisciplinary science has led to multiple working hypotheses for the interrelationships among the various proxy records.

From core to surface, our next major advances will likely arise through determining accurate ways to measure ancient geomagnetic field strength; obtaining more complete records of mantle plume activity; developing novel methods for estimating crustal growth and continental emergence; solving the palaeogeography of supercontinents and supercratons, with consequent 'ground-truthing' of proposed tectonic processes and mineral deposit evolution; creating new chemical and isotopic proxies for the evolution of mantle, crust, and surface; dating these records precisely with new analytical techniques; and integrating these strands of data into robust geodynamic models. With these continuing advancements the Archaean–Proterozoic transition is at last coming into focus.

We thank Brendan Murphy and Alan Vaughan for formal manuscript reviews; and Alan Collins, Paul Hoffman, Jim Kasting, Rajat Mazumder, and Bruce Simonson for informal discussions. This work is a contribution to UNESCO's International Geoscience Programme (IGCP) Project 509 and is TIGeR publication No. 165.

References

ABBOTT, D. H. & ISLEY, A. E. 2002. The intensity, occurrence, and duration of superplume events and eras over geological time. *Journal of Geodynamics*, **34**, 265–307.

ABBOTT, D., BURGESS, L., LONGHI, J. & SMITH, W. H. F. 1994. An empirical thermal history of the Earth's upper mantle. *Journal of Geophysical Research*, **99**, 13835–13850.

ADDISON, W. D., BRUMPTON, G. R. ET AL. 2005. Discovery of distal ejecta from the 1850 Ma Sudbury impact event. *Geology*, **33**, 193–196.

ALLEGRE, C. J. & ROUSSEAU, D. 1984. The growth of the continent through geological time studied by Nd isotope analysis of shales. *Earth and Planetary Science Letters*, **67**, 19–34.

ALLEGRE, C. J., STAUDACHER, T., SARDA, P. & KURZ, M. 1983. Constraints on evolution of Earth's mantle from rare gas systematics. *Nature*, **303**, 762–766.

ANBAR, A. D. & KNOLL, A. H. 2002. Proterozoic ocean chemistry and evolution: a bioinorganic bridge? *Science*, **297**, 1137–1142.

ANBAR, A. D., DUAN, Y. ET AL. 2007. A whiff of oxygen before the Great Oxidation Event? *Science*, **317**, 1903.

ANDERSON, D. L. 1994. Superplumes or supercontinents? *Geology*, **22**, 39–42.

ARMSTRONG, R. L. 1981. Radiogenic isotopes: the case for crustal recycling on a near-steady-state no-continental growth earth. *Royal Society of London, Philosophical Transactions, Series A*, **301**, 443–472.

ARMSTRONG, R. L. 1991. The persistent myth of crustal growth. *Australian Journal of Earth Sciences*, **38**, 613–630.

ARNDT, N. T. & GOLDSTEIN, S. L. 1989. An open boundary between lower continental crust and mantle: its role in crust formation and crustal recycling. *Tectonophysics*, **161**, 201–212.

ARNOLD, G. L., ANBAR, A. D., BARLING, J. & LYONS, T. W. 2004. Molybdenum isotope evidence for widespread anoxia in mid-Proterozoic oceans. *Science*, **304**, 87–90.

ASPLER, L. B. & CHIARENZELLI, J. R. 1998. Two Neoarchean supercontinents? Evidence from the Paleoproterozoic. *Sedimentary Geology*, **120**, 75–104.

BARLEY, M. E. & GROVES, D. I. 1992. Supercontinent cycles and the distribution of metal deposits through time. *Geology*, **20**, 291–294.

BARLEY, M. E., KRAPEZ, B., GROVES, D. I. & KERRICH, R. 1998. The Late Archaean bonanza: metallogenic and environmental consequences of the interaction between mantle plumes, lithospheric tectonics and global cyclicity. *Precambrian Research*, **91**, 65–90.

BARLEY, M. E., BEKKER, A. & KRAPEZ, B. 2005. Late Archean to Early Paleoproterozoic global tectonics, environmental change and the rise of atmospheric oxygen. *Earth and Planetary Science Letters*, **238**, 156–171.

BAU, M., ROMER, R. L., LUDERS, V. & BEUKES, N. J. 1999. Pb, O, and C isotopes in silicified Mooidraai dolomite (Transvaal Supergroup, South Africa): implications for the composition of Paleoproterozoic seawater and 'dating' the increase of oxygen in the Precambrian atmosphere. *Earth and Planetary Science Letters*, **174**, 43–57.

BEKKER, A., KAUFMAN, A. J., KARHU, J. A., BEUKES, N. J., SWART, Q. D., COETZEE, L. L. & ERIKSSON, K. A. 2001. Chemostratigraphy of the Paleoproterozoic Duitschland Formation, South Africa: implications for coupled climate change and carbon cycling. *American Journal of Science*, **301**, 261–285.

BEKKER, A., HOLLAND, H. D. ET AL. 2004. Dating the rise of atmospheric oxygen. *Nature*, **427**, 117–120.

BEKKER, A., KAUFMAN, A. J., KARHU, J. A. & ERIKSSON, K. A. 2005. Evidence for Paleoproterozoic cap carbonates in North America. *Precambrian Research*, **137**, 167–206.

BEKKER, A., KARHU, J. A. & KAUFMAN, A. J. 2006. Carbon isotope record for the onset of the Lomagundi carbon isotope excursion in the Great Lakes area, North America. *Precambrian Research*, **148**, 145–180.

BENGTSON, S., RASMUSSEN, B. & KRAPEZ, B. 2007. The Paleoproterozoic megascopic Stirling biota. *Paleobiology*, **33**, 351.

BENNETT, V. C. 2003. Compositional evolution of the mantle. *In*: CARLSON, R. W. (ed.) *The Mantle and Core*. Treatise on Geochemistry. Elsevier/Pergamon, Oxford, 493–519.

BERCOVICI, D. 2003. The generation of plate tectonics from mantle convection. *Earth and Planetary Science Letters*, **205**, 107–121.

BERCOVICI, D. & KARATO, S. 2003. Whole-mantle convection and the transition-zone water filter. *Nature*, **425**, 39–44.

BERNER, R. A. 2006. GEOCARBSULF: a combined model for Phanerozoic atmospheric O_2 and CO_2. *Geochimica et Cosmochimica Acta*, **70**, 5653–5664.

BERRY, A. J., DANYUSHEVSKY, L. V., ST. C. O'NEILL, H., NEWVILLE, M. & SUTTON, S. R. 2008. Oxidation state of iron in komatiitic melt inclusions indicates hot Archaean mantle. *Nature*, **455**, 960–963.

BETTS, P. G. & GILES, D. 2006. The 1800–1100 Ma tectonic evolution of Australia. *Precambrian Research*, **144**, 92–125.

BEUKES, N. J. & GUTZMER, J. 2008. Origin and paleoenvironmental significance of major iron formations at the Archean–Paleoproterozoic boundary. *Society of Economic Geologists Reviews*, **15**, 5–47.

BISPO-SANTOS, F., D'AGRELLA-FILHO, M. S. ET AL. 2008. Columbia revisited: paleomagnetic results from the 1790 Ma colider volcanics (SW Amazonian Craton, Brazil). *Precambrian Research*, **164**, 40–49.

BJERRUM, C. J. & CANFIELD, D. E. 2002. Ocean productivity before about 1.9 Gyr ago limited by phosphorus adsorption onto iron oxides. *Nature*, **417**, 159.

BLEEKER, W. 2003. The late Archean record: a puzzle in ca. 35 pieces. *Lithos*, **71**, 99–134.

BLEEKER, W. & ERNST, R. 2006. Short-lived mantle generated magmatic events and their dyke swarms: the key unlocking Earth's palaeogeographic record back to 2.6 Ga. *In*: HANSKI, E., MERTANEN, S., RÄMÖ, T. & VUOLLO, J. (eds) *Dyke Swarms–Time Markers of Crustal Evolution*. Taylor & Francis, London, 3–26.

BOUSSAU, B., BLANQUART, S., NECSULEA, A., LARTILLOT, N. & GOUY, M. 2008. Parallel adaptations to high temperatures in the Archaean eon. *Nature*, **456**, 942–945.

BRADLEY, D. C. 2008. Passive margins through Earth history. *Earth Science Reviews*, **91**, 1–26.

BRASIER, M. D., GREEN, O. R. ET AL. 2002. Questioning the evidence for Earth's oldest fossils. *Nature*, **416**, 76–81.

BRIMHALL, G. 1987. Preliminary fractionation patterns of ore metals through Earth history. *Chemical Geology*, **64**, 1–16.

BROCKS, J. J., LOGAN, G. A., BUICK, R. & SUMMONS, R. E. 1999. Archean molecular fossils and the early rise of eukaryotes. *Science*, **285**, 1033.

BROCKS, J. J., BUICK, R., LOGAN, G. A. & SUMMONS, R. E. 2003a. Composition and syngeneity of molecular fossils from the 2.78 to 2.45 billion-year-old Mount Bruce Supergroup, Pilbara Craton, Western Australia. *Geochimica et Cosmochimica Acta*, **67**, 4289–4319.

BROCKS, J. J., BUICK, R., SUMMONS, R. E. & LOGAN, G. A. 2003b. A reconstruction of Archean biological diversity based on molecular fossils from the 2.78 to 2.45 billion-year-old Mount Bruce Supergroup, Hamersley Basin, Western Australia. *Geochimica et Cosmochimica Acta*, **67**, 4321–4335.

BROCKS, J. J., LOVE, G. D., SUMMONS, R. E., KNOLL, A. H., LOGAN, G. A. & BOWDEN, S. A. 2005. Biomarker evidence for green and purple sulphur bacteria in a stratified Palaeoproterozoic sea. *Nature*, **437**, 866–870.

BROWN, M. 2007. Metamorphic conditions in orogenic belts: a record of secular change. *International Geology Review*, **49**, 193–234.

BUCHAN, K. L., MORTENSEN, J. K., CARD, K. D. & PERCIVAL, J. A. 1998. Paleomagnetism and U-Pb geochronology of diabase dyke swarms of Minto block, Superior Province, Quebec, Canada. *Canadian Journal of Earth Sciences*, **35**, 1054.

BUCHAN, K. L., GOUTIER, J., HAMILTON, M. A., ERNST, R. E. & MATTHEWS, W. A. 2007. Paleomagnetism, U-Pb geochronology, and geochemistry of Lac Esprit and other dyke swarms, James Bay area, Quebec, and implications for Paleoproterozoic deformation of the Superior Province. *Canadian Journal of Earth Sciences*, **44**, 643–664.

BUICK, I. S., UKEN, R., GIBSON, R. L. & WALLMACH, T. 1998. High-$\delta^{13}C$ Paleoproterozoic carbonates from the Transvaal Supergroup, South Africa. *Geology*, **26**, 875–878.

BUICK, R. 1992. The antiquity of oxygenic photosynthesis: evidence from stromatolites in sulphate-deficient Archaean lakes. *Science*, **255**, 74–77.

BUTLER, S. L., PELTIER, W. R. & COSTIN, S. O. 2005. Numerical models of the Earth's thermal history: effects of inner-core solidification and core potassium. *Physics of the Earth and Planetary Interiors*, **152**, 22–42.

BUTTERFIELD, N. J. 2000. Bangiomorpha pubescens n. gen., n. sp.: implications for the evolution of sex, multicellularity, and the Mesoproterozoic/Neoproterozoic radiation of eukaryotes. *Paleobiology*, **26**, 386–404.

CAMPBELL, I. H. & ALLEN, C. M. 2008. Formation of supercontinents linked to increases in atmospheric oxygen. *Nature Geoscience*, **1**, 554–558.

CANFIELD, D. E. 1998. A new model for Proterozoic ocean chemistry. *Nature a-z index*, **396**, 450–453.

CANFIELD, D. E. 2005. The early history of atmospheric oxygen: Homage to Robert M. Garrels. *Annual Reviews of Earth and Planetary Sciences*, **33**, 1–36.

CANFIELD, D. E., POULTON, S. W., KNOLL, A. H., NARBONNE, G. M., ROSS, G., GOLDBERG, T. & STRAUSS, H. 2008. Ferruginous conditions dominated later Neoproterozoic deep-water chemistry. *Science*, **321**, 949.

CARD, K. D. & POULSEN, K. H. 1998. Geology and mineral deposits of the Superior Province of the Canadian Shield. *In*: LUCAS, S. B. & ST. ONGE, M. R. (co-ordinates) *Geology of the Precambrian Superior and Grenville Provinces and Precambrian Fossils in North America*. Geological Survey of Canada, Geology of Canada, **7**, 13–194.

CATLING, D. C. & CLAIRE, M. W. 2005. How Earth's atmosphere evolved to an oxic state: a status report. *Earth and Planetary Science Letters*, **237**, 1–20.

CATLING, D. C., ZAHNLE, K. J. & MCKAY, C. P. 2001. Biogenic methane, hydrogen escape, and the irreversible oxidation of early Earth. *Science*, **293**, 839–843.

CAWOOD, P. A., KRÖNER, A. & PISAREVSKY, S. 2006. Precambrian plate tectonics: criteria and evidence. *GSA Today*, **16**, 4–11.

CHADWICK, B., CLAEYS, P. & SIMONSON, B. 2001. New evidence for a large Palaeoproterozoic impact: spherules in a dolomite layer in the Ketilidian orogen, South Greenland. *Journal of the Geological Society*, **158**, 331–340.

CHANDLER, F. W. & PARRISH, R. R. 1989. Age of the Richmond Gulf Group and implications for rifting in the Trans-Hudson Orogen, Canada. *Precambrian Research*, **44**, 277–288.

CHENEY, E. S. 1996. Sequence stratigraphy and plate tectonic significance of the Transvaal succession of southern Africa and its equivalent in Western Australia. *Precambrian Research*, **79**, 3–24.

CLOUD, P. E. 1968. Atmospheric and hydrospheric evolution on the primitive Earth. *Science*, **160**, 729–736.

COLLERSON, K. D. & KAMBER, B. S. 1999. Evolution of the continents and the atmosphere inferred from Th–U–Nb systematics of the depleted mantle. *Science*, **283**, 1519.

CONDIE, K. C. 1995. Episodic ages of greenstones: a key to mantle dynamics? *Geophysical Research Letters*, **22**, 2215–2218.

CONDIE, K. C. 1998. Episodic continental growth and supercontinents: a mantle avalanche connection? *Earth and Planetary Science Letters*, **163**, 97–108.

CONDIE, K. C. 2000. Episodic continental growth models: afterthoughts and extensions. *Tectonophysics*, **322**, 153–162.

CONDIE, K. C. 2004. Supercontinents and superplume events: distinguishing signals in the geologic record. *Physics of the Earth and Planetary Interiors*, **146**, 319–332.

CONDIE, K. C. & KRONER, A. 2008. When did plate tectonics begin? Evidence from the geologic record. *In*: CONDIE, K. C. & PEASE, V. (eds) *When Did Plate Tectonics Begin on Planet Earth?* Geological Society of America Special Paper, **440**, 281–294.

CONDIE, K. C. & PEASE, V. 2008. *When Did Plate Tectonics Begin on Planet Earth?* Geological Society of America Special Paper, **440**.

CONDIE, K. C., O'NEILL, C. & ASTER, R. C. 2009. Evidence and implications for a widespread magmatic shutdown for 250 My on Earth. *Earth and Planetary Science Letters*, **282**, 294–298.

CRUSE, T. & HARRIS, L. B. 1994. Ediacaran fossils from the Stirling Range Formation, Western Australia. *Precambrian Research*, **67**, 1–10.

DAVIES, G. F. 1980. Thermal histories of convective Earth models and constraints on radiogenic heat production in the Earth. *Journal of Geophysical Research*, **85**, 2517–2530.

DAVIES, G. F. 1988. Ocean bathymetry and mantle convection 1. Large-scale flow and hotspots. *Journal of Geophysical Research*, **93**, 10451–10466.

DAVIES, G. F. 1995. Penetration of plates and plumes through the mantle transition zone. *Earth and Planetary Science Letters*, **133**, 507–516.

DAVIES, G. F. 2007. Mantle regulation of core cooling: a geodynamo without core radioactivity? *Physics of the Earth and Planetary Interiors*, **160**, 215–229.

DE KOCK, M. O., EVANS, D. A. D. & BEUKES, N. J. 2009. Validating the existence of Vaalbara in the late Neoarchean. *Precambrian Research*, in press.

DEWEY, J. F. & WINDLEY, B. F. 1981. Growth and differentiation of the continental crust. *Philosophical Transactions of the Royal Society of London. Series A, Mathematical and Physical Sciences*, **301**, 189–206.

DEXTER-DYER, B., KRETZSCHMAR, M. & KRUMBEIN, W. E. 1984. Possible microbial pathways in the formation of Precambrian ore deposits. *Journal of the Geological Society*, **141**, 251–262.

DUTKIEWICZ, A., VOLK, H., GEORGE, S. C., RIDLEY, J. & BUICK, R. 2006. Biomarkers from Huronian oil-bearing fluid inclusions: an uncontaminated record of life before the Great Oxidation Event. *Geology*, **34**, 437–440.

DUTKIEWICZ, A., GEORGE, S. C., MOSSMAN, D. J., RIDLEY, J. & VOLK, H. 2007. Oil and its biomarkers associated with the Palaeoproterozoic Oklo natural fission reactors, Gabon. *Chemical Geology*, **244**, 130–154.

EARTH IMPACT DATABASE 2009. http://www.unb.ca/passc/ImpactDatabase/. Accessed, 16 Feb 2009.

EIGENBRODE, J. L., FREEMAN, K. H. & SUMMONS, R. E. 2008. Methylhopane biomarker hydrocarbons in Hamersley Province sediments provide evidence for Neoarchean aerobiosis. *Earth and Planetary Science Letters*, **273**, 323–331.

ERIKSSON, K. A. & SIMPSON, E. L. 1998. Controls on spatial and temporal distribution of Precambrian eolianites. *Sedimentary Geology*, **120**, 275–294.

ERIKSSON, K. A. & SIMPSON, E. L. 2000. Quantifying the oldest tidal record: The 3.2 Ga Moodies Group, Barberton Greenstone Belt, South Africa. *Geology*, **28**, 831–834.

ERIKSSON, K. A., SIMPSON, E. L., MASTER, S. & HENRY, G. 2005a. Neoarchaean (c. 2.58 Ga) halite casts: implications for palaeoceanic chemistry.

Journal of the Geological Society of London, **162**, 789–799.

ERIKSSON, P. G., CATUNEANU, O., SARKAR, S. & TIRSGAARD, H. 2005b. Patterns of sedimentation in the Precambrian. *Sedimentary Geology*, **176**, 17–42.

ERIKSSON, P. G., MAZUMDER, R., CATUNEANU, O., BUMBY, A. J. & ILONDO, B. O. 2006. Precambrian continental freeboard and geological evolution: a time perspective. *Earth-Science Reviews*, **79**, 165–204.

ERNST, R. E. & BUCHAN, K. L. 2001. Large mafic magmatic events through time and links to mantle-plume heads. *In*: ERNST, R. E. & BUCHAN, K. L. (eds) *Mantle Plumes: Their Identification Through Time*. Geological Society of America Special Paper **352**, Boulder, Colorado, 483–575.

ERNST, R. E. & BUCHAN, K. L. 2003. Recognizing mantle plumes in the geological record. *Annual Review of Earth and Planetary Sciences*, **31**, 469–523.

EVANS, D. A. D. 2003. A fundamental Precambrian-Phanerozoic shift in earth's glacial style? *Tectonophysics*, **375**, 353–385.

EVANS, D. A. D. 2006. Proterozoic low orbital obliquity and axial-dipolar geomagnetic field from evaporite palaeolatitudes. *Nature*, **444**, 51–55.

EVANS, D. A. D. 2009. The palaeomagnetically viable, long-lived and all-inclusive Rodinia supercontinent reconstruction. *In*: MURPHY, J. B., KEPPIE, J. D. & HYNES, A. (eds) *Ancient Orogens and Modern Analogues*. Geological Society, London, Special Publications, **327**, in press.

EVANS, D. A., BEUKES, N. J. & KIRSCHVINK, J. L. 1997. Low-latitude glaciation in the Palaeoproterozoic era. *Nature*, **386**, 262–266.

EVANS, D. A. D. & PISAREVSKY, S. A. 2008. Plate tectonics on early Earth? Weighing the paleomagnetic evidence. *In*: CONDIE, K. C. & PEASE, V. (eds) *When Did Plate Tectonics Begin on Planet Earth?* Geological Society of America, Special Paper, **440**, 249–263.

FARQUHAR, J., BAO, H. & THIEMENS, M. 2000. Atmospheric influence of Earth's earliest sulfur cycle. *Science*, **289**, 756–758.

FARQUHAR, J., PETERS, M., JOHNSTON, D. T., STRAUSS, H., MASTERSON, A., WIECHERT, U. & KAUFMAN, A. J. 2007. Isotopic evidence for Mesoarchaean anoxia and changing atmospheric sulphur chemistry. *Nature*, **449**, 706–709.

FEDONKIN, M. A. & YOCHELSON, E. L. 2002. Middle Proterozoic (1.5 Ga) Horodyskia moniliformis Yochelson and Fedonkin, the oldest known tissue-grade colonial eucaryote. *Smithsonian Contributions to Paleobiology*, **94**, 1–29.

FISCHER, A. G. 1984. The two Phanerozoic supercycles. *In*: BERGGREN, W. A. & VAN COUVERING, J. A. (eds) *Catastrophes and Earth History*. Princeton University Press, Princeton, 129–148.

FLAMENT, N., COLTICE, N. & REY, P. F. 2008. A case for late-Archaean continental emergence from thermal evolution models and hypsometry. *Earth and Planetary Science Letters*, **275**, 326–336.

FRALICK, P., DAVIS, D. W. & KISSIN, S. A. 2002. The age of the Gunflint Formation, Ontario, Canada: single zircon U-Pb age determinations from reworked volcanic ash. *Canadian Journal of Earth Sciences*, **39**, 1085–1091.

GARDE, A. A., CHADWICK, B., GROCOTT, J., HAMILTON, M. A., MCCAFFREY, K. J. W. & SWAGER, C. P. 2002. Mid-crustal partitioning and attachment during oblique convergence in an arc system, Palaeoproterozoic Ketilidian orogen, southern Greenland. *Journal of the Geological Society*, **159**, 247–261.

GASTIL, R. G. 1960. The distribution of mineral dates in time and space. *American Journal of Science*, **258**, 1.

GAUTHIER-LAFAYE, F. & WEBER, F. 2003. Natural nuclear fission reactors: time constraints for occurrence, and their relation to uranium and manganese deposits and to the evolution of the atmosphere. *Precambrian Research*, **120**, 81–100.

GEOFFREY, F. D. 1993. Cooling the core and mantle by plume and plate flows. *Geophysical Journal International*, **115**, 132–146.

GEORGE, S. C., DUTKIEWICZ, A., HERBERT, V., RIDLEY, J. & DAVID, J. 2008. Eukaryote-derived steranes in Precambrian oils and rocks: fact or fiction? *Nature*, **395**, 885–888.

GLIKSON, A. Y. 2005. Geochemical signatures of Archean to Early Proterozoic Maria-scale oceanic impact basins. *Geology*, **33**, 125–128.

GONNERMANN, H. M., MANGA, M. & JELLINEK, A. M. 2002. Dynamics and longevity of an initially stratified mantle. *Geophysical Research Letters*, **29**, 33–31.

GOWER, C. F., RYAN, A. B. & RIVERS, T. 1990. Mid-Proterozoic Laurentia-Baltica: an overview of its geological evoluation and summary of contributions made by this volume. *In*: GOWER, C. F., RYAN, A. B. & RIVERS, T. (eds) *Mid Proterozoic Laurentia-Baltica*. GAC Special Paper **38**. Geological Association of Canada, 1–22.

GRAHAM, D. W. 2002. Noble gas isotope geochemistry of mid-ocean ridge and ocean island basalts: characterization of mantle source reservoirs. *Reviews in Mineralogy and Geochemistry*, **47**, 247.

GREY, K., WILLIAMS, I. R., MARTIN, D. M. B., FEDONKIN, M. A., GEHLING, J. G., RUNNEGAR, B. N. & YOCHELSON, E. L. 2002. New occurrences of 'strings of beads' in the Bangemall Supergroup: a potential biostratigraphic marker horizon. *Annual Report of the Geological Survey of West Australia*, 69–73.

GRIEVE, R. & THERRIAULT, A. 2000. Vredefort, Sudbury, Chicxulub: Three of a Kind? *Annual Review of Earth and Planetary Sciences*, **28**, 305–338.

GROSS, G. A. 1965. Geology of iron deposits in Canada. 1, General geology and evaluation of iron deposits. Geological Survey of Canada.

GROSS, G. A. 1983. Tectonic systems and the deposition of iron formation. *Precambrian Research*, **20**, 171–187.

GROTZINGER, J. P. 1989. Facies and evolution of Precambrian carbonate depositional systems: emergence of the modern platform archetype. *In*: CREVELLO, P. D., WILSON, J. L., SARG, J. F. & READ, J. F. (eds) *Controls on Carbonate Platform and Basin Development*. Society of Economic Paleontologists and Mineralogists, **44**, Tulsa, 79–106.

GROTZINGER, J. P. & JAMES, N. P. 2000. Precambrian carbonates: evolution of understanding. In: GROTZINGER, J. P. & JAMES, N. P. (eds) Carbonate Sedimentation and Diagenesis in the Evolving Precambrian World. Society for Sedimentary Geology Special Publications, 67, 3–22.

GROTZINGER, J. P. & KASTING, J. F. 1993. New constraints on Precambrian ocean composition. Journal of Geology, 101, 235–243.

GROTZINGER, J. P. & MCCORMICK, D. S. 1988. Flexure of the Early Proterozoic lithosphere and the evolution of Kilohigok Basin (1.9 Ga), northwest Canadian Shield. In: KLEINSPEHN, K. L. & PAOLA, C. (eds) New Perspectives in Basin Analysis. Springer-Verlag, New York, 405–430.

GROVE, T. L. & PARMAN, S. W. 2004. Thermal evolution of the Earth as recorded by komatiites. Earth and Planetary Science Letters, 219, 173–187.

GROVES, D. I., VIELREICHER, R. M., GOLDFARB, R. J. & CONDIE, K. C. 2005. Controls on the heterogeneous distribution of mineral deposits through time. In: MCDONALD, I., BOYCE, A. J., BUTLER, I. B., HERRINGTON, R. J. & POLYA, D. A. (eds) Mineral Deposits and Earth Evolution. Geological Society, London, Special Publications, 248, 71–101.

GUBBINS, D., MASTERS, G. & NIMMO, F. 2008. A thermochemical boundary layer at the base of Earth's outer core and independent estimate of core heat flux. Geophysical Journal International, 174, 1007–1018.

GURNIS, M. 1988. Large-scale mantle convection and the aggregation and dispersal of supercontinents. Nature, 332, 695–699.

HALE, C. J. 1987. Palaeomagnetic data suggest link between the Archaean-Proterozoic boundary and inner-core nucleation. Nature, 329, 233–237.

HALLS, H. C., DAVIS, D. W., STOTT, G. M., ERNST, R. E. & HAMILTON, M. A. 2008. The Paleoproterozoic Marathon Large Igneous Province: new evidence for a 2.1 Ga long-lived mantle plume event along the southern margin of the North American Superior Province. Precambrian Research, 162, 327–353.

HAN, T. M. & RUNNEGAR, B. 1992. Megascopic eukaryotic algae from the 2.1-billion-year-old Negaunee iron-formation, Michigan. Science, 257, 232–235.

HANNAH, J. L., BEKKER, A., STEIN, H. J., MARKEY, R. J. & HOLLAND, H. D. 2004. Primitive Os and 2316 Ma age for marine shale: implications for Paleoproterozoic glacial events and the rise of atmospheric oxygen. Earth and Planetary Science Letters, 225, 43–52.

HARRISON, D. & BALLENTINE, C. J. 2005. Noble gas models of mantle convection and mass reservoir transfer. In: VAN DER HILST, R. D., BASS, J., MATAS, J. & TRAMPERT, J. (eds) Earth's Deep Mantle: Structure, Composition and Evolution. Geophysical Monograph, 160. American Geophysical Union, 9–26.

HAWKESWORTH, C. J. & KEMP, A. I. S. 2006a. Evolution of the continental crust. Nature, 443, 811–817.

HAWKESWORTH, C. J. & KEMP, A. I. S. 2006b. Using hafnium and oxygen isotopes in zircons to unravel the record of crustal evolution. Chemical Geology, 226, 144–162.

HAWKESWORTH, C., CAWOOD, P., KEMP, T., STOREY, C. & DHUIME, B. 2009. Geochemistry: A Matter of Preservation. Science, 323, 49.

HAZEN, R. M., PAPINEAU, D., BLEEKER, W. ET AL. 2008. Mineral evolution. American Mineralogist, 93, 1693.

HEAMAN, L. M. 1997. Global mafic magmatism at 2.45 Ga: remnants of an ancient large igneous province. Geology, 25, 299–302.

HERNLUND, J. W., THOMAS, C. & TACKLEY, P. J. 2005. A doubling of the post-perovskite phase boundary and structure of the Earth's lowermost mantle. Nature, 434, 882–886.

HIETALA, S., MOILANEN, J. & KIVELANTIE, B. 2004. Keurusselkä -a new impact structure in central Finland. In: 35th Lunar and Planetary Science Conference Abstract, March 15–19. League City, Texas.

HILTON, D. R. & PORCELLI, D. 2003. Noble gases as mantle tracers. Treatise on Geochemistry, 2.

HIROSE, K., FEI, Y., MA, Y. & MAO, H. K. 1999. The fate of subducted basaltic crust in the Earth's lower mantle. Nature, 397, 53–56.

HIROSE, K. & LAY, T. 2008. Discovery of post-perovskite and new views on the core-mantle boundary region. Elements, 4, 183–189.

HIRTH, G. & KOHLSTEDT, D. L. 1996. Water in the oceanic upper mantle: implications for rheology, melt extraction and the evolution of the lithosphere. Earth and Planetary Science Letters, 144, 93–108.

HOFFMAN, P. F. 1988. United plates of America, the birth of a craton: early proterozoic assembly and growth of laurentia. Annual Reviews of Earth and Planetary Sciences, 16, 543–603.

HOFFMAN, P. F. 1989. Speculations on Laurentia's first gigayear (2.0 to 1.0 Ga). Geology, 17, 135–138.

HOFFMAN, P. F. 1997. Tectonic genealogy of North America. In: VAN DER PLUIJM, B. A. & MARSHAK, S. (eds) Earth Structure: An Introduction to Structural Geology and Tectonics. New York, McGraw-Hill, 459–464.

HOFFMAN, P. F. & SCHRAG, D. P. 2002. The snowball Earth hypothesis: testing the limits of global change. Terra Nova, 14, 129–155.

HOFMANN, A. W. 2003. Sampling mantle heterogeneity through oceanic basalts: isotopes and trace elements. Treatise on Geochemistry, 2, 61–101.

HOFMANN, H. J. 1976. Precambrian microflora, Belcher Islands, Canada: significance and systematics. Journal of Paleontology, 50, 1040–1073.

HOLLAND, H. D. 2002. Volcanic gases, black smokers, and the Great Oxidation Event. Geochimica et Cosmochimica Acta, 66, 3811–3826.

HOLLAND, H. D. 2006. The oxygenation of the atmosphere and oceans. Philosophical Transactions of the Royal Society B: Biological Sciences, 361, 903–915.

HOU, G., SANTOSH, M., QIAN, X., LISTER, G. S. & LI, J. 2008a. Configuration of the Late Paleoproterozoic supercontinent Columbia: insights from radiating mafic dyke swarms. Gondwana Research, 14, 395–409.

HOU, G., SANTOSH, M., QIAN, X., LISTER, G. S. & LI, J. 2008b. Tectonic constraints on, 1.3–1.2 Ga final breakup of Columbia supercontinent from a giant radiating dyke swarm. Gondwana Research, 14, 561–566.

HURLEY, P. M. & RAND, J. R. 1969. Pre-drift continental nuclei. *Science*, **164**, 1229–1242.

HUSTON, D. L. & LOGAN, G. A. 2004. Barite, BIFs and bugs: evidence for the evolution of the Earth's early hydrosphere. *Earth and Planetary Science Letters*, **220**, 41–55.

HYNES, A. 2001. Freeboard revisited: continental growth, crustal thickness change and Earth's thermal efficiency. *Earth and Planetary Science Letters*, **185**, 161–172.

ISLEY, A. E. & ABBOTT, D. H. 1999 Plume-related mafic volcanism and the deposition of banded iron formation. *Journal of Geophysical Research*, **104**, 15461–15477.

JACOBS, J. A. 1953. The Earth's Inner Core. *Nature*, **172**, 297–298.

JACOBSEN, S. B. & WASSERBURG, G. J. 1979. Mean age of mantle and crustal reservoirs. *Journal of Geophysical Research*, **84**, 7411–7428.

JAUPART, C., LABROSSE, S., MARESCHAL, J. C. & GERALD, S. 2007. *Temperatures, Heat and Energy in the Mantle of the Earth*. Treatise on Geophysics. Elsevier, Amsterdam, 253–303.

JEANLOZ, R. 1990. The nature of the Earth's core. *Annual Review of Earth and Planetary Sciences*, **18**, 357–386.

KAH, L. C. & RIDING, R. 2007. Mesoproterozoic carbon dioxide levels inferred from calcified cyanobacteria. *Geology*, **35**, 799.

KAH, L. C., LYONS, T. W. & FRANK, T. D. 2004. Low marine sulphate and protracted oxygenation of the Proterozoic biosphere. *Nature*, **431**, 834–838.

KARHU, J. A. & HOLLAND, H. D. 1996. Carbon isotopes and the rise of atmospheric oxygen. *Geology*, **24**, 867–870.

KASTING, J. F. 2005. Methane and climate during the Precambrian era. *Precambrian Research*, **137**, 119–129.

KASTING, J. F., EGGLER, D. H. & RAEBURN, S. P. 1993. Mantle redox evolution and the oxidation state of the Archean atmosphere. *Journal of Geology*, **101**, 245–257.

KAUFMAN, A. J., JOHNSTON, D. T. ET AL. 2007. Late Archean biospheric oxygenation and atmospheric evolution. *Science*, **317**, 1900.

KELLOGG, L. H., HAGER, B. H. & VAN DER HILST, R. D. 1999. Compositional stratification in the deep mantle. *Science*, **283**, 1881–1884.

KEMP, A. I. S., HAWKESWORTH, C. J., PATERSON, B. A. & KINNY, P. D. 2006. Episodic growth of the Gondwana supercontinent from hafnium and oxygen isotopes in zircon. *Nature*, **439**, 580–583.

KEMP, A. I. S., HAWKESWORTH, C. J. ET AL. 2007. Magmatic and crustal differentiation history of granitic rocks from Hf-O isotopes in zircon. *Science*, **315**, 980.

KIRSCHVINK, J. L. 1992. Late Proterozoic low-latitude global glaciation: the Snowball Earth. *In*: SCHOPF, W. S. & KLEIN, C. (eds) *The Proterozoic Biosphere*. Cambridge University Press, Cambridge, 51–52.

KIRSCHVINK, J. L. & KOPP, R. E. 2008. Palaeoproterozoic ice houses and the evolution of oxygen-mediating enzymes: the case for a late origin of photosystem II. *Philosophical Transactions of the Royal Society B: Biological Sciences*, **363**, 2755–2765.

KIRSCHVINK, J. L., GAIDOS, E. J., BERTANI, L. E., BEUKES, N. J., GUTZMER, J., MAEPA, L. N. & STEINBERGER, R. E. 2000. Paleoproterozoic snowball Earth: extreme climatic and geochemical global change and its biological consequences. *Proceedings of the National Academy of Sciences*, **97**, 1400–1405.

KLEIN, C. 2005. Some Precambrian banded iron-formations (BIFs) from around the world: Their age, geologic setting, mineralogy, metamorphism, geochemistry, and origins. *American Mineralogist*, **90**, 1473–1499.

KLEIN, C. & BEUKES, N. J. 1992. Time distribution, stratigraphy, and sedimentologic setting, and geochemistry of Precambrian iron-formations. *In*: SCHOPF, J. W. & KLEIN, C. (eds) *The Proterozoic Biosphere: A Multidisciplinary Study*. Cambridge University Press, Cambridge, 139–146.

KNAUTH, L. P. 2005. Temperature and salinity history of the Precambrian ocean: implications for the course of microbial evolution. *Palaeogeography Palaeoclimatology Palaeoecology*, **219**, 53–69.

KNAUTH, L. P. & LOWE, D. R. 2003. High Archean climatic temperature inferred from oxygen isotope geochemistry of cherts in the 3.5 Ga Swaziland Supergroup, South Africa. *Bulletin of the Geological Society of America*, **115**, 566–580.

KNOLL, A. H., JAVAUX, E. J., HEWITT, D. & COHEN, P. 2006. Eukaryotic organisms in Proterozoic oceans. *Philosophical Transactions of the Royal Society B: Biological Sciences*, **361**, 1023–1038.

KOMIYA, T. 2004. Material circulation model including chemical differentiation within the mantle and secular variation of temperature and composition of the mantle. *Physics of the Earth and Planetary Interiors*, **146**, 333–367.

KONHAUSER, K. O., LALONDE, S. V., AMSKOLD, L. & HOLLAND, H. D. 2007. Was there really an Archean phosphate crisis? *Science*, **315**, 1234.

KOPP, R. E., KIRSCHVINK, J. L., HILBURN, I. A. & NASH, C. Z. 2005. The Paleoproterozoic snowball Earth: a climate disaster triggered by the evolution of oxygenic photosynthesis. *Proceedings of the National Academy of Sciences*, **102**, 11131–11136.

KORENAGA, J. 2006. Archean geodynamics and the thermal evolution of Earth. *In*: BENN, K., MARESCHAL, J.-C. & CONDIE, K. C. (eds) *Archean Geodynamics and Environments*. Geophysical Monograph 164, American Geophysical Union, Washington, D.C, 7–32.

KORENAGA, J. 2008a. Comment on 'Intermittent Plate Tectonics?' *Science*, **320**, 1291a.

KORENAGA, J. 2008b. Invited review Plate tectonics, flood basalts and the evolution of Earth's oceans. *Terra Nova*, **20**, 419–439.

KORENAGA, J. 2008c. Urey ratio and the structure and evolution of Earth's mantle. *Reviews of Geophysics*, **46**; doi: 10.1029/2007RG000241.

KRAPEZ, B. 1999. Stratigraphic record of an Atlantic-type global tectonic cycle in the Palaeoproterozoic Ashburton Province of Western Australia. *Australian Journal of Earth Sciences*, **46**, 71–87.

KUMP, L. R. & BARLEY, M. E. 2007. Increased subaerial volcanism and the rise of atmospheric oxygen 2.5 billion years ago. *Nature*, **448**, 1033–1036.

KUMP, L. R., KASTING, J. F. & BARLEY, M. E. 2001. Rise of atmospheric oxygen and the 'upside-down' Archean mantle. *Geochemistry Geophysics Geosystems*, **2**, U1–U10.

LABRENZ, M., DRUSCHEL, G. K., THOMSEN-EBERT, T. ET AL. 2000. Formation of sphalerite (ZnS) deposits in natural biofilms of sulfate-reducing bacteria. *Science*, **290**, 1744–1747.

LABROSSE, S. & JAUPART, C. 2007. Thermal evolution of the Earth: secular changes and fluctuations of plate characteristics. *Earth and Planetary Science Letters*, **260**, 465–481.

LABROSSE, S., POIRIER, J. P. & LE MOUËL, J. L. 2001. The age of the inner core. *Earth and Planetary Science Letters*, **190**, 111–123.

LABROSSE, S., HERNLUND, J. W. & COLTICE, N. 2007. A crystallizing dense magma ocean at the base of the Earth's mantle. *Nature*, **450**, 866–869.

LEDRU, P., JOHAN, V., MILESI, J. P. & TEGYEY, M. 1994. Markers of the last stages of the Palaeoproterozoic collision: evidence for a 2 Ga continent involving circum-South Atlantic provinces. *Precambrian Research*, **69**, 169–191.

LEPOT, K., BENZERARA, K., BROWN, G. E. & PHILIPPOT, P. 2008. Microbially influenced formation of, 2,724-million-year-old stromatolites. *Nature Geoscience*, **1**, 118–121.

LI, Z. X. 2000. Palaeomagnetic evidence for unification of the North and West Australian cratons by ca.1.7 Ga: new results from the Kimberley Basin of northwestern Australia. *Geophysical Journal International*, **142**, 173–180.

LI, Z. X., BOGDANOVA, S. V. ET AL. 2008. Assembly, configuration, and break-up history of Rodinia: A synthesis. *Precambrian Research*, **160**, 179–210.

LI, Z. X., BOGDANOVA, S. V. ET AL. 2009. How not to assemble a Precambrian supercontinent – Reply to comment by J. D. A. Piper on 'Assembly, configuration, and break-up history of Rodinia: A synthesis' by LI, Z. X., BOGDANOVA, S. V., ET AL. Precambrian Research, **160** (1–2), 179–210 (2008). *Precambrian Research* in press.

LIANG, M. C., HARTMAN, H., KOPP, R. E., KIRSCHVINK, J. L. & YUNG, Y. L. 2006. Production of hydrogen peroxide in the atmosphere of a snowball earth and the origin of oxygenic photosynthesis. *Proceedings of the National Academy of Sciences*, **103**, 18896.

LINDSAY, J. F. & BRASIER, M. D. 2002. Did global tectonics drive early biosphere evolution? Carbon isotope record from 2.6 to 1.9 Ga carbonates of Western Australian basins. *Precambrian Research*, **114**, 1–34.

LONG, D. G. F. 2006. Architecture of pre-vegetation sandy-braided perennial and ephemeral river deposits in the Paleoproterozoic Athabasca Group, northern Saskatchewan, Canada as indicators of Precambrian fluvial style. *Sedimentary Geology*, **190**, 71–95.

LOWE, D. R. & TICE, M. M. 2007. Tectonic controls on atmospheric, climatic, and biological evolution 3.5–2.4 Ga. *Precambrian Research*, **158**, 177–197.

LOWMAN, J. P. & JARVIS, G. T. 1999. Effects of mantle heat source distribution on supercontinent stability. *Journal of Geophysical Research*, **104**, 12,733–712,746.

MACDONALD, F. A., BUNTING, J. A. & CINA, S. E. 2003. Yarrabubba-a large, deeply eroded impact structure in the Yilgarn Craton, Western Australia. *Earth and Planetary Science Letters*, **213**, 235–247.

MACOUIN, M., VALET, J. P., BESSE, J., BUCHAN, K., ERNST, R., LEGOFF, M. & SCHARER, U. 2003. Low paleointensities recorded in 1 to 2.4 Ga Proterozoic dykes, Superior Province, Canada. *Earth and Planetary Science Letters*, **213**, 79–95.

MACOUIN, M., VALET, J. P. & BESSE, J. 2004. Long-term evolution of the geomagnetic dipole moment. *Physics of The Earth and Planetary Interiors*, **147**, 239–246.

MARSHALL, H., WALKER, J. & KUHN, W. 1988. Long-term climate change and the geochemical cycle of carbon. *Journal of Geophysical Research*, **93**.

MARTIN, D. M. B. 2004. Depositional environment and taphonomy of the 'strings of beads': mesoproterozoic multicellular fossils in the Bangemall Supergroup, Western Australia. *Australian Journal of Earth Sciences*, **51**, 555–561.

MATZ, M. V., FRANK, T. M., MARSHALL, N. J., WIDDER, E. A. & JOHNSEN, S. 2008. Giant deep-sea protist produces bilaterian-like traces. *Current Biology*, **18**, 1849–1854.

MAZUMDER, R. 2004. Implications of lunar orbital periodicity from, the Chaibasa tidal rhythmite (India) of late Paleoproterozoic age. *Geology*, **32**, 841–844.

MAZUMDER, R. & ARIMA, M. 2005. Tidal rhythmites and their implications. *Earth Science Reviews*, **69**, 79–95.

MCCULLOCH, M. T. & BENNETT, V. C. 1994. Progressive growth of the Earth's continental crust and depleted mantle: geochemical constraints. *Geochimica et Cosmochimica Acta*, **58**, 4717–4738.

MCLENNAN, S. M. 1989. Rare earth elements in sedimentary rocks; influence of provenance and sedimentary processes. *Reviews in Mineralogy and Geochemistry*, **21**, 169–200.

MEERT, J. G. 2002. Paleomagnetic evidence for a Paleo-Mesoproterozoic supercontinent Columbia. *Gondwana Research*, **5**, 207–215.

MEERT, J. G. & TORSVIK, T. H. 2003. The making and unmaking of a supercontinent: Rodinia revisited. *Tectonophysics*, **375**, 261–288.

MEERT, J. G. & TORSVIK, T. H. 2004. Paleomagnetic constraints on Neoproterozoic 'Snowball Earth' continental reconstructions. *In*: JENKINS, G. S., MCMENAMIN, M. A. S., MCKAY, C. P. & SOHL, L. (eds) *The Extreme Proterozoic: Geology, Geochemistry, and Climate*. AGU Geophysical Monograph Series, **146**, 5–11.

MELEZHIK, V. A. 2006. Multiple causes of Earth's earliest global glaciation. *Terra Nova*, **18**, 130–137.

MELEZHIK, V. A., FALLICK, A. E., MEDVEDEV, P. V. & MAKARIKHIN, V. V. 1999. Extreme $^{13}C_{carb}$ enrichment in ca. 2.0 Ga magnesite-stromatolite-dolomite-red beds' association in a global context: a case for the world-wide signal enhanced by a local environment. *Earth-Science Reviews*, **48**, 71–120.

MELEZHIK, V. A., FALLICK, A. E., HANSKI, E. J., KUMP, L. R., LEPLAND, A., PRAVE, A. R. & STRAUSS, H. 2005a. Emergence of an aerobic biosphere during the Archean-Proterozoic transition: challenges of future research. *GSA Today*, **15**, 4–11.

MELEZHIK, V. A., FALLICK, A. E., RYCHANCHIK, D. V. & KUZNETSOV, A. B. 2005b. Palaeoproterozoic evaporites in Fennoscandia: implications for seawater sulphate, the rise of atmospheric oxygen and local amplification of the delta C-13 excursion. *Terra Nova*, **17**, 141–148.

MELEZHIK, V. A., HUHMA, H., CONDON, D. J., FALLICK, A. E. & WHITEHOUSE, M. J. 2007. Temporal constraints on the Paleoproterozoic Lomagundi-Jatuli carbon isotopic event. *Geology*, **35**, 655.

MURAKAMI, M., HIROSE, K., KAWAMURA, K., SATA, N. & OHISHI, Y. 2004. Post-perovskite phase transition in $MgSiO_3$. *Science*, **304**, 855–858.

NANCE, R. D., WORSLEY, T. R. & MOODY, J. B. 1986. Post-Archean biogeochemical cycles and long-term episodicity in tectonic processes. *Geology*, **14**, 514–518.

NANCE, R., WORSLEY, T. & MOODY, J. 1988. The supercontinent cycle. *Scientific American*, **259**, 72–79.

NIMMO, F., PRICE, G. D., BRODHOLT, J. & GUBBINS, D. 2004. The influence of potassium on core and geodynamo evolution. *Geophysical Journal International*, **156**, 363–376.

NOFFKE, N., HAZEN, R. & NHLEKO, N. 2003. Earth's earliest microbial mats in a siliciclastic marine environment (2.9 Ga Mozaan Group, South Africa). *Geology*, **31**, 673–676.

NOFFKE, N., BEUKES, N., GUTZMER, J. & HAZEN, R. 2006a. Spatial and temporal distribution of microbially induced sedimentary structures: a case study from siliciclastic storm deposits of the 2.9 Ga Witwatersrand Supergroup, South Africa. *Precambrian Research*, **146**, 35–44.

NOFFKE, N., ERIKSSON, K. A., HAZEN, R. M. & SIMPSON, E. L. 2006b. A new window into Early Archean life: microbial mats in Earth's oldest siliciclastic tidal deposits (3.2 Ga Moodies Group, South Africa). *Geology*, **34**, 253–256.

NOFFKE, N., BEUKES, N., BOWER, D., HAZEN, R. M. & SWIFT, D. J. P. 2008. An actualistic perspective into Archean worlds-(cyano-) bacterially induced sedimentary structures in the siliciclastic Nhlazatse Section, 2.9 Ga Pongola Supergroup, South Africa. *Geobiology*, **6**, 5–20.

NOMADE, S., CHEN, Y. ET AL. 2003. The Guiana and the West African shield Palaeoproterozoic grouping: new palaeomagnetic data for French Guiana and the Ivory Coast. *Geophysical Journal International*, **154**, 677–694.

O'NEILL, C., LENARDIC, A., JELLINEK, A. M. & MORESI, L. 2009. Influence of supercontinents on deep mantle flow. *Gondwana Research*, **15**, 276–287.

O'NIONS, R. K. & OXBURGH, E. R. 1983. Heat and helium in the Earth. *Nature*, **306**, 429–431.

O'NIONS, R. K., EVENSEN, N. M. & HAMILTON, P. J. 1979. Geochemical modeling of mantle differentiation and crustal growth. *Journal of Geophysical Research*, **84**, 6091–6101.

OGAWA, M. 2008. Mantle convection: A review. *Fluid Dynamics Research*, **40**, 379–398.

OHMOTO, H., WATANABE, Y., IKEMI, H., POULSON, S. R. & TAYLOR, B. E. 2006. Sulphur isotope evidence for an oxic Archaean atmosphere. *Nature*, **442**, 908.

ONO, S., EIGENBRODE, J. L., PAVLOV, A. A., KHARECHA, P., RUMBLE, D., KASTING, J. F. & FREEMAN, K. H. 2003. New insights into Archean sulfur cycle from mass-independent sulfur isotope records from the Hamersley Basin, Australia. *Earth and Planetary Science Letters*, **213**, 15–30.

ONO, S., BEUKES, N. J. & RUMBLE, D. 2009a. Origin of two distinct multiple-sulfur isotope compositions of pyrite in the 2.5 Ga Klein Naute Formation, Griqualand West Basin, South Africa. *Precambrian Research*, **169**, 48–57.

ONO, S., KAUFMAN, A. J., FARQUHAR, J., SUMNER, D. Y. & BEUKES, N. J. 2009b. Lithofacies control on multiple-sulfur isotope records and Neoarchean sulfur cycles. *Precambrian Research*, **169**, 58–67.

ONSTOTT, T. C. & HARGRAVES, R. B. 1981. Proterozoic transcurrent tectonics: palaeomagnetic evidence from Venezuela and Africa. *Nature*, **289**, 131–136.

ONSTOTT, T. C., HARGRAVES, R. B., YORK, D. & HALL, C. 1984. Constraints on the motions of South American and African shields during the Proterozoic; I, 40 Ar/39 Ar and paleomagnetic correlations between Venezuela and Liberia. *Bulletin of the Geological Society of America*, **95**, 1045–1054.

PAPINEAU, D., MOJZSIS, S. J. & SCHMITT, A. K. 2007. Multiple sulfur isotopes from Paleoproterozoic Huronian interglacial sediments and the rise of atmospheric oxygen. *Earth and Planetary Science Letters*, **255**, 188–212.

PARMAN, S. W. 2007. Helium isotopic evidence for episodic mantle melting and crustal growth. *Nature*, **446**, 900–903.

PAVLOV, A. A. & KASTING, J. F. 2002. Mass-independent fractionation of sulfur isotopes in Archean sediments: strong evidence for an anoxic Archean atmosphere. *Astrobiology*, **2**, 27–41.

PAVLOV, A. A., BROWN, L. L. & KASTING, J. F. 2001. UV shielding of NH_3 and O_2 by organic hazes in the Archean atmosphere. *Journal of Geophysical Research*, **106**, 23267–23287.

PAVLOV, A. A., HURTGEN, M. T., KASTING, J. F. & ARTHUR, M. A. 2003. Methane-rich Proterozoic atmosphere? *Geology*, **31**, 87–90.

PEARSON, D. G., PARMAN, S. W. & NOWELL, G. M. 2007. A link between large mantle melting events and continent growth seen in osmium isotopes. *Nature*, **449**, 202–205.

PERCIVAL, J. A., SANBORN-BARRIE, M., SKULSKI, T., STOTT, G. M., HELMSTAEDT, H. & WHITE, D. J. 2006. Tectonic evolution of the western Superior Province from NATMAP and Lithoprobe studies. *Canadian Journal of Earth Sciences*, **43**, 1085.

PESONEN, L. J., ELMING, S. ET AL. 2003. Palaeomagnetic configuration of continents during the Proterozoic. *Tectonophysics*, **375**, 289–324.

PIPER, J. D. A. 2003. Consolidation of continental crust in Late Archaean – Early Proterozoic times: a palaeomagnetic test. *Gondwana Research*, **6**, 435–448.

PIPER, J. D. A. 2007. The Neoproterozoic supercontinent Palaeopangaea. *Gondwana Research*, **12**, 202–227.

PIRAJNO, F. 2004. Hotspots and mantle plumes: global intraplate tectonics, magmatism and ore deposits. *Mineralogy and Petrology*, **82**, 183–216.

PIRAJNO, F., HAWKE, P., GLIKSON, A. Y., HAINES, P. W. & UYSAL, T. 2003. Shoemaker impact structure, Western Australia. *Australian Journal of Earth Sciences*, **50**, 775–796.

PIRAJNO, F., HOCKING, R. M., REDDY, S. M. & JONES, A. J. 2009. A review of the geology and geodynamic evolution of the Palaeoproterozoic Earaheedy Basin, Western Australia. *Earth-Science Reviews*, **94**, 39–77.

PISAREVSKY, S. A. & NATAPOV, L. M. 2003. Siberia and Rodinia. *Tectonophysics*, **375**, 221–245.

PISAREVSKY, S. A., NATAPOV, L. M., DONSKAYA, T. V., GLADKOCHUB, D. P. & VERNIKOVSKY, V. A. 2008. Proterozoic Siberia: a promontory of Rodinia. *Precambrian Research*, **160**, 66–76.

PLUMB, K. A. 1991. New Precambrian time scale. *Episodes*, **14**, 139–140.

POLLER, U., GLADKOCHUB, D., DONSKAYA, T., MAZUKABZOV, A., SKLYAROV, E. & TODT, W. 2005. Multistage magmatic and metamorphic evolution in the Southern Siberian Craton: Archean and Palaeoproterozoic zircon ages revealed by SHRIMP and TIMS. *Precambrian Research*, **136**, 353–368.

POPE, M. C. & GROTZINGER, J. P. 2003. Paleoproterozoic Stark Formation, Athapuscow Basin, Northwest Canada: record of cratonic-scale salinity crisis. *Journal of Sedimentary Research*, **73**, 280–295.

PORCELLI, D. & BALLENTINE, C. J. 2002. Models for distribution of terrestrial noble gases and evolution of the atmosphere. *Reviews in Mineralogy and Geochemistry*, **47**, 411.

PORCELLI, D. & ELLIOTT, T. 2008. The evolution of He isotopes in the convecting mantle and the preservation of high ^{3}He/^{4}He ratios. *Earth and Planetary Science Letters*, **269**, 175–185.

PROKOPH, A., ERNST, R. E. & BUCHAN, K. L. 2004. Time-series analysis of large igneous provinces: 3500 Ma to present. *Journal of Geology*, **112**, 1–22.

PUFAHL, P. K., HIATT, E. E., STANLEY, C. R., MORROW, J. R., NELSON, G. J. & EDWARDS, C. T. 2007. Physical and chemical evidence of the 1850 Ma Sudbury impact event in the Baraga Group, Michigan. *Geology*, **35**, 827.

RASMUSSEN, B., BENGTSON, S., FLETCHER, I. R. & MCNAUGHTON, N. J. 2002a. Discoidal impressions and trace-like fossils more than 1200 million years old. *Science*, **296**, 1112–1115.

RASMUSSEN, B., BOSE, P. K., SARKAR, S., BANERJEE, S., FLETCHER, I. R. & MCNAUGHTON, N. J. 2002b. 1.6 Ga U–Pb zircon age for the Chorhat Sandstone, lower Vindhyan, India: possible implications for early evolution of animals. *Geology*, **30**, 103–106.

RASMUSSEN, B., FLETCHER, I. R., BENGTSON, S. & MCNAUGHTON, N. J. 2004. SHRIMP U–Pb dating of diagenetic xenotime in the Stirling Range Formation, Western Australia: 1.8 billion year minimum age for the Stirling biota. *Precambrian Research*, **133**, 329–337.

RASMUSSEN, B., FLETCHER, I. R., BROCKS, J. J. & KILBURN, M. R. 2008. Reassessing the first appearance of eukaryotes and cyanobacteria. *Nature*, **455**, 1101–1104.

RAY, J. S., MARTIN, M. W., VEIZER, J. & BOWRING, S. A. 2002. U–Pb zircon dating and Sr isotope systematics of the Vindhyan Supergroup, India. *Geology*, **30**, 131–134.

REYMER, A. & SCHUBERT, G. 1984. Phanerozoic addition rates to the continental crust and crustal growth. *Tectonics*, **3**, 63–77.

RICHTER, F. M. 1985. Models for the Archean thermal regime. *Earth and Planetary Science Letters*, **73**, 350–360.

RICHTER, F. M. 1988. A major change in the thermal state of the Earth at the Archean-Proterozoic boundary: consequences for the nature and preservation of continental lithosphere. *Journal of Petrology, Special Lithosphere Issue*, 39–52.

RINO, S., KOMIYA, T., WINDLEY, B. F., KATAYAMA, I., MOTOKI, A. & HIRATA, T. 2004. Major episodic increases of continental crustal growth determined from zircon ages of river sands; implications for mantle overturns in the Early Precambrian. *Physics of the Earth and Planetary Interiors*, **146**, 369–394.

RITTS, B. D. & GROTZINGER, J. P. 1994. Depositional facies and detrital composition of the Paleoproterozoic Et-Then Group, NWT, Canada: sedimentary response to intracratonic indentation. *Canadian Journal of Earth Sciences*, **31**, 1763–1778.

ROGERS, J. J. W. 1996. A history of continents in the past three billion years. *Journal of Geology*, **104**, 91–107.

ROGERS, J. J. W. & SANTOSH, M. 2002. Configuration of Columbia, a Mesoproterozoic supercontinent. *Gondwana Research*, **5**, 5–22.

ROLLINSON, H. 2006. Crustal Generation in the Archean. In: BROWN, M. & RUSHMER, T. (eds) *Evolution and Differentiation of the Continental Crust*. Cambridge University Press, 173.

ROSCOE, S. M. 1973. *The Huronian Supergroup, a paleoaphebian succession showing evidence of atmospheric evolution*. Geological Association of Canada Special Paper, **12**, 31–47.

ROSE, E. C., MCLOUGHLIN, N. & BRASIER, M. D. 2006. Ground truth: the epistemology of searching for the earliest life on Earth. In: SECKBACH, J. (eds) *Life as we know it. Cellular Origin, Life in Extreme Habitats and Astrobiology*, 10. Springer, Berlin, 259–285.

ROUXEL, O. J., BEKKER, A. & EDWARDS, K. J. 2005. Iron isotope constraints on the Archean and Paleoproterozoic ocean redox state. *Science*, **307**, 1088–1091.

RUDNICK, R. L. 1995. Making continental crust. *Nature*, **378**, 571–577.

RUDNICK, R. L., GAO, S., HEINRICH, D. H. & KARL, K. T. 2003. Composition of the Continental Crust. Treatise on Geochemistry 3. Pergamon, Oxford, 1–64.

SAGAN, C. & MULLEN, G. 1972. Earth and Mars: evolution of atmospheres and surface temperatures. *American Association for the Advancement of Science*, **177**, 52–56.

SAKURAI, R., ITO, M., UENO, Y., KITAJIMA, K. & MARUYAMA, S. 2005. Facies architecture and sequence-stratigraphic features of the Tumbiana Formation in the Pilbara Craton, northwestern Australia: implications for depositional environments of oxygenic stromatolites during the Late Archean. *Precambrian Research*, **138**, 255–273.

SALMINEN, J. & PESONEN, L. J. 2007. Paleomagnetic and rock magnetic study of the Mesoproterozoic sill, Valaam island, Russian Karelia. *Precambrian Research*, **159**, 212–230.

SCHIDLOWSKI, M., EICHMANN, R. & JUNGE, C. E. 1975. Precambrian sedimentary carbonates: carbon and oxygen isotope geochemistry and implications for the terrestrial oxygen budget. *Precambrian Research*, **2**, 1–69.

SCHNEIDER, D. A., BICKFORD, M. E., CANNON, W. F., SCHULZ, K. J. & HAMILTON, M. A. 2002. Age of volcanic rocks and syndepositional iron formations, Marquette Range Supergroup: implications for the tectonic setting of Paleoproterozoic, iron formations of the Lake Superior region. *Canadian Journal of Earth Sciences*, **39**, 999–1012.

SCHUBERT, G., STEVENSON, D. & CASSEN, P. 1980. Whole planet cooling and the radiogenic heat scource contents of the Earth and the Moon. *Journal of Geophysical Research*, **85**, 2531–2538.

SCHUBERT, G., TURCOTTE, D. L. & OLSON, P. 2001. *Mantle Convection in the Earth and Planets*. Cambridge University Press.

SEILACHER, A. 2007. The nature of vendobionts. *In*: VICKERS-RICH, P. & KOMAROWER, P. (eds) *The Rise and Fall of the Ediacaran Biota*. Geological Society, London, Special Publications, **286**, 387–398.

SEILACHER, A., BOSE, P. K. & PFLUGER, F. 1998. Triploblastic animals more than 1 billion years ago: trace fossil evidence from India. *Science*, **282**, 80–83.

SENGOR, A. M. C. & NATAL'IN, B. A. 2001. Rifts of the World. *In*: ERNST, R. E. & BUCHAN, K. L. (eds) *Mantle Plumes: Their Identification Through Time*. Geological Society of America Special Paper, **352**, Boulder, Colorado, 389–482.

SHCHERBAKOVA, V. V., LUBNINA, N. V., SHCHERBAKOVA, V. P., MERTANEN, S., ZHIDKOV, G. V., VASILIEVA, T. I. & TSEL'MOVICH, V. A. 2008. Palaeointensity and palaeodirectional studies of early Riphaean dyke complexes in the Lake Ladoga region (Northwestern Russia). *Geophysical Journal International*, **175**, 433–448.

SHEN, Y., CANFIELD, D. E. & KNOLL, A. H. 2002. Middle Proterozoic ocean chemistry: evidence from the McArthur Basin, northern Australia. *American Journal of Science*, **302**, 81.

SHEN, Y., KNOLL, A. H. & WALTER, M. R. 2003. Facies dependence of sulfur isotopes in a mid-Proterozoic basin. *Nature*, **423**, 632–635.

SHIELDS, G. A. 2007. A normalised seawater strontium isotope curve: possible implications for Neoproterozoic-Cambrian weathering rates and the further oxygenation of the Earth. *Earth Discussions*, **2**, 69–84.

SHIELDS, G. A. & KASTING, J. F. 2007. Palaeoclimatology: Evidence for hot early oceans? *Nature*, **447**, E1.

SHIELDS, G. & VEIZER, J. 2002. Precambrian marine carbonate isotope database: Version 1.1. *Geochemistry, Geophysics, Geosystems*, **3**, 1031; doi: 10.1029/2001GC000266.

SHIREY, S. B., KAMBER, B. S., WHITEHOUSE, M. J., MUELLER, P. A. & BASU, A. R. 2008. A review of the isotopic and trace element evidence for mantle and crustal processes in the Hadean and Archean: implications for the onset of plate tectonic subduction. *In*: CONDIE, K. C. & PEASE, V. (eds) *When Did Plate Tectonics Start on Earth*. Geological Society of America Special Paper, **440**, 1–29.

SILVER, P. G. & BEHN, M. D. 2008a. Intermittent plate tectonics? *Science*, **319**, 85.

SILVER, P. G. & BEHN, M. D. 2008b. Response to Comment on 'Intermittent Plate Tectonics?' *Science*, **320**, 1291b.

SIMONSON, B. M. & GLASS, B. P. 2004. Spherule layers – Records of ancient impacts. *Annual Review of Earth and Planetary Sciences*, **32**, 329–362.

SIMONSON, B. M., SUMNER, D. Y., BEUKES, N. J., JOHNSON, S. & GUTZMER, J. 2009. Correlating multiple Neoarchean-Paleoproterozoic impact spherule layers between South Africa and Western Australia. *Precambrian Research*, **169**, 100–111.

SMIRNOV, A. V., TARDUNO, J. A. & PISAKIN, B. N. 2003. Paleointensity of the early geodynamo (2.45 Ga) as recorded in Karelia: A single-crystal approach. *Geology*, **31**, 415–418.

STERN, R. J. 2005. Evidence from ophiolites, blueschists, and ultrahigh-pressure metamorphic terranes that the modern episode of subduction tectonics began in Neoproterozoic time. *Geology*, **33**, 557–560.

STEVENSON, D. J. 1981. Models of the Earth's core. *Science*, **214**, 611–619.

STEVENSON, D. J., SPOHN, T. & SCHUBERT, G. 1983. Magnetism and thermal evolution of the terrestrial planets. *Icarus*, **54**, 466–489.

STOCKWELL, C. H. 1982. *Proposals for Time Classification and Correlation of Precambrian Rocks and Events in Canada and Adjacent Areas of the Canadian Shield: Part 1, a Time Classification of Precambrian Rocks and Events*. Geological Survey of Canada.

STOREY, B. C., LEAT, P. T., WEAVER, S. D., PANKHURST, R. J., BRADSHAW, J. D. & KELLEY, S. 1999. Mantle plumes and Antarctica-New Zealand rifting: evidence from mid-Cretaceous mafic dykes. *Journal of the Geological Society*, **156**, 659–671.

STRIK, G., BLAKE, T. S., ZEGERS, T. E., WHITE, S. H. & LANGEREIS, C. G. 2003. Palaeomagnetism of flood basalts in the Pilbara Craton, Western Australia: late Archaean continental drift and the oldest known reversal of the geomagnetic field. *Journal of Geophysical Research*, **108**, 2551.

SUMNER, D. Y. & GROTZINGER, J. P. 1996. Were kinetics of Archean calcium carbonate precipitation related to oxygen concentration? *Geology*, **24**, 119–122.

TACKLEY, P. J. 2000. Mantle convection and plate tectonics: toward an integrated physical and chemical theory. *Science*, **288**, 2002–2007.

TACKLEY, P. J. 2008. Geodynamics: Layer cake or plum pudding? *Nature Geoscience*, **1**, 157–158.

TAJIKA, E. 2003. Faint young Sun and the carbon cycle: implication for the Proterozoic global glaciations. *Earth and Planetary Science Letters*, **214**, 443–453.

TAYLOR, S. R. 1964. Abundance of chemical elements in the continental crust: a new table. *Geochimica et Cosmochimica Acta*, **28**, 1273–1285.

TAYLOR, S. R. & MCLENNAN, S. M. 1985. *The Continental Crust: its Composition and Evolution: an Examination of the Geochemical Record Preserved in Sedimentary Rocks*. Blackwell Scientific Publications, Oxford.

TAYLOR, S. R. & MCLENNAN, S. 1995. The geochemical composition of the continental crust. *Reviews in Geophysics*, **33**, 241–265.

TRAMPERT, J. & VAN DER HILST, R. D. 2005. *Towards a Quantitative Interpretation of Global Seismic Tomography*. Geophysical Monograph, American Geophysical Union, **160**, 47.

TRENDALL, A. F. 2002. *The significance of iron-formation in the Precambrian stratigraphic record*. Special

Publication of the International Association of Sedimentologists, **33**, 33–66.

TRINDADE, R. I. F., D'AGRELLA-FILHO, M. S., EPOF, I. & BRITO NEVES, B. B. 2006. Paleomagnetism of Early Cambrian Itabaiana mafic dikes (NE Brazil) and the final assembly of Gondwana. *Earth and Planetary Science Letters*, **244**, 361–377.

TUREKIAN, K. K. & WEDEPOHL, K. H. 1961. Distribution of the Elements in Some Major Units of the Earth's Crust. *Geological Society of America Bulletin*, **72**, 175–192.

TYLER, S. A. & BARGHOORN, E. S. 1954. Occurrence of structurally preserved plants in Pre-cambrian rocks of the Canadian Shield. *Science*, **119**, 606–608.

VALENTINE, J. W. & MOORES, E. M. 1970. Plate-tectonic regulation of faunal diversity and sea level: a model. *Nature*, **228**, 657–659.

VALLEY, J. W., LACKEY, J. S. ET AL. 2005. 4.4 billion years of crustal maturation: oxygen isotope ratios of magmatic zircon. *Contributions to Mineralogy and Petrology*, **150**, 561–580.

VAN DEN BOORN, S. H. J. M., VAN BERGEN, M. J., NIJMAN, W. & VROON, P. Z. 2007. Dual role of seawater and hydrothermal fluids in Early Archean chert formation: evidence from silicon isotopes. *Geology*, **35**, 939–942.

VAN DER HILST, R. D., WIDIYANTORO, S. & ENGDAHL, E. R. 1997. Evidence for deep mantle circulation from global tomography. *Nature*, **386**, 578–584.

VAN HUNEN, J., VAN KEKEN, P. E., HYNES, A. & DAVIES, G. F. 2008. Tectonics of early Earth: some geodynamic considerations. *In*: CONDIE, K. C. & PEASE, V. (eds) *When Did Plate Tectonics Begin on Planet Earth?* The Geological Society of America, Special Paper, **440**, 157–172.

VAN KEKEN, P. E., HAURI, E. H. & BALLENTINE, C. J. 2002. Mantle mixing: the generation, preservation, and destruction of chemical heterogeneity. *Annual Review of Earth and Planetary Sciences*, **30**, 493–525.

VAUGHAN, A. P. M. & STOREY, B. C. 2007. A new supercontinent self-destruct mechanism: evidence from the Late Triassic-Early Jurassic. *Journal of the Geological Society*, **164**, 383.

VEIZER, J. & JANSEN, S. L. 1979. Basement and sedimentary recycling and continental evolution. *Journal of Geology*, **87**, 341–370.

VENTURA, G. T., KENIG, F. ET AL. 2007. Molecular evidence of Late Archean archaea and the presence of a subsurface hydrothermal biosphere. *Proceedings of the National Academy of Sciences*, **104**, 14260.

WALDBAUER, J. R., SHERMAN, L. S., SUMNER, D. Y. & SUMMONS, R. E. 2009. Late Archean molecular fossils from the Transvaal Supergroup record the antiquity of microbial diversity and aerobiosis. *Precambrian Research*, **169**, 28–47.

WALTER, M. R., DU, R. & HORODYSKI, R. J. 1990. Coiled carbonaceous megafossils from the Middle Proterozoic of Jixian (Tianjin) and Montana. *American Journal of Science*, **290**, 133–148.

WALZER, U. & HENDEL, R. 2008. Mantle convection and evolution with growing continents. *Journal of Geophysical Research-Solid Earth*, **113**, B09405.

WARDLE, R. J., JAMES, D. T., SCOTT, D. J. & HALL, J. 2002. The southeastern Churchill Province: synthesis of a Paleoproterozoic transpressional orogen. *Canadian Journal of Earth Sciences*, **39**, 639–663.

WESTALL, F. 2009. Life on an anaerobic planet. *Science*, **323**, 471.

WHITTINGTON, A. G., HOFMEISTER, A. M. & NABELEK, P. I. 2009. Temperature-dependent thermal diffusivity of the Earth's crust and implications for magmatism. *Nature*, **458**, 319–321.

WILLE, M., KRAMERS, J. D. ET AL. 2007. Evidence for a gradual rise of oxygen between 2.6 and 2.5 Ga from Mo isotopes and Re-PGE signatures in shales. *Geochimica et Cosmochimica Acta*, **71**, 2417–2435.

WILLIAMS, G. E. 2000. Geological constraints on the Precambrian history of Earth's rotation and the moon's orbit. *Reviews of Geophysics*, **38**, 37–59.

WILLIAMS, G. E. 2005. Subglacial meltwater channels and glaciofluvial deposits in the Kimberley Basin, Western Australia: 1.8 Ga low-latitude glaciation coeval with continental assembly. *Journal of the Geological Society*, **162**, 111–124.

WILLIAMS, H., HOFFMAN, P. H., LEWRY, J. F., MONGER, J. W. H. & RIVERS, T. 1991. Anatomy of North America: thematic geologic portrayals of the continents. *Tectonophysics*, **187**, 117–134.

WINGATE, M. T. D. 1998. A palaeomagnetic test of the Kaapvaal – Pilbara (Vaalbara) connection at 2.78 Ga. *South African Journal of Geology*, **101**, 257–274.

WINGATE, M. T. D., PISAREVSKY, S. A., GLADKOCHUB, D. P., DONSKAYA, T. V., KONSTANTINOV, K. M., MAZUKABZOV, A. M. & STANEVICH, A. M. 2009. Geochronology and paleomagnetism of mafic igneous rocks in the Olenek Uplift, northern Siberia: implications for Mesoproterozoic supercontinents and paleogeography. *Precambrian Research*, **170**, 256–266.

WISE, D. U. 1974. Continental margins, freeboard and the volumes of continents and oceans through time. *In*: BURK, C. A. & DRAKE, C. L. (eds) *The Geology of Continental Margins*. Springer, New York, 45–58.

WORSLEY, T. R., NANCE, D. & MOODY, J. B. 1984. Global tectonics and eustasy for the past 2 billion years. *Marine Geology*, **58**, 373–400.

ZAHNLE, K., CLAIRE, M. & CATLING, D. 2006. The loss of mass-independent fractionation in sulfur due to a Palaeoproterozoic collapse of atmospheric methane. *Geobiology*, **4**, 271–283.

ZANG, W. L. 2007. Deposition and deformation of late Archaean sediments and preservation of microfossils in the Harris Greenstone Domain, Gawler Craton, South Australia. *Precambrian Research*, **156**, 107–124.

ZEGERS, T. E., DE WIT, M. J., DANN, J. & WHITE, S. H. 1998. Vaalbara, Earth's oldest assembled continent? A combined structural, geochronological, and palaeomagnetic test. *Terra Nova*, **10**, 250–259.

ZHAO, G. C., CAWOOD, P. A., WILDE, S. A. & SUN, M. 2002. Review of global 2.1–1.8 Ga orogens: implications for a pre-Rodinia supercontinent. *Earth Science Reviews*, **59**, 125–162.

ZHAO, G., SUN, M., WILDE, S. A. & LI, S. 2004. A paleo-mesoproterozoic supercontinent: assembly, growth and breakup. *Earth Science Reviews*, **67**, 91–123.

The IGCP 509 database system: design and application of a tool to capture and illustrate litho- and chrono-stratigraphic information for Palaeoproterozoic tectonic domains, large igneous provinces and ore deposits; with examples from southern Africa

BRUCE M. EGLINGTON[1]*, STEVEN M. REDDY[2] & DAVID A. D. EVANS[3]

[1]*Saskatchewan Isotope Laboratory, University of Saskatchewan, 114 Science Place, Saskatoon, Saskatchewan, S7N 5E2, Canada*

[2]*The Institute for Geoscience Research, Department of Applied Geology, Curtin University of Technology, GPO Box U1987, Perth, WA 6845, Australia*

[3]*Department of Geology and Geophysics, Yale University, New Haven, CT 06520-8109, USA*

*Corresponding author (e-mail: bruce.eglington@usask.ca)

Abstract: The IGCP 509 project is collating global information for the Palaeoproterozoic era through the activities of numerous international collaborators. A database system (StratDB) and web interface has been designed to facilitate this process with links to an existing geochronology database (DateView). As a result, all information captured will remain available in a digital format for future researchers. The philosophy and design of the database and some of the outputs available from it are described. One of the principal features of the system is that it facilitates the construction of time–space correlation charts using an innovative application of GIS technology to non-geographic information, which permits users to query a variety of attribute information associated with lithostratigraphic units, metamorphic and deformation episodes associated with user-selected tectonic domains, large igneous provinces and major ore deposits. In the process, much of the manual labour normally associated with the construction of such charts in standard graphical or drafting packages is avoided. Associations between units, deformation, metamorphism, large igneous provinces and ore deposits may become more apparent once linked information is available for querying and investigation. Geochronological information from the DateView database may also be linked to entities stored in StratDB. GIS maps may be linked to the attribute information in StratDB and DateView to construct a variety of time-slice maps or palaeogeographic reconstructions with the same symbology as is used in the time–space correlation charts. This database system will facilitate the dissemination of lithostratigraphic information for many countries to a broader community and will help non-specialists to easily view information for various Palaeoproterozoic tectonic domains. The system is illustrated using a preliminary compilation of information for the Palaeoproterozoic of southern Africa. The correlation charts and time-slice maps provide insights to the geological evolution of this region which emphasize some aspects and correlations which have not previously been extensively considered; for instance, possible correlation of units in the central and western zones of the Limpopo Belt (South Africa, Zimbabwe and Botswana) with the Magondi Belt of Zimbabwe and its extension into northern Botswana.

Supplementary data is available at http://www.geolsoc.org.uk/SUP18352

The IGCP 509 project seeks to collate global information for the Palaeoproterozoic era, with the express objectives of developing a thorough database of the geological record with up-to-date geochronological constraints and to produce global time–space correlation charts. In order to meet these objectives, published and new information is being compiled by numerous researchers, coordinated by more than 20 regional experts. With so many individuals, drawn from a wide variety of sub-disciplines within the earth sciences, varied organizations, cultures and languages, it was decided to establish a database system to facilitate data capture, sharing and standardization and to provide standardized software for producing time–space correlation charts derived from information in the database. An added advantage of this approach is that all information captured will remain available in a digital format for future researchers. Here, we describe the philosophy and design of the principal database employed by participants in the IGCP 509 project, its links to other allied database systems, and illustrate some of the outputs available. Example outputs are based on

From: REDDY, S. M., MAZUMDER, R., EVANS, D. A. D. & COLLINS, A. S. (eds) *Palaeoproterozoic Supercontinents and Global Evolution.* Geological Society, London, Special Publications, **323**, 27–47.
DOI: 10.1144/SP323.2   0305-8719/09/$15.00 © Geological Society of London 2009.

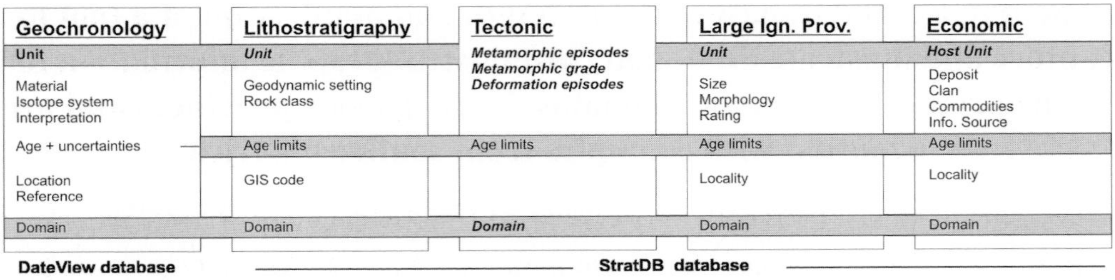

Geochronology	Lithostratigraphy	Tectonic	Large Ign. Prov.	Economic
Unit	Unit	Metamorphic episodes Metamorphic grade Deformation episodes	Unit	Host Unit
Material Isotope system Interpretation	Geodynamic setting Rock class		Size Morphology Rating	Deposit Clan Commodities Info. Source
Age + uncertainties	Age limits	Age limits	Age limits	Age limits
Location Reference	GIS code		Locality	Locality
Domain	Domain	Domain	Domain	Domain
DateView database		**StratDB database**		

Fig. 1. Schematic of the conceptual design of the StratDB and DateView databases, illustrating the main tables, fields and some typical values. Tables shown primarily relate to the lithostratigraphic aspects of the database design and do not represent a formal entity-relationship diagram.

an initial compilation of the geology and geochronology of Palaeoproterozoic southern Africa.

The principal database system used for the IGCP 509 project is the StratDB database which provides storage of lithostratigraphic, tectonic domain, large igneous province and ore deposit information. Geochronological information is drawn from the DateView database. Figure 1 provides a schematic view of the major components comprising the StratDB and DateView databases and their associated links. Both the StratDB and DateView databases are available on the web at http://sil.usask.ca/databases.htm.

We also describe the general methodology adopted for the construction of time–space correlation charts. The approach used is an innovative application of GIS technology to non-geographic information so as to benefit from its ability to query and portray attribute information associated with polygons. In the process, much of the manual labour normally associated with the construction of such charts in standard graphical or drafting packages is avoided. Although some compilations have produced flexible legends for GIS maps (i.e. Steinshouer et al. 1999; Raines et al. 2007), we are not aware of any equivalent use of GIS (geographic information system) technology in the manner instituted here. In addition to creating correlation charts, the system permits users with access to appropriate digital GIS data sources (via shapefiles, geodatabases, etc.) to produce maps using exactly the same graphical symbology as for the charts.

Origin of the database systems

StratDB is a web-enabled extension of an earlier desktop system which captured information for lithostratigraphic units recognized by the South African Committee for Stratigraphy (Eglington et al. 2001). Several enhancements were specifically developed for the IGCP 509 project, in particular to facilitate the capture of rock-type and geodynamic setting information for lithostratigraphic units; summary information for multiple metamorphic and deformation episodes within tectonic domains; and large igneous province information and summary attributes for ore deposits. StratDB also provides links to geochronological information stored in the DateView database. Aspects of this geochronology database system have previously been described by Eglington (2004) and Eglington & Armstrong (2004).

Many of the concepts adopted for outputs from the StratDB system are based on correlation charts produced by previous compilers, for example Hartzer et al. (1998), Wardle et al. (2002) and Ansdell et al. (2005), all of which were produced using standard commercial graphics packages.

Database design

The principal tables in the StratDB database contain essential information for each lithostratigraphic unit, structural domain, ore deposit and large igneous province (Fig. 2). Other tables provide referential, look-up values or linkage fields to draw together additional information from various of the database tables. Another group of tables within the database (Fig. 3) stores additional information for use in the construction of the time–space correlation charts.

The database utilizes a relational structure with full referential integrity and normalization. It is designed so that data integrity is constrained by a series of primary and foreign keys. Records (new or modified) may not contain values which do not already exist in master tables and these tables also act as 'look-up' sources to facilitate the construction of user queries. Any changes made to 'master' key field values are automatically propagated to all linked tables and records.

The principal table in the StratDB database contains information defining the lithostratigraphic units. Each unit has a unique integer ID and is associated with one country. If a unit with the

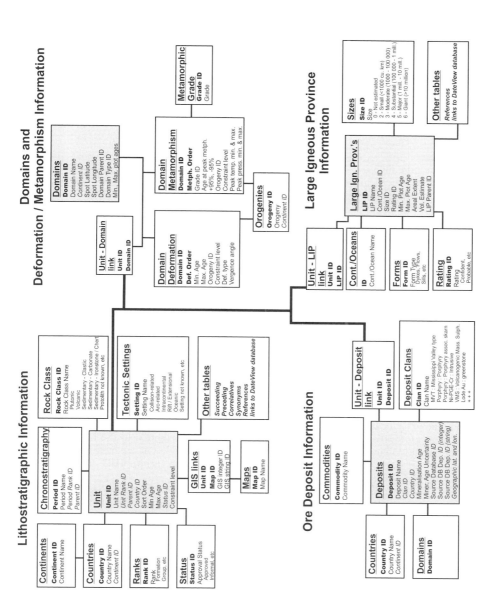

Fig. 2. Schematic of the conceptual design of the StratDB database, illustrating the major tables, fields and some typical values for the lithostratigraphic, domain (deformation/metamorphism), large igneous province and ore deposit components of the system. Tables and links shown do not represent a formal entity-relationship diagram. Bold field names are primary keys.

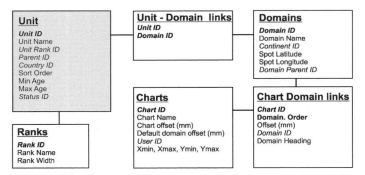

Fig. 3. Schematic of the conceptual design of the StratDB database, illustrating the main tables and fields required to construct time–space correlation charts.

same name occurs in more than one country, then additional unique records need to be created for each country. Unit names are stored in Unicode strings so as to handle non-Western characters. All non-numeric primary and foreign keys, however, require ASCII characters. Each unit is recursively linked to a 'parent' unit via a ParentID field so as to create hierarchical relationships between units. The relative order of subunits in these hierarchical relationships is maintained by a longer integer (4 byte) 'sort order' field. For maximum flexibility, this field expects a value representing the minimum age of the unit in years. Minimum and maximum estimates for the age of a unit (in Ma) are also stored in appropriate fields. Other fields in this table store information on the rank of the unit and the status of the unit, for instance, whether the name is officially recognized, informal, historical, etc. Descriptions of a unit, stored in a BLOB (Binary Large Object) field, are invaluable for users who do not know the local geology well.

Two important tables, for the IGCP 509 project, hold information on the dominant rock class and geodynamic settings for each unit. Some of the acceptable values for these two characteristics are shown in Figure 2. These values are instrumental in illustrating the nature of each unit in the time–space correlation charts. The depositional setting of units (e.g. fluviatile, shallow marine, deep marine, etc.) may also be captured and used as attributes for correlation charts and maps. Dominant palaeocurrent directions for sedimentary units may also be captured so as to facilitate comparisons across country boundaries and in palaeogeographic reconstructions.

All lithostratigraphic units are associated with hierarchical ranks. Each rank has an associated rank width which controls the width of the rectangular polygons for the time–space correlation charts. Higher level units are plotted as wider boxes. Each unit is also linked to chronostratigraphic time-scale intervals (stages, periods, etc.), utilizing the terms recommended by Gradstein et al. (2004). Links should be made at the lowest possible classification, for example stage rather than period if an appropriate stage exists. A hierarchical list of chronostratigraphic time-scale intervals with associated minimum and maximum ages may be downloaded as an Excel spreadsheet.

If GIS map attribute information is available for units, it is useful to also capture the unique GIS identifiers used in the GIS system. Some GIS coverages and shapefiles use unique integer values for units, whereas others use unique strings. Either can be stored in the StratDB database and associated with the unique unit record number used by the database. With this information in place, it is possible to link any of the attribute data from the database to GIS maps, as was done by Eglington & Armstrong (2004) for the ages of units in the Kaapvaal Craton. In future, once the OneGeology initiative (www.onegeology.org) provides a web file server (WFS) service in addition to the initial web map server (WMS) service, it ought to be possible to link attribute data in StratDB to the various OneGeology maps.

Other tables hold information specifying which units precede or succeed one another and details of which units are correlatives or synonyms for the current unit. If information is added for either preceding or succeeding units, the software can add the reverse relationship. The intention is to capture information for definitive relationships, not necessarily for every contact. Correlatives are equivalent units recognized as different units, whereas synonyms are alternative terms used for exactly the same unit and outcrop area. In most cases, synonyms are historical unit names which should no longer be used. In addition to providing useful insights into the lithostratigraphy of an area, these links facilitate navigation between units in the web browser interface.

Each tectonic domain also receives a unique integer ID and is associated with a hierarchical parent via a 'parent ID' number. Associated central latitude and longitude values facilitate queries to extract information from the database within user-selected geographic limits without implementing a WFS capability. Various episodes of deformation receive unique integer 'order' numbers and each episode of deformation has a minimum and maximum age. Each episode is also associated with an orogeny, which permits people who do not know the regional geology to select domains associated with more widespread orogenesis. Deformation style (e.g. ductile, brittle) and vergence direction (at $45°$ increments) may also be stored. Similarly, metamorphic episodes within the domain have unique integer IDs. The age of peak metamorphism and its $\pm 95\%$ uncertainties are also stored, together with the typical grade of metamorphism for the domain, plus estimates for minimum and maximum peak temperature and metamorphism and the associated orogeny.

In order to construct time–space correlation charts, units and domains need to be associated with unique charts which any user of the system may create. Charts are not shared by groups of users; each user creates their own by defining the order (from left to right across a chart) in which tectonic domains are to be drawn. However, the information associated with each domain (lithostratigraphy, metamorphic and deformation information, etc.) is common to all users. Figure 3 illustrates the principal features of the database tables and links used to construct the outputs which are subsequently imported to a GIS system for final compilation of the charts. All point locality geographic information stored in both StratDB and DateView should be relative to the WGS84 datum.

Hardware and software

The data are stored in a Firebird open-source client/server type relational database management system which provides full referential integrity. Although open-source, Firebird provides similar capabilities to major commercial database systems such as Oracle and is very easy to manage. The web interface to the database was programmed using the Delphi language, utilizing Intraweb, TeeChart and Flexcel components. The database server is currently located on a standard Pentium 4 computer with 512 Mb memory, running the Red Hat Linux operating system. Web interface programs for each database are standalone web servers, running as services on a standard Windows XP, Pentium 4 computer with 1 Gb of memory. Browsers currently supported by the web servers are Internet Explorer, Firefox, Netscape (versions 6 and 7), Opera and Safari.

All access to the database has been coded using dbExpress components within Delphi, which has the advantage that the data can be ported to other client-server database systems with minimal change to the software code, provided that these database systems support triggers, generators and BLOB fields. Intraweb was chosen for the web interface since it speeded up the programming considerably. Most web database connections are stateless, that is the database does not maintain any memory of previous requests from a user. Whilst effective for simple reads from a database, this approach introduces considerable programming and network overheads when inserting new information or updating existing information in a database. Intraweb provides a stateful interface. In addition, the programming environment is much closer to traditional graphical user interfaces (GUI), which also speeded up development of the software. As a result of this stateful control, the normal 'back' button in browsers is disabled and navigation has to be performed by appropriate links programmed on each web page. TeeChart is a rich graphing component which integrates well with Intraweb and is easily modified to achieve the various graphs required by both StratDB and DateView. It also provides the ability to display GIS shapefiles, which is important for illustrating the geographic location of samples in DateView. Flexcel is a component for importing and exporting spreadsheets in Excel format. It has the advantage that templates can be designed using Excel and stored outside of the program executable. As a result, spreadsheet formats for downloading information are easily changed without the need to recompile any software. Flexcel also provides access to most Excel capabilities, the most important of which for these database interfaces has been the use of outlines to facilitate grouping of data (in StratDB) and the use of pivot tables to organize data for different variables measured for many samples (in DateView).

All user access to the system is controlled via a separate Firebird database which contains user information such as user ID, passwords, e-mail addresses and access permissions. When users log in, this system is queried to determine the user's allowed access rights and to upload any stored information for the user, for instance the definition terms of the user's last query.

Administration of the database, including regular backups, sweeps, design changes, etc., is performed using a commercial GUI package, IBExpert.

Time–space correlation charts are created using GIS technology. Although the data in the charts are not geographic, the x (offsets from an arbitrary origin) and y (age) information may be cast as a

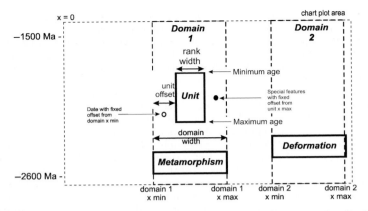

Fig. 4. Variables used, and stored in the database, to define the layout of polygons delimiting domains, lithostratigraphic units, metamorphic and deformation episodes.

Cartesian problem in which the node values permit the construction of rectangles (polygons in GIS terminology) with associated attribute information (Fig. 4). These attributes may be flexibly queried and used to construct the charts using any GIS package. We experimented with several products and have used ArcGIS from ESRI because it is available at most universities and geological organizations, and it provides style sheets which facilitate easy changes to the symbology used to portray different rock classes, geodynamic settings, etc.

Data are downloaded from the database as Microsoft Excel files, which are then imported to an ArcGIS personal geodatabase using Microsoft Access, ESRI ArcCatalog and ArcMap. Values for the vertices of the various polygons are derived from the minimum and maximum ages of each unit, LIP or episode of metamorphism or deformation, cast as negative values so that the youngest events will be at the top of a chart. The left side of each polygon is defined by an offset for each unit into the space allocated for each domain. The width of polygons is controlled by the 'rank width' parameter stored for each rank. Hierarchically higher level units have larger rank widths and thus plot as wider boxes. Various parameters which control the layout of the time–space correlation charts are illustrated in Figure 3. Creation of polygons and bounding polylines from initial point feature classes is performed using the free ET Geowizards plug-in for ArcMap.

Functionality

User-specified options

Users may modify and save various options which impact on the output of database queries. At present, this is limited to selecting whether to show associated geochronological records from the DateView database and whether to limit records to hierarchical ranks above those of seams (i.e. member rank and above). In part, these options are intended to reduce the time needed to run queries for situations where the extra information is not needed.

Querying the database

Queries against the database are defined in a series of up to three windows. The first requires users to select which continents are to be included in the query. The second window (for unit queries) shows all countries in the previously selected continents which have records in the database. If one wants all possible units from all available countries, one may omit checking the checkbox above the list of countries. The third window lists all lithostratigraphic units in the selected countries, plus lists of selectable values for variables such as: minimum age; maximum age; minimum and maximum sort order values; tectonic domains; reference sources; validation status; and chronostratigraphic periods.

In the case of domain queries, users have the option to select from a list of orogenies and domain types (province, terrane, etc.), in addition to a list of domain names associated with the user-selected continents.

A checkbox is associated with each query parameter so as to reduce the impact of changing which variables are selected during a session. For instance, assume one selected some specific units within an age range, ran the query and then decided to see all units within the same age range. For the first query, one would check the checkboxes for units, minimum age and maximum age and provide appropriate values for these fields. For the second query, all that is required is to uncheck the units checkbox. If one then wants to repeat the first query, it is not necessary to select each unit again; one only needs to re-check the unit's checkbox. All unchanged selections are remembered during a session and, if checked, are stored in the user database for when users next log in, provided that users logout by clicking on the 'log out' hot link.

Modifying the database

The default access permission for new users is read-only but users wishing to modify existing records or insert new records may contact the database administrator to request modify/insert rights. Users with modify permission for the StratDB database see a button with the caption 'Edit' on several of the detail forms. It is necessary to click on this button to go into edit mode, at which point editable fields are enclosed by box outlines or show drop-down boxes. Once changes have been made, one must click on the 'Save' button to update the information in the database. If one moves to another record without saving, all changes made are lost. If one realizes that a mistake has been made one may either correct the mistake or click on the 'Cancel' button. It is important to realize that every view of the database contents is a 'snapshot' of the situation when a query is run and is held in memory. Users never edit the database itself, only a copy of the information. Changes to the database are automatically managed by the relational database management system, based on what is changed and when it is changed. Thus multiple users may edit different parts of the same record and the software will update those that require changes when users activate the save process. Other users will see these changes the next time information is read from the database.

There is also a button on this form to export the lithostratigraphic unit name, unique database ID and some other information to the DateView database. Anybody wishing to add data to DateView is encouraged to first add the unit information to StratDB as this will ensure that the two databases are correctly linked without the need for manual editing of the DateView database. When appropriate links are in place, 'published' geochronological information added to DateView for a unit will automatically be visible in StratDB without the need to manually edit several tables in DateView.

For security, only a very limited number of individuals have permission to delete records from the database. When a master record is deleted, other 'detail' records which have referential integrity links are automatically also deleted. Incorrect deletion of records in the database could thus lead to substantial loss of information which can only be recovered from backups.

Adding data to the database

Only some fields for lithostratigraphic units may be inserted online. This is, in part, because of the complexity of adding records for all associated tables and the rather tedious, slow process involved in doing this one record at a time. Most initial data capture is performed offline, using either a Microsoft Access database template or a series of Microsoft Excel spreadsheet templates, as described at http://sil.usask.ca/sdb_compilations.htm. Once the basic data are in the database, additional attribute information such as GIS links, references and associations with tectonic domains, may be added online.

Raising an objection to a record

An important feature of the database software, which will facilitate peer review of information, is the ability for any user to 'raise an objection' to any record. Users doing so enter text describing their reasons for disagreeing with the information currently in the database. Various volunteers with regional or topic-specific expertise act as moderators (validators) for information in the database. They are tasked with adjudicating any objections raised so as to either achieve a compromise or add additional records to capture significantly different interpretations. The validation status of records may also be used as a search term.

Confidentiality of information

The DateView geochronology database is designed to allow storage of confidential information, for example personal, unpublished dates. Access to all information is controlled by user permissions and these permissions are controlled across both the DateView and StratDB database systems. Hence, users with personal information in linked fields in DateView will be able to see these data from StratDB, whilst most users will only see records marked as 'public'. Only those DateView records which have interpretations set as intrusion, extrusion, detrital or diagenesis (for units); or as metamorphism or cooling (for domains), are visible in StratDB. Other isotope data are not shown at present.

Outputs from the database

A number of outputs from the database are available, most of which are activated from menu links at the left of the main menu. In all cases, the results of a query are provided as a grid. From this grid, one may click on the unit (or other) ID hotlink to drill down for more information on the chosen unit, LIP, domain or deposit. Once one has selected a specific entity (Fig. 5), one may navigate down or up the hierarchy by either clicking on hotlinks associated with each sub-unit or by clicking the 'Go to parent' button. Links are also provided to information such as tectonic settings, rock class, chronostratigraphic period, references, GIS links, etc.

Geochronological information for the current unit and all units that are hierarchically one level lower may be extracted from the DateView database if link-fields exist. Some other outputs, designed for export to Excel spreadsheets, are described below.

Listings of look-up tables

Users may download Excel spreadsheets containing the contents of several of the master (look-up) tables. Contributors should check on up-to-date

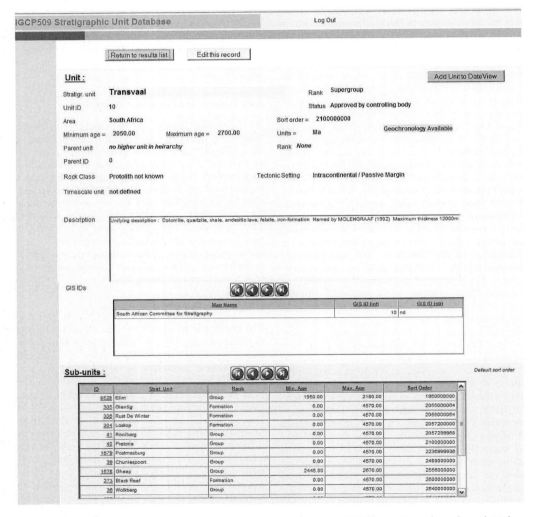

Fig. 5. View of the first part of detailed information for a lithostratigraphic unit. Users may navigate down the unit hierarchy by clicking on the sub-unit ID numbers or up the hierarchy by clicking on the 'Go to parent' button when it is visible. In the example illustrated (Transvaal Supergroup from South Africa), there is no hierarchically higher lithostratigraphic unit, hence the button is not shown. Note that the software indicates whether geochronological data are available in DateView for the current or for any of the units one level lower in the hierarchy. Actual data are provided lower on the form (not shown).

values for the various tables prior to compiling information in the Microsoft Access and Excel templates.

Lithostratigraphic hierarchy

The results of a query may be downloaded as an Excel spreadsheet, as illustrated in Figure 6. The contents of the spreadsheet are grouped using the outline capability of Excel and sorted according to values of the 'sort order' variable so as to illustrate the hierarchical relationship between units. Similar capabilities exist or are planned for tectonic domains and for large igneous provinces.

Time–space correlation charts

In addition to providing a long-term resource for researchers, StratDB and DateView also allow project participants to query the databases and produce outputs for use in the construction of time–space charts. The web interface provides data output in a format suitable for construction of time–space diagrams using standard GIS software such as ArcMap, MapInfo or UDig. The diagrams are queryable using various attribute information, such as Rock Class, Geodynamic Setting, Depositional Setting, etc. The system has been tested, using ESRI ARCGIS versions 9.1 and 9.2, and testing is ongoing as more 'real' data are added. More information on the specific procedures to follow when creating time–space correlation charts is provided in separate documentation available from http://sil.usask.ca/databases.htm.

The principal legend for the space-time charts is illustrated in Figure 7. This legend is based on the one used for previous compilations of Canadian Palaeoproterozoic terranes (Wardle *et al.* 2002; Ansdell *et al.* 2005), but with data shown as a matrix derived from two properties: rock class and geodynamic setting. With the pseudo-GIS approach utilized, each of the attributes associated with individual rectangles on the chart is selectable, making for a very flexible resource. Specific colours and symbology for the geodynamic setting – rock-class matrix are derived from an ESRI ArcGIS style sheet and can thus be easily changed to suit the requirements of all IGCP 509 project coordinators without any time-consuming recoding by participants. The level of certainty for the minimum and maximum age limits determine the outline style of bounding polygons.

An alternative time–space correlation chart, utilizing information on depositional environment may also be produced. The legend for this type of chart is illustrated in Figure 8, together with some other attribute information which may be plotted with either form of chart. Additional attribute information, such as special features of units (e.g. units exhibiting ^{13}C isotope excursions) or

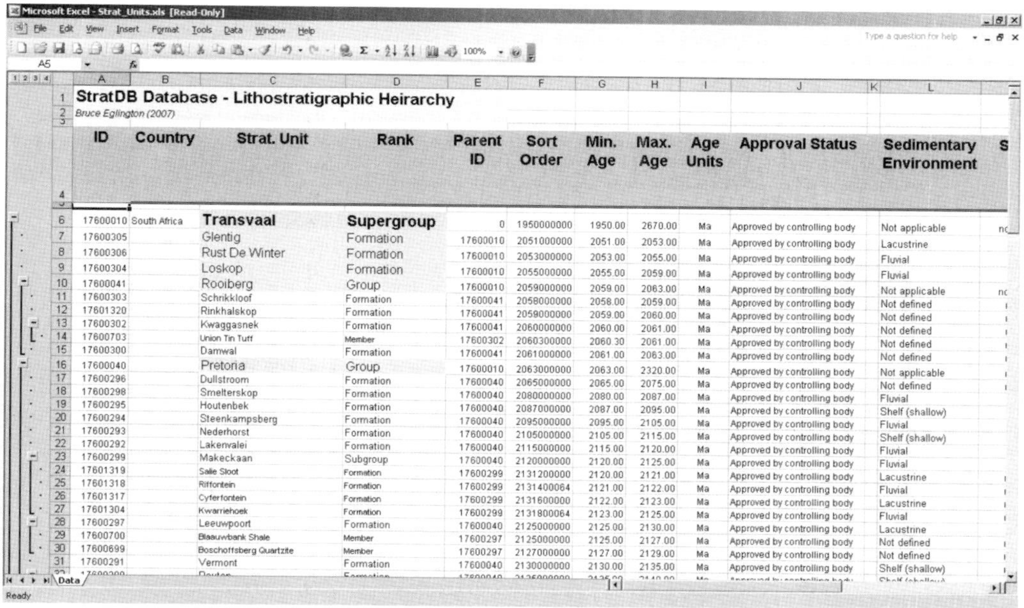

StratDB Database - Lithostratigraphic Heirarchy

Bruce Eglington (2007)

	ID	Country	Strat. Unit	Rank	Parent ID	Sort Order	Min. Age	Max. Age	Age Units	Approval Status	Sedimentary Environment
6	17600010	South Africa	Transvaal	Supergroup	0	1950000000	1950.00	2670.00	Ma	Approved by controlling body	Not applicable
7	17600305		Glentig	Formation	17600010	2051000000	2051.00	2053.00	Ma	Approved by controlling body	Lacustrine
8	17600306		Rust De Winter	Formation	17600010	2053000000	2053.00	2055.00	Ma	Approved by controlling body	Fluvial
9	17600304		Loskop	Formation	17600010	2055000000	2055.00	2059.00	Ma	Approved by controlling body	Fluvial
10	17600041		Rooiberg	Group	17600010	2059000000	2059.00	2063.00	Ma	Approved by controlling body	Not applicable
11	17600303		Schrikkloof	Formation	17600041	2058000000	2058.00	2059.00	Ma	Approved by controlling body	Not defined
12	17601320		Rinkhalskop	Formation	17600041	2059000000	2059.00	2060.00	Ma	Approved by controlling body	Not defined
13	17600302		Kwaggasnek	Formation	17600041	2060000000	2060.00	2061.00	Ma	Approved by controlling body	Not defined
14	17600703		Union Tin Tuff	Member	17600302	2060300000	2060.30	2061.00	Ma	Approved by controlling body	Not defined
15	17600300		Damwal	Formation	17600041	2061000000	2061.00	2063.00	Ma	Approved by controlling body	Not defined
16	17600040		Pretoria	Group	17600010	2063000000	2063.00	2320.00	Ma	Approved by controlling body	Not applicable
17	17600296		Dullstroom	Formation	17600040	2065000000	2065.00	2075.00	Ma	Approved by controlling body	Not defined
18	17600298		Smelterskop	Formation	17600040	2080000000	2080.00	2087.00	Ma	Approved by controlling body	Fluvial
19	17600295		Houtenbek	Formation	17600040	2087000000	2087.00	2095.00	Ma	Approved by controlling body	Shelf (shallow)
20	17600294		Steenkampsberg	Formation	17600040	2095000000	2095.00	2105.00	Ma	Approved by controlling body	Fluvial
21	17600293		Nederhorst	Formation	17600040	2105000000	2105.00	2115.00	Ma	Approved by controlling body	Shelf (shallow)
22	17600292		Lakenvalei	Formation	17600040	2115000000	2115.00	2120.00	Ma	Approved by controlling body	Fluvial
23	17600299		Makeckaan	Subgroup	17600040	2120000000	2120.00	2125.00	Ma	Approved by controlling body	Fluvial
24	17601319		Salie Sloot	Formation	17600299	2131200000	2120.00	2121.00	Ma	Approved by controlling body	Lacustrine
25	17601318		Riffontein	Formation	17600299	2131400064	2121.00	2122.00	Ma	Approved by controlling body	Fluvial
26	17601317		Cyferfontein	Formation	17600299	2131600000	2122.00	2123.00	Ma	Approved by controlling body	Lacustrine
27	17601304		Kwarriehoek	Formation	17600299	2131800064	2123.00	2125.00	Ma	Approved by controlling body	Fluvial
28	17600297		Leeuwpoort	Formation	17600040	2125000000	2125.00	2130.00	Ma	Approved by controlling body	Lacustrine
29	17600700		Blaauwbank Shale	Member	17600297	2125000000	2125.00	2127.00	Ma	Approved by controlling body	Not defined
30	17600699		Boschoffsberg Quartzite	Member	17600297	2127000000	2127.00	2129.00	Ma	Approved by controlling body	Not defined
31	17600291		Vermont	Formation	17600040	2130000000	2130.00	2135.00	Ma	Approved by controlling body	Shelf (shallow)

Fig. 6. Example download of query information for an hierarchical lithostratigraphic succession from southern Africa. The outline feature of Excel provides flexible control of which units are visible.

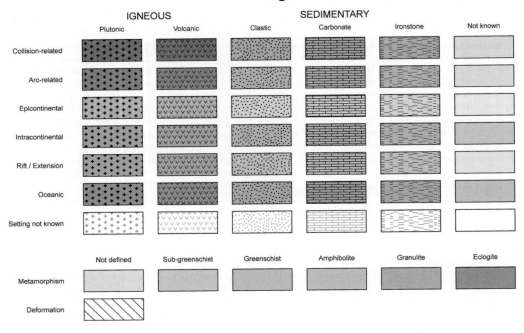

Fig. 7. Legend illustrating the matrix of geodynamic settings (left) and rock classes (top) used to classify all lithostratigraphic units. Dyke swarms and pre-existing crust are classified independent of tectonic setting and rock class. Legend was created as a series of labelled polygons using the ArcMap component of ArcGIS.

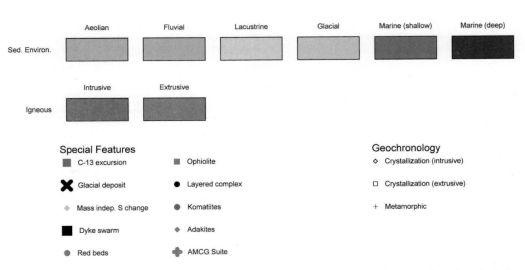

Fig. 8. Legend for environment of formation for lithostratigraphic units and of special features and geochronological method symbols. Legend was created as a series of labelled polygons and symbols using the ArcMap component of ArcGIS.

geochronological information, may be plotted as point symbols. Values and symbols for these attributes are shown in Figure 8. Figure 9 provides a miniature version of a time–space correlation chart constructed for the Palaeoproterozoic of southern Africa, to illustrate the general layout of the charts with space for titles, authors, locality maps, legend and the various tectonic domains. Full size versions of the Rock Class – Geodynamic Setting and of the Depositional Environment time–space correlation charts for southern Africa are available as supplementary files (SUP18352). In the charts, smaller polygons are plotted on top of larger ones based on the rank width of the polygons, utilizing queries within ArcGIS. Polygons for all plutonic and hypabyssal intrusions are offset to the right of sedimentary and volcanic units to emphasize their intrusive relationship. Labelling (not shown on the miniature version in Figure 9 because of size constraints) is automated, using default options in ArcMap. The label positions may, however, be modified manually if so desired.

Time–space correlation chart for southern Africa

A number of features of the time–space correlation charts which may be produced for the IGCP 509 project are illustrated in the following figures, based on examples drawn from a preliminary compilation of data for the Palaeoproterozoic of southern Africa. Capabilities and limitations of this approach to creating charts are also described. A more detailed assessment of the Palaeoproterozoic geology of southern Africa, together with more comprehensive cross-referencing of published literature sources, will be presented elsewhere towards the end of the IGCP 509 project.

The area covered by the Palaeoproterozoic compilation for southern Africa is illustrated in Figure 10, which also shows the extent of the various tectonic domains considered. The backdrop to this diagram is a compilation of aeromagnetic anomalies provided by the Council for Geoscience, South Africa, some features of which are important in defining domain boundaries and in regional correlations. The domain polygons are drawn from a separate GIS database or from individual shapefiles, not from the StratDB database.

Figure 11 illustrates the lithostratigraphy and metamorphism for the interval from c. 2300–1800 Ma. Here, we illustrate that it is possible to compile and present different correlation schemes, for instance that of the South Africa Committee for Stratigraphy (SACS) [as most recently summarized by Eriksson et al. (2006) and Moen (2006)] and more recent work based on dating of detrital zircons in sediments of the Kimberley domain. This recent work (Dorland 2004) has recognized that the Lucknow and Mapedi units, previously considered to be part of the Olifantshoek succession and coeval with the Waterberg Group, are much older and are actually coeval with units of the Segwagwa and Pretoria Groups (upper Transvaal Supergroup). The Lucknow and Mapedi units are therefore proposed to form an Elim Group which succeeds the Postmasburg Group in the Kimberley domain (Dorland 2004). This recent correlation places ^{13}C isotope excursions (blue squares in the figure) in the Lucknow Formation (upper Elim Group) and Silverton Shale (Pretoria Group) at similar ages (Bekker et al. 2009). Hartley basalt volcanism in the Kimberley domain (Cornell et al. 1998) is coeval with dykes in the Kanye domain but precedes other, younger post-Waterberg dykes in the same domains (Hanson et al. 2004). Thermal metamorphism associated with the Bushveld Complex (Witwatersrand domain) and coeval intrusions in the Kanye and Kimberley domains, is distinct from deformation, metamorphism and igneous activity associated with the Vredefort impact event (see DateView database for age information). At least two separate episodes of glacial activity (large X symbols) are evident in the lower part of the upper Transvaal succession (Postmasburg and Pretoria Groups).

The geodynamic setting of the units was initially epicratonic but was succeeded by collision-related activity during deposition of the Waterberg Group. Epicontinental sedimentation resumed along the western margin of the Kaapvaal Craton after about 1930 Ma, possibly with some faulting or thrusting subsequent to formation of the Neylan Formation (not labelled) and Hartley basalts, but prior to deposition of younger units of the Olifantshoek succession (Tinker et al. 2002). There is, however, no sign of significant (Kheis) orogenesis in the Kimberley domain during this interval of time, a point that has been made previously (Eglington & Armstrong 2004; Eglington 2006).

Depositional environments for the Transvaal Supergroup varied considerably (Fig. 12). The lower part of the succession was dominated by shallow to deeper marine environments with deeper facies predominating along the western edge of the Kaapvaal Craton (Coetzee 2001; Dorland 2004; Eriksson et al. 2006; Sumner & Beukes 2006). Recent dating (Dorland 2004) suggests that conventional correlation of the Black Reef Formation (Kanye, Witwatersrand and Pietersburg domains) with the Vryburg Formation (Kimberley domain) is incorrect. His work suggests that the Black Reef conglomerates are considerably younger than the Vryburg sediments and that a correlation with the Motiton Member, Monteville Formation is more

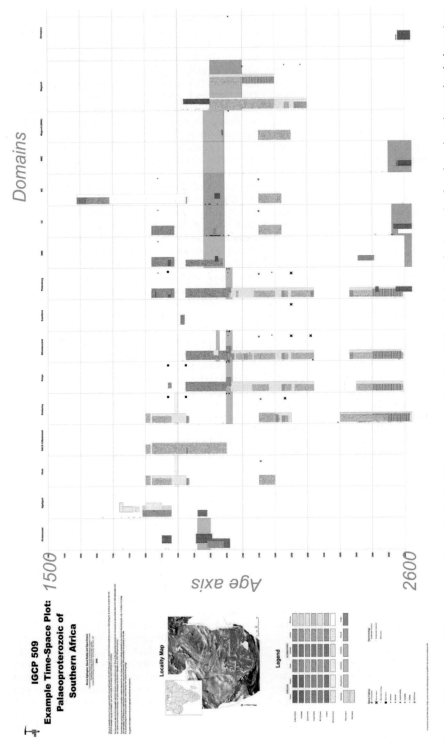

Fig. 9. Miniature view of time–space correlation chart for the Palaeoproterozoic of southern Africa. Symbology is for the matrix of geodynamic settings and rock classes (see Fig. 7). Lithostratigraphic unit labels have been omitted for clarity and a few labels (large grey text) have been added. Full version of this chart and of an equivalent for depositional setting are available as supplementary information.

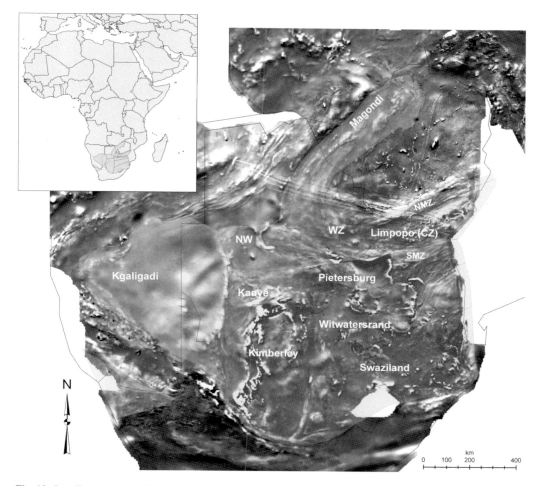

Fig. 10. Locality map (copied from the full time–space correlation chart) for southern Africa. Domain extents and boundaries are shown over an aeromagnetic compilation of the region (provided by the Council for Geoscience, South Africa).

likely. After an almost 100 Ma hiatus in sedimentation (or at least no preservation), sedimentation resumed with more proximal sediments dominated by fluvial and lacustrine settings, although marine environments continued along the western edge of the Kaapvaal Craton (Kimberley domain). A prominent episode of volcanism formed the Ongeluk, Tsatsu and Hekpoort units. Age constraints on the succession are provided by zircons from tuffs in the lower Transvaal succession in the Kimberley domain but very few direct geochronological data are available in the other domains.

Plutonic igneous activity is recorded in the Pietersburg domain (ages based on recalculation of multiple bulk zircon analyses National Physical Research Laboratory, Council for Scientific and Industrial Research; see DateView database) and was coeval with the deposition of carbonate sediments in the same domain and with the latter stages of plutonism in the central zone of the Limpopo Belt (Fig. 9 and full size charts). This association presents some conceptual problems and highlights the benefits of time–space correlation charts for emphasizing issues warranting further investigation. In this case, how does one juxtapose plutonic igneous activity and contemporaneous deposition of marine carbonates? Possibly this could be explained by considering the geographic distribution of the units concerned, an aspect that cannot be illustrated on time–space correlation charts alone. Another issue with the current approach to constructing the rectangles (polygons) for igneous units is that their

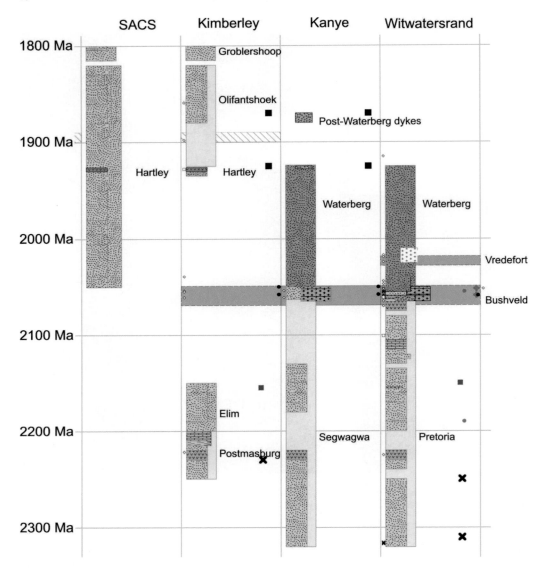

Fig. 11. Extract from the southern African time–space correlation chart to illustrate features relevant to correlation of the upper Transvaal succession and overlying lithostratigraphic units. Automatic labelling by the GIS system has been omitted (to reduce confusion due to the reduction in size of the image) and selected labels have been added manually in a graphics package. For legend see Figure 7.

vertical dimension is controlled by either real duration of activity or the uncertainty of the ages. In the case of the Moletsi granite, the range in age is due to age uncertainty and not due to a long duration of igneous activity. Some way to distinguish between, and illustrate, these two options would be useful but, other than to use different line styles for the polygon borders (as is currently provided), this issue is not resolved. In some cases, sedimentary units have similar limitations but usually the relationship between units within a hierarchical succession provides some limitation on possible minimum and maximum ages where geochronological constraints are absent or insufficient.

Igneous and metamorphic activity associated with the Bushveld igneous event and with the Limpopo Belt are two distinct events, as shown in Figure 13. Most dates for c. 2 Ga activity in the

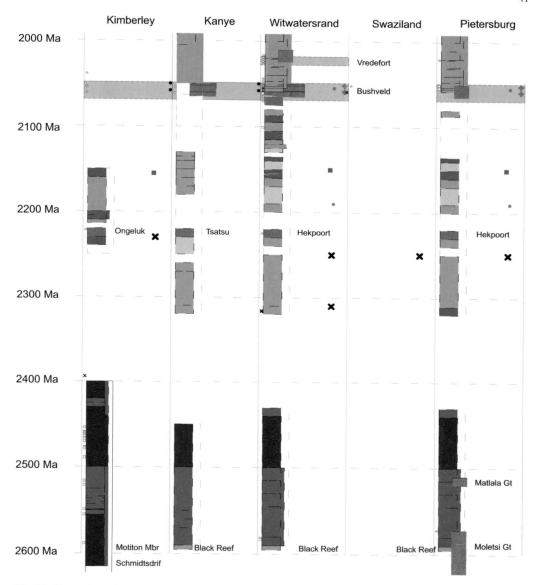

Fig. 12. Extract from the southern African time–space correlation chart to illustrate aspects of the environment of deposition view. Automatic labelling by the GIS system has been omitted (to reduce confusion due to the reduction in size of the image) and selected labels have been added manually in a graphics package. For legend see Figure 8.

Limpopo Belt are from the central and western zones of the belt. Dates associated with lower temperature closure of isotope systems are not shown but help define the age limits and grades of metamorphism in the various domains constituting the Limpopo Belt. The age span of metamorphism and deformation suffers from a similar problem to that of plutonic igneous activity. The interval for high grade metamorphism in the central zone of the Limpopo Belt is well constrained by numerous zircon and monazite ages but medium to high-grade metamorphism in the Magondi Belt is largely unconstrained. In this latter case, metamorphism must be younger than deposition of the Piriwiri and Lomagundi sediments (also poorly constrained) and older than post-tectonic granitoids. The exact duration and extent to which metamorphism and plutonism might be diachronous is not easily

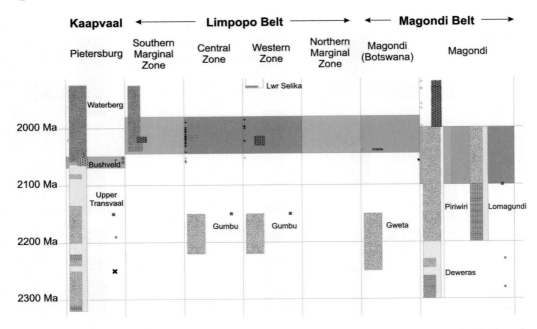

Fig. 13. Extract from the southern African time–space correlation chart to illustrate aspects of correlation for the northern Kaapvaal Craton, various domains in the Limpopo Belt and the Magondi Belt of northwestern Zimbabwe. Symbology is based on geodynamic setting and rock class. Automatic labelling by the GIS system has been omitted (to reduce confusion due to the reduction in size of the image) and selected labels have been added manually in a graphics package. For legend see Figure 7.

illustrated in time–space correlation charts intended to provide a broad, regional perspective.

Deposition of sediments along the northwestern margin of Zimbabwe (Deweras and younger Piriwiri and Lomagundi successions) must have been essentially contemporaneous with deposition of the upper Transvaal sediments on the Kaapvaal Craton (here illustrated only for the Pietersburg domain) and with sediments of the Gumbu Group (Buick et al. 2003) in the Limpopo Belt and of sediments encountered during drilling at Gweta and at Sua Pan, Botswana (Mapeo et al. 2001; Majaule et al. 2001). About 2.0 Ga to c. 1.93 Ga, post-tectonic igneous activity is recorded from the Magondi Belt of NW Zimbabwe (Treloar & Kramers 1989; Munyanyiwa et al. 1997; McCourt et al. 2000), effectively providing a minimum age for Lomagundi and Piriwiri sedimentation. Metamorphism in the Magondi Belt is broadly contemporaneous with that in the Limpopo Belt. Indeed, sediments with similar depositional and metamorphic ages follow the arcuate trend in aeromagnetic anomalies evident in Figure 10 (see also Fig. 14). Several of these sediments and time-equivalents on the Kaapvaal craton also exhibit major ^{13}C isotope excursions (note blue squares on figure) (Schidlowski et al. 1976; Buick et al. 2003; Bekker et al. 2009), and have been inferred to represent a single perturbation in the carbon isotope composition of marine carbonates. Traditionally, the Magondi Belt has been thought to extend southwards along the western margin of the Kaapvaal craton to join up with the Kheis Belt (Hartnady et al. 1985). Eglington & Armstrong (2004) and Eglington (2006) have emphasized, though, that there is no geochronological evidence for a major c. 2 Ga orogeny normally envisaged for the southwestern margin of the Kaapvaal Craton. Eglington & Armstrong (2004) also drew attention to the arcuate aeromagnetic anomaly pattern around the western, southwestern and southern margins of the Zimbabwe Craton. Several earlier studies have also commented on or illustrated possible links of the Magondi and Limpopo Belts (Hartnady et al. 1985; Mapeo et al. 2001; Ranganai et al. 2002). The time–space correlation charts (Figs 9 & 13), GIS maps of coeval units (Fig. 14), geophysics (Fig. 10) and geochronology (Fig. 13) all appear to support a link between these two belts, reflecting Palaeoproterozoic movement of the Zimbabwe Craton to the SW. The various zones of the Limpopo Belt have traditionally been considered in isolation from the possible broader regional picture (Barton et al. 2006; Kramers et al. 2006) which may, in part, have contributed to

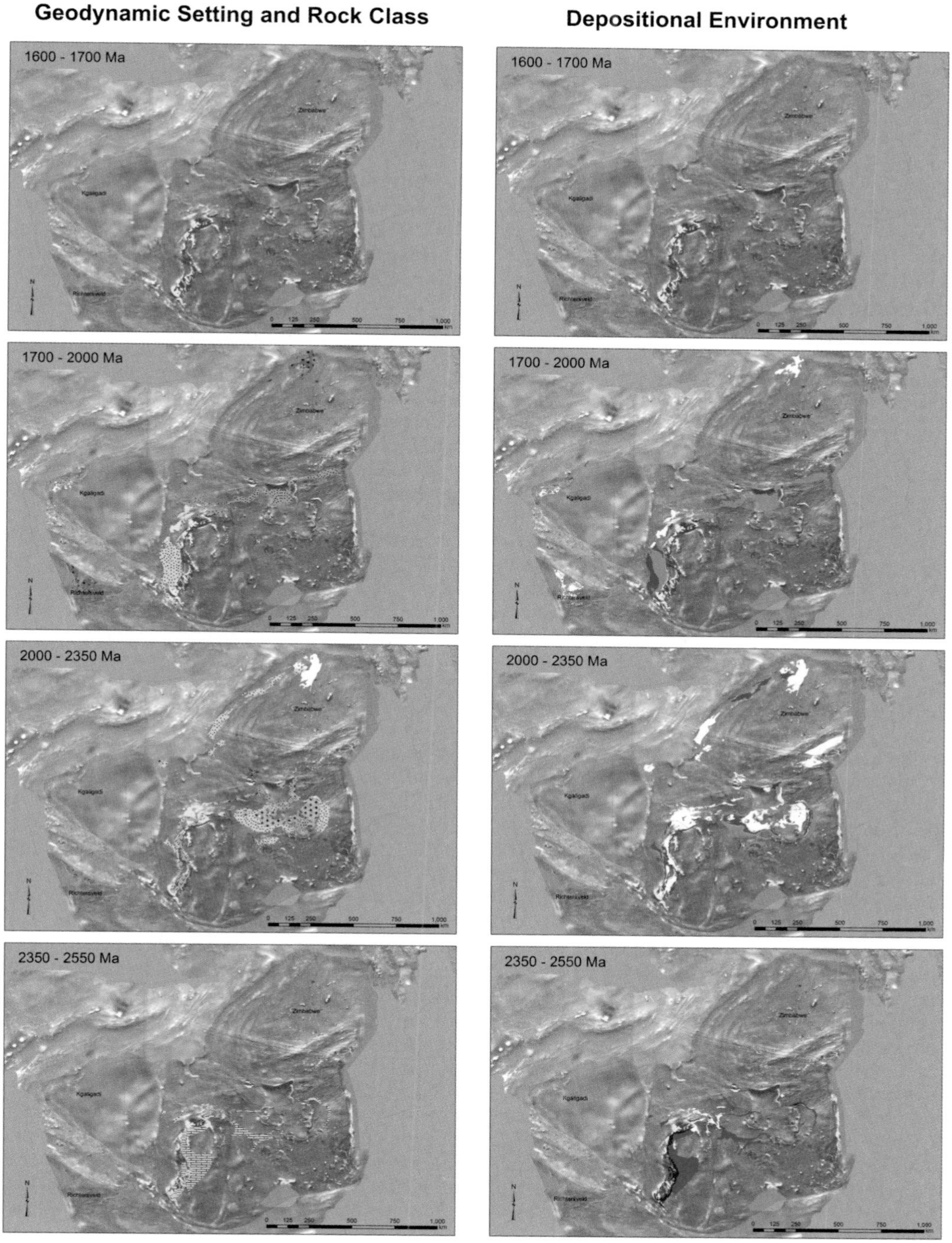

Fig. 14. Time slice maps for the Palaeoproterozoic evolution of southern Africa, using symbology for geodynamic setting v. rock class and for sedimentary depositional environment. Maps were constructed in ArcGIS using linked attribute information from the StratDB database and GIS maps for South Africa (provided by the Council for Geoscience, South Africa), Namibia (provided by the Geological Survey of Namibia), Botswana (Key & Ayres 2000), Zimbabwe (cut from the 1:5 000 000 geological map of Africa, Council for Geoscience, South Africa) and the 1:2 500 000 geological map of sub-Kalahari units (Council for Geoscience, South Africa). Time slices were selected to match logical breaks in sedimentation as identified in the time–space correlation chart of the area. For legend: left hand panels see Figure 7; right hand panels see Figure 8.

disagreements as to the regional significance of *c.* 2 Ga activity relative to Late Archaean activity (see Barton *et al.* & Kramers *et al.* for recent summaries). If the Zimbabwe Craton was an indenter into early Proterozoic sediments around its western, southwestern and southern margins, then this has significant implications for the palaeogeographic evolution of the region and for mineral exploration.

Time-slice maps illustrating lithostratigraphic evolution of southern Africa

Since the StratDB database permits one to store link-field values for individual units in different GIS maps, it is very easy to construct time-slice maps which illustrate the development of the domains included in the correlation charts. Example maps for southern Africa, produced from data exported from StratDB and linked to GIS information from five different map compilations, are illustrated in Figure 14. Here, the maps illustrate the Rock Class – Geodynamic Settings and the Depositional Environment attributes stored in StratDB but a similar approach could be used for any attribute information from the database or from offline compilations which use the unique unit IDs created for StratDB. These GIS compilations provide a useful insight into the progressive development of crust during the Palaeoproterozoic of southern Africa. Early epicontinental, mostly marine sedimentation dominated by carbonate and then ironstone lithologies spread across the Kaapvaal Craton with some plutonic and late metamorphic igneous activity continuing in parts of the Limpopo Belt. This was followed by continued epicontinental sedimentation in fluvial and lacustrine to shallow marine settings on the Kaapvaal Craton and along the northwestern and southwestern margins of the Zimbabwe craton. Marine facies were most common along the western margins of both cratons. Fault-bounded basins associated with sinistral collision and movement of the Zimbabwe Craton relative to the Kaapvaal Craton then provided repositories for collision-related sediments of the Waterberg Group and Soutpansberg Group subsequent to *c.* 2.05 Ga. Clastic, epicontinental sedimentation continued on the western margin of the Kaapvaal Craton. Sparse evidence for sedimentation from 1700–1600 Ma is preserved along the western margin of the Kaapvaal Craton and in Botswana.

The time–space correlation charts and time-slice maps highlight avenues for further research. Extension of the maps and correlation charts back in time may help elucidate the earlier (Late Archaean) evolution of the Limpopo Belt by providing inter-regional perspectives and data compilations otherwise not always available or freely accessible.

Future directions

StratDB and DateView are both a work in progress and are still under active development as suggestions for improvements are received from the user community. This development proceeds along several fronts.

First, the database design can be modified or extended. StratDB provides a very useful central system in which information for lithostratigraphic units, tectonic domains, large igneous provinces and ore deposits is captured. Each record receives a unique database ID which may be used as a common link-field value by external databases storing information for other purposes. For instance, palaeomagnetic information stored in an external database could include the StratDB unit ID and so facilitate linking of the two databases or StratDB could be modified to store unique ID values (as used by the external database) for each palaeomagnetic pole. Geochemical data for units may be stored in a similar way. The DateView geochronology database also utilizes the same unique unit ID and could be linked to a palaeomagnetic database to show all available ages associated with specific poles, in the same way that StratDB currently draws age information for lithostratigraphic units, tectonic domains, large igneous provinces and ore deposits from DateView. The database design also provides links to GIS shapefiles, geodatabases, coverages, etc., which may facilitate enhanced spatial interfacing of the data in future and could be used to plot the location of igneous or metamorphic activity, mineralization, etc. for plates after rotation about selected Euler poles. Another area of application for both DateView and StratDB is with efforts to capture information for magmatic activity associated with large igneous provinces. Ernst & Buchan (2001) have already compiled a database of many of these episodes and the geochronological data used by Prokoph *et al.* (2004) have been imported to DateView and to StratDB. Association of these records in StratDB with unique lithostratigraphic unit IDs will facilitate future linking of these data whilst also providing continued updating of the geochronology as new data become available for the various plutonic, hypabyssal and volcanic intrusions. Ore deposit information has been imported from the Geological Survey of Canada's (GSC) global ore deposit database and contains links to locality, clan type, mineralization age, host unit and major commodities. Addition of data from other economic databases and links to these systems are envisaged. An advantage of such linkages is that new data added to any one database are immediately visible to other systems such that users may concentrate on their own fields of expertise, yet still draw on information compiled by experts in other sub-disciplines.

Second, the user interface will also require updates as user requirements grow and change. At present, all software development is by one of us (BME) but, hopefully, others will play a role in future. The Delphi language versions originally used for development could not handle Unicode strings, but this has changed and full international language support is being fully implemented. The present interface is entirely via a browser interface but other web interfaces will no doubt become necessary. Possibilities include accessing data from other web portals and systems using web services such as the simple object access protocol (SOAP). Controlling user permissions will, however, need to be addressed before this protocol is implemented. Alternatively, web services could be set up to provide only basic functionality and access to public records.

As time permits, it will also be useful to provide options for online capture of information, both for individual records and for batches of records. Graphs will also be added to illustrate the variation of detrital zircon age data (extracted from raw data stored in the DateView database) relative to interpreted stratigraphic age (the minimum age field) for each unit. Direct output of the information required to produce the time–space correlation charts to either shapefiles or a personal geodatabase would also be an advantage so as to reduce the effort required to produce these charts.

The database design is intended to facilitate control of the contents of the database by an interested community of collaborators, more of whom will hopefully accept the role of moderators (validators) for information in the database. Future involvement of other international interest groups, beyond IGCP 509, is also likely and will help ensure that the database remains a long-term resource which facilitates international earth science research. As with any database endeavour, continued relevance of the data captured will require the ongoing interest and involvement of the user community to ensure that changes in knowledge are reflected in the contents of the database and that the database remains available for use.

Conclusions

The StratDB database system described here will play a significant role in the achievement of several of the objectives of the IGCP 509 project. It will enhance the dissemination of lithostratigraphic information for many countries to a broader community and will help non-specialists to easily view information for various Palaeoproterozoic tectonic domains. The database, its web interface and the outputs of the system will greatly facilitate the construction of standardized time–space correlation charts by leveraging other technologies such as GIS.

Although primarily intended at this stage to cater for the IGCP 509 project in its investigation of the Palaeoproterozoic, the design of the database is flexible and can handle data from all of Earth history. Rapid and easy comparison of domains, now widely separated on different continents, using the StratDB and DateView databases, coupled with the GIS charting and mapping technology described here, will greatly facilitate palaeogeographic assessments and enhance understanding of the Palaeoproterozoic development of the Earth. As the first and, to the best of our knowledge, only international databases of their kind, it is hoped that the design and implementation of StratDB and the associated DateView database will stimulate the collation of international data in a uniform structure which will benefit many Earth science researchers in years to come.

This is a contribution to IGCP 509. UNESCO and the International Geoscience Programme are thanked for providing some funding to facilitate a workshop from which design enhancements for the database system flowed. CodeGear (previously Borland), AtoZed, TMS, Steema, HK-Software and ESRI provided educational versions of their software. The Saskatchewan Isotope Laboratory, which hosts the databases, receives financial support for its infrastructure from the University of Saskatchewan. The Council for Geoscience, South Africa, and the Geological Survey of Namibia provided digital copies of various maps and geophysical images which were invaluable in developing the database system and in creating some of the diagrams, charts and maps. We also thank all the regional coordinators and other contributors to the IGCP 509 project for their assistance in populating the database and for suggestions for improvements. Two reviewers, S. Pisarevsky and N. Neumann, provided several suggestions for improvements to the manuscript. This is TIGeR Publication No. 166.

References

ANSDELL, K. M., HEAMAN, L. M. ET AL. 2005. Correlation chart of the evolution of the Trans-Hudson Orogen – Manitoba–Saskatchewan segment. *Canadian Journal of Earth Sciences*, **42**, 761.

BARTON, J. M., KLEMD, R. & ZEH, A. 2006. The Limpopo Belt: a result of Archean to Proterozoic, Turkic-type orogenesis? In: REIMOLD, W. U. & GIBSON, R. L. (eds) *Processes on the Early Earth*. Geological Society of America, Special Paper, **405**, 315–332.

BEKKER, A., BEUKES, N. J., HOLMDEN, C. H., KENIG, F., EGLINGTON, B. M. & PATTERSON, W. P. 2009. Fractionation between inorganic and organic carbon during the Lomagundi (2.22–2.1 Ga) carbon isotope excursion. *Earth and Planetary Science Letters*, **271**, 278–291.

BUICK, I. S., WILLIAMS, I. S., GIBSON, R. L., CARTWRIGHT, I. & MILLER, J. A. 2003. Carbon and U-Pb evidence for a Palaeoproterozoic crustal component in the Central Zone of the Limpopo Belt, South Africa. *Journal of the Geological Society, London*, **160**, 601–612.

COETZEE, L. L. 2001. *Genetic stratigraphy of the Palaeoproterozoic Pretoria Group in the western Transvaal.* MSc thesis (unpubl.), Rand Afrikaans University.

CORNELL, D. H., ARMSTRONG, R. A. & WALRAVEN, F. 1998. Geochronology of the Proterozoic Hartley Basalt Formation, South Africa: constraints on the Kheis tectogenesis and the Kaapvaal Craton's earliest Wilson Cycle. *Journal of African Earth Sciences*, **26**, 5–27.

DORLAND, H. C. 2004. *Provenance ages and timing of sedimentation of selected neoarchean and Palaeoproterozoic successions on the Kaapvaal Craton.* PhD thesis (unpubl.), Rand Afrikaans University.

EGLINGTON, B. M. 2004. DateView: a Windows geochronology database. *Computers and Geosciences*, **30**, 847–858.

EGLINGTON, B. M. 2006. Evolution of the Namaqua-Natal Belt, southern Africa – a geochronological and isotope geochemical review. *Journal of African Earth Sciences*, **46**, 93–111.

EGLINGTON, B. M. & ARMSTRONG, R. A. 2004. The Kaapvaal Craton and adjacent orogens, southern Africa: a geochronological database and overview of the geological development of the craton. *South African Journal of Geology*, **107**, 13–32.

EGLINGTON, B. M., TUCKER, S., WOLMARANS, L. G. & JOHNSON, M. R. 2001. Desktop implementation of South African lithostratigraphic unit database: version 2.0. *Council for Geoscience Open File Report*, 2001–0027, 1–21.

ERIKSSON, P. G., ALTERMANN, W. & HARTZER, F. J. 2006. The Transvaal Supergroup and its precursors. *In*: JOHNSON, M. R., ANHAEUSSER, C. R. & THOMAS, R. J. (eds) *The Geology of South Africa*, Geological Society of South Africa and the Council for Geoscience, Pretoria, 237–260.

ERNST, R. E. & BUCHAN, K. L. 2001. Large mafic magmatic events through time and links to mantle-plume heads. *In*: ERNST, R. E. & BUCHAN, K. L. (eds) *Mantle Plumes: their Identification through Time*. Geological Society of America, Special Paper, **352**, 483–575.

GRADSTEIN, F. M., OGG, J. G. & SMITH, A. G. 2004. Chronostratigraphy: linking time and rock. *In*: GRADSTEIN, F. M., OGG, J. G. & SMITH, A. G. (eds) *A Geologic Timescale 2004*. Cambridge University Press, Cambridge, 20–46.

HANSON, R. E., GOSE, W. A. *ET AL*. 2004. Palaeoproterozoic intraplate magmatism and basin development on the Kaapvaal Craton: age, paleomagnetism and geochemistry of ~1.93 to ~1.87 Ga post-Waterberg dolerites. *South African Journal of Geology*, **107**, 233–254.

HARTNADY, C. J. H., JOUBERT, P. & STOWE, C. W. 1985. Proterozoic crustal evolution in southwestern Africa. *Episodes*, **8**, 236–244.

HARTZER, F. J., JOHNSON, M. R. & EGLINGTON, B. M. 1998. *Stratigraphic Table of South Africa*. Council for Geoscience, Pretoria, South Africa.

KEY, R. M. & AYRES, N. 2000. The 1998 edition of the National Geological Map of Botswana. *Journal of African Earth Sciences*, **30**, 427–451.

KRAMERS, J. D., MCCOURT, S. & VAN REENEN, D. D. 2006. The Limpopo Belt. *In*: JOHNSON, M. R., ANHAEUSSER, C. R. & THOMAS, R. J. (eds) *The Geology of South Africa*. Geological Society of South Africa and the Council for Geoscience, Pretoria, South Africa, 209–236.

MAJAULE, T., HANSON, R. M., KEY, R. M., SINGLETARY, S. J., MARTIN, M. W. & BOWRING, S. A. 2001. The Magondi Belt in northeast Botswana: regional relations and new geochronological data from the Sua Pan area. *Journal of African Earth Sciences*, **32**, 257–267.

MAPEO, R. B. M., ARMSTRONG, R. A. & KAMPUNZU, A. B. 2001. SHRIMP U–Pb zircon geochronology of gneisses from the Gweta borehole, northeast Botswana: implications for the Palaeoproterozoic Magondi Belt in southern Africa. *Geological Magazine*, **138**, 299–308.

MCCOURT, S., HILLIARD, P., ARMSTRONG, R. A. & MUNYANYIWA, H. 2000. SHRIMP U–Pb zircon geochronology of the Hurungwe granite northwest Zimbabwe: age constraints on the timing of the Magondi orogeny and implications for the correlation between the Kheis and Magondi belts. *South African Journal of Geology*, **104**, 39–46.

MOEN, H. F. G. 2006. The Olifantshoek Supergroup. *In*: JOHNSON, M. R., ANHAEUSSER, C. R. & THOMAS, R. J. (eds) *The Geology of South Africa*. Geological Society of South Africa and the Council for Geoscience, Pretoria, South Africa, 319–324.

MUNYANYIWA, H., KRONER, A. & JAECKEL, P. 1997. U–Pb and Pb–Pb single ages for charnoenderbites from Magondi mobile belt, northwest Zimbabwe. *South African Journal of Geology*, **98**, 52–57.

PROKOPH, A., ERNST, R. E. & BUCHAN, K. L. 2004. Time-series analysis of large igneous provinces: 3500 Ma to present. *Journal of Geology*, **112**, 1–22.

RAINES, G. L., HASTINGS, J. T. & MOYER, L. A. 2007. ArcGeology (version 1): a geodatabase design for digital geologic maps using ArcGIS. http://support.esri.com/index.cfm?fa=downloads.dataModels.filteredGateway&dmid=30

RANGANAI, R. T., KAMPUNZU, A. B., ATEKWANA, E. A., PAYA, B. K., KING, J. G., KOOSIMILE, D. I. & STETTLER, E. H. 2002. Gravity evidence for a larger Limpopo Belt in southern Africa and geodynamic implications. *Geophysical Journal International*, **149**, F9–F14.

SCHIDLOWSKI, M., EICHMANN, R. & JUNGE, C. E. 1976. Carbon isotope geochemistry of Precambrian Lomagundi Carbonate Province, Rhodesia. *Geochimica et Cosmochimica Acta*, **40**, 449–455.

STEINSHOUER, W., QIANG, J., MCCABE, P. J. & RYDER, R. T. 1999. Maps showing geology, oil and gas fields, and geologic provinces of the Asia Pacific region. *USGS Open File Report*, 97–470F, 1–16.

SUMNER, D. Y. & BEUKES, N. J. 2006. Sequence stratigraphic development of the Neoarchean Transvaal carbonate platform, Kaapvaal Craton, South Africa. *South African Journal of Geology*, **109**, 11–22.

TINKER, J., DE WIT, M. J. & GROTZINGER, J. P. 2002. Seismic stratigraphic constraints on Neoarchaean – Palaeoproterozoic evolution of the western margin of the Kaapvaal Craton, South Africa. *South African Journal of Geology*, **105**, 107–134.

TRELOAR, P. J. & KRAMERS, J. D. 1989. Metamorphism and geochronology of granulites and migmatitic granulites from the Magondi Mobile Belt, Zimbabwe. *Precambrian Research*, **45**, 277–289.

WARDLE, R. J., GOWER, C. F. ET AL. 2002. Correlation chart of the Proterozoic assembly of the northeastern Canadian–Greenland Shield. *Canadian Journal of Earth Sciences*, **39**, 895.

The Columbia connection in North China

T. M. KUSKY[1]* & M. SANTOSH[2]

[1]*Department of Earth and Atmospheric Sciences, St. Louis University, MO, USA*
[2]*Faculty of Science, Kochi University, Akebono-cho 2-5-1, Kochi 780-8520, Japan*
Corresponding author (e-mail: kusky@eas.slu.edu)

Abstract: The Archaean and Proterozoic geology, structure and metamorphism of the North China Craton (NCC) reveal that the amalgamated Eastern and Western blocks of the craton collided as a single entity with the Columbia supercontinent at 1.93 Ga, and that the Northern Hebei orogen correlates with the Transamazonian-Eburnian belts of Africa and South America, and the Svecofennian of Baltica. This metamorphic belt preserves evidence for extreme crustal metamorphism with diagnostic ultrahigh-temperature (UHT) assemblages such as sapphirine + quartz, spinel + quartz, high alumina orthopyroxene + sillimanite + quartz and high temperature perthites which record temperatures exceeding 1000 °C and pressure above 12 kbar. The metamorphic $P-T$ trajectory is characterized by initial isobaric cooling followed by steep isothermal decompression, defining an overall anticlockwise exhumation history. Electron probe monazite geochronology and precise SHRIMP zircon dating of sapphirine-bearing granulites constrain the timing of the UHT event at c. 1.92 Ga, suggesting that the UHT metamorphism coincided with collisional orogenesis as the North China Craton joined the Columbia supercontinent amalgam along the Northern Hebei orogen during the Palaeoproterozoic. Fluid inclusion studies in the UHT rocks provide evidence for the involvement of synmetamorphic pure CO_2, linking the thermal anomaly and fluid flux to underplated mafic magmas during asthenospheric upwelling. Evidence for probable-plume related mafic magmatism is also provided by the extensive mafic dyke swarms cutting the region and elsewhere within the NCC, with geochemical characters testifying to emplacement within rifts that opened up during the extensional collapse of the orogen subsequent to the collisional event. Recognition of the Palaeoproterozoic Columbia suture in the Northern Hebei orogen represents a major paradigm shift, as one popular group of models for the NCC suggests that the Palaeoproterozoic suture resides in an older orogen, the Central Orogenic belt. However, either model for the location of the suture is able to explain metamorphic $P-T-t$ data for crustal thickening at 1.85 Ga, whereas only the model for a Late Archaean collision in the Central Orogenic belt and a Palaeoproterozoic collision in the Northern Hebei orogen can explain the structural, sedimentological, geochronological, and petrological data. The Central Orogenic belt contains several hundred fragments of a c. 2.505 Ga ophiolite suite, a contemporaneous 2.5–2.4 Ga foreland basin deposited on 2.7–2.5 Ga passive margin sediments on the Eastern Block, and contains rare evidence for c. 2.5 Ga granulite facies metamorphism that was largely overprinted by 1.92–1.85 Ga high-grade assemblages. East-directed 2.5 Ga fold-thrust structures are overprinted by 1.92–1.85 Ga south-directed thrusts associated with large-scale thickening of the craton, succeeded by strike-slip shear zones that slice the orogen into numerous fault-bounded terranes that preserve different levels of exhumation.

Understanding the distribution of continental blocks in pre-Rodinian supercontinents is hampered by a paucity of reliable palaeomagnetic data, by limited distribution of outcrop belts of orogens related to the amalgamation of these continents, and by incomplete knowledge of the tectonic histories of these belts. Global similarities of ages of high-grade metamorphism associated with continental collision suggests that there was a peak in the amalgamation of landmasses into a supercontinent in the late Palaeoproterozoic (Rogers & Santosh 2003; Condie 2002, 2003a, 2004; Zhao et al. 2001a, b, 2002a, b, 2006; Kusky et al. 2007a, b, c). Rogers & Santosh (2003, 2004) proposed the framework of a supercontinent, Columbia, that formed and was broken up during the period between 1.9 and 1.5 billion years ago (Fig. 1). They proposed that Columbia may have contained nearly all of the Earth's continental blocks during this period when eastern India, Australia and attached parts of Antarctica were apparently sutured to western North America, and the eastern margin of North America to the southern margin of Baltica. The Inner Mongolia–Northern Hebei orogen (NHO) of North China has recently been recognized as a major Palaeoproterozoic orogen that links Baltica, North China, Tarim, West Africa and the western margin of the Amazon shield in a continuous zone of continental outbuilding (Kusky

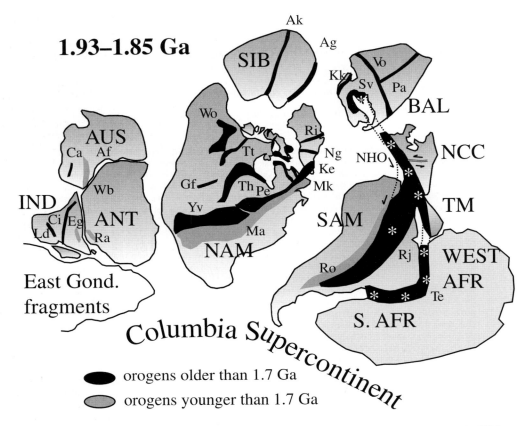

Fig. 1. Reconstruction of the Columbia Supercontinent at 1.93–1.85 Ga, modified after Rogers & Santosh (2004) and Kusky *et al.* (2007*a, b, c, d*). Palaeoproterozoic orogens are shown in black, younger Neoproterozoic ones in grey. Stars show the locations of *c.* 1.9–1.85 Ga UHT and related metamorphic assemblages that we relate to the amalgamation of Columbia. Note that the timing of events in the North China Craton coincides with those in adjacent cratons when the collisional orogen that brought the NCC into Columbia is recognized as the Northern Hebei orogen. This orogen continues for 3000 km across the NCC and through the Tarim block, and then connects with the Transamazonian-Eburnian and Rio Negro belts of Africa and South America, and the Svecofennian belt of Baltica. Abbreviations of cratons as follows: AFR, Africa; IND, India; AUS, Australia; ANT, Antarctica; NAM, North America; SIB, Siberia; BAL, Baltica; NCC, North China Craton; SAM, South America. Abbreviations of orogens as follows: Af, Albany-Fraser; Ad, Aravalli-Delhi; Ak, Akitkan; An, Angara; Ca, Capricorn; Ci, Central Indian; Eg, Eastern Ghats; Fo, Fox; Gf, Great Falls; Ke, Ketilidian; Kg, Konisbergian-Gothian; Kk, Kola-Karelia; La, Labradorian; Li, Limpopo; Ln, Lurio-Namama; Ma, Mazatzal; Mk, Makkovikian; Ng, Nagssugtoqidian; NHO, North Hebei Orogen; Nq, New Quebec; Pa, Pachemel; Pe, Penokian; Ra, Rayner; Ri, Rinkian; Rj, Rio Negro-Juruena; Ro, Rondonian; Sv, Sveckofennian; Te, Transamazonian-Eburnian; Th, Trans-Hudson; TM, Tarim; Tt, Thalston-Thelon; To, Torngat; Un, Ungava; Wb, Windmill Islands-Bunger Hills; Wo, Wopmay; Yv, Yavapai.

et al. 2007*a* and references therein), making the North China Craton (NCC) a keystone block in matching different elements of Columbia. This orogen contains a number of accreted arcs recording different magmatic pulses, ultrahigh-temperture (UHT) metamorphism, ophiolites and accretionary terranes. Fragmentation locally began at *c.* 1.8 Ga and in earnest by 1.6 Ga, when rifting occurred in North China and across Columbia. Rifting continued until about 1.4 Ga in most of Columbia, and a similar age of rifting north of the Zimbabwe craton of southern Africa suggests that an entire continental block stretching from Australia to South Africa separated from Columbia at this time. Further separation of North America from South America/Africa and rotation of the different blocks ultimately led to the fragmentation of Columbia and its reorganization to Rodinia.

Several models have previously been proposed with different scenarios about the location of the NCC in the Palaeoproterozoic supercontinent, and most of these were based on petrological and geochonological data (Wilde et al. 2002; Zhai et al. 2005), as well as the synthesis of regional palaeomagnetic data and distribution of major orogenic belts (Rogers & Santosh 2004). However, detailed structural studies as well as sequence stratigraphy are also critical for a complete understanding of orogenesis (e.g. Polat et al. 2007). In some of the recent models for the NCC, the data were largely derived from a geographically-limited outcrop area at the junction of a north–south striking Archaean orogen, overprinted by a east–west striking Palaeoproterozoic orogen, which was again overprinted by a north–south striking Mesozoic–Cenozoic extensional orogen. The Palaeoproterozoic $P-T-t$ paths were erroneously correlated with the north–south striking orogen in these works (e.g. Zhao et al. 2006a, b and references therein). The late Palaeoproterozoic to early Mesoproterozoic tectonic framework and tectonothermal episodes of the NCC have recently been re-interpreted (Kusky & Li 2003; Kusky et al. 2007a, b, c; Polat et al. 2005a, b, 2006, 2007; Li & Kusky 2007; Hou et al. 2007; Faure et al. 2007) and these new models necessitate a re-evaluation of the position of the NCC in the Columbia supercontinent, and its correlation with other cratons in the period 2.00–1.70 Ga.

Most importantly, the new tectonic models for the NCC have shown that the c. 1.93–1.85 Ga orogenic belt associated with incorporation of the NCC into the Columbia supercontinent is located along the present northern margin of the craton (Kusky & Li 2003; Kusky et al. 2007a, c, d), and not in the centre of the craton as assumed by earlier workers (Zhao et al. 2006a, b; Wilde et al. 1998). This belt of 1.93–1.85 Ga deformation can now be shown to extend another 2000 km from along the northern margin of the NCC through the adjacent Tarim block (Zhang et al. 2007), extending the known length of the North Hebei orogen in China to more than 3000 km. Deformation in the interior of the North China Craton at this time includes farfield north over south thrusting in the Jiao-Liao-Ji belt, and along the Zhuzhifang fault in the Hengshan, and strike-slip shear zones that merge with the orogen in the north (O'Brien et al. 2005; Kusky et al. 2007b; Faure et al. 2007). Granulite facies metamorphism at 1.9–1.8 Ga is not confined to the Central Orogenic belt (Trans-North China Orogen of Zhao et al. 2001a, b, 2005, 2006a, b), but strikes right across the Central Orogenic belt and affects much of the north half of the craton. Reconstructions of Columbia that assume that the eastern and western blocks of the NCC amalgamated at 1.85 Ga during Columbia assembly are therefore incorrect and need revision. In this contribution we present details of a new and more viable correlation of the NCC with Columbia, based on some of the new findings and a better understanding of the tectonics of the NCC.

The new recognition that North China accreted with Columbia at 1.92 Ga along the Northern Hebei orogen, and that this 1.9 Ga event extends for >5000 km along strike, demonstrates that the Palaeoproterozoic suture could not be in the Central Orogenic belt (or TNCO of Zhao et al. 2006a, b and references therein). The 1.92 Ga suture encloses the Eastern Block in the internal zones of the orogen, such that models that postulate a 1.85 Ga suture between the Eastern and Western blocks find the impossible situation of requiring a younger collision within blocks that had already accreted. The only major collisional events possible in the Central Orogenic belt must be older than 1.92 Ga.

Geological evolution of North China Craton

Tectonic divisions and Archaean history of the craton

Beneath an extensive Phanerozoic cover, the North China Craton (Fig. 2) is comprised dominantly of c. 3.8–2.5 Ga banded gneiss, tonalite-trondhjemite-granodiorite (TTG), granite, migmatite, amphibolite, ultramafite, mica schist and dolomitic marble, garnet-sillimanite- and graphite-bearing granulites (khondalites) interlayered with Mg–Al rich ultrahigh-temperature assemblages, calc-silicate rocks, banded iron formation (BIF) and metaarkose (Jahn & Zhang 1984a, b; Bai et al. 1992, 1996; Bai & Dai 1998; Wu et al. 1998; Jahn et al. 1987; He et al. 1991, 1992; Shen et al. 1992; Shen & Qian 1995; Wang et al. 1997; Kusky et al. 2007b; Santosh et al. 2007a). The most important period of crustal growth in the craton is between 2700 and 2520 Ma, based on a compilation of U–Pb and Sm–Nd ages, and Nd isotopic studies (Kusky et al. 2007a).

The North China Craton is divided into two major blocks (Fig. 2) separated by the Neoarchaean Central Orogenic belt in which virtually all U–Pb zircon ages (upper intercepts) fall between 2.55 and 2.50 Ga (Zhao et al. 1998, 1999a, b, 2000a, b, 2001a, b; Kröner et al. 1998, 2002, 2005; Li 2000a; Wilde et al. 1998, 2004; Zhao 2001; Kusky et al. 2001, 2007a, b; Kusky & Li 2003; Polat et al. 2006). The Western Block contains Archaean gneiss intruded by a narrow belt of 2.55–2.50 Ga arc plutons along its eastern margin. Much of the Archaean geology of the Western Block is poorly exposed as it is covered by thick platformal cover, ranging in age from Late Archaean to Cretaceous (e.g. Ritts et al. 2004).

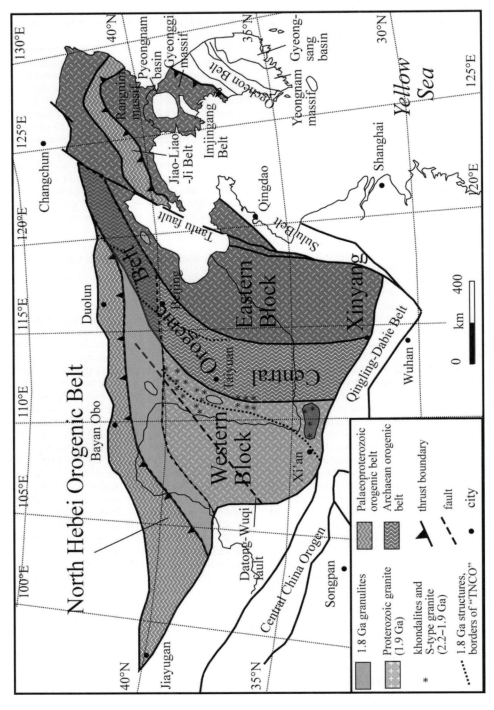

Fig. 2. Simplified tectonic map of the North China Craton, modified after Kusky et al. 2007a, b, c, d.

The Eastern Block contains a variety of c. 3.80–2.50 Ga gneissic rocks and greenstone belts locally overlain by 2.60–2.50 Ga sandstone and carbonate units. Deformation is complex, polyphase and indicates the complex collisional, rifting and underplating history of this block from the early Archaean through the Mesoproterozoic (Zhai et al. 1992, 1995, 2002, 2003; Zhou et al. 2002; Li et al. 2002; Kusky et al. 2001, 2003, 2004a, b; Polat et al. 2005a, b, 2006; Wan et al. 2006). Early Archaean basement (>3.4 Ga) is limited in extent to the Anshan and eastern Hebei areas, with c. 3.8 Ga trondhjemetic gneisses and 3.6–3.4 Ga supracrustal rocks (Song 1992; Li et al. 2001a, b). Middle Archaean (3.4–2.8 Ga) basement includes the Lower Anshan, Qianxi and Longgang Groups with related granite plutons. Late Archaean (2.8–2.5 Ga) basement, occupying 85% of the total exposure of Archaean basement of Eastern Block, is composed of tonalitic-trondhjemitic-granodioritic (TTG) gneisses with minor supracrustal rocks intruded by syntectonic granites/charnockites with greenschist to granulite facies metamorphism. The Archaean rocks are overlain by the 1.85–1.40 Ga Mesoproterozoic Changcheng (Great Wall) system (Li et al. 2000a, b; Kusky & Li 2003). In some areas in the central part of the NCC, 2.40–1.90 Ga Palaeoproterozoic sequences deposited in cratonic graben are preserved (Kusky & Li 2003; Kusky et al. 2007a, b).

Rocks of the North China Craton have traditionally been grouped into different Archaean to Palaeoproterozoic tectonic cycles, including the Qianxi (>3.0 Ga), Fuping (3.0–2.5 Ga), Wutai (2.5–2.4 Ga) and Lüliang (2.4–1.8 Ga) 'movements' or orogenies. Palaeoproterozoic rocks of the North China Craton are widely exposed in the North China Craton, especially in the Khondalite belt in the north-central to northwestern part of the craton, and also along eastern margin in the Jiao-Liao-Ji belt.

The Central Orogenic belt extends through the middle part of North China Craton, and includes Late Archaean–Palaeoproterozoic fragments of ancient oceanic crust, mélanges, high pressure granulites and retrograded eclogites (Zhao et al. 2000a, b; Kusky et al. 2001, 2004a, b, 2007a, b, c, d; Li et al. 2002). Based on available data, Zhao et al. (2001) suggested that Palaeoproterozoic Central Orogenic belt (COB) is the collisional orogen amalgamating both Eastern Block and Western Block together to form the North China Craton, and in a series of papers suggested that the time of the collision was c. 1.85 Ga (Zhao et al. 2006a, b and references therein). However, Kusky et al. (2001, 2004a, b, 2007a, b, c, d) provided evidence that the COB is a late Archaean–Palaeoproterozoic belt, reactivated in the Palaeoproterozoic along with much of the northern half of the craton. The Central Orogenic belt is bordered on the east by a major Late Archaean foreland basin that records events in this Archaean orogen (Li & Kusky 2007). Faure et al. (2007) present data that shows the COB may have a polyphase history, and that after the 2.5 Ga events, additional shortening may be recorded at 2.3 Ga. Kusky, Windley & Zhai (2007a, b) review the geology and history of recognition that the Central Orogenic belt is an Archaean, not Palaeoproterozoic orogen.

A long and wide c. 2.3–1.85 Ga east–west-striking Palaeoproterozoic orogen is located along the northern margin of the craton (Fig. 2). This orogen, named the Inner Mongolia/North Hebei orogen (NHO; Kusky & Li 2003; Kusky et al. 2007a, c, d), although poorly known, contains several belts of arc-related plutonic rocks and evidence for two phases of granulite facies metamorphism, including one UHT event at 1.92 Ga (Santosh et al. 2006a, 2007a, b). Mafic granulites of the Hengshan region are located geographically in both the Central Orogenic belt and the North Hebei orogen, which has confused petrologists about which orogenic belt their metamorphism may be related to. These granulites exhibit clockwise $P-T-t$ paths and isothermal decompression characteristic of the classic models of continental collisional environments. However, recent studies on rocks subjected to extreme thermal metamorphism in this orogen identify some terranes exhibiting an anticlockwise $P-T$ path with steep isothermal decompression suggesting deep subduction and rapid exhumation at the leading edge of the continental margin that was involved in the collisional assembly of the craton within the Columbia supercontinent (Santosh et al. 2007a). Earlier geologists (Zhao et al. 2001a, b) recognized the Late Archaean origin of the Central Orogenic belt, but related the collisional $P-T-t$ paths in the Central Orogenic belt to a presumed closure of this central orogen in the Palaeoproterozoic, requiring a missing orogenic accretionary history of nearly 700 million years (Polat et al. 2007). Only recently has the significance of the Palaeoproterozoic collisions along the North Hebei orogen been widely-enough appreciated to realize that the $P-T-t$ paths are related to a major collision on the north side of the craton, and not in its centre (Kusky et al. 2001, 2003, 2004a, b, 2007a, c, d; Santosh et al. 2007a, b).

Late Palaeoproterozoic tectonics (between 2.10 and 1.90 Ga)

The North China Craton is bordered by late Palaeoproterozoic accretionary orogens, and cut internally by belts of Palaeoproterozoic deformation.

The Inner Mongolia–North Hebei Orogenic belt (or more simply the North Hebei Orogenic belt: NHO) is elongated along the northern margin of the craton, characterized by east–west-striking composite folds and shear zones. It consists of belts of plutonic rocks (gabbro, diorite, granites, mostly 2.3 Ga and 2.10–1.90 Ga) and volcanic-sedimentary sequences. In the NHO, the metamorphic ages (2.00 Ga) are commonly older than the crystallization ages of granites (1.90 Ga) (Li et al. 2000b). They are interpreted to result from accretion and collision along the northern margin of the NCC (Li et al. 2000b, 2001a, b; Kusky & Li 2003; Kusky et al. 2007a, b). Another so-called orogenic belt is located in the Sino-Korea boundary area, and is named the Jiao-Liao-Ji Orogenic belt (Fig. 2). This belt forms a band around the North Korean block (Bai et al. 1992; Li et al. 2006), and separates the eastern part of NCC from the North Korean massif. Since this belt is only 30–300 km wide, 1000 km long, and does not separate basement blocks with any significant differences (Li et al. 2006), we prefer to call this a structural belt instead of an orogen in the classic usage of the term. Both of these belts, and their possible relationships, are discussed below.

The Palaeoproterozoic Northern Hebei orogen: Reworked cratonic sedimentary sequences, khondalites, arc magmatism and accreted terranes in the northwestern NCC. After the Eastern and Western blocks of the NCC were sutured at 2.5 Ga, the craton experienced a brief episode of extension until about 2.4 Ga. This extensional event is associated with the intrusion of the Hengshan mafic igneous province, the Miyuan dykes and the deposition of mixed clastic and volcanic successions in graben including the Zhorstiao rift, the Luliang rift, the Taihang rift and extensive rift-to-passive margin sequences along the northern edge of the craton (Kusky & Li 2003). Circa 2.4–1.85 Ga cratonic cover, passive margin and foreland basin deposits were strongly deformed, metamorphosed and intruded by magmas at 1.9–1.85 Ga along the northern margin of the craton. Various authors have named different parts of the Palaeoproterozoic orogen along the northern margin of the craton the Inner Mongolia–Northern Hebei orogen (IMNHO, Kusky & Li 2003) the Northern Hebei Orogenic belt (NHOB; Qian & Li 1999) and the Khondalite belt (Santosh et al. 2006a, 2007a, b), after the distinctive granulite facies metapelitic rocks that characterize this belt.

The NHO consists of granitoid gneisses (2.42–2.47 Ga), a gabbro-diorite complex, minor ultramafic rocks and voluminous supracrustal sequences (2.30–2.64 Ga), all of which are intruded by late granites (2.38–2.44 Ga) (Geng et al. 1997; Li et al. 1998) and exhibit amphibolite- to granulite-facies prograde metamorphism. It forms an east–west-striking belt along the northern margin of the NCC and was probably generated by accretion along an active continental margin at the end of the Neoarchaean and the beginning of the Palaeoproterozoic (Kusky & Li 2003). To the south, the NHO is thrust over the khondalites belt of the NCC.

Kusky & Li (2003) suggested that parts of the Northern Hebei orogen represent an arc/accretionary complex that initially grew in the ocean to the north of the NCC after rifting at 2.5–2.4 Ga, and was accreted to the North China Craton by 2.3 Ga. These rocks include the low- to intermediate-grade Guyang and Chifeng metamorphic terranes, 2.49–2.45 Ga tonalitic-granitic gneiss, 2.48–2.40 Ga diorite-gabbro, scattered ultramafic rocks, 2.393 Ma trondhjemite and several 2.45–2.33 Ga supracrustal sequences including BIF, turbidites, and biotite-hornblende gneiss, all intruded by 2.44–2.38 Ga granitoids. Multigrain analysis of zircons reveals that an early phase of metamorphism occurred at 2237 Ma (Kusky & Li 2003).

Oblique collision along the khondalite belt (NHO) was partitioned into left lateral strike-slip motion and southward thrusting in the area. The sedimentary cover of NCC underwent compressive to shearing deformation and high-grade metamorphism with clockwise $P-T-t$ paths, or anticlockwise $P-T-t$ paths at the leading edge. Especially voluminous S-type granites with emplacement ages ranging from 2.00–1.90 Ga, were generated through anatexis and metamorphism. Ages of 2.00–1.85 Ga are recorded from metasediments in the northwestern part of the craton (Liu et al. 1992; Guo et al. 1993).

The khondalite belt. The khondalite belt comprises a vast sequence of metasupracrustal rocks distributed in a large area along the northwestern segment of the North China Craton (e.g. Mei 1997; Kusky & Li 2003; Xia et al. 2006; Santosh et al. 2006a, 2007a, b) forming the southern part of the Northern Hebei orogen (Figs 2 & 3). The rocks occur broadly within three terranes, the Jining-Liangcheng-Fengzhen, Daqingshan-Wulashan and Helanshan (Kusky et al. 2007b). The main rock associations in Jining-Liangcheng terrane are khondalites and S-type granites with minor metagabbro. The S-type granite belts are largely controlled by ENE-striking sinistral strike-slip shear zones. To the west, the Daqingshan-Wulashan terrane is made up of reworked granulite-charnockite complex, khondalites and granites. The Daqingshan-Wulashan terrane is dominated by nearly east–west-trending structural grains. Further to the west, the Helanshan terrane also consists of khondalites with an S-type granite association. This terrane is characterized by ENE-striking foliation, associated with a nearly

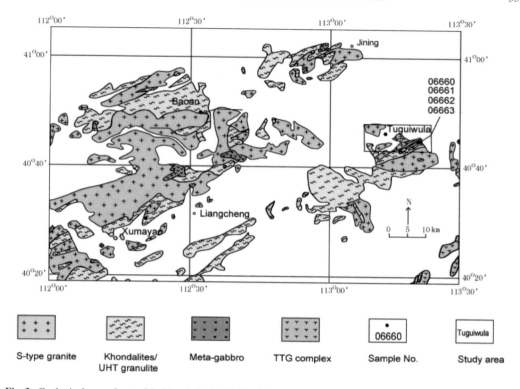

Fig. 3. Geological map of part of the khondalite belt in North China Craton after Santosh et al. (2007a). The location of sapphirine-bearing UHT granulites from Tuguiwula area are shown inside box. The sample numbers correspond to those studied by Santosh et al. (2007a, b) for petrology and zircon SHRIMP geochronology. Uncoloured areas are post-Proterozoic cover sequences.

east–west- to WNW-trending structural dome. The common association of the khondalite terranes with sinistral shear zones that host syntectonic emplacement of S-type granites have been correlated with Palaeoproterozoic reworking episodes along the northern margin of the North China Craton. A molasse basin, older than 1.90 Ga, lies along the northern margin of the belt. The khondalite belt is succeeded to the south by amphibolite-facies metasupracrustals and S-type granites. The Neoarchaean structural patterns and metamorphic lithology and greenstone belts have been locally preserved in the weakly-reworked basement. To the south, nearly east–west-striking strike-slip to thrust shear zones are uniformly superimposed on the Neoarchaean domains of the NCC. Mafic dyke intrusions of c. 1.80–1.76 Ga occur in all the three terranes (Hou et al. 2007).

The major rock types in the khondalite belt comprise garnet- and sillimanite-bearing granulite grade metapelites, garnet- and biotite-bearing leptynites, marbles and calc-silicate rocks, broadly similar to the lithological units in the khondalite belt of southern Kerala in India metamorphosed during the Late Neoproterozoic–Cambrian times (cf. Santosh et al. 2006b). Condie et al. (1992) and Li et al. (2000b) summarized the geochemical characteristics of the khondalites from North China Craton and described features typical of post-Archaean shales. These rocks are characterized by enrichment of large ion lithophile elements, low Zr–Hf–Sr contents, significant enrichment in LREE and negative Eu anomalies. The khondalite suite of lithologies is thought to be derived from the high-grade metamorphism of interlayered psammitic, psammopelitic, pelitic and carbonaceous sediments deposited in a shallow sea upon a cratonic shelf. Graphite deposits are common in many localities within the khondalite belt. The voluminous S-type granites containing garnet in most places may represent extensive partial melting during the high grade metamorphism of the thick sedimentary pile and melt movement. The metasedimentary sequence overlies a high-pressure granulite belt, the Hengshan belt, which extends over 700 km (cf. Kusky & Li 2003).

Zircons from the Jining Group khondalites have yielded two distinct age populations: 2.30–2.60 Ga and 1.85–2.00 Ga; the older ages are interpreted as the detrital zircon age and the younger ones as the metamorphic age (e.g. Kusky & Li 2003). A number of granite bodies with ages ranging from 2.2–1.9 Ga are emplaced within the Jining khondalite terrane, particularly the voluminous S-type granites which were possibly generated through anatexis of the pelitic sediments (Guo et al. 1999; Santosh et al. 2006a). These S-type granites from the khondalite terranes of southern Inner Mongolia at the northwestern part of the North China Craton have yielded zircon U–Pb ages of 1971 ± 60 Ma and 1837 ± 40 Ma, Pb–Pb ages of 1975 ± 23 Ma and 1893 ± 14 Ma, and Rb–Sr isochron ages between 1970–1840 Ma (Guo et al. 1999). Among the granites, two types have been recognized, including garnet granite and leucogranite. The formation age of the garnet granite from the Sanggan area has been estimated as 1836 ± 18 Ma from zircon U–Pb Concordia, while the leucogranites gave an upper intercept age of 1912 ± 98 Ma (Guo et al. 1999). Detrital zircons from the garnet-sillimanite-plagioclase gneiss (khondalite) yielded an age of 2130 Ma, while the metamorphic zircons are dated at 1872 Ma (Wu et al. 1998). Monazite dating of khondalites with spinel + quartz UHT assemblage yielded two age peaks at 1927 ± 11 Ma and 1819 ± 11 Ma correlated to distinct thermal perturbations during the prograde and peak stages of extreme crustal metamorphism in this belt (Santosh et al. 2006a). In another study, Santosh et al. (2007a) reported U–Pb–Th electron microprobe analysis of monazites in textural association with UHT granulites occurring within the khondalite belt and constrained the timing of metamorphism as 1917 ± 48 Ma. Since the khondalites as well as the Mg–Al granulites experienced extreme metamorphism at T >950–1000 °C, the monazite ages fix the timing of the UHT event in North China Craton to be of Palaeoproterozoic age.

A detailed zircon SHRIMP U–Pb study precisely constrains the timing of the UHT event in the NCC (Santosh et al. 2007b). The metamorphic zircons from the UHT rocks show ovoid shapes with numerous high-order crystal faces and low luminescence in cathodoluminescence (CL) images. As metamorphic temperatures increase, inferred from the presence of UHT indicator minerals, the relict oscillatory zones are progressively destroyed and the grains become uniformly dark in CL. The oldest detrital cores reveal $^{207}Pb/^{206}Pb$ ages up to 2090 ± 22 Ma, but most have ages of c. 1970 Ma. The highest grade sample with a sapphirine-bearing UHT assemblage contains only a single zircon population with a weighted mean $^{207}Pb/^{206}Pb$ age of 1919 ± 10 Ma. This is the same age, within error, as the youngest populations in the other samples and Santosh et al. (2007b) interpreted this age to record the time of UHT metamorphism in the North China Craton. The c. 1.92 Ga UHT metamorphic event in the North China Craton coincided with the assembly of the Palaeoproterozoic supercontinent Columbia.

In another recent study, Wan et al. (2006) reported U–Pb zircon ages from various Palaeoproterozoic metasedimentary units in the North China Craton. These rocks contain detrital igneous zircon grains with ages ranging from 3.0–2.1 Ga, with the majority displaying an age between 2.3 and 2.0 Ga. According to these authors, the latter age group represents a series of magmatic events that occurred between the formation of the Archaean basement and deposition of the sediments. The metasedimentary rocks analysed by Wan et al. (2006) occur as two associations; older volcano-sedimentary units that formed between 2.37 and 2.0 Ma, and younger sedimentary units that formed mainly between 2.0 and 1.88 Ga. All these rock associations were then subjected to metamorphism during the late Palaeoproterozoic (1.88 and 1.85 Ga) tectonothermal event. Wan et al. (2006) correlate this age with a continent–continent collision event that led to the amalgamation of the North China Craton. We suggest the continents were the amalgamated NCC and Columbia.

Liu et al. (2006) analysed U–Th–Pb in monazites from the low- to medium-grade Luliang and Wutai Complexes in the Central Orogenic belt, in the area affected by crustal thickening and collisional plateau formation (Kusky & Li 2003) from the 1.92 Ga collision in the NHO. The data revealed five age populations: (1) 1940–1938 Ma; (2) 1880–1847 Ma; (3) 1795–1755 Ma; (4) 1720–1703 Ma; and (5) c. 1648 Ma. Of these, the first four have also been recorded in the Wutai Complex. The oldest age range of 1940–1930 Ma is interpreted to record the widespread emplacement of mafic dykes in rocks of the Lüliang and Wutai Complexes. The age range of 1882–1822 Ma is interpreted by Liu et al. (2006) as the age of the major metamorphic event caused by amalgamation between the Eastern and the Western blocks. The age range of 1795–1755 Ma is consistent with emplacement of large-scale unmetamorphosed mafic swarms at 1800–1765 Ma, interpreted as the time of post-orogenic extension. The 1720–1703 Ma and c. 1648 Ma ages are considered to date later multiple stages of hydrothermal alteration, since monazites with such young ages occur only along fractures in older monazite grains and at the rims. Liu et al. (2006) considered their data to support a tectonic model for the evolution of the North China Craton involving the collision of discrete Eastern and Western blocks along the Trans-North China Orogen at c. 1.85 Ga and post-collisional extension in the period 1795–1755 Ma. Although Liu *et al.* (2006) did not study any UHT granulite assemblages, some of their monazite ages overlap with the age ranges from monazites in the UHT granulites of the North China Craton reported in Santosh *et al.* (2006a, 2007a, b). Liu *et al.* (2006) also did not consider whether their data may be consistent with a continent–continent collision along the Northern Hebei Orogen; instead they restricted their interpretation to possible events only within the Central Orogenic belt. Since the metamorphic P–T–t data cannot distinguish the orientation of major shear and collisional belts, but only the relative depth of rock units in the crust, we agree with their general conclusion of a collision at that time, but dispute the location of the collisional orogen.

The granulite terranes of the North China Craton show a common structural association. The oldest, Neoarchaean structures strike generally NNE to NE, and are cut by *c.* 1.85–1.80 Ga East–West-striking shear zones. One of the best examples of these cross-cutting relationship is in the Hengshan, where NE-striking foliations in *c.* 2.55 Ga gneisses are cut by the *c.* 1.8 Ga east–west-striking Zhujiafang shear zone that emplaces high pressure granulites in the north over medium pressure granulites to the south (O'Brien *et al.* 2005; Kusky *et al.* 2007a). The boundaries to the Hengshan uplift are controlled by the NE-striking faults of the Cenozoic Shaanxi graben system, perhaps re-activating older Mesozoic or Proterozoic structures. These crosscutting relationships show the complex structural history that these rocks have experienced, including orogenic events at 2.5 Ga, 1.9–1.85 Ga and Mesozoic to active tectonics related to Pacific subduction and the collision of India with Asia. Thus, the rock record does not support the contention of Zhao *et al.* (2000a, b, 2002, 2005) that the NE-striking shear zones are boundaries to an ocean that was open for 700 million years from 2.5–1.8 Ga, but rather strengthens the concept that the region records a complex history of several orogenic events with clear cross-cutting structural and superimposed stratigraphic relationships (e.g. Kusky & Li 2003; Kusky *et al.* 2007; Faure *et al.* 2007). At 1.9–1.85 Ga, the northern margin of the NCC was experiencing a major collisional and thickening event while the COB was behaving largely as a strike-slip belt in the orogenic foreland.

UHT metamorphism at 1.92 Ga in the NHO. High pressure (HP) and UHT metamorphic assemblages have been described from granulites in the Bao'an, Hongshaba, Xumayao and Tuguiwula areas within the khondalite belt, with temperatures exceeding 1000 °C and pressures above 12 kbar (Santosh *et al.* 2006a, 2007). In the Tuguiwula area (Fig. 3) Mg–Al granulites outcrop in association with garnet-sillimanite–spinel-bearing gneisses, garnet- and biotite-bearing felsic gneiss and S-type granite (Santosh *et al.* 2006a, 2007a). The rocks show prominent compositional layering with garnet-sillimanite ± cordierite ± spinel-rich dark bands and quartzofeldspathic leucosome layers indicating *in situ* partial melting. The khondalite layers commonly range in thickness from a few metres to several tens of metres, and show polyphase deformational structures. Early NW-trending and later SW-trending lineations are identified. The garnet-gneiss belts are strongly folded and the enclosing S-type granites also suffered strong ductile deformation with a sinistral sense of shear. At least four generations of deformation can be recognized with the khondalites, involving D_1 early thrusting, D_2 extension, D_3 strike-slip shear deformation (mainly sinistral) and D_4 intrusion of mafic dykes.

The Bao'an area (Fig. 3) is characterized mainly by S-type granites with abundant melanocratic residual bands of metapelites. At least four mafic dykes cut the S-type granites in this area, ranging in width from 1.5–5 m. The S-type granite-khondalite association is also exposed in the Honshaba area, where the khondalite band shows strong folding. Garnet-bearing granites and khondalites underwent strong ductile shear deformation and were partly mylonitized in the Xumayao area. The felsic bands show stretching, folding and development of S_2 foliation. Garnet-bearing dark grey coloured mylonitized gneiss appears in Zhaohaogou area, with large bands and relics of aluminous metapelites. Melanocratic residual bands rich in garnet and sillimanite occur within partially melted khondalites in the Tuguiwula area.

Santosh *et al.* (2007a) reported the occurrence of ultrahigh-temperature Mg–Al granulites from the Tuguiwula area, where these occur as several centimetre- to decimetre-sized concordant bands and layers within khondalites. The UHT indicator assemblages (Santosh *et al.* 2006a, 2007a) in these rocks include equilibrium sapphirine + quartz, high alumina orthopyroxene + sillimanite + quartz and high temperature perthites. The surrounding khondalites also carry spinel + quartz equilibrium assemblage. The history of prograde to peak metamorphism of these rocks is indicated by inclusions of sapphirine, spinel, quartz and sillimanite within garnet. Whereas the stability of sapphirine + quartz and orthopyroxene + sillimanite + quartz in the matrix assemblage suggests $T > 1000$ °C and $P > 10$ kbar peak metamorphism, conventional geothermobarometric estimates confirm the ultrahigh-temperature nature and also trace the retrograde conditions (930–970 °C at $P > 10$ kbar – Al in Opx; 900–1000 °C – perthite; 930–990 °C – Grt-Opx; and 910–940 °C – sapphirine-spinel). An evaluation of the assemblages and textures using appropriate petrogenetic grids indicates that following peak ultrahigh-temperature conditions, the rocks underwent initial isobaric cooling and subsequent isothermal decompression. Santosh et al. (2007a) interpreted these trajectories to be part of an overall anticlockwise $P-T$ evolution, indicating significant magmatic underplating may have played a role in the high temperatures. Dating of monazites in the UHT assemblage yielded ages of 1917 ± 48 Ma, comparing well with other ages of 1927 ± 11 Ma on monazites from surrounding terranes. Zircon SHRIMP dating of sapphirine-bearing UHT granulites also show similar ages 1919 ± 10 Ma (Santosh et al. 2007b) precisely constraining the timing of UHT metamorphism in NCC as c. 1.92 Ga. Santosh et al. (2007a, b) relate these features to collision of the northern margin of the NCC during assembly of the Palaeoproterozoic supercontinent Columbia.

Although the occurrence of sapphirine was noted in some parts of the NCC in some of the earlier studies published in the Chinese literature (e.g. Liu et al. 1992, 2000), these were from boulders with no free quartz or other diagnostic UHT assemblages. Guo et al. (2006) reported sapphirine-bearing granulites within in situ exposures from Huhhot, about 175 km west of the Tuguiwula locality from where Santosh et al. (2007a) reported sapphirine-bearing UHT granulites and retrieved temperatures of 920–970 °C, although again, characteristic UHT assemblages were not recorded. These reports, together with our ongoing studies, clearly indicate widespread UHT metamorphism all along the northern margin of the North China Craton, which has important tectonic implications for the collisional assembly of this craton with the Columbia supercontinent during the Palaeoproterozoic.

The dry mineral assemblage that characterizes the UHT rocks provides important constraints to evaluate granulite formation in the deep crust as well as crust–mantle interaction in relation to the dynamics of supercontinents. A detailed fluid inclusion study of the UHT rocks from the NCC revealed the common occurrence of CO_2 rich fluid inclusions in peak UHT minerals (Santosh et al. 2008). Microthermometric and laser Raman spectroscopic studies revealed that the synmetamorphic fluid in the UHT rocks is pure CO_2. Although very high density carbonic fluids are preserved in some cases, corresponding to entrapment at peak conditions, most of the inclusions underwent density modification during rapid and near-isothermal exhumation of the rocks following peak UHT metamorphism. The present study area within the northern margin of the North China Craton defines a collisional orogenic belt where continental collision and deep subduction were possibly followed by extensional collapse of the orogen. Evidence for the extensional phase is provided by the several mafic magmatic intrusives of Palaeoproterozoic age that are seen emplaced within the granulite facies supracrustals (Hou et al. 2007). The thermal anomaly displayed by the UHT granulites may thus correlate with asthenospheric upwelling and emplacement of mafic magmas, possibly associated with plume activity, which provided the heat and CO_2 input to achieve extreme crustal metamorphism and generation of dry granulite facies assemblages. The fluid inclusion study by Santosh et al. (2008) is the first report of synmetamorphic CO_2-rich fluids in UHT granulites from a Palaeoproterozoic terrane where the fluid densities were partly modified during rapid decompression when extensional collapse of the orogen occurred after continental collision. Geodynamic models for CO_2 advection during extreme crustal metamorphism should take into consideration the heat source to account for the high T/P conditions that characterize the ultrahigh-temperature assemblages.

In a recent evaluation of the thermal anomaly models to explain UHT metamorphism Santosh & Omori (2007) suggested that lower crustal heating is the one model that is most consistent with the episodic formation of high- to ultrahigh-temperature granulites at different times in Earth history, which also correlate with the episodic assembly and disruption of supercontinents, or plume activity during various periods in Earth history (cf. Condie 2003b). From thermodynamic computations and quantitative estimations, Santosh & Omori (2007) illustrated that both decarbonation of calc-silicates and sub-lithospheric mantle are suitable sources for CO_2 involved in the petrogenesis of these rocks. The sub-lithospheric mantle decarbonation process is coeval with the thermal anomaly induced by plumes or arc magmas, and the estimated temperatures satisfy the computed phase equilibria conditions for the release of CO_2, as well as the observed fluid inclusion characteristics in granulite minerals. The formation of the high- and ultrahigh-temperature rocks and multiple CO_2 infiltration events have close petrogenetic relation, if the sub-lithospheric mantle consists of considerable amounts of carbonates. An evaluation of the Archaean plate tectonic system with hydrothermal carbonation of oceanic crust, and its subduction and decarbonation during subduction zone metamorphism suggests that a substantial amount of CO_2 was fixed in the mantle from the Archaean atmosphere (Santosh & Omori 2007). Such carbonated oceanic crust could have released CO_2-bearing fluids during various thermotectonic processes associated with the formation of UHT granulites in the Proterozoic.

The present study area within the northern margin of the North China Craton defines a collisional orogenic belt where continental collision and deep subduction were possibly followed by extensional collapse of the orogen (Kusky & Li 2003; Kusky et al. 2007a). Evidence for the extensional phase is provided by the several mafic magmatic intrusives of Palaeoproterozoic age that are seen emplaced within the granulite facies supracrustals (Peng et al. 2007; Hou et al. 2007). These late Palaeoproterozoic mafic dyke swarms are widespread in the Shanxi Province in the Western Block and Central Orogenic Zone, and also in the Shandong Province of the Eastern Block of the North China Craton. Their geochemical features indicate a continental rift environment and suggest that the magma originated from partial melting of the subcontinental lithospheric mantle. Hou et al. (2007) correlated the mafic dyke swarms with a continuous extension event from 1.84–1.77 Ga in response to the breakup of the supercontinent Columbia. The extensive mafic dyke swarms therefore signal a major anomalous thermal event that correlates with asthenospheric upwelling and emplacement of mafic magmas, which provided the heat and CO_2 input to achieve extreme crustal metamorphism and generation of dry granulite facies assemblages. A cartoon sketch is shown in Figure 4 to illustrate the link between mantle upwelling, decarbonation and UHT metamorphism in the deep continental crust. In order to explain the ultrahigh temperatures, along with the CO_2 rich fluids, we suggest that the initial geotherms were elevated in an arc or back arc region in the NHO, and that when this active arc collided with South American and other blocks in the Columbia amalgam, the high-temperature lower crust was rapidly thickened, further elevating the temperatures in the lower crust. These rocks were then exhumed during post-orogenic collapse and isostatic adjustment to the thickened crust.

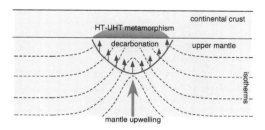

Fig. 4. Cartoon model illustrating the formation of 'dry' ultrahigh-temperature granulite facies assemblages through heat and CO_2 input at the base of the crust from mantle upwelling (after Santosh & Omori 2007). See text for discussion.

Events in the foreland: intermontane structural/metamorphic belts, foreland basin sedimentation

Chang Cheng Series and foreland basin

The northern part of the North China Craton was covered by thick predominantly clastic sedimentary successions of the Chang Cheng Series deposited between 1.85 and 1.4 Ga, which Kusky & Li (2003) interpreted as a foreland basin related to a continental collision along the NHO. Most rocks of this series are relatively little deformed and metamorphosed, but show shallow-level fold-thrust belt style structures. However, possibly correlative rocks of the Jiao-Liao-Ji belt are strongly deformed and metamorphosed, which we interpret to mean that rocks in that belt define a zone of intracontinental deformation at 1.85 Ga, similar to deep levels of the Tarim and Tsaidam basins north of the active India–Asia collision.

The Chang Cheng Series is divided into five formations; the Changzhougou, Chuanlingguo, Tuanshanzi, Dahongyu and Gaoyuzhuang Formations (Hebei Bureau 1989; Sun et al. 1985; Zhu & Chen 1992; Fang et al. 1998). The Changzhougou Formation consists of 450 m of sandstone and conglomerate, beginning with a basal boulder and cobble conglomerate containing well-rounded 45 cm clasts of the underlying basement, vein quartz and metamorphic rocks. These are interbedded with graded beds of coarse-grained sandstone and grade up into quartz sandstone with low-angle trough cross beds and then into fluvial sandstone. In the central and southern parts of the craton, the sandstones are interbedded with and grade laterally into dolostones, suggesting that the source of the sandstone and conglomerate was in the north, and that the Changzhougou Basin was relatively free of clastic influx a few hundred kilometres from the North Hebei Orogen. Such relationships are characteristic of younger foreland basins, where clastic rocks grade laterally into shallow water carbonates from the orogen toward the craton. Yu et al. (1994) report Pb–Pb ages of 1848 ± 39 Ma for clay minerals in the Changzhougou Formation.

The Chuanlingguo Formation rests conformably on the Changzhougou Formation and ranges from 900 m thick in the east to 50 m thick in the west (Zhu & Chen 1992). It consists mainly of thinly bedded siltstone, interbedded with micrite and stromatolitic dolostone. To the south, the formation consists mainly of thinly-bedded siltstone, shale and fine-grained sandstone. Yu et al. (1994) report a Pb/Pb age of 1785 ± 19 Ma for illite in the upper part of the formation. The lack of conglomerates and coarse-grained sandstone in the

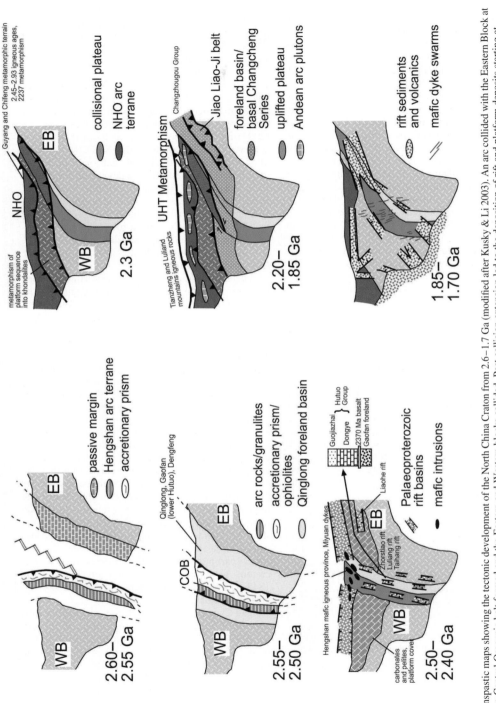

Fig. 5. Palinspastic maps showing the tectonic development of the North China Craton from 2.6–1.7 Ga (modified after Kusky & Li 2003). An arc collided with the Eastern Block at 2.5 Ga as the Central Orogenic belt formed and the Eastern and Western blocks collided. Post collision extension led to the deposition of rift and platform deposits starting at 2.5 Ga, and a passive margin developed on the north margin of the craton. From 2.5 Ga onwards, the amalgamated orogen that links Western and Eastern blocks is represented by the solid green area of the COB. The margin converted to an active margin through an arc collision at 2.3 Ga, and experienced UHT metamorphism, and strong deformation at 1.92 Ga as the North China Craton collided with Columbia. Sediments of the Chang Cheng Series were deposited in a large foreland basin that propagated across the craton, and the Jiao-Liao Ji belt formed as an intermontane basin south of the collisional plateau. Following this collision along the northern margin of the craton at 1.92–1.85 Ga, extensional tectonics formed a

Chuanlingguo Formation suggests that the main denudation of the mountains to the north had been completed by 1785 Ma. Kusky & Li (2003) interpreted the Chuanlingguo Formation to mark the transition of the northern margin of the craton from a foreland basin at 1850–1790 Ma, to a rift and cratonic cover sequence at 1785–1600 Ma.

The upper three formations of the Chang Cheng System, the Tuanshanzi, Dahongyu and Gaoyuzhuang Formations, include dominantly micrites, dolostones, stromatolitic dolostone, quartzite and shale. Kusky & Li (2003) interpreted these formations to represent a continuation of cratonic sedimentation. The Tuanshanzi Formation has yielded a U–Pb whole rock isochron of 1776 Ma (no error reported, Bai et al. 1998), and a Rb–Sr whole rock isochron age of 1606 ± 19 Ma for a volcanic flow in the unit (Lu & Li 1991). The Dahongyu Formation has several interbedded potassic lava flows, one of which has yielded a U–Pb zircon age of 1626 ± 6 Ma (Lu & Li 1991).

Jiao-Liao-Ji belt: Palaeoproterozoic intermontane basins on collisional plateau

South of the NHO, and apparently on the southern edge of the collisional plateau, are a couple of belts of multiply deformed metasedimentary rocks, with metamorphic ages of c. 1.85–1.8 Ga. Although rocks in these belts were strongly deformed and metamorphosed at 1.85–1.8 Ga, the belts do not separate different types of basement (Li et al. 2006), and are therefore regarded as zones of Palaeoproterozoic intracontinental deformation and not sutures or orogens (for an opposite interpretation, see Zhao et al. 2005). These belts resemble the Tarim and Tsaidam basins north of Tibet. The largest of these is the Palaeoproterozoic Jiao-Liao-Ji belt (Fig. 2) located in the Eastern Block of the craton, extending from North Korea through the southern Anshan-Benxi, eastern Hebei, southern Jilin, Liaoning, through Bohai Sea, and into Miyun-Chengde, western Shandong and eastern Shandong Provinces. It is up to 1000 km long and 50–300 km wide (Li et al. 2005).

The belt is composed of Palaeoproterozoic metamorphic rocks derived from the Eastern Block. Most of this belt is characterized by greenschist facies to lower amphibolite facies metamorphic rocks, rare volcanic successions, as well as associated granitic and mafic intrusions (Zhao et al. 2005), which include the Macheonayeong Group in North Korea, Ji'an and Liaoling groups in southern Jilin, the Liaohe Group of eastern Liaoning, as well as the Jingshan and Fenzishan groups of eastern Shandong (Fig. 2). Among these sedimentary and bimodal volcanic rocks, there are large amounts of A-type granites and mafic layered intrusions (Luo et al. 2004; Zhao et al. 2005; Li et al. 2006).

Jiao-Liao-Ji belt rock types, ages and depositional environment. Palaeoproterozoic rocks of the Jiao-Liao-Ji belt include the Fengzishan and Jingshan groups in eastern Shandong Province, Liaohe Group in the Liaodong Peninsula of Liaoning Province, Guanghua, Ji'an and Liaoling groups in Southern Jilin Province, and possibly the Macheonayeong Group in North Korea. To date there is no consensus on the depositional environment, time of deposition or tectonic setting of these rocks. However, we note that there appears to be no significant difference between the metasedimentary sequences preserved in the Jiao-Liao-Ji belt and the Palaeoproterozoic cover sequences such as the Chang Cheng Series recognized across the NCC (e.g. Kusky & Li 2003; Kusky et al. 2007a), except that the metamorphic grade and extent of deformation are higher in the Jiao-Lioa-Ji belt than is typical of elsewhere on the craton. Although this belt remains poorly known, some studies have been undertaken on the granitoids and low-grade sedimentary basins in attempts to understand their tectonic setting (e.g. Zhao et al. 2005; Lu et al. 2004; Li et al. 2006).

The Guanghua and Ji'an groups are the only units that include volcanic rocks in the Jiao-Liao-Ji belt; the Liaoling Group is composed of clastic sediments and carbonates (Lu et al. 2004; Luo et al. 2004; Wan et al. 2006; Wu et al. 2005). For the stratigraphic sequences, the Fenzishan Group of eastern Shandong is correlated with North Liaohe Group of Liaoning and Laoling Group in southern Jilin; Jingshan Group of eastern Shandong also exhibits similar stratigraphic features as the South Liaohe Group of Liaoning Province and Ji'an Group in

---

Fig. 5. (*Continued*) series of aulacogens, rifts and mafic dykes that propagated across the craton. Rifting led to the development of an ocean along the southwestern margin of the craton, where passive margin and cratonic sedimentation continued until 1.5 Ga. Throughout the rest of the Precambrian and until the end of the Palaeozoic, the NCC behaved as a coherent, stable continental block, as evidenced by deposition of shallow marine carbonate platform sediments, with the only breaks in sedimentation being associated with deformation and orogeny along the margins of the craton, and a regional disconformity between the Upper Ordovician and Upper Carboniferous. The latter may have resulted from the global eustatic low stand of sea level following the Caledonian orogeny or from double-vergent subduction beneath the north and south margin of the craton (the Qaidam plate was subducted beneath the southern margin of the craton, and several oceanic plates subducted beneath the north margin of the craton (Kusky et al. 2007a, c, d).

southern Jilin (Zhao et al. 2005). Lu et al. (2004) carried out a REE and zircon U–Pb study on the Liaohe Group of Liaodong Peninsula of Liaoning Province and the Guanghua, Ji'an and Liaoling groups in the Tonghua area of southern Jilin Province. They show that the Guanghua Group consists of metabasalt containing the assemblage plagioclase, hornblende, garnet and minor quartz. Older gneiss consists of plagioclase, biotite, hornblende, quartz and garnet, and contains layers of marble. The Ji'an Group consists of diopside-bearing gneiss, graphite-bearing gneiss, garnet-bearing gneiss and amphibolite. The diopside-gneiss is interlayered with amphibolite, graphitic gneiss and marble, taken to indicate a mixed volcanic and sedimentary protolith for the gneiss. The Liaoling Group mainly consists of coarse-grained quartz sandstone, marble and fine-grained quartz sandstone.

Lu et al. (2004) classified the Guanghua, Ji'an and Liaoling groups as volcano-sedimentary rocks, among which the Guanghua and Ji'an groups also consist of abundant volcanic rocks. Zircon U–Pb isotopic dating shows that they range in age between 2.13 Ga and 1.98 Ga, whereas the Liaoling Group is characterized by clastic sediments and carbonates, and the dating shows that they exhibit two depositional ages, the first at 2.5–2.2 Ga and the second at 2.0–1.9 Ga. The older group corresponds to the age of carbonate and clastic sediments deposited across much of the northern and central part of the craton following post-orogenic extension at 2.5–2.4 Ga, after the collision in the Central Orogenic belt (Kusky et al. 2007a, c, d). The younger age group corresponds to the foreland basin deposits of the Chang Cheng Series deposited across the northern part of the craton following the 1.92 Ga collision in the Northern Hebei orogen (Kusky & Li 2003). Overall, Lu et al. (2004) describes the Palaeoproterozoic tectonic evolution of Tonghua area in Eastern Block of NCC as the event of initial extension from 2.12–1.98 Ga, and subsequent collision. In the Liaonan Block, the timing of extension was dated as around 2.16 Ga.

Jiao-Liao-Ji belt metamorphism. The metamorphic history of the Jiao-Liao-Ji belt has raised considerable interest, although detailed petrological studies are yet to be done. In the Jiao Bei section, the rock succession includes garnet biotite gneiss, garnet-sillimanite/kyanite gneiss intercalated with mica-rich schistose rocks as well as calc-silicates. Towards the south, in the Jiao Nan area, garnet-, spinel- and sillimanite-bearing metapelites along with garnet–spinel- and orthopyroxene-bearing charnockites are exposed. The garnet–biotite gneiss, garnet–cordierite–biotite–tourmaline granulite with occasional sillimanite (fibrolite) and inclusions of kyanite and the calcite + diopside + phlogopite + plagioclase + quartz bearing calc-silicate rock sequences resemble typical shallow marine sediments metamorphosed at upper amphibolite to granulite facies conditions. In most cases, the matrix mineral assemblage is lower grade (amphibolite facies) although rare kyanite inclusions in garnet indicate traces of high pressures. Some of the rocks here also contain tourmaline, thought to indicate high boron content in the sediment, which has prompted some workers to suggest an intracontinental rift-basin within which the sediments were deposited (Zhao et al. 2005; Lu et al. 2004; Li et al. 2006; Luo et al. 2004). However, the presence of tourmaline alone cannot be taken as a criterion to infer a rift environment, since tourmaline is known from many supracrustal belts, and even tectonic settings in accretionary wedges. For instance, in the amphibolite-facies metapelites in the Yanai area, Ryoke metamorphic belt, SW Japan. Kawakami & Ikeda (2003) reported the common occurrence of tourmaline, where the mineral formed by the breakdown of muscovite + quartz. The size of the tourmaline grains increases drastically where the muscovite + quartz breakdown reaction is advanced. Tourmaline is abundant in pegmatites and granite stocks cutting the southern Alaska accretionary wedge (Kusky et al. 2003), also a non-rift setting. A similar situation is envisaged for the Jiao Bei section, with the escape of the boron-bearing melt or the fluid resulting in the formation of melt pockets, veins and pegmatites carrying coarse crystals of tourmaline. Thus, the occurrence of tourmaline in these supracrustals would bear little if any relationship to the original tectonic environment. Also, the Jia Bei section does not preserve any evidence for UHT metamorphism.

Previous tectonic models for the Jiao-Liao-Ji belt. Based on the geochemical analyses of the volcanic and sedimentary successions of the Jiao-Liao-Ji belt, the Palaeoproterozoic tectonic evolution of this belt has been suggested to be an intracontinental rift that opened then closed, undergoing both collisional and extensional activites (Zhao et al. 2005; Lu et al. 2004; Li et al. 2006; Luo et al. 2004). We contend that this model is not the best model for the belt, but rather is inconsistent with the data used to form the model. Present data also indicate that the Laoling, Ji'an and Liaohe groups are all strongly deformed sequences of different ages, including two main depositional ages of 2.5–2.2 and 2.0–1.9 Ga. These ages correspond to the ages of the platform and foreland cover sequences deposited on the NCC, as described by Kusky & Li (2003). All of them have suffered from the intense deformation without showing the common stratigraphy (Lu et al. 2004). It is difficult to identify the original layering and the contact relationships.

The opening and closing of the intra-continental rifting model (Zhao et al. 2005; Lu et al. 2004; Li et al. 2006; Luo et al. 2004) is based on the following major evidence:

- The Jiao-Liao-Ji orogenic belt contains rare meta-mafic volcanic rocks with greenschist facies and amphibolites facies, and meta-rhyolites;
- The existence of the large amounts of A-type granites and mafic layered intrusions (minor post-tectonic alkaline syenites and rapakivi granites);
- On the opposite side of the orogenic belt, the geochemical and geochronological characteristics are similar with each other. Both have the Late Archaean TTG basement gneisses and mafic dyke swarms;
- The Ji'an, South Liaohe and Jingshan groups, preserve Palaeoproterozoic rocks with low-pressure features and anticlockwise $P-T$ paths. All of these features are related with the intrusion and underplating of mantle magmas.

In these models, the Western and Eastern blocks of North China Craton were amalgamated during the Late Archaean to Palaeoproterozoic, then the Eastern Block underwent rifting along its eastern margin during the Early Palaeoproterozoic, accompanied by the formation of the Fenzishan and Jingshan groups in eastern Shandong, South and North Liaohe groups in Liaoning and Laoling and Ji'an groups in south Jilin. After that, the model rift is supposed to have closed upon itself in the late Palaeoproterozoic.

Re-interpretation of the Jiao-Liao-Ji belt as an intermontane basin within a collisional plateau.
Rocks of the Jiao-Liao-Ji belt are the same age as other similar rocks deposited across the North China Craton in response to post-orogenic extension at 2.5–2.4 Ga, and in a foreland basin developed on the craton after a 1.92 Ga collision in the North Hebei orogen. The only major differences between the Jiao-Liao-Ji belt and these other sequences is that the Jiao-Liao-Ji belt is strongly deformed and metamorphosed, while most of the other sequences are not. We suggest that this is a result of the Jiao-Liao-Ji belt having experienced intracontinental deformation in a collision-related basin on the edge of the uplifted plateau that formed during the 1.92 Ga collision of the NCC with Columbia. This setting is similar to the Tarim and Tsaidam basins that are presently forming on the north side of the Tibet plateau. Rocks at depth in these basins are presently being deformed and metamorphosed, and when exhumed will likely resemble those of the Jiao-Liao-Ji belt.

Origin of rifting and breakup of NCC at 1.80 Ga.
There has not been much discussion about what initiated rifting of the NCC at 1.85–1.80 Ga, although a plume model has been postulated (Zhai et al. 2005; Peng et al. 2005). A mantle plume can account for the domal uplift and subsequent continental extension and fragmentation within the North China Craton documented in this chapter. Prior to continental rifting at 1.80 Ga, uplift during this stage extended across a dome hundreds of kilometres in diameter, beyond the Yanliao-Taihang aulacogen, in response to underlying mantle upwelling. It resulted in erosion of the subcontinental lithospheric mantle. The domal uplift and cracking caused radial fracturing with the lithosphere, and initiated the breakup of the NCC, with the formation of three branches of Yanliao-Taihang aulacogen, Xionger-Zhongtiao aulacogen and emplacement of mafic dyke swarms. However, beyond the uplift, rift-aulacogen systems, such as BR, ZA, were also controlled by regional (i.e. plate-boundary) stresses (e.g. see Kusky & Li 2003).

Evidence for domal uplift and branched rifting is compatible with mantle plume or mantle diapir models (Condie 2002). The plume model is also supported by anorogenic magmatism within rift-aulacogen systems and emplacement of related mafic dyke swarms. The total duration of magmatism and associated rifting is up to 100–150 Ma (between 1.85 and 1.70 Ga), comparable to those of long-lived plumes (Condie 2003b). We suggest therefore that a mantle plume initiated the uplift, extension and subsidence that formed the extensive passive margin along the southwestern side of the North China Craton, accompanied by the development of oceanic basins at about 1.80 Ga.

Correlation of the Proterozoic tectonics of the NCC with the Columbia Supercontinent

Global correlation of a 2.0–1.9 Ga UHT extreme metamorphic event related to continental collision

Collisions of the NCC with other blocks to the north at about 2.00–1.90 Ga resulted in reworking of the foreland in the northwestern part of the NCC. Thrust belts and intermontane basins were formed, flysch basins propagated southward across the craton, and the northern edge of the craton experienced strong deformation, granulite, HT and UHT metamorphism related to the collision. Comparable orogenic-reworking events at the same time have been widely documented in the foreland and hinterland of other cratonic blocks, including the Aldan shield, Laurentia, Baltica, Greenland, the Kalahari and Sao Francisco cratons (Teixeira & Figueiredo 1991; Nutman et al. 1992; Gorbatschev & Bogdanova 1993; Trompette & Carozzi 1994;

Frost et al. 1998; Jahn et al. 1998; Whitehouse 1998; Orrell et al. 1999; Scott 1999; Teixeira et al. 1999). Most continental collisions during this period were related with the assembly of the supercontinent Columbia.

At about 1.85 Ga, the NCC switched from collision and reworking to cratonic extension, culminating with the development of rift-aulacogen systems and intrusion of mafic dyke swarms at about 1.80–1.70 Ma (Kusky & Li 2003; Peng et al. 2007; Hou et al. 2007). The present southwestern and northern margins of the NCC began to form by the breakup of a larger craton, with development of oceanic basins, as extension continued at 1.80 Ga. The onset of extension in the North China Craton is older than in most other cratons of Columbia, with the exceptions including similar-aged rifts on the margins of the Kalahari Craton.

As indicated by its two distinct rifted margins and extensional regime at about 1.80 Ga, the major part of the NCC was far from a collisional setting during that period. With development of rifts and oceanic basins along both its northern and southwestern margins, the NCC was broken from the Columbia supercontinent, and then drifted to form widely separated fragments. The age of rifting-breakup (1.80 Ga) of the NCC was clearly earlier than those of the Laurentia, and slightly younger than post-orogenic magmatism in Baltica (c. 1.77 Ga).

The northern and the southwestern rifted margins of the NCC formed an acute angle on its northwestern corner at about 1.80 Ga. This distinct geometry of the cratonic block and relevant rifted margins is similar to RRR (rift-rift-rift) triple junction margins commonly associated with supercontinent fragmentation, consistent with the observation that the NCC split from the hinterland of the Columbia supercontinent at 1.80 Ga. Similar acute angles and paired rifted margins have been preserved on the southern ends of several cratonic fragments, such as Africa and India, associated with fragmentation of Gondwana (e.g. Rogers & Santosh 2004).

Moreover, the timing of extensional-anorogenic magmatism of the NCC mainly shows a range from 1.80–1.70 Ga. The Taihang mafic dyke swarms, and the late rapakivi granite of North China can be correlated with those of the Aldan shield, Amazonian craton and Urayualagua shield (Fig. 1). In the Aldan shield and Amazonian craton, rapakivi granites with ages of 1.90–1.70 Ga are reported (Larin et al. 2006; Ramo et al. 2002). The Svecofennian orogen saw the intrusion of post-orogenic rapakivi granites, gabbros and anorthosites after collisional processes ended and the orogen was under extension by 1.77 Ga (Korja & Heikkinen 2005). In the Amazonian craton and Urayualagua shield, a mafic dyke swarm of 1.75–1.72 Ga has been identified (Chaves & Neves 2005). The Archaean basement was also intruded by both rapakivi granite and mafic dyke swarms in the Aldan shield, as has the eastern part of the Amazonian craton. All the evidence suggests a possible connection of the NCC with the Aldan shield and Amazonian craton in the late Palaeoproterozoic. Figure 1 shows our proposed reconstruction of Columbia based on these data.

Correlation of c. 1.9 Ga UHT metamorphism in the NHO with the Tarim-Transamazonian/Eburnian–Rodonian-Rio Negro–Svecofennian Belts

Recognition of c. 1.9 Ga UHT metamorphism at several places, hundreds of kilometres apart, in the Northern Hebei orogen, and thousands of kilometres apart along correlative belts, suggests that these extreme metamorphic conditions may be a product of tectonic environment along the length of the orogen. Thus, future studies should reveal more locations with c. 1.9 Ga UHT assemblages in the Tarim-Transamazonian/Eburnian–Rodonian-Rio Negro–Svecofennian Belts, that Kusky et al. (2007b) suggested were correlative with the North Hebei orogen.

The Tarim block of western China was located closer to the NCC before movement on Altyn Tagh fault in response to the India–Asia collision (Yin & Niu 1996). Most of the Precambrian rocks of the Tarim basin are covered beneath younger deposits, even more extensively than the Western Block of the NCC. Nevertheless, a sequence of deformation and metamorphism has been worked out for the Tarim Block from studies along its margins. Zhang et al. (2007) outline a tectonic evolution for the Tarim Basin area that is similar to that proposed by Kusky et al. (2007a, b) for the once-contiguous North China Craton. Crustal formation is regarded to have been largely complete by 2.5 Ga, followed by rifting at 2.41 and 2.35–2.34 Ga, subsequent to which the rocks were affected by a major metamorphic event at 1.9 Ga. Thus, the 1.9 Ga metamorphic event can be shown to extend thousands of kilometres east–west along the northern margin of the NCC and Tarim cratons, and probably linking with the Rio Negro-Juruena, Transamazonian and Eburnian belts as suggested by Kusky et al. (2007b). The 1.9 Ga event in the North China Craton and Tarim block is clearly not associated with the north–south-striking Central Orogenic belt of the North China Craton, more than 2000 kilometres to the east. The c. 2.1 Ga Birimian terranes of West Africa (Abouchami et al. 1990) appear to represent accretion and imbrication of an oceanic plateau terrane in the pre-Columbian ocean to the margin of the West African craton, adjacent to the Tarim block (Abocuhami et al. 1990; Kusky 1998). Similar-aged accretion followed by UHT metamorphism at 2.1–1.9 Ga extends down the length of the Eburnian and Transamazonian belts, as discussed below. In addition, the geological history of the Svecofennian Belt shows a history of accretion and convergent activity from 1.9–1.84 Ga (e.g. Windley 1995), and represents the westward (in our reconstruction) correlative of the North Hebei orogen.

Several c. 2.1–1.9 Ga UHT metamorphic rocks are known from the Transamazonia–Eburnian belts of South America and Africa, supporting the correlation of these belts with the NHO, and extending the known length of the orogen an additional 5000 km. Other granulites that appear to show a peak of c. 1.3 Ga are known from the Rio Negro-Juruena and related belts of Rondonia, although some data indicate an older event in this belt as well, perhaps near 1.9 Ga (Moraes et al. 2006a, b, c). Palaeoproterozoic granulites were formed in Roraima, Amapa, Bahia, Tocantis, northern Goias, Parana, Santa Catarina and Rio Grande do Sul, and are likely related to the formation of the Columbia supercontinent. The metamorphic history of many of these rocks is only recently coming to light, and preliminary results suggest that two events may be most significant, including a HT–UHT event at 2.0–1.9 Ga, and a later granulite overprinting at c. 630 Ma, related to the amalgamation of Gondwana (e.g. Bendaoud et al. 2006; Moller et al. 2006; Moares et al. 2006a, b). We highlight a few of the examples below to demonstrate the similarity with the NHO.

A c. 2.0 Ga granulite facies event is recorded in the Transamazonian belt of the northern Mantiqueira Province, Brazil. Jordt-Evangelista et al. (2006) report opx–cpx–hbl–plag mafic and gt–bio–plag–or–qtz felsic granulites with c. 2.0 Ga ages determined for the metamorphism (Rb–Sr and Pb/Pb ages). Just to the north in the Salvador-Curaca belt, Carlson et al. (2006) have documented 2.08–2.05 Ga UHT sapphirine-bearing granulites, formed during the transpressional Rhyacian orogeny. P–T conditions are estimated at 900–1000 °C at 5–8 kbar, falling within the realm of UHT rocks. In southern Cameroon, Toteu et al. (1994) and Feybesse et al. (1998) have described 2.05 Ga granulites with lenses of 2.09 Ga retrogressed eclogites from the Eburnian–Transamazonian orogen, formed during the collision of the Sao Francisco and Congo cratons during the assembly of Columbia. Temperatures reached 800 °C and pressures in the eclogites reached 16 kbar (equivalent to 60 km). Diagnostic UHT mineral assemblages have also been recorded from several other localities in Brazil (e.g. Moraes & Fuck 2000), although precise age constraints on the timing of the extreme metamorphic event are yet to be reported.

Recently Brandt et al. (2007) reported UHT metamorphic rocks from the Epupa Complex situated at the eastern margin of the Kaoko belt in northwestern Namibia. The latter is part of the Pan-African mobile belt system of West Gondwana and is thought to have formed by the collision of the Congo Craton with the Rio de la Plata Craton in the late Neoproterozoic. Based on radiogenic age data, the Epupa Complex has been considered to constitute the southwestern margin of the Archaean to Mesoproterozoic Congo Craton. The Palaeoproterozoic zircon ages (c. 1.8 Ga) from the eastern margin of the Epupa Complex are within the age range (2.0–1.6 Ga) known from other basement complexes in north Namibia and south Angola, signifying a major Eburnian tectonothermal event in the entire region. Kroner et al. (2004) reported radiometric age data from the Kaoko belt which belongs to the Neoproterozoic mobile belt system of western Gondwana, and whose geodynamic evolution is assumed to have resulted from the collision between the Congo Craton (present Africa) and the Rio de la Plata Craton (present South America). They identified prominent Palaeoproterozoic U–Pb magmatic emplacement ages from this belt between c. 2.03 and 1.96 Ga, and correlated these with the Eburnian event (c. 1.8–2.0 Ga), which is widespread in Africa. In addition, two distinct thermal events were also traced in the southwestern Congo Craton in late Palaeoproterozoic and Mesoproterozoic times, indicated by magmatic ages around 1.77 Ga and between c. 1.52 and c. 1.45 Ga.

Moving to the other side of the NCC in Columbia, the Svecofennian orogen is known to contain a >500 km-long belt of high-grade migmatites and gneiss deformed and metamorphosed at 1.84–1.83 Ga (e.g. Ehlers et al. 1993). The Svecofennian Orogen is up to 1200 km wide and extends from Poland and southern Belarus to central Sweden, and is characterized mainly by accretion of juvenile island arcs at 1.9 Ga, along with associated accretionary prisms and ophiolites, including the Jourma ophiolite (Windley 1995; Peltonin et al. 2004). The Jourma ophiolite was thrust over rift and passive margin deposits that rest in the same structural position as the post-2.4 Ga rift to passive margin sequences of the northern margin of the North China Craton (Kusky & Li 2003). Many of the metasedimentary rocks have been transformed to biotite-bearing granitic gneisses and schists, separated by belts of arc plutons. Most deformation in the Svecofennian orogen ended by 1.77 Ga, and was followed after a hiatus of about 70 Ma by the intrusion of epizonal rapakivi granites, gabbros, anorthosites, and mafic dyke swarms. Windley (1993) related these late intrusions to post-orogenic collapse and extension of the orogen, which continued to about 1.55 Ga. In Belarus the late rapakivi granites intruded c. 1.8–1.9 Ga shear zones and granites between 1.8 and 1.5 Ga.

Similarly, in the NCC post-orogenic events include intrusion of anorogenic magmas around 1.7 Ga (rapakivi granite, gabbro, anorthosite) (Qian & Chen 1987; Li et al. 2000a; Peng et al. 2005, 2007; Halls et al. 2000; Kusky & Li 2003; Hou et al. 2003, 2005; Xie 2005; Song 1992; Hu et al. 1990; Sun & Hu 1993; Li & Wang 1997; Wang et al. 1996; Wang 1991). In the Aldan shield and Amazonian craton, the rapakivi granites with ages of 1.90–1.70 Ga are reported. In the Amazonian craton and Urayualagua shield, a mafic dyke swarm of 1.75–1.72 Ga has been identified (Chaves & Neves 2005). The Archaean basement was also intruded by both rapakivi granite and mafic dyke swarms in the Aldan shield, as has the eastern part of the Amazonian craton. The Taihang mafic dyke swarms, and the rapakivi granite of North China are thus correlated with those of Aldan shield, Amazonian craton and Urayualagua shield. All the evidence suggests a possible connection of the NCC with the Aldan shield and Amazonian craton in the late Palaeoproterozoic.

Summary

The NCC records a sequence of comparable tectonothermal episodes (2.00–1.70 Ga) with those of Baltic shield, Amazonian and Sao Francisco cratons, and has a close tectonic relation with these cratons. The tectono-magmatic and metamorphic evolution of the Northern Hebei orogen is remarkably similar to the Svecofennian of Baltica, and the Transamazonian and Eburnian belts of Africa and South America. These continental blocks and orogens may have belonged to a single supercontinent or continent at 1.80–1.70 Ga. All these blocks were associated with extensional tectonics and breakup of a possible supercontinent over a time interval from 1.80–1.70 Ga, in contrast to the dominant accretion and collision tectonics of Laurentia and North Atlantic craton during the same period. Therefore, we suggest that the NCC in the latest Palaeoproterozoic was far from Laurentia and Greenland, and located between the Baltic Shield, and the Amazonian and Sao Francisco cratons (Fig. 1). Interestingly, the age of collision in the NHO (1.93 Ga) correlates remarkably well with the events in the Transamazonian orogens, which experienced sinistral transpression from 2.1–2.0 Ga, and late magmatic stages from 2.0–1.9 Ga (Rogers & Santosh 2004). Activity continued in the older part of the Rio Negro-Juruena Belt from 1.8–1.6 Ga (Rogers & Santosh 2004) and may be related to the extensional event recoded in the Baiyun Obo rift on the north side of the NCC at that time (Fig. 4).

This work was funded by the US National Science Foundation (Grant 02–07886 awarded to T. Kusky) and by St. Louis University. Research of M. Santosh was funded by Grant-in-aid No. 17403013 from the Japanese Ministry of Education, Culture, Sports, Science and Technology. We would like to thank J. Li and M. Zhai for valuable discussion during various stages of our study in the North China Craton. Valuable reviews of this manuscript were provided by J. J. W. Rogers and W. Xiao.

References

ABOUCHAMI, W., BOHER, M., MICHARD, A. & ALBAREDE, F. 1990. A major 2.1 Ga event of mafic magmatism in West Africa: an early stage of crustal accretion. *Journal of Geophysical Research*, **95, B11**, 17 605–17 629.

BAI, J. & DAI, F. Y. 1998. Archaean crust of China. *In*: MA, X. Y. & BAI, J. (eds) *Precambrian Crustal Evolution of China*. Springer-Geological Publishing, Beijing, 15–86.

BAI, J., HUANG, X. & WANG, H. 1996. *The Precambrian Crustal Evolution of China* (2nd edn). Geological Publishing House, Beijing, 80–90 (in Chinese with English abstract).

BAI, J., WANG, R. Z. & GUO, J. J. 1992. *The Major Geological Events of Early Precambrian and Their Dating in Wutaishan Region*. Geological Publishing House, Beijing, 1–60.

BENDADOUDE, A., OUZEGANE, K., DJEMAI, S. & KIENAST, J. R. 2006. The Eburnean ultrahigh-temperature metamorphism of the Amessmessa area (south in Ouzzal Terrane, Western Hoggar, Algeria). *In*: BROWN, M. & PICCOLI, M. (eds) *Granulites and Granulites*. Abstracts of Meeting, Brasilia, 10–12 July 2006, 15.

BRANDT, S., WILL, T. H. & KLEMD, R. 2007. Magmatic loading in the Proterozoic Epupa Complex, NW Namibia, as evidenced by ultrahigh-temperature sapphirine-bearing orthopyroxene–sillimanite–quartz granulites. *Precambrian Research*, **153**, 143–178.

CARLSON, L., BARBOSA, J., NICOLLET, C. & SABATE, P. 2006. Petrological evolution of silica-undersaturated sapphirine-bearing granulite at ultrahigh temperature conditions in the Salvador-Curaca Belt, Bahia, Brazil. *In*: BROWN, M. & PICCOLI, M. (eds) *Granulites and Granulites*. Abstracts of Meeting, Brasilia, 10–12 June 2006, 42.

CHAVES, A. O. & NEVES, J. M. C. 2005. Magmatism, rifting and sedimentation related to late Palaeoproterozoic mantle plume events of central and Southeastern Brazil. *Journal of Geodynamics*, **39**, 197–208.

CONDIE, K. C. 2002. Continental growth during a 1.9-Ga superplume event. *Journal of Geodynamics*, **34**, 249–264.

CONDIE, K. C. 2003a. *Plate Tectonics and Crustal Evolution*. Butterworth, Oxford, 69–83.

CONDIE, K. C. 2003b. *Mantle Plumes and Their Record in Earth History*. Cambridge University Press, Cambridge, 54–115.

CONDIE, K. C. 2004. Supercontinents and superplume events: distinguishing signals in the geologic record. *Physics of the Earth and Planetary Interiors*, **146**, 319–332.

CONDIE, K. C., BORYTA, M. D., LIU, J. Z. & QIAN, X. L. 1992. The origin of khondalites – geochemical evidence from the Archaean to early Proterozoic granulite belt in the North China Craton. *Precambrian Research*, **59**, 207–223.

EHLERS, C., LINDROOS, A. & SELONEN, O. 1993. The Late Svecofennian granite-migmatite zone of southern Finland: a belt of transpressive deformation and granite emplacement. *Precambrian Research*, **64**, 295–309.

FANG, L., FRIEND, C. R. L. ET AL. 1998. *Geology of the Santunying area of Eastern Hebei Province*. Geological Survey Publishing House, Beijing, 134.

FAURE, M., TRAP, P., LIN, W., MOINE, P. & BRUGUIER, O. 2007. Polyorogenic evolution of the Paleo-Proterozoic Trans-North China Belt, new insights from the Luliangshan-Hengshan-Wutaishan and Fuping massifs. *Episodes*, **30**, 95–106.

FEYBESSE, J. L., TRIBOULET, C., GUERROT, C., MAYAGA-MIKOLO, F., BOUCHOT, V. & EKO N'DONG, J. 1998. The West Central African Belt: a model of 2.5–2.0 Ga accretion and two-phase orogenic evolution, *Precambrian Research*, **87**, 161–216.

FROST, B. R., AVCHENKO, O. V., CHAMBERLAIN, K. R. & FROST, C. D. 1998. Evidence for extensive Proterozoic remobilization of the Aldan shield and implications for Proterozoic plate tectonic reconstructions of Siberia and Laurentia. *Precambrian Research*, **89**, 1–23.

GENG, Y., LIU, D. & SONG, B. 1997. Chronological framework of the early Precambrian important events of the Northwestern Hebei granulite terrain. *Acta Geologica Sinica*, **71**, 316–327. (in Chinese with English abstract).

GORBATSCHEV, R. & BOGDANOVA, S. 1993. Frontiers in the Baltic shield. *Precambrian Research*, **64**, 3–21.

GUO, J. H., CHEN, Y., PENG, P., LIU, F., CHEN, L. & ZHANG, L. Q. 2006. Saphirine granulites from Daqingshan, western block of the North China Craton: 1.8 Ga ultrahigh–high temperature metamorphism. Abstract, in *Proceedings of Annual Symposium on Petrology and Geodynamics in China*, Nanjing, 205–218.

GUO, J. H., ZHAI, M. G. & ZHANG, Y. G. 1993. Early Precambrian Manjinggou high-pressure granulites melange belt on the southern edge of the Huaian Complex, North China Craton: geological features, petrology and isotopic geochronology, *Acta Petrologica Sinica*, **9**, 329–341.

GUO, J. H., SHI, X., BIAN, A. G., XU, R. H., ZHAI, M. G. & LI, Y. G. 1999. Pb isotopic compositions of feldspar and U–Pb ages of zircons from early Proterozoic granites in the Sanggan area, North China Craton: metamorphism, crustal melting, and tectonothermal events. *Acta Petrological Sinica*, **15**, 199–207.

HALLS, H. C., LI, J. H., DAVIS, D., HOU, G., ZHANG, B. X. & QIAN, X. L. 2000. A precisely dated Proterozoic palaeomagnetic pole from the North China Craton, and its relevance to palaeocontinental reconstructions. *Geophysical Journal International*, **143**, 185–203.

HE, T. X., LIN, Q. & FANG, Z. R. 1992. *The Petrogenesis of Granitic Rocks in Eastern Hebei*. Jiling Science and Technology Press. Changchun, 1–4 (in Chinese with English abstract).

HE, G. P., LU, L. Z. & YE, H. W. 1991. *The Early Precambrian Metamorphic Evolution of the Eastern Heibei and the Southeastern Inner Mongolia*. Jilin University Press, Changchun, 1–17 (in Chinese with English abstract).

HEBEI GEOLOGICAL BUREAU 1989. *Regional Geology of Hebei Province, Beijing Municipality and Tianjing Municipality*. Geological Memoir, **15**, Geological Publishing House, Beijing.

HOU, G. T., LI, J. H. & HALLS, H. C. 2003. The flow structures and mechanics of Later Precambrian mafic dyke swarms in North China Craton. *Acta Geologica Sinica*, **77**, 210–216 (in Chinese with English abstract).

HOU, G. T., LI, J. H., LIU, Y. L. & QIAN, X. L. 2005. The latest Palaeoproterozoic extensional episodes: aulacogen and mafic dyke warm in North China Craton. *Progress in Natural Sciences*, **15**, 1366–1373.

HOU, G., LI, J. H., YANG, M., YAO, W., WANG, C. & WANG, Y. 2007. Geochemical constraints on the tectonic environment of the Late Proterozoic mafic dyke swarms in the North China Craton. *Gondwana Research*, **13**, 103–116.

HU, S. L., WANG, S. S. & SANG, H. Q. 1990. The isotopic age dating, rare earth element geochemistry of Damiao anorthosite, and their geological implication. *Scientia Geologica Sinica*, **4**, 332–343 (in Chinese with English abstract).

JAHN, B. M., AUVRAY, B., CORNICHET, J., BAI, Y.-L., SHEN, Q.-H. & LIU, D.-Y. 1987. 3.5 Ga amphibolites from Eastern Hebei Province, China; field occurrence, petrography, Sm–Nd isochron age and REE geochemistry. *Precambrian Research*, **34**, 311–346.

JAHN, B. M., GRUAU, G., CAPDEVILA, R., NEMCHIN, A., PIDGEON, R. & RUDNIK, V. A. 1998. Archaean crustal evolution of the Aldan Shield, Siberia: geochemical and isotopic constraints. *Precambrian Research*, **91**, 333–363.

JAHN, B. M. & ZHANG, Z. Q. 1984a. Archaean granulite gneisses from Eastern Hebei Province, China: rare earth geochemistry and tectonic implications. *Contributions to Mineralogy and Petrology*, **85**, 225–243.

JAHN, B. M. & ZHANG, Z. Q. 1984b. Radiometric ages (Rb–Sr, Sm–Nd, U–Pb) and REE geochemistry of Archaean granulite gneisses from Eastern Hebei Province, China. *In*: KRÖNER, A., HANSON, G. & GOODWIN, A. (eds) *Archaean Geochemistry*. Springer-Verlag, Berlin, 204–244.

JORDT-EVANGILISTA, H. & DELGADO, C. E. R. 2006. Kyanite-cordierite-bearing rocks from the Acaica granulite complex, southeastern Minas Gerais, Brazil. *In*: BROWN, M. & PICCOLI, M. (eds) *Granulites and Granulites*. Abstracts of Meeting, Brasilia, 10–12 July 2006, 37.

KAWAKAMI, T. & IKEDA, T. 2003. Boron in metapelites controlled by the breakdown of tourmaline and retrograde formation of borosilicates in the Yanai area, Ryoke metamorphic belt, SW Japan. *Contributions to Mineralogy and Petrology*, **145**, 131–150.

KORJA, A. & HEIKKINEN, P. 2005. The accretionary Svecofennian orogen-insight from the BABEL profiles. *Precambrian Research*, **136**, 241–268.

KRÖNER, A., CUI, W. Y., WANG, S. Q., WANG, C. Q. & NEMCHIN, A. A. 1998. Single zircon ages from highgrade rocks of the Jianping complex, Liaoning province, NE China. *Journal of Asia Earth Sciences*, **16**, 519–532.

KRÖNER, A., WILDE, S. & LI, J. H. 2005. Age and evolution of a late Archaean to early Palaeoproterozoic upper to lower crustal section in the Wutaishan/ Hengshan/Fuping terrane of northern China. *Journal of Asian Earth Sciences*, **24**, 577–595.

KRÖNER, A., WILDE, S., WANG, K. & ZHAO, G. C. 2002. Age and evolution of a late Archaean to early Proterozoic upper to lower crustal section in the Wutaishan/ Hengshan/Fuping terrane of northern China: a field guide. *Geological Society of America Penrose Conference*. Beijing, China, September 2002.

KRONER, S., KONOPASEK, J., KRONER, A., PASSCHIER, C. W., POLLER, U., WINGATE, M. T. D. & HOFFMANN, K. H. 2004. U–Pb and Pb–Pb zircon ages for metamorphic rocks in the Kaoko Belt of Northwestern Namibia: a Palaeo- to Mesoproterozoic basement reworked during the Pan-African orogeny. *South African Journal of Geology*, **107**, 455–476.

KUSKY, T. M. 1998. Tectonic setting and terrane accretion of the Archaean Zimbabwe craton. *Geology*, **26**, 163–166.

KUSKY, T. M. 2004a. Precambrian Ophiolites and Related Rocks, Introduction. *In*: KUSKY, T. M. (ed.) *Precambrian Ophiolites and Related Rocks*. Developments in Precambrian Geology **13**, Elsevier Amsterdam, 1–35.

KUSKY, T. M. 2004b. What, if anything, have we learned about precambrian ophiolites and early earth processes. *In*: KUSKY, T. M. (ed.) *Precambrian Ophiolites and Related Rocks*. Developments in Precambrian Geology **13**, Elsevier Amsterdam, 727–737.

KUSKY, T. M. & LI, J. H. 2003. Palaeoproterozoic Tectonic Evolution of the North China Craton. *Journal of Asian Earth Sciences*, **22**, 383–397.

KUSKY, T. M., LI, J. H. & TUCKER, R. T. 2001. The Archean Dongwanzi ophiolite complex, North China Craton: 2.505 billion year old oceanic crust and mantle. *Science*, **292**, 1142–1145.

KUSKY, T. M., BRADLEY, D. C., DONLEY, D. T., ROWLEY, D. & HAEUSSLER, P. 2003. Controls on intrusion of near-trench magmas of the Sanak-Baranof belt, Alaska, during Paleogene ridge subduction, and consequences for forearc evolution. *In*: SISSON, V. B., ROESKE, S. & PAVLIS, T. L. (eds) *Geology of a Transpressional Orogen Developed During a Ridge – Trench Interaction Along the North Pacific Margin*. Geological Society of America, Special Paper, **371**, 269–292.

KUSKY, T. M., LI, Z. H., GLASS, A. & HUANG, H. A. 2004. Archaean ophiolites and ophiolite fragments of the North China Craton. *In*: KUSKY, T. M. (ed.) *Precambrian Ophiolites and Related Rocks*. Developments in Precambrian Geology **13**, Elsevier, Amsterdam, 223–274.

KUSKY, T. M., LI, J. H. & SANTOSH, M. 2007a. The Palaeoproterozoic North Hebei Orogen: North China Craton's collisional suture with Columbia Supercontinent. *In*: ZHAI, M. G., XIAO, W. J., KUSKY, T. M. & SANTOSH, M. (eds) *Tectonic Evolution of China and Adjacent Crustal Fragments*. Special Issue of Gondwana Research, **12**, 4–28.

KUSKY, T. M., ZHI, X. C., LI, J. H., XIA, Q. X., RAHARIMAHEFA, T. & HUANG, X. N. 2007b. Chondritic Osmium isotopic composition of Archean ophiolitic mantle. *In*: ZHAI, M. G., XIAO, W. J., KUSKY, T. M. & SANTOSH, M. (eds) *Tectonic Evolution of China and Adjacent Crustal Fragments*. Special Issue of Gondwana Research, **12**, 67–76.

KUSKY, T. M., WINDLEY, B. F. & ZHAI, M. G. 2007c. Tectonic Evolution of the North China Block: from Orogen to Craton to Orogen. *In*: ZHAI, M. G., WINDLEY, B. F., KUSKY, T. M. & MENG, Q. R. (eds) *Mesozoic Sub-Continental Lithospheric Thinning Under Eastern Asia*. Geological Society of London, Special Publication, **280**, 1–34.

KUSKY, T. M., WINDLEY, B. F. & ZHAI, M. G. 2007d. Lithospheric thinning in eastern Asia; constraints, evolution, and tests of models. *In*: ZHAI, M. G., WINDLEY, B. F., KUSKY, T. M. & MENG, Q. R. (eds) *Mesozoic Sub-Continental Lithospheric Thinning Under Eastern Asia*. Geological Society of London, Special Publication, **280**, 331–343.

LARIN, A. M., KOTOV, A. B. *ET AL*. 2006. The Kalar Complex, Aldan-Stanovoi shield, and ancient anorthosite-mangerite-charnockite-granite association: Geochronologic, geochemical, and isotopic-geochemical characteristics. *Petrology*, **14**, 2–20.

LI, H. M. & WANG, R. Z. 1997. The single-grain zircon U–Pb age dating of Fenghuangshan granite. *Progress of Precambrian Research*, **20**(3), 56–62 (in Chinese with English abstract).

LI, J. H., QIAN, X. L. & GU, Y. C. 1998. Outline of Palaeoproterozoic tectonic division and plate tectonic evolution of North China Craton. *Earth Science*, **23**, 230–235.

LI, J. H., QIAN, X. L. & HOU, G. T. 2000a. New interpretation of the 'Luliang Movement'. *Earth Science*, **25**, 15–20 (in Chinese with English abstract).

LI, J. H., KRÖNER, A., QIAN, X. L. & O'BRIEN, P. 2000b. The tectonic evolution of early Precambrian highpressure granulite belt, North China Craton (NCC). *Acta Geologica Sinica*, **274**, 246–256.

LI, J. H., HOU, G. T. & QIAN, X. L. 2001a. Single-Zircon U–Pb age of the initial Mesoproterozoic basic dyke swarms in Hengshan Mountain and its implication for the tectonic evolution of the North China Craton. *Geological Review* **47**, 234–238 (in Chinese with English abstract).

LI, J. H., HOU, G. T. & HUANG, X. N. 2001b. The constraint for the supercontinental cycles: evidence from Precambrian geology of North China Block. *Acta Petrologica Sinica*, **17**, 177–186 (in Chinese with English abstract).

LI, J. H., KUSKY, T. M. & HUANG, X. N. 2002. Neoarchaean podiform chromitites and harzburgite tectonite in ophiolitic melange. North China Craton: remnants of Archaean oceanic mantle. *GSA Today*, **12**, 4–11.

LI, S. Z., ZHAO, G. C., SUN, M., HAN, Z. Z., LUO, Y., HAO, D. F. & XIA, X. P. 2005. Deformation history of the Palaeoproterozoic Liaohe assemblage in the Eastern Block of the North China Craton. *Journal of Asian Earth Sciences*, **24**, 659–674.

LI, S. Z., ZHAO, G. C., SUN, M., HAN, Z. Z., ZHAO, G. T. & HAO, D. 2006. Are the south and north Liaohe Groups of North China Craton different exotic terranes? Nd isotope constraints. *Gondwana Research*, **9**, 198–208.

LI, J. H. & KUSKY, T. M. 2007. A Late Archean foreland fold and thrust belt in the North China Craton: implications for early collisional tectonics. *In*: ZHAI, M.-G., XIAO, W. J., KUSKY, T. M. & SANTOSH, M. (eds) *Tectonic Evolution of China and Adjacent Crustal Fragments*. Special Issue of Gondwana Research, **12**, 47–66.

LIU, J., QIANK, K., LIU, X. & OUYANG, Z. 2000. Dynamics and genetic grids of sapphirine-bearing spinel gneiss in Daqing Mountain orogen zone, Inner Mongolia. *Acta Petrologica Sinica*, **16**, 245–255 (in Chinese with English Abstract).

LIU, S., ZHAO, G. *ET AL.* 2006. Th–U–Pb monazite geochronology of the Lüliang and Wutai Complexes: constraints on the tectonothermal evolution of the Trans-North China Orogen. *Precambrian Research*, **148**, 205–224.

LIU, X., WEI, J. & LI, S. 1992. Low pressure metamorphism of granulite facies in an early Proterozoic orogenic event in central Inner Mongolia. *Acta Geologica Sinica* **66**, 244–256 (in Chinese with English abstract).

LU, S. N. & LI, H. M. 1991. A precise U–Pb single zircon age determination for the volcanics of the Dahongyu Formation, Changcheng System in Jixain. *Bulletin of the Chinese Academy of Geological Sciences*, **22** (in Chinese).

LU, X. P., WU, F. Y., LIN, J. Q., SUN, D. Y., ZHANG, Y. B. & GUO, C. L. 2004. Geochronological successions of the Early Precambrian granitic magmatism in Southern Liaodong peninsula and its constraints on tectonic evolution of the North China Craton. *Scientia Geologica Sinica*, **39**, 123–138.

LUO, Y., SUN, M., ZHAO, G. C., LI, S. Z., XU, P., YE, K. & XIA, X. P. 2004. LA-ICP-MS U–Pb zircon ages of the Liaohe Group in the Eastern Block of the North China Craton: constraints on the evolution of the Jiao-Liao-Ji Belt. *Precambrian Research*, **134**, 349–371.

MEI, H. L. 1997. Tectonic division of the early rocks in the Yinshan area and the Palaeoproterozoic crustal collision lateral extension model. *Progress in Precambrian Research*, **20**, 51–56 (in Chinese with English abstract).

MOLLER, A., MORAES, R., HELLEBRAND, E., KENNEDY, A. & FUCK, R. A. 2006. Age and duration of the UHT event in the Brasilia fold belt: *in-situ* dating of zircon and rutile and equilibrium REE distribution between zircon and orthopyroxene. *In*: BROWN, M. & PICCOLI, M. (eds) *Granulites and Granulites*. Abstracts of Meeting, Brasilia, 10–12 July 2006, 54.

MORAES, R. & FUCK, R. A. 2000. Ultra-high-temperature metamorphism in Central Brazil: the Barro Alto complex. *Journal of Metamorphic Geology*, **18**, 345–358.

MORAES, R., NETO, C., DA COSTA, M. & FUCK, R. A. 2006a. More evidence of UHT mineral assemblages in the Anapolis-Itaucu complex, Gois, Brazil. *In*: BROWN, M. & PICCOLI, M. (eds) *Granulites and Granulites*. Abstracts of Meeting, Brasilia, 10–12 July 2006, 55.

MORAES, R., FUCK, R. A., BALDWIN, M., DANTAS, J. A., LAUX, E. L. & JUNGES, S. L. 2006b. UHT wollastonite + scapolite calc-silicate rocks from Goianira, Anapolis-Itaucu Complex, Gois, Brazil. *In*: BROWN, M. & PICCOLI, M. (eds) *Granulites and Granulites*. Abstracts of Meeting, Brasilia, 10–12 July 2006, 56.

MORAES, R., FUCK, R. A., DUARTE, R. A., PASCHOAL, B., BARBOSA, J. S. F. & LEITE, C. DE. M. M. 2006c. Granulite Facies rocks in Brazil. *In*: BROWN, M. & PICCOLI, M. (eds) *Granulites and Granulites*. Abstracts of Meeting, Brasilia, 10–12 July 2006, 57.

NUTMAN, A. P., CHERNYSHEV, I. V., BAADSGAARD, H. & SMELOV, A. P. 1992. The Aldan shield of Siberia, USSR: the age of its Archaean components and evidence for widespread reworking in the mid-Proterozoic. *Precambrian Research*, **54**, 195–210.

O'BRIEN, P. J., WALTE, N. & LI, J. H. 2005. The petrology of two distinct granulite types in the Hengshan Mts, China, and tectonic implications. *Journal of Asian Earth Sciences*, **24**, 615–627.

ORRELL, S. E., BICKFORD, M. E. & LEWRY, J. F. 1999. Crustal evolution and age of thermal tectonic reworking in the western hinterland of the Trans-Hudson orogen, northern Saskatchewan. *Precambrian Research*, **95**, 187–223.

PELTONEN, P. & KONTINEN, A. 2004. The Jourma ophiolite: a mafic-ultramafic complex from an ancient ocean-continent transition zone. *In*: KUSKY, T. M. (ed.) *Precambrian Ophiolites and Related Rocks*. Developments in Precambrian Geology **13**, 35–72.

PENG, P., ZHAI, M. G., ZHANG, H. F. & GUO, J. H. 2005. Geochronological constraints on the Palaeoproterozoic evolution of the North China Craton: SHRIMP zircon ages of different types of mafic dykes. *International Geology Review*, **47**, 492–508.

PENG, P., ZHAI, M. G., GUO, J. H., KUSKY, T. M. & ZHAO, T. P. 2007. Nature of mantle source contributions and crystal differentiation in the petrogenesis of the 1.78 Ga mafic dykes in the central North China Craton. *Gondwana Research*, **12**, 29–46.

POLAT, A., KUSKY, T. M., LI, J. H., FRYER, B. & PATRICK, K. 2005a. Geochemistry of the Late Archaean (c. 2.55–2.50 Ga) volcanic and ophiolitic rocks in the Wutaishan Greenstone Belt, Central Orogenic Belt, North China Craton: implications for geodynamic setting and continental growth. *Bulletin of the Geological Society of America*, **117**, 1387–1399.

POLAT, A., LI, J. H., FRYER, B., KUSKY, T., GAGNON, J. M. & ZHANG, S. 2005b. Geochemical characteristics of the Neoarchaean (2800–2700 Ma) Taishan Greenstone Belt, North China Craton: evidence for plume-craton interaction. *Chemical Geology*, **230**, 60–87.

POLAT, A., HERZBERG, C. *ET AL.* 2006. Geochemical and petrological evidence for a suprasubduction zone origin of Neoarchaean (c. 2.5 Ga) peridotites, central orogenic belt, North China Craton. *Bulletin of the Geological Society of America*, **118**, 771–784.

POLAT, A., KUSKY, T. M. & LI, J. H. 2007. Geochemistry of the Late Archaean (c. 2.55–2.50 Ga) volcanic and ophiolitic rocks in the Wutaishan Greenstone Belt, Central Orogenic Belt, North China

Craton: implications for geodynamic setting and continental growth, Reply. *Geological Society of America Bulletin*, **119**, 490–492.

QIAN, X. L. & CHEN, Y. P. 1987. Late Precambrian mafic swarms of the North China Craton. *In*: HALLS, H. C. ET AL. (eds) *Mafic Dike Swarms*. Geological Association of Canada, Special Papers, **34**, 385–391.

QIAN, X. L. & LI, J. H. 1999. Discovery of Neoarchaean unconformity and its implication for continental cratonization of North China Craton. *Science in China* **42**, 399–407.

RAMO, O. T., DALL'AGNOL, R., MACAMBRIA, M. J. B., LEITE, A. A. S. & DE OLIVEIRA, D. C. 2002. 1.88 Ga oxidized A-Type granites of the Rio Maria region, eastern Amazonian Craton, Brazil: positively Anorogenic! *Journal of Geology*, **110**, 603–610.

ROGERS, J. J. W. & SANTOSH, M. 2003. Supercontinents in Earth history. *Gondwana Research*, **6**, 357–368.

ROGERS, J. J. W. & SANTOSH, M. 2004. *Continents and Supercontinents*. Oxford University Press, New York.

RITTS, B. D., HANSON, A. D., DARBY, B. J., NANSON, L. & BERRY, A. 2004. Sedimentary record of Triassic intraplate extension in North China: evidence from the nonmarine NW Ordos Basin, Helan Shan and Zhuozi Shan. *Tectonophysics*, **386**, 177–202.

SANTOSH, M. & OMORI, S. 2008. CO_2 flushing: fact or fiction? *Gondwana Research*, **13**, 86–102.

SANTOSH, M., SAJEEV, K. & LI, J. H. 2006a. Extreme crustal metamorphism during Columbia supercontinent assembly: evidence from North China Craton. *Gondwana Research*, **10**, 256–266.

SANTOSH, M., MORIMOTO, T. & TSUTSUMI, Y. 2006b. Geochronology of the khondalite belt of Trivandrum Block, southern India: electron probe ages and implications for Gondwana tectonics. *Gondwana Research*, **9**, 261–278.

SANTOSH, M., TSUNOGAE, T., LI, J. H. & LIU, S. J. 2007a. Discovery of sapphirine-bearing Mg–Al granulites in the North China Craton: implications for Palaeoproterozoic ultrahigh-temperature metamorphism. *Gondwana Research*, **11**, 263–285.

SANTOSH, M., WILDE, S. A. & LI, J. H. 2007b. Timing of Palaeoproterozoic ultrahigh-temperature metamorphism in the North China Craton: evidence from SHRIMP U–Pb zircon geochronology. *Precambrian Research*, **159**, 178–196.

SANTOSH, M., TSUNOGAE, T., OHYAMA, H., SATO, K., LI, J. H. & LIU, S. J. 2008. Carbonic metamorphism at ultrahigh temperatures: Evidence from North China Craton. *Earth and Planetary Science Letters*, **266**, 149–165.

SCOTT, D. J. 1999. U–Pb geochronology of the eastern Hall Peninsula, south Baffin Island, Canada: a northern link between the Archaean of West Greenland and the Palaeoproterozoic Torngat Orogen of northern Labrador. *Precambrian Research*, **93**, 5–26.

SHEN, Q. & QIAN, X. 1995. Assemblages, episodes and tectonic evolution in the Archaean of China. *Episodes*, **18**, 44–48.

SHEN, Q. H., XU, H. F., ZHANG, Z. Q., GAO, J. F., WU, J. S. & JI, C. L. 1992. *Early Precambrian granulites in China*. Geological Publishing House, Beijing.

SONG, B. 1992. The isochronological age dating and rare earth element geochemistry of Miyun rapakivi granite, its implication for origin. *The Bulletin of Geological Institute. China Academy of Geological Science*, **25**, 137–157 (in Chinese with English abstract).

SUN, S., ZHANG, G. W. & CHEN, Z. M. 1985. *The Precambrian Geological Evolution of the Southern Part of North China Fault-block*. Beijing: Metallurgical Industry Publishing House, 176–185 (in Chinese with English abstract).

SUN, D. & HU, W. (eds) 1993. *Precambrian Chronotectonic framework and Chronological Crustal Structure of Zhongtiao Mt*. Geological Publishing House, Beijing, 77–117 (in Chinese with English abstract).

TEIXEIRA, W. & FIGUEIREDO, M. C. H. 1991. An outline of early Proterozoic crustal evolution in the Sao Francisco craton, Brazil: a review. *Precambrian Research*, **53**, 1–22.

TEIXEIRA, W., RENNE, P. R. & BOSSI, J. 1999. $^{40}Ar-^{39}Ar$ and Rb–Sr geochronology of the Uruguayan dyke swarms, Rio de la Plata Craton and implications for Proterozoic intraplate activity in western Gondwana. *Precambrian Research*, **93**, 153–180.

TOTEU, S. F., VAN SCHMUS, W. R. & NYOBE, J. B. 1994. U–Pb and Sm–Nd evidence for Eburnian and Pan-African high-grade metamorphism in cratonic rocks of Cameroon. *Precambrian Research*, **67**, 321–347.

TROMPETTE, R. & CAROZZI, A. V. 1994. *Geology of Western Gondwana (2000–500 Ma), Pan-African-Brasiliano Aggregation of South America and Africa*. Balkema, Rotterdam.

WAN, Y., SONG, B. ET AL. 2006. SHRIMP U–Pb zircon geochronology of Palaeoproterozoic metasedimentary rocks in the North China Craton: evidence for a major Late Palaeoproterozoic tectonothermal event. *Precambrian Research*, **149**, 249–271.

WANG, K.-Y. 1991. The preliminary study of Luyashang Palaeoproterozoic quartz-monzonite. *Shanxi Bulletin of Science*, **36**, 685–687.

WANG, K., LI, J. & HAO, J. 1996. The Wutaishan orogenic belt within the Shanxi Province, Northern China: a record of late Archaean collision tectonics. *Precambrian Research*, **78**, 95–103.

WANG, K., LI, J. & HAO, J. 1997. Late Archaean mafic-ultramafic rocks from the Wutaishan, Shanxi Province: a possible ophiolite melange. *Acta Petrologica Sinica*, **13**, 139–151.

WHITEHOUSE, M. J., KALSBEEK, F. & NUTMAN, A. P. 1998. Crustal growth and crustal recycling in the Nagssugtoqidian orogen of West Greenland: constraints from radiogenic isotope systematics and U–Pb zircon geochronology. *Precambrian Research*, 365–381.

WILDE, S. A., CAWOOD, P., WANG, K. Y. & NEMCHIN, A. 1998. SHRIMP U–Pb zircon dating of granites and gneisses in the Taihangshan-Wutaishan area: implications for the timing of crustal growth in the North China Craton. *Chinese Science Bulletin*, **43**, 1.

WILDE, S. A., ZHAO, G. C. & SUN, M. 2002. Development of the North China Craton during the Late Archaean and its final amalgamation at 1.8 Ga: some speculations on its position within a global Palaeoproterozoic supercontinent. *Gondwana Research*, **5**, 85–94.

WILDE, S. A., ZHAO, G. C., WANG, K. Y. & SUN, M. 2004. First precise SHRIMP U–Pb zircon ages for the Hutuo Group, Wutaishan: further evidence for the Palaeoproterozoic amalgamation of the North China Craton. *Chinese Science Bulletin*, **49**, 83–90.

WINDLEY, B. F. 1995. *The Evolving Continents* (3rd edn). John Wiley, New York.

WU, F. Y., ZHAO, G. C., WILDE, S. A. & SUN, D. Y. 2005. Nd isotopic constraints on the crustal formation of the North China Craton. *Journal of Asian Earth Sciences* **24**, 523–545.

WU, J., GENG, Y. S. & SHEN, Q. H. 1998. *Archaean Geology Characteristics and Tectonic Evolution of Sino-Korea Palaeocontinent*. Geological Publishing House, Beijing, pp. 1–104.

XIA, X. P., SUN, M., ZHAO, G. C., WU, F. Y., XU, P., ZHANG, J. & LUO, Y. 2006. U–Pb and Hf isotopic study of detrital zircons from the Wulushan khondalites: constraints on the evolution of the Ordos terrane, Western Block of the North China Craton. *Earth and Planetary Science Letters*, **241**, 581–593.

XIE, G. H. 2005. Petrology and geochemistry of the anorthosite in Damiao and the rapakivi granite in Miyun: a review on the global distribution and signification of the rock mass type anorthosite and rapakivi granite on the time and space. *Science Press, Beijing*, 1–155 (in Chinese).

YIN, A. & NIE, S. Y. 1996. A Phanerozoic palinspastic reconstruction of China and its neighboring regions. *In*: YIN, A. & HARRISON, T. M. (eds) *The Tectonic Evolution of Asia*. Cambridge University Press, Cambridge UK, 442–485.

YU, J. H., FU, H. Q. & ZHANG, F. L. 1994. Petrogenesis of potassic alkaline volcanism and plutonism in a Proterozoic rift trough near Beijing. *Regional Geology of China*, **2**, 115–122 (in Chinese with English abstract).

ZHAI, M. G. & LIU, W. J. 2003. Palaeoproterozoic tectonic history of the North China Craton: a review. *Precambrian Research*, **122**, 183–199.

ZHAI, M. G., GUO, J. H. & YAN, Y. H. 1992. Discovery and preliminary study of Archaean high-pressure granulites in the North China Craton. *Science in China*, **12**, 28–50 (in Chinese).

ZHAI, M. G., GUO, J. H., LI, Y. G. & YAN, Y. H. 1995. Discovery of Archaean retrograded eclogites in the North China Craton and their tectonic implications. *Bulletin of Science in China*, **40**, 706–721.

ZHAI, M. G., GUO, J. H. & LIU, W. J. 2005. Neoarchaean to Palaeoproterozoic continental evolution and tectonic history of the North China Craton: a review. *Journal of Asian Earth Science*, **24**, 547–561.

ZHAI, M. G., YANG, J. H., FAN, H. R., MIAO, L. C. & LI, Y. G. 2002. A large-scale cluster of gold deposits and metallogenesis in the eastern North China Craton. *International Geology Reviews*, **44**, 458–476.

ZHANG, C. L., LI, Z. X., YU, H. P. & YE, H. M. 2007. An early Palaeoproterozoic high-K intrusive complex in the southwestern Tarim Basin, NW China: age, geochemistry, and implications for tectonic evolution of the Tarim Basin. *In*: ZHAI, M. G., XIAO, W. J., KUSKY, T. M. & SANTOSH, M. (eds) *Special Issue of Gondwana Research on the Tectonic Evolution of China and Adjacent Crustal Fragments*, **12**, 101–112.

ZHAO, G. C. 2001. Palaeoproterozoic assembly of the North China Craton. *Geological Magazine*, **138**, 87–91.

ZHAO, G. C., WILDE, S. A., CAWOOD, P. A. & LU, L. Z. 1998. Thermal evolution of basement rocks from the eastern part of the North China Craton and its bearing on tectonic setting. *International Geology Reviews*, **40**, 706–721.

ZHAO, G. C., CAWOOD, P. A., WILDE, S. A., SUN, M. & LU, L. Z. 2000a. Metamorphism of basement rocks in the Central Zone of the North China Craton: implications for Palaeoproterozoic tectonic evolution. *Precambrian Research*, **103**, 55–88.

ZHAO, G. C., WILDE, S. A., CAWOOD, P. A. & LU, L. Z. 2000b. Petrology and P–T–t path of the Fuping mafic granulites: Implications for tectonic evolution of the central zone of the North China Craton. *Journal of Metamorphic Petrology*, **18**, 375–391.

ZHAO, G. C., WILDE, S. A., CAWOOD, P. A. & SUN, M. 2001a. Archaean blocks and their boundaries in the North China Craton: lithological, geochemical, structural and P–T path constraints and tectonic evolution. *Precambrian Research*, **107**, 45–73.

ZHAO, G. C., WILDE, S. A., CAWOOD, P. A. & LU, L. Z. 2001b. High-pressure granulites (retrograded eclogites) from the Hengshan Complex, North China Craton: petrology and tectonic implications. *Journal of Petrology*, **42**, 1141–1170.

ZHAO, G. C., CAWOOD, P. A., WILDE, S. A. & LU, L. Z. 2002. Review of global 2.1–1.8 Ga orogens: implications for a pre-Rodinia supercontinent. *Earth Science Reviews*, **59**, 125–162.

ZHAO, G. C., SUN, M. & WILDE, S. A. 2005. Late Archaean to Palaeoproterozoic evolution of the North China Craton: key issues revisited. *Precambrian Research*, **136**, 77–202.

ZHAO, G. C., SUN, M., WILDE, S. A., LI, S. Z., LIU, S. W. & ZHANG, J. 2006a. Composite nature of the North China Granulite-Facies Belt: tectonothermal and geochronological constraints. *Gondwana Research*, **9**, 337–348.

ZHAO, G. C., SUN, M., WILDE, S. A., LI, S. Z. & ZHANG, J. 2006b. Some key issues in reconstruction of Proterozoic supercontinents. *Journal of Asian Earth Sciences*, **28**, 3–19.

ZHU, S. X. & CHEN, H. 1992. Characteristics of Palaeoproterozoic stromatolites in China. *Precambrian Research*, **57**, 135–163.

The Precambrian Khondalite Belt in the Daqingshan area, North China Craton: evidence for multiple metamorphic events in the Palaeoproterozoic era

YUSHENG WAN[1,2]*, DUNYI LIU[1,2], CHUNYAN DONG[1,2], ZHONGYUAN XU[3], ZHEJIU WANG[4], SIMON A. WILDE[5], YUEHENG YANG[6], ZHENGHONG LIU[3] & HONGYING ZHOU[1,2]

[1]*Institute of Geology, Chinese Academy of Geological Sciences, Beijing 100037, China*
[2]*Beijing SHRIMP Centre, Beijing 100037, China*
[3]*College of Earth Sciences, Jilin University, Changchun 130061, China*
[4]*Chinese Academy of Geological Sciences, Beijing 100037, China*
[5]*Department of Applied Geology, Curtin University of Technology, Perth, WA 6845, Australia*
[6]*Institute of Geology and Geophysics, Chinese Academy of Sciences, Beijing 100029, China*
*Corresponding author (e-mail: wanyusheng@bjshrimp.cn)

Abstract: High-grade pelitic metasedimentary rocks (khondalites) are widely distributed in the northwestern part of the North China Craton and were named the 'Khondalite Belt'. Prior to the application of zircon geochronology, a stratigraphic division of the supracrustal rocks into several groups was established using interpretative field geology. We report here SHRIMP U–Pb zircon ages and Hf-isotope data on metamorphosed sedimentary and magmatic rocks at Daqingshan, a typical area of the Khondalite Belt. The main conclusions are as follows: (1) The early Precambrian supracrustal rocks belong to three sequences: a 2.56–2.51 Ga supracrustal unit (the previous Sanggan 'group'), a 2.51–2.45 Ga supracrustal unit (a portion of the previous upper Wulashan 'group') and a 2.0–1.95 Ga supracrustal unit (including the previous lower Wulashan 'group', a portion of original upper Wulashan 'group' and the original Meidaizhao 'group') the units thus do not represent a true stratigraphy; (2) Strong tectono-thermal events occurred during the late Neoarchaean to late Palaeoproterozoic, with four episodes recognized: 2.6–2.5, 2.45–2.37, 2.3–2.0 and 1.95–1.85 Ga, with the latest event being consistent with the assembly of the Palaeoproterozoic supercontinent Columbia; (3) During the late Neoarchaean to late Palaeoproterozoic (2.55–2.5, 2.37 and 2.06 Ga) juvenile, mantle-derived material was added to the crust.

Supplementary Material: U–Pb and Lu–Hf data for zircons (Tables EA1 & EA2) are available at http://www.geolsoc.org.uk/SUP18358

Knowledge of the assembly and breakup of Precambrian supercontinents (e.g. Bleeker 2003) is important for understanding the early evolution of continental crust and thereby global geodynamics. Prior to the Rodinia supercontinent, there may have been a Palaeoproterozoic supercontinent, commonly referred to as 'Columbia' (Rogers & Santosh 2002; Zhao et al. 2002, 2004). However, the timing of the assembly and breakup of Columbia is still debatable. A major reason is that in many cratonic blocks where tectono-thermal events relating to the assembly and breakup are well-developed, detailed geochronological studies have not been carried out. This chapter addresses this problem by presenting the results of an isotopic study of metasediments close to proposed Columbia suture zones in north China.

The North China Craton (NCC) is a cratonic block that contains some of the oldest rocks in the world. Major advancements in understanding the geological history of the NCC have been made in the past few years. Zhao et al. (1998, 1999, 2000, 2001, 2005) recognized two major Palaeoproterozoic collisional belts: the Khondalite Belt and Trans-North China Orogen, in the western and central parts of the craton, respectively (Fig. 1). The Trans-North China Orogen divides the NCC into two discrete blocks, named the Eastern and Western blocks, whereas the Palaeoproterozoic Khondalite Belt divides the Western Block into the Yinshan Terrane in the north and the Ordos Terrane in the south (Fig. 1). On the other hand, Li et al. (2002), Kusky & Li (2003) and Kusky et al. (2007) proposed collision in the Neoarchaean,

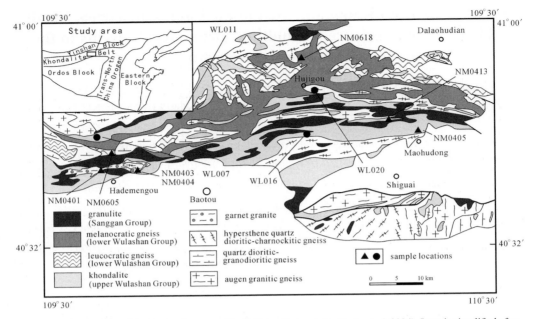

Fig. 1. Geological map of the Daqingshan area, North China Craton (after Yang *et al.* 2004). Inset is simplified after Zhao *et al.* (2005). Also shown are sample locations in this study (triangles) and by Xia *et al.* (2006*a*) (circles).

with the collisional belt similar in spatial distribution to that of the Palaeoproterozoic one proposed by Zhao *et al.* (1998, 1999, 2000, 2001, 2005), except that the northern portion of the belt turns eastwards into the western Liaoning–southern Jilin provinces. These authors proposed that the Western and Eastern blocks collided and amalgamated as early as the end of the Archaean, forming the unified NCC at this time. They also identified an Inner Mongolia–North Hebei Orogenic Belt (or more simply the North Hebei Orogenic Belt: NHOB) in the north and the Hengshan Plateau (including the Khondalite Belt) in the south along the northwestern part of the NCC (Kusky & Li 2003; Kusky *et al.* 2007). These authors interpreted the NHOB as probably being generated by accretion along an active continental margin at the end of Neoarchaean, whereas to the south the NHOB was thrust over the Khondalite Belt of the NCC. There are now coherent outlines for the timing and tectonic processes involved in the Palaeoproterozoic amalgamation and extensive knowledge concerning the pre-collisional history of the Trans-North China Orogen (Wu & Zhong 1998; Zhao *et al.* 1999, 2000, 2001, 2005, Wilde *et al.* 2002; Kröner *et al.* 2005*a*, *b*, 2006). However, the evolution of the Palaeoproterozoic Khondalite Belt still remains largely unknown.

The Khondalite Belt extends in an east–west direction (Fig. 1) and is dominated by graphite-bearing sillimanite-garnet gneiss, garnet quartzite, felsic paragneiss, calc-silicate rock and marble, which have previously been referred to informally as the 'khondalite series' in the Chinese literature (following the Indian usage) and were considered to represent stable continental margin deposits and to be Archaean in age, mainly based on their high metamorphic grade (Shen *et al.* 1990; Hu *et al.* 1994; Lu *et al.* 1996; Yang *et al.* 2000). Qian & Li (1999) suggested that the khondalites were deposited unconformably on TTG rocks, both of Archaean age. However, these metasedimentary rocks contain Palaeoproterozoic detrital zircons and give younger whole-rock Nd-isotope model ages, supporting recent field-based geological evidence for a post-Archaean age (Wu *et al.* 1997, 1998; Wan *et al.* 2000*a*, *b*; Xia *et al.* 2006*a*, *b*; Dong *et al.* 2007), The latter interpretation supports the view of Zhao *et al.* (2005) that the Western Block of the NCC consists of the Yinshan Block in the north and the Ordos Block in the south, separated by the Palaeoproterozoic Khondalite Belt. Recently, Guo *et al.* (2006) and Santosh *et al.* (2006, 2007*a*, *b*) reported the important discovery in the belt of diagnostic mineral assemblages indicating extreme crustal metamorphism at ultrahigh-temperature conditions. However, it is unclear whether all the metasedimentary rocks in the Khondalite Belt formed at the same time. Furthermore, besides the metasedimentary rocks, there are many other rock types present, such as TTG gneisses, mafic granulites, charnockites, as well as I-type and S-type granitic rocks. It has recently been established that some of these granitic rocks formed earlier than the khondalites (in the Palaeoproterozoic), as indicated by their emplacement ages of 2.3–2.0 Ga (Li *et al.* 2004; Zhong *et al.* 2006*a, b*; Dong *et al.* 2007).

In order to further understand the evolution of the Khondalite Belt during the Palaeoproterozoic, we report SHRIMP U–Pb zircon ages and Hf-isotope data on metamorphosed sedimentary and magmatic rocks in the Daqingshan area (Fig. 1). Our data show that the Palaeoproterozoic history of the Khondalite Belt is more complex than previously thought, possibly related to tectonic slices of different ages and origins being interdigitated with each other during the late Palaeoproterozoic continent–continent collision.

Geological background

The Daqingshan area is a representative part of the Khondalite Belt in the NCC (Fig. 1). The following summary of the lithologies is largely based on Yang *et al.* (2003, 2004), although their age interpretations are questioned in light of the new zircon geochronology presented in this study. At Daqingshan and adjacent areas, the Precambrian metamorphic rocks have been divided into the PalaeoArchaean Sanggan (Xinghe) Group, the MesoArchaean Wulashan Group, the Neoarchaean Seertenshan Group, the Palaeoproterozoic Meidaizhao Group and the Mesoproterozoic Zhaertai Group (Yang *et al.* 2004). The subdivisions are mainly based on geological field studies, with only a few conventional U–Pb zircon dates to support them. The Seertenshan and Zhaertai groups are distributed north of our study area, so they are not considered in detail here. The sequence of metamorphic rocks within these groups is unclear in many cases, so we are not necessarily dealing with a sedimentary sequence, rather a series of 'complexes' whose stratigraphy remains unknown. However, we will use the term 'group' in an informal way in this chapter, so that readers can compare our results with the existing Chinese literature.

The Sanggan group occurs mainly in the southern part of the study area (Fig. 1) and has been subdivided into a mesocratic granulite unit and a leucocratic granulite unit. The former is mainly composed of hypersthene-biotite-plagioclase gneiss, magnetite-hypersthene-plagioclase gneiss, two-pyroxene-plagioclase gneiss (with some retrogressed hypersthene-two-feldspar granulite), hypersthene-alkali-feldspar granulite, (retrogressed) two-pyroxene granulite and hypersthene-magnetite quartzite. The leucocratic unit is mainly composed of pyroxene-two-feldspar gneiss, with some two-pyroxene-alkali-feldspar gneiss and amphibole-hypersthene granulite. The Sanggan group underwent granulite-facies metamorphism and was locally retrograded to amphibolite facies. The protoliths of the mesocratic and leucocratic granulite units have been interpreted as basic–intermediate and intermediate–acid volcano-sedimentary rocks, respectively (Yang *et al.* 2004).

The Wulashan group has been subdivided into lower and upper subgroups. The lower subgroup is mainly distributed in the northern part of the study area (Fig. 1) and is further subdivided into melanocratic gneiss and leucocratic gneiss subunits. The former is mainly composed of quartz-bearing pyroxene amphibolite, quartz-alkali-feldspar amphibolite, amphibole-plagioclase gneiss, biotite- amphibole-two-feldspar gneiss and pyroxene-magnetite quartzite. The leucocratic gneiss is felsic in composition and mainly composed of biotite-alkali-feldspar gneiss, biotite-two-feldspar gneiss, biotite-plagioclase gneiss and amphibole-feldspar gneiss. Both sub-units underwent upper amphibolite- to granulite-facies metamorphism and have been regarded as having similar protoliths to rocks of the Sanggan group (Yang *et al.* 2004). There are some associated metamorphosed plutons in both the Sanggan group and lower Wulashan subgroup, but more work is required to define their spatial distribution and size.

The upper subgroup of the Wulashan group is similar in rock association and metamorphism to the khondalite rock series in southern India (Chacko *et al.* 1992) and was named the khondalite series in China. It is distributed widely in the study area (Fig. 1) and has been subdivided into three main components: garnet-biotite-quartzo-feldspathic gneiss, diopside gneiss and marble (Xu *et al.* 2005). The garnet-biotite gneiss sub-unit contains sillimanite-cordierite-garnet-biotite gneiss, graphite-bearing gneiss, fine-grained garnet leucogneiss, garnet quartzite and banded iron formation (BIF). The diopside gneiss sub-unit is mainly composed of feldspar-diopside gneiss, but also contains diopsidite and diopside marble. Some of the diopside marbles have been considered as crustally derived carbonate-rich magmatic rocks (Wan *et al.* 2008). Finally, the marble sub-unit is mainly composed of dolomitic marble, with thin interlayers of diopside quartzite and tremolite schist. These rocks underwent upper amphibolite- to granulite-facies metamorphism.

Based on petrographical studies and *P–T* calculations, Jin *et al.* (1991) established that the Sanggan and Wulashan groups had different *P–T–t* paths, showing anticlockwise and clockwise paths, respectively. They further suggested that the

Sanggan group was Neoarchaean and the Wulashan group was Palaeoproterozoic in age. The idea that the khondalite unit of the Wulashan group formed during the Palaeoproterozoic was also supported by the age distribution of detrital zircons (Xia et al. 2006b; Wan et al. 2006). In addition, the Sanggan and Wulashan groups were extensively modified by anatexis, locally resulting in the formation of granites, whose chemistry is consistent with derivation from the adjacent metasedimentary rocks (Song et al. 2005).

The Meidaizhao group occurs SE of the area shown in Figure 1 and consists mainly of fine-grained biotite gneiss, fine-grained quartzo-feldspathic gneiss and quartzite. The main minerals are plagioclase, K-feldspar, quartz, biotite, muscovite, chlorite and epidote, showing mineral associations typical of greenschist-facies metamorphism. It has been considered to be Palaeoproterozoic in age and to unconformably overlie the khondalite unit of the Wulashan group (Xu et al. 2003).

Analytical techniques

Zircon crystals were obtained using standard crushing and separation techniques and U–Pb dating was carried out using the SHRIMP II ion microprobe at the Beijing SHRIMP Centre, Chinese Academy of Geological Sciences (CAGS). The hand-picked crystals, together with the TEMORA standard (with a conventionally determined ^{206}Pb–^{238}U age of 417 Ma – Black et al. 2003), were cast in epoxy resin discs and polished. All grains were photographed in both transmitted and reflected light and then imaged using cathodoluminescence (CL) in order to reveal the internal structure and to identify preferred locations for SHRIMP analysis. Mineral and CO_2 inclusions were identified using a Jasco nitrogen Raman spectroscope with a Renishaw 1000 laser using the 514.5 nm line at the Institute of Geology, CAGS. Zircon dating procedures were similar to those described by Williams (1998). The intensity of the primary O^{2-} ion beam was 8 nA and the spot size was c. 30 μm, with each site rastered for 120–180 s before analysis. Five scans through the nine mass stations were made for each spot analysis. Data processing was carried out using the Squid and Isoplot programs (Ludwig 2001), and the measured ^{204}Pb was applied for the common lead correction, assuming an isotopic composition of Broken Hill lead. The uncertainties given in Table EA-1 (SUP18358) and figures for individual analyses are quoted at the 1σ level, whereas those for weighted mean ages are quoted at the 95% confidence level.

The in-situ Lu–Hf isotopic composition of zircon was measured with a Geolas-193 laser-ablation microprobe, attached to a Neptune multi-collector ICPMS at the State Key Laboratory of Lithospheric Evolution, Institute of Geology and Geophysics, Chinese Academy of Sciences in Beijing. A 193 nm UVArF excimer laser ablation system was used for laser ablation analysis. Instrumental conditions and analytical procedures are described in Wu et al. (2006). Ablation times were about 26 s for 200 cycles of each measurement, with a 6 Hz repetition rate, a laser power of 100 mJ/pulse and a spot size of 63 μm. Zircon standard 91500 (^{176}Hf/^{177}Hf = 0.282306; Woodhead et al. 2004) was analysed ($n = 13$) before and after the unknowns and yielded average ^{176}Hf/^{177}Hf$_{(c)}$ and ^{176}Lu/^{177}Hf$_{(c)}$ values of 0.282287 and 0.000286, respectively. The errors for the Lu–Hf isotope results are quoted at the 2σ level. The calculation of Hf-model ages was based on a depleted-mantle source with a present-day ^{176}Hf/^{177}Hf = 0.28325, using the ^{176}Lu decay constant 1.865×10^{-11} year^{-1} (Scherer et al. 2001). The calculation of εHf$_{(T)}$ values was based on zircon SHRIMP U–Pb ages and the chondritic values (^{176}Hf/^{177}Hf = 0.282772, ^{176}Lu/^{177}Hf = 0.0332; Blichert-Toft & Albarède 1997).

SHRIMP dating of zircons

In order to determine the formation ages of the main units and understand the tectono-thermal histories in the Daqingshan area, a total of eight rock samples, including metamorphic rocks from the Sanggan, Wulashan and Meidaizhao groups, garnet-biotite granite derived from khondalite and metamorphosed basic dyke and gabbro, were dated using the SHRIMP U–Pb zircon dating technique. The results are presented below.

Sanggan group

A two-pyroxene-bearing retrogressed granulite (sample NM0403) was collected from the mesocratic granulite unit of the Sanggan group NE of Hademengou (N40°41′55″, E109°41′52″) (Fig. 1). It comes from an homogenous outcrop (Fig. 2a) and is composed of hornblende, plagioclase, clinopyroxene and orthopyroxene, with accessory quartz and microcline. Some orthopyroxene grains show exsolution to clinopyroxene and amphibole is retrogressed from pyroxene; therefore, the rock is a partly retrogressed granulite, consistent with granulite-facies rocks with a clinopyroxene-orthopyroxene-plagioclase assemblage occurring adjacent to the sample location. The protolith of the basic granulite was of basaltic composition (Yang et al. 2004).

Zircons are commonly 200–300 μm in diameter, stubby in shape and show sector zoning in

Fig. 2. Field photographs of Precambrian rocks in the Daqingshan area, North China Craton. (**a**) Two-pyroxene-bearing retrogressed granulite (NM0403) of the Sanggan group, cut by meta-basic dyke (NM0404), NE of Hademengou; (**b**) Felsic gneiss (NM0605) of the leucocratic gneiss unit of the lower Wulashan group, north of Hademengou, showing anatectic features; (**c**) and (**d**) Meta-gabbro (NM0618), north of Hujigou; (c) shows evidence of partial melting; (**e**) Garnet-biotite gneiss (NM0413) of the upper Wulashan group, NW of Maohudong, showing local melt structures; (**f**) garnet-biotite granite (NM0401) at Hademengou, showing banded structure; (**g**) Garnet-biotite gneiss (NM0405) of the upper Wulashan group, north of Maohudong; (**h**) Feldspathic quartzite (NM0414) of the Meidaizhao group, c. 30 km SE of Shiguai.

CL image (grain 2 in Fig. 3a), being similar to grains commonly formed under granulite-facies conditions (Vavra et al. 1999). Some grains show a banded structure with evidence of marginal recrystallization (grain 3 in Fig. 3b), but have the same age as those with sector zoning. Inclusions of pyroxene, apatite (grain 2 in Fig. 3a), feldspar, quartz and CO_2 were identified in these zircons. Some calcite was also identified, but this occurs along cracks. These zircons are considered to have formed during granulite-facies metamorphism and are designated here as Meta I. These zircons show overgrowth rims (named Meta II), which are dark in CL and usually narrow in width, so that only a few were wide enough for analysis (Fig. 3a, b, c). A total of 23 analyses were made on 14 zircon grains. Except for spot 11.1 (a Meta I analysis which shows lead loss), the other 13 Meta I zircon domains give a $^{207}Pb/^{206}Pb$ weighted mean age of 2510 ± 8 Ma (MSWD = 3.8) (Fig. 4a). This age is interpreted as the time of granulite-facies metamorphism. The U and Th contents and Th/U ratios show large variations, being 65–4262 ppm, 4–1725 ppm and 0.07–1.10, respectively (Table EA-1). Some analyses show high Th/U ratios, which may be a feature of zircons from high-grade metamorphic rocks, especially ultrahigh-temperature ones (Zhou et al. 2004; Wan et al. 2006; Santosh et al. 2007b). Nine spots were analysed on the Meta II rims. The U and Th contents and Th/U ratios range from 221–966 ppm, 101–610 ppm and 0.14–1.37, respectively (Table EA-1). Six of the spots give a $^{207}Pb/^{206}Pb$ weighted mean age of 2419 ± 10 Ma (MSWD = 1.7) (Fig. 4a) and are considered to represent the time of a subsequent metamorphic event. Two spots (7.1 Meta II and 10.1 Meta II) on rims give $^{207}Pb/^{206}Pb$ ages of 2340 and 2354 Ma (Fig. 4a). This may record a still later metamorphic event or reflect disturbance by the c. 1.9 Ga event, although all rims are similar in CL image and Th/U ratio. There appears to be an inherited core in one grain (grain 2 in Fig. 3a) which gives a $^{207}Pb/^{206}Pb$ age of 2567 ± 10 Ma (analysis 2.2 in Fig. 4a, Table EA-1).

A meta-basic dyke (sample NM0404) that cuts the two-pyroxene retrogressed granulite (NM0403) described above (Fig. 2a), also experienced the high-grade metamorphic event. It is composed of hornblende, plagioclase, clinopyroxene and orthopyroxene, being similar in mineralogy to the country rock but without quartz and microcline and showing a more homogenous structure. Zircons are rare and occur as small, round or stubby crystals that have a uniform structure in CL images (Fig. 3d), and only apatite and CO_2 inclusions have been identified. Six spots were analysed on six zircons. The U and Th contents and Th/U ratios range from 695–865 ppm, 118–212 ppm and 0.18–0.25, respectively (Table EA-1). It is not certain whether the zircons are metamorphic in origin, although they are different from those from the two-pyroxene-bearing retrogressed granulite which the dyke cuts. All spots, with the exception of 2.1 which is younger in age (1854 Ma) but shows reverse discordance, give a $^{207}Pb/^{206}Pb$ weighted mean age of 1924 ± 11 Ma (MSWD = 1.8) (Fig. 4b). This age is the same as that obtained by Santosh et al. (2007b) for ultrahigh-temperature metamorphic rocks in the Khondalite Belt, so it probably records a metamorphic event, although it is difficult to understand why no metamorphic zircons with this age formed in the adjacent granulite (NM0403).

Wulashan group

A felsic gneiss (sample NM0605) was collected from the leucocratic gneiss sub-unit of the lower Wulashan group, north of Hademengou (N40°44′08″, E109°38′06″) (Fig. 1). It is composed of quartz, K-feldspar, plagioclase and biotite, but shows anatectic features, with reddish veins and patches composed mainly of coarser-grained K-feldspar, plagioclase and quartz (Fig. 2b). Nearby, meta-basic rocks are interlayered with, or occur as pods in the felsic gneiss. Zircon crystals are stubby or tabular in shape and can be divided into two main components: detrital cores and metamorphic (anatectic) rims. Many detrital cores themselves also show a 'core-mantle-rim' structure, with the cores displaying oscillatory zoning (Fig. 3e). It appears likely that both the rim and mantle are of metamorphic/anatectic origin (Vavra et al. 1999; Corfu et al. 2003) (so the ages should represent the metamorphic/anatectic time of their source region) but the mantles are too narrow for SHRIMP analysis and so we have not been able to establish this. Seven analyses of cores have U and Th contents and Th/U ratios ranging from 60–540 ppm, 57–538 ppm and 0.34–1.13, respectively (Table EA-1). They vary in apparent age from 2.58–2.20 Ga (Fig. 4c), with the oldest being 2582 ± 17 Ma in age. Eleven analyses of the rims have U and Th contents and Th/U ratios ranging from 310–1036 ppm, 14–72 ppm and 0.02–0.1, respectively (Table EA-1), and vary from 2.48–1.78 Ga in age, with 7 analyses having a $^{207}Pb/^{206}Pb$ weighted mean age of 2438 ± 8 Ma (MSWD = 1.2) (Fig. 4c). A few cores show a banded structure in CL image (grain 7 in Fig. 3f), and may be detrital magmatic grains. Three analyses of these give U and Th contents and Th/U ratios of 240–448 ppm, 139–190 ppm and 0.38–0.60, and show a large variation in $^{207}Pb/^{206}Pb$ age from 2398–2130 Ga (Fig. 4c), probably suggesting that

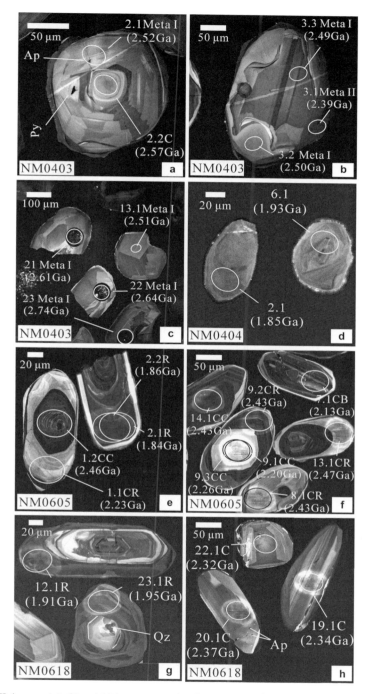

Fig. 3. Zircon CL images: (a), (b) and (c) two-pyroxene-bearing retrogressed granulite (NM0403) from the Sanggan group; (d) meta-basic dyke (NM0404) which cuts the two-pyroxene-bearing retrogressed granulite (NM0403); (e) and (f) anatectic felsic gneiss (NM0605) from the lower Wulashan group; (g) and (h) meta-gabbro (NM0618). Ellipse (c. 30 μm) and circle (c. 60 μm) show positions of SHRIMP U–Pb and ICP-MS Hf-analytical sites, respectively, with their identification numbers, as in Tables EA-1 and EA-2. Py, pyroxene; Ap, apatite; Qz, quartz, Meta I-zircon formed during the first metamorphic event (c. 2.5 Ga), Meta II-zircon formed during the second metamorphic event (c. 2.42 Ga).

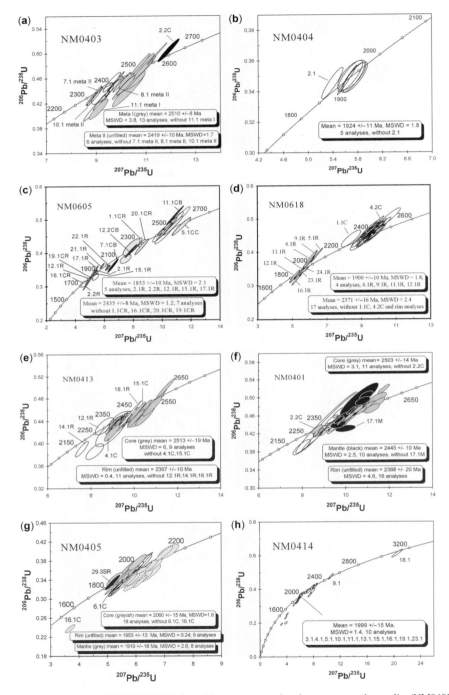

Fig. 4. Concordia diagrams of SHRIMP U–Pb data: (**a**) two-pyroxene-bearing retrogressed granulite (NM0403) from the Sanggan group: black, magmatic core; grey: Meta I; unfilled, Meta II; (**b**) meta-basic dyke (NM0404) cutting the two-pyroxene-bearing retrogressed granulite (NM0403); (**c**) anatectic felsic gneiss (NM0605) from the lower Wulashan group: light-grey, core of detrital zircon; unfilled, rim of detrital zircon; black, detrital zircon with banded texture; dark-grey, metamorphic rim; (**d**) meta-gabbro (NM0618): unfilled, magmatic core; grey, metamorphic rim; (**e**) garnet-biotite gneiss (NM0413) from the upper Wulashan group: grey, detrital core; unfilled, metamorphic rim;

they came from different sources. Five rim analyses, with the exception of 21.1R and 22.1R which overlap onto cores, give a $^{207}Pb/^{206}Pb$ weighted mean age of 1853 ± 10 Ma (MSWD = 2.1) (Fig. 4c). This age is interpreted as the age of late Palaeoproterozoic metamorphism and anatexis.

A meta-gabbro (sample NM0618) was collected from north of Hujigou (N40°53'21", E110°04'26") (Fig. 1). Its cutting relationship with the melanocratic gneiss unit of the lower Wulashan group can be observed. The gabbro has undergone anatexis with local development of leucocratic neosome (Fig. 2c). Away from the neosome, the gabbro is more homogeneous and medium- to coarse-grained (Fig. 2d) and is composed of plagioclase, hornblende, biotite and minor quartz. Zircons are prismatic or stubby in shape and show core-rim structures in CL images. Most of the cores are banded (Fig. 3g, h) or show oscillatory zoning (Fig. 3g). Some cores contain feldspar, apatite and quartz inclusions (Fig. 3g, h). Nineteen analyses on the cores have U and Th contents and Th/U ratios ranging from 72–545 ppm, 64–809 ppm and 0.91–1.53, respectively (Table EA-1, SUP18358). All analyses, except 1.1C and 4.2C, give a $^{207}Pb/^{206}Pb$ weighted mean age of 2371 ± 16 Ma (MSWD = 2.4) (Fig. 4d). The large error and MSWD value may partly reflect disturbance due to later tectono-thermal event(s). Metamorphic rims are common, but only a few are wide enough to analyse. They are homogenous or show weak zoning (Fig. 3g), and no mineral inclusions have been identified. Eight spots were analysed on the rims and the U and Th contents and Th/U ratios range from 261–1476 ppm, 18–338 ppm and 0.01–0.58, respectively (Table EA-1). They vary widely in $^{207}Pb/^{206}Pb$ age, with the four youngest analyses giving a $^{207}Pb/^{206}Pb$ weighted mean age of 1900 ± 10 Ma (MSWD = 1.8) (Fig. 4d). The slightly older rim ages may be partly due to analytical sites overlapping onto the cores.

A garnet-biotite gneiss (sample NM0413) was collected from a wide homogeneous layer NE of Maohudong (N40°48'32", E110°15'27") (Fig. 1), where the high-grade metasedimentary rocks are considered to represent typical khondalites of the upper Wulashan group. The rock shows weak anatexis, giving rise to a quartzo-feldspathic neosome (Fig. 2e). The sample consists of plagioclase, quartz, biotite and garnet; sillimanite was identified in rocks nearby, but was not present in this sample. Zircons are round or long–prismatic grains with rounded terminations and show a core-mantle-rim structure, though the mantle is usually narrow (Fig. 5a, b). The round, detrital grains mostly have wider rims than the elongated ones. Some cores contain apatite, quartz and feldspar inclusions, but the mantles and rims are inclusion-free. Ten spots were analysed on the cores and all analyses, except 15.1C, have U and Th contents and Th/U ratios ranging from 105–1882 ppm, 67–359 ppm and 0.09–1.13, respectively (Table EA-1). They have $^{207}Pb/^{206}Pb$ ages ranging from 2557 to 2477 Ma, with a weighted mean $^{207}Pb/^{206}Pb$ age of 2513 ± 19 Ma (MSWD = 6.1) (Fig. 4e), suggesting a late Archaean magmatic source region. Fourteen spots were analysed on the rims and the U and Th contents and Th/U ratios range from 279–1208 ppm, 5–210 ppm and 0.01–0.27, respectively, with Th/U ratio being mostly <0.1 (Table EA-1). Except for 12.1R, 14.1R and 18.1R, the remaining 11 analyses give a $^{207}Pb/^{206}Pb$ weighted mean age of 2397 ± 10 Ma (MSWD = 0.4) (Fig. 4e), which is interpreted as dating a subsequent metamorphic event.

A garnet-biotite granite (sample NM0401) from Hademengou (N40°42'29", E109°38'34") (Fig. 1) varies in texture, with some localities showing a distinct foliation (Fig. 2f). There are many metasedimentary and some mafic granulite enclaves in the granite, and geological and geochemical studies indicate it formed through anatexis of upper Wulashan group metasedimentary rocks (Yang et al. 2004; Song et al. 2005). The biotite is dark-brown in colour and contains many needle-shaped titanite inclusions. Zircons are round in shape, 150–200 μm in diameter and show complex structures in CL images, generally exhibiting core-mantle-rim structures (Fig. 5c, d). Most of the cores are characterized by concentric and patchy zoning in CL and are embayed by metamorphic mantles (Fig. 5c, d). The mantles are dark in CL and show a more uniform structure, with most of them wide enough to analyse. In some zircon grains, the boundary between mantle and core shows an intergrowth of bright and dark features in CL (Fig. 5c, d), interpreted to be the result of recrystallization, probably under the influence of fluids. The rims are bright in CL and show transitional boundaries with the mantle. Some rims show oscillatory zoning (Fig. 5c), indicating a magmatic origin. Apatite, feldspar and quartz inclusions are present in some cores, suggesting that they formed originally from a granitic magma: no inclusions were found in the mantles and rims. Twelve analyses of the cores have U and Th contents and Th/U ratios ranging

---

Fig. 4. (*Continued*) (**f**) garnet-biotite granite (NM0401) intruding the upper Wulashan group: grey, inherited core; dark, metamorphic mantle; unfilled, magmatic rim; (**g**) garnet-biotite gneiss (NM0405) of the upper Wulashan group: pale-grey, detrital core; darker grey, metamorphic mantle; unfilled, metamorphic rim; black, outer metamorphic rim; (**h**) feldspathic quartzite (NM0414) of the Meidaizhao group. Error ellipses are at the 1σ level.

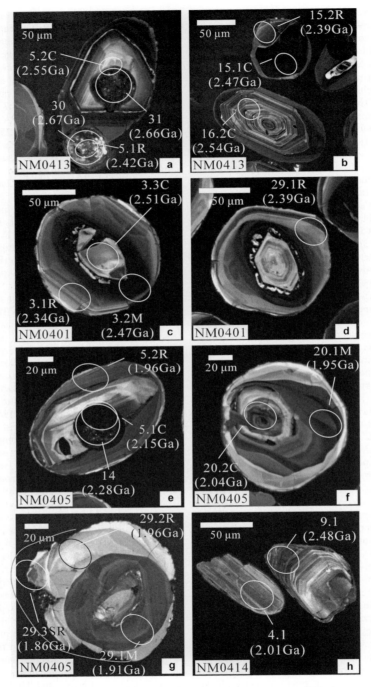

Fig. 5. Zircon CL images: (**a**) and (**b**) garnet-biotite gneiss (NM0413) from the upper Wulashan group; (**c**) and (**d**) garnet-biotite granite (NM0401) intruding the upper Wulashan group; (**e**), (**f**) and (**g**) garnet-biotite gneiss (NM0405) of the upper Wulashan group; curve in (g) represents the boundary of the outer rim of the zircon grain which has been partly destroyed during polishing; (**h**) feldspathic quartzite (NM0414) of the Meidaizhao group. Ellipse (c. 30 μm) and circle (c. 60 μm) show position of SHRIMP U–Pb and ICP-MS Hf-analytical sites, respectively, with their identification numbers as in Tables EA-1 and EA-2.

from 50–605 ppm, 33–489 ppm and 0.58–1.07, respectively (Table EA-1). All core analyses, with the exception of 2.2C which is much younger in age (2302 Ma), give a $^{207}Pb/^{206}Pb$ weighted mean age of 2503 ± 14 Ma (MSWD = 3.1) (Fig. 4f). Eleven analyses of mantles show higher U contents (329–1690 ppm), lower Th contents (8–49 ppm) and, therefore, very low Th/U ratios (0.01–0.08); with the exception of 11.1 M, which has a Th/U ratio of 0.31. All spots, except outlier 17.1, give a $^{207}Pb/^{206}Pb$ weighted mean age of 2445 ± 10 Ma (MSWD = 2.5) (Fig. 4f). Sixteen analyses of the rims have U and Th contents and Th/U ratios ranging from 89–175 ppm, 44–80 ppm and 0.26–0.61, respectively, and give a $^{207}Pb/^{206}Pb$ weighted mean age of 2388 ± 20 Ma (MSWD = 4.6) (Fig. 4f). The large error and MSWD may be partly due to some sites overlapping onto mantles. The core ages are interpreted as the formation time of the detrital source rocks (2.5 Ga) to the khondalitic protolith, the mantle ages as the time of high-grade metamorphism of the original khondalitic sedimentary rock (2.45 Ga) and, finally, the rims record anatexis of the khondalite, leading to formation of the late to post-tectonic granite (2.39 Ga).

Another garnet-biotite gneiss (sample NM0405) was collected from a unit previously assigned to the upper Wulashan group, north of Maohudong (N40°45′37″, E110°19′02″) (Fig. 1). It is crudely banded (Fig. 2g) and composed of quartz, plagioclase, biotite, garnet and some K-feldspar. The garnets contain central mineral inclusions of fine-grained quartz, feldspar and light-brown biotite; the biotite in the matrix is darker brown in colour. Quartz, plagioclase, K-feldspar and biotite occur as aggregates, possibly a result of strong deformation and recrystallization. The zircons are round or long-prismatic grains and commonly show core-mantle-rim structures (Fig. 5e, f, g). Grain 29 in Figure 5g shows a complex structure with a mantle exhibiting weak zoning, possibly suggesting an anatectic origin. The cores contain apatite, quartz, feldspar and anatase (TiO_2 polymorph) inclusions, but no inclusions were identified in the mantles or rims. Based on 20 analyses, the cores have U and Th contents and Th/U ratios of 57–763 ppm, 18–485 ppm and 0.08–0.94, with most of the ratios being >0.25 (Table EA-1). Eighteen core analyses give a $^{207}Pb/^{206}Pb$ weighted mean age of 2060 ± 15 Ma (MSWD = 1.8) (Fig. 4g). Eight analyses of the mantles show large variations of U and Th contents (141–572 ppm and 6–98 ppm) and Th/U ratios (0.01–0.72) and define a $^{207}Pb/^{206}Pb$ weighted mean age of 1913 ± 18 Ma (MSWD = 2.6) (Fig. 4g). Nine analyses of the rims have U and Th contents and Th/U ratios of 57–288 ppm, 1–60 ppm and 0.02–0.34, respectively, and give a $^{207}Pb/^{206}Pb$ weighted mean age of 1953 ± 13 Ma (MSWD = 0.24) (Fig. 4g). The mantles are thus apparently younger than the rims. It may be that the mantles and rims formed during the same metamorphic event, although it is not possible to constrain which, if either, is the more correct estimate of the timing of this event. There are also narrow outer rims in some zircons and one of these (29.3SR in Fig. 5g) gives a $^{207}Pb/^{206}Pb$ age of 1856 ± 17 Ma, with U and Th contents and Th/U ratios being 177 ppm, 10 ppm and 0.06, respectively (Table EA-1). This age may represent the time of a later metamorphic event, since it is similar to the main event recorded in the Trans-North China Orogen (Zhao et al. 2005).

Meidaizhao group

A feldspathic quartzite (NM0414) sample was collected from the Meidaizhao group c. 30 km SE of Shiguai (N40°37′19″, E109°35′02″). It is composed of plagioclase, microcline, quartz, biotite and muscovite. The plagioclase is commonly altered to sericite, and the biotite is chloritized. The mineral assemblage defines greenschist-facies metamorphism, and the rock only shows a weak deformation fabric (Fig. 2h); in this respect it is different from the high-grade metamorphic rocks of the Sanggan and Wulashan groups. The detrital zircons are commonly small and stubby to prismatic in shape, and oscillatory zoning is clearly observed in CL images (Fig. 5h), indicating that they are of magmatic origin. Narrow metamorphic rims are present in some zircons (grain 9 in Fig. 5h). Twenty-three sites were analysed on 23 zircons and the U and Th contents and Th/U ratios range from 88–590 ppm, 26–894 ppm and 0.21–1.63, respectively. Two analyses have $^{207}Pb/^{206}Pb$ ages of 3.15 and 2.48 Ga (18.1 and 9.1), but the remainder are concentrated between 2.3 to 2.1 and at c. 2.0-Ga (Fig. 4h). The 10 youngest analyses on or close to concordia give a $^{207}Pb/^{206}Pb$ weighted mean age of 1999 ± 15 Ma (MSWD = 1.4). This defines the maximum age of deposition of the quartzite.

Hf isotope compositions of zircons

Sanggan group

The two-pyroxene-bearing retrogressed granulite (NM0403) was investigated for Lu–Hf systematics and a total of 25 analyses were made on 22 zircon grains (Table EA-2). Twenty-two analyses on Meta I domains have $\varepsilon Hf_{(T)}$ (T = 2510 Ma, the weighted mean $^{207}Pb/^{206}Pb$ age of the Meta I domains) ranging from +2.00 to +5.43 and T_{DM} model ages of 2743–2615 Ma (Fig. 6). Three analyses of Meta II domains have $\varepsilon Hf_{(T)}$ (T = 2419 Ma, the weighted mean $^{207}Pb/^{206}Pb$ age

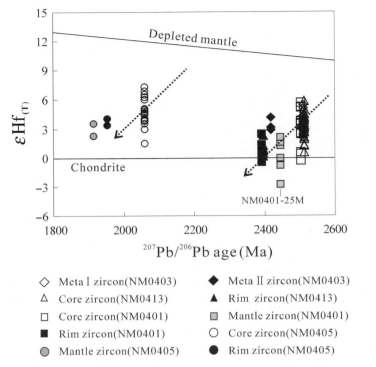

Fig. 6. εHf$_{(T)}$ v. ^{207}Pb/^{206}Pb age diagram of zircons from early Precambrian rocks in the Daqingshan area.

of the Meta II domains) and T$_{DM}$ age ranging from +2.92 to +4.19 and 2632–2584 Ma, respectively, similar to the Meta I domains.

Wulashan group

The garnet-biotite gneiss (NM0413) was investigated and a total of 35 analyses were made on 29 zircon grains (Table EA-2, SUP18358). Twenty-six analyses on detrital cores have εHf$_{(T)}$ (T = 2513 Ma, the weighted mean ^{207}Pb/^{206}Pb age of the detrital cores) ranging from +0.55 to +5.83 and T$_{DM}$ model ages of 2803–2603 Ma (Fig. 6), thus showing a larger Hf-isotope variation than zircons from the two-pyroxene retrogressed granulite of the Sanggan group. Nine analyses on metamorphic rims have εHf$_{(T)}$ (T = 2397 Ma, the weighted mean ^{207}Pb/^{206}Pb age of the metamorphic rims) ranging from +0.86 to +2.28 and T$_{DM}$ model ages of 2719–2642 Ma (Fig. 6), being similar to those of the detrital cores. This demonstrates that the Lu–Hf isotopic system has not been affected by high-grade metamorphism (Dong et al. 2007).

The garnet-biotite granite (NM0401), considered to be the result of anatexis of upper Wulashan metasedimentary rocks, was investigated and a total of 28 analyses were made on 21 zircon grains (Table EA-2). Twelve analyses on inherited cores have εHf$_{(T)}$ (T = 2503 Ma, the weighted mean ^{207}Pb/^{206}Pb age of the inherited cores) ranging from +0.27 to +5.73 and T$_{DM}$ model ages of 2833–2598 Ma (Fig. 6), being similar to the detrital cores of zircons from the garnet-biotite gneiss (NM0413) and therefore consistent with the view that the granite formed by anatexis of the metasedimentary rocks. Six analyses on metamorphic mantles have εHf$_{(T)}$ (T = 2445 Ma, the weighted mean ^{207}Pb/^{206}Pb age of the metamorphic mantles) of −0.80 to +2.10 and T$_{DM}$ model ages 2802–2691 Ma, except for analysis 25M, which has a lower εHf$_{(T)}$ (−2.76) and an older T$_{DM}$ age (2875 Ma) (Table EA-2, Fig. 6). Nine analyses on anatectic rims have εHf$_{(T)}$ (T = 2388 Ma, the weighted mean ^{207}Pb/^{206}Pb age of the anatectic rims) and T$_{DM}$ ages of −0.37 to +2.50 and 2728–2619 Ma, respectively. The metamorphic mantles and anatectic rims are therefore similar to the inherited cores, suggesting that the Hf-isotope composition of the sources from which the zircons formed was not changed during subsequent metamorphism and anatexis.

Another garnet-biotite gneiss (sample NM0405) was also investigated and a total of 31 analyses were made on 29 zircon grains (Table EA-2).

Twenty-seven analyses on detrital cores have $\varepsilon Hf_{(T)}$ (T = 2060 Ma, the weighted mean $^{207}Pb/^{206}Pb$ age of the core) ranging from +1.50 to +7.35 and T_{DM} model ages of 2380–2153 Ma. The large positive $\varepsilon Hf_{(T)}$ values show that the source region has no relationship to the older rocks in the area (Fig. 6). A few analyses on metamorphic mantles and rims have similar Hf-isotope compositions to the detrital cores (Fig. 6), indicating that the Lu–Hf isotope systematics were not affected by metamorphism. The protolith of this garnet-bearing gneiss thus sampled an entirely different source region to garnet gneiss sample NM0413.

Discussion

The new SHRIMP U–Pb zircon dating, combined with the zircon Hf-isotope study, indicates a thorough reassessment of the Khondalite Belt is required. Our data show new complexities in the depositional and metamorphic history, as summarized in Table 1, and indicates that the previous stratigraphic division needs to be revised.

Sanggan group. The two-pyroxene-bearing retrogressed granulite (NM0403) has an age of 2.51 ± 0.01 Ga for the Meta I zircon domains and this is interpreted as the time of granulite-facies metamorphism of the Sanggan group. An inherited core-like domain in one zircon grain (grain 2 in Fig. 3a) gives a $^{207}Pb/^{206}Pb$ age of 2567 ± 10 Ma, which may record the magmatic age of the rock, although more data are required to establish this. The Meta I zircon domains show large variations in Lu/Hf, but with $\varepsilon Hf_{(T)}$ being generally positive and ranging from +2.00 to +5.43. These results indicate that the Sanggan group in the Daqingshan area formed during the Neoarchaean, not the Palaeoarchaean, as previously thought.

Wulashan group. Zircons from anatectic felsic gneiss sample (NM0605) from the leucocratic gneiss unit of the lower Wulashan group show complex variations in texture and age, with most detrital cores themselves showing core-mantle-rim structures. The homogenous or weakly zoned rims give a $^{207}Pb/^{206}Pb$ weighted mean age of 2.44 ± 0.01 Ga, which may represent the time of metamorphism of the source region. The youngest detrital core gives a $^{207}Pb/^{206}Pb$ age of 2.13 ± 0.02 Ga, which limits the time of deposition to between 2.13 and 1.85 Ga, the latter being the age of the metamorphic rims. No ages have been determined for the mesocratic gneiss unit of the lower Wulashan group.

Detrital and metamorphic zircons from garnet-biotite gneiss (NM0405) give $^{207}Pb/^{206}Pb$ weighted

Table 1. *Zircon U–Pb data of early Precambrian rocks in the Daqingshan area, Khondalite Belt*

Sample No.	Rock type	2.6–2.55	2.5	2.45	2.4	2.37	2.3	2.2–2.0	1.95	1.90	1.85
NM0403	Two-pyroxene-bearing retrogressed granulite	V (2.57)	M (2.51)		M (2.42)						
NM0605	Felsic gneiss	D (2.58)	D (2.50)	D (2.44)							
NM0618	Meta-gabbro						D (2.30)	D (2.13)		M (1.90)	M (1.85)
NM0413	Garnet-biotite gneiss		D (2.51)		M (2.40)	E (2.37)					
NM0401	Garnet-biotite granite		I (2.50)	M (2.45)	E (2.39)						
NM0405	Garnet-biotite gneiss		D (2.48)					D (2.06)	M (1.95)	M (1.91)	M (1.85)
NM0414	Feldspathic quartzite							D (2.00)		M (1.90)	
NM0409	Crustal carbonatite (formed by anatexis of impure marble)						D (2.30)		E (1.95)		
NM0407-5	Diopsidite								M (1.95)		
NM0404	Meta-basic dyke									M (1.92)	

Note: (1) V, D, I, E and M represent volcanic, detrital, inherited, emplacement and metamorphic zircons, respectively. (2) Age in Ga. (3) Data of NM0409 and NM0407-5 are from Wan *et al.* (2008).

mean ages of 2.06 ± 0.02 Ga and 1.95 ± 0.01 Ga, respectively, limiting the time of deposition of the khondalite precursors of the upper Wulashan group to late Palaeoproterozoic, not Mesoarchaean as previously suggested. This is consistent with the geochronological study of Santosh et al. (2007b), who obtained detrital core and metamorphic rim ages of zircons from UHT rocks in the Zhuozi-Tuguiwula area in the eastern segment of the Khondalite Belt, of c. 1.97 and 1.92 Ga, respectively. Based on detrital zircon ages, Xia et al. (2006a, b) drew a similar conclusion from central and eastern segments of the Khondalite Belt. However, these latter authors considered that the protolith of the khondalites was deposited after 1.84 Ga. This is not consistent with evidence that metamorphism occurred between 1.95 and 1.85 Ga (Wan et al. 2006; Santosh et al. 2007a, b; Dong et al. 2007; this study) and that the khondalite was intruded by diorite with a zircon age of 1.92 Ga (Gou et al. 2002).

It is therefore evident that most khondalites were deposited during the late Palaeoproterozoic (Fig. 1, Xia et al. 2006a, b; Santosh et al. 2007a, b; Dong et al. 2007; this study). However, garnet-biotite gneiss sample NM0413 has detrital and metamorphic zircon ages of 2.51 ± 0.02 Ga and 2.40 ± 0.01 Ga, respectively, suggesting that the protolith of some khondalites was deposited during the earliest Palaeoproterozoic. This is further supported by dating of the garnet-biotite granite (NM0401), which intrudes some metasedimentary rocks of the upper Wulashan group at c. 2.39 Ga (the age of zircon rims with some showing magmatic zoning, Fig. 5c, d). The metamorphic zircon mantles in this granite record a slightly older age of 2.46 ± 0.01 Ga, further limiting the deposition time of the sedimentary rocks to between 2.51 and 2.46 Ga. Since the garnet-biotite gneiss (NM0413) is interlayered with other high-grade rocks assigned to the Sanggan and Wulashan groups, it appears that gneisses of different age have been juxtaposed during later deformation; a feature commonly observed in Precambrian terrains worldwide.

Meidaizhao group. Xu et al. (2003) suggested that the Meidaizhao group formed during the late Palaeoproterozoic. Although metamorphic zircon ages could not be obtained from the sample of feldspathic quartzite (NM0414), because there was no new zircon growth owing to the low grade of metamorphism, the youngest detrital zircon age (2.00 ± 0.02 Ga) may support this view. In addition: (1) no regional metamorphism occurred after the Palaeoproterozoic in the study area; and (2) the Meidaizhao group metasediments were intruded by post-orogenic granite at 1.82 ± 0.01 Ga (Xu et al. 2003). On the basis of geological mapping, Xu et al. (2003) indicated that the Meidaizhao group rests unconformably on the khondalites. This implies that the age of the underlying khondalite was >2.0 Ga.

The Precambrian supracrustal rocks in the Daqingshan area can now be divided into at least three units in terms of their formation age: (1) The 2.56–2.51 Ga supracrustal unit which is equivalent to the original Sanggan group and is mainly composed of basic-intermediate-acid volcano-sedimentary rocks at granulite facies. Rocks of the same age are widely distributed in the Jining area, east of Daqingshan (Guo et al. 2005) and in the Yinshan Block, north of the Khondalite Belt (Jian et al. 2005; Chen 2007); (2) The 2.51–2.45 Ga supracrustal unit was originally considered to belong to the khondalite sequence of the upper Wulashan group (Yang et al. 2004) but this has now been shown to be incorrect. The detritus may have come from c. 2.6–2.5 Ga basement rocks which underwent high-grade metamorphism prior to erosion. More work is required to identify their spatial distribution and to distinguish them from the late Palaeoproterozoic khondalites; (3) The 2.0–1.95 Ga supracrustal unit which includes the original lower Wulashan group, most of the khondalites of the upper Wulashan group and the Meidaizhao group. It is still necessary to investigate whether all the lower Wulashan group rocks belong to this unit, since we can only limit the age of formation to the leucocratic gneisses within an age range of 2.13–1.85 Ga. The late Palaeoproterozoic rocks of the upper Wulashan and Meidaizhao groups formed during the same period, but the former are high grade and strongly deformed whereas the latter are low grade and only weakly deformed. This phenomenon has also been observed in other orogenic belts, such as the Jingshan and Fenzhishan groups in eastern Shandong province (Wan et al. 2006).

Neoarchaean and Palaeoproterozoic tectonothermal events in the Daqingshan area

Strong and periodical tectono-thermal (magmatic or/and metamorphic) events occurred during the late Neoarchaean and Palaeoproterozoic in the Daqingshan area (Table 1) and can be divided into four phases, namely, at 2.6–2.5, 2.45–2.37, 2.3–2.0 and 1.95–1.85 Ga.

2.6–2.5 Ga event

Only a few zircon grains with ages of 2.6–2.55 Ga have been identified in this study. However, zircons from different rocks in the Daqingshan area record ages of *c*. 2.5 Ga, and most of them are of detrital or inherited origin (Table 1). The most important is the metamorphic age of 2.51 Ga determined from two-pyroxene-bearing retrogressed granulite sample NM0403, which indicates that high-grade metamorphism occurred at the end of the Archaean. A strong *c*. 2.5 Ga event has also been recorded in the Yinshan Block, north of the Khondalite Belt, including the *c*. 2.5 Ga Seertershan (group) greenstone belt, which was considered to have formed by a combination of subduction and mantle plumes (Chen 2007) and TTG granites that show island-arc geochemical features and contain old inherited zircons (Jian *et al*. 2005). However, it is still debatable whether the *c*. 2.5 Ga tectonothermal event, that occurred widely in the NCC, represents arc magmatism that culminated in collisional orogeny or reflects underplating from mantle plumes (Liu *et al*. 2007).

2.45–2.37 Ga event

This event can be subdivided into three stages: high-grade metamorphism (2.45 Ga); anatexis (2.40 Ga); and basic magmatism (2.37 Ga). The occurrence of *c*. 2.4 Ga U–Pb zircon ages is rare in the NCC, being recognized only in a few areas such as in northern Hebei (Liu *et al*. 2007). Both the Archaean basement (represented by sample NM0403 from the Sanggan group) and the early Palaeoproterozoic supracrustal unit (represented by sample NM0413 from the original upper Wulashan group) were involved in the 2.45–2.37 Ga event. However, based on CL images and U–Pb dating of detrital zircons, the metamorphic event at *c*. 2.45 Ga occurred in the source region from which some late Palaeoproterozoic sediments were derived (represented by sample NM0605 from the original lower Wulashan group) and so pre-dated deposition. Metasedimentary rocks (sample NM0413) from the early Palaeoproterozoic supracrustal unit show relatively high maturity. How-ever, the age gap between detrital and metamorphic zircons is only *c*. 50 Ma, possibly suggesting that these rocks were deposited in an active continental margin or back-arc environment (Kimbrough *et al*. 2001). The garnet biotite granite (NM0401) is the first early Palaeoproterozoic crustally-derived granite identified in the area and formed at 2.39 Ga (age of magmatic, anatectic zircon rims). The metamorphic and magmatic zircons give ages of 2.45 and 2.39 Ga, respectively. We suggest that the magmatic zircon age of *c*. 2.4 Ga reflects the time when the continental crust underwent regional extension because of crustal uplift resulting from mantle underplating. This is supported by the geochemical features of the 2.37 Ga meta-gabbro (NM0618) which is high in SiO_2 and low in MgO, characterized by high REE contents (TREE = 351 ppm) and fractionation of LREE and HREE [$(La/Yb)n$ = 19.7], with a weak Eu anomaly (Eu/Eu* = 0.77) and negative Nb anomalies (Wan *et al*. unpublished data). These compositional features suggest that the basic magma underwent strong fractionation and probably continental contamination before crystallization of the gabbro. Therefore, it is possible that the mantle-derived magma may have resided for a long time at the base of the crust before it was emplaced into the upper crust. We therefore suggest that magmatic underplating probably caused the high-temperature metamorphism and anatexis at 2.45–2.4 Ga.

2.3–2.0 Ga event

Although rocks with 2.3–2.0 Ga ages have not been identified in the Daqingshan area, many metasedimentary rocks of late Palaeoproterozoic age contain detrital zircons with these ages, and most are *c*. 2.0 Ga, similar to results obtained by Xia *et al*. (2006*b*). There are also abundant 2.3–2.0 Ga detrital zircons in Palaeoproterozoic metasedimentary rocks in other parts of the Khondalite Belt, and the proportion is commonly higher than that of Archaean detrital zircons (Wan *et al*. 2006; Xia *et al*. 2006*a*, *b*; Dong *et al*. 2007). In the Qianlishan area, north of Helanshan, detrital zircons from Neoproterozoic to Mesozoic sedimentary rocks reveal an important age peak around 2.00–2.06 Ga (Darby & Gehrels 2006). Therefore, there must have been an extensive source composed of Palaeoproterozoic igneous rocks which provided the detritus. It remains unclear where the 2.3–2.0 Ga detritus came from: the Ordos massif, Khondalite Belt or somewhere else (Xia *et al*. 2006*a*, *b*; Dong *et al*. 2007)? Zhong *et al*. (2005*a*) identified 2.33–2.13 Ga MgO-rich granites in the Khondalite Belt and suggested an island-arc environment, similar to that of the Closepet granite in southern India (Moyen *et al*. 2001).

1.95–1.85 Ga event

A major, complex tectono-thermal event occurred in the Daqingshan area during the late Palaeoproterozoic. Three metamorphic (including anatectic) ages have been identified in this study, namely *c*. 1.95, *c*. 1.90 and *c*. 1.85 Ga. Interestingly, these ages have not been recorded in the Neoarchaean mafic volcanic precursor to the two-pyroxene retrogressed granulite (sample NM0403), but this does not necessarily mean that the rock was not involved in the event since, in the Jining area east of Daqingshan, Archaean rocks were affected by the late Palaeoproterozoic event (Guo *et al*. 2005). The

1.95 Ga metamorphism in the Khondalite Belt resulted in anatexis of carbonates of the upper Wulashan group to form crustally derived carbonate-rich magmatic rocks (Wan et al. 2008). A 1.90 Ga metamorphic event also occurred in the Khondalite Belt, with the 1.92–1.90 ages interpreted as the time of continent–continent collision (Santosh et al. 2007b). This is slightly earlier than collision of the Trans-North China Orogen (1.88–1.85 Ga) (Zhao et al. 2000, 2002; Guan et al. 2002; Kröner et al. 2005a, b, 2006; Wan et al. 2006). This supports the idea that the three blocks of the NCC were assembled during a single complex event, with the Yinshan and Ordos blocks colliding first, followed slightly later by juxtaposition of the Eastern and Western blocks along the Trans-North China Orogen (Zhao et al. 2005; Wan et al. 2006; Dong et al. 2007).

Although different groups in the Khondalite Belt are distributed separately in space, they commonly appear to be interlayered, especially near the boundaries between the different groups (Yang et al. 2003). The rocks have commonly undergone strong metamorphism and deformation and it is difficult to identify a stratigraphy in many cases, implying they occur as tectonic slices. It appears likely that the late Palaeoproterozoic collisional event led to the structural interdigitation of these units, since no strong tectono-thermal event has been recognized in the area after the Palaeoproterozoic. More work is required to test this hypothesis.

Protoliths of khondalites are commonly considered to have been deposited in stable environments, such as at passive continental margins or in cratonic basins, based on their high maturity (Condie et al. 1992; Wan et al. 2000b). However, the high-grade metamorphism indicates that they must once have been at a deep crustal level. The sequence of events may be summarized as follows: protoliths of khondalites were deposited at a passive continental margin, plate motion drove it to interact with another continent with an active continental margin and, finally, the passive margin was subducted below the active margin, resulting in high-grade metamorphism of the sediments. In this scenario, the detritus making up the khondalite protolith would have come from an old continent, therefore most of the detrital zircons should be much older than the depositional age of the sediments. In the Daqingshan area, however, the late Palaeoproterozoic khondalites of the upper Wulashan group contain detrital zircons as young as c. 2.06 Ga, but were deposited between c. 2.0 and 1.95 Ga, with only a relatively short time gap of c. 50 Ma. This suggests that the sediments were deposited in an active continental margin or back-arc setting (Kimbrough et al. 2001), consistent with other features of the total rock assemblage of the Wulashan group.

Ultrahigh-temperature (UHT) metamorphism occurs in many orogens and is considered to result from underplating/intraplating of mafic magmas or upwelling of asthenospheric mantle (Santosh et al. 2007b). Recently, Guo et al. (2006) carried out petrological studies on the Daqingshan UHT sapphirine granulites. The metamorphic peak was determined to be at 910–980 °C and 7.1–9.2 kbar (Guo et al. 2006) with a clockwise $P–T$ path (Jin et al. 1991; Liu et al. 2000) and a metamorphic age of 1.85 Ga (Guo et al. 2006). In the Khondalite Belt, metamorphism of this age has been identified by Dong et al. (2007) and in this study. At Jining, Santosh et al. (2006, 2007a, b) reported UHT (>900 °C and 9–12 kbar) metamorphic rocks, including sapphirine-bearing and spinel-bearing assemblages. Based on detailed petrographical and geochronological studies, they determined the time of metamorphism as 1.92 Ga (three zircon samples and one monazite sample), being 70 Ma older than the metamorphic age obtained by Guo et al. (2006). However, Santosh et al. (2006) had previously obtained monazite ages of 1819 ± 11 Ma from the 1.92 Ga UHT rocks, suggesting that they had been influenced by a later tectono-thermal event: this may have also affected rocks in the present study.

The Palaeoproterozoic geological evolution of the Daqingshan area can therefore be divided into two main phases: at 2.5–2.37 and 2.3–1.85 Ga. These probably represent two collision-extensional events during the late Archaean to late Palaeoproterozoic. The 1.95–1.85 Ga tectono-thermal event widely recorded in the Daqingshan area is similar to tectono-thermal events recognized in other areas of the NCC and considered to lead to the formation of the NCC (Zhao et al. 2002, 2004, 2005; Guo et al. 2005; Kröner et al. 2005b; Wan et al. 2006; Dong et al. 2007). These events have also been identified in other parts of the world (Hoffman 1988; Rosen et al. 1994; Clowes et al. 2000; Lee et al. 2000; Hartmann 2002; Nutman et al. 2007), being consistent with the assembly of the Palaeoproterozoic supercontinent Columbia (Rogers & Santosh 2002; Zhao et al. 2002; Santosh et al. 2007a, b).

An age histogram (Fig. 7) based on published zircon data from the Khondalite Belt and those from this study (Table 2) allows us to make the following comments: (1) There were strong tectono-thermal events between 2.5 and 2.3 Ga in the late Neoarchaean to early Palaeoproterozoic; (2) there is no evidence of tectono-thermal activity between 2.3 and 2.15 Ga, although a few low-quality ages have been reported (Table 2); (3) there are some 2.4–2.0 Ga magmatic zircons,

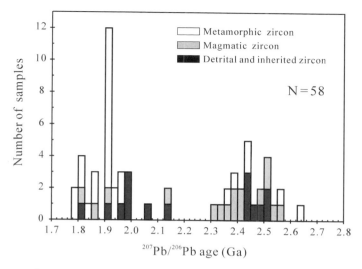

Fig. 7. Age histogram of early Precambrian zircons from the Khondalite Belt, North China Craton (1) Only the data marked with an asterisk in Table 2 were used; (2) dating methods include TIMS isotope dilution, TIMS evaporation, SHRIMP and LAM-ICPMS; (3) the bin width is 25 Ma; (4) age distribution of magmatic, metamorphic and detrital (plus inherited) zircons is shown separately.

whose igneous hosts may possibly be the source of the late Palaeoproterozoic sedimentary rocks; and (4) strong metamorphic and weaker magmatic events occurred during the late Palaeoproterozoic. The Palaeoproterozoic age distribution for the Khondalite Belt, as shown in Figure 7, is different from that outside the Khondalite Belt, as summarized by Wan et al. (2006), with the latter showing a strong peak between 2.2 and 2.05 Ga. However, both are similar in that there is an age gap between 2.3 and 2.2 Ga.

Crustal addition during the late Neoarchaean to late Palaeoproterozoic

The Hf-isotope data of zircons from four samples in this study can broadly be divided into two groups in terms of formation ages. One includes samples NM0403, NM0413 and NM0401 with Archaean and early Palaeoproterozoic ages, and the second is represented by sample NM0405 of late Palaeoproterozoic age (Fig. 6). In the first group, Meta I zircons from the two-pyroxene-bearing retrogressed granulite from the Sanggan group are metamorphic in origin and their $\varepsilon Hf_{(T)}$ and Hf model ages are 2.0–5.4 and 2.74–2.62 Ga, respectively, possibly reflecting the Hf-isotopic feature of the precursor volcanic rock. Detrital and inherited zircons from garnet-biotite gneiss from the Wulashan group, and garnet-biotite granite which was derived from this group, have $^{207}Pb/^{206}Pb$ ages of c. 2.5 Ga and show very similar Hf-isotope compositions (Table EA-2, Fig. 6). This suggests that: (1) the detrital and inherited zircons were originally from a source region similar in Hf-isotopic composition to the Archaean basement; therefore (2) a large amount of material in the Daqingshan and adjacent areas was derived from a mantle source at the end of the Archaean; and (3) the mantle-derived material might be partly contaminated by continental crust because their zircon $\varepsilon Hf_{(T)}$ values vary largely, with some being close to 0 (Fig. 6).

The c. 2.06 Ga detrital zircon cores from the late Palaeoproterozoic metasedimentary rock (NM0405) have $\varepsilon Hf_{(T)}$ of 1.5–7.4 and T_{DM} of 2.38–2.15 Ga (Table EA-2, Fig. 6). The large variation in positive $\varepsilon Hf_{(T)}$ suggests there was strong mantle addition in the source region at about 2.0 Ga, with some continental contamination, a conclusion also drawn by Xia et al. (2006a). Some c. 2.0 Ga detrital zircons analysed by Xia et al. (2006a) gave very negative $\varepsilon Hf_{(T)}$ values, suggesting they were derived from a source region characterized by recycling of continental crust. The Hf-isotopic composition of the meta-gabbro sample (NM0618) has not been analysed in this study; however, the geochemical features of this rock indicate that mantle addition occurred at 2.37 Ga. Therefore, there were at least three periods of mantle addition recorded in the Daqingshan area during the late Neoarchaean to late Palaeoproterozoic (2.55–2.5, c. 2.37 and c. 2.06 Ga).

Table 2. *Zircon U–Pb data of early Precambrian rocks from the Khondalite Belt, North China Craton*

Sample No.	Rocks	Occurrences	Locations
AD115TW1	Gneissic granite	Meta-intrusion	Bayan Ul, Inner Mongolia
HD01TW1	Garnet-mica two-feldspar gneiss	Helanshan group	Helanshan, Ningxia
$14Z_{333-1}$	Garnet granite	Crustally derived granite	Helanshan, Ningxia
$11Z_{71-1}$	Sillimanite cordierite gneiss	Helanshan group	Helanshan, Ningxia
$11Z_{232-1}$	Fine-grained biotite gneiss	Helanshan group	Helanshan, Ningxia
01ZSPt	Sandstone	Neoproterozoic stratum	Qianlishan, Inner Mongolia
01ZSC	Sandstone	Cambrian stratum	Qianlishan, Inner Mongolia
01ZSOr	Sandstone	Ordovician stratum	Qianlishan, Inner Mongolia
WL007	Garnet-bearing metapelitic gneiss	Wulashan group	Daqingshan, Inner Mongolia
WL011	Sillimanite-garnet-biotite gneiss		Daqingshan, Inner Mongolia
WL016	Quartzite		Daqingshan, Inner Mongolia
WL020	Garnet-bearing metapelitic gneiss		Daqingshan, Inner Mongolia
NM03-1	Quartz dioritic gneiss	Meta-intrusion	Daqingshan, Inner Mongolia
NM03-2	Granodioritic gneiss	Meta-intrusion	Daqingshan, Inner Mongolia
NM03-3	Biotite hornblende granite	Intrusion	Daqingshan, Inner Mongolia
Guo	Sapphirine granulite	Wulashan group	Daqingshan, Inner Mongolia
NM0403	Two-pyroxene-bearing retrogressed granulite granulite	Sanggan group	Daqingshan, Inner Mongolia
NM0605	Felsic gneiss	Wulashan group	Daqingshan, Inner Mongolia
NM0618	Meta-gabbro	Meta-intrusion	Daqingshan, Inner Mongolia
NM0413	Garnet-biotite gneiss	Wulashan group	Daqingshan, Inner Mongolia
NM0401	Garnet-biotite granite	Intrusion	Daqingshan, Inner Mongolia
NM0405	Garnet-biotite gneiss	Wulashan group	Daqingshan, Inner Mongolia
NM0414	Feldspathic quartzite	Meidaizhao group	Daqingshan, Inner Mongolia
NM0404	Meta-basic dyke	Meta-intrusion	Daqingshan, Inner Mongolia
NM0409	Crustal carbonatite	Intrusion	Daqingshan, Inner Mongolia
NM0407-5	Diopsidite	Wulashan group	Daqingshan, Inner Mongolia
HL-5	Banded basic-intermediate granulite	Xinghe group	Guyang, Inner Mongolia
P7JD4-25	Pyroxene garnet quartzite	Xinghe group	Guyang, Inner Mongolia
GY32	Hornblende granite	Intrusion	Guyang, Inner Mongolia
GY49	Diorite	Intrusion	Guyang, Inner Mongolia
03ZT-10	Quartzite	Mesoproterozoic Zhaertai	Guyang, Inner Mongolia
03ZT-14	Migmatitic granite	Intrusion	Guyang, Inner Mongolia
Hz3-2	Adakitic granite	Intrusion	Daqinshan, Inner Mongolia

Ages (Ma)	Analysis No.	Interpretation	Methods	References
2323 ± 20*	Weighted mean age (7)	Emplacement age	SHRIMP	Dong et al. (2007)
1923 ± 28*	Weighted mean age (7)	Metamorphic age		
1856 ± 12*	Weighted mean age (5)	Metamorphic age		
2871–2469	8	Detrital zircon age		
1978 ± 17*	Weighted mean age (9)	Detrital zircon age		
1893–1975	2	Metamorphic age	Zircon evaporation	Hu et al. (1994)
1898–1853	3	Metamorphic age		
1902–2102	4	Metamorphic age		
1879–2785	99	Detrital zircon age	ICPMS	Brian J. D. et al. (2006)
1928–2951	99	Detrital zircon age		
1848–2880	52	Detrital zircon age		
1801–2562	59	Detrital zircon age	LA-ICPMS	Xia et al. (2006a)
1858–2097	51	Detrital zircon age		
1930–2095	57	Detrital zircon age		
1880–2502	67	Detrital zircon age		
2410 ± 6*	Intercept age	Emplacement age	TIMS dilution	Xu et al. (2003)
2540 ± 23*	Intercept age	Emplacement age		
1820 ± 8*	Intercept age (5)	Emplacement age		
1850*	?	Metamorphic age	SHRIMP	Guo et al. (2006)
2567 ± 10	1	Volcanic age	SHRIMP	In this study
2518–2501	8	Metamorphic age		
2429–2416	4	Metamorphic age		
2582 ± 17	1	Detrital zircon age		
2494–2456	4	Detrital zircon age		
2435 ± 8*	7	Detrital zircon age		
2349–2201	6	Detrital zircon age		
2135 ± 15	1	Detrital zircon age		
1853 ± 10*	Weighted mean age (5)	Metamorphic age		
2371 ± 16*	Weighted mean age (17)	Emplacement age		
1900 ± 10*	Weighted mean age (4)	Metamorphic age		
2513 ± 19*	Weighted mean age (9)	Detrital zircon age		
2397 ± 10*	Weighted mean age (11)	Metamorphic age		
2503 ± 14*	Weighted mean age (11)	Inherited zircon age		
2445 ± 10*	Weighted mean age (10)	Metamorphic age		
2388 ± 20*	Weighted mean age (16)	Emplacement age		
2060 ± 15*	Weighted mean age (18)	Detrital zircon age		
1953 ± 11*	Weighted mean age (10)	Metamorphic age		
1913 ± 17*	Weighted mean age (6)	Metamorphic age		
1859–1849	3	Metamorphic age		
2478 ± 14	1	Detrital zircon age		
2318–2256	4	Detrital zircon age		
1999 ± 15*	Weighted mean age (10)	Detrital zircon age		
1924 ± 11*	Weighted mean age (5)	Metamorphic age		
1951 ± 5*	Weighted mean age (14)	Emplacement age		
1899–1882	2	Metamorphic age		
1957 ± 38	Weighted mean age (14)	Metamorphic age		
2649 ± 5*	Intercept age (4)	Metamorphic age	TIMS dilution	Wang et al. (2001)
2503 ± 20*	Intercept age (3)	Detrital zircon age		
2911–3484	2	Inherited zircon age	SHRIMP	Jian et al. (2005)
2520 ± 9*	Weighted mean age (12)	Emplacement age		
2556 ± 14*	Weighted mean age (11)	Emplacement age		
2464 ± 15*	Weighted mean age (11)	Detrital zircon age	Zircon U–Pb dilution	Li et al. (2007)
2480 ± 13*	Weighted mean age (7)	Emplacement age	SHRIMP	
2564 ± 19*	Weighted mean age (5)	Metamorphic age	SHRIMP	
2435 ± 12*	Weighted mean age (9)	Emplacement age	SHRIMP	Zhong et al. (2006a)
1909 ± 28*	Weighted mean age (6)	Metamorphic age		

(*Continued*)

Table 2. *(Cont.) Zircon U–Pb data of early Precambrian rocks from the Khondalite Belt, North China Craton*

Sample No.	Rocks	Occurrences	Locations
Hz2-1	Sanukitoid	Intrusion	Daqinshan, Inner Mongolia
C1001	Closepet	Intrusion	Daqinshan, Inner Mongolia
BSH5-2	Strongly peraluminous granite	Crustally derived granite	Liangcheng, Inner Mongolia
HL401	Strongly peraluminous granite	?	Liangcheng, Inner Mongolia
LC401	Strongly peraluminous granite	?	Liangcheng, Inner Mongolia
96JL06	Garnet granite	Intrusion	NW of Liangcheng, Inner Mongolia
96JL11	Leucogranite	Crustally derived granite	Liangcheng, Inner Mongolia
Santosh	UHT metapelite	Jining group	East of Tuguiwula, Inner
1DIM$_1$	Biotite-bearing hornblende plagioclase gneiss	Jining group	Xinghe, Inner Mongolia
Nm84-10A	Hornblende two-pyroxene plagioclase granulite	Jining group	Xinghe, Inner Mongolia
Nm84-10B	Hornblende two-pyroxene plagioclase granulite	Jining group	Xinghe, Inner Mongolia
Nm84-20A	Garnet sillimanite K-feldspar gneiss	Jining group	Xinghe, Inner Mongolia
Nm84-20B	Garnet sillimanite K-feldspar gneiss	Jining group	Xinghe, Inner Mongolia
1DIM$_3$	Two-pyroxene-bearing amphibolite	Jining group	Xinghe, Inner Mongolia
01M020	Sillimanite-garnet-feldspar gneiss	Jining group	Jining, Inner Mongolia
01M038	Sillimanite-garnet-feldspar gneiss	Jining group	Jining, Inner Mongolia
01M041	Sillimanite-garnet gneiss	Jining group	Jining, Inner Mongolia
01M053	Cordierite-garnet-sillimanite gneiss	Jining group	Jining, Inner Mongolia
6662	Sapphirine-orthopyroxene-sillimanite granulite	Jining group	Jining, Inner Mongolia
6614	garnet-orthopyroxene-spinel granulite	Jining group	Jining, Inner Mongolia
6631	garnet-sillimanite khondalite	Jining group	Jining, Inner Mongolia
XYS01	HP granulite	Meta-basic vein	Northwest of Yangyuan, Hebei
DST02	HP granulite	Meta-basic vein	Northeast of Xuanhua, Hebei
G084	Garnet-sillimanite K-feldspar gneiss	Jining group	Huangtuyao, Inner Mongolia
N9306	Biotite horblende schist	Erdaowa group	Hohhot, Inner Mongolia

Ages (Ma)	Analysis No.	Interpretation	Methods	References
2416 ± 8*	Weighted mean age (6)	Emplacement age		
1917 ± 17*	Weighted mean age (7)	Metamorphic age		
2494 ± 16*	Weighted mean age (4)	Inherited zircon age		
2426 ± 41	Weighted mean age (2)	Inherited zircon age		
2330 ± 54	Weighted mean age (3)	Inherited zircon age		
2133 ± 21*	Weighted mean age (3)	Emplacement age		
1825 ± 19*	Weighted mean age (3)	Metamorphic age		
1916 ± 10*	Weighted mean age (15)	Emplacement age	SHRIMP	Zhong et al. (2006b)
1904 ± 12*	Weighted mean age (5)	Emplacement age		
1980–2120	2	Inherited zircon age		
1933 ± 7*	Weighted mean age (4)	Metamorphic age		
2140*	Intercept age (5)	Inherited zircon age		
1836 ± 18	1	Emplacement age	TIMS dilution	Guo et al. (1999)
1912 ± 98	Intercept age (7)	Emplacement age		
1927 ± 11		Metamorphic age	Monazite electron	Santosh M. et al. (2006)
1819 ± 11		Metamorphic age		
2467 + 54/−35	Intercept age (6)	Metamorphic age	TIMS dilution	Shen et al. (1987)
2339 ± 13*	Intercept age (5)	Emplacement age		
2382 + 16/−15*	Intercept age (6)	Emplacement age		
1962 + 69/−64	Intercept age (5)	Metamorphic age		
1821 + 12/−28*	Intercept age (5)	Metamorphic age		
1958 + 38/−23	Intercept age (5)	Metamorphic age		
1902 ± 16	Intercept age (46)	Detrital zircon age	LA-ICP-MS	Xia et al. (2006b)
1811 ± 23	Weighted mean age (5)	Detrital zircon age		
1882–2031	48	Detrital zircon age		
2175 ± 42	Weighted mean age (2)	Detrital zircon age		
1717–2196	54	Detrital zircon age		
1838–2384	51	Detrital zircon age		
1919 ± 10*	Weighted mean age (26)	Metamorphic age	SHRIMP	Santosh M. et al. (2007b)
1975 ± 24*	Intercept age (6)	Detrital zircon age		
1922 ± 11*	Intercept age (9)	Metamorphic age		
1970 ± 24*	Weighted mean age (4)	Detrital zircon age		
1923 ± 11*	Weighted mean age (5)	Metamorphic age		
2034 ± 80	Weighted mean age (6)	Emplacement age	SHRIMP	Zhang et al. (2006)
1748 ± 82	Weighted mean age (16)	Metamorphic age		
1792 ± 14*	Weighted mean age (8)	Metamorphic age		
2310 ± 46	Intercept age (6)	Detrital zircon age	TIMS dilution	Wu et al. (1998)
1873 ± 32	Intercept age (2)	Metamorphic age		
1793 ± 10*	Weighted mean age (3)	Metamorphic age	Rutile TIMS dilution	
2372 ± 14*	Intercept age (3)	Metamorphic age	TIMS dilution	Wang et al. (1996)
2185 ± 47	Intercept age (3)	Emplacement age		
2096 ± 38	Intercept age (4)	Emplacement age		
2237 ± 64	Intercept age (2)	Metamorphic age		
1995 ± 15	1	Metamorphic age		
2404 ± 24*	Weighted mean age (18)	Metamorphic age		

Note: (1) Age interpretations follow original authors. (2) Data with asterisk were used for Figure 7.

Conclusions

The early Precambrian supracrustal rocks of the Daqingshan area can be divided into three main groups: the 2.56–2.51, 2.50–2.45 and 2.0–1.95 Ga supracrustal units. However, these do not correlate with the previous interpretation of the geology that subdivided the rocks into a stratigraphic succession that was composed in part of, from the base upward, the Sanggan, Wulashan and Meidaizhao groups. Our new results show that the Sanggan group formed during the Neoarchaean (not Palaeoarchaean) and the Wulashan group during the Palaeoproterozoic (not Mesoarchaean), as previously believed. We have also identified, for the first time, early Palaeoproterozoic metasedimentary rocks forming a unit within the Wulashan group. This study does, however, support the view of Xu et al. (2003) that the Meidaizhao group formed during the Palaeoproterozoic and not in the Archaean.

Some zircons show complex internal structures and record different ages, indicating that they underwent several stages of growth/recrystallization. They enable four tectono-thermal events to be recognized in the Daqingshan area, each recording an important stage in the evolution of the Precambrian crust. The events are at 2.6–2.5, 2.45–2.37, 2.3–2.0 and 1.95–1.85 Ga, with the latest event being consistent with the formation of the NCC. Zircon data obtained by others from the Khondalite Belt show a similar age distribution to the Daqingshan area in this study but with age records at c. 1.8 Ga. The ages of 1.9–1.85 and 1.85–1.80 Ga reflect collision and extension of the Khondalite Belt, respectively, being consistent with the assembly and breakup of the Palaeoproterozoic supercontinent of Columbia.

Furthermore, there were at least three periods of mantle addition to the crust, based on the hafnium zircon data from the Daqingshan area: at 2.55–2.5, 2.37 and 2.06 Ga. Whereas most samples record recycling of Neoarchaean crust, garnet-biotite gneiss NM0405 contains 2.06 Ga zircons that have $\varepsilon Hf_{(T)}$ values of 1.5–7.4 and Hf-model ages of 2.38–2.15 Ga, indicating a later mantle extraction event. Although no Hf-isotope data were obtained for the 2.37 Ga meta-gabbro (NM0618), it is likewise interpreted as being derived from the mantle.

We thank H. Tao, Q. D. Zhang and Q. Ye for making the zircon mounts; Y. H. Zhang and Z. Q. Yang for help with SHRIMP U–Pb dating; F. Y. Wu and L. W. Xie for help with LA-ICP-MS Hf-data collection; and H. Yong and L. Yan for help with Raman spectroscopy. We also thank A. Nutman, A. Kröner, Y. Rojas-Agramonte, G. C. Zhao, M. Sun, X. P. Xia, J. S. Wu, Q. H. Shen, J. S. Ren, Y. S. Geng, C. H. Yang, X. Y. Yin, L. L. Du, S. W. Liu, F. Y. Wu, J. H. Gou, M. G. Zhai, S. N. Lu, C. T. Zhong, J. J. Li, C. S. Liu and H. T. Xin for their valuable discussions and assistance during the research. We thank G. C. Zhao and M. Santosh for their valuable comments. This manuscript is a contribution to IGCP project 509 (Palaeoproterozoic Supercontinents and Global Evolution). The study was supported financially by the Key Program of the Land and Resource Ministry of China (1212010711815), the Scientific Research Program of the Ministry of Science and Technology of China (J0901), the Programs of the Beijing SHRIMP Centre and the State Key Laboratory of Lithospheric Evolution, Institute of Geology and Geophysics, Chinese Academy of Sciences.

References

BLACK, L. P., KAMO, S. L., ALLEN, C. M., ALEINIKOFF, J. N., DAVIS, D. W., KORSCH, R. J. & FOUDOULIS, C. 2003. Temora 1: a new zircon standard for Phanerozoic U–Pb geochronology. *Chemical Geology*, **200**, 155–170.

BLEEKER, W. 2003. The late Archaean record: a puzzle in c. 35 pieces. *Lithos*, **71**, 99–134.

BLICHERT-TOFT, J. & ALBAREDE, F. 1997. The Lu–Hf geochemistry of chondrites and the evolution of the mantle-crust system. *Earth and Planetary Science Letters*, **148**, 243–258.

BRIAN, J. D. & GEORGE, G. 2006. Detrital zircon reference for the North China Block. *Journal of Asian Earth Sciences*, **26**, 637–648.

CHACKO, T., KUMAR, G. R., MEEN, J. K. & ROGERS, J. J. W. 1992. Geochemistry of high-grade supracrustal rocks from the Kerala Khondalite Belt and adjacent massif charnockites, South India. *In*: VAN REENEN, D. D., ROERING, C. & ASHWAL, L. D. (eds) *The Archaean Limpopo Granulite Belt: Tectonics and Deep Crustal Processes. Precambrian Research*, **55**, 469–489.

CHEN, L. 2007. *Geochronology and geochemistry of the Guyang Greenstone Belt*. Post-Doctorate Report. Institute of Geology and Geophysics, Chinese Academy of Sciences, Beijing, 1–40.

CLOWES, R., COOK, F., HAJNAL, Z., HALL, J., LEWRY, J., LUCAS, S. & WARDLE, R. 2000. Canada's LITHOPROBE project (collaborative, multidisplinary geoscience research leads to new understanding of continental evolution). *Episodes*, **22**, 4–20.

CONDIE, K. C., BORYTA, M. D., LIU, J. Z. & QIAN, X. L. 1992. The origin of khondalites: geochemical evidence from the Archaean to Early Proterozoic granulite belt in the North China Craton. *Precambrian Research*, **59**, 207–223.

CORFU, F., HANCHAR, J. M., HOSKIN, P. W. O. & KINNY, P. 2003. Atlas of zircon textures. *In*: HANCHAR, J. M. & HOSKIN, P. W. O. (eds) *Zircon*. Mineralogical Society of America, Washington, 469–500.

DARBY, B. J. & GEHRELS, G. 2006. Detrital zircon reference for the North China Block. *Journal of Asian Earth Sciences*, **26**, 637–648.

DONG, C. Y., LIU, D. Y. ET AL. 2007. Palaeoproterozoic Khondalite Belt in the western North China

Craton: new evidence from SHRIMP dating and Hf-isotope composition of zircons from metamorphic rocks in the Bayanwula-Helanshan area. *Chinese Science Bulletin*, **52**, 1913–1922.

GUAN, H., SUN, M., WILDE, S. A., ZHOU, X. H. & ZHAI, M. G. 2002. SHRIMP U–Pb zircon geochronology of the Fuping Complex: implications for formation and assembly of the North China Craton. *Precambrian Research*, **113**, 1–18.

GUO, J. H., SHI, X., BIAN, A. G., XU, R. H., ZHAI, M. G. & LI, Y. G. 1999. Pb-isotopic composition of feldspar and U–Pb age of zircon from early Proterozoic granite in Sanggan area North China Craton: metamorphism crustal melting and tectono-thermal event. *Acta Petrologica Sinica*, **15**, 199–207 (in Chinese with English abstract).

GUO, J. H., ZHAI, M. G. & XU, R. H. 2002. Timing of granulite facies metamorphism occurring widely in the Sanggan area, North China: zircon U–Pb geochronology. *Science in China*, **32**, 10–18.

GUO, J. H., SUN, M., CHEN, F. K. & ZHAI, M. G. 2005. Sm–Nd and SHRIMP U–Pb zircon geochronology of high-pressure granulites in the Sanggan area, North China Craton: timing of Palaeoproterozoic continental collision. *Journal of Asian Earth Sciences*, **24**, 629–642.

GUO, J. H., CHEN, Y., PENG, P., LIU, F., CHEN, L. & ZHANG, L. Q. 2006. Sapphirine granulite in the Daqingshan area, Inner Mongolia: 1.8 Ga ultra-high temperature (UHT) metamorphism. Abstract, *Conference on Petrology and Geodynamics*, 215–218.

HARTMANN, L. A. 2002. The Mesoproterozoic supercontinental Atlantica in the Brazilian shield-review of geological and U–Pb zircon and Sm–Nd isotopic evidence. *Gondwana Research*, **5**, 157–163.

HOFFMAN, P. F. 1988. United plates of America, the birth of a craton: early Proterozoic assembly and growth of Laurentia. *Annual Review of Earth Planetary Sciences*, **16**, 543–603.

HU, N. G., YANG, J. X., WANG, Z. B., WANG, T. & LI, W. P. 1994. *The composition and evolution of complex in Helanshan*. Xi'an Map Publishing House, Xi'an, 1–59 (in Chinese).

JIAN, P., ZHANG, Q., LIU, D. Y., JIN, W. L., JIA, X. Q. & QIAN, Q. 2005. SHRIMP dating and geological significance of late Archaean high-Mg diorite (sanukite) and hornblende-granite at Guyang of Inner Mongolia. *Acta Petrologica Sinica*, **21**, 151–157 (in Chinese with English abstract).

JIN, W., LI, S. X. & LIU, X. S. 1991. A study on characteristics of early Precambrian high-grade metamorphic rock series and their metamorphic dynamics. *Acta Petrologica Sinica*, **4**, 27–35 (in Chinese with English abstract).

KIMBROUGH, D. L., SMITH, D. P. ET AL. 2001. Fore-arc-basin sedimentary response to Rapid Late Cretaceous batholith emplacement in the Peninsular Ranges of southern and Baja California. *Geology*, **29**, 491–494.

KRÖNER, A., WILDE, S. A., LI, J. H. & WANG, K. Y. 2005a. Age and evolution of a late Archaean to Palaeoproterozoic upper to lower crustal section in the Wutaishan/Hengshan/Fuping terrain of northern China. *Journal of Asian Earth Science*, **24**, 577–595.

KRÖNER, A., WILDE, S. A., O'BRIEN, P. J., LI, J. H., PASSCHIER, C. W., WALLE, N. P. & LIU, D. Y. 2005b. Field relationships, geochemistry, zircon ages and evolution of a Late Archaean to Palaeoproterozoic lower crustal section in the Hengshan Terrain of Northern China. *Acta Geologica Sinica*, **79**, 605–629.

KRÖNER, A., WILDE, S. A. ET AL. 2006. Zircon geochronology and metamorphic evolution of mafic dykes in the Hengshan Complex of northern China: evidence for late Palaeoproterozoic extension and subsequent high-pressure metamorphism in the North China Craton. *Precambrian Research*, **146**, 45–67.

KUSKY, T. M. & LI, J. H. 2003. Palaeoproterozoic tectonic evolution of the North China Craton. *Journal of Asian Earth Sciences* **22**, 383–397.

KUSKY, T., LI, J. H. & SANTOSH, M. 2007. The Palaeoproterozoic North Hebei Orogen: North China Craton's collisional suture with the Columbia supercontinent. *Gondwana Research*, **12**, 4–28.

LEE, S. R., CHO, M., YI, K. & STERN, R. A. 2000. Early Proterozoic granulites in Central Korea: tectonic correlation with Chinese cratons. *Journal of Geology*, **108**, 729–738.

LI, J. H., KUSKY, T. M. & HUANG, X. 2002. NeoArchaean podiform chromitites and harzburgite tectonite in ophiolitic melange, North China Craton, remnants of Archaean oceanic mantle. *GSA Today*, **12**, 4–11.

LI, J. J., SHEN, B. F., LI, H. M., ZHOU, H. Y., GUO, L. J. & LI, C. Y. 2004. Single-zircon U–Pb age of granodioritic gneiss in the Bayan Ul area, western Inner Mongolia. *Regional Geology of China*, **23**, 1243–1245 (in Chinese with English abstract).

LI, Q. L., CHEN, F. K., GUO, J. H., LI, X. H., YANG, Y. H. & SIEBEL, W. 2007. Zircon ages and Nd–Hf isotopic composition of the Zhaertai Group (Inner Mongolia): evidence for early Proterozoic evolution of the northern North China Craton. *Journal of Asian Earth Sciences*, **30**, 573–590.

LIU, J. Z., QIANG, X. K., LIU, X. S. & OUYANG, Z. Y. 2000. Dynamics and genetic grids of sapphirine-bearing spinel gneiss in Daqing Mountain orogen zone, Inner Mongolia. *Acta Petrologica Sinica*, **16**, 245–255 (in Chinese with English abstract).

LIU, D. Y., WAN, Y. S., WU, J. S., WILDE, S. A., ZHOU, H. Y., DONG, C. Y. & YIN, X. Y. 2007. EoArchaean rocks and zircons in the North China Craton. *In*: VAN KRANENDONK, M. J., SMITHIES, R. H. & BENNETT, V. C. (eds) *Developments in Precambrian Geology*, Elsevier, **15**, 251–274.

LIU, S. W., LÜ, Y. J., FENG, Y. G., ZHANG, C., TIAN, W. & LIU, X. M. 2007. Geology and zircon U–Pb isotopic chronology of Dantazi Complex, Northern Hebei province. *Geological Journal of China Universities*, **13**, 484–497 (in Chinese with English abstract).

LU, L. Z., XU, X. C. & LIU, F. L. 1996. *Early Precambrian Khondalite Series of North China*. Changchun Publishing House, Changchun, 1–272 (in Chinese).

LUDWIG, K. R. 2001. *SQUID 1.02, a User's Manual*. Berkeley Geochronology Centre Special Publications, **2**.

MOYEN, J. F., MARTIN, H. & JAYANANDA, M. 2001. Multi-element geochemical modelling of crust–mantle interactions during late-Archaean crustal growth: the Closepet granite (South India). *Precambrian Research*, **112**, 87–105.

NUTMAN, A. P., DAWES, P. R., KALSBEEK, F. & HAMILTON, M. A. 2007. Palaeoproterozoic and Archaean gneiss complexes in northern Greenland: Palaeoproterozoic terrane assembly in the High Arctic. *Precambrian Research*. doi: 10.1016/j.precamres.2007.09.006.

QIAN, X. L. & LI, J. H. 1999. The discovery of NeoArchaean unconformity and its implication for continental cratonization of the North China Craton. *Sciences in China, Series. D*, **42**, 401–407.

ROGERS, J. J. W. & SANTOSH, M. 2002. Configuration of Columbia, a Mesoproterozoic supercontinent. *Gondwana Research*, **5**, 5–22.

ROSEN, O. M., CONDIE, K. C. & NATOPOV, L. M. 1994. Archaean and early Proterozoic evolution of the Siberian Craton: a preliminary assessment. *In:* CONDIE, K. C. (ed.) *Archaean Crustal Evolution*. Elsevier, Amsterdam, 411–459.

SANTOSH, M., SAJEEV, K. & LI, J. H. 2006. Extreme crustal metamorphism during Columbia supercontinent assembly: evidence from North China Craton. *Gondwana Research*, **10**, 256–266.

SANTOSH, M., TSUNOGAE, T., LI, J. H. & LIU, S. J. 2007a. Discovery of sapphirine-bearing Mg–Al granulites in the North China Craton: implications for Palaeoproterozoic ultrahigh-temperature metamorphism. *Gondwana Research*. **11**, 263–285.

SANTOSH, M., WILDE, S. A. & LI, J. H. 2007b. Timing of Palaeoproterozoic ultrahigh-temperature metamorphism in the North China Craton: evidence from SHRIMP U–Pb zircon geochronology. *Precambrian Research*. doi: 10.1016/j.precamres.2007.06.006.

SCHERER, E., MUNKER, C. & MEZGER, K. 2001. Calibration of the Lutetium–Hafnium clock. *Science*, **293**, 683–687.

SHEN, Q. H., LIU, D. Y., WANG, P., GAO, J. F. & ZHANG, Y. F. 1987. U–Pb and Rb–Sr isotopic age study of the Jining group from NEL Mongol of China. *Bulletin of the Chinese Academy of Geological Sciences*, **16**, 165–178 (in Chinese with English abstract).

SHEN, Q. H., ZHANG, Y. F., GAO, J. F. & WANG, P. 1990. *Study on Archaean Metamorphic Rocks in mid-southern Nei Mongol of China*. Geological Publishing House, Beijing, 1–192 (in Chinese).

SONG, H. F., XU, Z. Y. & LIU, Z. H. 2005. Geochemical characteristics and origin of garnet migmatitic granites in Daqingshan area, Inner Mongolia. *Acta Petrologica et Mineralogica*, **24**, 489–495 (in Chinese with English abstract).

VAVRA, G., SCHMID, R. & GEBAUER, D. 1999. Internal morphology, habit and U–Th–Pb microanalysis of amphibolite-to-granulite facies zircons: geochronology of the Ivrea Zone (Southern Alps). *Contribution to Mineralogy and Petrology*, **134**, 380–404.

WAN, Y. S., GENG, Y. S., LIU, F. L., SHEN, Q. H., LIU, D. Y. & SONG, B. 2000a. Age and composition of the khondalites of the North China Craton and its adjacent area. *Progress in Precambrian Research*, **23**, 221–235 (in Chinese with English abstract).

WAN, Y. S., GENG, Y. S., SHEN, Q. H. & ZHANG, R. X. 2000b. Khondalite series-geochronology and geochemistry of the Jihekou Group in Luliang area, Shanxi Province. *Acta Petrologica Sinica*, **16**, 49–58 (in Chinese with English abstract).

WAN, Y. S., SONG, B. *ET AL*. 2006. SHRIMP U–Pb zircon geochronology of Palaeoproterozoic metasedimentary rocks in the North China Craton: evidence for a major Late Palaeoproterozoic tectono-thermal event, *Precambrian Research*, **149**, 249–271.

WAN, Y. S., LIU, D. Y. *ET AL*. 2008. Palaeoproterozoic crustally derived carbonatite from the Daqinshan area, North China Craton: Geological, petrographical, geochronological and geochemical (Hf, Nd, O and C) evidence. *American Journal of Science*, **308**, 351–378.

WANG, H. C. & XIU, Q. Y. 1996. Single zircon U–Pb ages of Erdaowa group in Northern Hohhot, Inner Mongolia. *Geology of Inner Mongolia*, **1**, 13–17 (in Chinese with English abstract).

WANG, H. C., YUAN, G. B. & XIN, H. T. 2001. U–Pb single zircon ages for granulites in Cunkongshan area, Guyang Inner Mongolia and enlightenment for its geological signification, China. *Progress in Precambrian Research*, **24**, 28–34 (in Chinese with English abstract).

WILDE, S. A., ZHAO, G. C. & SUN, M. 2002. Development of the North China Craton during the Late Archaean and its final amalgamation at 1.8 Ga: some speculation on its position within a global Palaeoproterozoic Supercontinent. *Gondwana Research*, **5**, 85–94.

WILLIAMS, I. S. 1998. U–Th–Pb geochronology by ion microprobe, applications of microanalytical techniques to understanding mineralizing processes. *In:* MCKIBBEN, M. A., SHANKS, W. C. & RIDLEY, W. I. (eds) *Reviews in Economic Geology*, **7**, 1–35.

WOODHEAD, J., HERGT, J., SHELLEY, M., EGGINS, S. & KEMP, R. 2004. Zircon Hf-isotope analysis with an excimer laser, depth profiling, ablation of complex geometries, and concomitant age estimation. *Chemical Geology*, **209**, 121–135.

WU, C. H. & ZHONG, C. T. 1998. Early Proterozoic SW–NE collision model for the central part of the North China Craton. *Progress in Precambrian Research*, **21**, 28–50 (in Chinese with English abstract).

WU, C. H., LI, H. M., ZHONG, C. T. & CHEN, A. Q. 1998. The ages of zircon and rutile (cooling) from khondalite in Huangtuyao, Inner Mongolia. *Geological Review*, **44**, 618–626 (in Chinese with English abstract).

WU, C. H., ZHONG, C. T. & CHEN, Q. G. 1997. Discussion on the age of khondalite in Jin-Meng (Shanxi-Nei Mongol) high-grade terrain. *Acta Petrologica Sinica*, **13**, 289–302 (in Chinese).

WU, F. Y., YANG, Y. H., XIE, L. W., YANG, J. H. & XU, P. 2006. Hf-isotopic compositions of the standard zircons and baddeleyites used in U–Pb geochronology. *Chemical Geology*, **44**, 105–126.

XIA, X. P., SUN, M., ZHAO, G. C. & LUO, Y. 2006a. LA-ICP-MS U–Pb geochronology of detrital zircons from the Jining Complex, North China Craton and its tectonic significance. *Precambrian Research*, **144**, 199–212.

XIA, X. P., SUN, M., ZHAO, G. C., WU, F. Y., XU, P., ZHANG, J. H. & LUO, Y. 2006b. U–Pb and Hf-isotopic study of detrital zircons from the Wulashan khondalites: constraints on the evolution of the Ordos Terrane, Western Block of the North China Craton. *Earth and Planetary Science Letters*, **241**, 581–593.

XU, Z. Y., LIU, Z. H. & YANG, Z. S. 2003. The discovery of Zaoergou angular unconformity and establishment of Meidaizhao Group-complex in Daqingshan Mountains, Inner Mongolia: Palaeoproterozoic low-grade metamorphic strata on the khondalite series. *Geological Bulletin of China*, **22**, 480–486 (in Chinese with English abstract).

XU, Z. Y., LIU, Z. H., HU, F. X. & YANG, Z. S. 2005. Geochemical characteristics of the calc-silicate rocks in khondalite series in Daqingshan area, Inner Mongolia. *Journal of Jilin University (Earth Science Edition)*, **35**, 681–689 (in Chinese with English abstract).

YANG, Z. S., XU, Z. Y. & LIU, Z. H. 2000. Khondalite event and Archaean crust structure evolution. *Progress in Precambrian Research*, **23**, 206–211 (in Chinese with English abstract).

YANG, Z. S., XU, Z. Y. & LIU, Z. H. 2003. Consideration and practice of the construction of lithostratigraphic system in high-grade metamorphic terrains: a case study in the Daqingshan-Wulashan area. *Geology in China*, **30**, 343–351 (in Chinese with English abstract).

YANG, Z. S., LIU, M. X., LIU, Z. H. & XU, Z. Y. 2004. *Explanatory text of the geological map of Baotou sheet, 1:250 000.* 1–572 (in Chinese).

ZHANG, H. F., ZHAI, M. G. & PENG, P. 2006. Zircon SHRIMP U–Pb age of the Palaeoproterozoic high-pressure granulites from the Sanggan area, the North China Craton and its geologic implications. *Earth Science Frontiers*, **13**, 190–2199 (in Chinese with English abstract).

ZHAO, G. C., WILDE, S. A., CAWOOD, P. A. & LU, L. Z. 1998. Thermal evolution of Archaean basement rocks from the eastern part of the North China craton and its bearing on tectonic setting. *International Geology Review*, **40**, 705–721.

ZHAO, G. C., WILDE, S. A., CAWOOD, P. A. & LU, L. Z. 1999. Tectono-thermal history of the basement rocks in the western zone of the North China Craton and its tectonic implications. *Tectonophysics*, **310**, 37–53.

ZHAO, G. C., CAWOOD, P. A., WILDE, S. A., SUN, M. & LU, L. Z. 2000. Metamorphism of basement rocks in the central zone of the North China Craton: implications for Palaeoproterozoic tectonic evolution. *Precambrian Research*, **103**, 55–88.

ZHAO, G. C., WILDE, S. A., CAWOOD, P. A. & SUN, M, 2001. Archaean blocks and their boundaries in the North China Craton: lithological geochemical, structural and $P–T$ path constraints and tectonic evolution. *Precambrian Research*, **107**, 45–73.

ZHAO, G. C., CAWOOD, P. A., WILDE, S. A. & SUN, M. 2002. Review of global 2.1–1.8 Ga orogens: implications for a pre-Rodinia supercontinent. *Earth Science Reviews*, **59**, 125–162.

ZHAO, G. C., SUN, M., WILDE, S. A. & LI, S. H. 2004. A Palaeo-Mesoproterozoic supercontinent: assembly, growth and breakup. *Earth Science Reviews*, **67**, 91–123.

ZHAO, G. C., SUN, M., WILDE, S. A. & LI, S. Z. 2005. Late Archaean to Palaeoproterozoic evolution of the North China Craton: key issues revisited. *Precambrian Research*, **136**, 177–202.

ZHONG, C. T., DENG, J. F., WAN, Y. S., MAO, D. B., LI, H. M., CHEN, B. & ZHOU, H. Y. 2006a. Zircon U–Pb SHRIMP dating and tectonic significations of adakite granite-sanukite-Closepet granite in Daqingshan of Inner Mongol, *Beijing SHRIMP Centre annals (2005)*. Beijing Geological Publishing House, Beijing, 27–29 (in Chinese).

ZHONG, C. T., DENG, J. F., WAN, Y. S., MAO, D. B., XI, Z. & CHEN, B. 2006b. Magma record of Palaeoproterozoic Orogen: Rock geochemical characteristics and zircon SHRIMP dating of strongly peraluminous granitoids in the northern margin of the North China Craton, *Beijing SHRIMP Centre annals (2005)*. Beijing Geological Publishing House, Beijing, 30–31 (in Chinese).

ZHOU, D. W., SU, L., JIAN, P., WANG, R. S., LIU, X. M., LU, G. X. & WANG, J. L. 2004. Zircon U–Pb SHRIMP ages of high-pressure granulite in Yushugou ophiolitic terrane in southern Tianshan and their tectonic implications. *Chinese Science Bulletin*, **49**, 1415–1419.

The Lüliang Massif: a key area for the understanding of the Palaeoproterozoic Trans-North China Belt, North China Craton

P. TRAP[1]*, M. FAURE[1], W. LIN[2] & S. MEFFRE[3]

[1]*Institut des Sciences de la Terre d'Orléans, CNRS–Université d'Orléans (UMR 6113), 45067 Orléans Cedex 2, France*

[2]*State Key Laboratory of Lithospheric Evolution, Institute of Geology and Geophysics, Chinese Academy of Sciences, Beijing 100029, China*

[3]*CODES ARC Centre of Excellence in Ore Deposits, University of Tasmania, Private Bag 79, Hobart 7001, Australia*

*Corresponding author (e-mail: Pierre.Trap@univ-orleans.fr)

Abstract: This paper documents the first detailed structural analysis of the Lüliang Massif in the Trans-North China Belt, North China Craton. A nappe, made up of a Terrigeneous and Mafic Unit (TMU) derived from an oceanic basin thrust over gneisses and volcanic-sedimentary rocks, is interpreted as a magmatic arc deposited upon a TTG basement. The nappe is rooted to the west in the Trans-North China Suture that separates the Fuping Block from the Western Block. Nappe stacking, coeval with a top-to-the-SE synmetamorphic D_1 event, is dated around 1890–1870 Ma using chemical U–Th/Pb EPMA datings on monazite and U–Pb LA-ICP-MS dating on zircon. A second D_2 ductile event, characterized by SE-verging folds, reworks the D_1 structures. D_2 is the first event recorded in the late-orogenic sedimentary series that unconformably covers the metamorphic units formed during D_1. These lithological, structural and geochronological results are correlated with those described in the eastern massifs of Hengshan, Wutaishan and Fuping. The Trans-North China Belt resulted from the collision of the Fuping Block and the Western Block after a westward-directed subduction and subsequent closure of an oceanic basin where the TMU was deposited.

The NNE–SSW trending Trans-North China Belt (TNCB) is a collisional orogen representing the final amalgamation of the North China Craton. Several aspects of the tectonic evolution of the TNCB remain disputed. These are the timing of the collision, the bulk architecture of the belt, the location of suture zone and the succession of the tectonic metamorphic and plutonic evolution of the TNCB. The tectonic evolution of the TNCB can be summarized by three basic models. The first model considers the collision of the Archaean Western and Eastern blocks to have occurred during the late Palaeoproterozoic (*c.* 1900-Ma) following southeastward directed subduction (Fig. 1; Zhao *et al.* 2001a, 2004, 2005, 2007; Kröner *et al.* 2005). In a second model, the collision occurred in the Neoarchaean and resulted in the emplacement of several eastward or southeastward directed thrust sheets (Kusky & Li 2003; Li & Kusky 2007; Polat *et al.* 2005). A third model proposed a collision at around 1880 Ma after a northwestward subduction of an intervening micro-continental block, called the Fuping Block, beneath the Western (or Ordos) Block (Faure *et al.* 2007; Trap *et al.* 2007).

In the Western Central part of Shanxi Province, the Lüliang Massif is a key area for the understanding of the TNCB as it exposes the westernmost part of the orogen, and as argued below, the suture zone between the Western Block and the Fuping Block. Hence, study of the Lüliang Massif may resolve many of the fundamental questions regarding the timing and nature of the amalgamation of the North China Craton. Geographically, the Lüliang Massif is formed by two topographic highs, called Lüliangshan and Yunzhongshan from west to east, respectively (Fig. 1).

Previous geological work in the Lüliang Massif has concentrated on geological and geochronological investigations (Yu *et al.* 1997a, b; Geng *et al.* 2000, 2003, 2004; Wan *et al.* 2000). Although the previous study provided important new insights, further understanding of the geodynamic evolution of the Trans-North China Belt is hindered by the lack of reliable structural investigations which are reflected in the incomplete and poorly constrained nature of available geological maps for the TNCB. In particular, tectonic information on the Lüliang Massif is not available. We present here the results of field and laboratory studies, leading to the reappraisal of lithological and tectonic units that formed the Lüliang area. Several features of this area provide keys for the understanding of the

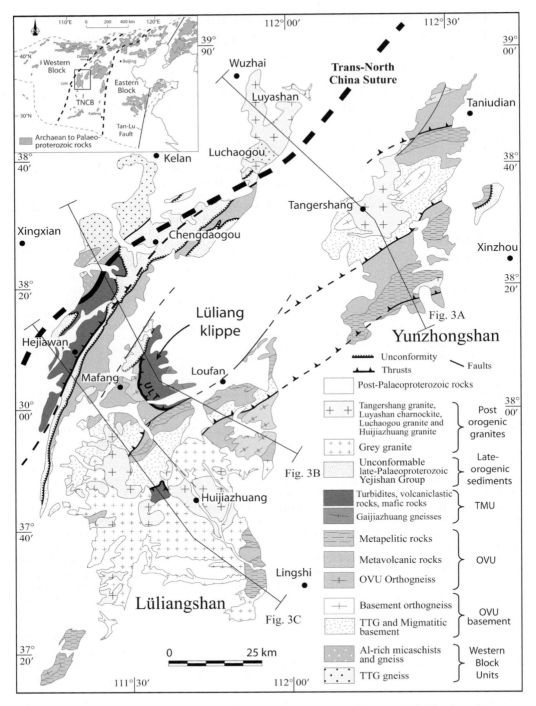

Fig. 1. Geological map of the Lüliang Massif composed of the Lüliangshan to the west and the Yunzhongshan to the NE. Insert: Location of the Lüliang Massif in the westernmost part of the TNCB, in the North China Craton (modified after Zhao *et al.* 2005).

tectonic history of the Trans-North China Belt. Our structural study allows us to recognize that the Lüliang Massif was built through at least two main ductile deformations, D_1 and D_2. In addition, we undertook detailed geochronological work, via U–Th/Pb EPMA dating on polygenetic monazite grains and LA-ICP-MS dating on zircon grains. These geochronological results complement other studies (Yu et al. 1997a; Geng et al. 2000, 2004; Zhao et al. 2008) and support the polyphase tectonic evolution of the Lüliang Massif and the North China Craton from c. 2.5–1.8 Ga. A revised tectonic evolution of the Lüliang Massif is proposed and incorporated into the general framework of the Trans-North China Belt.

Main lithotectonic units of the Lüliang Massif

In the Lüliang Massif, six main lithotectonic units arranged along NNE–SSW trending stripes have been recognized (Fig. 1). The succession from NW to SE is: (1) TTG gneiss and Al-rich metasedimentary rocks; (2) a mafic, turbiditic and volcanic-sedimentary unit; (3) an orthogneiss and metavolcanic unit; (4) a lower gneissic TTG and migmatitic unit; (5) an unconformable late-orogenic weakly metamorphosed sedimentary series; and (6) several generations of post-tectonic granites.

The TTG and Al-rich metasedimentary rocks

The northwestern part of the Lüliang Massif (Fig. 1) consists of two lithological series, namely a gneissic TTG basement made up of tonalite and granodiorite, covered to the SE by Al-rich metasedimentary rocks. Both are metamorphosed under amphibolite facies conditions. The Al-rich metasedimentary rocks consist chiefly of biotite-sillimanite-garnet bearing micaschists and gneiss that are here ascribed to a part of the Jiehekou formation (SBGMR 1989; Yu et al. 1997a; Geng et al. 2000; Wan et al. 2000). The Al-rich metasedimentary rock of the Jiehekou formation crop out northwestward in the Jining, Daqingshan and Helanshan complexes; it belongs to the 'Khondalite series' of the Chinese literature, which is considered to have developed along the passive continental margin of the Ordos Terrane of the Western Block (Wan et al. 2000, 2006; Zhao et al. 2004, 2005; Xia et al. 2006). Together with these previous works we consider that the TTG gneisses and the Al-rich metasedimentary rocks that crop out in the western part of the Lüliang Massif are the basement and the supracrustal cover, respectively, of the eastern margin of the Western Block.

The Terrigeneous and Mafic Unit

The turbiditic rocks. This unit that crops out along a 5–10 km-wide belt lying to the SE of the western metamorphic rocks is commonly included in the Jiehekou group that belongs to the Khondalites series (SBGMR 1989; Geng et al. 2000; Wan et al. 2000; Liu et al. 2006; Zhao et al. 2008). However, the series of well-bedded, fine- to coarse-grained, centimetre to decimetre-thick beds of sandstone-mudstone alternations is quite distinct from those of the Western Block. In spite of a widespread amphibolite facies metamorphism, characterized by the development of biotite, garnet and sillimanite in the pelitic facies, the regular arrangement of sandstone and pelite suggests a turbiditic series. This turbiditic unit is well-exposed, for instance along the Qiushuihe valley, in the vicinity of the village of Hejiawan (Figs 1 & 2a). Although rather homogeneous, the turbidites contain tremolite marble boudins of several metres to hundreds of metres long that might represent olistoliths (SBGMR 1989). As previously described by Wu & Zhong (1998), ultramafic, mafic rocks and greenschists slices are intercalated within the graded sandstones. In some places, heterogenous zones are composed of interleaved marble, meta-sandstone, amphibolites and biotite gneiss. Due to the intense ductile deformation, the primary relationships between these lithologically contrasted fragments within the turbiditic matrix are not settled. They either represent a tectonic imbrication or a block-in-matrix formation. These rocks cannot be compared with those of the Western Block unit and thus they must be regarded as a separate unit. This formation is interpreted here as a subduction complex.

The Lüliang group. NE of Mafang, the following succession is observed from west to east is: (1) basalt, gabbro and quartz-green schists intercalated within greenschist facies metapelites (Fig. 2b); (2) the same turbiditic series as observed to the east, with quartzite and pale-yellow mudstone alternations; (3) black-bluish banded siltites and black wacke interpreted as metatuffs; (4) biotite-bearing gneiss sheets, named as the Gaijiazhuang gneisses (Zhao et al. 2008). Minor andesite, rhyolite, dacite and BIF are also described (Yu et al. 1999; Wan et al. 2000). This lithological assemblage is called the Lüliang Group in the Chinese literature (SBGMR 1989). The available geochemical studies indicate that some of the volcanic rocks of the Lüliang Group were erupted in an oceanic rift-type tectonic setting (Yu et al. 1997b; Geng et al. 2003). Rocks from the Lüliang Group are metamorphosed under greenschist facies conditions. Lithological similarity between the turbiditic series and

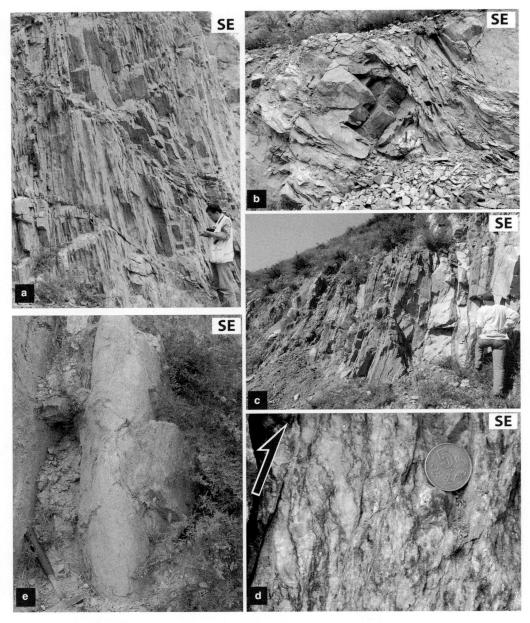

Fig. 2. (**a**) Turbiditic series made of meta-sandstone and meta-pelitic alternation, Lüliangshan, Qiushuihe Valley (N38°09.034′/E111°18.516′). (**b**) Metric rounded gabbro-blocks scattered in the greenschist matrix. The surrounding greenschists are well-foliated with S_1 foliation that warps around gabbro blocks, Lüliang Group, NE of Mafang (N38°05.556′/E111°31.679′). (**c**) Typical alternation of decimetre- to metre-sized amphibolite-leptynite formations that form one of the typical elements of the (OVU), Yunzhongshan (N38°45.766′/E112°33.320′). (**d**) Ductile shear zone within an OVU orthogneiss, S_1 strikes N60E and dips 60° towards the NW, shear bands and asymmetric porphyroclasts indicate a top-to-the-SE shearing, SE of Loufan (N37°57.831′/E111°57.184′). (**e**) Synfolial F_1 fold within turbiditic rocks showing a steep dipping fold axis parallel to L_1 (N38°30.110′/E111°24.424′).

the Lüliang Group leads us to group these rocks into a single unit called the Terrigeneous and Mafic Unit (TMU) (Fig. 1).

The Orthogneiss-and-Volcanite Unit (OVU)

This litho-tectonic unit covers more than half of the surface of Lüliangshan and Yunzhongshan (Fig. 1). Three lithologies are recognized in this unit: (1) a centimetre to metre-thick alternation of acidic gneiss and amphibolite; (2) a metasedimentary series of Al-rich gneisses and micaschists with some marble layers; and (3) several K-feldspar augen gneisses derived from porphyritic dioritic and tonalitic plutons. The protoliths of the amphibolites are basalt or diabase and those of the gneisses are rhyolite, dacite and felsic volcaniclastic rocks (Fig. 2c). As a whole, this characteristic sequence of acidic-gneiss and amphibolite corresponds to a volcaniclastic suite. The relationship between orthogneiss and volcaniclastic rocks is not always clear. Some of those orthogneisses are derived from granitoids that intrude the series, but other orthogneisses seem to belong to an underlying basement upon which the volcaniclastic suite deposited. As stated above, detailed petrological or geochemical studies are rare, nevertheless, this litho-tectonic unit resembles the Orthogneiss-and-Volcanic Unit (OVU) defined in the Hengshan and Wutaishan Massifs to the east (Trap et al. 2007; see Discussion). For this reason, this above described unit of the Lüliang Massif is also attributed to the OVU.

The TTG migmatitic basement

Within the Lüliang Massif, TTG gneiss and migmatite crop out in several places in the core of anticlines. TTG gneiss represents the basement upon which the volcaniclastic and Al-rich sedimentary rocks deposited. The structural relationships between the migmatite and TTG gneiss is somewhat difficult to settle in the field. However, in some places migmatite is likely to derive from partial melting of TTG gneiss and OVU rocks, as both occur as xenoliths within the migmatite. However, the age of migmatization is not constrained yet. TTG gneiss and migmatite might represent the continental basement upon which the OVU deposited, as was suggested more to the east in the Wutaishan massif (Faure et al. 2007; Trap et al. 2007).

The unconformable syn- to late orogenic sedimentary series

All previously described lithologies are unconformably covered by sub-greenschist facies metamorphosed or unmetamorphosed conglomerates, sandstones, quartz wackes and minor carbonates, known as the late Palaeoproterozoic Yejishan group (SGBMR 1989). These rocks mainly crop out along a narrow, NE–SW-trending strip that extends in the western part of the Lüliangshan (Fig. 1). Since these weakly metamorphosed sedimentary rocks are also deformed (see Structural analysis below) but unconformably deposited over amphibolite rocks, they represent syn- to late orogenic deposits. At the scale of the TNCB, this terrigeneous series is correlated with the Hutuo Supergroup that also corresponds to a late orogenic deposit (Yang et al. 1986; Faure et al. 2007).

The post-tectonic granites

Several generations of undeformed granites intrude the above-described lithological units. The pink fine- to coarse-grained Tangershang granite crosscuts the OVU in Yunzhongshan. The Luyashan porphyritic charnockite and the Luchaogou porphyritic granite are both encountered in northern Lüliangshan (Fig. 1). In addition, a large granitic complex of >2000 km^2 develops in the southern part of Lüliangshan. It consists of a large body of fine- to medium-grained grey granite, with a magmatic foliation, intruded by coarser-grained granites such as the Huijiazhuang granite (Fig. 1). Numerous xenoliths of TTG and migmatite enclosed within the post-tectonic granite provide additional evidence for the existence of a gneissic basement underlying the OVU.

Structural analysis

Bulk architecture

The bulk structure of the Lüliang Massif appears as a stack of nappes (Figs 1 & 3). The uppermost unit of the Lüliang Massif is represented by the mafic volcaniclastic rocks that crop out between Loufan and Mafang. As described in the next section, this unit is a klippe, called the 'Lüliang klippe', which has been displaced from NW–SE. The same unit can be also recognized in the central part of Lüliangshan. More to the west, the amount of mafic rocks decreases and metasandstone-mudstone alternations predominate. East and below the TMU, several thrust sheets of the TTG, migmatitic basement and OVU are bounded by secondary thrust faults (Figs 1 & 3). The Lüliang klippe is rooted along the eastern boundary of the Western Block that corresponds to the Trans-North China Suture (TNCS).

The bulk crustal architecture of the Lüliang Massif, namely nappe formation and thrusting (top-to-the SE), is a result of D_1 deformation

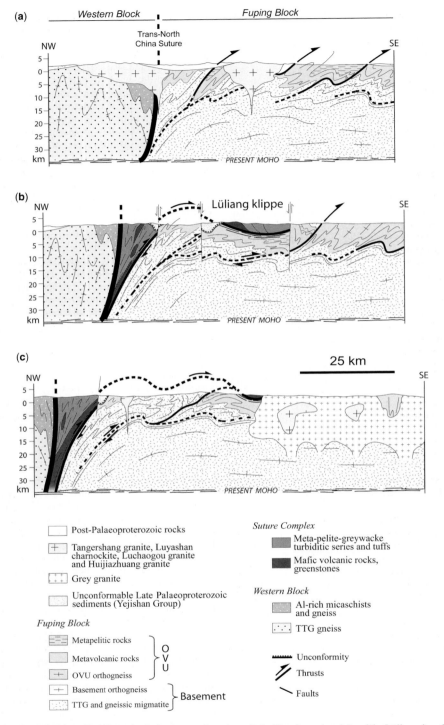

Fig. 3. Interpretative crustal-scale geological cross-sections through the Yunzhongshan (a) and the Lüliangshan (b & c).

which is defined mesoscopically by a penetrative, originally flat-lying penetrative foliation (S_1), NW–SE trending lineation (L_1) and syn-folial folds (F_1). A second ductile deformation D_2 is responsible for the folding of early D_1 structures. The stack of nappe is crosscut by post-tectonic granitic plutons. In the following, we describe the mesoscopic and microscopic structural features of D_1 and D_2 as well as the overprinting relationships between the two deformations, observed at several key locations.

D_1 deformation

S_1, L_1 and F_1. The S_1 foliation is the most widespread deformation fabric observed in rocks of the Lüliang Massif. The preferred orientation of metamorphic hornblende, biotite, sillimanite, kyanite and aggregates of recrystallized quartz defines the S_1 foliation surface which is also marked by a layering segregation. The structural investigation revealed three main S_1 trends at around N40E, N90E and N150E, among which the N40E general strike is the dominant one (Fig. 4). Conversely, west of the Trans-North China Suture, within the Western Block unit, the N90E trend predominates but it becomes NE–SW in the vicinity of the TNCS. A similar east–west trend occurs in a small area enclosed within the post-tectonic granites SE of the Lüliangshan (Fig. 4). The N150E trend is restricted to the middle part of the Lüliangshan, within and south of the Lüliang klippe. In general, the foliation dips steeply preferentially towards the NW, with >50% of dip measurements lying between 70° and 90°. A flat lying foliation is commonly observed in the central part of Lüliangshan in the surroundings of the Lüliang klippe. The rapid change in S_1 trend and dip reflects a later folding.

The S_1 foliation holds a prominent mineral and stretching lineation (L_1) defined by the parallel alignment of euhedral or subhedral grains with an elongated crystal shape such as hornblende, sillimanite and kyanite, elongated segregations of felsic and mafic minerals, plagioclase shadows and elongated grain aggregates (Fig. 5). L_1 also occurs as finely coloured stripes of strongly sheared rocks, as commonly observed within the metasandstone of the turbiditic series. Within the OVU metavolcanics, L_1 is also marked by boudins with irregular shape characterized by strongly elongated necks affected by pinch-and-swell structures (Fig. 6e). The structural study reveals two dominant trends for the stretching lineation, namely N120E and N150E. The observed orientation variation of L_1 is a consequence of D_2 deformation. L_1 trends of N120E and N150E correlate to the orientation of L_1 on either limb of F_2 folds. Along the L_1 lineation, several shear criteria such as shear bands and sigmoidal-shaped minerals argue for a top-to-the SE sense of shear. The kinematic indicators are well-developed in coarse-grained rocks such as, for instance, in the OVU augen orthogneisses (Fig. 2d).

Synfolial isoclinal folds (F_1) are observed in all metamorphosed lihologies. For instance, metre-scale intrafolial folds are delineated by meta-sandstone layers within the turbiditic sequence (Fig. 2e). Parallelism between fold axes and stretching lineation suggests that those folds formed during the D_1 shearing. Such folds can be viewed as early folds subsequently stretched out and re-oriented during D_1 (e.g. Carrerras et al. 2005). Other minor folds with oblique axes with respect to the direction of L_1 can be interpreted as imperfectly re-oriented folds during D_1.

Thrust faults. The S_1 foliation planes locally grade into ductile shear zones marked by the development of a penetrative mylonitic fabric associated with a well-pronounced L_1 stretching lineation. The shear zones thickness varies from several decimetres to several decametres. Here we describe the main mylonitic shear zones and distinguish their structural significance since they separate distinct lithological and metamorphic units.

The most significant shear zone is the Upper Lüliang Thrust (ULT, Fig. 1) that places the mafic and turbiditic rocks over the magmatic arc series of the OVU. This tectonic contact is well-exposed along the Suiyu Valley, NE of Mafang, where it appears as a flat lying (S_1 dip <20°), hectometre-thick shear zone made of mylonitized or ultra-mylonitized augen gneiss derived from the OVU (Figs 6a, b). Top-to-the SE shearing is clearly demonstrated by asymmetric feldspar porphyroclasts (Fig. 6b). Similarly, gabbros lying near the base of the Lüliang klippe underwent intense shearing, illustrated by a well-developed mylonitic fabric defined by elongated pyroxene grains with length-wide-ratio up to 1:6. Away from the mylonitic sole of the nappe, deformation decreases to the extent that, in the uppermost Lüliang Nappe, sedimentary bedding is well-preserved. West of the Lüliang klippe, a similar tectonic contact separates the OVU rocks from the TMU. There, we did not observe the mylonitic basal shear zone, as it is hidden below the late-orogenic sedimentary rocks of the Yejishan Group. However, the existence of this tectonic contact is likely since an increasing deformation can be observed when approaching the westward dipping contact. In spite of a later recrystallization due to intrusive granitic plutons, just before touching the Yejishan discontinuity, fine-grained gneisses show a strong ductile deformation marked by a well-developed foliation formed by

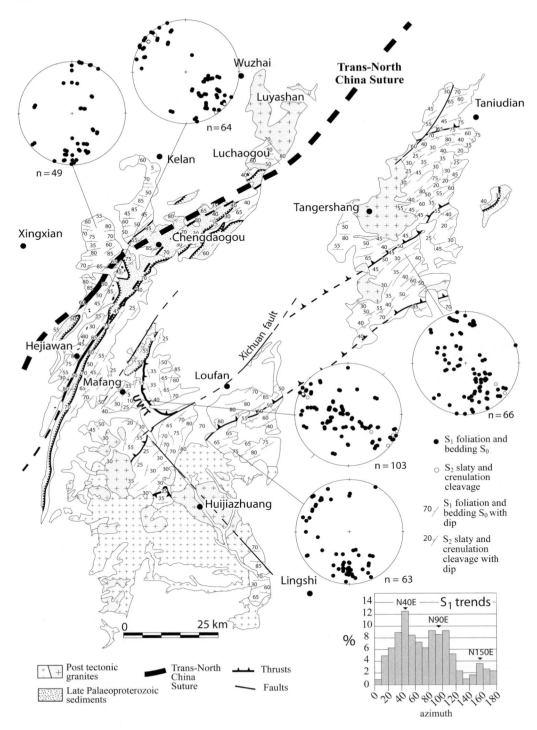

Fig. 4. Structural map showing the foliation pattern within the Lüliang Massif.

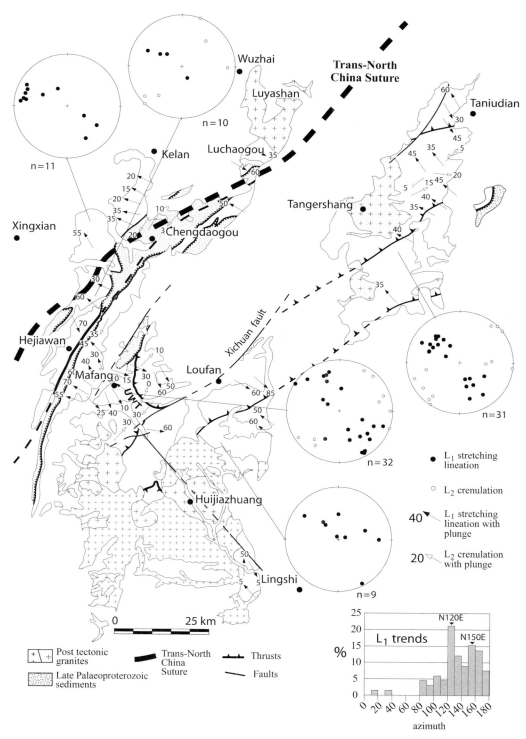

Fig. 5. Structural map showing the lineation pattern within the Lüliang Massif.

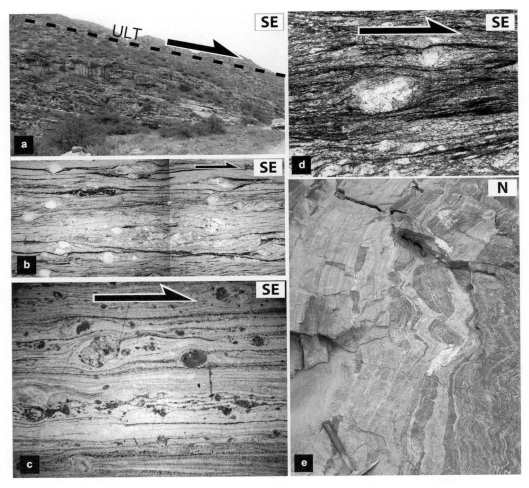

Fig. 6. (**a**) Hectometre-thick flat-lying mylonitic augen gneiss that define the ULT at the base of the Lüliang klippe, north of Mafang (N38°02.079′/E111°29.649′). (**b**) Thin section from the same rock showing the well-developed mylonitic fabric with top-to-the SE shear criteria. (**c**) Typical mylonitic fabric in a gneiss within one of the metre-scale shear zone observed along the TNCS (N38°28.838′/E111°25.427′). (**d**) Strongly sheared sillimanite-bearing micaschists within the TNCS with shear band and asymmetric pressure shadows around FK porphyroclasts showing a top-to-the SE shearing (N38°21.528′/E111°17.987′). (**e**) Overprinting relationships showing the L_1 marked by boudinage of amphibolite layers refolded by F_2 folds, Yunzhongshan Massif (N38°42.905′/E112°33.812′).

alignment of biotite grain, flattened quartz and sub-grain development. The planar and linear fabric is again associated with a top-to-the SE shearing. In the cross-section in Figure 3, the ULT is drawn as a folded, but initially flat lying, shear zone along which the Lüliang Nappe emplaced from the NW to the SE, upon the OVU. The Lüliang Nappe is rooted in the western part of the Lüliangshan along the suture zone (Figs 1 & 3).

The second main shear zone, the Trans-North China Suture (Faure *et al.* 2007) is not exposed as a single thrust fault like the ULT. Conversely, the Trans-North China Suture is made of several decimetre- to metre-scale mylonitic to ultramylonitic shear zones (Fig. 6c) forming a *c.* 1 km-thick band that strikes N30–40E and dips steeply westward. The micaschists that crop out within this kilometre-scale thick band also exhibit consistent top-to-the SE kinematics developed along L_1 (Fig. 6d).

Three metre to decametre-scale steeply NW dipping mylonitic shear zones show the same top-to-the SE kinematic pattern as the thrust zones described above. These shear zones developed in the meta-volcano-sedimentary rocks and orthogneisses as well as at the cover–basement interface.

D_2 deformation

Within the whole Lüliang Massif, the bedding observed in the late-Palaeoproterozoic sedimentary rocks and the D_1-related structures are deformed by a D_2 deformation. F_2 folds occur throughout the whole area and represent the most pervasive structural element related to D_2 deformation. The F_2 folds are characterized by a steeply to moderately northwestward dipping axial plane ($50°-80°$) striking around N20E–N60E. F_2 open to tight folds ranging in scale from several metres to decametres (Fig. 7). In addition, kilometric-scale F_2 folds are sometimes observed in outcrop (Fig. 7a) or inferred from S_1 attitude, for instance the Lüliang klippe preserved within a D_2 synform. The F_2 folds are consistently overturned to the SE. Such a vergence complies with the top-to-the SE shearing developed during D_1. As stated above, the F_2 folds are responsible for the scattering of L_1 in two directional sets. The observation of the two sets of lineation is in good agreement with the F_2 fold axes measurements and the homogeneous distribution of both F_2 folds and L_1 across the whole area (Fig. 8).

The F_2 folding is coeval with the development of an S_2 cleavage. Within the Lüliang klippe and the upper part of the OVU, S_2 appears as a crenulation cleavage overprinting the S_1 foliation. At the outcrop scale, a S_2 crenulation cleavage is observed within F_2 fold hinges (Fig. 7b). A crenulation lineation L_2 parallel to the fold hinges is associated with the crenulation cleavage (Figs 5 & 7e). At the thin-section scale, S_2 is defined by the homogeneous distribution of well-aligned biotite minerals (Fig. 7c). The development of biotite within the S_2 crenulation cleavage shows that a second metamorphic event (M_2) is coeval with the D_2 deformation.

D_2 is also well-developed in the late-orogenic late Palaeoproterozoic sedimentary rocks that did not experience the early D_1 deformation. The S_2 surface is a NE–SW trending and steeply dipping slaty cleavage that overprints and reworks the S_0 bedding. Indeed, the S_2 cleavage cuts across bedding at a high angle in the gently-dipping limbs of the metric-scale F_2 folds, and refracted at the lithological interface (Fig. 7d). Near the F_2 fold hinge, S_0 underwent a strong transposition into S_2. The sandstone beds flow in the S_2 planes, well-developed within less competent pelitic layers. Due to the D_2 event, the primary unconformable contact between the late Palaeoproterozoic Yejishan Group and the underlying metamorphic unit is difficult to observe and sometimes sheared.

Relationship between polyphase deformation and metamorphism

As stated above, the Lüliang Massif is a stack of nappes thrust towards the SE. The main part of the transported material underwent a metamorphism under amphibolite facies conditions (Liu et al. 2006). As the amphibolite facies metamorphic mineral assemblage defines the D_1 planar and linear fabric, it is reasonable to consider that the D_1 deformation is coeval with the amphibolite facies metamorphism. However, within the Lüliang Group rocks, D_1 structural elements are defined by the observed greenschist facies mineral assemblage. Therefore, D_1 is likely contemporaneous to the greenschist grade and the amphibolite grade metamorphisms (both represent M_1) that affected the Lüliang Group and OVU rocks, respectively.

Within the greenschist facies Lüliang klippe, a second metamorphic event is mainly characterized by the development of biotite, almandine and staurolite (Li & Zhai 1984; SBGMR 1989; Yu et al. 1999). According to Yu et al. (1999), these isogrades are at high angle to the S_1 foliation. Mineral assemblages associated with M_2 define the D_2 fabric as evidenced by biotite crystallization along the S_2 crenulation cleavage plane (Fig. 7c). Hence M_2 is interpreted to be coeval with the D_2 deformation event. This second metamorphism is well-recognized in the Lüliang klippe because there the pre-D_2 metamorphic grade was low, that is greenschist facies. Nevertheless, further studies are required to constrain the pressure and temperature conditions and $P-T-t$ path for each of the metamorphic events.

Geochronological study

Previous studies

Figure 9 synthesizes the principal geochronological works performed in the Lüliang Massif. Conversely to other Trans-North China massifs, such as Hengshan, Wutaishan or Fuping massifs, ages reported within the Lüliang Massif range mostly within the 2.3–1.8 Ga period (Geng et al. 2000, 2003, 2004; Wan et al. 2000; Zhao et al. 2008). Just a few Archaean components are reported there (Wan et al. 2000; Zhao et al. 2008). The 2800 Ma age comes from inherited zircons in the metasedimentary rocks of the Western Block (Wan et al. 2000). In the Yunzhongshan, a TTG gneiss yields an age at c. 2499 ± 9 Ma interpreted as that of the earliest arc-related magmatic event (Zhao et al. 2008).

Ages around c. 2150–2030 Ma reported in several granitoid rocks with a calc-alkaline

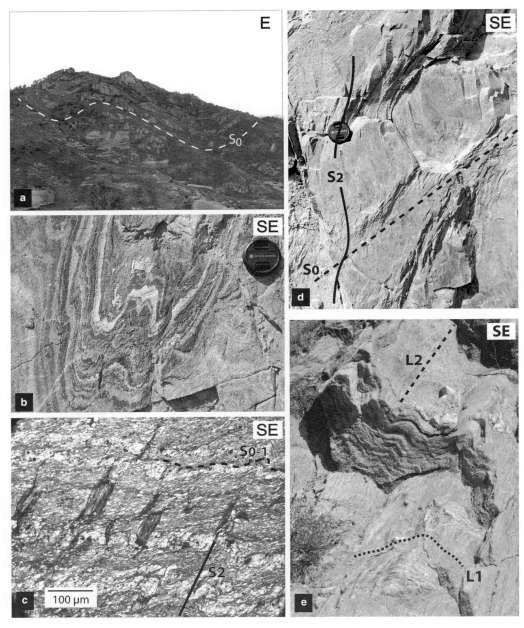

Fig. 7. (**a**) Hectometre-sized F_2 folds within the Yejishan group, north of Chengdaogou village (N38°29.095′/E111°36.478′). (**b**) F_2 fold hinge showing subvertical S_2 crenulation cleavage within metavolcanic rocks (OVU), F_2 is overturned towards the SE, Yunzhongshan (N°38°45.465/E112°33.820). (**c**) S_2 crenulation cleavage overprinting the S_1 foliation; note the preferred orientation of biotites that grow along the S_2 planes, Lüliang group (N38°00.286′/E111°35.463′). (**d**) Refraction of S_2 slaty cleavage at lithological boundaries (S_0) within sandstones of the unmetamorphosed Yejishan terrigeneous rocks, south of Chengdaogou village (N38°29.393′/E111°35.206′). (**e**) L_1 refolded by F_2 folds marked by the L_2 crenulation lineation, within metavolcanic rocks of the OVU, Yunzhongshan (N38°43.509′/E112°33.431′).

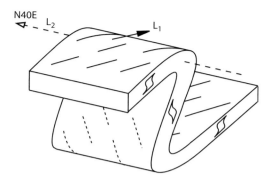

Fig. 8. Reorientation of the early stretching lineation (L_1).

geochemical signature are interpreted as igneous ages related to the most widespread arc-related magmatic event in the Lüliang Massif (Geng *et al.* 2000, 2004; Zhao *et al.* 2008). In the Western units, the same Palaeoproterozoic ages are interpreted to reflect an amphibolite facies metamorphism (Geng *et al.* 2000; Wan *et al.* 2000). Yu *et al.* (1997*a*) suggest that meta-basalt and meta-rhyolite of the Lüliang group erupted at about 2100 Ma.

Based on U–Pb zircon and U–Th/Pb monazite ages, the regional metamorphism of the Lüliang Massif is thought to occur in the period 1880–1820 Ma (Liu *et al.* 2006; Zhao *et al.* 2008). However, Liu *et al.* (2006) performed a U–Th–Pb EPMA dating of two metapelite samples from the southeastern part of the Lüliang Massif (Fig. 9) that revealed an age group at c. 1940 Ma interpreted as the crystallization age of an igneous monazite resedimented in the protoliths of the metapelites. Thus, the age of the syn-D_1 amphibolite facies metamorphism is not clearly settled yet. Finally, post-collisional granites, dated around 1800 Ma, are widespread across the whole area (Yu *et al.* 1997*a*; Geng *et al.* 2004; Zhao *et al.* 2008).

U–Th/Pb EPMA dating

Analytical procedure. EPM analysis were performed using a Cameca SX 50 electron probe microanalyser, cooperated by the BRGM and ISTO in Orléans (France), equipped with five wave-length-dispersive spectrometers using an acceleration voltage of 20 kV and a beam current of 100 nA. Counting times (peak + background) were 240 s for Pb, 200 s for U, and 40 s for all other elements. Details of the U–Th/Pb EPMA dating on monazite are to be found in Cocherie *et al.* (1998) and Cocherie & Albarede (2001) so is not repeated here. U–Th/Pb monazite chemical dating of all samples was undertaken using *in situ* analysis on polished thin sections. The use of *in situ* geochronological analysis allows for monazite growth and subsequently age data to be related to the growth of the metamorphic mineral assemblage. In addition, investigations about the zonation of monazite were performed using back-scattered electron images.

EPMA results. Back-scattered electron images of all analysed monazite grains reveal a well-defined core and rim zonation. Therefore, for each rock, EPM analyses carried out on the totality of the monazite grains were separated in two datasets, one corresponding to the monazite core and one to the rim. Analytical data are summarized in Table 1. For each core and rim age calculation, analytical results show a large range in Th/U ratio and thus the chemical composition of the core and rims of monazite grains is favourable for using the Th–Pb v. U–Pb diagram (Cocherie & Albarede 2001). A good data spread is obtained for every computed age. Details on geochronological results are given in the following.

Rock FP276 was sampled in Yunzhongshan, SW of Taniudian. It consists of kyanite-rich gneiss from the OVU with a main assemblage of quartz + muscovite + kyanite + tourmaline that mark the M_1 metamorphism developed during D_1. Monazite grains are located in the matrix of quartz + muscovite and range in size from 50 μm–200 μm (Fig. 10a). Grains are prismatic-shaped with a long axis parallel to the foliation marked by preferred orientation of muscovite and kyanite. Cores analyses (n = 67) revealed a greater amount of Pb (2500 ppm) than in the rims (c. 2000 ppm, Table 1). The individual U–Pb and Th–Pb of 2360 +109/−120 Ma and 2450 +72/−68 Ma, respectively are concordant within error. The regression line, close to the theoretical isochron, suggests that the dataset is in good agreement with a single age. A mean age of 2401 ± 15 Ma was calculated at the centroid of the population (Fig. 10b).

For the rims age calculation, a total of 124 analyses was used. The variation of Th/U ratio is similar to that for the monazite cores, that is 9.9 ± 3.5. The two intercepts ages are similar within error, that is U–Pb age: 1970 +115/−129 Ma and Th–Pb age: 1835 +67/−62 Ma. The regression line, within the errors envelope suggests that the dataset is in good agreement with a single age. A mean age of 1881 ± 10 Ma was calculated at the centroid of the population (Fig. 10c).

Sample FP313 is a sillimanite-bearing gneiss sampled nearby the TNCS. It contains a quartz + biotite + sillimanite + muscovite + microcline main assemblage. The rock is intensely sheared with a mylonitic fabric marked by alignment of biotite

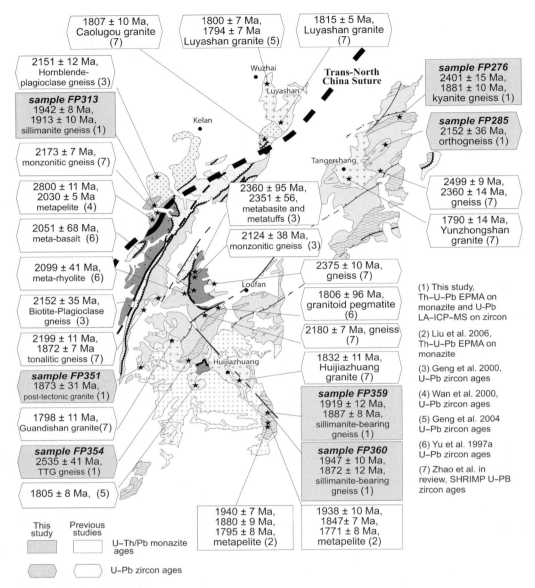

Fig. 9. Synthesis of the principal geochronological results obtained within the Lüliang Massif. Each sample is located by a black star. One box represents one sample that could have produced one, or several ages in the case of polygenetic grains.

and fibrolite aggregate and especially strongly elongated quartz grains with shape ratio up to 10:1 showing wide subgrain development. Monazite occurs in the matrix or as inclusion within biotite. Monazite grains are euhedral or subhedral with no shape orientation and range in size from 40–60 μm (Fig. 10d).

A total of 43 analyses was used for the core age calculation. The intercept ages are well defined and similar within error (U–Pb age: 1905 +47/−51 Ma and Th–Pb age: 1985 +57/−54 Ma). The regression line lies between the errors envelope and an isochron age of 1942 ± 8 Ma is calculated at the centroid of the population (Fig. 10b).

Rims analyses show a very different composition from the core, with a much lower U content, 7300 v. 9600 ppm (Table 1). The two intercept ages are similar within error, that is

Table 1. Summary of electron micropobe data for the polygenetic monazites

U (ppm) ± σ S.D.	Th (ppm) ±	Pb (ppm) ±	Th/U ±	Isochron age ± 2σ Ma	No. of data
Kyanite-rich gneiss (FP276) of the OVU, Yunzhongshan massif (N38°43.879'/E112°33.203')					
Cores of 9 monazite grains					
1733 ± 170	15243 ± 400	2534 ± 100	9.95 ± 0.72	2401 ± 15	67
Rims of 9 monazite grains					
1782 ± 86	15893 ± 190	1975 ± 34	9.69 ± 0.60	1881 ± 10	124
Sillimanite-bearing micaschist (FP313) of the Western sedimentary cover, (N38°21.897'/E111°17.856')					
Cores of 7 monazite grains					
9630 ± 950	42470 ± 1400	7479 ± 260	3.86 ± 0.57	1942 ± 8	43
Rims of 7 monazite grains					
7328 ± 290	41728 ± 980	6286 ± 120	5.64 ± 0.56	1913 ± 10	84
Sillimanite-garnet gneiss (FP359) of the OVU, Lüliang Massif (N37°34.335'/E111°53.165')					
Cores of 5 monazite grains					
3903 ± 290	19144 ± 1500	3394 ± 240	5.42 ± 0.46	1919 ± 12	67
Rims of 6 monazite grains					
2295 ± 300	42449 ± 2000	4704 ± 130	21.6 ± 2.9	1887 ± 8	60
Sillimanite-garnet gneiss (FP360) of the OVU, Lüliang Massif (N37°34.335'/E111°53.165')					
Cores of 3 monazite grains					
3066 ± 540	31640 ± 4200	4618 ± 330	12.0 ± 2.3	1941 ± 10	33
Rims of 3 monazite grains					
3184 ± 510	33527 ± 2900	4275 ± 270	11.6 ± 1.8	1872 ± 12	33

U–Pb age: 1895 +48/−46 Ma and Th/Pb age: 1935 +56/−60 Ma. The regression line is close to the theoretical isochron. The mean age calculated at the centroid of the population obtained on 6 grain rims is 1913 ± 10 Ma (Fig. 9e).

Sample FP359 is a biotite + plagioclase + sillimanite + garnet metasedimentary rock exposed in the southeasternmost part of Lüliangshan, west of Lingshi village (Fig. 9). The S_1 foliation is marked by preferred orientation of biotite and sillimanite. Some centimetre-scale leucocratic pockets of undeformed quartz-feldspar assemblage and chlorite growth after biotite are evidence for fluid relocation. Monazite is homogeneously distributed in the rock but monazite grains within leucocratic pockets are much altered and EPM dating is therefore precluded. Age computations were performed only on clean monazite grains that occur within the biotite-sillimanite layers. Monazite grains range in size from 100–200 μm and lie at the interface between biotite and sillimanite or as inclusions within biotite. As for the two previous samples, monazite appears with a pronounced core-and-rim zonation on BSE images (Fig. 11a). This zonation is also well-represented by the difference in Th/U ratio between core (6.3 ± 3.3) and rim (24.5 ± 11.1) (Table 1).

The two intercepts ages computed from monazite cores are similar within error: 1890 +77/−84 Ma and 1944 +63/−58 Ma for U–Pb and Th–Pb ages, respectively. The regression line is close to the theoretical isochron and a mean age of 1919 ± 12 Ma was calculated at the centroid of the data population (Fig. 11b).

The two intercept ages computed from monazite rims are similar within error, that is U–Pb age: 1898 +70/−72 Ma and Th/Pb age: 1884 +19/−18 Ma. The regression is almost parallel to the theoretical isochron and yields a mean age of 1887 ± 8 Ma, calculated at the centroid of the data population (Fig. 11c).

The sample FP360 was collected in the same outcrop as the previous sample (FP359). It consists of a paragneiss with a main assemblage of quartz + garnet + biotite + sillimanite + plagioclase that mark the M_1 metamorphism developed during D_1. Biotite and sillimanite are concentrated along millimetre-thick shear bands that anastomose around 1 cm-sized garnet porphyroblasts. Monazite grains (Fig. 11d) are rare and only three large grains (c. 60 μm) that occur in the matrix or around garnet porphyroblasts were analysed.

For monazite cores, the computed regression line fits almost perfectly the theoretical isochron and the two U–Pb and Th–Pb intercept ages of 1936 +69/−73 Ma and 1943 +29/−28 Ma, respectively, are similar within errors. The mean age calculated at the population centroid is 1941 ± 10 Ma (Fig. 11e).

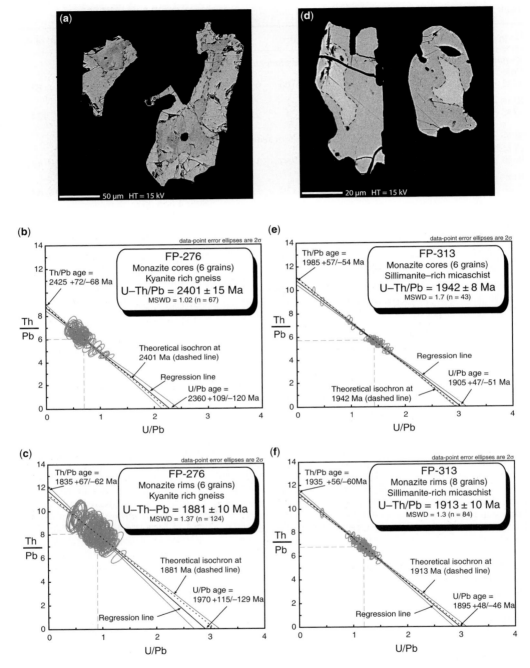

Fig. 10. Back-scattered electron pictures and Th–Pb v. U–Pb diagrams for monazites from samples FP276 (a, b and c) and FP313 (d, e and f). All errors are quoted at 95% confidence level and error ellipses are plotted at 2σ. See text for explanations.

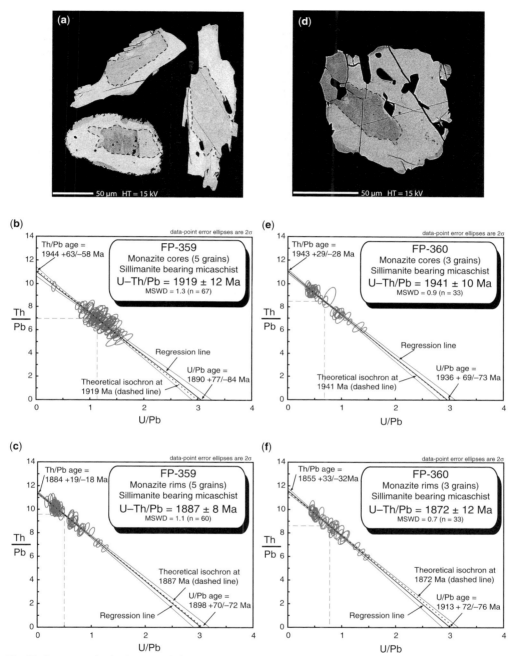

Fig. 11. Representative back-scattered electron pictures and Th–Pb v. U–Pb diagrams for monazites from samples FP359 (**a**, **b** and **c**) and FP360 (**d**, **e** and **f**). All errors are quoted at 95% confidence level and error ellipses are plotted at 2σ. See text for explanations.

Monazite rims show a slightly different Th/U ratio than monazite cores. The two intercepts ages computed from monazite rims are similar within error: 1895 +48/−46 Ma and 1935 +56/−60 Ma for U–Pb and Th–Pb ages, respectively. The regression line is close to the theoretical isochron and a mean age of 1872 ± 12 Ma has been calculated at the centroid of the population (n = 33) (Fig. 11f).

U–Pb zircon ages

LA-ICP-MS zircon analyses were performed on a leucosome of migmatite (FP354), a basement orthogneiss (FP285) and a post-tectonic granite (FP351). Sample locations are shown in Figure 9. Sample FP354 is a leucosome of migmatite made of a quartz-feldspath assemblage with minor biotite and hornblende grains. The migmatitic layering marks the S_1 foliation. Sample FP285 is an augen gneiss with a quartz + plagioclase + microcline + biotite + magnetite assemblage and a D_1 fabrics. Sample FP351 is a biotite-bearing granite that does not show any metamorphic or deformation features and is interpreted to be post-tectonic.

Method. The rocks were crushed in a ring mill and sieved (400 micron). Non-magnetic and slightly magnetic heavy minerals were separated from the <400 micron fraction using a plastic gold pan and an iron-boron-rare earth magnet. Large clear crystals were then picked from the heavy mineral separate and mounted in epoxy. All three samples contained large zircon, monazite and apatite crystals.

The samples were then analysed using a Hewlet Packard 4500 quadrupole ICPMS and a New Wave UP 213 nm laser at CODES, University of Tasmania. Ablation was performed in a custom-designed chamber in an He atmosphere using a laser pulse rate of 5 Hz on a beam 30 micron in size delivering about 13 mJ cm^{-2}. A total of 11 masses were analysed (Zr, Hf, Nd, Hg, Pb, Th, U) with longer counting times on the Pb and U isotopes. Each analysis began with a 30 s gas blank followed by 30 s with the laser switched on. Four primary (Temora zircons of Black *et al.* 2003) and two secondary standards (91500 zircons of Wiedenbeck *et al.* 1995) were analysed before and after each 12 zircons to correct for mass bias, machine drift and downhole fractionation. Data reduction was performed using the methods outlined by Black *et al.* (2004). Weighted averages and Concordia plots were calculated using the isoplot software of Ludwig (2003).

U–Pb zircon results. U–Pb analytical results are summarized in Table 2. Most of the zircon analyses from leucosome of migmatite, sample FP354, plot along a tightly constrained isochron intersecting Concordia at 874 ± 75 and 2535 ± 41 Ma with analyses with the highest U contents having the lowest Pb/U and $^{207}Pb/^{206}Pb$ ratios (Fig. 12a). This suggests that the Archaean or earliest Proterozoic zircons within this sample were extensively reset or overgrown in the Neoproterozoic. The analysis with the highest U content (0.3%) plots to the left of the isochron, suggesting recent Pb loss associated with metamictization.

Zircons analysed from orthogneiss, sample FP285, have medium to low U contents with simple U–Pb isotopic composition. The zircon data clusters tightly around Concordia at 2152 ± 36 Ma (Fig. 12b). The $^{207}Pb/^{206}Pb$ age was chosen as the most reliable for this sample as it gives slightly better precision compared to the $^{206}Pb-^{238}U$ age (2150 ± 38 Ma). The zircon with the highest U content shows evidence of slight Pb loss, which should not affect the $^{207}Pb/^{206}Pb$ age.

Most of the zircons analysed from a post-tectonic granite, sample FP351, show some evidence of recent Pb loss (low $^{206}Pb-^{238}U$ but unaffected $^{207}Pb/^{206}Pb$) within high U zones (Fig. 12c). These high U zones were omitted from the calculations. One of the zircons has high ^{204}Pb and anomalously young $^{206}Pb-^{238}U$ ages, indicating both radiogenic Pb loss and common Pb gain. This zircon was omitted from the calculations. To minimize problems associated with Pb loss, only the $^{207}Pb/^{206}Pb$ age was used to calculate a Palaeoproterozoic age of 1873 ± 31 Ma for the zircons.

Interpretation of EPMA monazite and LA-ICP-MS zircon ages

The eleven new monazite and zircon ages presented in this study are plotted together with the most recent and reliable geochronological results obtained from the Lüliang and Hengshan-Wutaishan-Fuping massifs. Four main geochronological trends are revealed (Fig. 13).

The older ages of 2401 ± 15 and 2535 ± 41 Ma recorded in a kyanite-bearing gneiss (sample FP276, core of monazite grains) and a migmatite (sample FP354 zircon), respectively, are within the range of the majority of zircon ages reported from the adjacent Hengshan–Wutaishan–Fuping massifs (Fig. 13). These Late Archaean to Early Palaeoproterozoic ages preserved in relic grains represent the time of formation of an early volcanic arc of the TNCB. The migmatite (sample FP354) belongs to the basement upon which the metavolcanic rocks and metasedimentary rocks (sample FP276) were deposited.

The age of 2152 ± 36 Ma obtained on zircon from an orthogneiss (sample FP285) that intrudes

Table 2. *LA-ICP-MS data for zircon from samples FP285, FP351 and FP354*

	Pb	Th	U	Th/U	$^{207}Pb/^{235}U$	1σ	$^{206}Pb/^{238}U$	1σ	$^{207}Pb/^{206}Pb$	1σ	$^{206}Pb/^{238}U$ age (Ma)	1σ
Basement migmatite leucosome (FP354), Lüliangshan Massif (N37°45.254'/E111°26.342')											
H1	349	86	1669	0.05	3.78	0.22	0.231	0.003	0.118	0.004	1285	18.7
H2	207	50	715	0.07	6.18	0.28	0.312	0.005	0.144	0.004	1664	25.7
H3	43	47	87	0.54	10.42	0.48	0.474	0.007	0.165	0.006	2498	47.2
H4	48	39	87	0.45	12.58	0.46	0.526	0.006	0.177	0.004	2783	53.0
H5	177	193	429	0.45	8.47	0.38	0.404	0.005	0.154	0.004	2132	31.3
H6	270	30	756	0.04	8.77	0.54	0.395	0.007	0.158	0.006	2071	42.2
H7	173	113	543	0.21	6.42	0.20	0.325	0.003	0.147	0.003	1727	18.6
H8	116	65	280	0.23	10.14	0.52	0.441	0.007	0.163	0.005	2307	46.3
H9	62	49	119	0.41	12.09	0.41	0.504	0.006	0.178	0.004	2626	43.0
H10	154	97	393	0.25	8.63	0.27	0.415	0.005	0.156	0.003	2189	29.0
H11	358	42	2032	0.02	2.96	0.16	0.197	0.003	0.110	0.004	1115	18.2
H12	411	1102	2960	0.37	2.22	0.10	0.140	0.002	0.121	0.003	789	8.9
OVU augen orthogneiss (FP285), Yunzhongshan Massif (N38°371598'/E112°27.847')											
F1	30	38	74	0.51	7.34	0.35	0.411	0.008	0.134	0.005	2240	47.1
F2	48	79	114	0.69	7.14	0.43	0.382	0.007	0.131	0.006	2083	43.8
F3	20	35	45	0.77	7.47	0.78	0.384	0.011	0.142	0.012	2061	71.0
F4	41	44	93	0.48	7.21	0.29	0.413	0.006	0.131	0.004	2261	38.1
F5	22	30	52	0.58	7.06	0.34	0.387	0.007	0.137	0.005	2092	43.1
F6	95	116	254	0.46	6.30	0.41	0.357	0.008	0.131	0.006	1940	44.6
F7	14	22	31	0.72	7.10	0.37	0.393	0.007	0.136	0.006	2125	42.4
F8	21	31	48	0.65	6.88	0.33	0.388	0.007	0.131	0.005	2113	40.6
F9	19	27	47	0.57	6.80	0.37	0.376	0.008	0.132	0.006	2042	49.3
F10	56	47	130	0.36	7.32	0.27	0.393	0.006	0.141	0.004	2113	33.3
F11	23	27	54	0.50	7.79	0.52	0.415	0.010	0.131	0.006	2269	60.7
F12	45	53	100	0.53	7.03	0.29	0.402	0.006	0.133	0.004	2192	37.8
Biotite-bearing granite (FP351), Lüliangshan Massif (N38°44.314'/E111°15.560')											
G1	126	450	285	1.58	5.06	0.16	0.332	0.003	0.113	0.002	1850	18.5
G2	97	91	347	0.26	4.76	0.37	0.276	0.007	0.119	0.006	1524	39.6
G3	72	287	153	1.87	5.71	0.55	0.345	0.011	0.119	0.008	1906	64.5
G4	145	534	361	1.48	4.41	0.24	0.289	0.005	0.118	0.005	1601	27.9
G5	180	352	621	0.57	3.81	0.24	0.253	0.006	0.111	0.005	1418	31.9
G6	80	122	212	0.58	5.39	0.39	0.271	0.007	0.150	0.007	1440	39.7
G7	97	366	207	1.76	4.85	0.19	0.331	0.004	0.112	0.003	1843	25.0
G8	88	266	193	1.38	5.46	0.34	0.347	0.007	0.122	0.005	1909	40.3
G9	154	405	364	1.11	5.32	0.18	0.349	0.004	0.114	0.002	1941	21.2
G10	104	406	235	1.73	4.63	0.18	0.314	0.004	0.112	0.003	1748	21.1
G11	125	368	292	1.26	5.29	0.19	0.343	0.004	0.114	0.003	1910	23.7
G12	86	339	188	1.80	5.34	0.23	0.336	0.004	0.120	0.003	1851	25.5

the TTG basement corresponds to the crystallization age of the porphyric granite before its deformation. This age is consistent with numerous ages at c. 2100 Ma obtained from similar intrusive plutonic bodies within the metavolcano-sedimentary series and its TTG gneissic and migmatitic basement in the Hengshan–Wutaishan–Fuping massifs (Fig. 13) (Wilde *et al.* 1998, 2004a, 2005; Zhao *et al.* 2002).

Several geochronological results provide ages around 1880 Ma: the 1887 ± 8, 1881 ± 10 and 1872 ± 12 Ma U–Th/Pb monazite ages and the 1873 ± 31 Ma U–Pb zircon age. Within Al-rich metasedimentary rocks, monazite growth is commonly linked to staurolite-in or aluminosilicate-in reactions because of allanite breakdown during prograde metamorphism (Smith & Barreiro 1990; Kingsbury *et al.* 1993; Bingen *et al.* 1996; Ferry 2000; Wing *et al.* 2003; Khon & Malloy 2004). The U–Th/Pb monazite ages are reported from metamorphic overgrowths within rocks that experienced the amphibolite grade metamorphism. Hence, we suggest that they correspond to the time of the main metamorphic event (M_1) that is coeval with the nappe stacking and thickening during the D_1 event. These results are consistent with metamorphic zircon overgrowth rims reported from a tonalitic gneiss that yielded a weighted mean $^{207}Pb/^{206}Pb$ age of 1872 ± 7 Ma interpreted as the approximate time of peak metamorphism in the Lüliang Complex (Zhao *et al.* 2008).

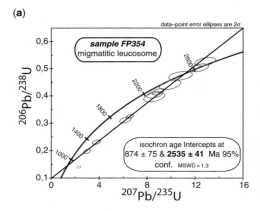
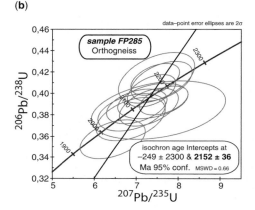
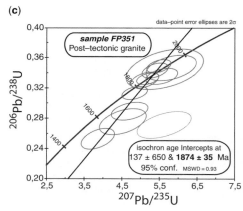

Fig. 12. Concordia plots for U–Pb LA-ICP-MS analyses of zircons extracted from (a) a leucosome within a basement migmatite, NW of Huijiazhuang (N37°45.254′/111°26.342′), (b) an OVU augen orthogneiss, Yunzhongshan (N38°37.598′/E112°27.847′), (c) a biotite-bearing granite, Lüliangshan (N38°44.314′/E111°15.560′). All errors are quoted at 95% confidence level and error ellipses are plotted at 2σ. See text for explanations.

Similarly, the U–Th/Pb EPMA monazite age of 1880 ± 9 Ma reported by Liu et al. (2006) (Fig. 9) is in agreement with our data. The amphibolite facies metamorphism has also been dated at 1880–1890 Ma in the adjacent Hengshan and Wutaishan massifs (Trap et al. 2007). As a further example, along the Longquanguan Thrust, a strongly sheared orthogneiss yield a U–Th–Pb EPMA dating monazite age of 1877 ± 11 Ma, interpreted as the age of the ductile shearing (Zhao et al. 2006). The U–Pb zircon age of 1873 ± 31 Ma reported for a post-tectonic granite (sample FP351) is difficult to interpret as there is no textural constraint. The sample FP351 is a post-tectonic granite that does not show D_1 or D_2 deformation and the age at 1873 ± 31 Ma might be an inherited age of the main metamorphic event. However, the error in the age is ±31 Ma, so the granite can statistically be as young as 1842 Ma, which is much younger than the interpreted age range for peak metamorphism. If so, it would mean that the D_2 event is older than 1842 Ma. Therefore the age of 1873 ± 31 Ma remains difficult to interpret since all the post-tectonic granites are well-constrained around 1790–1815 Ma (Zhao et al. 2008). In spite of this, our results do not allow us to constrain the age of the D_2 deformation. However, it is likely that D_2 occurred between 1873 and 1815 Ma, which is the youngest D_1 estimate and the oldest post-tectonic granite, respectively.

In the sample FP313, monazite cores are dated at 1942 ± 8 Ma. Similar ages, around 1940 Ma, are reported in the Khondalite Belt that represents an important tectonic feature within the Western Block of the North China Craton (Zhao et al. 2005; Santosh et al. 2007). It is thus likely that the ages of c. 1940 Ma recorded by the monazite cores are that of igneous grains reworked as detrital minerals in the protoliths of the khondalites. This result is consistent with those of Wan et al. (2006) who identified SHRIMP U–Pb ages of 1940–1930 Ma in detrital zircon grains from the Khondalite Belt located c. 150 km north of the Lüliang Massif. An age of 1913 ± 10 Ma has been calculated from monazite rims of the sample FP313. It can be interpreted as that of a metamorphic event since it was calculated from overgrowths. Santosh et al. (2007) also reported a SHRIMP U–Pb zircon age at 1919 ± 10 Ma from a sapphirine-bearing Mg–Al granulite in the Khondalite Belt near the northern margin of the NCC and assigned it as the time of an UHT metamorphism.

The sample FP359 yields an age of 1919 ± 12 Ma calculated for monazites cores. Similarly, a high thermal event is associated with the widespread emplacement of mafic dykes at c. 1915 Ma within the Henghan Massif (Peng et al. 2005; Kröner

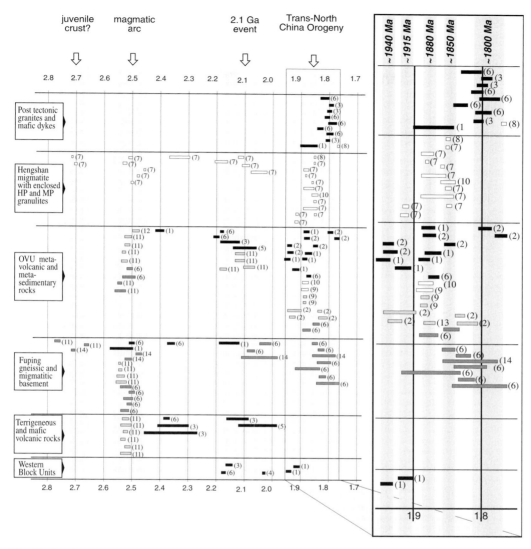

Fig. 13. Geochronological dataset of the Lüliang Massif (black boxes), Hengshan Massif (white boxes), Wutaishan Massif (pale-grey boxes), Fuping Massif (dark-grey boxes). (1): This study; (2) Liu et al. 2006; (3) Geng et al. 2003, 2004; (4) Wan et al. 2000; (5) Yu et al. 1997a; (6) Zhao et al. 2002, 2008; (7) Kröner et al. 2005, 2006; (8) Chang et al. 1999; (9) Trap et al. 2007; (10) Faure et al. 2007; (11) Wilde et al. 1997, 1998, 2004a, b, 2005; (12) Cawood et al. 1998; (13) Zhao et al. 2006; (14) Guan et al. 2002.

et al. 2006). Due to the scarcity of such ages within the Trans-North China Belt (Fig. 13), they are difficult to interpret in terms of a tectono-metamorphic event. Nevertheless, the c. 1915 Ma ages reported in two metapelites of the Lüliang Massif represent a thermal or tectonic-metamorphic event older than the main regional metamorphism dated at 1880 Ma.

In sample FP360, monazite cores yield an U–Th/Pb age of 1947 ± 10 Ma (Figs 9 & 13). Liu et al. (2006) reported U–Th/Pb monazite ages of 1942 ± 8 Ma and 1945 ± 6 Ma, from metapelites of the same locality in the southeastern part of the Lüliang Massif and interpreted these ages as recording the crystallization of igneous monazite. Such ages at c. 1945 Ma are still not documented in the Hengshan-Wutaishan-Fuping area. The c. 1940 Ma grains probably come from the Khondalite belt, and were incorporated in the sedimentary protoliths of the metapelites before the development of the Trans-North China orogeny.

Discussion

Correlation between the Lüliang Massif and the Hengshan–Wutaishan–Fuping massifs

Although the Lüliang Massif is considered to belong to the TNCB, the recent models that try to explain the tectonic evolution of the TNCB are based only on the results obtained in the Hengshan–Wutaishan–Fuping massifs, ignoring the Lüliang Massif (Zhao et al. 2001a, 2004; Kusky & Li 2003). However, many of the problems in understanding the tectonic evolution of the TNCB might be partly resolved by studying the Lüliang Massif. In particular, the comparison of lithologies, deformation sequence and age constraints between the Lüliang Massif and the Hengshan–Wutaishan–Fuping massifs presents numerous similarities (Fig. 14).

The Lüliang Massif exhibits the same stack of nappes, as previously recognized in the Hengshan–Wutaishan domain (Trap et al. 2007). The greenschist facies Lüliang klippe has its counterpart within the Wutaishan, represented by the Low Grade Mafic Unit (LGMU) that also contains mafic magmatic, volcanic-clastic, serpentinites and sedimentary rocks (Faure et al. 2007; Trap et al. 2007; Fig. 14). These klippes are the remnants of a larger greenschist facies nappe that roots along the Trans-North China Suture where the same lithologies crop out. The MORB-like geochemical signature of some metabasalts of the LGMU, within the Wutai klippe, is considered as evidence for oceanic crust (Wang et al. 2004; Polat et al. 2005; Polat & Kusky 2007). This oceanic area that separated the Western Block from the Fuping Block has been called the Lüliang Ocean (Faure et al. 2007; Trap et al. 2006, 2007). The structural analysis documents the southeastward thrust of the TMU and LGMU along a synmetamorphic shear zone represented by the Upper Lüliang Thrust (ULT) and the Upper Wutai Thrust (UWT) in the Lüliang Massif and Wutaishan, respectively (Trap et al. 2007; this study).

The underlying para-autochonous domain overthrust by the LGMU thrust is represented by the OVU and its underlying basement interpreted as the westward extension of the Fuping Massif, called the Fuping Block. This domain can be subdivided into several crustal slices separated by low-angle ductile shear zones with a top-to-the SE kinematics. The extension of these tectonic contacts at depth is speculative, thus they are represented as dotted-lines in the general cross-sections (Fig. 14). The Fuping Block appears as deformed by several crustal slices. However, we propose that the thrust faults observed within the Lüliang and Wutaishan massifs may branch into a single basal thrust fault, that crops out to the east as the Longquanguan Thrust (LQGT; Fig. 14).

Al-rich metapelites located at the southeastern edge of Lüliangshan, near Lingshi, do not belong to the Western Block as commonly documented in the literature (Zhao et al. 2008; Liu et al. 2006; Xia et al. 2006) but belong to the OVU. According to the general NE–SW trend of structures, we can easily correlate the metapelite and marble that crop out in the southeastern part of the Lüliang Massif, near Lingshi, with the same rocks that crop out in the southernmost part of the Wutaishan Massif (Fig. 14). In this view, the Longquanguan Thrust might extend southwestward east of Lingshi. Zhao et al. (2008) consider that the Yunzhongshan is made only of granitic gneiss, named the Yunzhongshan TTG gneiss, intruded by the post-tectonic Tangershang granite. On the contrary, our study demonstrates that the Yunzhongshan exhibits the same lithologies and structural pattern as those viewed in the Lüliangshan but also in the Wutaishan.

The OVU is thrust over the Fuping Massif through a main ductile shear zone named the Longquanguan Thrust that represents the frontal thrust of the Trans-North China Belt in the study area. By comparison with the Lüliang, Hengshan and Wutaishan massifs, the Fuping Massif displays a very different architecture made of east–west trending dome and basins developed during an earlier tectonic and thermal event at c. 2100 Ma (Trap et al. 2008).

Within the Lüliang Massif, migmatites are older than 2.1 Ga since they are intruded by the 2150 ± 34 Ma orthogneiss, in agreement with the migmatitic leucosome dated at 2535 ± 41 Ma, and are thus considered to represent the basement upon which the OVU volcano-sedimentary rocks deposited, before both were subsequently thrust and folded southeastward, during the collision at 1890–1870 Ma. Conversely, in the Hengshan, migmatites are widespread but as they are dated around 1850 Ma (Faure et al. 2007; Trap et al. 2007), they are coeval with a thermal doming only visible in this massif, responsible for the exhumation of the HP granulites (O'Brien et al. 2005; Zhao et al. 2001b), not observed in the Lüliang Massif.

In agreement with previous work (SBGMR 1989), we correlate the late Palaeoproterozoic Yejishan group of the Lüliang Massif with the Hutuo Group of the Wutaishan Massif. They both consist of siliciclastic sedimentary rocks derived from the erosion of rocks formed during the early Trans-North China Belt tectonic phase and accumulated in a series of sedimentary basins. In the Wutaishan domain, the sediments are interpreted as foreland basins (Bai et al. 1992; Wu & Zhong 1998; Wilde et al. 2004b). In the Lüliang Massif, the Yejishan group is preserved in isolated and

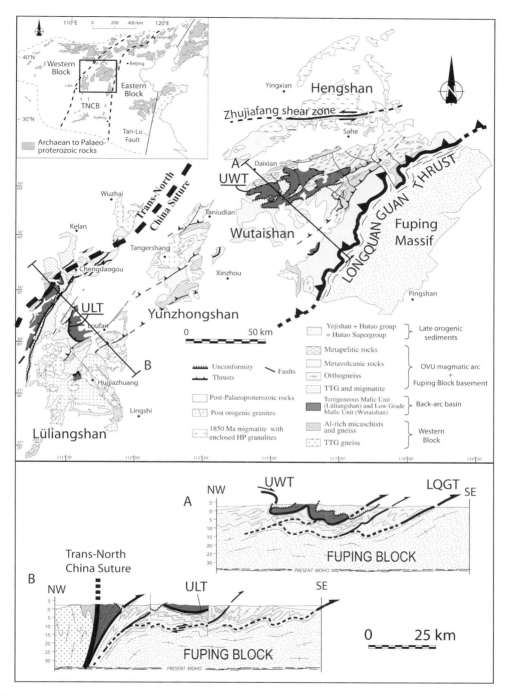

Fig. 14. Structural map showing the correlation between the Lüliang, Yunzhongshan, Hengshan, Wutaishan and Fuping massifs. Crustal-scale interpretative cross-sections of the Trans-North China Belt showing its bulk architecture and the dominant flat-lying structures. Note that the structure of the Fuping Massif is not represented here.

often faulted strips that could be interpreted as intra-mountain basins. Both in the Lüliang and Wutaishan massifs, the Hutuo sediments were subsequently deformed by folds associated with an axial planar cleavage and thrust during D_2 (Trap et al. 2007; this study). The age of the Hutuo Supergroup is presently unknown but since it unconformably covers the amphibolite facies OVU rocks in which metamorphism is dated at 1880 Ma and because it is intruded by the Luchaogou pluton dated at 1807 ± 10 Ma, the age of the Hutuo Supergroup must be bracketed between these two dates.

Tectonic evolution

The c. 2500 Ma ages represent the first and most widely developed arc magmatism in the North China Craton that resulted in large TTG intrusion within the Fuping Block and the deposition of the OVU. However, this Neoarchaean–Early-Palaeoproterozoic event is still poorly understood.

The c. 2100 Ma event is largely reported within the Lüliang Massif and within the other massifs of the North China Craton as well. These ages reflect a second arc-related magmatism represented by the granitoids that intrudes both the OVU and its underlying TTG and gneissic migmatite basement (Faure et al. 2007; Zhao et al. 2008; this study). Within the Lüliang Massif, some inherited ages are also related to the 2.1 Ga Khondalite belt of the Western Block. The formation of the Khondalite belt, the second-arc magmatism, and the intense partial melting in the Fuping Massif are certainly linked but not yet understood. The significance of the 2.1 Ga event in the Trans-North China Belt is still speculative. Faure et al. (2007) propose that the 2.1 Ga event recorded in the Fuping–Wutaishan–Hengshan massifs could be interpreted as an older west-directed subduction of the Eastern Block below the Fuping Block.

Around 1890–1870 Ma, regional amphibolite facies metamorphism developed during the D_1 event. The nappe stacking associated with eastward or southeastward displacement is the consequence of the Western and Fuping block collision after the westward subduction and subsequent closure of the Lüliang Ocean. The last deformation event is the D_2 event that might represent the younger increment of the same deformation continuum, although D_1 and D_2 are separated by the deposition of the Yejishan Group. The syn-D_1 fabric remains well-preserved but F_2 folds are responsible for the final architecture of the Lüliang Massif.

Concluding remarks

The structural investigation of the Lüliang Massif documents polyphase, SE-directed synmetamorphic thrust tectonics, and the geochronological results, in agreement with previous ones, support a Palaeoproterozoic age, not a Neoarchaean one for the formation of the TNCB. The Lüliang Massif exhibits a peculiar TMU that can be compared with the LGMU defined to the NE in Wutaishan (Faure et al. 2007; Trap et al. 2007). These rocks, presently preserved as thrust sheets in the Wutai klippe and Lüliang klippe, can be compared with the ophiolitic nappes that characterize the modern collisional orogens. The Trans-North China Suture, located in the western part of the Lüliang Massif, exposes the root zone of the TMU and corresponds to the remnant of an oceanic basin called the Lüliang Ocean. In agreement with previous works (e.g. Faure et al. 2007; Trap et al. 2007, 2008), the underlying eastern continental rocks are attributed to the Fuping Block, which crops out more to the east in the Fuping Massif. Thus, a west-directed oceanic subduction of the Lüliang Ocean, followed by the continental subduction of the Fuping Block and finally its collision with the Western (or Ordos) Block accounts well for the bulk architecture and the structural and metamorphic evolution of the Lüliang, Massif which appears to represent the innermost massif of the TNCB. A complete discussion of the entire TNCB is beyond the scope of this chapter.

This work was supported by National Science Foundation of China grant no 40472116 and the contribution of the UMR 6113 (CNRS). G. Zhao is thanked for allowing us to use his unpublished geochronological work, in review during the drafting of the manuscript.

References

BAI, J., WANG, R. Z. & GUO, J. J. 1992. *The Major Geologic Events of Early Precambrian and their Dating in Wutaishan Region*. Geological Publishing House, Beijing.

BINGEN, B., DEMAIFFE, D. & HERTOGEN, J. 1996. Redistribution of rare earth elements, thorium, and uranium over accessory minerals in the course of amphibolite to granulite facies metamorphism: the role of apatite and monazite in orthogneisses from southwestern Norway. *Geochimica et Cosmochimica Acta*, **60**, 1341–1354.

BLACK, L. P., KAMO, S. L., ALLEN, C. M., ALEINIKOFF, J. N., DAVIS, D. W., KORSCH, R. J. & FOUDOULIS, C. 2003. TEMORA 1: a new zircon standard for Phanerozoic U–Pb geochronology. *Chemical Geology*, **200**, 155–170.

BLACK, L. P., KAMO, S. L. ET AL. 2004. Improved $^{206}Pb/^{238}U$ microprobe geochronology by the monitoring of a trace-element related matrix effect; SHRIMP, ID-TIMS, ELA-ICP-MS, and oxygen isotope documentation for a series of zircon standards. *Chemical Geology*, **205**, 115–140.

CARRERAS, J., DRUGUET, E. & GRIERA, A. 2005. Shear zone-related folds. *Journal of Structural Geology*, **27**, 1229–1251.

CAWOOD, P. A., WILDE, S. A., WANG, K. & NEMCHIN, A. A. 1998. Application of integrated field mapping and SHRIMP U/Pb geochronology to subdivision of the Precambrian in China: constraints from the Wutaishan. Abstract-vol. ICOG-9, Beijing 1998. *Chinese Science Bulletin*, **43**, 17.

CHANG, X. Y., CHEN, Y. D. & ZHU, B. Q. 1999. U–Pb zircon isotope age of metabasites from Hengshan grey gneiss. *Acta Mineralogica Sinica*, **19**, 263–266 (in Chinese with English abstract).

COCHERIE, A. & ALBAREDE, F. 2001. An improved U–Th–Pb age calculation for electron microprobe dating of monazite. *Geochimica et Cosmochimica Acta*, **65**, 4509–4522.

COCHERIE, A., LEGENDRE, O. & PEUCAT, J. J. 1998. Geochronology of polygenetic monazites constrained by *in situ* electron microprobe Th–U–total lead determination: implications for lead behaviour in monazite. *Geochimica et Cosmochimica Acta*, **62**, 2475–2497.

FAURE, M., TRAP, P., LIN, W., MONIÉ, P. & BRUGUIER, O. 2007. The formation of the North China Craton by two Palaeoproterozoic continental collisions in Lüliang–Hengshan–Wutaishan–Fuping massifs. *Episodes*, **30**, 1–12.

FERRY, J. M. 2000. Patterns of mineral occurrence in metamorphic rocks. *American Mineralogist*, **85**, 1573–1588.

GENG, Y. S., WAN, Y. S., SHEN, Q. H., LI, H. M. & ZHANG, R. X. 2000. Chronological framework of the early Precambrian important events in the Lüliang area, Shanxi Province. *Acta Geologica Sinica*, **74**, 216–223 (in Chinese with English abstract).

GENG, Y. S., WAN, Y. S. & YANG, C. H. 2003. The Palaeoproterozoic rift-type volcanism in Lüliangshan Area, Shanxi Province, and its geological significance. *Acta Geoscientica Sinica*, **24**, 97–104 (in Chinese with English abstract).

GENG, Y. S., YANG, C. H., SONG, B. & WANG, Y. S. 2004. Post-orogenic granites with an age of 1800 Ma in Lüliang area, North China Craton: constraints from isotopic geochronology and geochemistry. *Geological Journal of Chinese University*, **10**, 477–487 (in Chinese with English abstract).

GUAN, H., SUN, M., WILDE, S. A., ZHOU, X. H. & ZHAI, M. G. 2002. SHRIMP U–Pb zircon geochronology of the Fuping Complex: implications for formation and assembly of the North China craton. *Precambrian Research*, **113**, 1–18.

KHON, M. J. & MALLOY, M. A. 2004. Formation of monazite via prograde metamorphic reactions among common silicates: implications for age determinations. *Geochimica et Cosmochimica Acta*, **68**, 101–113.

KINGSBURY, J. A., MILLER, C. F., WOODEN, J. L. & HARRISON, T. M. 1993. Monazite paragenesis and U–Pb systematics in rocks of the eastern Mojave Desert, California, USA; implications for thermochronometry. *Chemical Geology*, **110**, 147–167.

KRÖNER, A., WILDE, S. A., LI, J. H. & WANG, K. Y. 2005. Age and evolution of a late Archaean to early Palaeoproterozoic upper to lower crustal section in the Wutaishan/Hengshan/Fuping terrain of northern China. *Journal of Asian Earth Sciences*, **24**, 577–596.

KRÖNER, A., WILDE, S. A. ET AL. 2006. Zircon geochronology and metamorphic evolution of mafic dykes in the Hengshan Complex of northern China: evidence for late Palaeoproterozoic extension and subsequent high-pressure metamorphism in the North China Craton. *Precambrian Research*, **146**, 45–67.

KUSKY, T. M. & LI, J. H. 2003. Palaeoproterozoic tectonic evolution of the North China Craton. *Journal of Asian Earth Sciences*, **22**, 23–40.

LI, Z. H. & ZHAI, A. M. 1984. Regional metamorphism of Palaeoproterozoic metapelitic series, Lüliang Areas. *Tianjin Institute of Geology and Mineral Resources Report*, **11**, 83–108.

LI, J. H. & KUSKY, T. M. 2007. A Late Archaean foreland fold and thrust belt in the North China Craton: implications for early collisional tectonics. *Gondwana Research*, **12**, 47–66.

LIU, S. W., ZHAO, G. C. ET AL. 2006. Th–U–Pb monazite geochronology of the Lüliang and Wutai Complexes: constraints on the tectonothermal evolution of the Trans-North China Orogen. *Precambrian Research*, **148**, 205–225.

LUDWIG, K. R. 2003. *Users manual for ISOPLOT/EX, version 3. A geochronological toolkit for Microsoft Excel*. Berkeley Geochronology Center, Special Publication, **Vol. 4**.

O'BRIEN, P. J., WALTE, N. & LI, J. H. 2005. The petrology of two distinct Palaeoproterozoic granulite types in the Hengshan Mts., North China Craton, and tectonic implications. *Journal of Asian Earth Sciences*, **24**, 615–627.

PENG, P., ZHAI, M. G., ZHANG, H. F. & GUO, J. H. 2005. Geochronological constraints on the palaeoproterozoic evolution of the North China craton: SHRIMP zircon ages of different types of mafic dykes. *International Geological Revue*, **47**, 492–508.

POLAT, A. & KUSKY, T. M. 2007. Discussion of geochemistry of the Late Archean (c. 2.55–2.50 Ga) volcanic and ophiolitic rocks in the Wutaishan Greenstone Belt, Central Orogenic Belt, North China Craton: implications for geodynamic setting and continental growth, reply to G. ZHAO & A. KRÖNER. *Geological Society American Bulletin*, **119**, 490–492.

POLAT, A., KUSKY, T. M., LI, J. H., FRYER, B., KERRICH, R. & PATRICK, K. 2005. Geochemistry of NeoArchaean (c. 2.55–2.50) volcanic and ophiolitic rocks in the Wutaishan greenstone belt, central orogenic belt, North China craton: implications for geodynamic setting and continental growth. *Geological Society American Bulletin*, **117**, 1387–1399.

SANTOSH, M., WILDE, S. A. & LI, J. H. 2007. Timing of Palaeoproterozoic ultrahigh-temperature metamorphism in the North China Craton: evidence from SHRIMP U–Pb zircon geochronology. *Precambrian Research*, **159**, 178–196.

SBGMR (Shanxi Bureau of Geology and Mineral Resources) 1989. *Regional Geology of Shanxi Province*. Geological Publishing House, Beijing.

SMITH, H. A. & BARREIRO, B. 1990. Monazite U–Pb dating of staurolite grade metamorphism in pelitic schists. *Contribution to Mineralogy and Petrology*, **105**, 602–615.

TRAP, P., FAURE, M., LIN, W., LE BRETON, N., BRUGUIER, O. & MONIÉ, P. 2006. Structural, metamorphic and geochronological works in the Hengshan-Wutaishan-Fuping massifs and correlation with the Lüliang massif: implication for the tectonic evolution of the Trans-North China Belt. *2006 IAGR Annual Convention and International Conference, Hong Kong, Abstracts*, pp. 55–56.

TRAP, P., FAURE, M., LIN, W. & MONIÉ, P. 2007. Late Palaeoproterozoic (1900–1800 Ma) nappe stacking and polyphase deformation in the Hengshan-Wuatishan area: implication for the understanding of the Trans-North China Belt, North China Craton. *Precambrian Research*, **156**, 85–106.

TRAP, P., FAURE, M., LIN, W., BRUGUIER, O. & MONIÉ, P. 2008. Contrasted tectonic styles for the paleoproterozoic evolution of the North China Craton. Evidence for a c. 2.1 Ga thermal and tectonic event in the Fuping Massif. *Journal of Structural Geology*, **30**, 1109–1125.

WAN, Y. S., GENG, Y. S., SHEN, Q. H. & ZHANG, R. X. 2000. Khondalite series-geochronology and geochemistry of the Jiehekou Group in the Lüliang area, Shanxi Province. *Acta Petrologica Sinica*, **16**, 49–58.

WAN, Y. S., SONG, B. *ET AL*. 2006. SHRIMP U–Pb zircon geochronology of Palaeoproterozoic metasedimentary rocks in the North China Craton: evidence for a major Late Palaeoproterozoic tectono-thermal event. *Precambrian Research*, **149**, 249–271.

WANG, Z., WILDE, S. A., WANG, K. & YU, L. 2004. A MORB-arc basalt-adakite association in the 2.5 Ga Wutai greenstone belt: late Archaean magmatism and crustal growth in the North China Craton. *Precambrian Research*, **131**, 323–343.

WIEDENBECK, M., ALLE, P. *ET AL*. 1995. 3 Natural Zircon Standards for U–Th–Pb, Lu–Hf, Trace-Element and REE Analyses. *Geostandards Newsletter*, **19**, 1–23.

WILDE, S. A., CAWOOD, P., WANG, K. Y. & NEMCHIN, A. 1997. The relationship and timing of granitoid evolution with respect to felsic volcanism in the Wutai Complex, North China Craton. Proceedings of the 30th International Geological Congress, Beijing. *Precambrian Geology Metamorphism and Petrology*, **17**, 75–88.

WILDE, S. A., CAWOOD, P. A., WANG, K. Y. & NEMCHIN, A. 1998. SHRIMP U–Pb zircon dating of granites and gneisses in the Taihangshan–Wutaishan area: implications for the timing of crustal growth in the North China Craton. *Chinese Sciences Bulletin*, **43**, 144–145.

WILDE, S. A., CAWOOD, P. A., WANG, K. Y. & NEMCHIN, A. A. 2005. Granitoid evolution in the Late Archaean Wutai Complex, North China Craton. *Journal of Asian Earth Sciences*, **24**, 597–613.

WILDE, S. A., CAWOOD, P. A., WANG, K. Y., NEMCHIN, A. & ZHAO, G. C. 2004a. Determining Precambrian crustal evolution in China: a case study from Wutaishan, Shanxi Province, demonstrating the application of precise SHRIMP U–Pb geochronology. *In*: MALPAS, J., FLETCHER, C. J. N., ALI, J. R. & AITCHISON, J. C. (eds) *Aspects of the Tectonic Evolution of China*, **Vol. 226**. Geological Society of London, 5–25 (special publication).

WILDE, S. A., ZHAO, G. C., WANG, K. Y. & SUN, M. 2004b. First SHRIMP zircon U–Pb ages for the Hutuo Group in Wutaishan: further evidence for amalgamation of North China Craton. *Chinese Sciences Bulletin*, **49**, 83–90.

WING, B. N., FERRY, J. M. & HARRISON, T. M. 2003. Prograde destruction and formation of monazite and allanite during contact and regional metamorphism of pelites: petrology and geochronology. *Contribution to Mineralogy and Petrology*, **145**, 228–250.

WU, C. H. & ZHONG, C. T. 1998. Early Proterozoic SW–NE collision model for the central part of the North China Craton: implications for tectonic regime of the Khondalite downward into lower crust in Jin-Meng high grade region. *Progress in Precambrian Research*, **21**, 28–50 (in Chinese with English abstract).

XIA, X. P., SUN, M., ZHAO, G. C. & LUO, Y. 2006. LA-ICP-MS U–Pb geochronology of detrital zircons from the Jining Complex, North China Craton and its tectonic significance. *Precambrian Research*, **144**, 199–212.

YANG, Z., CHENG, Y. & WANG, H. 1986. *The Geology of China*. Clarendon Press, Oxford.

YU, J. H. & LI, H. M. 1997a. Ages of the Lüliang Group and its main metamorphism in the Lüliang Mountains, Shanxi: evidence from single-grain zircon U–Pb ages. *Geological Revue*, **43**, 403–408.

YU, J. H., WANG, D. Z. & WANG, C. Y. 1997b. Geochemical characteristics and petrogenesis of the early Proterozoic bimodal volcanic rocks from the Lüliang Group. *Acta Petroligica Sinica*, **13**, 59–70.

YU, J. H., WANG, C. Y., LAI, M. Y., CHEN, S. Q. & LU, B. Q. 1999. Re-division of the metamorphic facies zonation of Lüliang Group in the Shanxi province and its significance. *Geological Journal of China Universities*, **5**, 65–74.

ZHAO, G. C., WILDE, S. A., CAWOOD, P. A. & SUN, M. 2001a. Archaean blocks and their boundaries in the North China Craton: lithological, geochemical, structural and P–T path constraints and tectonic evolution. *Precambrian Research*, **107**, 45–73.

ZHAO, G. C., CAWOOD, P. A., WILDE, S. A. & LU, L. Z. 2001b. High-pressure granulites (retrograded eclogites) from the Hengshan Complex North China Craton: petrology and tectonic implications. *Journal of Petrology*, **42**, 1141–1170.

ZHAO, G. C., WILDE, S. A., CAWOOD, P. A. & SUN, M. 2002. SHRIMP U–Pb zircon ages of the Fuping Complex: implications for late Archaean to Palaeoproterozoic accretion and assembly of the North China Craton. *American Journal of Sciences*, **302**, 191–226.

ZHAO, G. C., SUN, M., WILDE, S. A. & GUO, J. H. 2004. Late Archaean to Palaeoproterozoic evolution of the Trans-North China Orogen: insights from synthesis of existing data from the Hengshan–Wutai–Fuping belt. *In*: MALPAS, J., FLETCHER, C. J. N., ALI, J. R. & AITCHISON, J. C. (eds) *Aspects of the Tectonic Evolution of China*. Geological Society, London, Special Publications, **226**, 27–55.

ZHAO, G. C., SUN, M., WILDE, S. A. & LI, S. Z. 2005. Late Archaean to Palaeoproterozoic evolution of the North China Craton: key issues revisited. *Precambrian Research*, **136**, 177–202.

ZHAO, L., ZHANG, J. J. & LIU, S. W. 2006. Syn-deformational granites of the Longquanguan ductile shear zone and their monazite electronic microprobe dating. *Acta Petrologica et Mineralogica*, **25**, 210–218.

Zhao, G. C., Kröner, A. *et al.* 2007. Lithotectonic elements and geological events in the Hengshan–Wutai–Fuping belt: a synthesis and implications for the evolution of the Trans-North China Orogen. *Geological Magazine*, **144**, 753–775.

Zhao, G. C., Wilde, S. A., Sun, M., Li, S. Z., Li, X. P. & Zhang, J. 2008. SHRIMP U–Pb zircon ages of granitoid rocks in the Lüliang Complex: implications for the accretion and evolution of the Trans-North China Orogen. *Precambrian Research*, **160**, 213–226.

Palaeoproterozoic to Eoarchaean crustal growth in southern Siberia: a Nd-isotope synthesis

DMITRY P. GLADKOCHUB[1]*, TATIANA V. DONSKAYA[1], STEVEN M. REDDY[2], ULRIKE POLLER[3], TAMARA B. BAYANOVA[4], ANATOLIY M. MAZUKABZOV[1], SERGEI DRIL[5], WOLFGANG TODT[3] & SERGEI A. PISAREVSKY[6]

[1]*Institute of the Earth's Crust, Siberian Branch of the Russian Academy of Sciences, 128 Lermontov Street, 664033 Irkutsk, Russia*

[2]*The Institute for Geoscience Research, Department of Applied Geology, Curtin University of Technology Box U 1987, Perth, WA 6845, Australia*

[3]*Max-Plank Institute für Chemie, Postfach 3060, 55020 Mainz, Germany*

[4]*Geological Institute of Kola Scientific Centre of the Russian Academy of Sciences, 14 Fersmana, 184209, Apatity, Russia*

[5]*Institute of Geochemistry, Siberian Branch of the Russian Academy of Sciences, 1 Favorsky Street, 664033 Irkutsk, Russia*

[6]*School of Geosciences, University of Edinburgh, King's Buildings, West Mains Road, Edinburgh EH9 3JW, UK*

Corresponding author (e-mail: dima@crust.irk.ru)

Abstract: Nd-isotope analyses from 114 rock samples are reported from the southern part of the Siberian craton to establish a first-order crustal formation scheme for the region. The Nd-isotope data show considerable variability within and among different cratonic units. In many cases this variability reflects differing degrees of mixing between juvenile and older (up to Eoarchaean) crustal components. The fragments of Palaeoproterozoic juvenile crust within the studied segment of the Siberian craton margin have Nd-model ages of c. 2.0–2.3 Ga. Voluminous Palaeoproterozoic granites (c. 1.85 Ga) were intruded into cratonic fragments and suture zones. These granites mark the stabilization of the southern Siberian craton. The complexity in the Nd data indicate a long history of crustal development, extending from the Eoarchaean to the Palaeoproterozoic eras, which is interpreted to reflect the amalgamation of distinct Archaean crustal fragments, with differing histories, during Palaeoproterozoic accretion at 1.9–2.0 Ga and subsequent cratonic stabilization at 1.85 Ga. Such a model temporally coincides with important orogenic events on nearly every continent and suggests that the Siberian craton participated in the formation of a Palaeoproterozoic supercontinent at around 1.9 Ga.

The Siberian craton is one of the largest Early Precambrian lithospheric units of Northern Eurasia, extending for a distance of over 2500 km from the Arctic Ocean in the north to Lake Baikal in the south. The craton comprises five major provinces (superterranes) (Aldan, Anabar, Olenek, Stanovoy and Tungus) and two large-scale orogenic belts (Angara & Akitkan) (Fig. 1a) that are defined by lithological, structural and geophysical (gravity and magnetic anomaly) data (Rosen *et al.* 1994, 2005). The amalgamation of these blocks to form the Siberian craton was initially assumed to have taken place from c. 1.9–2.1 Ga (Condie & Rosen 1994). This has recently been confirmed by radiometric dating of granulite-facies metamorphic rocks and collision-related granites located at the margins of the different blocks (Nutman *et al.* 1992; Donskaya *et al.* 2002; Turkina *et al.* 2003, 2006; Poller *et al.* 2004, 2005; Rosen *et al.* 2005; Gladkochub *et al.* 2006*a*; Larin *et al.* 2006; Sal'nikova *et al.* 2007).

Recently, northern and eastern parts of the Siberian craton have been extensively studied to investigate the crustal-formation ages of terranes that make up these areas of the craton (Parfenov *et al.* 2001 and references therein). Geochronological data from igneous rocks and metamorphic complexes of southern Siberia (Fig. 1a) indicate tectonic

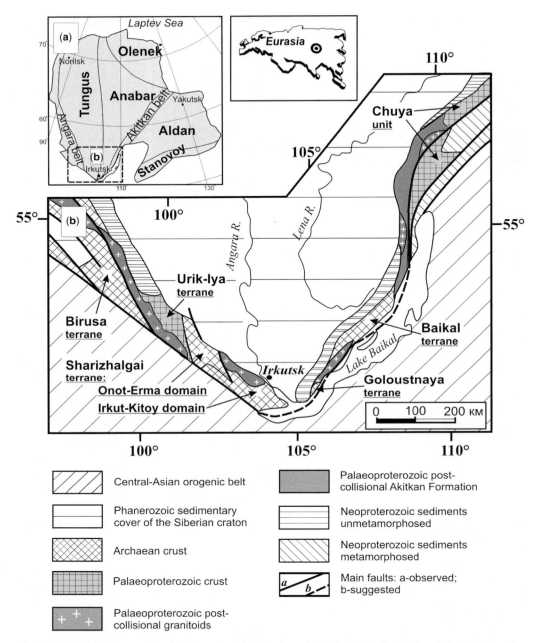

Fig. 1. The tectonic scheme of the Siberian craton after Rosen *et al.* (2005) (**a**) The distribution of the major tectonic units of the Siberian craton, (**b**) with detail of these tectonic units.

activity spanning 3.39–1.85 Ga (Neymark *et al.* 1998; Poller *et al.* 2005; Turkina 2005; Bibikova *et al.* 2006; Sal'nikova *et al.* 2007). Magmatic zircons as old as 3.39 Ga (Bibikova *et al.* 2006) and 3.39 Ga inherited zircon cores from Palaeoproterozoic granulites (Poller *et al.* 2005) indicate the likely presence of Early Archaean crust in the southern part of the Siberian craton. However, the interior structure and the variation in crustal-formation age in the different crustal blocks that form the southern part of the craton are currently poorly known. In this chapter, a synthesis of published geochronological data is combined with 114 new Sm–Nd analyses from the southern Siberian craton. These new data record the variation in Nd-isotopic compositions within the main crustal blocks, establish variations in crustal-formation age and allow us to assess whether the different crustal components were derived from isotopically juvenile material or older continental crustal components. This is the most comprehensive synthesis of crustal age distribution in southern Siberia so far, and provides an assessment of the continental growth record of the region during the Archaean and Palaeoproterozoic.

Geological setting and main unit descriptions

The two largest cratonic blocks of the Siberian craton (Anabar and Tungus) and the Angara and Akitkan orogenic belts are all found within the southern Siberian craton in the proximity of Irkutsk (Fig. 1a). However, the majority of the Siberian craton is overlain by Mesoproterozoic–Neoproterozoic and Phanerozoic sediments such that only a few small areas of basement crop out. These areas are principally the Birusa, Sharizhalgai, Goloustnaya, Baikal and Urik-Iya terranes (Fig. 1b). According to recent studies, the Birusa terrane is considered as the southern termination of the Angara orogenic belt; the Sharizhalgai terrane represents the southern part of the Tungus Block; and the Goloustnaya and Baikal terranes are the southern building blocks of the Akitkan orogenic belt (Fig. 1a) (Rosen et al. 2005; Gladkochub et al. 2006a). The Urik-Iya terrane is thought to represent a graben comprising Palaeoproterozoic sediments that now separates the Birusa and Sharizhalgai terranes (Gladkochub et al. 2002).

The Birusa terrane comprises amphibolite facies gneisses, migmatites, schists and rare amphibolites. Within the metamorphic complex are numerous thick layers of quartzites, marbles and high-Al gneisses. Although the dominant metamorphic grade in the Birusa terrane is amphibolite facies, relicts of retrograde granulites are locally preserved and indicate a complex metamorphic history in the terrane basement (Turkina et al. 2003). ^{207}Pb/^{206}Pb zircon ages of c. 1.90 Ga from biotite gneiss are interpreted as the age of the amphibolite facies metamorphic event (Turkina et al. 2006). The ages of post-tectonic tonalites, granites and diorites record a restricted age range to 1.86–1.87 Ga (Turkina et al. 2006). The youngest anorogenic granite intrusions within the Birusa terrane yield a zircon crystallization age of 1.75 Ga (Turkina et al. 2003). The Early Precambrian basement of the Birusa terrane is also intruded by Neoproterozoic (741 Ma) mafic dykes and sills thought to have formed during the breakup of Rodinia (Gladkochub et al. 2006b). Early Palaeozoic (500 Ma) mafic dykes and sills are related to accretion, and subsequent extension, of several microcontinents during the earliest stages of Palaeoasian ocean closure within the Central Asian orogenic belt to the south (Donskaya et al. 2000; Gladkochub et al. 2008).

The Sharizhalgai terrane is located to the SE of the Birusa terrane (Fig. 1b) and is well-exposed for up to 400 km from the Urik River to the southern part of Lake Baikal. The Sharizhalgai terrane consists of two major tectonic units, the Onot-Erma domain and Irkut-Kitoy domain. Both of these domains comprise orthogneisses and migmatites of felsic and mafic composition interlayered with sillimanite-bearing paragneiss, marbles and schists. These rocks record amphibolite and granulite facies metamorphic conditions.

At the western margin of the Sharizhalgai terrane, the Onot-Erma domain (Fig. 1b) comprises a lower metamorphic sequence composed of TTG-orthogneisses, relicts of amphibolites, orthogneiss (metadacite), marbles and quartzites. The U–Pb zircon age of TTG-orthogneisses of the Onot-Erma basement yields one of the oldest ages reported for the Siberian craton at 3.39 Ga (Bibikova et al. 2006). This is similar to the U–Pb age of inherited zircon cores determined from granulites exposed in the Irkut-Kitoy domain (Poller et al. 2005). On the basis of U–Pb zircon data, both domains comprise two main periods of metamorphism at c. 2.6 and 1.9 Ga (Aftalion et al. 1991; Poller et al. 2005; Sal'nikova et al. 2007).

The metamorphic rocks of the Sharizhalgai terrane are intruded by the Neoarchaean (2.53 Ga) Kitoy granites (Gladkochub et al. 2005), Palaeoproterozoic (1.85–1.86 Ga) Sayan and Shumikha granites (Donskaya et al. 2002; Levitskii et al. 2002) and Neoproterozoic (0.74–0.76 Ma) mafic dykes (Sklyarov et al. 2003; Gladkochub et al. 2001b, 2006b).

The Birusa and Sharizhalgai terranes are separated by the amphibolite facies volcanic-sedimentary rocks of the Palaeoproterozoic Urik-Iya terrane (Fig. 1b). The western contact of the Urik-Iya terrane with the Birusa terrane is marked by 1.86 Ga post-collisional granites (Levitskii et al. 2002) that intrude the Urik-Iya volcano-sedimentary sequence. To the east, the contact with the Sharizhalgai terrane unit is marked by a Palaeoproterozoic fault. This eastern part of the Urik-Iya terrane contains relicts of eclogite facies mafic-ultramafic rocks and associated metavolcanics and metasediments (pelites, quartzites and marbles). Based on field and geochemical characteristics, this association of rocks has been interpreted as a dismembered and metamorphosed ophiolite sequence of Palaeoproterozoic (c. 1.9 Ga) age that separated the Birusa and Sharizhalgai domains (Gladkochub et al. 2001a). Recently, numerous strongly tectonized fragments of this potential ophiolite have been found in the marginal part of the Sharizhalgai terrane (Gladkochub et al. 2001a).

In the central (axial) zone of the Urik-Iya terrane, a younger sequence of unmetamorphosed coarse-grained terrigenous rocks (sandstones and conglomerates) are intruded by dykes of anorogenic porphyritic granite. The age of this granite has been constrained by $^{40}Ar/^{39}Ar$ biotite dating and yields an age of 1.53 Ga (Gladkochub et al. 2002). This date provides a minimum age for sedimentation related to intracratonic extension within the Urik-Iya unit.

To the NE of the Sharizhalgai terrane lie the Goloustnaya and Baikal terranes. According to the modern tectonic interpretation of the Siberian craton (Rosen et al. 2005; Gladkochub et al. 2006a) both of these terranes belong to the Akitkan orogenic belt (Fig. 1a). The Goloustnaya terrane is well-exposed on the western shore of Lake Baikal (Fig. 1b). This terrane comprises gneisses, foliated granites and amphibolites. The age of foliated granites have been determined by SHRIMP and TIMS U–Pb zircon methods which yield ages of 2.02 and 2.04 Ga, respectively (Poller et al. 2005). The basement complex of the Goloustnaya terrane is intruded by 1.86 Ga post-collisional granite of the Primorsk complex (Donskaya et al. 2003). The Baikal terrane (Fig. 1b) is exposed along the Lake Baikal shore and comprises Neoarchaean (2.88 Ga) (Donskaya et al. 2005a) and Palaeoproterozoic (1.91–2.02 Ga) (Bibikova et al. 1987; Neymark et al. 1998; Larin et al. 2006) foliated granites and Palaeoproterozoic volcanic-sedimentary sequences (Sarma and Chuya groups). The Sarma Group rocks are interpreted to be fragments of a dismembered back-arc basin, while the Chuya Group represents arc material (Neymark et al. 1998). Consequently, the Sarma and Chuya groups are considered to have formed at the Palaeoproterozoic active-margin of the Siberian craton.

The basement complex rocks of the Baikal terrane record a wide range of metamorphic pressure–temperature (PT) conditions from predominantly lower greenschist facies to locally preserved granulite facies mineral assemblages. Granulite facies metamorphism has been dated by U–Pb dating of zircon at 1.88 Ga (Poller et al. 2005) and has been interpreted to reflect a collisional event along this segment of the craton's southern margin (Gladkochub et al. 2006a).

The basement rocks of the Baikal terrane are unconformably overlain by a younger, unmetamorphosed Palaeoproterozoic volcanic-sedimentary sequence (Akitkan Formation). The lower part of this formation contains mafic volcanic rocks while its upper parts are more felsic. The sedimentary rocks in this formation are sandstones and siltstones. U–Pb zircon dating of felsic volcanic rocks of the Akitkan Formation, sampled at different parts of the Baikal terrane, is reported at 1.85–1.88 Ga (Larin et al. 2003; Donskaya et al. 2007; Didenko et al. 2009).

Within the Baikal terrane there are numerous massifs of undeformed and unmetamorphosed post-collisional 1.85–1.87 Ga granitoids (Neymark et al. 1991; Poller et al. 2005; Larin et al. 2006). Some of these Palaeoproterozoic granitoids (e.g. the Irel Complex) are considered to be co-magmatic intrusions associated with the felsic volcanics of the Akitkan Formation (Neymark et al. 1991). According to recent constraints on the tectonic evolution of the southern Siberian craton, the Akitkan volcanic-sedimentary sequence, as well as granites mentioned above, are considered to have formed during post-collision extension immediately after the main collision event reported for the area at c. 1.9 Ga (Larin et al. 2003; Donskaya et al. 2005b; Poller et al. 2005; Gladkochub et al. 2006b). In contrast to c. 1.85–1.87 Ga post-collisional complexes, the Chuya granite complex intruding the metamorphosed volcanic-sedimentary sequence of the Chuya Group (eastern flank of the Baikal terrane, see Fig. 1b) records a slightly older age (2.02 Ga) and geochemical affinities typical for island-arc granitoids (Neymark et al. 1998; Larin et al. 2006).

Three intracratonic mafic dyke complexes have a local distribution within the Baikal terrane. The age of the oldest mafic dykes is reportedly similar to the mafic volcanics of the Akitkan Formation (Gladkochub et al. unpublished U–Pb zircon SHRIMP data). Younger dykes yield a Sm–Nd mineral-whole rock isochron age of 1.67 Ga (Gladkochub et al. 2007). The youngest dyke intruded into the Archaean and Palaeoproterozoic complexes of the Baikal terrane have Neoproterozoic ages (c. 0.79 Ga) and are associated with rifting related to the Rodinia breakup (Gladkochub et al. 2007).

Analytical approach and methodology

Sm–Nd-isotope data are commonly used to determine the timing of continental crust formation, and identify possible rock sources (DePaolo & Wasserburg 1976). The Nd-model age calculation for igneous rocks produced by partial melting of a mantle source is based on the assumption that the major chemical fractionation of Sm and Nd took place when the source material was derived from depleted mantle (DM), that is prior to its incorporation into the crust (e.g. McCullogh 1987; Jahn & Condie 1995). This requires that intra-crustal processes, such as crystal fractionation, do not affect the initial Sm–Nd ratios (Taylor & McLennan 1985). There is empirical support for this assumption since Sm–Nd systems appear to remain undisturbed during multiple deformation events (Barovich & Patchett 1992), and even granulite-facies metamorphism (Whitehouse 1988). Consequently, a 'single-stage' evolution model produced by a single (uniform) magma source yields positive values of εNd(T) that indicate short-lived juvenile sources, whereas negative values of εNd(T) are typical for rocks produced by reworking of long-lived crustal materials.

It should be noted that caution is warranted where multiple sources and magma mixing may be present. In the cases where mixed sources are likely, for example, some igneous felsic rocks close in chemical composition to I- and A-type granites (hereafter granite abbreviations are given according to Chappell & White 1974; Whalen et al. 1987; Liew et al. 1989), Nd-model ages do not reflect the correct age of crust-forming events and demonstrate Nd-model ages older than the true age of melting (Arndt & Goldstein 1987). However, these data provide indirect evidence for the existence of older crustal components within the source region and are therefore useful for tectonic interpretations.

Sedimentary rocks composed mainly of material derived from a single source (uniform type of sediments, according to Dickin 2005) may yield useful constraints on the model age of the source. However, Nd-model ages of sedimentary rocks produced by several sources (mixed sediments) only yield the minimum Nd-model age of their crustal provenances (Dickin 2005). Consequently, better interpretation of Nd data from sediments requires isotopic analysis to be coupled with petrological analysis that allow potential source components to be identified. However, in both cases Sm–Nd characteristics of sedimentary rocks may provide useful provenance information (Dickin 2005).

The Nd-isotope system of meta-sedimentary rocks may be disturbed during diagenesis and metamorphic processes (Stille & Clauer 1986; Bros et al. 1992; Bock et al. 1994; Cullers et al. 1997). However, for such samples it is possible to use a 'two-stage' model age calculation (Keto & Jacobsen 1987).

In this study, the Sm–Nd systematics of numerous mafic intrusions (dykes/sills), widespread within the studied area, have been investigated. All of these are strongly contaminated by crustal components (Gladkochub et al. 2001a, 2006b). Consequently, the Nd-isotope characteristics can only be used to provide constraints on the Nd-model age of the crustal components.

For this study, unweathered geochemical samples were collected from sample locations given in Table 1. The samples were crushed, aliquoted, and approximately 300 g of the resulting material was pulverized using an agate mill. For mineral separation, the crushed material was ground using a rotary mill to a grain size of <500 μm. For Sm–Nd, XRF and ICP-MS analyses, the fine-grained (<50 μm) powder from the agate mill was used.

Isotopic measurements were carried out at: the Max-Planck Institut (MPI) für Chemie in Mainz, Germany; the Geological Institute of the Kola Science Centre RAS (GI KSC RAS), Apatity, Russia; the Institute of Precambrian Geology and Geochronology of the RAS (IPGG RAS), St Petersburg, Russia; the Institute of Geochemistry of the Siberian Branch of the Russian Academy of Sciences (IGC SB RAS), Irkutsk, Russia; and at Hokkaido University (HUni), Sapporo, Japan. In MPI, IPGG RAS and HUni, the measurements were carried out using Finnigan MAT 261 mass spectrometers. In the case of GI KSC RAS and IGC SB RAS, Finnigan MAT 262 mass spectrometers were used. All systems employ multiple collectors operating in static mode. The analytical Sm–Nd procedures are described in detail by Poller et al. (2001) for MPI, White & Patchett (1984) for HUni, Neymark et al. (1993) for IPGG RAS and IGC SB RAS and in Bayanova (2004) for GI KSC RAS.

The total laboratory blanks were: 0.06 ng for Sm and 0.3 ng for Nd in GI KSC RAS, IGS SB RAS and Hokkaido Uni; <100 pg for Nd and Sm in MPI, 0.03 ng for Sm and 0.1–0.5 for Nd in IPGG RAS. All $^{147}Sm/^{144}Nd$ and $^{143}Nd/^{144}Nd$ ratios were normalized to $^{146}Nd/^{144}Nd$ values of 0.7219 and $^{143}Nd/^{144}Nd$ values of 0.511860 or 0.512100 reported for the La Jolla and JNdi-1 Nd standards, respectively. The weighted mean $^{143}Nd/^{144}Nd$ values for the La Jolla Nd standard during the period of measurements were 0.511839 ± 7 (n = 13) in IPGG RAS; 0.511833 ± 6 (n = 11) in GI KSC RAS; 0.511835 ± 9 (n = 20) in MPI; and 0.511831 ± 7 (n = 12) in Hokkaido Uni; values for the JNdi-1 Nd standard were 0.512092 (n = 10) in IGC SB RAS. The εNd(T) values and model ages T_{DM} were calculated using the currently accepted parameters of CHUR ($^{143}Nd/^{144}Nd$ = 0.512638 and $^{147}Sm/^{144}Nd$ = 0.1967) (Jacobsen & Wasserburg 1984) and depleted mantle ($^{143}Nd/^{144}Nd$ = 0.513151 and $^{147}Sm/^{144}Nd$ = 0.2136) (Goldstein & Jacobsen 1988). Errors on isotopic ratios are quoted at the 2σ level.

Nd isotope data for the southern Siberian craton

Birusa terrane

The studied gneisses of the Birusa terrane basement show a narrow range of Sm and Nd concentrations

Table 1. *Nd isotope data from southern Siberian craton*

Sample #	Lithological unit	abbr.	T (Ma)	Error (Ma)	Method	Age interpr. Ref.	Sm (ppm)	Nd (ppm)	$^{147}Sm/^{144}Nd$	$^{143}Nd/^{144}Nd$	2σ error	εNd (T)	T (DM) 1-step	T (DM-2) 2-step	Lab	Ref. for Nd data
Birusa terrane																
9-00	gneiss	GN	1900	30	Pb–Pb	M, 1	6.47	31.18	0.1052	0.511311	8	−3.6	2574	2680	IGC	1
38-81	gneiss	GN	1900	30	Pb–Pb	M, 1	5.87	32.68	0.1086	0.511336	11	−3.9	2620	2708	IGC	1
33-81	gneiss	GN	1900	30	Pb–Pb	M, 1	5.44	30.13	0.1091	0.511212	13	−6.5	2811	2917	IGC	2
01097*	S-granite (Birusa)	S-GR	1879	8	U–Pb	I, 8	8.60	49.48	0.1046	0.511226	14	−5.4	2677	2808	IGS	
02100*	S-granite (Birusa)	S-GR	1879	8	U–Pb	I, 8	5.60	32.42	0.1045	0.511175	24	−6.3	2744	2887	MPI	
20-00	I-granite (Podpor)	I-GR	1869	10	U–Pb	I, 9	7.04	40.47	0.1051	0.511355	3	−3.1	2510	2612	IGC	1
21-00	I-granite (Podpor)	I-GR	1869	10	U–Pb	I, 9	1.77	9.04	0.1183	0.511510	11	−3.2	2611	2624	IGC	1
17-01	I-granite (Uda)	I-GR	1859	10	U–Pb	I, 1	4.06	2.41	0.1049	0.511436	18	−1.6	2394	2480	IGC	1
17-00	A-granite (Podpor)	A-GR	1747	4	U–Pb	I, 9	12.52	71.74	0.1055	0.511321	4	−5.3	2567	2693	IGC	1
14-00	A-granite (Podpor)	A-GR	1747	4	U–Pb	I, 9	12.00	68.35	0.1062	0.511355	17	−4.8	2536	2652	IGC	1
33-01	A-granite (Uda)	A-GR	1859	10	U–Pb	I, 1	3.76	24.22	0.0938	0.511256	24	−2.4	2400	2551	IGC	1
01064	diabase	DB	741	2	Ar–Ar	I, 10	3.38	15.58	0.1310	0.511673	8	−12.6	2712	2482	IPG	
03200*	diabase	DB	741	2	Ar–Ar	I, 10	3.97	18.26	0.1309	0.511696	22	−12.2	2666	2443	MPI	
03201*	diabase	DB	511	5	U–Pb	I, 10	10.84	51.20	0.1274	0.511787	21	−12.1	2402	2252	MPI	
Urik-Iya terrane																
1538	I-granite (Oka)	I-GR	1533	22	Ar–Ar	I, 11	3.20	21.11	0.0912	0.511273	13	−5.9	2329	2573	MPI	
1517b	aleurolite	MS	1860		Age of main igneous event		1.57	7.00	0.1350	0.512021	53	2.7	2183	2131	MPI	
7008g	amphibolite	MB	1880	90	Sm–Nd	M, 12	1.97	6.10	0.1944	0.512598	54	−0.2	4339	2388	MPI	
Sharizhalai terrane																
Onot-Erma domain																
157-5	amphibolite	MB	3386		Age of host-rock		3.06	9.99	0.1850	0.512490	14	2.2	3494	3419	IGC	3
63-95	TTG-orthogneiss	TTG	3386	14	U–Pb	I, 4	3.08	18.56	0.1002	0.510476	10	0.0	3565	3603	IGC	1
48-95	TTG-orthogneiss	TTG	3386	14	U–Pb	I, 4	1.29	9.97	0.0783	0.509955	15	−0.6	3570	3652	IGC	1
48-03	TTG-orthogneiss	TTG	3386	14	U–Pb	I, 4	1.85	10.29	0.1087	0.510695	27	0.5	3539	3557	IGC	1
173-95	TTG-orthogneiss	TTG	3386	14	U–Pb	I, 4	3.19	19.98	0.0964	0.510390	4	0.0	3560	3604	IGC	1
P-40-3	TTG-orthogneiss	TTG	3386	14	U–Pb	I, 4	1.85	11.72	0.0954	0.510396	9	0.5	3524	3560	IGC	4
35-04	TTG-orthogneiss	TTG	3386	14	U–Pb	I, 4	4.55	27.00	0.1018	0.510599	15	1.7	3451	3464	GIK	5
60-04	TTG-orthogneiss	TTG	3386	14	U–Pb	I, 4	6.97	42.40	0.0993	0.510497	19	0.8	3510	3537	GIK	5
2-03	orthogneiss	HMV	2600		Age of metamorphism		13.11	68.83	0.1151	0.511019	16	−4.3	3276	3311	IGC	4
6-03	orthogneiss	HMV	2600		Age of metamorphism		7.95	36.24	0.1326	0.511227	25	−6.1	3591	3458	IGC	4
23-04	orthogneiss	HMV	2600		Age of metamorphism		13.30	96.10	0.0834	0.510297	10	−7.8	3315	3594	GIK	5
3-03	A-granite (Shumikha)	A-GR	1861	1	U–Pb	I,13	16.45	89.92	0.1106	0.511159	19	−8.3	2929	3036	IGC	1
6-03-Sh	A-granite (Shumikha)	A-GR	1861	1	U–Pb	I,13	15.27	88.65	0.1041	0.511121	24	−7.5	2809	2969	IGC	1
91614	A-granite (Shumikha)	A-GR	1861	1	U–Pb	I,13	9.25	50.0	0.1119	0.511220	6	−7.4	2876	2963	IPG	
9060Б	A-granite (Shumikha)	A-GR	1861	1	U–Pb	I,13	18.40	109.8	0.1013	0.511092	4	−7.4	2778	2961	IPG	

Sample	Rock	Code	Age	n	Method	I/M	Sm	Nd	147Sm/144Nd	143Nd/144Nd	n	εNd	T_DM	T_DM2	Lab	Ref
Irkut-Kitoy domain																
0329*	mafic granulite	MB	2623	32	U–Pb	M, 14	1.79	5.35	0.2014	0.512806	34	1.7	4259	2838	MPI	
02162*	metaandesite	HMV	2623	32	U–Pb	M, 14	6.40	31.52	0.1222	0.511133	16	−4.2	3340	3323	MPI	
02216*	mafic granulite	MB	2623	32	U–Pb	M, 14	8.21	31.55	0.1421	0.511453	24	−4.7	3589	3361	MPI	
9305b*	amphibolite-xenolithe	MB	2623	32	U–Pb	M, 14	4.20	22.55	0.1121	0.510504	12	−13.2	3937	4047	MPI	
9309b*	granulite after sediment	HMS	2623	32	U–Pb	M, 14	1.62	9.53	0.1023	0.510995	33	−0.2	2934	2993	MPI	
0365*	granulite after sediment	HMS	2623	32	U–Pb	M, 14	8.43	47.95	0.1058	0.511053	4	−0.2	2948	2997	MPI	
01001*	granulite after sediment	HMS	2623	32	U–Pb	M, 14	3.64	17.41	0.1258	0.511361	30	−1.0	3088	3059	MPI	
02250*	granulite after sediment	HMS	2623	32	U–Pb	M, 14	4.69	33.90	0.0833	0.510573	6	−2.0	2995	3141	MPI	
8922	granulite after sediment	HMS	2623	32	U–Pb	M, 14	3.52	17.02	0.1245	0.511162	101	−4.4	3375	3339	MPI	
9311a*	granulite after sediment	HMS	2623	32	U–Pb	M, 14	3.12	15.39	0.1220	0.510851	26	−9.7	3793	3768	MPI	
81605	I-S-granite (Kitoy)	I-S-GR	2532	12	U–Pb	I, 15	5.18	25.90	0.1209	0.511208	7	−3.2	3172	3166	IPG	
0354*	I-S-granite (Sayan)	I-S-GR	1844	14	U–Pb	I, 16	5.64	46.80	0.0725	0.510614	9	−10.1	2726	3171	MPI	
02161*	I-S-granite (Sayan)	I-S-GR	1870	6	U–Pb	I, 14	6.78	55.60	0.0734	0.510574	11	−10.7	2785	3239	MPI	
02172*	I-granite (Sayan)	I-GR	1866	3	U–Pb	I, 14	1.82	14.50	0.0756	0.510761	25	−7.6	2625	2983	MPI	
01011*	I-granite (Sayan)	I-GR	1866	3	U–Pb	I, 14	9.50	48.79	0.1172	0.511308	17	−6.9	2896	2926	MPI	
02171*	diabase	DB	758	4	Ar–Ar	I, 17	1.69	5.95	0.1627	0.511749	27	−14.1			MPI	
81609	diabase	DB	758	4	Ar–Ar	I, 17	1.60	6.42	0.1508	0.512011	19	−7.8	2751	2098	HUn	
81603	diabase	DB	758	4	Ar–Ar	I, 17	0.95	3.84	0.1504	0.511845	10	−11.0	3127	2362	HUn	
8324	diabase	DB	758	4	Ar–Ar	I, 17	1.23	4.76	0.1563	0.511860	9	−11.3	3407	2385	HUn	
8330	diabase	DB	758	4	Ar–Ar	I, 17	0.95	3.83	0.1500	0.511477	20	−18.2	3973	2950	HUn	
Goloustnaya terrane				Age of main igneous event												
0660	gneiss	GN	1860				3.85	21.62	0.1072	0.511309	19	−4.6	2624	2728	MPI	
0355	I-granite	I-GR	2018	28	U–Pb	I, 14	3.55	22.82	0.0936	0.511224	37	−0.9	2437	2554	MPI	
0356	I-granite	I-GR	2018	28	U–Pb	I, 14	2.47	16.07	0.0931	0.511168	17	−1.8	2495	2632	GIK	
5836	A-granite (Primorsk)	A-GR	1859	16	U–Pb	I, 18	9.72	55.07	0.1067	0.511228	5	−6.1	2726	2849	IPG	
Baikal terrane																
0265	TTG-type granitoide	TTG	2884	12	U–Pb	I, 19	4.10	34.57	0.0717	0.510270	24	0.2	3074	3172	GIK	
05104	TTG-type granitoide	TTG	2884	12	U–Pb	I, 19	2.57	16.91	0.0917	0.510679	20	0.8	3070	3126	GIK	
0266	TTG-type granitoide	TTG	2884	12	U–Pb	I, 19	2.03	13.36	0.0917	0.510581	21	−1.1	3190	3282	GIK	
0253	TTG-type granitoide	TTG	2884	12	U–Pb	I, 19	2.72	14.20	0.0967	0.510652	23	−1.6	3234	3322	GIK	
03128*	metapelite (Sarma)	MS	1876	6	U–Pb	M, 14	6.19	30.89	0.1206	0.511728	32	0.6	2322	2319	MPI	
05028*	metasandstone (Sarma)	MS	1876	6	U–Pb	M, 14	3.25	20.53	0.0953	0.511279	7	−2.1	2400	2539	MPI	
PO-2375	metasandstone (Sarma)	MS	1876	6	U–Pb	M, 14	5.96	31.1	0.1157	0.511467	7	−3.4	2608	2641	IPG	
05078*	metasandstone (Sarma)	MS	1876	6	U–Pb	M, 14	1.10	5.73	0.1155	0.511322	22	−6.2	2826	2871	MPI	
PO-2309	metasandstone (Sarma)	MS	1876	6	U–Pb	M, 14	2.44	14.4	0.1020	0.511077	9	−7.7	2816	2996	IPG	
05006*	metasandstone (Sarma)	MS	1876	6	U–Pb	M, 14	3.83	19.11	0.1206	0.511252	13	−8.8	3091	3083	IPG	
05008*	metasandstone (Sarma)	MS	1876	6	U–Pb	M, 14	9.09	49.44	0.1107	0.511376	14	−3.9	2615	2688	IGS	
8714	A-granite (Irel)	A-GR	1854	5	U–Pb	I, 21	28.53	157.18	0.1097	0.511504	9	−1.4	2405	2465	IPG	7
A-67	A-granite (Irel)	A-GR	1854	5	U–Pb	I, 21	14.61	78.73	0.1122	0.511520	14	−1.7	2440	2489	IPG	7

(Continued)

Table 1. Continued

Sample #	Lithological unit	abbr.	T (Ma)	Error (Ma)	Method	Age interpr. Ref.	Sm (ppm)	Nd (ppm)	^{147}Sm/^{144}Nd	^{143}Nd/^{144}Nd	2σ error	εNd (T)	T (DM) 1-step	T (DM-2) 2-step	Lab	Ref. for Nd data
0292/1	A-granite (Irel)	A-GR	1854	5	U–Pb	I, 21	2.61	12.88	0.1220	0.511509	20	−4.3	2716	2698	MPI	
05122	A-granite (Irel)	A-GR	1854	5	U–Pb	I, 21	9.90	53.69	0.1115	0.511279	15	−6.3	2777	2861	GIK	
06360	A-granite (Irel)	A-GR	1854	5	U–Pb	I, 21	12.81	71.25	0.1087	0.511404	17	−3.1	2524	2606	GIK	
0292	A-granite (Irel)	A-GR	1854	5	U–Pb	I, 21	18.43	99.24	0.1123	0.511526	29	−1.6	2433	2481	GIK	
0371*	A-granite (Irel)	A-GR	1854	5	U–Pb	I, 21	10.44	62.21	0.1010	0.511065	21	−8.0	2807	3000	MPI	
0268	A-granite (Tatarnik)	A-GR	1854	5	U–Pb	I, 21	11.99	62.90	0.1152	0.511575	31	−1.4	2430	2459	GIK	
0270	A-granite (Tatarnik)	A-GR	1854	5	U–Pb	I, 21	11.53	58.74	0.1186	0.511510	56	−3.4	2620	2631	GIK	
05123*	diabase	DB	1844	11	U–Pb	I, un.	5.46	25.33	0.1298	0.511542	14	−5.6	2908	2797	MPI	
03101*	diabase	DB	1844	11	U–Pb	I, un.	8.23	43.83	0.1130	0.511332	39	−5.7	2741	2807	MPI	
03115*	diabase	DB	1844	11	U–Pb	I, un.	7.88	39.88	0.1189	0.511400	21	−5.8	2802	2813	MPI	
05036*	diabase	DB	1844	11	U–Pb	I, un.	8.02	37.30	0.1294	0.511411	22	−8.0	3128	2999	IGS	
05038*	diabase	DB	1844	11	U–Pb	I, un.	5.86	27.73	0.1272	0.511463	15	−6.5	2959	2873	IGS	
0392*	diabase	DB	1844	11	U–Pb	I, un.	6.05	26.62	0.1368	0.511463	12	−8.8	3324	3060	IGS	
2250	diabase	DB	1674	29	Sm–Nd	I, 20	3.97	19.11	0.1258	0.511663	35	−3.8	2569	2514	MPI	
01055	diabase	DB	787	21	Ar–Ar	I, 20	3.06	13.59	0.1324	0.511324	15	−19.2	3400	3058	HUn	
01056	diabase	DB	787	21	Ar–Ar	I, 20	3.81	17.31	0.1294	0.511223	15	−20.9	3460	3195	HUn	

Akitkan Formation of the Baikal terrane

Lower part

Sample #	Lithological unit	abbr.	T (Ma)	Error (Ma)	Method	Age interpr. Ref.	Sm (ppm)	Nd (ppm)	^{147}Sm/^{144}Nd	^{143}Nd/^{144}Nd	2σ error	εNd (T)	T (DM) 1-step	T (DM-2) 2-step	Lab	Ref. for Nd data
875	andesite-basalt	AMV	1854	5	U–Pb	I, 21	11.98	65.11	0.1110	0.511354	13	−4.7	2655	2732	IPG	7
871	andesite-basalt	AMV	1854	5	U–Pb	I, 21	7.52	41.46	0.1090	0.511324	11	−4.8	2648	2741	IPG	7
883	andesite-basalt	AMV	1854	5	U–Pb	I, 21	8.25	45.11	0.1103	0.511333	9	−4.9	2668	2752	IPG	7
886	andesite-basalt	AMV	1854	5	U–Pb	I, 21	9.22	49.83	0.1116	0.511338	7	−5.1	2694	2769	IPG	7
884	andesite-basalt	AMV	1854	5	U–Pb	I, 21	7.99	43.87	0.1099	0.511314	8	−5.2	2685	2774	IPG	7
887	andesite-basalt	AMV	1854	5	U–Pb	I, 21	8.73	45.44	0.1159	0.511370	7	−5.5	2762	2802	IPG	7
05050*	basalt	AMV	1854	5	U–Pb	I, 21	4.97	25.56	0.1170	0.511280	11	−7.6	2933	2968	IGS	
05060*	basalt	AMV	1854	5	U–Pb	I, 21	5.48	27.77	0.1188	0.511332	9	−7.0	2907	2920	MPI	
05048*	basalt	AMV	1854	5	U–Pb	I, 21	5.85	30.44	0.1156	0.511291	8	−7.0	2875	2923	MPI	

Sample	Rock type		Age (Ma)		Method	Refs	Sm	Nd	$^{147}Sm/^{144}Nd$	$^{143}Nd/^{144}Nd$		$\varepsilon Nd(T)$	$T(DM)$	$T(DM)-2$	Lab	Ref
Middle part																
A-104	rhyolite-dacite	AFV	1854	5	U–Pb	1, 21	14.12	83.32	0.1025	0.511408	17	−1.6	2380	2478	IPG	7
05120*	rhyolite	AFV	1854	5	U–Pb	1, 21	9.81	66.33	0.0890	0.511226	9	−1.9	2345	2507	MPI	
05121	rhyolite	AFV	1854	5	U–Pb	1, 21	11.27	65.04	0.1048	0.511155	22	−7.1	2779	2929	GIK	
05114*	rhyolite	AFV	1854	5	U–Pb	1, 21	8.03	48.41	0.0998	0.511092	17	−7.1	2742	2933	IGS	
38329	rhyolite-dacite	AFV	1854	5	U–Pb	1, 21	14.80	84.70	0.1056	0.511405	10	−2.4	2452	2544	IPG	7
0285	rhyolite	AFV	1854	5	U–Pb	1, 21	14.99	85.42	0.1061	0.511394	23	−2.7	2478	2571	GIK	
0219	rhyolite	AFV	1854	5	U–Pb	1, 21	18.33	103.81	0.1067	0.511352	25	−3.7	2552	2651	GIK	
05100*	rhyolite	AFV	1854	5	U–Pb	1, 21	12.01	108.53	0.0666	0.510790	39	−5.1	2437	2768	MPI	
0234	rhyolite-dacite	AFV	1854	5	U–Pb	1, 21	11.23	92.34	0.0736	0.510839	28	−5.8	2504	2826	GIK	
860	rhyolite-dacite	AFV	1854	5	U–Pb	1, 21	11.30	66.60	0.1025	0.511172	9	−6.2	2700	2857	IPG	7
01022	rhyolite	AFV	1854	5	U–Pb	1, 21	16.97	96.30	0.1065	0.511218	16	−6.3	2735	2862	GIK	
0376	rhyolite	AFV	1854	5	U–Pb	1, 21	6.92	40.65	0.1030	0.511027	19	−9.2	2907	3098	GIK	
Upper part																
0490	rhyolite	AFV	1854	5	U–Pb	1, 21	10.04	56.71	0.1066	0.511295	62	−4.8	2629	2739	MPI	
06377	rhyolite	AFV	1854	5	U–Pb	1, 21	15.74	86.17	0.1104	0.511539	19	−0.9	2370	2423	GIK	
04100	basalt	AMV	1854	5	U–Pb	1, 21	2.66	11.37	0.1408	0.511796	23	−3.1	2820	2606	MPI	
Chuya unit of the Baikal terrane																
35-89	I-granite (Chuya)	I-GR	2020	12	U–Pb	1, 7	1.38	7.15	0.1163	0.511660	7	1.8	2325	2337	IPG	7
31/10-29	I-granite (Chuya)	I-GR	2020	12	U–Pb	1, 7	5.62	38.00	0.0894	0.511276	8	1.3	2291	2380	IPG	7
Ab-4	A-granite (Abchada)	A-GR	1861	8	U–Pb	1, 22	11.01	64.36	0.1034	0.511489	11	−0.1	2289	2365	IPG	7
Ab-3	A-granite (Abchada)	A-GR	1861	8	U–Pb	1, 22	31.39	172.45	0.1100	0.511512	14	−1.3	2400	2458	IPG	7

T (Ma) – data age for calculation εNd (T) and T(DM) and T(DM)-2 step.

Age interpr. Ref., Age interpretation, Reference; M, metamorphic age; I, intrusion age; Ref., data sources.

* = $^{147}Sm/^{144}Nd$ ratios are based on ICP-MS measured element concentrations.

Rock types abbreviation: AFV, Akitkan Formation Felsic volcanics; AMV, Akitkan Formation mafic volcanics; DB, diabase; GN, gneiss; HMS, High metamorphosed sediment; HMV, High metamorphosed volcanics; MB, metabasite; MS, metasediment; TTG, TTG association; A-GR; I-GR; I-S-GR; S, GR, granite types.

Labs abbreviation: GIK, Geological Inst. of the Kola Science Centre RAS; HUn, Hokkaido University; IGC, Inst. of Geochemistry RAS; IGS, Inst. Of Geochemisty SB RAS; IPG, Institute of Precambrian Geology and Geochronology RAS; MPI, Max-Planck-Institut für Chemie.

Data Sources common for whole Table: 1, Turkina et al. 2006; 2, Turkina, 2005; 3, Turkina, 2004; 4, Bibikova et al. 2006; 5, Turkina et al. 2007; 6, Makrigina et al. 2005; 7, Neymark et al. 1998; 8, Donskaya et al. 2005c; 9, Turkina et al. 2003; 10, Gladkochub et al. 2006b; 11, Gladkochub et al. 2002; 12, Gladkochub et al. 2001a; 13, Donskaya et al. 2002; 14, Poller et al. 2005; 15, Gladkochub et al. 2005; 16, Poller et al. 2004; 17, Sklyarov et al. 2003; 18, Donskaya et al. 2003; 19, Donskaya et al. 2005a; 20, Gladkochub et al. 2007; 21, Larin et al. 2003; 22, Neymark et al. 1990.

(5.44–6.47 and 30.13–32.68 ppm, respectively; Table 1). $^{147}Sm/^{144}Nd$ ratios display typical crustal values (0.1052–0.1091), indicating that the whole-rock Sm–Nd system remained undisturbed during metamorphism. As the U–Pb zircon age of the gneisses is unknown, initial εNd values were calculated for t = 1900 Ma, which is considered to be a minimum for the main metamorphic event in the region (Turkina et al. 2006). The analysed gneisses are characterized by εNd(T) values between −3.6 and −6.5 (Fig. 2) and have corresponding Nd-model ages ranging between 2.57 and 2.81 Ga (Table 1, Fig. 3).

The Nd-isotope characteristics of seven granites (S-, I- and A-types) of the Birusa terrane have been investigated. Among these rocks, S-type granites (based on mineral composition and chemical affinities, see Donskaya et al. 2005c) display mostly negative εNd(T) = up to −6.3 and a maximum Nd-model age of 2.74 Ga. Both of these values are similar to εNd(T) and the Nd-model age of the Birusa terrane gneiss (sample 33-81, Table 1).

Palaeoproterozoic I- and A-type granites demonstrate variable values of εNd(T) (from −1.6 to −5.3) and Nd-model ages range from 2.39–2.61 Ga (Table 1, Figs 2 & 3). These variations in Nd-model age reflect the variable mixing proportion of the Palaeoproterozoic mantle-derived material and Archaean crustal source represented by the nearby Birusa gneisses (Fig. 2). The mantle-derived material input into the Podporog A-type granite source (εNd(T) = −4.8 to −5.3) is estimated at about 35%, while for the Uda granite (εNd(T) = −1.6 to −2.4) it is estimated at around 40–50% (Turkina et al. 2006).

Two main groups of dolerite intrusions are widespread in the Biryusa terrane. The first group is represented by Neoproterozoic dolerite sills (741 Ma) that are found in the passive margin volcanic-sedimentary sequence (Gladkochub et al. 2006b). Younger mafic dolerite dykes (511 Ma) intrude basement rock and the Neoproterozoic sediments (Gladkochub et al. 2006b). Both groups of dolerite have enriched LREE profiles, with

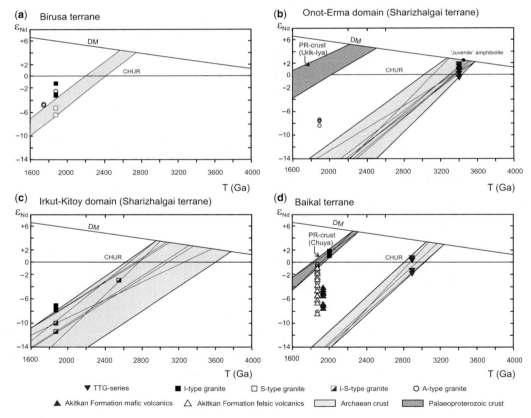

Fig. 2. εNd(T) v. primary age diagram for rocks of major terranes and units of the southern Siberian craton. Shaded areas correspond to Nd-isotope evolution fields. The symbols are given for samples provided by U–Pb zircon emplacement age.

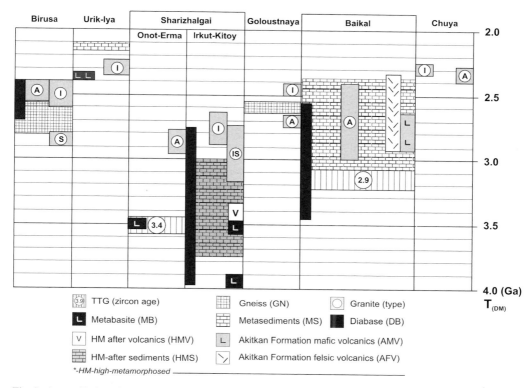

Fig. 3. A compilation of Nd-model ages of major terranes and domains within the southern Siberian craton.

$^{147}Sm/^{144}Nd = 0.1274-0.1310$, very low $\varepsilon Nd(T)$ values (-12.1 up to -12.6) and Nd-model ages of 2.40–2.71 Ga (Table 1, Fig. 3). The $\varepsilon Nd(T)$ values of dolerites extend to very negative values and Nd-model age variations are consistent with varied degrees of the original melts being contaminated by much older (at least Neoarchaean) continental crust (Fig. 2). Other geochemical characteristics, reported by Gladkochub et al. (2006b), support the inference that the dolerites were sourced from mantle-derived melts contaminated by crustal components.

Sharizhalgai Terrane

Onot-Erma Domain. Onot-Erma TTG-orthogneisses of the Onot-Erma basement yield a narrow range of Nd-model ages, with a mean T_{DM} age of 3.55 Ga. This model age is 160 Ma older than the 3.39 Ga U–Pb zircon age of Onot-Erma TTG-orthogneiss (Bibikova et al. 2006). The Nd-isotope data of Onot-Erma TTG-orthogneisses ($\varepsilon Nd(T)$) range from $+1.7$ to -0.6) suggest an older evolved crustal component in this region (Turkina 2004; Turkina et al. 2007) (Fig. 2, Table 1).

The youngest Nd-model age of orthogneisses (metadacite) of the Onot-Erma unit is 3.28 Ga (Fig. 3) while the Nd-model age (3.49 Ga) of amphibolite relicts among TTG-orthogneisses of the Onot-Erma basement is similar to the TTG-orthogneisses model age. Positive value of $\varepsilon Nd(T) = 2.2$ estimated for the amphibolite corresponds to a DM value at 3.57 Ga (Fig. 2). Thus, the Nd-isotope composition of this amphibolite indicates a signature of juvenile Palaeoarchaean crust in the Onot-Erma unit (Turkina 2004).

Four samples of unmetamorphosed postcollisional A-type granitoids of Palaeoproterozoic age (1.86 Ga; Donskaya et al. 2002; Turkina et al. 2006) from the Onot-Erma unit show a narrow range in $^{147}Sm/^{144}Nd$ (0.1013–0.1119) (Table 1). Their $\varepsilon Nd(T)$ (at T = 1861 Ma) values are remarkably consistent, ranging from -7.4 to -8.3. Because of the insignificant variation in $^{147}Sm/^{144}Nd$, the depleted-mantle model ages from these granitoids show minimal variation (2.78–2.93 Ga; Table 1), suggesting a relatively homogenous source.

The difference between the Nd-model age (2.78–2.93 Ga) and U–Pb-zircon age (1.86 Ga) of A-type granitoids exposed in the Onot-Erma unit indicates that the granitoids were derived from sources with long crustal-residence time. The Nd-model ages of A-type granitoids are distinctly younger than the model age of their possible crustal source (Palaeoarchaean TTG-orthogneisses with Nd-model age 3.55 Ga; Fig. 3) and indicate a mixture of Palaeoproterozoic juvenile mantle-derived material with an older crustal source (Fig. 2; Donskaya et al. 2005b; Turkina et al. 2006).

Irkut-Kitoy Domain. Metamorphic rocks of the Irkut-Kitoy Domain attained peak high-grade conditions around 2.6 Ga (Poller et al. 2005) and are characterized by the following features: mafic rocks with chondritic REE profiles, $^{147}Sm/^{144}Nd = 0.2014$, and εNd(T) of +1.7 (at T = 2.62 Ga); mafic and intermediate rocks with LREE enrichment, $^{147}Sm/^{144}Nd = 0.1421-0.1121$, and εNd(T) of −4.2 to −13.2; and metasediments, some with significant LREE enrichment, $^{147}Sm/^{144}Nd$ up to 0.0833, and εNd(T) of −0.2 to −9.7 (at 2.62 Ga). The εNd depleted mantle value of +1.7 in the Irkut-Kitoy mafic granulite (sample 0329) indicates a juvenile depleted-mantle Palaeoarchaean source. The rest of the metamorphic rocks of the Irkut-Kitoy domain can be considered as reworked Neoarchaean (c. 2.6 Ga, after Poller et al. 2004, 2005) relicts of older Archaean lower (granulites after mafic and intermediate rocks) and upper (granulite after sediments) crust. Mean Nd-model ages of lower crust relicts vary from 3.34–3.94 Ga (Figs 2 & 3), and fit to the range 2.94–3.79 Ga for upper crust relicts.

Three analysed granites (81605, 02161 and 0354), which are typically transitional I-S type according to their chemical composition, have been derived through melting of mainly volcanic-sedimentary sources (Gladkochub et al. 2005). Nd-model ages of these rocks (2.73–3.17 Ga) are fairly consistent with Nd-model ages of upper crustal gneisses of the Sharizhalgai terrane (Table 1, Fig. 2).

Palaeoproterozoic I-type granites (Table 1) formed through melting of an igneous or intracrustal source yield lower εNd values, reflecting possibly the mixing of Palaeoproterozoic mantle-derived material and an Archaean crustal source. Neodymium-isotope studies provide support that the change in Palaeoproterozoic granite geochemistry from I-S-type (εNd(T) = −10.7) to I-type (minimum εNd(T) = −6.9), reflects an addition of mantle-derived magmatic source material (Fig. 2).

The Sm–Nd whole rock isotope systematics from five Neoproterozoic mafic dykes intruding the basement of the Sharizhalgai terrane have been investigated. All analysed samples are calc-alkaline in composition and have geochemical characteristics indicating they were sourced from mantle-derived melts contaminated by crustal components (Gladkochub et al. 2007). These dykes are slightly LREE-enriched ($^{147}Sm/^{144}Nd = 0.1710–0.1500$) and have intrusion ages of 0.76 Ga (Sklyarov et al. 2003).

The initial εNd values, calculated for the intrusion age of the dykes, range from −7.8 to −18.2 and yield linear correlations with estimated Nd-model ages (from 2.75 Ga at εNd(T) = −7.8 to 3.97 Ga at εNd(T) = −18.2, Table 1). These correlations can be explained as reflecting different degrees of contamination of mantle-derived magma by early Precambrian crustal components of the Sharizhalgai terrane. In general, the lowest εNd value (dolerite sample 8330) may indicate more crustal involvement in its genesis. Nevertheless, as indicated by the Nd-model ages, the crustal component required for mixing must at least be of Archaean age.

Urik-Iya terrane

Metamorphosed under eclogite facies conditions (Sklyarov et al. 1998) mafic rocks within the Urik-Iya terrane (7008 g in Table 1) yield close to chondritic REE ratios, $^{147}Sm/^{144}Nd = 0.1944$, and εNd value −0.2 at 1.88 Ga (Table 1), which is a reasonable minimum age of eclogite metamorphism in the area (Gladkochub et al. 2001a). This εNd value is interpreted to represent mixing of juvenile and older crustal components, both of Palaeoproterozoic age.

To the NW, away from the suture zone, mature sandstone of the Urik-Iya terrane (sample 1517b, in Table 1) is slightly enriched in LREE ($^{147}Sm/^{144}Nd = 0.1350$) and has an initial εNd(T) value of 2.7 at 1.86 Ga. We used this age value as time of possible disturbance of the Sm–Nd system in the rock studied, related to voluminous post-collisional granite intrusions in the area (Levitskii et al. 2002; Donskaya et al. 2002). A two-step Nd-model age of the mature sediment of 2.13 Ga indicates Palaeoproterozoic crust as a possible source for this sandstone.

Mesoproterozoic (1.53 Ga) anorogenic I-type granite (Gladkochub et al. 2002) intrudes basement and sedimentary sequence of the Urik-Iya terrane and records LREE enrichment ($^{147}Sm/^{144}Nd = 0.0912$) and initial εNd value −5.9. The Nd-model age of the granite is 2.33 Ga (Table 1). The geochemical affinities of the granites indicate their generation by melting of tonalitic crustal basement of the Urik-Iya terrane (Gladkochub et al. 2002).

Goloustnaya terrane

The basement of the Goloustnaya terrane comprises gneisses which are LREE enriched ($^{147}Sm/^{144}Nd = 0.1072$) and record initial εNd(T) value of −4.6 for 1.86 Ga, which is the time of possible Sm–Nd system disturbance by large-scale granite intrusions (Donskaya et al. 2003). The Nd-model age of the Gouloustnaya terrane gneisses is 2.62 Ga and is younger than the Nd-model age of the nearby basement rocks of the Sharizhalgai terrane (Table 1). As we used the only Nd data for the Goloustnaya basement complex, we did not show it on Figure 2.

Two I-type granitoids demonstrate a narrow range in $^{147}Sm/^{144}Nd$ (0.0931–0.0936). Their εNd values are similar and range from −0.9 to −1.8 (at 2.0 Ga, the time of their intrusion; see Poller et al. 2005) with depleted-mantle model ages showing little variation among 2.44–2.50 Ga (Table 1). In contrast to the I-type granites, A-type granite (Primorsk complex) intruding the Goloustnaya gneiss basement displays more 'crustal' values of εNd(T) (−6.1) and an older Nd-model age (2.7 Ga). The data for A- and I-granites indicate a more significant crustal contribution of a Neoarchaean crustal source in A-type granites than in the I-type granitoids of the Goloustnaya terrane.

Baikal terrane

The lower crustal metamorphic rocks of the Baikal-terrane are not exposed, so only the Mesoarchaean foliated TTG-type granitoids of the Baikal terrane are available for Nd-isotopic study (Donskaya et al. 2005a). These rocks record typical crustal values for $^{147}Sm/^{144}Nd$ (0.0717–0.0967) and display no major Sm–Nd fractionation during the Palaeoproterozoic metamorphic event reported in the region (Poller et al. 2005; Gladkochub et al. 2006a). TTG-type granitoids show εNd(T) values estimated for time of their emplacement (2.88 Ga, Donskaya et al. 2005a) between +0.8 and −1.6. The Nd-model ages of the granitoids range between 3.07 and 3.23 Ga, which are below the depleted mantle line (Table 1, Figs 2 & 3).

The Sm–Nd isotopic systematics of seven whole-rock samples from the Palaeoproterozoic metamorphosed sedimentary rocks of the Sarma Group have been investigated (Table 1). For all of these samples, the εNd(T) and two-stage Nd-model ages were calculated for T = 1.88 Ga, which is considered to be a minimum age of the main metamorphic event capable to affect the Sm–Nd system of the samples. The metamorphosed mature sandstones display a narrow range in $^{147}Sm/^{144}Nd$ (0.0953–0.1206), negative values of εNd(T) (from −2.1 to −8.8) and maximum two-stage T_{DM} values of 3.08 Ga. Such values of Nd-model age are similar to the Nd-model age of the TTG-type granitoids, indicating the possibility that these may be the major source for Sarma Group sandstones. The metamorphosed pelite (03128, Table 1) has positive (0.6) values of εNd(T) and provides a relatively young two-stage T_{DM} of 2.32 Ga.

The Baikal terrane basement complex is unconformably overlain by undeformed and unmetamorphosed post-collisional volcanic-sedimentary rocks of the Akitkan Formation. The age of the final eruption/igneous event for Akitkan Formation is 1.85 Ga (Larin et al. 2003). This age is considered reasonable for εNd(T) calculations for all of the Akitkan Formation volcanic rocks and co-magmatic granitoids. The analysed basalts and andesite-basalts of the lower part of the formation demonstrate moderate LREE enrichment, ($^{147}Sm/^{144}Nd$ = 0.1090–0.1188), and εNd(T) of −4.7 to −7.6. The Nd-model age of the samples from the lower part of the Akitkan Formation ranges within 2.65–2.93 Ga (Table 1). The rhyolite and rhyolite-dacite from the middle part of the Akitkan Formation (all close to A-type granite in their compositions) display stronger LREE enrichment ($^{147}Sm/^{144}Nd$ up to 0.066), and more variable values of εNd(T) from −1.6 to −9.2, and Nd-model ages from 2.38–2.91 Ga. Neodymium-model ages of rocks with the lowest εNd(T) values represent minimal mixing of young-juvenile and older crustal components; hence the basement of the Akitkan Formation is likely to be at least of Mesoarchaean age (Fig. 2). The Nd-isotope data of samples from the upper part of the Akitkan Formation (εNd(T) = −0.9 to −4.8, and Nd-model age ranging from 2.37–2.82 Ga; Figs 2 & 3) confirm the significant involvement of Mesoarchaean crustal material as a component in felsic and mafic volcanic sources.

Some A-type granites of the Baikal terrane are co-magmatic to Akitkan felsic volcanics and display clear similarities in Nd-isotope characteristics (Table 1). The Palaeoproterozoic A-type granitoids are also enriched in LREE ($^{147}Sm/^{144}Nd$ = 0.1087–0.1212), show wide variations of εNd(T) values (−1.4 to −8.0) and Nd-model ages from 2.40–2.81 Ga (Figs 2 & 3). Such variation could be explained by mixing Palaeoproterozoic juvenile material and an older crustal component of the Baikal terrane basement.

Juvenile-type relicts, described in the Chuya Unit (Fig. 1b, Neymark et al. 1998) are represented by 2.02 Ga arc-related Chuya granites (Table 1), with εNd(T) values ranging from +1.3 to +1.8, and Nd-model ages between 2.29 and 2.33 Ga.

Small massifs of A-type granite in the Chuya Unit have εNd(T) values from −0.1 to −1.3, and Nd-model ages from 2.29–2.40 Ga. However, in contrast to A-type granitoids which are co-magmatic to Akitkan felsic volcanics, the source of the Chuya Unit A-type granites (Abchada complex) appears to be Palaeoproterozoic juvenile crust (Fig. 2) (Neymark et al. 1998).

Three different groups of dolerite dykes intruding the Baikal terrane were investigated for Sm–Nd systematics (Table 1). The oldest dykes are thought to be synchronous with felsic volcanic rocks of the Akikan Formation (c. 1.85 Ga, our U–Pb zircon SHRIMP data, unpublished). The dolerites show a narrow range of εNd(T) values (from −5.6 to −8.8) and Nd-model ages between 2.74 and 3.33 Ga (Table 1, Fig. 3). Nd-isotope data for 1.67 Ga dolerite shows εNd(T) = −3.8 and Nd-model age of 2.57. The εNd(T) estimated for Neoproterozoic (Gladkochub et al. 2006b) dolerites range between −19.2 and −20.9 and demonstrate significant contamination of basaltic magma by very older crustal material. Nd-model ages obtained for these dolerites are 3.40–3.46 Ga and indicate the likely presence of older crust deep in the Baikal terrane. Such a result is consistent with geochemical characteristics reported by Gladkochub et al. (2007).

Discussion

Increasing evidence, particularly from recent studies on Nd-model ages, supports the recycling of Palaeo- and Eoarchaean crust during Palaeoproterozoic granite intrusion in the southern Siberian craton. In the Sharizhalgai terrane, the period 3.4–3.6 Ga represents early crust formation as determined from Nd-isotope compositions of basement granulites and TTG-orthogneisses. The mafic xenoliths from these basement rocks provide evidence of the existence of juvenile crust in the area, perhaps as old as 3.9 Ga (Table 1). Nd-isotopic data indicate that the Birusa and Goloustnaya terranes, separated by the Sharizhalgai and Urik-Iya terranes, are underlain by younger crust (2.8 and 2.6 Ga, respectively). Farther NE of the Goloustnaya terrane, early crust is locally present in the Baikal terrane, as indicated by Nd ages of 3.0–3.2 Ga. Much older Nd-model ages of 3.4–3.5 Ga from the dolerite dykes indicate that older crust has been also involved in the generation of mafic magma.

The distribution of different Nd-model ages throughout the southern Siberian craton indicates a significant heterogeneity of the crustal protoliths (Fig. 3).

The Urik-Iya and Chuya units are relicts of Palaeoproterozoic (c. 2.0–2.3 Ga) juvenile crust within the studied segment of the Siberian craton margin. The Chuya Unit is interpreted as a Palaeoproterozoic island-arc reworked during major 1.9–2.0 Ga orogenesis. The juvenile nature of the Urik-Iya Palaeoproterozoic crust suggests that at 2.0 Ga the Birusa terrane was not in direct contact with the older Sharizhalgai terrane. Moreover, the Nd-isotopic characteristics of Archaean rocks from the Sharizhalgai terrane contrast with those of the Birusa terrane, confirming that, although their Proterozoic evolution has been parallel, the two terranes did not share a common Archaean history. High-grade (up to eclogite facies) metamorphosed mafic-ultramafic relicts in the eastern side of the Urik-Iya terrane mark the boundary of these two terranes whose assembly probably occurred around 1.9 Ga during a Palaeoproterozoic accretion-collision event.

The Nd-isotope data of Neoarchaean collisional-type granites and numerous Palaeoproterozoic post-collisional granite plutons (excluding the Abchada granite of the Chuya Unit) indicate that all of these intrusions inherited their Nd-isotope composition from the Palaeo- to Neoarchaean basement from which they were derived and through which they passed. Moreover the Palaeo- and Neoproterozoic mafic dykes in the studied terranes also show a contamination by older basement in their Nd-isotope record.

This synthesis of the new Nd-isotope highlights the significance of Nd-isotope data for constraining the crustal growth processes responsible for the evolution of the Siberian craton. The data provides evidence that the southern part of the Siberian craton preserves a long history of crustal development extending from the Neoproterozoic as far back as the Eoarchaean. The complicated and heterogeneous structure of the southern Siberian craton indicates that the craton formed from a series of distinct Archaean crustal fragments with different histories (Fig. 3) that were amalgamated by Palaeoproterozoic accretion between 2.0 and 1.9 Ga. Following amalgamation, voluminous granites were intruded at c. 1.85 Ga into the Archaean crustal units and the suture zones separating them. These granites mark the final stage of the stabilization of the southern Siberian craton. The assemblage of Siberia broadly coincides with important orogenic events on nearly every Precambrian continent (Windley 1998; Zhao et al. 2002), suggesting that the evolution of the southern part of the Siberian craton reflects its involvement in the amalgamation of a Palaeoproterozoic supercontinent (Condie 2002; Zhao et al. 2002).

The authors are grateful to V. P. Kovach (IPGG RAS), Yu. A. Paholchenko (IGC SB RAS) and A. M. Agashev (IGM SB RAS) for help in sample preparation and analysis. This research was supported in part by grants of Russian Ministry of Education (MD 242.2007.5, NSH 3082.2008.5), Russian Foundation for Basic Research (08-05-00245, 08-05-98070) and Research-Education Centre 'Baikal' of Irkutsk State University (RNP 2.2.1.1.7334). This is TIGeR manuscript 167.

References

AFTALION, M., BIBIKOVA, E. V., BOWES, D. R., HOPWOOD, A. M. & PERCHUK, L. L. 1991. Timing of Early Proterozoic collisional and extensional events in the granulite-gneiss-charnokite-granite complex, Lake Baikal, USSR: A U–Pb, Rb–Sr and Sm–Nd isotopic study. *The Journal of Geology*, **99**, 851–861.

ARNDT, N. T. & GOLDSTEIN, S. I. 1987. Use and abuse of crustal formation ages. *Geology*, **15**, 893–895.

BAROVICH, K. M. & PATCHETT, P. J. 1992. Behaviour of isotopic systematics during deformation and metamorphism: a Hf, Nd and Sr isotope study of milonitized granite. *Contribution to Mineralogy and Petrology*, **109**, 386–393.

BAYANOVA, T. B. 2004. *The Age of the Key Geological Complex of the Kola Region and Duration of the Magmatic Processes*. Nauka, Saint-Petersburg (in Russian).

BIBIKOVA, E. V., KORIKOVSKY, S. P., KIRNOZOVA, T. I., SUMIN, L. V., ARAKELYANTS, M. M., FEDOROVSKY, V. S. & PETROVA, Z. I. 1987. Age determinations of the rocks of the Baikal-Vitim greenstone belt by isotopic geochronological methods. *In*: SHULYUKOV, YU. A. (ed.) *Isotopic dating of the metamorphic and metasomatic processes*. Nauka, Moscow, 154–164 (in Russian).

BIBIKOVA, E. V., TURKINA, O. M., KIRNOZOVA, T. I. & FUGZAN, M. M. 2006. Ancient Plagiogneisses of the Onot Block of the Sharyzhalgai Metamorphic Massif: isotopic geochronology. *Geochemistry International*, **44**, 310–322.

BOCK, B., MCLENNAN, & HANSON, G. N. 1994. Rare earth element redistribution and its effects on the Nd isotope system in the Austin Glen Member of the Normanskill Formation. New York, USA. *Geochimica et Cosmochimica Acta*, **58**, 5245–5253.

BROS, R., STILLE, P., GAUTHIER-LAFAYE, F., WEBER, F. & CLAUER, N. 1992. Sm–Nd isotope dating of Proterozoic clay mineral: an example from the Francevillian sedimentary series, Gabon. *Earth Planetary Science Letters*, **113**, 207–218.

CHAPPELL, B. W. & WHITE, A. J. R. 1974. Two contrasting granite types. *Pacific Geology*, **8**, 173–174.

CONDIE, K. C. & ROSEN, O. M. 1994. Laurentia-Siberia connection revisited. *Geology*, **22**, 168–170.

CONDIE, K. C. 2002. Breakup of a Palaeoproterozoic supercontinent. *Gondwana Research*, **5**, 41–43.

CULLERS, R. L., BOCK, B. & GUIDOTTI, C. 1997. Elemental distributions and Nd isotopic compositions of Silurian metasediments, western Maine, USA: redistribution of rare earth elements. *Geochimica et Cosmochimica Acta*, **61**, 1847–1861.

DEPAOLO, D. J. & WASSERBURG, G. J. 1976. Nd isotopic variations and petrogenetic models. *Geophysics Research Letters*, **3**, 249–252.

DICKIN, A. P. 2005. *Radiogenic Isotope Geology* (2nd edn) Cambridge University Press, Cambridge.

DIDENKO, A. N., VODOVOZOV, V. YU. ET AL. 2009. Palaeomagnetism and U–Pb dates of the Palaeoproterozoic Akitkan Group (South Siberia) and implications for pre-Neoproterozoic tectonics. *In*: REDDY, S. M. ET AL. (eds) *Palaeoproterozoic Supercontinents and Global Evolution*. The Geological Society, London. Special Publications, **323**, 145–163.

DONSKAYA, T. V., SKLYAROV, E. V. ET AL. 2000. The Baikal collisional metamorphic belt. *Doklady Earth Sciences*, **374**, 1075–1079.

DONSKAYA, T. V., SAL'NIKOVA, E. B. ET AL. 2002. Early Proterozoic postcollision magmatism at the southern flank of the Siberian craton: new geochronological data and geodynamic implications. *Doklady Earth Sciences*, **383**, 125–128.

DONSKAYA, T. V., BIBIKOVA, E. V. ET AL. 2003. The Primorsky granitoid complex of western Cisbaikalia: geochronology and geodynamic typification. *Russian Geology and Geophysics*, **44**, 968–979.

DONSKAYA, T. V., GLADKOCHUB, D. P., MAZUKABZOV, A. M., POLLER, U. & TODT, W. 2005a. Archean of Cisbaikalia: new geochronological, geochemical and isotopic data. *In*: GLEBOVITSKY, V. A. (ed.) *Archean Geology and Geodynamic*, abstract volume. CIC, Sankt-Peterburg, 108–113 (in Russian).

DONSKAYA, T. V., GLADKOCHUB, D. P., KOVACH, V. P. & MAZUKABZOV, A. M. 2005b. Petrogenesis of Early Proterozoic postcollisional granitoids in the southern Siberian craton. *Petrology*, **13**, 229–252.

DONSKAYA, T. V., GLADKOCHUB, D. P. & MAZUKABZOV, A. M. 2005c. Petrogenesis of Early Proterozoic two-mica granite of the Birusa terrane. *In*: MITROFANOV, F. P. (ed.) *The Origin of Igneous Rocks*, abstract volume. KSC RAN, Apatite, 73–74 (in Russian).

DONSKAYA, T. V., MAZUKABZOV, A. M. ET AL. 2007. Stratotype of the Chaya Formation of the Akitkan Group in the North Baikal volcanoplutonic belt: age and time of sedimentation. *Russian Geology and Geophysics*, **48**, 707–710.

GLADKOCHUB, D. P., SKLYAROV, E. V., MEN'SHAGIN, YU. V. & MAZUKABZOV, A. M. 2001a. Geochemistry of ancient ophiolites of the Sharyzhalgai Uplift. *Geochemical International*, **39**(10), 947–959.

GLADKOCHUB, D. P., SKLYAROV, E. V., DONSKAYA, T. V., MAZUKABZOV, A. M., MENSHAGIN, YU. V. & PANTEEVA, S. V. 2001b. Petrology of gabbro-dolerites from Neoproterozoic dike swarms in the Sharyzhalgai Block with reference to the problem of breakup of the Rodinia supercontinent. *Petrology*, **9**(6), 560–577.

GLADKOCHUB, D. P., DONSKAYA, T. V., MAZUKABZOV, A. M., SKLYAROV, E. V., PONOMARCHUK, V. A. & STANEVICH, A. M. 2002. The Urik-Iya Graben of the Sayan Inlier of the Siberian Craton: new geochronological data and geodynamic implications. *Doklady Earth Sciences*, **386**, 74–78.

GLADKOCHUB, D. P., DONSKAYA, T. V., MAZUKABZOV, A. M., SALNIKOVA, E. B., SKLYAROV, E. V. & YAKOVLEVA, S. Z. 2005. Kitoy granite complex (southern part of the Siberian craton): composition, age, geodynamic setting. *Russian Geology and Geophysics*, **46**(11), 1121–1133.

GLADKOCHUB, D., PISAREVSKY, S., DONSKAYA, T., NATAPOV, L. M., MAZUKABZOV, A., STANEVICH, A. M. & SLKYAROV, E. 2006a. Siberian Craton and its evolution in terms of Rodinia hypothesis. *Episodes*, **29**(3), 169–174.

GLADKOCHUB, D. P., WINGATE, M. T. D., PISAREVSKY, S. A., DONSKAYA, T. V., MAZUKABZOV, A. M.,

PONOMARCHUK, V. A. & STANEVICH, A. M. 2006b. Mafic intrusions in southwestern Siberia and implications for a Neoproterozoic connection with Laurentia. *Precambrian Research*, **147**, 260–278.

GLADKOCHUB, D. P., DONSKAYA, T. V., MAZUKABZOV, A. M., STANEVICH, A. M., SKLYAROV, E. V. & PONOMARCHUK, V. A. 2007. Signature of Precambrian extension events in the southern Siberian craton. *Russian Geology and Geophysics*, **48**(1), 17–41.

GLADKOCHUB, D. P., DONSKAYA, T. V. ET AL. 2008. Petrology, geochronology, and tectonic implications of c. 500 Ma metamorphic and igneous rocks along the northern margin of the Central-Asian Orogen (Olkhon terrane, Lake Baikal, Siberia). *Journal of the Geological Society, London*, **165**(1), 235–246.

GOLDSTEIN, S. J. & JACOBSEN, S. B. 1988. Nd and Sr isotopic systematics of rivers water suspended material: implications for crustal evolution. *Earth and Planetary Science Letter*, **87**, 249–265.

JACOBSEN, S. B. & WASSERBURG, G. J. 1984. Sm–Nd evolution of chondrites and a chondrites, II. *Earth and Planetary Science Letters*, **67**, 137–150.

JAHN, B. M. & CONDIE, K. C. 1995. Evolution of the Kaapvaal craton as viewed from geochemical and Sm–Nd isotopic analyses of intracratonic pelites. *Geochimica et Cosmochimica Acta*, **59**, 2239–2258.

KETO, L. S. & JACOBSEN, S. B. 1987. Nd and Sr isotope variations of Early Palaeozoic oceans. *Earth and Planetary Science Letter*, **84**, 27–41.

LARIN, A. M., SAL'NIKOVA, E. B. ET AL. 2003. The North Baikal Volcanoplutonic Belt: age, formation duration, and tectonic setting. *Doklady Earth Sciences*, **392**, 963–967.

LARIN, A. M., SAL'NIKOVA, E. B., KOTOV, A. B., MAKAR'EV, L. B., YAKOVLEVA, S. Z. & KOVACH, V. P. 2006. Early Proterozoic syn- and postcollision granites in the northern part of the Baikal Fold Area. *Stratigraphy and Geological Correlation*, **14**(5), 463–474.

LEVITSKII, V. I., MEL'NIKOV, A. I. ET AL. 2002. Early Proterozoic postcollisional granitoids in southwestern Siberian craton. *Russian Geology and Geophysics*, **43**, 679–692.

LIEW, T. C., FINGER, F. & HÖCK, V. 1989. The Moldanubian granitoid plutons of Austria: chemical and isotopic studies bearing on their environmental setting. *Chemical Geology*, **76**, 41–55.

MAKRIGINA, V. A., PETROVA, Z. I., SANDIMIROVA, G. P. & PAKHOLCHENKO, YU. A. 2005. New data on the age of the strata framing the Chuya and Cisbaikalian uplifts (northern and western Baikal areas). *Russian Geology and Geophysics*, **46**, 714–722.

MCCULLOGH, M. T. 1987. Sm–Nd isotopic constraints on the evolution of Precambrian crust in the Australian continent. *In*: KRONER, A. (ed.) *Proterozoic Lithospheric Evolution*. Geodynamic Series, 17. American Geophysical Union, Washington, DC, 115–130.

NEYMARK, L. A., LARIN, A. M., YAKOVLEVA, S. Z., SRYVTSEV, N. A. & BULDYGEROV, V. V. 1991. New data on the age of the Akitkan Group, Baikal – Patom foldbelt: constraints from U–Pb dating of zircons. *Dokladi Akademii Nauk SSSR*, **320**, 182–186 (in Russian).

NEYMARK, L. A., KOVACH, V. P & NEMCHIN, A. A. 1993. Late Archean intrusive complexes in the Olekma granite-greenstone terrane (Eastern Siberia): geochemical and isotopic study. *Precambrian Research*, **62**, 453–472.

NEYMARK, L. A., LARIN, A. M., NEMCHIN, A. A., OVCHINNIKOVA, G. V. & RYTSK, E. Yu. 1998. Anorogenic nature of magmatism in the Northern Baikal volcanic belt: evidence from geochemical, geochronological (U–Pb), and isotopic (Pb, Nd) data. *Petrology*, **6**, 124–148.

NUTMAN, A. P., CHERNYSHEV, I. V., BAADSGAARD, H. & SMELOV, A. P. 1992. The Aldan shield of Siberia, USSR: the age of its Archaean components and evidence for widespread reworking in the Mid-Proterozoic. *Precambrian Research*, **54**, 195–210.

PARFENOV, L. M. & KUZMIN, M. I. 2001. *Tectonics, Geodynamics, and Metallogeny of the Sakha Republic (Yakutia)*. MAIK Nauka, Interperiodica, Moscow (in Russian).

POLLER, U., KOHUT, M., TODT, W. & JANAK, M. 2001. Nd, Sr, Pb isotope study of the Western Carpathians: implications for Palaeozoic evolution. *Schweizerische Mineralogisch-Petrographische Mitteilungen*, **81**, 159–174.

POLLER, U., GLADKOCHUB, D. P., DONSKAYA, T. V., MAZUKABZOV, A. M., SKLYAROV, E. V. & TODT, W. 2004. Early Proterozoic collisional magmatism along the Southern Siberian craton – Constrains from U–Pb single zircon data. *Transactions of the Royal Society Edinburgh*, **152**, 1116–1127.

POLLER, U., GLADKOCHUB, D., DONSKAYA, T., MAZUKABZOV, A., SKLYAROV, E. & TODT, W. 2005. Multistage magmatic and metamorphic evolution in the Southern Siberian Craton: archean and Palaeoproterozoic zircon ages revealed by SHRIMP and TIMS. *Precambrian Research*, **136**, 353–368.

ROSEN, O. M., CONDIE, K. C, NATAPOV, L. M. & NOZHKIN, A. D. 1994. Archean and Early Proterozoic evolution of the Siberian craton: a preliminary assessment. *In*: CONDIE, K. C. (ed.) *Archean Crustal Evolution*. Elsevier, Amsterdam, 411–459.

ROSEN, O. M., MANAKOV, A. V. & SERENKO, V. P. 2005. Palaeoproterozoic collisional system and diamondiferous lithospheric keel of the Yakutian kimberlite province. *Russian Geology and Geophysics*, **46**, 1259–1272.

SAL'NIKOVA, E. B., KOTOV, A. B. ET AL. 2007. Age constraints of high-temperature metamorphic events in crystalline complexes of the Irkut block, the Sharyzhalgai ledge of the Siberian platform basement: results of the U–Pb single zircon dating. *Stratigraphy and Geological Correlation*, **15**, 343–358.

SKLYAROV, E. V., GLADKOCHUB, D. P., MAZUKABZOV, A. M. & MENSHAGIN, YU.V. 1998. Metamorphism of the ancient ophiolites of the Sharyzhalgai. *Russian Geology and Geophysics*, **39**, 1722–1739.

SKLYAROV, E. V., GLADKOCHUB, D. P., MAZUKABZOV, A. M., MENSHAGIN, Y. V., WATANABE, T. & PISAREVSKY, S. A. 2003. Neoproterozoic mafic dike swarms of the Sharyzhalgai metamorphic massif

(southern Siberian craton). *Precambrian Research*, **122**, 359–376.
STILLE, P. & CLAUER, N. 1986. Sm-Ns isochron-age and provenance of the argillites of the Gunflint Iron Formation in Ontario, Canada. *Geochimica et Cosmochimica Acta*, **50**, 1141–1146.
TAYLOR, S. R. & MCLENNAN, S. M. 1985. *The Continental Crust: its Composition and Evolution*. Blackwell, Oxford.
TURKINA, O. M. 2004. The amphibolite plagiogneiss complex of the Onot Block, Sharyzhalgai Uplift: isotopic geochemical evidence for the Early Archean evolution of the continental crust. *Doklady Earth Sciences*, **399**, 1296–1300.
TURKINA, O. M. 2005. Proterozoic tonalites and trondhjemites of the southwestern margin of the siberian craton: isotope geochemical evidence for the lower crustal sources and conditions of melt formation in collisional settings. *Petrology*, **13**, 35–48.
TURKINA, O. M., BIBIKOVA, E. V. & NOZHKIN, A. D. 2003. Stages and geodynamic settings of Early Proterozoic granite formation on the southwestern margin of the Siberian Craton. *Doklady Earth Sciences*, **388**, 159–163.
TURKINA, O. M., NOZHKIN, A. D. & BAYANOVA, T. B. 2006. Sources and formation conditions of Early Proterozoic granitoids from the southwestern margin of the Siberian Craton. *Petrology*, **14**, 262–283.
TURKINA, O. M., NOZHKIN, A. D., BAYANOVA, T. B. & DMITRIEVA, N. V. 2007. Isotopic provinces and evolution stages of the Precambrian crust at the southwestern margin of the Siberian Craton and its folded framing. *Doklady Earth Sciences*, **413**, 481–486.
WHALEN, J. B., CURRIE, K. L. & CHAPPELL, B. W. 1987. A-type granites: geochemical characteristics, discrimination and petrogenesis. *Contribution to Mineralogy and Petrology*, **95**, 407–419.
WHITE, W. M. & PATCHETT, J. 1984. Hf–Nd–Sr isotopes and incompatible element abundances in island arcs: implications for magma origins and crust-mantle evolution. *Earth and Planetary Science Letter*, **67**, 167–185.
WHITEHOUSE, M. J. 1988. Granulite facies Nd-isotopic homogenisation in the Lewisian complex of northwest Scotland. *Nature*, **331**, 705–707.
WINDLEY, B. F. 1998. *The Evolving Continents*. Wiley & Sons, New York.
ZHAO, G., CAWOOD, P. A., WILDE, S. A. & SUN, M. 2002. Review of global 2.1–1.8 Ga orogens: implications for a pre-Rodinia supercontinent. *Earth Science Reviews*, **59**, 125–162.

Palaeomagnetism and U–Pb dates of the Palaeoproterozoic Akitkan Group (South Siberia) and implications for pre-Neoproterozoic tectonics

ALEXEI N. DIDENKO[1,6]*, VLADIMIR Y. VODOVOZOV[2,7], SERGEI A. PISAREVSKY[3], DMITRY P. GLADKOCHUB[4], TATYANA V. DONSKAYA[4], ANATOLY M. MAZUKABZOV[4], ARKADY M. STANEVICH[4], ELENA V. BIBIKOVA[5] & TATYANA I. KIRNOZOVA[5]

[1]*Geological Institute of the Russian Academy of Sciences, 7, Pyzhevsky Lane, 119017 Moscow, Russia*

[2]*Institute of Physics of the Earth of the Russian Academy of Sciences, 10 Bol. Gruzinskaya Street, 123995 Moscow, Russia*

[3]*School of Geosciences of the University of Edinburgh, the King's Buildings, West Mains Road, Edinburgh EH9 3JW, UK*

[4]*Institute of the Earth's Crust of the Siberian Branch of the Russian Academy of Sciences, 128, Lermontova Avenue, 664033 Irkutsk, Russia*

[5]*Vernadsky Institute of geochemistry and analytical chemistry of the Russian Academy of Sciences, 19, Kosygina Street, 117975 Moscow, Russia*

[6]*Institute of Tectonics and Geophysics of the Eastern Branch of the Russian Academy of Sciences, 65, Kim Yu. Chen Street, 680063 Khabarovsk, Russia*

[7]*Lomonosov Moscow State University, Faculty of Geology, 1 Leninskiye Gory, 119899 Moscow, Russia*

*Corresponding author (e-mail: alexei_didenko@mail.ru)

Abstract: We present new geochronological and palaeomagnetic results from the late Palaeoproterozoic Akitkan Group in South Siberia. The zircon U–Pb conventional age of the rhyodacite from the upper part of the group is 1863 ± 9 Ma and the age of the dacite from the lower part of the group is 1878 ± 4 Ma. Palaeomagnetic study of sedimentary and some igneous rocks from the upper part of the group isolated a high-temperature characteristic component ($D = 193°$, $I = 19°$, $k = 51$, $\alpha_{95} = 7°$) which is supported by two of three applied conglomerate tests. However, the third intra-formational conglomerate test demonstrates a contaminating overprint of uncertain nature for a part of our collection. The analysis of data suggests that this overprint occurred at time when the geomagnetic field's direction was similar to that at the time of the deposition. Therefore the corresponding palaeomagnetic pole (22.5 °S, 97.4 °E, $dp = 1.5°$, $dm = 2.8°$) may be considered as representative for the deposition time. Palaeomagnetic study of the sediments in the lower part of the Akitkan Group isolated a stable primary remanence ($D = 189°$, $I = 8°$, $k = 111$, $\alpha_{95} = 5°$) supported by positive intra-formational conglomerate and fold tests. The palaeomagnetic pole (30.8 °S, 98.7 °E, $dp = 2.5°$, $dm = 5.0°$) is nearly coeval with the 1879 Ma Molson B pole from the Superior craton. We used these two poles to compare the relative position of Siberia and the Superior craton in the late Palaeoproterozoic. It is different from their reconstruction around 1000 Ma. This demonstrates their relative movements in the Mesoproterozoic.

Pre-Neoproterozoic history of the Earth has been a matter of increased interest in the last decade. Conceptions vary from denying the very existence of plate tectonics before 1 Ga (Stern 2005), to considering a single supercontinent during the Precambrian (Piper 2000), to suggesting assemblages and breakups of one or more supercontinents (Borukaev 1985; Khain & Bozhko 1988; Hoffman

From: REDDY, S. M., MAZUMDER, R., EVANS, D. A. D. & COLLINS, A. S. (eds) *Palaeoproterozoic Supercontinents and Global Evolution*. Geological Society, London, Special Publications, **323**, 145–163.
DOI: 10.1144/SP323.7 0305-8719/09/$15.00 © Geological Society of London 2009.

1991; Windley 1995; Rogers 1996; Dalziel 1997; Condie 2002; Rogers & Santosh 2002). The variety of opinions results (at least in part) from a paucity of pre-1000 Ma palaeomagnetic data. The latest version of the IAGA Global Palaeomagnetic Database (GPMDB, Pisarevsky 2005) contains about 1100 pre-Neoproterozoic results compared to about 8000 Phanerozoic poles. Moreover, a great majority of these data fail to pass reasonable reliability criteria, because many palaeomagnetic poles are very poorly dated and only a few are supported by field tests. In addition, many results were obtained without proper demagnetizing. As a result, there are only 45 very reliable pre-Neoproterozoic poles (Evans & Pisarevsky 2008).

Until recently, Precambrian drift of Siberia was poorly constrained due to lack of precise geochronology and few reliable palaeomagnetic data. For example, at least six different Siberia–Laurentia reconstructions in the youngest Precambrian supercontinent Rodinia (equally poorly constrained) have been published (Pisarevsky et al. 2003, 2007; Pisarevsky & Natapov 2003). This situation improved greatly due to recent palaeomagnetic (Gallet et al. 2000; Pavlov et al. 2000, 2002) and geochronological (Rainbird et al. 1998; Sklyarov et al. 2003; Gladkochub et al. 2006a) studies. However, the Siberian role in pre-Rodinian palaeogeography is still practically unknown. There are only three relatively reliable (i.e. well-dated and properly demagnetized) Siberian palaeopoles of a pre-Rodinian age, namely 1503 Ma Kuonamka dykes, 1384 Ma Chieress dykes (Ernst et al. 2000) and 1850 Ma Shumikhin granites (Didenko et al. 2003, 2005). Moreover, Kuonamka and Chieress poles, as admitted by Ernst et al. (2000), are based on a small number of dykes, so the palaeosecular variation is not adequately averaged. Hence there are no reliable (s. s.) Siberian palaeopoles for the 800 Ma time interval between the c. 1045 Ma Malgina pole of Gallet et al. (2000) and the 1850 Ma Shumikhin pole of Didenko et al. (2003, 2005). As a result, most of pre-Rodinian global palaeogeographic reconstructions (Rogers 1996; Rogers & Santosh 2002; Zhao et al. 2002) place Siberia in the position next to north Laurentia by analogy with early Rodinian reconstructions (Hoffman 1991), or based on vague geological suggestions, some of which were made many years ago and have since proved to be incorrect. On the other hand, Siberia is very important for the pre-Rodinian palaeogeography, because it was almost surrounded by Mesoproterozoic passive margins (Pisarevsky & Natapov 2003; Gladkochub et al. 2006b), which means that Siberia could be a core of some late Palaeoproterozoic–Mesoproterozoic supercontinent in the same way as Laurentia apparently was the core of Rodinia (Dalziel 1997). Consequently, the need of precisely dated and highly reliable Siberian Meso- and Palaeoproterozoic palaeomagnetic poles is obvious. Here we present results of palaeomagnetic and geochronological studies of two sections of the late Palaeoproterozoic Akitkan Group in South Siberia.

Geology and sampling

The Siberian craton (Fig. 1a) is a Palaeoproterozoic collage of mostly Archaean superterranes, assembled at 2.1–1.8 Ga (Rosen et al. 2005). Collision of two major superterranes, Aldan and Anabar, resulted in the formation of the Akitkan orogenic belt (Fig. 1a; Rosen et al. 1994; Condie & Rosen 1994; Rosen et al. 2005) at c. 1.88 Ga (Poller et al. 2005). Post-collisional extension caused emplacement of voluminous 1.87–1.84 Ga granites (Neymark et al. 1991; Donskaya et al. 2002; Larin et al. 2003; Poller et al. 2005) and a development of the 550 km-long and 60 km-wide North-Baikal volcano-plutonic belt (Fig. 1b; Larin et al. 2003; Donskaya et al. 2005) composed of c. 4500 m of volcanic and volcano-sedimentary strata of the Akitkan Group and their co-magmatic granitoids (Bukharov 1987). Red sediments of the Akitkan Group were the primary target for our palaeomagnetic study.

Mazukabzov et al. (2006) proposed a new stratigraphic scheme of the Akitkan Group, based on the synthesis of previous schemes (Mats 1965; Salop 1967; Mats et al. 1968; Bukharov 1987) and on the new geochronological data (Larin et al. 2003; Poller et al. 2005). According to this scheme, the Akitkan Group unconformably overlies the Archaean foliated granites and schists and metavolcanicas of the Palaeoproterozoic Sarma Group in the southern part of the belt (Donskaya et al. 2005, 2007). The Sarma Group is cut by the 1910 Ma Kocherikovo granite (Bibikova et al. 1987). Direct contacts of the Akitkan Group with the Archaean basement and with the Sarma Group have been observed near the Malaya Kosa cape on the shores of Lake Baikal (Fig. 1d).

The lower part of the Akitkan Group is composed mostly of the clastic Malaya Kosa suite overlain by felsic volcanics and minor sediments of the Khibelen suite (Fig. 1d). In the northern part of the North-Baikal belt, the Akitkan Group is underlain by Palaeoproterozoic metamorphic rocks and granitoids. The Khibelen suite is also exposed here (Fig. 1c) and is conformably overlain by the volcano-sedimentary upper part of the Akitkan Group (Chaya suite). In the study area the Akitkan Group is conformably overlain by the sandstones of the Okun Group, which is, in turn, unconformably overlain by the Neoproterozoic Baikal Group. The Baikal Group directly overlies the Akitkan Group

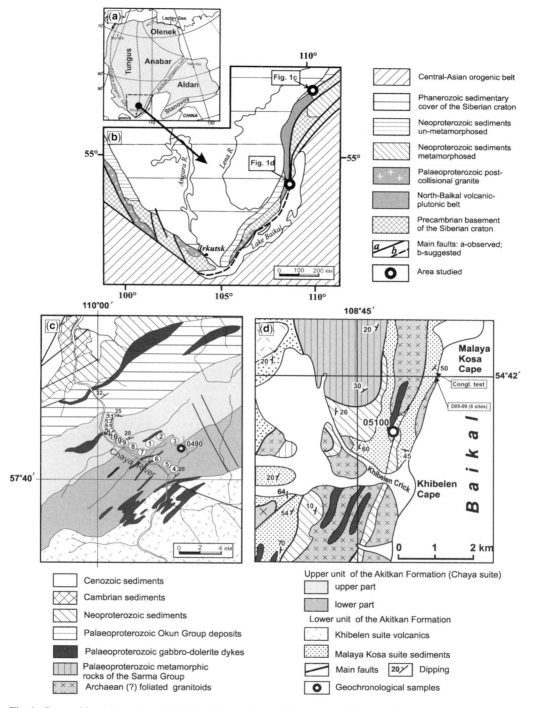

Fig. 1. Geographic setting and geological sketch map of the study area: (**a**) major tectonic elements of the Siberian craton (after Pisarevsky *et al.* 2007); (**b**) geology of the southern part of the Siberian craton (modified after Donskaya *et al.* 2007); (**c**) geological map of Chaya area (northern part of the Akitkan orogenic belt), the prefix D04 is omitted in the numbers of sample sites; (**d**) geological map of the Khibelen area (southern part of the Akitkan orogenic belt).

in most other areas. All of these rocks are weakly metamorphosed up to a low-temperature greenschist grade. The whole stratigraphy is folded; the age of this folding has been considered as Riphean (Meso- to Neoproterozoic) until recently. However, Zorin *et al.* (2008) reported 420 Ma syn-kinematic granites in the Chaya river section, so it is more likely that the pulses of folding continued at least until the Silurian.

The age of the Akitkan Group is constrained by the U–Pb zircon dates of 1866 ± 6 Ma (Neymark *et al.* 1991) and 1869 ± 6 Ma (Larin *et al.* 2003) from two distinct horizons of felsic volcanics in the lower to middle part of the Akitkan Group. The 1854 ± 5 Ma U–Pb zircon age on rhyolite of the upper part of the Akitkan Group (Larin *et al.* 2003) is probably close to the end of its deposition. The whole Akitkan Group (and the Okun Group in one of our sampling localities) is intruded by thick gabbro-diabase intrusions of the Chaya complex with the Sm–Nd age of 1674 ± 29 Ma (Fig. 1c; Gladkochub *et al.* 2007). The correlatives of these intrusions with similar mineral composition, geochemistry, isotope data and field relationship (Gladkochub *et al.* 2007) exist near the Khibelen Cape in the southern part of the belt (Fig. 1d).

The North-Baikal volcano-plutonic belt is a typical post-collisional belt formed under extension. The collision stage within the 'Akitkan orogenic belt' is marked by the Kaltygey Cape granulites (Poller *et al.* 2005) and the 1910 Ma Kocherikovo granite (Bibikova *et al.* 1987). All of these collision-related complexes are uncomfortably overlain by unmetamorphosed volcanics and sediments of the post-collisional North-Baikal volcano-plutonic belt (the Akitkan Group).

We chose two areas for our study: outcrops along the Chaya river in the northern part of the North-Baikal belt (Fig. 1c) and coastal outcrops near the Malaya Kosa Cape of Baikal Lake in the southern part of the belt (Fig. 1d).

In the first area, felsic volcanics of the lower part of the Akitkan Group (Khibelen suite) are conformably overlain by volcano-sedimentary rocks of the upper part of the Akitkan Group (Chaya suite). The latter was our primary target. It is gently (up to 30°) dipping to the NW and has a combined thickness of c. 3100 m (Fig. 1c). The Chaya suite in its lower part consists of grey-green and reddish sandstones with thin layers of tuffs, tuffites, siltstones, conglomerates and lenses of felsic volcanics. Typical thickness of this part is about 1600 m. The c. 1500 m upper part of the suite is composed of violet, reddish and green arkoses, polymictic sandstones, and conglomerates. Minor rhyolite flows and small bodies of porphyric andesites also exist. We collected 128 oriented block samples in both parts of the suite from 9 sedimentary sites (cherry-red siltstones and silty sandstones) and from two igneous units of rhyolites (site D04-12) and porphyric andesites (site D04-13). Two sets of pebbles from conglomerate layers (sites D04-3 and D04-11, Fig. 1c) were collected for the conglomerate test. These pebbles were derived from felsic volcanics of the underlying Khibelen suite and/or from the sedimentary rocks of the Chaya suite itself (intra-formational conglomerates). The geochronological sample 0490 has been collected from the rhyodacite flow in the middle part of the section (Fig. 1c).

In the second locality (Fig. 1d), we collected 41 oriented block samples from 8 sites of the c. 160 m-thick cherry-red and grey siltstones and fine-grained sandstones of the lower part of the Akitkan Group (Malaya Kosa suite) near its direct contact with Archaean tonalite, 3.3 km south of the Malaya Cosa Cape. These sediments are underlain by a 7 m-thick basal conglomerate and are interbedded with layers of tuffs, tuffites and intra-formational conglomerates. Twenty-nine oriented pebbles from one of these intra-formational conglomerates were collected for the conglomerate test. This is the southeastern limb of an anticline; the strata are dipping to the SE with angles of dip varying from $40°–50°$. The geochronological sample 05100 has been collected from felsic volcanics of the overlying Khibelen suite (Fig. 1d).

A magnetic compass was used for orientation; declinations measured directly against the outcrop and at c. 1 m away gave consistent results, suggesting that no local magnetic anomalies are present to bias our results (e.g. lightning or strong outcrop magnetizations). Two to four cubic specimens with sides of 2 cm were trimmed from each oriented block sample.

Methods and techniques

Samples were analysed in the palaeomagnetic laboratories at the Geological Institute (GIN RAS, Moscow), at the Institute of Physics of Solid Earth (IFZ RAS, Moscow) and at the Geophysical Observatory (GO, Borok, Russia). Remanence composition was determined by detailed stepwise thermal demagnetization (12–18 steps, to 680 °C), using a TD-48 furnace (ASC Scientific) and home-made furnaces (residual field of c. 10–20 nT), and the JR-4 (AGICO) spinner-magnetometer. To monitor possible mineralogical changes during heating and for the magnetic fabric study, magnetic susceptibility was measured in selected samples after each heating step using a KLY-2 (AGICO) kappa-bridge. Magnetic mineralogy was investigated from detailed spectra of the unblocking temperatures and Curie temperatures obtained using the

Curie balance and the 2-component thermomagnetometer of GO Borok (IFZ RAS). Magnetization vectors were isolated using Principal Component Analysis (Kirschvink 1980).

The U–Pb isotope study has been performed in the geochronological laboratory of the Vernadsky Institute of Geochemistry and Analytical Chemistry (RAS, Moscow) by conventional methods. Decomposition of zircon microsamples (0.5–0.1 mg) and chemical separation of U and Pb for isotopic analyses has been done using the method of Krogh (1973). U and Pb concentrations were determined by the isotopic delution method using a mixed spike ^{208}Pb + ^{235}U, blanks being 0.1 ng of Pb and 0.005 ng of U. Isotopic compositions were measured on a multi-collector mass spectrometer TRITON. The ISOPLOT program (Ludwig 1999) was used for processing of experimental data. Common-lead correction was introduced for the age 1850 Ma by the model of Stacey & Kramers (1975). The precision in U–Pb isotopic ratios was 0.5%. All errors are given at the 2σ level. Preliminary selective decomposition (SD) was undertaken for some zircon fractions (Mattinson 1994) to improve the concordance of U–Pb ratios in zircons. According to this method, the preliminary treatment of zircons was done in concentrate HF for 8 hours at 150 °C. After that, the crystalline residue was treated twice with 3.1 N HCl at 180 and at 200 °C during 6 and 10 hours. The crystalline residue was washed twice with 1N HNO$_3$ and decomposition has been done conventionally.

Upper part of the Akitkan Group

Geochronology. The geochronological sample 0490 has been collected from the rhyodacite lens within coarse sediments of the lower part of the Chaya suite in the River Suslinka (right Chaya's tributary, Fig. 1c). The rock composition corresponds to that of quartz-porphyry. Elements of fluidal texture are locally visible. Geochronological U–Pb analyses were done for two grain size fractions of the accessory zircon and for the residue after selective decomposition (Mattinson 1994). One of the points (+100 μm, Table 1) was not used for discordia calculations. The lower intercept of discordia reflecting the time of lead loss is 154 ± 170 Ma. The upper intercept age is 1863.2 ± 8.7 Ma (Fig. 2a). The morphology of zircon is magmatic and permits us to regard the age of 1863 ± 9 Ma as the time of crystallization of the melt parent for rhyodacites. This age is similar to the earlier reported result of 1854 ± 5 Ma by Larin et al. (2003) from another outcrop of felsic volcanics of the upper part of the Akitkan Group. We conclude that the age of the upper part of the Akitkan Group is close to 1860 Ma.

Magnetic minerals and rock magnetism. The natural remanent magnetization (NRM) of red sediments ranges from $0.8-5.0 \times 10^{-5}$ A m^{-1}, and their magnetic susceptibility from about $1-3 \times 10^{-4}$ SI units. The ratio of remanent magnetization to induced magnetization (Koenigsberger's Q parameter) usually reflects the grain size distribution of the magnetic material and, therefore, how magnetically 'hard' or 'soft' the particles are. Q varies from 2–5 (Fig. 3a) for the studied red beds, indicating their high palaeomagnetic stability, which is typical for most red beds (Khramov 1987). NRM of igneous rocks vary from $0.9-1.0 \times 10^{-5}$ A m^{-1}, and their magnetic susceptibility from about $1-7 \times 10^{-4}$ SI units. Q varies mostly from 2–5 (Fig. 3a), but two rhyolite samples have Q close to 1.

Magnetic saturation v. temperature (J_s-T) curves are mainly of Q- and R-types (see fig. 1–24 of Nagata 1961). The J_s-T curves (both the first and the second heating) of the porphyritic andesite (Fig. 3b) indicate two bends corresponding to Curie temperatures of 560 °C and 670 °C. Saturation remanence (J_{rs}-T) curves for the same rock indicate two blocking temperatures of 560 °C and 660 °C (Fig. 3b). These data suggest that the studied porphyritic andesite crystallized in subaerial conditions producing low-titanium titanomagnetite. Its deuteric oxidation during initial cooling caused generation of magnetite and hematite (Ade-Hall et al. 1971; Pechersky et al. 1975; McElhinny & McFadden 2000).

Both J_s-T and J_{rs}-T curves for red beds indicate that hematite carry the bulk of their remanence, but some minor magnetite is also indicated by small bends near 560–570 °C on the J_s-T curves (Fig. 3c).

Palaeomagnetism. Some typical examples of thermal demagnetization are shown in Figure 4. The majority of samples exhibit an interpretable demagnetization plot with the exception of the weak and unstable rhyolites (site D04-12). The present-day field (PDF) has been destroyed in most samples at 200–300 °C. In the rest of the samples this component is practically absent (Fig. 4b, d, e, f).

The shallow SSW high-temperature component was isolated in almost all samples and in some cases this is the only component carried by both magnetite and hematite (Fig. 4a, f, h). However, in some cases the hematite is its only carrier (Fig. 4b, c, d). Note that the direction of the high-temperature component and the thermal demagnetization behaviour is similar in porphyritic andesites (Fig. 4e, f) and in sedimentary rocks (other diagrams in Fig. 4). Stereoplots of the high-temperature component are shown in Figure 4i and j (sample

Table 1. U–Pb isotopic data for zircons from sample No. 0490 (dacite-porphyry from Upper unit of the Akitkan Formation) and from sample No. 05100 (dacite from Lower unit of the Akitkan Formation)

No.	Fraction (μm)	Weight (g)	Contents (ppm)		Isotopic composition of Pb				Isotopic ratio and age, Ma		
			U	Pb	$^{206}Pb/^{204}Pb$	$^{206}Pb/^{207}Pb$	$^{206}Pb/^{208}Pb$	$^{206}Pb/^{238}U$	$^{207}Pb/^{235}U$	$^{207}Pb/^{206}Pb$
					Sample 0490 (rhyodacite from the upper part of the Akitkan Formation, Chaya River area)						
1	+100	0.00102	381.47	125.86	3170	8.6038	5.9930	0.2976	4.5979	1833.1 ± 1.0
2	−100 + 75	0.00100	258.13	81.20	7551	8.7110	6.1440	0.2860	4.4607	1849.9 ± 1.0
3	−75	0.00760	471.05	177.09	8170	8.7360	6.1590	0.3321	5.2200	1846.8 ± 1.0
4	+75, SD				2866	8.4660	6.6340	0.3191	4.9939	1856.4 ± 1.1
5	+75, SD				18710	8.7525	7.0203	0.3144	4.9262	1858.2 ± 1.0
					Sample 05100 (dacite from lower part of the Akitkan Formation, the Khibelen Cape area)						
1	+100	0.00114	218.59	82.43	1416	7.3264	2.8220	0.2911	5.1093	2060.8 ± 1.0
2	+75	0.00110	258.29	91.02	534	7.2304	2.2207	0.2504	3.9077	1850.9 ± 1.2
3	−75	0.00100	225.50	82.96	1050	7.9031	2.4694	0.2755	4.3238	1861.0 ± 2.0
4	−75, SD				4610	8.4841	3.1101	0.3209	5.0358	1861.2 ± 0.9

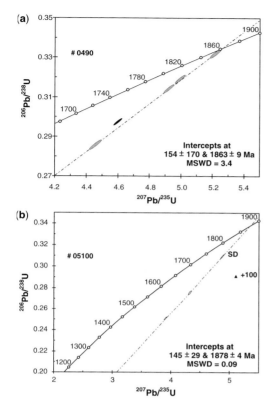

Fig. 2. U–Pb Concordia diagram and analysis of zircon crystal from (a) rhyodacite from Upper unit of the Akitkan Formation and (b) dacite from Lower unit of the Akitkan Formation.

and site means correspondingly). Table 2 shows the corresponding sample and site means and statistics, palaeomagnetic pole calculate from means after the tilt correction. The fold test of Enkin (2003) is indeterminate. Interestingly, our result (D = 192.5°, I = 18.6°, k = 50.5, α_{95} = 6.9°) is close to the old result of Gurevich from the same rocks (D = 191°, I = 14°, k = 30, α_{95} = 2°) based on a bulk demagnetization of samples without vector diagrams (Gurevich 1982, in Pisarevsky 2005, entry 5595). This proves once more that not all old results are bad results. Importantly, Gurevich reported the presence of both polarities.

In the medium-temperature interval (somewhere between 300 and 580 °C) many Zijderveld plots show some evidence of the possible presence of an additional remanence component (Fig. 4b, c, d, g). In some cases this 'medium-temperature component' may be artificial due to the overlapping of the blocking temperature spectra of the low- and high-temperature components. However, it is probably real in 22 samples in the upper part of the section. Its mean direction (after tilt correction) is D = 188°, I = −29°, k = 6.3, α_{95} = 13.5°. This is very close to the remanence directions of 1674 Ma mafic intrusions of the Chaya complex (Vodovozov et al. pers. comm.).

The medium-temperature component was also found in 29 samples of the lower part of the section, but it is very scattered there. However, suggesting the same direction of remagnetization (D = 188°, I = −29°) we applied the test for a random distribution of Shipunov et al. (1998). It revealed $\rho = 0.244$ ($\rho = n^{-1}\Sigma \cos \phi_i$, where n is the number of samples and ϕ_i is the angle between the ith unit vector and the suggested direction of the remagnetization). The critical value $\rho_c = 0.176$ is less than the dataset value of 0.244, so the partial influence of a 1674 Ma remagnetization is suggested.

Conglomerate test. We collected oriented pebbles from two conglomerate layers (sites D04-3 and D04-11, Fig. 1c) to constrain the age of our characteristic remanence by the conglomerate test of Graham (1949). The first set of pebbles includes 17 porphyritic clasts of the underlying Khibelen suite (circles in Fig. 5c, e). Their thermal demagnetization in most cases isolated just one high-temperature remanence component (Fig. 5a) with a chaotic direction (Fig. 5e), and a few samples also contain a low-temperature component with a PDF direction (Fig. 5c). A random distribution of the high-temperature component is indicated with $r/r_c = 0.161/0.388$, where $r = R/n$ and r_c is its critical value for n = 17 at the 95% confidence level (Mardia 1972). This confidence level has been chosen for all tests mentioned hereafter. The conglomerate test of Shipunov et al. (1998), with suggested direction of remagnetization of D = 192.5°, I = 18.6° (our site mean direction for the Chaya suite) is also positive: $\rho/\rho_c = 0.112/0.230$.

Seven pebbles from the same set (site D04-3) are sandstones and siltstones of the Chaya suite (triangles in Fig. 5e). Their thermal demagnetization also revealed a randomly distributed high-temperature component, according to both Rayleigh ($r/r_c = 0.371/0.597$) and Shipunov's ($\rho/\rho_c = 0.141/0.362$) tests. Hence this intra-formational conglomerate test is also positive.

Twenty-nine pebbles of siltstones and fine sandstones of the Chaya suite were collected in site D04-11 (Fig. 1c). Most of their thermal demagnetization plots show low- and high-temperature components (Fig. 5b). The low-temperature component is generally distributed around the PDF (Fig. 5d), probably being a recent viscous remanence. The distribution of the high-temperature component

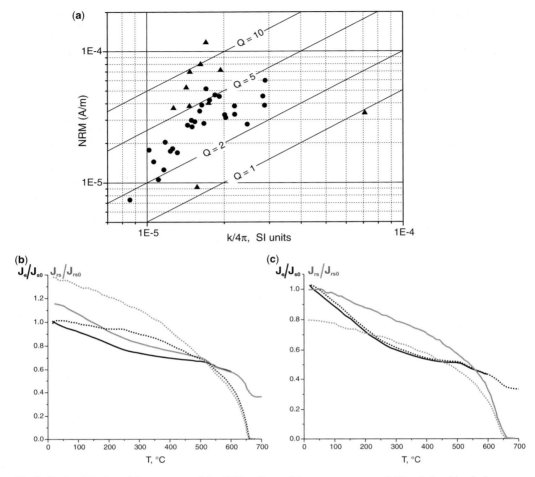

Fig. 3. Rock magnetism of the upper part of the Akitkan Group: (**a**) remanence–susceptibility relationships for igneous (triangles) and sedimentary (circles) rocks; (**b–c**) J_s-T (black) and J_{rs}-T (grey) curves (dotted lines – second heating): (**b**) site D04-13 (porphyric andesite); (**c**) site D04-15 (siltstone.)

(Fig. 5f) is not entirely random. The Rayleigh uniformity test (Mardia 1972) is negative ($r/r_c = 0.499/0.296$) and the mean direction (D = 172°, I = 20°, k = 1.93, $\alpha_{95} = 27.4°$, after the tilt correction) is close to the characteristic remanence direction for the Chaya suite (Table 2).

Thus we have three conglomerate tests in the Chaya suite. Two of them, including one intra-formational test (site D04-3) are positive, but the second intra-formational test (site D04-11) is negative. This probably means that the characteristic remanence of igneous and some sedimentary rocks of the Chaya suite is primary, but that of some other sediments may be partly remagnetized. However, there is no systematic difference between the remanence directions of igneous and sedimentary parts of our collection (Table 2, Fig. 4j). In addition, the calculated palaeopole (Table 2) is close to the c. 1850 Ma Shumikhin granites pole of Didenko et al. (2003, 2005) that is supported by the presence of two polarities (Table 4, Fig. 9) and to our c. 1880 Ma Malaya Kosa pole that proved to be primary by positive conglomerate and fold tests (see below). This puzzle could be explained if some minor remagnetization event occurred after the deposition of the Chaya suite and the direction of this overprint was similar to the characteristic remanence direction. The 1674 Ma mafic intrusions intruding the Chaya succession could cause such an overprint. We should also note that our Chaya pole lies close to the late Ordovician–Silurian part of the Siberian APWP (Didenko & Pechersky 1993; Smethurst et al. 1998). As mentioned above there is evidence of the Early–Middle Palaeozoic

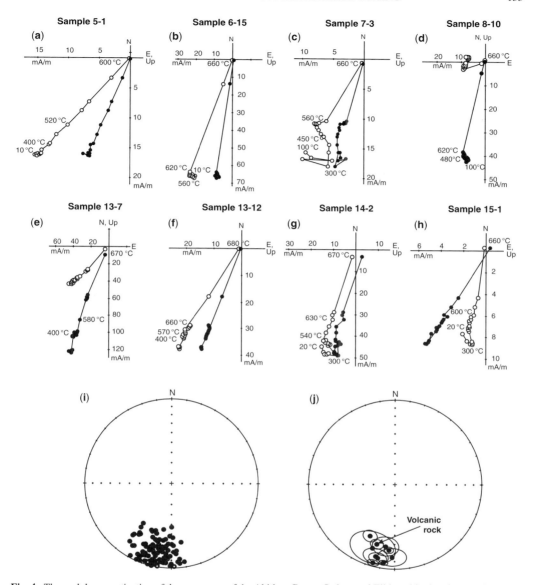

Fig. 4. Thermal demagnetization of the upper part of the Akitkan Group. Orthogonal Zijderveldt plots for: (**a–d, g–h**) sedimentary rocks; (**e–f**) igneous rocks (site D04-13); solid and open symbols denote horizontal and vertical projections, respectively. Stereographic projections of the high-temperature component: sample means (**i**) and site means (**j**); solid and open symbols represent downward and upward directions.

tectonic activity in the region and the 420 Ma syn-kinematic granites have been found in the Chaya river section (Zorin et al. 2008), so some Silurian remagnetization is also possible. The nature of remagnetization in both scenarios is not clear, because igneous and some sedimentary rocks were not affected. We think that this rules out a thermal remagnetization, but leaves a possibility for some chemical process selectively influencing some parts of the studied section.

Lower unit of the Akitkan Group

Geochronology. The geochronological sample 05100 has been collected from Khibelen dacite (Fig. 1d). This dacite conformably rests on the

Table 2. *Mean palaeomagnetic directions for high temperature component from the upper part of the Akitkan Formation (Chaya River, 57.6°N; 110.8°E)*

Site	N/n	Dip dir./Dip	In situ				Tilt corrected				Palaeomagnetic pole				
			D, °	I, °	k	α_{95}, °	D, °	I, °	K	α_{95}, °	Plat, °N	Plong, °E	d_p, °	d_m, °	φ_a, °
D04-5	6/8	348/22	185.3	7.0	40.4	10.7	188.2	26.9	40.4	10.7					
D04-6	15/15	348/28	183.0	−2.7	115.7	3.6	185.0	23.3	115.4	3.6					
D04-7	14/15	345/60	197.2	−14.0	43.3	6.1	205.1	33.4	44.2	6.0					
D04-8	15/15	335/22	188.3	−9.3	150.5	3.1	188.1	7.5	157.0	3.1					
D04-1	6/11	359/42	198.0	−16.4	14.6	18.2	198.9	21.7	14.6	18.2					
D04-2	8/12	327/39	182.0	−14.9	25.7	11.1	182.0	14.8	27.4	10.8					
D04-13	10/12	355/30	193.0	0.4	83.5	5.3	196.4	27.5	83.7	5.3					
D04-14	13/13	340/25	192.3	−9.2	40.6	6.6	192.8	11.8	41.3	6.5					
D04-15	5/5	348/27	194.1	−9.4	67.6	9.4	194.6	13.4	67.6	9.4					
D04-16	16/16	333/25	197.2	−14.9	89.0	4.7	195.3	4.7	86.7	4.0					
Sample mean:	108/128		191.1	−8.8	35.0	2.3	192.5	17.8	26.6	2.7	22.5	277.4	1.5	2.8	9.1
Site mean:	10/11		191.0	−8.4	70.2	5.8	192.5	18.6	50.5	6.9	22.1	277.5	3.7	7.2	9.6

Note: N/n, number of samples or sites used/collected; D/I, mean declination/inclination; k, Fisher's precision parameter; α_{95}, the semi-angle of the 95% cone of confidence; Plat/Plong, latitude and longitude of palaeomagnetic poles; d_p/d_m, semi-axes of the cone of confidence about the pole at the 95%; φ_a, palaeolatitude.

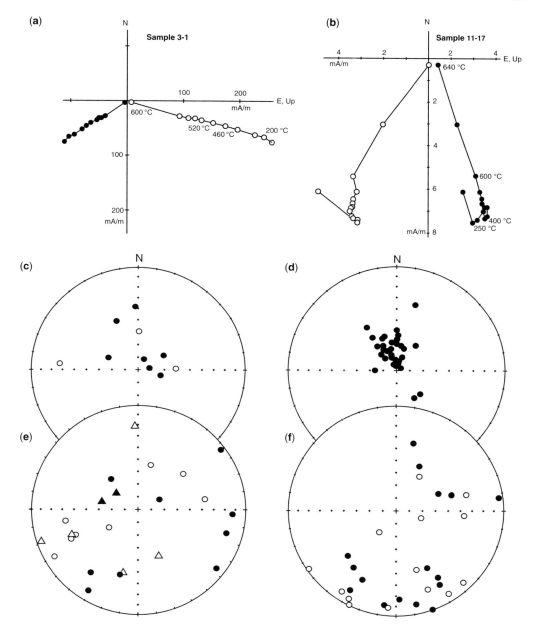

Fig. 5. Conglomerate tests for the upper part of the Akitkan Group. Left column, site D04-3; right column, site D04-11; (**a–b**) orthogonal plots; (**c–d**) stereoplots for the low-temperature component; (**e–f**) stereoplots for the high-temperature component. Triangles denote seven intra-formational pebbles of the Chaya suite. See also caption to Figure 4.

sedimentary sequence (Malaya Kosa suite) sampled for our palaeomagnetic study. The separated accessory zircons are short prismatic semi-transparent yellow crystals with smoothed terminations and sometimes visible inner zonation. The sample for analysis was handpicked avoiding grains containing inclusions or cracks. The U–Pb analyses have been done for three size fractions and one residue after selective decomposition (SD). One point containing cores (Table 1; +100) was not used in age calculations. The lower intercept age of 145 ± 29 Ma points to relatively recent lead loss. The upper intercept age is 1877.7 ± 3.8 Ma (Fig. 2b). Morphology and inner structure of zircons suggest their magmatic genesis, so we interpret the age of 1878 ± 4 Ma as the most reliable estimate for the time of the parent melt crystallization. The age of granulites near the Kaltygey Cape of 1876 ± 6 Ma (Poller *et al.* 2005) is overlapping with our new date. These granulites are overlain by the Akitkan Group, suggesting that its lower part (unmetamorphosed volcanics and sediments) has been accumulated just after the end of the collisional event accompanying the basement metamorphism. According to geological observations, the analysed dacite lies directly on the earliest sediments of the Akitkan Group. The age of this dacite could therefore be interpreted as the earliest volcanic event within the Akitkan Group. The latest volcanic activity of the lower and middle part of the Akitkan Group could be constrained by a U–Pb age of 1866 ± 6 Ma (Neymark *et al.* 1991).

Magnetic minerals and rock magnetism. The natural remanent magnetization (NRM) of red sediments ranges from $0.07–2.0 \times 10^{-5}$ A m^{-1}, and their magnetic susceptibility from about $0.5–2 \times 10^{-5}$ SI units with $Q > 1$ for the most part of the collection (Fig. 6a), indicating high palaeomagnetic stability. The J_s-T curves (both the first and the second heating) indicate Curie temperatures between 660 and 680 °C typical for the hematite

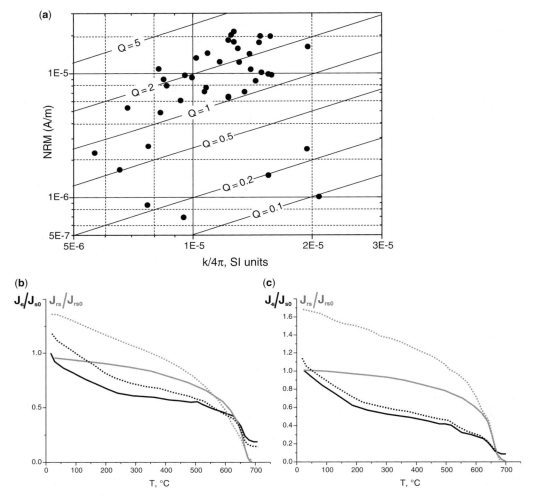

Fig. 6. Rock magnetism of the lower part of the Akitkan Group: (**a**) remanence-susceptibility relationships; (**b–c**) J_s-T (black) and J_{rs}-T (grey) curves (dotted lines–second heating): (**b**) site D05-9 (siltstone); (**c**) site D05-9 (fine sandstone).

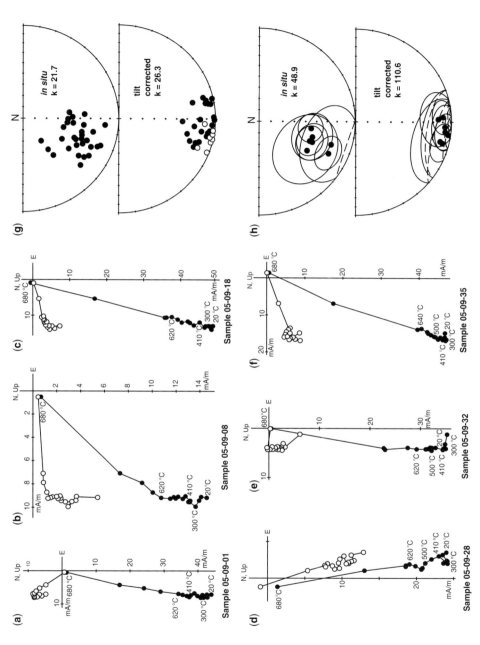

Fig. 7. Thermal demagnetization results of the lower part of the Akitkan Group: (a–f) orthogonal Zijderveldt plots; (g–h) stereographic projections of the high temperature component before and after the tilt correction: sample means (g) and site means (h). See caption to Figure 4.

(Fig. 6b, c). J_{rs}-T curves show that unblocking temperatures are mostly between 600 and 680 °C, confirming the dominance of hematite.

Palaeomagnetism. Some typical examples of the thermal demagnetization are shown in Figure 7. The random 'laboratory' viscous component and the present-day field (PDF) component have been destroyed in most samples at 200–300 °C.

The shallow SSW high-temperature hematite component was isolated in all samples (Fig. 7). The stereoplots of the high-temperature component are shown in Figure 7g (sample mean – D = 188.0°, I = 8.2°, k = 26.5, α_{95} = 5.2°) and 7 h (site mean – D = 188.7°, I = 8.1°, k = 110.6, α_{95} = 5.0°). Table 3 shows the corresponding means, palaeomagnetic pole and statistics.

The medium-temperature component has been isolated between 400 and 520–540 °C (sometimes up to 600 °C). Its direction after tilt correction (D = 183°, I = 19°, k = 12, α_{95} = 7.5°) is close to the direction of the high-temperature component in the Chaya suite (Table 2), but it is also close to the characteristic component for the Malaya Kosa suite (Table 3). We suggest that in part it may be artificial due to the overlapping of the blocking temperature spectra of the low- and high-temperature components. However, this component may be related to an overprint (c. 1674 Ma or c. 420 Ma, see above).

Field tests. Figure 7g and h demonstrate a better grouping of the high-temperature component directions after the tilt correction. The fold test of Enkin (2003) is positive with maximum grouping after 99.2% of unfolding. This test demonstrates that the characteristic high-temperature remanence of the Malaya Kosa sediments is pre-folding, that is at least pre-Silurian, as the latest folding event in the area occurred at c. 420 Ma (Zorin et al. 2008).

Twenty-five out of twenty-nine oriented pebbles of red sediments collected from the intraformational conglomerate near the base of the sampled section (Fig. 1d) revealed interpretable thermal demagnetization plots (Fig. 8a, b). The low-temperature component, with unblocking temperatures <350 °C, is distributed near the PDF direction (Fig. 8c), and the Rayleigh uniformity test (Mardia 1972) is negative (r/r_c = 0.953/0.296). The high-temperature component has been isolated in 18 pebbles. It is chaotically distributed (Fig. 8d) and the Rayleigh uniformity test (Mardia 1972) is positive (r/r_c = 0.117/0.377). Hence we conclude that the characteristic remanence of the Malaya Kosa sediments is primary. As mentioned above, the time of deposition of these sediments was very short, so the 1878 ± 4 Ma age is a good approximation of this new palaeopole.

Table 3. *Mean palaeomagnetic directions for high temperature component from the lower part of the Akitkan Formation (Khibelen Cape, 54.7°N; 108.8°E)*

Site	N/n	Dip dir./Dip	In situ			Tilt corrected			Palaeomagnetic pole			
			D, °	I, °	k	α_{95}, °	D, °	I, °	k	α_{95}, °	Plat, °N	Plong, °E	d_p, °	d_m, °	φ_a, °
D05-9(1–5)	5/5	145/50	204.6	40.3	28.3	14.6	186.5	6.9	28.3	14.6					
D05-9(6–11)	5/6	155/40	190.5	44.9	45.6	11.4	179.7	9.7	45.6	11.4					
D05-9(12–17)	4/6	155/47	214.7	41.2	11.1	28.8	196.3	9.8	11.1	28.8					
D05-6(18–23)	6/6	156/47	204.3	42.6	32.3	5.9	189.6	5.9	32.3	5.9					
D05-9(24–28)	4/5	138/42	199.4	41.3	24.8	18.8	181.0	14.5	24.8	18.8					
D05-9(29–34)	3/6	129/48	201.2	22.0	19.8	28.5	191.0	2.2	19.8	28.5					
D05-9(35–38)	3/4	130/43	208.1	21.3	96.6	12.6	196.9	7.7	96.6	12.6					
Sample mean:	30/41		203.0	38.2	21.7	5.8	188.0	8.2	26.5	5.2	30.8	279.5	2.6	5.2	4.1
Site mean:	7/8		203.4	36.4	48.9	7.6	188.7	8.1	110.6	5.0	30.8	278.7	2.5	5.0	4.0

Note: see Table 2.

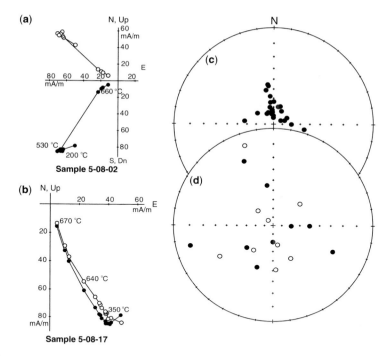

Fig. 8. Conglomerate test for the lower part of the Akitkan Group (Malaya Kosa). (**a–b**) orthogonal Zijderveldt plots. Stereoplots: (**c**) low-temperature component; (**d**) high-temperature component. See caption to Figure 4.

Discussion

The results of this study provide two reliable Siberian Palaeoproterozoic palaeopoles. The reliability index Q of Van der Voo (1990) for the 1878 Ma Malaya Kosa pole is 6, but we are more cautious about the Chaya pole due to less convincing conglomerate tests. These two poles, together with the Shumikhin pole (Didenko et al. 2003, 2005), provide robust constraints on the palaeopositions of Siberia at 1850–1880 Ma. Lack of reliable Mesoproterozoic palaeopoles prevents a construction of the pre-Neoproterozoic Siberian APWP, and we must consider two polarity options for any reconstruction. However, there are some direct tectonic applications of our new results.

The age of the Malaya Kosa pole is remarkably close to the age of the highly reliable 1880 Ma Molson dykes B-pole (Halls & Heaman 2000). Such closeness in age of two reliable poles from different continents is rare for pre-Neoproterozoic times (Cawood et al. 2006; Evans & Pisarevsky 2008). As the latest Mesoproterozoic Laurentia-Siberia reconstruction is palaeomagnetically constrained (Gallet et al. 2000; Pavlov et al. 2000, 2002; Pisarevsky & Natapov 2003), our study provides an important test for the differential movements of continental blocks in the late Palaeoproterozoic–Mesoproterozoic, challenged by Stern (2005). Stern argued that the modern-style subduction-related plate tectonics did not exist before 1 Ga, as in his opinion such evidence for an ancient subduction as blueschists, high-pressure metamorphics and ophiolite complexes are apparently absent in pre-Neoproterozoic. High-quality palaeomagnetic data are crucial to test this suggestion, because they may indicate the differential movements of different continents. Two pairs of coeval poles from two continental blocks allow a comparison of their mutual positions at two different time slices. Cawood et al. (2006) and Evans & Pisarevsky (2008) found only a few examples of such evidence of differential movements between Superior and Kalahari, Baltica and Australia, Superior and Karelia. Now we have an opportunity to test late Palaeoproterozoic–Mesoproterozoic movements between Siberia and Superior/Laurentia. Figure 9a and b demonstrate possible Superior–Siberia reconstructions at 1880 Ma in two polarity options based on the Malaya Kosa and Molson B poles. Multiple images of Siberia for each option highlight longitudinal uncertainty in craton position.

Pisarevsky & Natapov (2003) made a Siberia–Laurentia late Mesoproterozoic reconstruction using a best fit between the c. 950–1050 Ma set of

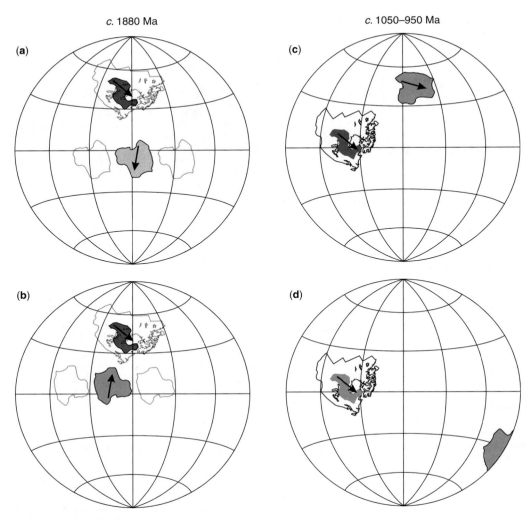

Fig. 9. Superior–Siberia reconstructions in two polarity options at (**a–b**) 1880 Ma; (**c–d**) 1050–950 Ma (after Pisarevsky & Natapov 2003). The arrows indicate to the modern 'North'.

poles from the Uchur-Maya area (Gallet et al. 2000; Pavlov et al. 2000, 2002) and the coeval fragment of the Laurentian APWP (Pisarevsky & Natapov 2003 for reference). Two polarity options are shown in Figure 9c and d. In this case we do not have a longitudinal uncertainty, because APWP fragments were used rather than individual poles. Figure 9 demonstrates a significant difference between relative palaeopositions of Superior/Laurentia and Siberia between c. 1880 and c. 1000 Ma, suggesting independent drift of two continents in the Mesoproterozoic, which is possible only in the present-day style tectonic environments with generation and consumption of lithosphere between these blocks on a constant-radius Earth. However, this should be considered just as a reasonable suggestion, because most of 950–1050 Ma Siberian and Laurentian poles cannot be considered as 'key-poles' for various reasons, mainly their poor age constraints (Evans & Pisarevsky 2008).

Conclusions

(1) Two parts of the Palaeoproterozoic Akitkan Group of south Siberia are dated at 1863 ± 9 and 1878 ± 4 Ma by the U–Pb zircon conventional method, confirming the earlier published ages (Neymark et al. 1991; Larin et al. 2003).

(2) A stable high-temperature remanence has been isolated in sedimentary rocks of the lower part of the Akitkan Group. Positive conglomerate and fold tests confirm that this magnetization is primary. The corresponding Malaya Kosa pole is highly reliable and may be considered as the 1878 Ma key-pole for Siberia.

(3) A stable high-temperature remanence has been isolated in sedimentary rocks of the upper part of the Akitkan Group. Conglomerate tests in two conglomerate layers gave contrasting results, suggesting that this remanence is primary in some studied rocks, but it might be contaminated in other rocks by a secondary remagnetization of uncertain nature. However, all remanence directions are coherent and resemble the direction of coeval rocks in a distal area. We suggest that the overprint direction could be close to the primary direction and the palaeopole is reliable.

(4) A comparison of coeval Malaya Kosa pole and Laurentian Molson B pole demonstrate that the Siberian craton changed its position with respect to the Superior craton during the Mesoproterozoic.

The authors thank E. Sklyarov, I. Kozakov, A. Bukharov, S. Shipunov, A. Dvorova and N. Dvorova for discussion and assistance during this work. This research was supported by the Branch of the Earth Sciences and Siberian Branch of Russian Academy Sciences (Program 10), and Russian Foundation for Basic Research (Project nos. 06-05-64352, 06-05-64458). We thank K. Buchan, R. Ernst and D. Evans for their constructive reviews.

References

ADE-HALL, J. M., PALMER, H. C. & HUBBARD, T. P. 1971. The magnetic and opaque petrological response of basalts to regional hydrothermal alteration. *Geophysical Journal International*, 24, 137–174.

BIBIKOVA, E. V., KORIKOVSKY, S. P., KIRNOZOVA, T. I., SUMIN, L. V., ARAKELYANTS, M. M., FEDOROVSKY, V. S. & PETROVA, Z. I. 1987. Age determinations of the rocks of the Baikal-Vitim greenstone belt by isotopic geochronological methods. In: SHULYUKOV, YU. A. (ed.) *Isotopic Dating of the Metamorphic and Metasomatic Processes*. Nauka, Moscow, 154–164 (in Russian).

BORUKAEV, ZH. B. 1985. *Structures of the Precambian and Plate Tectonics*. Nauka, Novosibirsk (in Russian).

BUKHAROV, A. A. 1987. *Protoactivated Zones of Ancient Platforms*. Nauka, Novosibirsk (in Russian).

CAWOOD, P. A., KRÖNER, A. & PISAREVSKY, S. A. 2006. Precambrian plate tectonics: criteria and evidence. *GSA Today*, 16, 4–11.

CONDIE, K. C. 2002. Breakup of a Palaeoproterozoic supercontinent. *Gondwana Research*, 5, 41–43.

CONDIE, K. C. & ROSEN, O. M. 1994. Laurentia-Siberia connection revised. *Geology*, 22, 168–170.

DALZIEL, I. W. D. 1997. Neoproterozoic-Paleozoic geography and tectonics: review, hypothesis, environmental speculation. *Geological Society of America Bulletin*, 109, 16–42.

DIDENKO, A. N. & PECHERSKY, D. M. 1993. Revised Paleozoic apparent polar wander paths for E. Europe, Siberia, N. China and Tarim plates. *L.P. Zonenshain Memorial conference on Plate Tectonics*. Moscow, 47–48 (in Russian).

DIDENKO, A. N., KOZAKOV, I. K. ET AL. 2003. Palaeoproterozoic granites of the Sharyzhalgai block, Siberian craton: palaeomagnetism and geodynamic inferences. *Doklady Earth Sciences*, 390, 510.

DIDENKO, A. N., VODOVOZOV, V. YU., KOZAKOV, I. K. & BIBIKOVA, E. V. 2005. Palaeomagnetic and geochronological study of post-collisional Early Proterozoic granitoids in the Southern Siberian Platform: methodological and geodynamic aspects. *Izvestiya, Physics of the Solid Earth*, 41, 156–172.

DONSKAYA, T. V., SAL'NIKOVA, E. B. ET AL. 2002. Early Proterozoic postcollision magmatism at the southern flank of the Siberian craton: new geochronological data and geodynamic implications. *Doklady Earth Sciences*, 383, 125–128.

DONSKAYA, T. V., GLADKOCHUB, D. P., KOVACH, V. P. & MAZUKABZOV, A. M. 2005. Petrogenesis of Early Proterozoic postcollisional granitoids of the southern Siberian craton. *Petrology*, 13, 229–252.

DONSKAYA, T. V., MAZUKABZOV, A. M. ET AL. 2007. Stratotype of the Chaya Formation of the Akitkan Group in the North Baikal volcanoplutonic belt: age and time of sedimentation. *Russian Geology and Geophysics*, 48, 707–710.

ENKIN, R. J. 2003. The direction-correction tilt test: an all-purpose tilt/fold test for palaeomagnetic studies. *Earth and Planetary Science Letters*, 212, 151–166.

ERNST, R. E., BUCHAN, K. L., HAMILTON, M. A., OKRUGIN, A. V. & TOMSHIN, M. D. 2000. Integrated palaeomagnetism and U–Pb geochronology of mafic dikes of the Eastern Anabar shield region, Siberia: implications for Mesoproterozoic paleolatitude of Siberia and comparison with Laurentia. *Journal of Geology*, 108, 381–401.

EVANS, D. A. D. & PISAREVSKY, S. A. 2008. Plate tectonics on the early Earth? – weighing the palaeomagnetic evidence. In: CONDIE, K. & PEASE, V. (eds) *When Did Plate Tectonics Begin?* Geological Society of America Special Paper 440, 249–263.

GALLET, Y., PAVLOV, V. E., SEMIKHATOV, M. A. & PETROV, P. Y. 2000. Late Mesoproterozoic magnetostratigraphic results from Siberia: paleogeographic implications and magnetic field behavior. *Journal of Geophysical Research*, 105, 16 481–16 500.

GLADKOCHUB, D. P., WINGATE, M. T. D., PISAREVSKY, S. A., DONSKAYA, T. V., MAZUKABZOV, A. M., PONOMARCHUK, V. A. & STANEVICH, A. M. 2006a. Mafic intrusions in southwestern Siberia and implications for a Neoproterozoic connection with Laurentia. *Precambrian Research*, 147, 260–278.

GLADKOCHUB, D., PISAREVSKY, S. A., DONSKAYA, T., NATAPOV, L. M., MAZUKABZOV, A., STANEVICH, A. M. & SLKYAROV, E. 2006b. The Siberian Craton

and its evolution in terms of the Rodinia hypothesis. *Episodes*, **29**, 169–174.
GLADKOCHUB, D. P., DONSKAYA, T. V., MAZUKABZOV, A. M., STANEVICH, A. M., SKLYAROV, E. V. & PONOMARCHUK, V. A. 2007. Signature of Precambrian extension events in the southern Siberian craton. *Russian Geology and Geophysics*, **48**, 17–31.
GRAHAM, J. W. 1949. The stability and significance of magnetism in sedimentary rocks. *Journal of Geophysical Research*, **54**, 131–167.
HALLS, H. C. & HEAMAN, L. M. 2000. The palaeomagnetic significance of new U–Pb age data from the Molson dyke swarms, Cauchon Lake area, Manitoba. *Canadian Journal of Earth Sciences*, **37**, 957–966.
HOFFMAN, P. F. 1991. Did the breakout of Laurentia turn Gondwana inside out? *Science*, **252**, 1409–1412.
KHAIN, V. E. & BOZHKO, N. A. 1988. *Historical Geotectonics Precambrian*. Nedra, Moscow (in Russian).
KHRAMOV, A. N. 1987. *Paleomagnetology*. Springer-Verlag, Berlin.
KIRSCHVINK, J. L. 1980. The least-square line and plane and the analysis of palaeomagnetic data. *Geophysical Journal of the Royal Astronomical Society*, **62**, 699–718.
KROGH, T. E. 1973. A low-contamination method for hydrothermal decomposition of zircon and extraction of U and Pb for isotopic age determination. *Geochimica et Cosmochimica Acta*, **37**, 485–494.
LARIN, A. M., SAL'NIKOVA, E. B. ET AL. 2003. The North Baikal volcanoplutonic belt: age, formation duration, and tectonic setting. *Doklady Earth Sciences*, **392**, 963.
LUDWIG, K. R. 1999. *ISOPLOT/Ex.Version 2.06. A Geochronological Toolkit for Microsoft Excel*. Berkley Geochronology Center, Special Publications, 1a.
MARDIA, K. V. 1972. *Statistics of Directional Data*. Academic Press, London.
MATS, V. D. 1965. *Upper Precambrian of the West Gisbaikalia and North-Baikalian upland (stratigraphy and historic development)*. PhD Thesis, Geological Institute SB RAS, Novosibirsk (in Russian).
MATS, V. D., BUKHAROV, A. A. & EGOROVA, O. P. 1968. Stratigraphy and some lithological affinities of volcanic-sedimentary sequences of the North-Baikal volcanic-plutonic belt. *In*: ODINTSOV, M. M. (eds) *Geology and Geophysics of the Siberian Platform*. VSKI, Irkutsk, 116–139 (in Russian).
MATTINSON, J. M. 1994. A study of complex discordance in zircons using step-wise dissolution techniques. *Contributions to Mineralogy and Petrology*, **116**, 117–129.
MAZUKABZOV, A. M., GLADKOCHUB, D. P., DONSKAYA, T. V., BUKHAROV, A. A. & STANEVICH, A. M. 2006. North-Baikal volcanic-plutonic belt: the analysis of geochronological data of volcanic-sedimentary sequences. *In*: GORDIENKO, I. V. (eds) *Volcanism and Geodynamics*. BNC, Ulan-Ude, **2**, 473–476 (in Russian).
MCELHINNY, M. W. & MCFADDEN, P. L. 2000. *Palaeomagnetism: Continents and Oceans*. Academic Press, San Diego.
NAGATA, T. 1961. *Rock Magnetism*. Marusen Company Ltd, Tokyo.

NEYMARK, L. A., LARIN, A. M., YAKOVLEVA, S. Z., SRYVTSEV, N. A. & BULDUGEROV, V. V. 1991. The new dates about age (U–Pb zircons) Akitkan Group of the Baikal-Patom Fold Belt. *Doklady Earth Sciences*, **320**, 182–186 (in Russian).
PAVLOV, V. E., GALLET, I., PETROV, P. Yu., ZHURAVLEV, D. Z. & SHATSILLO, A. V. 2002. The Ui Group and Late Riphean sills in the Uchur-Maya region: Isotopic and palaeomagnetic data and the problem of the Rodinia supercontinent. *Geotektonika*, **4**, 26–41.
PAVLOV, V. E., GALLET, Y. & SHATSILLO, A. V. 2000. Palaeomagnetism of the upper Riphean Lakhanda Group of the Uchur–Maya area and the hypothesis of the late Proterozoic supercontinent. *Fiz. Zemli* **8**, 23–34 (in Russian).
PECHERSKY, D. M., BAGIN, V. I., BRODSKAYA, S. YU. & SHARONOVA, Z. V. 1975. *Magnetism and Conditions of Origin in Extrusive Rocks*. Nauka, Moscow, (in Russian).
PIPER, J. D. A. 2000. The Neoproterozoic Supercontinent: Rodinia or Palaeopangaea? *Earth and Planetary Science Letters*, **176**, 131–146.
PISAREVSKY, S. A. 2005. New edition of the global palaeomagnetic database. *EOS transactions*, **86**, 170.
PISAREVSKY, S. A. & NATAPOV, L. M. 2003. Siberia and Rodinia. *Tectonophysics*, **375**, 221–245.
PISAREVSKY, S. A., WINGATE, M. T. D., POWELL, C. MCA., JOHNSON, S. & EVANS, D. A. D. 2003. Models of Rodinia assembly and fragmentation. *In*: YOSHIDA, M., WINDLEY, B. & DASGUPTA, S. (eds) *Proterozoic East Gondwana: Supercontinent Assembly and Breakup*. Geological Society, London, Special Publications, **206**, 35–55.
PISAREVSKY, S. A., NATAPOV, L. M., DONSKAYA, T. V., GLADKOCHUB, D. P. & VERNIKOVSKY, V. A. 2007. Proterozoic Siberia: a promontory of Rodinia. *Precambrian Research* doi:10.1016/j.precamres.2007.04.016.
POLLER, U., GLADKOCHUB, D., DONSKAYA, T., MAZUKABZOV, A., SKLYAROV, E. & TODT, W. 2005. Multistage magmatic and metamorphic evolution in the Southern Siberian Craton: Archean and Paleoproterozoic zircon ages revealed by SHRIMP and TIMS. *Precambrian Research*, **136**, 353–368.
RAINBIRD, R. H., STERN, R. A., KHUDOLEY, A. K., KROPACHEV, A. P., HEAMAN, L. M. & SUKHORUKOV, V. I. 1998. U–Pb geochronology of Riphean sandstone and gabbro from southeast Siberia and its bearing on the Laurentia – Siberia connection. *Earth and Planetary Science Letters*, **164**, 409–420.
ROGERS, J. J. W. 1996. A history of continents in the past three billion years. *Journal of Geology*, **104**, 91–107.
ROGERS, J. J. W. & SANTOSH, M. 2002. Configuration of Columbia, a Mesoproterozoic supercontinent. *Gondwana Research*, **5**, 5–22.
ROSEN, O. M., CONDIE, K. C., NATAPOV, L. M. & NOZHKIN, A. D. 1994. Archean and Early Proterozoic evolution of the Siberian craton: a preliminary assessment. *In*: CONDIE, K. C. (ed.) *Archean Crustal Evolution*. Elsevier, Amsterdam, 411–459.
ROSEN, O. M., MANAKOV, A. V. & SERENKO, V. P. 2005. Palaeoproterozoic collisional system and diamondiferous lithospheric keel of the Yakutian

kimberlite province. *Russian Geology and Geophysics*, **46**, 1259–1272.

SALOP, L. I. 1967. *Geology of Baykal Mountain Area*. Book 1. Nedra, Moscow (in Russian).

SHIPUNOV, S. V., MURAVIEV, A. A. & BAZHENOV, M. L. 1998. A new conglomerate test in palaeomagnetism. *Geophysical Journal International*, **133**, 721–725.

SKLYAROV, E. V., GLADKOCHUB, D. P., MAZUKABZOV, A. M., MEN'SHAGIN, YU. V., WATANABE, T. & PISAREVSKY, S. A. 2003. Neoproterozoic mafic dike swarms of the Sharyzhalgai metamorphic massif (southern Siberian craton). *Precambrian Research*, **122**, 359–376.

SMETHURST, M. A., KHRAMOV, A. N. & TORSVIK, T. H. 1998. The Neoproterozoic and Paleozoic palaeomagnetic data for the Siberian platform: from Rodinia to Pangea. *Earth Science Reviews*, **43(1)**, 1–24.

STACEY, J. S. & KRAMERS, I. D. 1975. Approximation of terrestrial lead isotope evolution by a two-stage model. *Earth and Planetary Science Letters*, **26**, 207–221.

STERN, R. J. 2005. Evidence from ophiolites, blueschists, and ultrahigh-pressure metamorphic terranes that the modern episode of subduction tectonics began in Neoproterozoic time. *Geology*, **33**, 557–560.

VAN DER VOO, R. 1990. The reliability of palaeomagnetic data. *Tectonophysics*, **184**, 1–9.

WINDLEY, B. F. 1995. *The Evolving Continents*, 3rd edn. Wiley, Chichester, 52.

ZHAO, G., CAWOOD, P. A., WILDE, S. A. & SUN, M. 2002. Review of global 2.1–1.8 Ga orogens: implications for a pre-Rodinia supercontinent. *Earth-Science Reviews*, **59**, 125–162.

ZORIN, YU. A., MAZUKABZOV, A. M., GLADKOCHUB, D. P., DONSKAYA, T. V., PRESNYAKOV, S. L. & SERGEEV, S. A. 2008. Silurian age of main deformation in the Riphean strata of the Baikal-Patom zone. *Doklady Earth Sciences*, **423**, 1–6.

Timing and duration of Palaeoproterozoic events producing ore-bearing layered intrusions of the Baltic Shield: metallogenic, petrological and geodynamic implications

T. BAYANOVA[1]*, J. LUDDEN[2] & F. MITROFANOV[1]

[1]*Geological Institute, Kola Science Centre RAS, 14 Fersman Street, Apatity, Russia*

[2]*British Geological Survey, Keyworth, Nottingham NG12 5GG, UK*

Corresponding author (e-mail: tamara@geoksc.apatity.ru)

Abstract: There are two 300–500 km long belts of Palaeoproterozoic layered intrusions in the Baltic (Fennoscandian) Shield; the Northern (Kola) Belt and the Southern (Fenno-Karelian) Belt. New U–Pb (TIMS) ages and radiogenic isotopic (Nd–Sr–He) data have been determined for mafic-ultramafic Cu–Ni–Ti–Cr and PGE-bearing layered intrusions of the Kola Belt. U–Pb zircon and baddeleyite data from gabbronorite and anorthosite bodies of the Fedorovo-Pansky, Monchepluton, Main Ridge (Monchetundra and Chunatundra) and Mt Generalskaya intrusions, and from gabbronorite intrusions and dykes associated with the Imandra lopolith, yield ages from c. 2.52–2.39 Ga. The age range of 130 Ma recorded in the Kola Belt samples, associated with at least four intrusive phases (three PGE-bearing and one barren), is significantly greater than that for intrusions of the Southern (Fenno-Karelian) Belt which clusters at 2.44 Ga. Nd isotopic values for the Kola Belt range from -1.1 to -2.4 and indicate an enriched mantle 'EM-1 type' reservoir for these layered intrusions. Initial Sr isotopic data for the Kola intrusions are radiogenic relative to bulk mantle, with I_{Sr} values from 0.703 to 0.704, but geochemical data and $^4He/^3He$ isotopic ratios of various minerals record a significant contribution from a mantle source rather than simply crustal melting. The geological and geochronological data indicate that in the eastern part of the Baltic Shield, mafic-ultramafic intrusive magmatism was active over a protracted period and was related to plume magmatism associated with continental breakup that also involved the Superior and Wyoming provinces.

There are about 20 Palaeoproterozoic layered mafic-ultramafic bodies in Finland, most of which occur in a roughly east–west-trending, 300 km-long belt known as the Tornio-Näränkävaara Belt (Alapieti *et al.* 1990; Vogel *et al.* 1998; Iljina & Hanski 2005). The belt (Fig. 1) extends for a few kilometres into Sweden (Tornio intrusion), and for several tens of kilometres into the Russian Karelia (Olanga complex). Together the intrusions make up the Southern, or Fenno-Karelian Belt, FKB (Mitrofanov *et al.* 1997).

In the NE of the province, the Northern, or Kola Belt (KB) strikes northwestwards for about 500 km (Fig. 1). It includes more than ten isolated layered mafic-ultramafic bodies that are mostly ore-bearing (Mitrofanov *et al.* 1997). The central part of the Kola Belt has been suggested to be part of a triple junction typical of intraplate rifting (Pirajno 2007) and is occupied by the Monchegorsk Layered Complex with a fairly complete range of ore types (Cr, Cu, Ni, Co, Ti, V, Pt, Pd, Rh). The western and eastern arms of the triple junction are composed of large anorthosite-troctolite (Main Ridge, Pyrshin,

Kolvitsa) intrusions (Fig. 1). The most typical PGE-bearing layered pyroxenite-norite-gabbro-anorthosite intrusions of the Kola Belt (e.g. Mt Generalskaya, Monchegorsk Layered Complex, Fedorovo-Pansky) are confined to boundaries between early Proterozoic rifts which were in-filled with volcano-sedimentary rocks overlying the Archaean basement (Schissel *et al.* 2002; Mitrofanov *et al.* 2005). In these cases, similar to those in Finland, the intrusive rocks underwent relatively low-grade local metamorphism and preserve cumulus and intracumulus minerals.

Convincing arguments in support of the mantle plume hypothesis, either as 'shallow plumes' (from c. 670 km) or 'deep plumes' (from the core – mantle boundary) have been put forward for relatively young well-preserved Palaeozoic and recent large igneous provinces (LIPs) (Coffin & Eldhom 1994; Heaman 1997; Ernst & Buchan 2003; French *et al.* 2008). Voluminous magmatism is considered to be related to mantle plumes that occurred throughout the Precambrian (Condie 2001). The best records of a plume source are

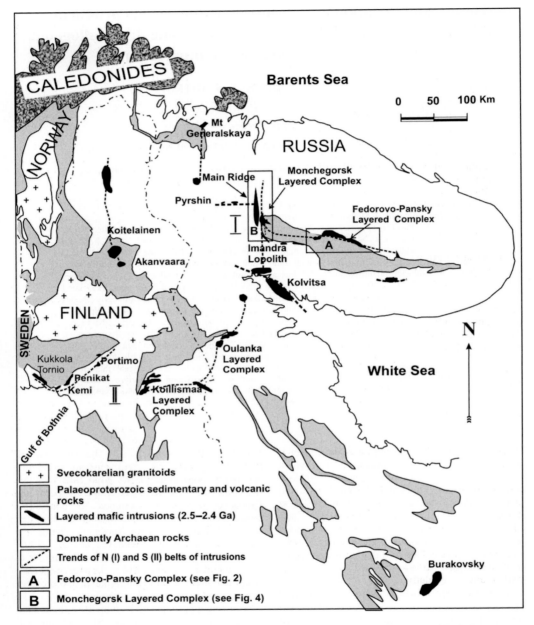

Fig. 1. Generalized geological map of the northeastern part of the Baltic Shield and the location of Early Proterozoic mafic layered intrusions.

evident in Neoarchaean and Palaeoproterozoic magmatic events and often have been associated with mineralization (Campbell 2001; Pirajno 2007). Many of these magmas have generated layered mafic-ultramafic complexes with exploitable PGE deposits (Pt, Pd and Rh), and Cr, Ni, Cu, Co, Ti and V mineralization (Li & Naldrett 1993; Mitrofanov et al. 1997, 2005). Magmatic igneous ore systems are present in the Windimurra (2.8 Ga), Stillwater (2.7 Ga), Bushveld (2.06 Ga) and Duluth (1.1 Ga) complexes (Li & Naldrett 1993; Eules & Cawthorn 1995; Pirajno 2007). Many of these intrusions include massif-type anorthosites and troctolites enriched in Cr, Ni,

Cu, PGE, Ti, V and Fe. All the above-mentioned rock types, metals, host rift-associated volcanic rocks and mafic dykes are found in the relatively recently described East Scandinavian Palaeoproterozoic Large Igneous Province (ESCLIP) (Iljina & Hanski 2005) with a total area of more than 200 000 km^2 (Fig. 1). In Finland, these geological complexes have been widely studied (Alapieti et al. 1990; Huhma et al. 1990; Vogel et al. 1998; Hanski et al. 2001) and the data were summarized by Iljina & Hanski (2005). Only a few publications on similar Russian complexes of the Baltic Shield have been written or translated into English (Papunen & Gorbunov et al. 1985; Balashov et al. 1993; Bayanova & Balashov 1995; Amelin et al. 1995; Sharkov et al. 1995; Mitrofanov et al. 1997; Mitrofanov & Bayanova 1999; Chashchin et al. 2002; Schissel et al. 2002; Mitrofanov et al. 2002, 2005).

This chapter presents a brief geological description of the Russian mafic-ultramafic intrusions of the Baltic Shield and associated mineralization. It focuses on new U–Pb (TIMS) and Sm–Nd geochronological data which constrain timing of magmatic pulses and the duration of the emplacement of Cr, Cu, Ni, Ti and PGE-bearing layered intrusions of the Kola Belt. Nd, Sr and He-isotope data help define geodynamic models for a long-lived early Precambrian mantle source expressed either in a large mantle diapir or multiple plume processes for one of the earliest clearly identifiable old intraplate LIPs and its metallogeny.

Geological setting of ore-bearing intrusions of the Baltic Shield

Palaeoproterozoic layered pyroxenite-peridotite-gabbronorite-anorthosite ore-bearing intrusions form two belts in the eastern part of the Baltic (Fennoscandian) Shield (Fig. 1). The Northern Belt strikes northwestward for over 500 km and is confined to the southwestern edge of the Archaean Kola-Norwegian Block and to the northern and southern borders of the Palaeoproterozoic Pechenga-Imandra-Varzuga rift. The Northern Belt includes such intrusions as Mt Generalskaya, Fedorovo-Pansky, Monchegorsk Layered Complex and Imandra lopolith, which are described in this chapter. The Southern Belt contains intrusions of the Olanga group (Kivakka, Tsipringa and Lukkulaisvaara) and the Burakovsky intrusion situated in Karelia, and some Finnish intrusions (e.g. Koilismaa, Näränkävaara, Koitilainen, Penikat, Akanvaara, Kemi). The Southern Belt stretches approximately from east to west over 350 km and occupies the northern margin of the Karelian craton. The intrusions of the Southern Belt formed in a similar pre-rifting geodynamic setting, and are generally confined to the margins of the intracontinental rift structures and are located at contacts with the Archaean basement.

Geology and petrology of the layered intrusions of the Kola Belt

Mt Generalskaya

This intrusion (Fig. 1) has a prominent cross-cutting contact with the Archaean gneiss complex in the west. The basal conglomerates associated with the Palaeoproterozoic Pechenga rift contain abundant gabbronorite pebbles in the Luostari area. The intrusion is cut by quartz dolerite dykes, which are similar in composition to the lower volcanogenic Sariola Majarvi Suite of the Pechenga structure. The intrusion crops out over an area of 3.5×1.5 km, but its actual size is assumed to be greater. It gently dips southwestward at an angle of $30°-35°$ underlying the conglomerates of the Televi Suite and volcanic andesite–basaltic flows of the Majarvi Suite, both of which constitute the lower part of the Pechenga section. The whole rock Rb–Sr age of the volcanic sequence is 2324 ± 28 Ma (Bayanova et al. 1999). The NE-trending intrusion ($10°-20°$) has a keel-like shape and a well-defined internal structure. The eastern and western contacts dip towards each other at angles of $60-65°$ and $30-50°$, respectively. The intrusion is dissected by younger faults of northeastern and northwestern strike; one of the faults separates the intrusion into two large blocks. The total thickness of the intrusive sequence increases from $200-1700$ m from NE to SW. The section is dominated by gabbronorite, olivine gabbronorite, gabbro, anorthosite or leucogabbro, norite, bronzitite, whereas serpentinized peridotite is minor (Hanski et al. 1990; Bayanova et al. 1999; Skuf'in & Bayanova 2006). Low sulphide disseminated PGE mineralization is now being explored by a mining company.

Geological observations and two zircon U–Pb ages (2496 ± 10 and 2446 ± 10 Ma, see below) for the rocks of the intrusion show that anorthositic injections took place later than gabbronoritic ones (Bayanova et al. 1999).

The Fedorovo-Pansky Complex

The Fedorovo-Pansky Layered Complex (Fig. 1) outcrops over an area of >400 km^2. It strikes northwestwards for >60 km and dips southwestwards at an angle of $30-35°$. The total rock sequence is about $3-4$ km thick. Tectonic faults divide the complex into several blocks. The major blocks from west to east (Fig. 2) are known as the Fedorov, the Lastjavr, the Western Pansky and the Eastern Pansky

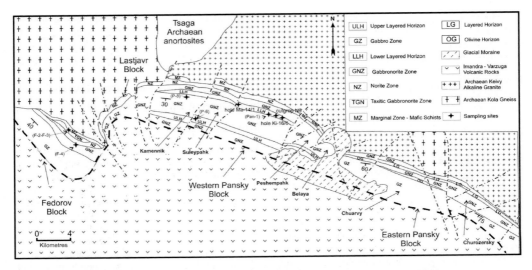

Fig. 2. General geological map of the Fedorovo-Pansky Layered Complex (Mitrofanov *et al.* 2005).

(Mitrofanov *et al.* 2005). The Fedorovo-Pansky complex is bordered by the Archaean Keivy terrane and the Palaeoproterozoic Imandra-Varzuga rift. The rocks of the complex crop out close to the Archaean gneisses only in the northwestern extremities, but their contacts cannot be established due to poor exposure. In the north, the complex borders with the alkaline granites of the White Tundra intrusion. The alkaline granites were recently proved to be Archaean with a U–Pb zircon age of 2654 ± 15 Ma (Bayanova 2004; Zozulya *et al.* 2005). The contact of the Western Pansky Block with the Imandra-Varzuga volcano-sedimentary sequence is mostly covered by Quaternary deposits. However, drilling and excavations in the south of Mt Kamennik reveal a strongly sheared and metamorphosed contact between the intrusion and overlying Palaeoproterozoic volcano-sedimentary rocks that we interpret to be tectonic in origin.

The Fedorovo-Pansky Complex comprises predominantly gabbronorites with varying proportions of mafic minerals and different structural features (Fig. 3). From bottom up, the composite layered sequence is as follows (Fig. 3):

- Marginal Zone (50–100 m) of plagioclase–amphibole schists with relicts of massive fine-grained norite and gabbronorite, which are referred to as chilled margin rocks;
- Taxitic Zone (30–300 m) that contains ore-bearing gabbronoritic matrix (2485 Ma, see below) and early xenoliths of plagioclase-bearing pyroxenite and norite (2526–2516 Ma, see below). Syngenetic and magmatic ores are represented by Cu and Ni sulphides with Pt, Pd and Au, and Pt and Pd sulphides, bismuthotellurides and arsenides;
- Norite Zone (50–200 m) with cumulus interlayers of harzburgite and plagioclase-bearing pyroxenite that includes an intergranular injection Cu–Ni–PGE mineralization in the lower part. The rocks of the zone are enriched in chromium (up to 1000 ppm) and contain chromite that is also typical of the rocks of the Penikat and Kemi intrusions (Finland) derived from the earliest magma portion (Iljina & Hanski 2005). Basal Cu–Ni–PGE deposits of the Fedorov Block have been explored and prepared for licensing (Schissel *et al.* 2002; Mitrofanov *et al.* 2005).
- Main Gabbronorite Zone (*c.* 1000 m) that is a thickly layered 'stratified' rock series (Fig. 3) with a 40–80 m thinly layered lower horizon (LLH) at the upper part. The LLH consists of contrasting alteration of gabbronorite, norite, pyroxenite and interlayers of leucocratic gabbro and anorthosite. The LLH contains a reef-type PGE deposit poor in base-metal sulphides. The deposit is now being extensively explored (Mitrofanov *et al.* 2005). According to the field investigations (Latypov & Chistyakova 2000), the LLH anorthositic layers have been intruded later, as shown by cutting injection contacts. This is confirmed by a zircon U–Pb age for the anorthosite of 2470 ± 9 Ma (see below).
- Upper Layered Horizon (ULH) between the Lower and Upper Gabbro Zones. The ULH consists of olivine-bearing troctolite, norite,

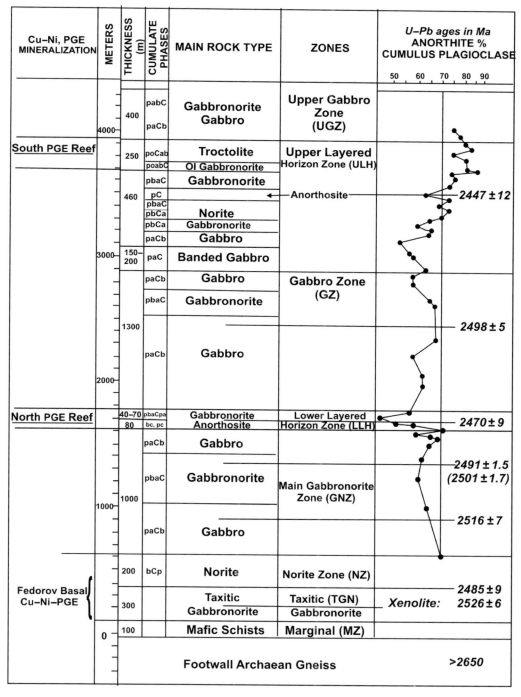

Fig. 3. Composite 'stratigraphic' section of the Fedorovo-Pansky Complex with Cu–Ni and PGE mineralization (modified after Schissel et al. 2002). The cumulate mineral terminology used in this paper is that of cumulate phase minerals in small letters, in order of volume percent, preceding the capital C for cumulate, with postcumulate mineral phases following. Major mineral abbreviations are: a, augite; b, bronzite; c, chromite; o, olivine; p, plagioclase (see Table 5 for references). Modified after Schissel et al. 2002.

gabbronorite and anorthosite (Fig. 3). It comprises several layers of rich PGE (Pd ≫ Pt) ore poor in base-metal sulphides (Mitrofanov et al. 2005). The U–Pb age on zircon and baddeleyite of the ULH rocks of 2447 ± 12 Ma (see below) is the youngest among those obtained for the rocks of the Fedorovo-Pansky Complex.

The Monchepluton, intrusions of the Main Ridge (Monchetundra and Chunatundra) and adjacent intrusions – Monchegorsk Layered Complex

The Monchegorsk Layered Complex (Fig. 4) has long been the subject of detailed investigation due to the exploitation of rich Cu–Ni ores of the Monchepluton (Papunen & Gorbunov 1985; Chashchin et al. 2002; Smolkin et al. 2004). The complex is located at a triple junction (Fig. 1) where weakly metamorphosed early Proterozoic rift-related rocks and deep-seated Archaean rocks metamorphosed at granulite to amphibolite facies become contiguous at the modem erosion level. The Monchepluton is an S-shaped body with an area of c. 65 km^2. It consists of two parts which probably represent independent magma chambers.

The northwestern and central parts of the Monchepluton (NKT: Mts Nittis, Kumuzhya and Travyanaya and Mt Sopcha) are mainly composed of non-metamorphosed ultramafic rocks, which from bottom-up are represented by a 10–100 m-thick basal zone of quartz-bearing norite and gabbronorite, harzburgite (100–200 m), alternating harzburgite and orthopyroxenite (250–400 m), orthopyroxenite (300–700 m) with chromitite lenses (Mt Kumuzhya) and 1–5 m-thick Cu–Ni-bearing dunite-harzburgite layers (Mt Sopcha, '330 horizon'). The total thickness of the NKT intrusion expands southwards from 200–1000 m and culminates at Mt Sopcha (1600 m).

The southeastern part of the Monchepluton (NPV: Mts Nyud, Poaz and Vurechuaivench) consists mainly of 100–600 m-thick mafic rocks: basal quartz-bearing gabbronorite and norite (up to 50 m), melanocratic norite with lenses of olivine-bearing harzburgite and norite, ore-bearing 'critical

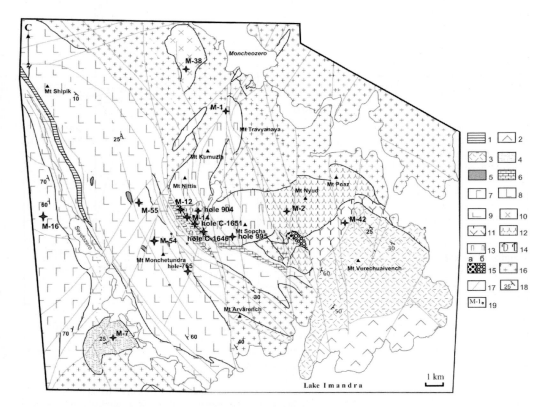

Fig. 4. Geological map of the Monchegorsk layered complex (Smolkin et al. 2004).

horizon' with xenoliths, olivine-free mesocratic and leucocratic norite and gabbronorite, gabbronorite, leucogabbro, anorthosite with PGE mineralization (Mt Vurechuaivench).

Both parts of the Monchepluton (NKT and NPV chambers) have a trough-like shape with a near-horizontal floor and flanks dipping southwestwards at an angle of 20–40°. The complex is underlain by the Archaean gneiss and migmatite, and overlain by the Sumi rocks of the Imandra-Varzuga rift (near Mt Vurechuaivench). The intrusive rocks of the Monchepluton are cut by veins of basic to intermediate pegmatites and diorite, and by dolerite and lamprophyre dykes.

The syngenetic disseminated Cu–Ni ore occurs in layers and is usually spatially confined to the layers of olivine-bearing rocks. The ore location is controlled by the primary structural elements of the intrusions. It also may be found in the upper and basal parts of the intrusions. The mineralization is related to the coarse-grained pegmatoid rocks. Occurrences of syngenetic and nest-disseminated ore with bedded, lens-shaped and stock-like forms are locally confined to the parts of the intrusions where fine-grained and irregular-grained rocks, pegmatoids and rocks related to the intrusion ('critical horizon') are widely developed. The distribution of the two last-mentioned rock varieties may in some cases serve to reveal ore-controlling zones. Exploitable Cu–Ni–PGE deposits of veined epigenetic ores in the Monchepluton are confined to the systems of steeply dipping shear fractures trending NNE and dipping SSE, which trace the primary structural elements of the intrusion (geometry of intrusive blocks, primary jointing, etc.). The main ore-controlling elements in the occurrences of epigenetic stringer-disseminated ores are the zones of tectonic dislocations marked by schistose and blastomylonitized rocks. Most favourable for the concentration of injected stringer-disseminated ores are the places where the tectonic zones pass along the bend of the contact between rocks sharply different in physico-mechanical properties, for example between ultramafic rocks and Archaean granite-gneiss. The epigenetic sulphide Ni–Cu ores of the complex tend to occur in bodies with a mainly NE strike and SW plunge.

The rocks of the Monchepluton were dated earlier by U–Pb methods on zircons and baddeleyite at the Geological Institute KSC RAS (Bayanova 2004) and at the Royal Ontario Museum laboratory in Canada (Amelin et al. 1995) with a good convergence of results (see below). These ages fall in the range of 2507–2490 Ma and favour the correlation of the Monchepluton mafic-ultramafic layered series with the mafic layered series of the second intrusive phase of the Fedorovo-Pansky massif. In both intrusions, the main phase melts have produced Cu–Ni–PGE economic mineralization where base metals predominate, but the portion of platinum in the PGE disseminated occurrences is at least 20%. The ore bodies within the ultramafic rocks of the Monchepluton (Papunen & Gorbunov et al. 1985) are considerably richer than those of the Fedorov block deposit (Schissel et al. 2002). However, the deposits of the Monchegorsk region have already been mined out, while the Fedorovo-Pansky Complex is now being carefully investigated for future development.

Extensive areas of the Monchegorsk ore region are occupied by amphibolite-facies high-pressure garnet-bearing gabbronorite-anorthosite and anorthosite with numerous conformable and cutting veins of leucogabbro and pegmatoid rocks. These are the intrusions of the Main Ridge and Lapland-Kolvitsa granulite belts (Pyrchin, etc.) located within strongly metamorphosed country rocks.

The rocks of the intrusions are insufficiently studied by modern geological and petrological methods, but have been investigated by mining companies because of the presence of high PGE and V–Ti concentrations. The Monchetundra intrusion is separated from the Monchepluton by a thick (a few hundreds of metres) blastomylonite zone with a garnet-amphibole mineral association (Smolkin et al. 2004). Regional shear zones cut and transform the primary monolith-like shape of the intrusion composed of roughly layered leucocratic mafic rocks. This results in the lens-like morphology of the intrusions.

Available U–Pb isotope ages of these anorthosites fall in a wide time interval (Mitrofanov & Nerovich 2003; Bayanova 2004). The zircons derived from magmatic plagioclase yield an age varying from 2500–2460 Ma for different intrusions. A few generations of metamorphic zircons yield an age of multistage metamorphism that took place 2420, 1940 and 1900 Ma (Mitrofanov & Nerovich 2003).

The Imandra lopolith

The Imandra lopolith is represented by a chain of six isolated sheet-like bodies with a thickness of up to 3 km that extend for tens of kilometres. These are Mt Devich'ja, Yagel'naya and Bol'shaya Varaka, Monchepoluostrov, Prikhibin'je and Umbarechka.

The Imandra lopolith cuts the Archaean biotite-amphibole gneiss and amphibolite basement rocks, as well as the volcano-sedimentary rocks of the Palaeoproterozoic Strelna Series of the Imandra-Varzuga palaeorift. The northern part of the lopolith occurs among the Strelna Series rocks, nearly conformably to the horizon of the Seidorechka felsic volcanic rocks. The southern part is sandwiched between the Palaeoarchaean amphibolites and felsic volcanic rocks of the Proterozoic Tominga Series. The intrusive relationships between the lopolith and country rocks are manifested in the recrystallization and formation of hornfels at the exocontact zone, and in the presence of schist and effusive xenoliths in the lopolith rocks. The Imandra lopolith is cut by dykes of varying composition.

The composite section of the lopolith includes the lower, the main, the upper and the roof zones. The lower layered zone has a thickness of 100–200 m. At the contact with the country rocks, there is a 5–7 m-thick layer of fine-grained mesocratic metamorphosed gabbroids that evidently represent the marginal part of the lopolith. Above this there are alternating plagioclase-bearing orthopyroxenite and poikilitic melanocratic norite and gabbronorite with a thickness of 45–60 m. The rocks contain up to 1 m-thick seams composed of mesocratic gabbro associated with chromite mineralization. The upper part of the lower layered zone (45–55 m) consists of gabbronorite.

The main zone is dominated by mesocratic gabbronorite with a thickness of about 2 km. In the lower part of the main zone there are thin (0.1–0.4 m) seams of olivine-bearing gabbronorite, and in the upper, seams of leucocratic gabbronorite of varying thickness.

The upper layered zone (c. 300 m) displays rhythmic layering resulted from the alternation of mesocratic and leucocratic gabbro.

The roof zone ranges from 150–500 m in thickness and is composed of mesocratic quartz gabbro and gabbro-diorite. Disseminated Ti-magnetite mineralization is confined to the bottom of the zone. The upper contact of the lopolith has not been established (Bayanova & Balashov 1995; Bayanova *et al.* 2001).

In modal composition (pyroxenite, norite, gabbro, quartz gabbro-diorite and anorthosite) the V–Ti–Fe mineralization of the Imandra lopolith is similar to that of the Porttivaara block of the Koillismaa layered igneous complex in Finland (Iljina & Hanski 2005).

Analytical methods

U–Pb (TIMS) method. Following the method proposed by Krogh (1973), the samples were dissolved in strong (48%) hydrofluoric acid at a temperature of 205–210 °C over 1–10 days. In order to dissolve fluorides, the samples were reacted with 3.1 N HCl at a temperature of 130 °C for 8–10 hours. To determine the isotope composition of lead and concentrations of lead and uranium, the sample was divided into two aliquots in 3.1 N HCl, and a mixed ^{208}Pb + ^{235}U tracer was added. Pb and U were separated on an AG 1 × 8, 200–400 mesh anion exchanger in Teflon columns. The laboratory blank for the whole analysis was <0.1–0.08 ng for Pb and 0.01–0.04 ng for U. All isotopic determinations for zircon and baddeleyite were made on Finnigan MAT-262 and MI 1201-T mass spectrometers and the Pb isotopic composition was analysed on a secondary-ion multiplier on a Finnigan MAT-262 in ion counting mode. The measurements of the Pb isotopic composition are accurate to 0.025% (Finnigan MAT-262) and 0.15% (MI 1201-T) when calibrated against NBS SRM-981 and SRM-982 standards, respectively. The U and Pb concentrations were measured in single-filament mode with the addition of H_3PO_4 and silica gel using the method (Scharer & Gower 1988; Scharer *et al.* 1996). Pb and U concentrations were measured within the temperature ranges of 1350–1450 and 1450–1550 °C, respectively. All of the isotopic ratios were corrected for mass discrimination during the static processing of replicate analyses of the SRM-981 and SRM-982 standards (0.12 ± 0.04% for the Finnigan MAT-262 and 0.17 ± 0.05% per a.m.u.). The errors in the U–Pb ratios were calculated during the statistical treatment of replicate analyses of the IGFM-87 standard and were assumed equal to 0.5% for Finnigan MAT-262 and 0.7% for MI 1201-T. If the actual analytical errors were higher, they are reported in the table of isotopic data. Isochrons and sample points were calculated Squid and Isoplot programs (Ludwig 1991, 1999). The age values were calculated with the conventional decay constants for U (Steiger & Jager 1977), all errors are reported for a 2 sigma level. Corrections for common Pb were made according to Stacey & Kramers (1975). Corrections were also made for the composition of Pb separated from syngenetic plagioclase or microcline if the admixture of common Pb was >10% of the overall Pb concentration and the ^{206}Pb/^{204}Pb ratios were <1000.

^{143}Sm–^{144}Nd *method.* In order to define concentrations of samarium and neodymium, the sample was mixed with a compound tracer ^{149}Sm/^{150}Nd prior to dissolution. It was then diluted with a mixture of HF + HNO$_3$ (or +HClO$_4$) in teflon sample bottles at a temperature of 100 °C until complete dissolution. Further extraction of Sm and Nd was carried out using standard procedures with two-stage ion-exchange and extraction-chromatographic separation using ion-exchange tar «Dowex» 50 × 8 in chromatographic columns employing 2.3 N and 4.5 N HCl as an eluent. The separated Sm and Nd fractions were transferred into nitrate form, whereupon the samples (preparations) were ready for mass-spectrometric analysis. Measurements of Nd-isotope composition and Sm and Nd concentrations by isotope dilution were performed using a multicollector mass-spectrometer in a Finnigan

MAT 262 (RPQ) in a static mode using Re + Re and Ta + Re filament. The measured reproducibility for ten parallel analysis of Nd-isotope composition for the standard La Jolla = 0.511833 ± 6 was <0.0024% (2σ). The same reproducibility was obtained from 11 parallel analyses of the Japanese standard: Ji Nd1 = 0.512078 ± 5. The error in ^{147}Sm/^{144}Nd ratios of 0.2% (2σ), the average of seven measures, was accepted for statistic calculations of Sm and Nd concentrations using the BCR standard. The blanks for laboratory contamination for Nd and Sm are 0.3 and 0.06 ng, respectively. Isochron parameters were developed from programs of Ludwig (1991, 1999). The reproducibility of measurements was ±0.2% (2σ) for Sm/Nd ratios and ±0.003% (2σ) for Nd-isotope analyses. All ^{147}Sm/^{144}Nd and ^{143}Nd/^{144}Nd ratios were normalized to ^{146}Nd/^{144}Nd = 0.7219 and adjusted to ^{143}Nd/^{144}Nd = 0.511860 using the La Jolla Nd standard. The ε_{Nd} (T) values and model T_{DM} ages were calculated using the currently accepted parameters of CHUR (Jacobsen & Wasserburg 1984): ^{143}Nd/^{144}Nd = 0.512638 and ^{147}Sm/^{144}Nd = 0.1967 and DM (Goldstein & Jacobsen 1988): ^{143}Nd/^{144}Nd = 0.513151 and ^{147}Sm/^{144}Nd = 0.2136.

^{87}Rb–^{86}Sr method. The samples and minerals were all treated with double distilled acids (HCl, HF and HNO$_3$) and H$_2$O distillate. A sample of 20–100 mg (depending on Rb and Sr contents) was dissolved with 4 ml of mixed HF and HNO$_3$ (5:1) in corked teflon sample bottles and left at a temperature of about 200 °C for one day. The solution was then divided into three aliquots in order to determine Rb and Sr isotope compositions and concentrations. These were measured by isotope dilution using separate ^{85}Rb and ^{84}Sr tracers. Rb and Sr extraction was performed by eluent chromatography with «Dowex» tar 50 × 8 (200–400 mesh); 1.5 N and 2.3 N HCl served as an eluent. Tar volumes in the columns were c. 7 and c. 4 sm^3. The separated Rb and Sr fractions were evaporated until dryness, followed by treatment with a few drops of HNO$_3$. Sr isotope compositions and Rb and Sr contents were measured by a MI-1201-T (Ukraine) mass spectrometer in the two-ribbon mode using Re filaments. The prepared samples were deposited on the ribbons in the form of nitrate. Sr isotope composition in all the measured samples was normalized to a value of 0.710235 recommended by NBS SRM-987. Errors on Sr isotope analysis (confidence interval of 95%) do not exceed 0.04%, and those of Rb–Sr ratio determination are of 1.5%. Blank laboratory contamination for Rb is 2.5 ng and for Sr 1.2 ng. The adopted Rb decay constant of Steiger & Jager (1977) was used for age calculations.

Results of age determinations for the layered intrusions of the Kola Belt

Mt Generalskaya

Samples for U–Pb dating were taken both from outcrops and from drill cores of the Mt Generalskaya intrusion. Two samples (SA-416/1 and SA-416/2) of 150 kg were collected at different times from the upper gabbronorite zone at a roadside outcrop at Pechenga-Luostari which runs across the northern flank of the intrusion. The samples are taken from the lower gabbronorite zone and consist of medium-grained, massive gabbronorite, and contain relict ortho- and clinopyroxenes, pseudomorphs and aggregates of hornblende and actinolite replacing pyroxene, large laths of saussuritized plagioclase and accessory Ti-magnetite. Sulphide and apatite are minor.

An 8 mg grain concentrate separated from the gabbronorite sample (SA 416/1) yielded two morphological types of zircons. In addition, baddeleyite was found for the first time in the layered intrusions of the Kola Peninsula (Bayanova et al. 1999).

Sample S-3464 of 200 kg was collected from a borehole at a depth between 600 and 800 m which penetrated the lower part of the layered unit. The sample is composed of slightly amphibolized meso- and leucocratic gabbronorite and of less melanocratic gabbronorite. The rocks are medium-grained and contain bronzite, augite (rare pigeonite), andesine-labradorite, hornblende, actinolite, chlorite, biotite, apatite and Ti-magnetite.

The U–Pb age obtained on magmatic zircon (Fig. 5) from the gabbronorite is 2496 ± 10 Ma, MSWD = 1.5; the lower intersection is at zero, indicative of modern lead loss. The zircons from the outcrop sample are more discordant than those from the drill core (Fig. 6a, Table 1).

Magmatic zircons from sample SA-416/1 (type 1) were sent to the Royal Ontario Museum laboratory in Canada to be reanalysed in order to compare the results. The U–Pb age obtained with the use of the abrasion technique was 2505 ± 1.5 Ma (Amelin et al. 1995).

Zircons xenocrysts are stubby columnar and brown-pinkish crystals up to 100 μm in size. They show well-defined broad zones. They contain more uranium than magmatic zircons. The ^{207}Pb/^{206}Pb ages on the zircons xenocrysts ranges from 2660–2606 Ma.

Sample SA-443 of 80 kg is taken from the middle part of the layered unit represented by thin (50 m) micro-rhythms of anorthosite (or leucogabbro). The latest phase has a trough-like shape and consists of rhythmically alternating layers of melanocratic and mesocratic gabbronorites which are 3–20 cm and 30–80 cm thick, respectively.

Mt Generalskaya intrusion

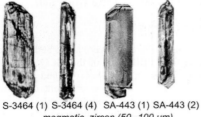 

S-3464 (1)　S-3464 (4)　SA-443 (1)　SA-443 (2)　　　　S-3464 (2)　　S-3464 (3)　　SA-443 (4)
magmatic zircon (50–100 μm)　　　　　　　　　　*zircon xenocrystic (70–80 μm)*

Fedorovo-Pansky Complex

Pan-1 (2)　　　Pan-1 (3)　　　　　　　　　　P-6 (4)　　　　　P-6 (5)
magmatic zircon (100–120 μm)　　　　　　　　*baddeleyite (50–70 μm)*

Monchepluton and Monchetundra

M-2 (3)　　　M-1 (1)　　　M-42 (3)　　　M-2 (1)　　M-42 (1)　　　　M-55 (1)　　　M-54 (1)
magmatic zircon from Monchepluton　　*baddeleyite from Monchepluton*　　*magmatic zircon from Monchetundra*
　　　　(175 μm)　　　　　　　　　　　(80–120 μm)　　　　　　　　　　　　(175 μm)

Imandra lopolith

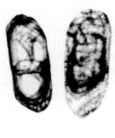    

6-57 (2)　7-57 (1)　　3-57 (4)　　5-77 (4)　　　6-57 (1)　　7-57 (4)　　5-77 (2)
magmatic zircon (125m)　zircon xenocrystic (225 m)　　baddeleyite (80–120 m)

Fig. 5. Photomicrographs of analysed zircon and baddeleyite.

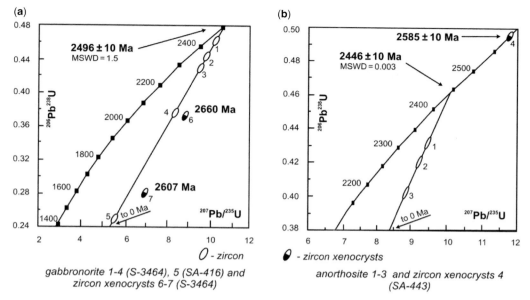

Fig. 6. U–Pb concordia diagrams for zircon gabbronorite (**a**) and from anorthosite (**b**) of the Mt Generalskaya intrusion (see Table 1).

The base of the micro-rhythmic zone contains an up to 4.0 m-thick lens-like anorthosite layer. The anorthosite is replaced by slightly amphibolized leucocratic gabbronorite along strike. The layer has sharp boundaries and includes pseudobrecciated fragments of mesocratic gabbronorite at its base. Thin anorthosite apophyses penetrate the underlain gabbronorite and indicate later crystallization of the anorthositic melt.

The anorthosite is dominated by tabular plagioclase, with minor amphibole, relict intercumulus pyroxene, chlorite, biotite and apatite. Co-existing ortho- and clinopyroxenes, pigeonite and plagioclase of labradorite composition are common.

Two zircon types (magmatic and xenocrysts) were separated from the anorthosite sample. Magmatic zircons are transparent bipyramidal-prismatic grains up to 150 μm in size. The terminations of bipyramids have simple faces. In immersion view, all the grains display narrow zoning towards the crystal edge.

Three zircon populations differing mainly in size were separated for U–Pb dating. The U–Pb age on magmatic zircons from the anorthosite is 2446 ± 10 Ma, MSWD = 0.003 (Fig. 6b, Table 1).

Fedorovo-Pansky Complex

Several large samples were selected for the U–Pb dating of the Fedorovo-Pansky Complex.

A 60 kg sample of medium- and coarse-grained gabbronorite was collected from the Lower Layered Horizon in the Eastern Kievey area. The separated zircons are transparent with a vitreous lustre. All the grains were divided into three types: Pan-1–regular bipyramidal-prismatic crystals of up to 120 μm; Pan-2–fragments of prismatic crystals; Pan-3–pyramidal apices of crystals of 80–100 μm. In immersion view, all the zircons display a simple structure with fine zoning and cross jointing (Fig. 5).

The discordia plotted on three points yields the upper intersection with the concordia and the U–Pb age at 2491 ± 1.5 Ma, MSWD = 0.05. The lower intersection of the discordia with the concordia is at zero and reflects modern lead losses (Fig. 7a, Table 2). The same zircon sample was analysed in the Royal Ontario Museum laboratory in Canada; the obtained U–Pb zircon age is 2501.5 ± 1.7 Ma (Amelin et al. 1995) that is somewhat older than ours. The age obtained is interpreted as the time of crystallization of the main gabbronorite phase rock (Mitrofanov et al. 1997; Mitrofanov & Bayanova 1999).

Sm–Nd dating on ortho- and clinopyroxene, plagioclase and whole-rock minerals extracted from the same gabbronorite gave an age of 2487 ± 51 Ma, MSWD = 1.5 (Balashov et al. 1993).

Three zircon populations of prismatic habit and light-yellow colour were separated from PGE-bearing gabbro-pegmatite (LLH). The

Table 1. *U–Pb zircon isotope data for the rocks of the Mt Generalskaya intrusion (from Bayanova et al. 1999)*

Sample No.	Weight (mg)	Concentration (ppm)		Pb isotopic composition[1]			Isotopic ratios[2]		Age[2] (Ma)
		Pb	U	$^{206}Pb/^{204}Pb$	$^{206}Pb/^{207}Pb$	$^{206}Pb/^{208}Pb$	$^{207}Pb/^{235}U$	$^{206}Pb/^{238}U$	$^{207}Pb/^{206}Pb$
gabbronorite (S-3464)									
1	0.7	68.0	112.1	2700	5.931	2.783	10.4131	0.45964	2501
2	0.9	157.1	275.5	5200	6.046	2.972	9.8889	0.43927	2490
3	0.8	107.5	194.5	4200	6.020	2.970	9.5879	0.42551	2491
4	0.8	121.3	246.3	1585	5.835	2.881	8.4272	0.37318	2495
gabbronorite (SA-416)									
5	1.3	76.4	227.5	2880	6.010	2.512	5.5480	0.24787	2480
zircon xenocrysts (S-3464)									
6	0.9	133.9	296.1	1000	5.311	5.050	8.9624	0.37118	2607
7	0.6	113.9	343.5	1400	5.281	5.621	6.9341	0.27819	2660
anorthosite (SA-443)									
1	0.8	109.4	199.8	4600	6.170	3.212	9.4737	0.43046	2452
2	0.8	115.9	208.6	1100	5.851	2.870	9.2448	0.41956	2454
3	0.5	52.6	101.9	1750	5.991	3.131	8.8624	0.40153	2456
zircon xenocrysts (SA-443)									
4	0.5	369.9	400.1	2600	5.641	1.023	11.7529	0.49322	2585

[1] All ratios are corrected for blanks of 0.1 ng for Pb and 0.04 ng for U and for mass discrimination of $0.17 \pm 0.05\%$.
[2] Correction for common Pb was determined for the age according to Stacey & Kramers (1975).

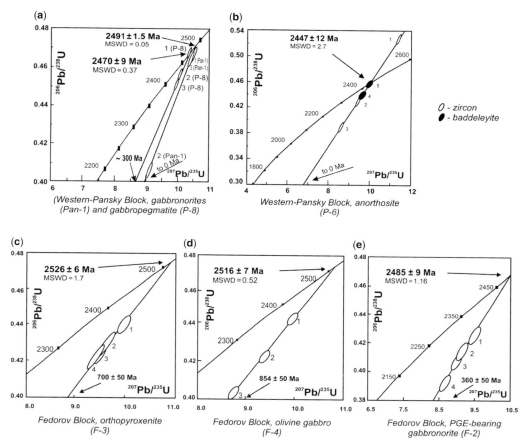

Fig. 7. U–Pb concordia diagrams for the Western-Pansky (**a, b**) and Fedorov (**c, d, e**) Blocks of the Fedorovo-Pansky Complex (see Table 2 for U–Pb data and references).

zircons from sample P-8 are stubby prismatic crystals with sharp outlines, about 100 μm in size. The crystals show cross-cracks and apparent zoning in immersion view. The zircons from samples D-15 and D-18 are multi-zoned pinkish fragments of prismatic crystals with adamantine lustre and 80 and 100 μm in size. The U–Pb zircon age of 2470 ± 9 Ma, MSWD = 0.37 (Fig. 7a, Table 2) was obtained from three points: one concordant and two lying in the upper part of the isochron. The lower intersection of the discordia with the concordia (c. 300 Ma) indicates lead loss associated with the Palaeozoic tectonic activation of the eastern Baltic Shield and the development of the giant Khibina and Lovozero intrusions of nepheline syenites (Kramm et al. 1993). Zircons from the gabbro-pegmatite are found to have higher U and Pb concentrations than those from the gabbronorite.

Three zircon and two baddeleyite populations were separated from a sample collected from the Upper Layered Horizon in the Southern Suleypahk area. All the zircons from anorthosite are prismatic, light-pink-coloured with vitreous lustre. In immersion view, they are zoned and fractured. A population of bipyramidal-prismatic zircons (Pb-1) is made up of elongate (3:1) crystals. Sample Pb-2 contains zircons of round-ellipsoidal habit and sample Pb-3 contains transparent flattened crystal fragments of up to 0.75 μm in size.

The separated baddeleyite crystals (first recorded in the anorthosite) were subdivided into two varieties, deep-brown and brown. All the grains are fragments of transparent baddeleyite crystals of 50 μm in size, without selvages and inclusions.

A U–Pb isochron plotted from three zircons and two baddeleyites intersects the concordia with an age of 2447 ± 12 Ma, MSWD = 2.7 (Fig. 7b, Table 2). The lower intersection of the discordia with the concordia records recent lead loss. The position of the baddeleyite points is near-concordant,

Table 2. U–Pb baddeleyite (bd) and zircon isotope data from the Western-Pansky and Fedorov Blocks of the Fedorovo–Pansky Complex

Sample No.	Weight (mg)	Concentration (ppm)		Pb isotopic composition[1]				Isotopic ratios[2]		Age[2] (Ma)
		Pb	U	$^{206}Pb/^{204}Pb$	$^{206}Pb/^{207}Pb$	$^{206}Pb/^{208}Pb$	$^{207}Pb/^{235}U$	$^{206}Pb/^{238}U$		$^{207}Pb/^{206}Pb$

Western-Pansky Block, gabbronorites (Pan-1); from Bayanova 2004

Sample No.	Weight (mg)	Pb	U	$^{206}Pb/^{204}Pb$	$^{206}Pb/^{207}Pb$	$^{206}Pb/^{208}Pb$	$^{207}Pb/^{235}U$	$^{206}Pb/^{238}U$	$^{207}Pb/^{206}Pb$
1	3.30	95.0	144	11740	6.091	3.551	10.510	0.4666	2491
2	1.90	70.0	142	10300	6.100	4.220	9.135	0.4061	2489
3	1.60	84.0	144	6720	6.062	3.552	10.473	0.4650	2491

Western-Pansky Block, gabbropegmatite (P-8); from Balashov et al. 1993

1	5.90	95.0	158	3240	5.991	3.081	10.435	0.4681	2471
2	7.30	181.0	287	8870	6.161	2.260	10.092	0.4554	2465
3	1.25	125.0	200	3400	6.012	2.312	10.082	0.4532	2468

Western-Pansky Block, anorthosite (P-6); from Bayanova 2004

1	0.75	218.0	322	5740	6.230	3.263	11.682	0.5352	2438
2	0.10	743.0	1331	3960	6.191	3.151	9.588	0.4393	2438
3	0.20	286.0	577	2980	6.021	3.192	8.643	0.3874	2474
4 (bd)	1.00	176.0	396	14780	6.290	63.610	9.548	0.4380	2435
5 (bd)	0.26	259.0	560	3360	6.132	54.950	9.956	0.4533	2443

Fedorov Block, orthopyroxenite (F-3); from Nitkina 2006

1	0.75	48.0	60.9	825	4.9191	1.3039	10.0461	0.44249	2504
2	0.80	374.0	598.6	4588	6.0459	1.9650	9.6782	0.43153	2484
3	0.85	410.2	630.2	4521	6.0281	1.6592	9.5667	0.42539	2488
4	1.00	271.0	373.1	2552	5.9916	1.2393	9.4700	0.42406	2476

Fedorov Block, olivine gabbro (F-4); from Nitkina 2006

1	1.80	725.3	1322.8	14649	6.1121	3.8177	10.0132	0.44622	2484
2	2.00	731.3	1382.8	8781	6.1522	3.5517	9.4306	0.42454	2467
3	1.95	680.9	1374.0	7155	6.2645	3.6939	8.7401	0.40155	2433

Fedorov block, PGE-bearing gabbronorite (F-2); from Nitkina 2006

1	0.30	498.0	833.4	2081	5.9502	2.2111	9.49201	0.42493	2477
2	0.65	513.8	932.2	5274	6.1519	2.6371	9.1373	0.41378	2458
3	0.55	583.2	999.3	3194	6.1132	2.0528	8.9869	0.40832	2452
4	0.80	622.5	1134.5	4114	6.1161	2.1914	8.6638	0.39165	2460

[1]All ratios are corrected for blanks of 0.1 ng for Pb and 0.04 ng for U and for mass discrimination of $0.17 \pm 0.05\%$.
[2]Correction for common Pb was determined for the age according to Stacey & Kramers (1975).

while zircon points (Sample P6-1) are above the concordia due to uranium loss. This age (2447 ± 12 Ma) is considered to constrain the origin of late-phase anorthosite because, as shown by Heaman & LeCheminant (1993), baddeleyite is commonly generated in residual melts.

The U–Pb zircon age of the early barren orthopyroxenite from the Fedorov Block, 2526 ± 6 Ma, is believed to be the time of emplacement (Fig. 7c, Table 2). The U–Pb age of 2516 ± 7 Ma (Fig. 7d, Table 2), obtained from zircon from barren olivine gabbro, is interpreted as the time of crystallization. The last Cu–Ni–PGE-bearing taxitic gabbronorite from the Fedorov Block (Fig. 7e, Table 2) yielded a U–Pb zircon age of 2485 ± 9 Ma (Nitkina 2006).

Monchegorsk Layered Complex

Ten samples of 50–120 kg were collected for U–Pb dating. Accessory baddeleyite and zircon were better preserved in drill core samples than in outcrops (Fig. 5).

The oldest rocks studied are pegmatites of gabbronorite composition, which are associated with the ore-bearing sulphide veins from the basal zone of Mt Travyanaya and the 'critical horizon' (Mt Hyud, Terassa deposit). Two baddeleyite and three zircon populations were examined from these rocks. All the crystals were unaltered. Baddeleyite grains are up to 80 μm long and light brown in colour. Zircons are prismatic and isometric, up to 150 μm in size, and feature narrow igneous zoning and various hues of brown. U and Pb concentrations are high, which is typical of pegmatite. A U–Pb age obtained on the five zircon and baddeleyite populations is 2500 ± 5 Ma, MSWD = 1.7; the lower intersection of the discordia and the concordia is at 349 ± 81 Ma, indicating Palaeozoic lead losses (Fig. 8a, Table 3). This age is comparable with that of 2493 ± 7 Ma obtained for gabbronorite of Mt Nyud, and with a zircon age for the norite of Mt Travyanaya (Fig. 8b, Table 3). A U–Pb age on baddeleyite and zircon recently obtained for the coarse-grained gabbronorite of Mt Vurechuaivench foothills (now considered as a PGE-bearing reef) is 2497 ± 21 Ma, being very similar to that for the Fedorovo-Pansky gabbronorite (Fig. 8c, Table 3).

To determine the age of the Sopcheozero chromite deposit located within the Dunite Block of the Monchepluton, cross-cutting dykes were analysed. The Dunite Block is composed of rocks poor in accessory minerals. The dykes are assumed to be associated with intrusive mafic rocks of the Monchepluton and are thought to have intruded the Dunite Block rocks before they had cooled. Thus the age of the dykes would constrain the minimum age limit of the Dunite Block and Sopcheozero deposit formation. For U–Pb dating, a sample was collected from Borehole 1586 at a depth of 63–125 m, from a coarse-grained gabbronorite dyke cutting the ultramafic rocks of the Dunite Block. Baddeleyite, two zircon populations and rutile were used for dating. Brown transparent plate-like baddeleyite grains of up to 70–80 μm in size are well preserved. Light-pink zircons of up to 150 μm in size have good outlines and thin zoning. The U–Pb age on zircon and baddeleyite is 2496 ± 14 Ma, MSWD = 0.011; the lower intersection of the discordia with the concordia is at 313 ± 271 Ma (Fig. 8d, Table 3). The point for the rutile has a near concordant value of c. 1.84 Ga that reflects the time of its formation. A similar U–Pb age (2506 ± 10 Ma) has also been obtained on zircon from a coarse-grained gabbronorite dyke from Borehole 1518 (Fig. 8e, Table 3). The gabbronorite dyke cuts the ultramafic rocks of the Dunite Block, therefore the Dunite Block must be older than the Monchepluton.

Small intrusions and dykes of the Monchegorsk Layered Complex were considered by most geologists to have the same age as the Monchepluton. In order to verify these relationships, diorite of the Yarva-Varaka intrusion was studied. Three zircon types and baddeleyite were selected from a sample of quartz diorite and granophyric hypersthene diorite collected in the upper part of the Yarva-Varaka section. Stubby prismatic, pink-brown zircons of up to 150 μm in size were divided by their colour hues into three populations. In immersion view, they are multi-zoned. Baddeleyite grains and fragments are prismatic in habit, light-brown coloured and up to 80 μm in size. A U–Pb age obtained on four points is 2496 ± 9 Ma, MSWD = 0.93; the lower intersection is at zero, indicating recent lead losses (Fig. 8f, Table 3).

The Ostrovsky intrusion also belongs to the series of small mafic-ultramafic intrusions of the Monchegorsk Layered Complex. It was considered to correlate in age with the Monchepluton and was interesting as a target for Cu–Ni prospecting. A sample for U–Pb dating was taken from mafic pegmatite veins in the middle part of the upper gabbronorite zone (Mt Ostrovskaya). The pegmatite body is >1 m-thick, up to 2 m long and has a complex morphology, with sinuous contacts with the coarse-grained slightly amphibolized host pigeonite gabbronorite. The sample is dominated by coarse-grained to pegmatoid gabbronorites with a poikilitic texture, made up mostly of calcic plagioclase and amphibolized clinopyroxene. The 60 kg sample produced two types of baddeleyite and two types of zircon. Baddeleyite grains of type 1 are up to 80 μm in size, with a deep-brown colour and flattened and tabular structure. Larger, up to 120 μm

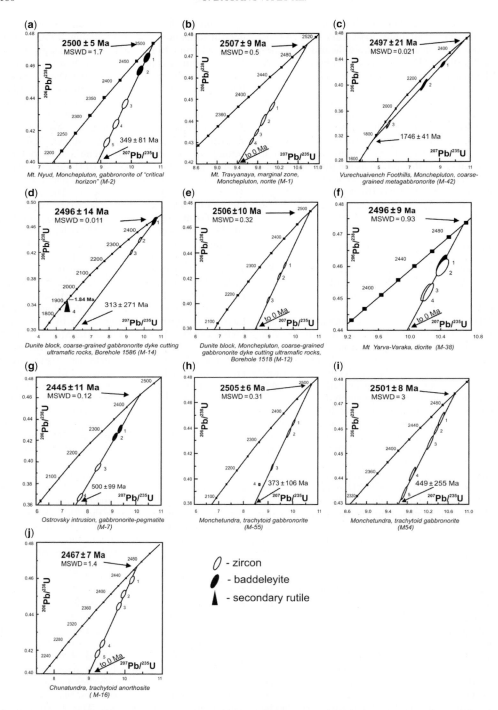

Fig. 8. U–Pb concordia diagrams for zircon, baddeleyite and rutile from different rocks of the Monchegorsk Layered Complex (see Table 3).

Table 3. U–Pb baddeleyite (bd), zircon and rutile (ru) isotope data from the Monchegorsk Layered Complex

Sample No.	Weight (mg)	Concentration (ppm)		Pb isotopic composition[1]				Isotopic ratios[2]		Age[2] (Ma)
		Pb	U	$^{206}Pb/^{204}Pb$	$^{206}Pb/^{207}Pb$	$^{206}Pb/^{208}Pb$	$^{207}Pb/^{235}U$	$^{206}Pb/^{238}U$		$^{207}Pb/^{206}Pb$

'Critical horizon', Mt Nyud Monchepluton, gabbronorite (M-2); from Bayanova 2004

Sample No.	Weight (mg)	Pb	U	$^{206}Pb/^{204}Pb$	$^{206}Pb/^{207}Pb$	$^{206}Pb/^{208}Pb$	$^{207}Pb/^{235}U$	$^{206}Pb/^{238}U$	$^{207}Pb/^{206}Pb$
1 (bd)	0.70	93.1	198.8	9432	6.0586	103.8300	10.4643	0.4636	2495
2 (bd)	0.40	170.8	364.4	3589	5.9833	50.7070	10.3199	0.4574	2494
3	0.40	117.4	183.1	4590	6.0153	1.8703	9.8274	0.4359	2492
4	0.80	187.9	308.4	13664	6.1169	1.9750	9.4740	0.4227	2483
5	0.50	152.2	252.5	5300	6.0842	1.8994	9.2129	0.4125	2477

marginal zone, Mt Travyanaya Monchepluton, norite (M-1); from Smolkin et al. 2004

1	0.30	308.3	504.8	5778	6.0202	2.3805	10.0760	0.4458	2497
2	0.35	185.4	319.8	8358	6.0582	2.7721	9.9277	0.4402	2493
3	0.40	264.5	441.6	23762	6.1006	2.2929	9.7814	0.4342	2491
4	0.40	434.8	793.1	6273	6.0541	3.2613	9.7060	0.4314	2489

Vurechuaivench Foothills Monchepluton, coarse-grained metagabbronorite (M-42); present study

1 (bd)	0.80	150.1	271.4	2982	6.3099	3.2054	9.23762	0.43446	2393
2 (bd)	0.65	65.1	122.6	2080	6.5863	2.7920	8.13574	0.40516	2295
3	0.75	137.4	288.5	911	6.4805	2.3018	5.75090	0.34208	2228

Dunite block, Monchepluton, coarse-grained gabbronorite dyke cutting ultramafic rocks, hole 1586 (M-14); from Bayanova 2004

1 (bd)	0.50	5.3	10.3	1307	5.7748	12.4320	10.5720	0.4684	2494
2	0.80	358.7	309.2	13360	6.1029	0.5312	9.8622	0.4391	2486
3	0.60	321.8	362.1	3791	6.0407	0.7838	9.3919	0.4199	2479
4 (ru)[3]	1.30	7.5	4.5	28	1.7085	0.8077	5.7139	0.3328	2022

Dunite block, Monchepluton, coarse-grained gabbronorite dyke cutting ultramafic rocks, hole 518 (M-12); from Smolkin et al. 2004

1	0.45	221.4	409.6	2152	5.9237	3.5776	9.6682	0.4303	2487
2	0.30	321.7	542.6	11260	6.1264	2.2049	9.4976	0.4249	2478
3	0.50	164.9	302.4	1952	5.9508	2.5602	8.9806	0.4031	2472

Mt Yarva-Varaka, diorite (M-38); from Bayanova 2004

1 (bd)	0.50	32.9	70.0	5615	6.0158	53.309	10.419	0.4608	2497
2	0.80	310.5	515.4	2587	5.9089	2.9118	10.420	0.4597	2501
3	1.40	151.8	262.7	4840	5.9895	3.1873	10.242	0.4519	2501
4	0.70	273.2	472.5	4590	5.9970	3.1828	10.217	0.4518	2497

(Continued)

Table 3. Continued

Sample No.	Weight (mg)	Concentration (ppm)		Pb isotopic composition[1]				Isotopic ratios[2]		Age[2] (Ma)
		Pb	U	$^{206}Pb/^{204}Pb$	$^{206}Pb/^{207}Pb$	$^{206}Pb/^{208}Pb$	$^{207}Pb/^{235}U$	$^{206}Pb/^{238}U$		$^{207}Pb/^{206}Pb$

Ostrovsky intrusion, gabbronorite-pegmatite (M-7); from Bayanova 2004

Sample	W	Pb	U	206/204	206/207	206/208	207/235	206/238	Age
1 (bd)	0.45	28.6	63.9	1820	6.162	29.610	9.350	0.4311	2405
2 (bd)	0.55	36.3	83.7	3380	6.346	42.038	9.210	0.4248	2389
3	0.45	336.4	694.8	9420	6.378	3.797	8.471	0.3953	2407
4	0.35	89.1	187.6	3700	6.375	2.942	7.756	0.3667	2384

Monchetundra, trachytoid gabbronorite (M-55); from Smolkin et al. 2004

Sample	W	Pb	U	206/204	206/207	206/208	207/235	206/238	Age
1	0.50	110.9	172.9	8690	6.0591	1.9737	10.0171	0.444092	2493
2	0.35	37.7	61.0	1122	5.7260	2.2170	9.82794	0.436123	2492
3	0.25	168.3	277.7	6350	6.0914	1.8221	9.15540	0.409417	2479
4	0.30	122.6	213.5	6159	6.2294	1.9243	8.64155	0.395489	2439

Monchetundra, trachytoid gabbronorite (M-54); from Smolkin et al. 2004

Sample	W	Pb	U	206/204	206/207	206/208	207/235	206/238	Age
1	0.50	308.9	494.5	9172	6.0283	2.4881	10.47020	0.46359	2503
2	0.35	374.3	587.5	18868	6.0791	2.2742	10.40220	0.46050	2496
3	0.40	72.8	118.6	6833	6.0271	2.5023	10.25210	0.45405	2498
4	0.25	206.3	333.1	7831	6.0324	2.1148	9.90668	0.44177	2499
5	0.45	196.6	311.9	14844	6.1123	1.9269	9.74196	0.43412	2484

Chunatundra, trachytoid anorthosite (M-16); from Bayanova 2004

Sample	W	Pb	U	206/204	206/207	206/208	207/235	206/238	Age
1	0.20	80.3	138.4	2710	6.020	3.404	10.216	0.4589	2471
2	2.05	122.83	214.1	4420	6.144	3.293	9.999	0.4527	2455
3	0.30	141.3	251.0	5140	6.137	3.291	9.831	0.4443	2461
4	0.60	92.7	169.4	7860	6.148	2.970	9.388	0.4228	2467
5	0.20	46.5	86.1	1090	5.805	3.242	9.262	0.4180	2463

[1]All ratios are corrected for blanks of 0.08 ng for Pb and 0.04 ng for U and for mass discrimination of 0.12 ± 0.04%.
[2]Correction for common Pb was determined for the age according to Stacey & Kramers (1975).
[3]Corrected for isotope composition of light cogenetic plagioclase: $^{206}Pb/^{204}Pb = 14.041 \pm 0.005$, $^{207}Pb/^{204}Pb = 14.581 \pm 0.007$, $^{208}Pb/^{204}Pb = 35.58 \pm 0.02$.

baddeleyite grains of type 2 were found within a fringe of metamict zircon and were exposed to aero-abrasion for 15 minutes in order to remove the metamict fringe. Zircons are prismatic, up to 125 μm in size, and are subdivided into light brown and brown varieties. Zircons show well-developed joints and thin zoning in immersion view. The U–Pb isochron age on two baddeleyite and two zircon points is 2445 ± 11 Ma, MSWD = 0.12 and the lower intersection of the discordia with the concordia is at 500 ± 99 Ma (Fig. 8g, Table 3).

To establish age correlations between the gabbronorite of the Monchepluton and the anorthosite of the Main Ridge intrusion, rock samples of the Monchetundra and Chunatundra intrusions were studied.

The Monchetundra intrusion has a complex structure and an overview of geological and geochronological investigations is given by Smolkin et al. (2004). It includes the upper zone comprised mainly of amphibolized gabbronorite and gabbro-anorthosite, and the lower zone, which consists of gabbronorite, norite and plagiopyroxenite (drilled by the deep borehole M-1).

The middle part of the upper zone, which contains a prominent horizon of slightly-altered medium to coarse-grained gabbronorite with trachytoid texture, was sampled for U–Pb dating. The sample yielded three zircon types. Prismatic acicular crystals up to 200 μm in size and their brown fragments were divided into three types by colour. In immersion view, multi-zoning, mineral inclusions, strong jointing, corrosion of the surface and spotted uneven grain colour are observed. The U–Pb ages (Fig. 8h, i, Table 3) on zircon from trachytoid gabbronorite are 2505 ± 6 Ma, MSWD = 0.31 and 2501 ± 8 Ma, MSWD = 3 (Bayanova & Mitrofanov 2005).

A sample was also taken from the rocks of the differentiated series of the Chunatundra intrusion. Zircons from medium-grained leucogabbro with trachytoid texture were divided into five types. Four types are up to 150 μm isometric fragments of brown and pink colour, whereas the last fraction is represented by up to 120 μm twinned pinkish-brown zircons with adamantine lustre. The U–Pb isochron plotted on five points has the upper intersection with the concordia at 2467 ± 7 Ma, MSWD = 1.4 and the lower intersection is at zero (Fig. 8j, Table 3). This age is close to the age obtained on magmatic zircon from anorthosite of the Pyrshin intrusion (Mitrofanov & Nerovich 2003) and on zircons from later anorthositic injections of the LLH (Fedorovo-Pansky Complex).

Imandra lopolith

Several large samples were taken from the different parts of the lopolith for dating. Baddeleyite was found in a chromite horizon and underlying norite of the Bol'shaya Varaka area. A sample of up to 350 kg, collected in order to study chromite mineralization, produced about 10 mg of zircon-baddeleyite concentrate. Angular and prismatic baddeleyite fragments were up to 80 μm in size; they are black and nearly opaque and have gouges on their faces (Fig. 5). The U–Pb isochron plotted on three points has the upper intersection with the concordia at 2446 ± 39 Ma, MSWD = 5.1. The lower intersection is at zero, reflecting recent lead loss (Fig. 9a, Table 4). The best-preserved and undisturbed U–Pb system was displayed by a zircon grain with a coordinate lying on the concordia within the determination error. The coordinate of one of the baddeleyite points is strongly discordant, while the other baddeleyite point lies above the concordia due to lead removal. The baddeleyite fraction was exposed to acid treatment in order to remove a thin white coating, following a similar technique to that applied to baddeleyite from carbonatites of the Phalaborwa intrusion in South Africa (Reischmann 1995). The U–Pb age on zircon and baddeleyite of 2446 ± 39 Ma is the oldest for the Imandra lopolith (Table 5).

A gabbronorite sample was taken from Borehole No. 6 within the depth interval of 57.8–75.5 m in the Umbarechka Block of the intrusion. The mineral Sm–Nd isochron corresponds to an age of 2444 ± 77 Ma at ε_{Nd} = −2.0 ± 0.6, MSWD = 0.31. Zircons separated from the sample were analysed at the Royal Ontario Museum Laboratory in Canada (Amelin et al. 1995), where a U–Pb age of 2441 ± 1.6 Ma was obtained (Table 5). The two ages determined by different systems give a good correlation and reflect crystallization time for the early rocks of the lower part of the lopolith – c. 2.46 Ga, which we refer to the main phase intrusion.

A close U–Pb age on zircon and baddeleyite was obtained for norite of the main phase of the Umbarechka intrusion. Three zircon points and baddeleyite lie on the isochron and give an age of 2437 ± 7 Ma (Fig. 9b, Table 4). Thus, the analysed rocks of the main phase of the Imandra lopolith formed within the interval of 10 Ma, that is from 2446–2437 Ma.

The second phase of the lopolith represented by leucogabbro-anorthosite and ferrogabbro with Ti-magnetite mineralization was studied in the Prikhibin'je and Bol'shaya Varaka Blocks at the upper contact of the lopolith, where one may observe a 400 m-thick ferrogabbro zone with 10 m anorthosite and Ti-magnetite gabbro, and a 400 m granophyric zone. The ferrogranophyre contains up to 64% SiO_2 and up to 15% of total iron oxide, and has high contents of P_2O_5, Co and Cu. The granophyre is close to the roof rhyodacite in silica

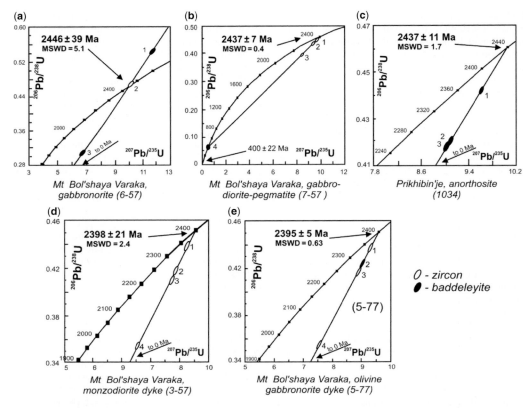

Fig. 9. U–Pb concordia diagrams for zircon and baddeleyite from rocks of the Imandra lopolith (see Table 4).

content, total iron oxide, Zr and REE (Bayanova et al. 2001).

Baddeleyite was found in thin sections of granophyre for the first time. Baddeleyite grains are brown, up to 100 μm in size and form single crystals and growths with ilmenite in plagioclase, pyroxene and later quartz and biotite interstices. Single crystals are confined to aggregates of biotite, which is genetically related to an early magmatic phase in the granophyric rocks. In all cases the baddeleyite and ilmenite were earlier minerals than quartz and micropegmatite. This suggests that they originated during the early stages of crystallization of the intercumulus melt under conditions of low silica activity. Zircon forms individual crystals and thin fringes (≤10 μm) around baddeleyite, which together with titanite, develop after ilmenite and are affected by subsolidus alteration. Multi-zoned light-coloured stubby prismatic zircon crystals up to 100–200 μm in size form aggregates in the micropegmatite matrix (Bayanova 2006).

Three baddeleyite populations of flattened, transparent brown, about 100 μm large crystals were separated from a 20 kg anorthosite sample.

The U–Pb age obtained on these baddeleyites is 2437 ± 11 Ma, MSWD = 1.7 (Fig. 9c, Table 4).

We consider that the granophyric rocks resulted from remelting of host volcano-sedimentary rocks along with a contamination by a felsic material. A 35 m-thick layer of newly-formed hybrid ferrodiorite melt gave rise to ferrogranophyric rocks. This conclusion is supported by similar U–Pb crystallization ages obtained for the rocks from the apex of the Imandra lopolith (2437 ± 11 Ma) and for the granophyres from its endocontact zone (2434 ± 15 Ma) (Table 5) (Bayanova & Balashov 1995). The roof granophyres of the Bushveld complex that originated from reworking of the host felsic rocks show similar relationships (Eules & Cawthorn 1995).

The rocks of the lower layered part of the Imandra lopolith are cut by monzodiorite dykes, which are part of a third phase of emplacement (Bayanova et al. 2001). A 50 kg sample from a monzodiorite dyke cutting the Umbarechka Block yielded well-preserved accessory zircons. Light, pale-pink and pinkish transparent zircon grains have prismatic habit, are about 100 μm in size and

Table 4. *U–Pb baddeleyite (bd) and zircon isotope data from the rocks of the Imandra lopolith (from Bayanova 2004)*

Sample No.	Weight (mg)	Concentration (ppm)		Pb isotopic composition[1]				Isotopic ratios[2]			Age (Ma)
		Pb	U	$^{206}Pb/^{204}Pb$	$^{206}Pb/^{207}Pb$	$^{206}Pb/^{208}Pb$		$^{207}Pb/^{235}U$	$^{206}Pb/^{238}U$		$^{207}Pb/^{206}Pb$
Mt Bol'shaya Varaka, gabbronorite (6–57)											
1 (bd)	0.60	257	463	5500	6.141	46.2300		12.0200	0.5435		2461
2	1.10	123	183	8040	6.231	2.0930		10.3600	0.4730		2445
3 (bd)	0.70	170	534	2570	6.162	31.0400		6.7081	0.3088		2430
Mt Bol'shaya Varaka, gabbro-diorite-pegmatite (7–57)											
1	0.90	274	474	8100	6.248	3.1196		9.9036	0.4532		2440
2	0.50	71	129	2310	6.108	3.4370		9.5718	0.4387		2437
3	0.60	125	232	2770	6.132	3.4848		9.4215	0.4311		2440
4 (bd)	0.30	60	128	1220	6.034	24.2930		9.4746	0.4425		2405
(Prikhibin'je, anorthosite (1034)											
1 (bd)	0.90	457.9	1032.6	28560	6.334	117.5		9.6090	0.4419		2431
2 (bd)	1.00	174.6	410.7	6280	6.316	87.0		9.1080	0.4219		2419
3 (bd)	0.80	330.6	790.5	14210	6.351	93.5		9.0000	0.4162		2422
Mt Bol'shaya Varaka, monzodiorite dyke (3–57)											
1	0.50	119	102	17330	5.821	0.5701		9.6611	0.4530		2397
2	0.71	114	104	1910	5.933	0.5801		9.2121	0.4322		2398
3	1.30	103	96	1530	5.612	0.5601		8.6314	0.4060		2393
4	0.50	152	184	1960	5.971	0.5812		6.9612	0.3260		2396
Mt Bol'shaya Varaka, olivine gabbronorite dyke (5–77)											
1	0.82	127	241	1630	6.173	4.5011		9.2951	0.4372		2393
2 (bd)	0.60	104	212	1490	5.547	8.2507		9.0262	0.4237		2397
3	0.63	332	607	1280	6.072	2.9501		8.9004	0.4171		2399
4	0.68	286	626	1590	6.144	3.1284		7.5613	0.3543		2399

[1] The ratios are corrected for blanks of 0.08 ng for Pb and 0.04 ng for U and for mass discrimination 0.17 ± 0.05%.
[2] Correction for common Pb was determined for the age according to Stacey & Kramers (1975).

display a thin faint zoning; they were subdivided into four types by colour. The U and Pb contents in the accessory minerals from the dyke are higher than those in zircon and baddeleyite from the Imandra lopolith.

The U–Pb isochron plotted on four varieties of zircon from the monzodiorite dyke gives an age of 2398 ± 21 Ma, MSWD = 2.4; the lower intersection is at zero, suggesting modern lead losses (Fig. 9d, Table 4). A close U–Pb age was obtained for zircon and baddeleyite from a pegmatoid olivine gabbronorite dyke. Three zircon types and baddeleyite yield a U–Pb age of 2395 ± 5 Ma, MSWD = 0.63 (Fig. 9e, Table 4).

Hence, dating of different parts of the Imandra lopolith makes it possible to determine the duration of crystallization processes and define three pulses of the Imandra lopolith evolution lasting for about 50 Ma. The interval agrees with the available data for the older Fedorovo-Pansky, Mt Generalskaya and Monchegorsk layered complexes.

Discussion

Specific features of the isotope investigation of the intrusions

The U–Pb concordia and isochron method has been used to define the age of crystallization of the rocks. The values obtained on zircons and baddeleyites from the same sample usually lie at the isochron (Figs 7, 8 & 9), indicating a similar age of magma crystallization and subsequent transformations. Coordinates of baddeleyites are near the concordia line. However, the method encounters the obstacle that mafic rocks contain very few zircon and baddeleyite grains. Samples of tens of kilograms yield only a few milligrams of these minerals.

The samples of gabbronorite and anorthosite taken from Mt Generalskaya and from the underlying granite-gneiss contain both magmatic and zircon xenocrysts. Refractory zircon xenocrysts reflect local processes of crustal contamination and lie outside the isochron (Fig. 6) showing a $^{207}Pb/^{206}Pb$ age of >2.6 Ga.

In order to compare our U–Pb results (23 isochron points), some samples were sent to the Royal Ontario Museum Laboratory in Canada. The data obtained there (Amelin et al. 1995) agree with ours within error (Table 5).

The Sm–Nd system is not an accurate geochronometer (c. 2–5%). However, the Sm–Nd isochron method can allow the establishment of crystallization times for mafic rocks on major rock-forming minerals (olivine, orthopyroxene, clinopyroxene and plagioclase). It is especially important for dating rocks with syngenetic ore minerals.

For example, this method has been used to determine for the first time the age (2482 ± 36 Ma) of the early ore body in the Fedorovo-Pansky Cu–Ni–PGE deposit of the Kola Peninsula that has economic importance (Serov et al. 2007).

For the mafic-ultramafic intrusions of the Kola belt, the Sm–Nd ages (16 points, Tables 5, 6) overlap because of high errors, but are commonly close to the U–Pb (TIMS) data on zircon and baddeleyite. They are especially valid for the marginal fast-crystallizing rocks of the Taxitic Zone of the Fedorovo-Pansky Complex, where the early barren orthopyroxenite and gabbro have the following ages: 2521 ± 42 Ma and 2516 ± 35 Ma (Sm–Nd method) and 2526 ± 6 Ma and 2516 ± 7 Ma (U–Pb method), respectively. The ore-bearing norite of the Fedorov Block yielded an age of 2482 ± 36 Ma (Sm–Nd method) and 2485 ± 9 Ma (U–Pb method) according to Nitkina (2006) and Serov et al. (2007) (Figs 7 & 10).

It is also important to stress that the Sm–Nd method provides valuable petrological and geochemical markers: $\varepsilon_{Nd}(T)$ and T_{DM}. The ε_{Nd} shows the degree of mantle magma source depletion, while T_{DM} indicates an approximate age of the mantle protolith (Faure 1986).

This chapter presents 37 ε_{Nd} values (Tables 5, 6, 8 & 9) and 24 T_{DM} measurements (Tables 6, 8 & 9) calculated from new data collected, and from previously published material (Tolstikhin et al. 1992; Amelin & Semenov 1990; Huhma et al. 1990; Hanski et al. 2001).

The Rb–Sr whole rock and mineral isochron method is mostly valuable for dating unaltered felsic igneous rocks and metamorphic amphibolite-facies associations (Faure 1986). In our work, 10 Rb–Sr isotope values for the rocks (Table 9) are considered to have only a petrological implication. Together with specific trace elements (Cu, Ni, Ti, V and LREE), ε_{Nd} (2.5 Ga), REE (Fig. 11, Table 7) and $^4He/^3He$ (Table 10) data, the values of initial $^{87}Sr/^{86}Sr$ (I_{Sr} [2.5 Ga]) indicate an enriched mantle reservoir 2.5 billion years ago which is comparable with the modern EM-I.

The timing, pulsation and total duration of magmatic activity

The largest and richest ore deposits of the Monchepluton and Fedorovo-Pansky complexes have been carefully studied by geochronological methods.

The layered or differentiated series of mafic-ultramafic rocks, from troctolite to leucogabbro-anorthosite, and syngenetic Cu–Ni–PGE ores of the Monchepluton formed within the time interval of 2516 (max)–2476 (min) Ma. Without analytical errors, the time interval is from 2507–2493 Ma

Table 5. *Summary of U–Pb and Sm–Nd geochronology for layered intrusions located in the eastern Baltic Shield*

Layered intrusions	Age (Ma) U–Pb	Age (Ma) Sm–Nd	$\varepsilon_{Nd(T)}$ @U/Pb age
Northern belt			
Mt Generalskaya			
Gabbronorite	2496 ± 10^1 $(2505 \pm 1.6)^2$	2453 ± 42^1	−2.3
Anorthosite	2446 ± 10^1		
Monchepluton			
Mt Travyanaya, norite	2507 ± 9^{15};		
Dunite block, gabbronorite dyke	2506 ± 10^{15}; 2496 ± 14^{15}		
Nyud Terrace, gabbronorite	2500 ± 5^{14}		
Nyud Terrace, gabbronorite	2493 ± 7^1 $(2504 \pm 1.5)^2$	2492 ± 31^3	−1.4
Vurechuaivench Foothills, metagabbronorite	2497 ± 21^{15}		
Main Ridge			
Monchetundra, gabbro	2463 ± 25^4; 2453 ± 4^5		
Monchetundra, gabbronorite	2505 ± 6^{14}; 2501 ± 8^{14}		
Chunatundra, anorthosite	2467 ± 7^{15}		
Ostrovsky intrusion			
Gabbronorite-pegmatite	2445 ± 11^{15}		
Fedorov-Pansky Complex			
Orthopyroxenite	2526 ± 6^{12}	2521 ± 42^{13}	−1.7
Olivine gabbro	2516 ± 7^{12}	2516 ± 35^{13}	−1.4
Magnetite gabbro	2498 ± 5^6		
Gabbronorite	2491 ± 1.5^7 $(2501 \pm 1.7)^2$	2487 ± 51^7	−2.1
Cu–Ni PGE-bearing gabbronorite	2485 ± 9^{12}	2482 ± 36^{13}	−2.4
PGE-gabbro-pegmatite	2470 ± 9^7		
PGE-anorthosite	2447 ± 12^7		
Imandra lopolith			
Gabbronorite	2446 ± 39^7 $(2441 \pm 1.6)^2$	2444 ± 77^7	−2.0
Gabbro-diorite-pegmatite	2440 ± 4^6		
Norite	2437 ± 7^6		
Leucogabbro-anorthosite	2437 ± 11^6		
Granophyre	2434 ± 15^6		
Olivine gabbronorite (dyke)	2395 ± 5^6		
Monzodiorite dyke	2398 ± 21^6		
Southern belt			
Kivakka			
Olivine gabbronorite	2445 ± 2^7	2439 ± 29^8	−1.2
Lukkulaisvaara			
Pyroxenite	2439 ± 11^7 $(2442 \pm 1.9)^2$	2388 ± 59^8	−2.4
Tsipringa			
Gabbro	2441 ± 1.2^2	2430 ± 26^8	−1.1
Burakovskaya intrusion			
Gabbronorite	2449 ± 1.1^2	2365 ± 90^8	−2.0
Kovdozero intrusion			
Pegmatoid gabbronorite	2436 ± 9^6		
Finnish group			
Koitelainen	2433 ± 8^9	2437 ± 49^{11}	−2.0
Koilismaa	2436 ± 5^{10}		
Nyaryankavaara	2440 ± 16^{10}		
Penikat		2410 ± 64^9	−1.6
Akanvaara	2437 ± 7^{11}	2423 ± 49^{11}	−2.1

1. Bayanova *et al.* 1999; 2. Amelin *et al.* 1995; 3. Tolstikhin *et al.* 1992; 4. Vrevsky & Levchenkov 1992; 5. Mitrofanov *et al.* 1993; 6. Bayanova 2004; 7. Balashov *et al.* 1993; 8. Amelin & Semenov 1996; 9. Huhma *et al.* 1990; 10. Alapieti *et al.* 1990; 11. Hanski *et al.* 2001; 12. Nitkina 2006; 13. Serov *et al.* 2007; 14. Bayanova & Mitrofanov 2005; 15. present study.

Table 6. *Sm–Nd isotope data on whole rock and mineral separates of the Fedorov Block of the Fedorovo-Pansky Complex (from Serov et al. 2007)*

Sample No.	Concentration (ppm)		Isotopic ratios		T_{DM} (Ga)	Sm–Nd (Ma)	ε_{Nd} (2.5 Ga)
	Sm	Nd	$^{147}Sm/^{144}Nd$	$^{143}Nd/^{144}Nd$			
orthopyroxenite (F-3)							
WR	0.32	1.17	0.1648	0.512196 ± 12	3.05	2521 ± 42	−1.73
Opx	0.12	0.38	0.2228	0.513182 ± 16			
Cpx	2.21	7.67	0.1745	0.512349 ± 17			
Pl	0.26	1.62	0.0960	0.511071 ± 29			
olivine gabbro (F-4)							
WR	0.63	2.80	0.1357	0.511548 ± 8	2.94	2516 ± 35	−1.53
Opx	0.23	0.72	0.1951	0.512555 ± 15			
Cpx	0.83	2.28	0.2187	0.512947 ± 16			
Pl	0.24	1.77	0.0815	0.510677 ± 14			
PGE-bearing gabbronorite (F-2)							
WR	0.42	1.66	0.1537	0.511807 ± 20	3.18	2482 ± 36	−2.50
Pl	0.41	2.88	0.0865	0.510709 ± 14			
Cpx	1.78	5.73	0.1876	0.512387 ± 8			
Opx	0.13	0.33	0.2323	0.513088 ± 40			

(Table 5). Some researchers (Smolkin *et al.* 2004) suggest that the Vurechuaivench part of the pluton, composed of gabbroids and anorthosites containing PGE deposits, is an independent magma chamber and that the age of rock and syngenetic PGE ore emplacement is 2497 ± 21 Ma.

The Fedorov Block of the Fedorovo-Pansky Complex represents an independent magma chamber, the rocks and ores of which differ significantly from those of the Western Pansky Block (Schissel *et al.* 2002). The 2 km-thick rock sequence, from the Marginal Zone to the Lower Gabbro Zone (Figs 2 & 3), is a layered or differentiated syngenetic series of relatively melanocratic pyroxenite-norite-gabbronorite-gabbro dated at 2526 ± 6 and 2516 ± 7 Ma. The Taxitic Zone is penetrated by concordant and cutting Cu–Ni–PGE-bearing gabbronorite (Fedorovo deposit) of the second pulse of magmatic injection, which is slightly younger (2485 ± 9 Ma; Table 5, Fig. 7).

The Western Pansky Block from the Main Gabbronorite Zone, without the Lower Layered Horizon and probably without the upper part (above 3000 m, Fig. 3), can also be considered a single syngenetic series of relatively leucocratic, mainly olivine-free gabbronorite-gabbro crystallized within the interval of 2503–2498–2491 ± 5 Ma (Fig. 3). In the lower part of the Block there are Norite and Marginal zones (Fig. 2). The Marginal zone contains poor disseminated Cu–Ni–PGE mineralization. This rock series can be correlated with certain parts of the Monchepluton and the Fedorov Block. The 40–80 m-thick Lower Layered Horizon (LLH) is prominent because of its contrasting structure with predominant leucocratic anorthositic rocks. The exposed part of the horizon strikes for almost 15 km (Fig. 2) and can be traced in boreholes down to a depth of 500 m (Mitrofanov *et al.* 2005). By its morphology, the horizon seems to be part of a single layered series. Nevertheless, there are anorthositic bodies that in outcrop show cutting contacts and apophyses (Latypov & Chistyakova 2000); the cumulus plagioclase compositions in the rocks of the horizon are different from those in the surrounding rocks (Fig. 3); and the age of the PGE-bearing leucogabbro-pegmatite, which is precisely defined by concordant and near-concordant U–Pb data on zircon as 2470 ± 9 Ma (Fig. 7a, Table 2), is slightly younger than the ages of the surrounding rocks (e.g. 2491 ± 1.5 Ma). The LLH rocks, especially the anorthosite and the PGE mineralization, probably represent an independent magmatic pulse.

The upper part and olivine-bearing rocks of the Western Pansky Block and the anorthosite of the Upper Layered Horizon (ULH) with the Southern PGE Reef (Figs 2 & 3) have been poorly explored. They differ from the main layered units of the Block in rock, mineral and PGE mineralization composition (Mitrofanov *et al.* 2005). Until now, only one reliable U–Pb age (2447 ± 12 Ma) has been obtained for the PGE-bearing anorthosite of the block, which may represent another PGE-bearing magmatic pulse.

The early magmatic activity of about 2.5 Ga manifested itself in the gabbronorite of the Monchetundra (2505 ± 6 and 2501 ± 8 Ma; Fig. 8h, j) and

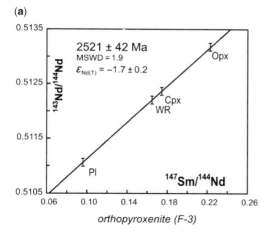
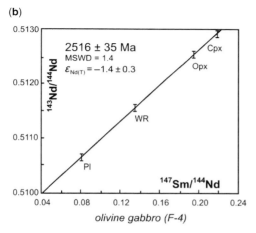
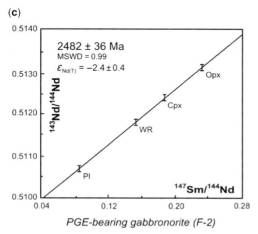

Fig. 10. Mineral Sm–Nd isochrons for rocks and rock-forming minerals of the Fedorov Block of the Fedorovo-Pansky Complex (Serov et al. 2007; see Table 6).

Mt Generalskaya (2496 ± 10 Ma; Fig. 6a). The magmatic activity that resulted in the formation of anorthosite took place about 2470 and 2450 Ma. It also contributed to the layered series of the Chunatundra (2467 ± 7 Ma; Fig. 8j) and Mt Generalskaya (2446 ± 10 Ma; Fig. 6b), Monchetundra gabbro (2453 ± 4 Ma; Table 5, Mitrofanov et al. 1993) and pegmatoid gabbronorite of the Ostrovsky intrusion (2445 ± 11 Ma; Fig. 8g).

The Imandra lopolith is the youngest large layered intrusion within the Kola Belt. It varies from the other intrusions of the Kola Belt both in its emplacement age and its metallogeny. There are five U–Pb zircon and baddeleyite ages for the rocks of the main magmatic pulse represented by norite, gabbronorite, leucogabbro-anorthosite, gabbrodiorite and granophyre; all formed within the interval from 2445–2434 Ma (Table 5).

Thus, several eruptive pulses of magmatic activity have been established in the complex intrusions of the Kola Belt, including at least four pulses (or phases) in the Fedorovo-Pansky Complex: a 2526–2516 Ma barren pulse and three ore-bearing of 2505–2485, 2470 and 2450 Ma. For similar intrusions of the Fenno-Karelian Belt, for example, the Penikat intrusion in Finland, five magmatic pulses varying only in geochemistry have been distinguished from the same deep chamber (Iljina & Hanski 2005).

A total duration for magmatic processes of over 80 Ma in the Kola Belt intrusions is unexpected for many researchers. The multi-phase magmatic duration of the Fenno-Karelian Belt intrusions was short-term and took place about 2.44 Ga years ago. However, there are only a few U–Pb precise age estimations for the Fenno-Karelian Belt intrusions (Iljina & Hanski 2005). A joint Russian–Finnish research collaboration intended for dating the intrusions of the both belts has recently been initiated. It is expected that the research will result in updating the knowledge about the timing and duration of the Palaeoproterozoic ore-forming intrusions on the Baltic Shield.

The Kola results underline that the layering of the intrusions with thinly-differentiated horizons and PGE reefs was not contemporaneous (or syngenetic), with each intrusion defining its own metallogentic trends in time and space.

Metallogenic implications

The Palaeoproterozoic magmatic activity in the eastern Baltic Shield is associated with the formation of widespread ore deposits: Cu–Ni (±PGE), Pt–Pd (+Rh, ±Cu, Ni, Au), Cr, Ti–V (Mitrofanov & Golubev 2008; Richardson & Shirey 2008).

On the Kola Peninsula, economic Cu–Ni (+PGE) deposits are known in the Monchegorsk

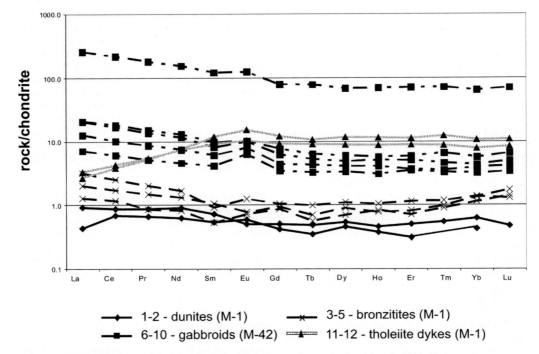

Fig. 11. REE data for rocks of the Monchepluton (plotted according to the data given in Table 7).

Table 7. *REE data for rocks of the Monchepluton*

Sample No.	1	2	3	4	5	6	7	8	9	10	11	12
	dunites (M-1)		*bronzitites (M-1)*			*gabbroids (M-42)*					*tholeiite dykes (M-1)*	
La	0.216	0.102	0.754	0.479	0.302	5.020	3.030	61.30	4.910	1.700	0.635	0.780
Ce	0.532	0.419	1.540	1.050	0.716	11.400	6.290	134.00	10.500	3.740	2.340	2.614
Pr	0.082	0.063	0.192	0.140	0.081	1.450	0.825	17.40	1.310	0.489	0.495	0.520
Nd	0.420	0.294	0.784	0.614	0.383	6.160	3.550	73.10	5.510	2.150	3.540	3.437
Sm	0.110	0.083	0.140	0.161	0.081	1.500	0.938	18.80	1.250	0.637	1.820	1.415
Eu	0.029	0.034	0.072	0.045	0.042	0.596	0.464	7.32	0.585	0.368	0.896	0.605
Gd	0.104	0.086	0.216	0.192	0.178	1.590	0.911	16.40	1.290	0.713	2.500	1.892
Tb	0.018	0.013	0.037	0.026	0.021	0.238	0.161	2.97	0.198	0.123	0.399	0.339
Dy	0.137	0.117	0.278	0.227	0.178	1.530	1.040	17.50	1.240	0.847	2.970	2.266
Ho	0.026	0.021	0.059	0.044	0.047	0.325	0.231	3.98	0.289	0.175	0.657	0.497
Er	0.083	0.051	0.190	0.135	0.118	0.980	0.607	11.80	0.842	0.578	1.880	1.477
Tm	0.014	<l.d.*	0.030	0.025	0.023	0.169	0.091	1.84	0.118	0.084	0.315	0.224
Yb	0.107	0.072	0.232	0.230	0.193	0.968	0.670	11.10	0.742	0.548	1.810	1.325
Lu	0.012	<l.d.*	0.044	0.033	0.036	0.168	0.107	1.81	0.127	0.086	0.279	0.213

REE determinations were measured by Ion probe Cameca 4F analyses by E. Deloule at the Centre de Recherches Petrographiques et Chimiques – CRPG-CNRS, Nancy. The data are provided by the Russian-French project RFFI-CNRS 01–05–22001 that was headed by the authors of the present study.
*<l.d. below limit of determination.

(c. 2500 Ma) and Pechenga (c. 1980 Ma) type intrusions. In the Monchepluton (the Monchegorsk type), syngenetic disseminated Cu–Ni (+PGE) ore bodies of magmatic origin are confined to basal parts of magmatic chambers (Papunen & Gorbunov 1985), while massive rich redeposited ores in the veined bodies of the Monchepluton bottom as well as beyond it (offset bodies) also contain a relatively high portion of platinum among PGE. They are associated mainly with c. 2500 Ma magnesium-rich mafic-ultramafic rocks with ε_{Nd} (2.5 Ga) values varying from -1 to -2. In comparison, Cu–Ni ($\pm$PGE) ores of the Pechenga type intrusions, that are not discussed, are related to the 1980 Ma gabbro-wehrlite rocks with ε_{Nd} (1.98 Ga) values varying from $+1$ to $+3$ (Hanski et al. 1990; Mitrofanov & Golubev 2008). The basal ores of the Fedorovo deposit are first of all valuable for platinum-group elements (Pt, Pd, Rh), but nickel, copper and gold are also of economic importance here (Schissel et al. 2002). The ore-forming magmatic and post-magmatic processes are closely related to the Taxitic Zone gabbronorite of 2485 $\pm$ 9 Ma magmatic pulse.

Pt–Pd ($\pm$Cu, Ni, Rh, Au) reef-type deposits and ore occurrences of the Vurechuaivench Foothills (Monchepluton) and Western Pansky Block (Fedorovo-Pansky Complex) seem, in terms of genesis, to be associated with pegmatoid leucogabbro and anorthosite rocks enriched in late-stage fluids. Portions of this magma produce additional injections of c. 2500 Ma (Vurechuaivench), c. 2470 Ma (the Lower, Northern PGE reef) and c. 2450 Ma (the Upper, Southern PGE reef of the Western Pansky Block and PGE-bearing mineralization of the Mt Generalskaya intrusion). These non-simultaneous injections are quite close in terms of composition, prevalence of Pd over Pt, ore mineral composition (Mitrofanov et al. 2005), and isotope geochemistry of Sm–Nd and Rb–Sr systems. The ε_{Nd} values for the rocks under consideration vary from -1 to -3, which probably indicates a single long-lived magmatic hearth.

Chromium concentration (>1000 ppm) is a typical geochemical feature of the lower mafic-ultramafic rocks of the layered intrusions of the Baltic Shield (Alapieti 1982; Iljina & Hanski 2005). The chromite mineralization is known in the basal series of the Monchepluton, Fedorovo-Pansky Complex, Imandra lopolith (Russia), Penikat and Narkaus intrusions (Finland) and in chromite deposits of the Kemi intrusion (Finland) and Dunite Block (Monchepluton, Russia). On the contrary, Fe–Ti–V mineralization of the Mustavaara intrusion (Finland) tends to most leucocratic parts of the layered series, and to leucogabbro-anorthosite and gabbro-diorite of the Imandra lopolith (Russia) and Koillismaa Complex (Finland).

Thus, PGE-bearing deposits of the region are represented by two types: the basal and the reef-like ones. According to modern economic estimations, the basal type of deposits is nowadays more preferable for mining, even if the PGE concentration (1–3 ppm) is lower compared to the reef-type deposits (>5 ppm). Basal deposits are thicker and contain more platinum, copper and, especially, nickel. These deposits are accessible to open pit mining.

Petrological and geodynamic implications

Magmatic processes since the Palaeoproterozoic (2.53 Ga) have affected almost the whole region of the East Scandinavian (Kola-Lapland-Karelian) province and a mature continental crust formed (2.55 Ga) in the Neoarchaean (Gorbatschev & Bogdanova 1993). Thick (up to 3 km) basaltic volcanites of the Sumian age (2.53–2.40 Ga) in Karelia, Kola and NE Finland cover an area of $>200 000$ km^2. In the north, magmatic analogues of these volcanic rocks are represented by two belts of layered intrusions and numerous dyke swarms (Vuollo et al. 2002; Vuollo & Huhma 2005). This together composes a single time- and space-related megacyclic association, the East Scandinavian Large Igneous Province (ESCLIP). All the magmatic units of the province covering a huge area show similar geological, compositional and metallogenic features.

Regional geological settings indicate anorogenic rift-like intraplate arrangements involving volcano-plutonic belts connecting different domains of the Palaeoarchaean Kola-Lapland-Karelia protocontinent. This resembles early advection extensional geodynamics of passive rifting that is typical of intraplate plume processes (Pirajno 2007).

Geochemical and isotope-geochemical data shed light on features of deep magma source for the ESCLIP rocks. T_{DM} values (Faure 1986) are approximately the age of the depleted mantle reservoir (DM) with slightly enriched Sm–Nd ratios. The T_{DM} values (Tables 6, 8 & 9) lie within the interval of 3.1–2.8 Ga. The ε_{Nd} values (Tables 5, 6, 8 & 9) vary from -1.1 to -2.4 and similar I_{Sr} values (0.703–0.704) obtained for discrete layered intrusions form a narrow range of enriched compositions. It is difficult to argue for a local crustal contamination and we suggest that the magmas producing different rocks of the ESCLIP layered intrusions were derived from a single homogenous mantle source enriched both with typically magmatic ore elements (Ni, TI, V and Pt) and lithophile elements including light REE (Fig. 11). To some extent, this reservoir is comparable with the modern EM-1 (Fig. 12) source (Hofmann 1997).

^{4}He/^{3}He ratio is also a reliable isotope tracer of mantle plume processes (Tolstikhin & Marty 1998;

Table 8. *Sm–Nd isotope analyses and model ages for the rocks of the Monchegorsk Layered Complex*

Sample No.	Rock type	Concentration (ppm)		Isotope ratios		T_{DM} (Ga)	ε_{Nd} (2.5 Ga)
		Sm	Nd	$^{147}Sm/^{144}Nd$	$^{143}Nd/^{144}Nd \pm 2\sigma$		
(Monchepluton)							
M-1	Mineralized norite	1.750	8.04	0.131957	0.511493 ± 3	2.91	−1.51
M-2	Gabbronorite-pegmatite	0.920	4.15	0.134055	0.511537 ± 4	2.90	−1.37
M-42	Metagabbronorite	1.480	6.67	0.134404	0.511462 ± 3	2.85	−2.98
C-1	Orthopyroxenite	0.564	2.56	0.133100	0.511477 ± 6	3.09	−2.30
H-4	Olivine pyroxenite	0.584	2.60	0.136000	0.511577 ± 8	3.01	−1.20
H-7	Gabbronorite	0.703	3.06	0.138900	0.511680 ± 5	3.07	−1.50
(Monchetundra)							
M-54	Trachytoid gabbronorite	0.940	3.54	0.159479	0.511963 ± 5	3.08	−1.23
(Chunatundra)							
M-16	Trachytoid anorthosite	0.730	3.08	0.143365	0.511740 ± 5	2.84	−0.63
(Mt Yarva-Varaka)							
M-38	Granophyric quartz diorite	7.380	38.48	0.115923	0.511184 ± 2	2.91	−2.48
(Ostrovsky intrusion)							
M-7	Gabbronorite-pegmatite	3.820	18.69	0.123446	0.511333 ± 2	2.90	−2.47
(dyke rocks of the Monchepluton)							
M-9	Coarse-grained melanonorite	0.469	2.24	0.126386	0.511354 ± 5	2.96	−2.60
M-12	Gabbronorite dyke	0.581	2.35	0.149344	0.511795 ± 11	2.98	−1.22
M-14	Gabbronorite dyke	1.770	8.86	0.120709	0.511573 ± 4	2.40	3.60

Average standard values: N = 11 (La Jolla: = 0.511833 ± 6); N = 100 (JNdi1: = 0.512098 ± 15).

Table 9. Sm–Nd and Rb–Sr isotope data for rocks of the Mt Generalskaya, Fedorovo-Pansky, Imandra, Monchepluton and Monchetundra intrusions

Sample No.	Concentration (ppm)		Isotopic ratios		ε_{Nd} (2.5 Ga)	T_{DM} (Ga)	$^{87}Rb/^{86}Sr$	$^{87}Sr/^{86}Sr$ ($\pm 2\sigma$) @2.5 Ga
	Sm	Nd	$^{147}Sm/^{144}Nd$	$^{143}Nd/^{144}Nd$ ($\pm 2\sigma$)				
(Mt Generalskaya)								
S-3464, gabbronorite	1.147	5.362	0.129320	0.511449 ± 14	−2.30	2.91	0.00534	0.70421 ± 22
(Fedorov-Pansky intrusion)								
Pan-1, gabbronorite	0.762	3.293	0.139980	0.511669 ± 7	−2.00	2.98	0.00135	0.70315 ± 10
Pan-2, gabbronorite	0.423	1.662	0.153714	0.511807 ± 20	−2.50	3.18	0.00174	0.70295 ± 17
F-4, olivine gabbro	0.629	2.801	0.135695	0.511548 ± 8	−1.53	2.94	0.00144	0.70288 ± 22
F-3, orthopyroxenite	0.318	1.166	0.164803	0.512196 ± 12	−1.73	3.05	0.00205	0.70333 ± 15
(Imandra lopolith)								
6–57, gabbronorite	2.156	10.910	0.119130	0.511380 ± 3	−2.00	2.88	0.00339	0.70455 ± 27
(Monchepluton)								
M-1, quartz norite	1.750	8.040	0.131957	0.511493 ± 3	−1.51	2.91	0.01053	0.70341 ± 9
H-7, gabbronorite	0.920	4.150	0.134055	0.511537 ± 4	−1.37	2.90	0.00227	0.70368 ± 24
(Monchetundra)								
MT-10, medium-grained pyroxenite	0.483	1.913	0.152689	0.511925 ± 33	−0.36	2.81	0.00495	0.70399 ± 17

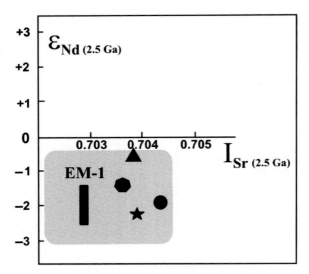

Fig. 12. ε_{Nd}–Sr plot of rocks from the Northern (Kola) Belt layered intrusions. Grey colour in the diagram shows EM-1 reservoir plotted for the layered intrusions of the Kola Peninsula based on the Sm–Nd and Rb–Sr isotope data given in Table 9.

Table 10. *Isotopic $^4He/^3He$ ratios of PGE layered intrusions of the Baltic Shield (data are provided by Kamensky Novikov, Geological Institute KSC RAS, unpublished)*

Hole No/sampling depth (m)	Rock, mineral	$^4He \times 10^{-6}$ $\mu cm^3/g$	$^4He/^3He \times 10^6$	Low Mantle contribution**%
(Fedorovo-Pansky intrusion)				
hole, Ki-16/6	Amphibole	81.00	9.10	0.11
hole, Ma-14/1	Orthopyroxene	9.90	12.80	0.08
outcrop, No 9	Ilmenite	43.90	16.50	0.06
(Monchepluton, Mt Sopcha)				
hole, 995/315	Olivinite, rock	17.00	6.25	0.16
hole, 995/315	Olivine	25.00	5.88	0.17
hole, 995/315	Orthopyroxene	31.00	6.25	0.16
hole, 995/315	Plagioclase	47.00	5.56	0.18
hole, 995/315	Magnetite	132.00	4.35	0.23
(Main Ridge, Monchetundra)				
hole, 765/905.9	Clinopyroxene	163.00	4.76	0.21
hole, 765/905.9	Orthopyroxene	21.00	4.76	0.21
hole, 765/985.3	Amphibole	97.00	4.76	0.21
hole, 765/985.3	Clinopyroxene	115.00	5.00	0.20
outcrop, MT-5	Gabbro	1.30	2.00	0.41
(Dunite block)				
hole, 904/102	Dunite, rock	218.00	1.47	0.68
hole, 904/102	Olivine	115.00	1.35	0.74
hole, 1651/244.9	Chromitite, ore	56.00	1.43	0.70
hole, C-1651/373.5*	Dunite-Bronzitite	28.00	0.83	1.20
hole, C-1622/7*	Chromitite, ore	2.80	0.69	1.44
hole, C-1646/450*	Dunite	2.20	1.29	0.77
hole, C-1651/373.5*	Dunite-Bronzitite-contact	0.13	0.60	1.68

Note: errors are according to the calculation method (Tolstikhin & Marty 1998)
*Step wise heating experiment (fraction under the temperature 1300 °C).
**Mantle components are given from value $^4He/^3He$ 0.55 × 10^4 (solar helium from lower mantle reservoir), Tolstikhin & Marty (1998).

Bayanova et al. 2006; Pirajno 2007). Their use in studying Precambrian rocks requires special care. Table 10 shows recent helium isotope data for the rocks and minerals of the Kola Belt intrusions. The data indicate that the $^4He/^3He$ isotope ratios of $n \times 10^{6-5}$ correspond to those of the upper mantle and differ from those of the crust ($n \times 10^8$) and lower mantle ($n \times 10^4$) (Tolstikhin & Marty 1998). The helium isotope data tend to favour a source dominated by mantle-derived magmas with only local crustal contamination.

According to the available data (Campbell 2001; Condie 2001; Vuollo et al. 2002; Bleeker 2003; Ernst & Buchan 2003; present study), the peak of the mafic-ultramafic magmatic activity of the Kola-Karelian, Superior and Wyoming provinces has been estimated at c. 2.45 Ga. Figure 13 presents an attempt to demonstrate some reconstruction of the Archaean supercontinent embodying these three provinces of Europe and North America (Heaman 1997). Insert (A) shows trends of the Kola and Fenno-Karelian Belts of 2.52–2.44 Ga

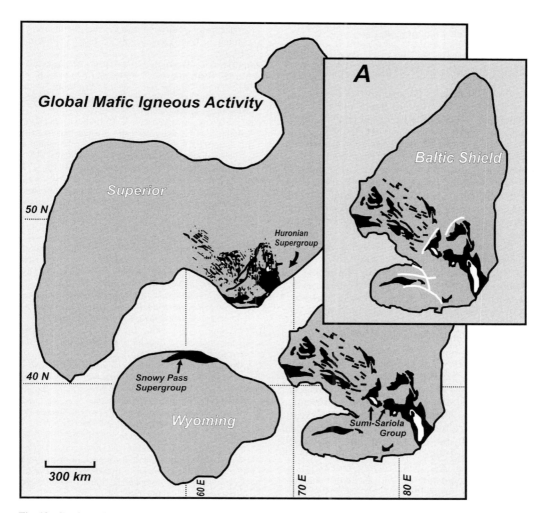

Fig. 13. Continental reconstruction at 2.45 Ga involving the Superior, Wyoming, and Karelian cratons (Heaman 1997). Patterned regions correspond to 2.45 Ga supracrustal rocks, including flood basalts, such as Huronian Supergroup (Superior), Snowy Pass Supergroup (Wyoming), and Suml-Sarlola-Strelna Supergroups (Karelia). Orientation of Karelia is based on alignment of Hearst and Karelian dike swarms and interpretation that they represent rift-parallel dykes. Estimated palaeolatitudes are determined from palaeomagnetic studies; palaeolongitudes are arbitrary. Black colour depicts Palaeoproterozoic mafic magmatism (layered intrusions and mafic dykes). **A** - shows trends (▬▬) of 2.52–2.44 Ga layered intrusions located on of the NE Baltic Shield.

layered intrusions with the intraplate nature interpreted from the results of the present study.

The ESCLIP layered intrusions are directly related to the Baltic Shield metallogeny (Mitrofanov & Golubev 2008). The >80 Ma duration and multiphase history of the Kola Belt layered mafic intrusions (i.e. 2.53–2.45 Ga) has been shown here. It has also been underlined that the younger intrusions of the Fenno-Karelian Belt (Fig. 1) clustre at 2.44 Ga (Iljina & Hanski 2005). The partially asynchronous evolution of these two belts, that are thought to be arms of a mantle plume, is now being examined in more detail as a follow-up to this study within the framework of Russian-Finnish research collaboration.

Summary and conclusions

A number of new U–Pb and Sm–Nd isotope data were obtained for various rocks of the mafic layered intrusions of the Kola Belt (Baltic Shield), including those which bear PGE, Ni–Cu and Ti–V mineralization. A surprisingly long period of multiphase magmatic activity, from 2530–2450 Ma (about 80 Ma), resulted in the intrusion of large-scale ore-bearing intrusions of the Kola Belt. Magmatism continued until about 2400 Ma and generated widespread dykes and small-scale intrusions. These results contrast with the published data, indicating short-term evolution interval (c. 2440 Ma) for similar intrusions of the Fenno-Karelian Belt (Iljina & Hanski 2005).

The two belts of mafic layered intrusions of the Baltic Shield (the Kola and Fenno-Karelian belts), together with the surrounding volcanic rocks and dyke swarms, compose the Palaeoproterozoic East Scandinavian Large Igneous Province (ESCLIP) with an area of >200 000 km². The petrological-geodynamic interpretation proposed by this chapter of the ESCLIP is a product of a vast long-lived plume is based on the homogenous and enriched isotope characteristics of the magmas and also the large volume and widespread distribution of the magmas. It is quite possible, and fully consistent with our observations, that the geochemical signatures of the ESCLIP magmas may well have been in part inherited from the subcontinental lithosphere, as described recently based on Os-isotope characteristics for the Bushveld magmas (Richardson & Shirey 2008).

The authors thank the following: L. Koval for baddeleyite and zircon separation from rock samples; E. Savchenko for baddeleyite and zircon analyses using a Cameca MS-46 and for taking images of baddeleyite crystals; N. Levkovich for the chromatographic separation of U and Pb for analyses by mass spectrometry at the Geological Institute, Kola Science Center, Russian Academy of Sciences.

The study was supported by the Russian Foundation of Fundamental Investigations, project no. 07-05-00956, 08-05-00324, 0F1-M-09-05-12028, Program no. 8 of the Division of Earth Science, Russian Academy of Sciences, and Program for Leading Research Schools NSH-1413.2006.5 (headed by F. P. Mitrofanov) and Interreg-Tacis K-0193. We thank S. Reddy who kindly provided a review of this chapter and also three anonymous reviewers.

References

ALAPIETI, T. T. 1982. The Koillismaa layered igneous complex, Finland: its structure, mineralogy and geochemistry, with emphasis on the distribution of chromium. *Geological Survey of Finland, Bulletin*, **319**, 116.

ALAPIETI, T. T., FILEN, B. A., LAHTINEN, J. J., LAVROV, M. M., SMOLKIN, V. F. & VOITEKHOVSKY, S. N. 1990. Early Proterozoic layered intrusions in the Northeastern part of the Fennoscandian Shield. *Mineralogy and Petrology*, **42**, 1–22.

AMELIN, Yu. V. & SEMENOV, V. S. 1996. U–Nd and Sr isotopic geochemistry of mafic layered intrusions in the eastern Baltic Shield: implications for the evolution of Palaeoproterozoic continental mafic magmas. *Contributions of Mineralogy and Petrology*, **124**, 255–272.

AMELIN, Yu. V., HEAMAN, L. M. & SEMENOV, V. S. 1995. U–Pb geochronology of layered mafic intrusions in the eastern Baltic Shield: implications for the timing and duration of Palaeoproterozoic continental rifting. *Precambrian Research*, **75**, 31–46.

BALASHOV, Y. A., BAYANOVA, T. B. & MITROFANOV, F. P. 1993. Isotope data on the age and genesis of layered basic-ultrabasic intrusions in the Kola Peninsula and northern Karelia, northeastern Baltic Shield. *Precambrian Research*, **64**, 197–205.

BAYANOVA, T. B. 2004. *Age of reference Geological complexes of the Kola region and the duration of igneous processes*. Nauka, Saint Petersburg, 174.

BAYANOVA, T. B. 2006. Baddeleyite: a Promising Geochronometer for Alkaline and Basic Magmatism. *Petrology*, **14**, 187–200.

BAYANOVA, T. B. & BALASHOV, Yu. A. 1995. Geochronology of Palaeoproterozoic layered intrusions and volcanites of the Baltic Shield: proceed. of the 1st International Barents Symposium 'Geology and minerals in the Barents Region'. Norges Geologiske Undersøkelse Special Publication **7**, 75–80.

BAYANOVA, T. B. & MITROFANOV, F. P. 2005. Layered Proterozoic PGE intrusions in Kola region: new isotope data. X international symposium of platinum 'Platinum-Group Elements – from Genesis to Beneficiation and Environmental Impact': extended abstracts. Oulu, Finland, 289–291.

BAYANOVA, T. B., SMOLKIN, V. F. & LEVKOVICH, N. V. 1999. U–Pb geochronological study of Mount Generalskaya layered intrusion, northwestern Kola Peninsula, Russia. *Transactions of the Institution of Mining and Metallurgy*, **108**, B83–B90.

BAYANOVA, T. B., GALIMZYANOVA, R. M. & FEDOTOV, G. A. 2001. Evidence of the multiphase complex history of the Imandra lopolith. Svekalapko.

Europrobe project. 6th Workshop. Abstracts. Lammi, Finland. University of Oulu, 7.

BAYANOVA, T. B., NOVIKOV, D. D., NITKINA, E. A., SEROV, P. A. & MITROFANOV, F. P. 2006. *Polychronic and long-time interval of the Formation Proterozoic PGE-bearing Fedorovo-Pansky intrusion.* Understanding the genesis of ore deposits to meet the demands of the 21st century. 12th quadrennial IAGOD symposium 2006: Abstract. Moscow. (fill №106).

BLEEKER, W. 2003. The late Archaean record: a puzzle in c. 35 pieces. *Lithos*, **71**, 99–134.

CAMPBELL, I. H. 2001. Identification of ancient mantle plumes. *In*: ERNST, R. E. & BUCHAN, K. L. (eds) *Mantle Plumes: their Identification through Time.* Geological Society of America, Special Papers, **352**, 5–22.

CHASHCHIN, V. V., BAYANOVA, T. B. & APANASEVICH, E. A. 2002. The Monchegorsk ore district as an example of the Palaeoproterozoic ore-bearing chamber structure (Kola, Russia). *Geology of Ore Deposits*, **44**, 142–149.

COFFIN, M. F. & ELDHOLM, O. 1994. Large igneous provinces: crustal structure, dimensions and external consequences. *Reviewes in Geophysics*, **32**, 1–36.

CONDIE, K. C. 2001. *Mantle Plumes and Their Record in Earth History.* Cambridge University Press, Cambridge.

ERNST, R. E. & BUCHAN, K. L. 2003. Recognizing mantle plumes in the geological record. *Annual Reviews in Earth Planetary Science*, **31**, 469–523.

EULES, H. V. & CAWTHORN, R. G. 1995. *The Bushveld Complex. Layered Intrusions.* Developments in Petrology 15. Elsevier, Amsterdam, 181–229.

FAURE, G. 1986. *Principles of Isotope Geology.* 2nd edn. Wiley, New York.

FRENCH, I. E., HEAMAN, L. M., CHACKO, T. & SRISTAVA, R. K. 2008. 1891–1883 Ma Southern Bastar-Cuddapah mafic igneous events, Judia: a newly recognized large igneous province. *Precambrian Research*, **160**, 308–322.

GOLDSTEIN, S. J. & JACOBSEN, S. B. 1988. Nd and Sr isotopic systematics of river water suspended material implications for crystal evolution. *Earth and Planetary Science Letters*, **87**, 249–265.

GORBATSCHEV, R. & BOGDANOVA, S. 1993. Frontiers in the Baltic Shield. *Precambrian Research*, **64**, 3–21.

HANSKI, E., HUHMA, H. ET AL. 1990. The age of the ferropicric volcanics and comagmatic Ni-bearing intrusions at Pechenga, Kola Peninsula, USSR. *Bulletin of the Geological Society of Finland*, 1990, **62**, 123–133.

HANSKI, E., WALKER, R. J., HUHMA, H. & SUOMINEN, I. 2001. The Os and Nd isotopic systematics of c. 2.44 Ga Akanvaara and Koitelainen mafic layered intrusions in northern Finland. *Precambrian Research*, **109**, 73–102.

HEAMAN, L. M. 1997. Global mafic magmatism at 2.45 Ga: remnants of an ancient large igneous province? *Geology*, **25**, 299–302.

HEAMAN, L. M. & LECHEMINANT, A. N. 1993. Paragenesis and U–Pb systematics of baddeleyite (ZrO). *Chemical Geology*, **110**, 95–126.

HOFMANN, A. W. 1997. Mantle geochemistry: the message from oceanic volcanism. *Nature*, **385**, 219–229.

HUHMA, H., CLIFT, R. A., PERTTUNEN, V. & SAKKO, M. 1990. Sm–Nd and Pb isotopic study of mafic rocks associated with early Proterozoic continental rifting: the Perapohja schist belt in Northern Finland. *Contributions to Mineralogy Petrology*, **104**, 369–379.

ILJINA, M. & HANSKI, E. 2005. Layered mafic intrusions of the Tornio-Näränkävaara belt. *In*: LEHTINEN, M., NURMI, P. A. & RÄMO, O. T. (eds) *Precambrian Geology of Finland – Key to the Evolution of the Fennoscandian Shield.* Elsevier B. V., Amsterdam. 101–138.

JACOBSEN, S. B. & WASSERBURG, G. J. 1984. Sm–Nd isotopic evolution of chondrites and achondrites, II. *Earth and Planetary Science Letters*, **67**, 137–150.

KRAMM, U. 1993. Mantle components of carbonatites from the Kola Alkaline Province, Russia and Finland: a Nd–Sr study. *European Journal of Mineralogy*, **5**, 985–989.

KROGH, T. E. 1973. A low-contamination method for hydrothermal dissolution of zircon and extraction of U and Pb for isotopic age determinations. *Geochimica et Cosmochimica Acta*, **37**, 485–494.

LATYPOV, R. M. & CHISTYAKOVA, S. Yu. 2000. Mechanism for differentiation of the Western-Pana layered intrusion. *Apatity: KSC RAS*, 315 (in Russian).

LI, C. & NALDRETT, A. J. 1993. Sulfide capacity of magma: a quantitative model and its applition to the formation of sulfide ores at Sudbury, Ontario. *Economic Geology*, **88**, 1253–1260.

LUDWIG, K. R. 1991. *PBDAT – A Computer Program for Processing Pb–U–Th isotope Data.* Version 1.22. Open-file report **88–542**. US Geological Survey, 38.

LUDWIG, K. R. 1999. ISOPLOT/Ex – A geochronological toolkit for Microsoft Excel, Version 2.05. *Berkeley Geochronology Center Special Publication*, **1a**, 49.

MITROFANOV, F. P. & BAYANOVA, T. B. 1999. Duration and timing of ore-bearing Palaeoproterozoic intrusions of Kola province. *In*: STANLEY, C. J. ET AL. (eds) *Mineral Deposits: Processes to Processing.* Balkema, Rotterdam, 1275–1278.

MITROFANOV, F. P. & GOLUBEV, A. 2008. *Russian Fennoscandia metallogeny.* Abstract to 33 IGC. Oslo, Norway.

MITROFANOV, F. P. & NEROVICH, L. I. 2003. Timing of magmatic crystallization and metamorphic transformation in the Pyrshin and Abvar Autonomous anorthosite massifs, Lapland granulate belt. *Contributions to Mineralogy and Petrology*, **11**, 343–351.

MITROFANOV, F. P., BALAGANSKY, V. V. ET AL. 1993. U–Pb age for gabbro-anorthosite of the Kola Peninsula. *Doklady RAN*, **331**, 95–98.

MITROFANOV, F. P., BALABONIN, N. L. ET AL. 1997. Main results from the study of the Kola PGE-bearing province, Russia. *In*: PAPUNEN, H. (ed.) *Mineral Deposits.* Balkema, Rotterdam, 483–486.

MITROFANOV, F. P., SMOLKIN, V. F., BAYANOVA, T. B., NERADOVSKY, YU. N., OHNENSTETTER, D., OHNENSTETTER, M. & LUDDEN, J. 2002. *Palaeoproterozoic (2.5–2.4 Ga) Plume Magmatism in the North-Eastern Baltic Shield and Origin of the PGE, Sulphide and Chromite Ore Deposit.* Extended abstracts: 9th International Platinum Symposium. Billings, Montana, USA, 309–311.

MITROFANOV, F. P., KORCHAGIN, A. U., DUDKIN, K. O. & RUNDKVIST, T. V. 2005. *Fedorovo–Pana layered mafic intrusion (Kola Peninsula, Russia): approaches, methods, and criteria for prospecting PGEs*. Exploration for platinum-group elements deposits. Short Course delivered on behalf of the Mineralogical Association of Canada in Oulu, Finland, **35**, 343–358.

NITKINA, E. A. 2006. U–Pb zircon dating of rocks of the platiniferous Fedorova-Pana layered massif, Kola Peninsula. *Doklady Earth Sciences*, **408**, 551–554.

PAPUNEN, H. & GORBUNOV, G. I. (eds) 1985. Nickel-copper deposits of the Baltic Shield and Scandinavian Caledonides. *Geological Survey of Finland, Bulletin*, **333**, 394.

PIRAJNO, F. 2007. Mantle plumes, associated intraplate tectono-magmatic processes and ore systems. *Episodes*, **30**, 6–19.

RICHARDSON, S. H. & SHIREY, S. B. 2008. Continental mantle signature of Bushveld magmas and coeval diamonds. *Nature*, **453**, 910–913.

REISCHMANN, T. 1995. Precise U–Pb age determination with baddeleyite (ZrO_2), a case study from the Phalaborwa Igneous Complex, South Africa. *South African Journal Geology*, **1**, 1–4.

SCHARER, U. & GOWER, C. F. 1988. Crustal evolution in Eastern Labrador: constraints from precise U–Pb Ages. *Precambrian Research*, **38**, 405–421.

SCHARER, U., WILMART, E. & DUCHESNE, J.-C. 1996. The short duration and anorogenic character of Anorthosite Magmatism: U–Pb Dating of the Rogaland Complex, Norway. *Earth and Planetary Science Letters*, **139**, 335–350.

SCHISSEL, D., TSVETKOV, A. A., MITROFANOV, F. P. & KORCHAGIN, A. U. 2002. Basal platinum-group element mineralization in the Fedorov Pansky layered mafic intrusion, Kola Peninsula. *Russian Economic Geology*, **97**, 1657–1677.

SEROV, P. A., NITKINA, E. A., BAYANOVA, T. B. & MITROFANOV, F. P. 2007. Comparison of the new data on dating using U–Pb and Sm–Nd isotope methods of early barren phase rocks and basal ore-hosting rocks of the Pt-bearing Fedorovo-Pansky layered intrusion (Kola peninsula). *Doklady Earth Sciences*, **415**, 1–3 (in Russian).

SHARKOV, E. V., BOGATIKOV, O. A., GROKHOVSKAYA, T. L., SNYDER, G. A. & TAYLOR, L. A. 1995. Petrology and Ni–Cu–Cr–PGE mineralization of the largest mafic pluton in Europe: the early Proterozoic Burakovsky Layered Intrusion, Karelia, Rossia. *International Geology Review*. **37**, 509–525.

SKUF'IN, P. K. & BAYANOVA, T. B. 2006. Early Proterozoic central-type volcano in the Pechenga Structure and its relation to the ore-bearing Gabbro-Wehrlite Complex of the Kola Peninsula. *Petrology*, **14**, 609–627.

SMOLKIN, V. F., FEDOTOV, ZH. A. *ET AL*. 2004. Layered intrusions of the Monchegorsk ore region: petrology, mineralization, isotope features and deep structure. *In*: MITROFANOV, F. P. & SMOLKIN, V. F. (eds) *Part 1. Apatity*, Kola Science Centre RAS, 177 (in Russian).

STACEY, J. S. & KRAMERS, J. D. 1975. Approximation of terrestrial lead isotope evolution by a two-stage model. *Earth and Planetary Science Letters*, **26**, 207–221.

STEIGER, R. H. & JAGER, E. 1977. Subcommission on geo-chronology: convention on the use of decay constants in geo- and cosmochronology. *Earth and Planetary Science Letters*, **36**, 359–362.

TOLSTIKHIN, I. N. & MARTY, B. 1998. The evolution of terrestrial volatiles: a view from helium, neon, argon and nitrogen isotope modeling. *Chemical Geology*, **147**, 27–52.

TOLSTIKHIN, I. N., DOKUCHAEVA, V. S., KAMENSKY, I. L. & AMELIN, YU. V. 1992. Juvenile helium in ancient rocks: II. U–He, K–Ar, Sm–Nd, and Rb–Sr systematics in the Monchepluton. $^3He/^4He$ rations frozen in uranium-free ultramafic rocks. *Geochimica et Cosmochimica Acta*, **56**, 987–999.

VOGEL, D. C., VUOLLO, J. I., ALAPIETI, T. T. & JAMES, R. S. 1998. Tectonic, stratigraphic, and geochemical comparison between *c.* 2500–2440 Ma mafic igneous events in the Canadian and Fennoscandian Shields. *Precambrian Research*, **92**, 89–116.

VREVSKY, A. B. & LEVCHENKOV, O. A. 1992. Geological-geochronological scale of the endogenous processes operated within the Precambrian complexes of the central part of the Kola Peninsula. *In*: MITROFANOV, F. P. & BOLOTOV, V. I. (eds) *Geodynamics and Deep Structure of the Soviet Baltic Schield*. Apatity, 150 (in Russian).

VUOLLO, J. I. & HUHMA, H. 2005. Palaeoproterozoic mafic dykes in NE Finland. *In*: LEHTINEN, M., NURMI, P. A. & RÄMÖ, O. T. (eds) *Precambrian Geology of Finland*. Elsevier B.V., Amsterdam, 195–236.

VUOLLO, J. I., HUHMA, H., STEPANA, V. & FEDOTOV, ZH. A. 2002. Geochemistry and Sm–Nd isotope studies of a 2.45 Ga dyke swarm: hints at parental magma compositions and PGE potential to Fennoscandian layered intrusions. *In*: BOUDREAU, A. (ed.) *9th International Platinum Symposium*, 21–25 July 2002, Billings, Montana USA, 469–470.

ZOZULYA, D. R., BAYANOVA, T. B. & NELSON, E. G. 2005. Geology and age of the Late Archaean Keivy Alkaline Province, Northeastern Baltic Shield. *Geology*, **113**, 601–608.

Palaeomagnetism of the Salla Diabase Dyke, northeastern Finland, and its implication for the Baltica–Laurentia entity during the Mesoproterozoic

J. SALMINEN[1]*, L. J. PESONEN[1], S. MERTANEN[2], J. VUOLLO[3] & M.-L. AIRO[2]

[1]*The Solid Earth Geophysics Laboratory, Physics Department, PO Box 64, University of Helsinki, FIN-00014 Helsinki, Finland*
[2]*Geological Survey of Finland, PO Box 96, FIN-02151 Espoo, Finland*
[3]*Geological Survey of Finland, PO Box 77, FIN-96101 Rovaniemi, Finland*
*Corresponding author (e-mail: johanna.m.salminen@helsinki.fi)

Abstract: New palaeomagnetic and rock magnetic results are presented for the 1122 ± 7 Ma Salla Diabase Dyke in NE Finland. A positive baked-contact test proves that the dyke has a primary natural remanent magnetization carried by magnetite. The characteristic remanent magnetization direction (D = 42.2°, I = 73.9°, k = 75.7°, α_{95} = 4.8°) of 13 sites along the large single dyke provides a virtual geomagnetic pole (VGP) position of Plat = 71°N, Plon = 113°E (A_{95} = 8.1°). Although secular variation may not have been fully averaged out, the new VGP provides an important result to define the late Mesoproterozoic position of Baltica. The VGP is not close to any known Proterozoic palaeopoles of Baltica, and therefore the pre-Sveconorwegian apparent polar wander path (APWP) of Baltica must be modified. The pre-Sveconorwegian (c. 1.3–1.0 Ga) APW swathes of Baltica, Laurentia (including the Logan Loop) and Kalahari cratons show similar shape, but new well-dated palaeomagnetic poles for c. 1.25–1.12 Ga interval from these continents are required to test the similarity. The Salla dyke VGP provides hints that the Mesoproterozoic Baltica–Laurentia connection in the Hudsonland supercontinent assembly lasted until 1.12 Ga.

Mafic dyke swarms are useful rocks for palaeomagnetism since they generally carry a stable and primary remanent magnetization and they can be accurately dated with high precision using the U–Pb method (Buchan & Halls 1990). Coupled with studies of dyke orientations and information of the anisotropy of magnetic susceptibility, the mafic dyke swarms yield valuable constraints for supercontinent research as well as for identification of large igneous provinces (Ernst & Buchan 1993). The WNW–ESE striking c. 150 km long and 60–100 m-wide Salla dyke of NE Finland (Fig. 1) represents a key target for palaeomagnetic study for several reasons. First, it has been precisely dated at 1122 ± 7 Ma (U–Pb, zircon; Lauerma 1995). Second, its pole position is important in refining the pre-Sveconorwegian plate movement of Baltica during 1.27–1.0 Ga, which then can be compared with the Laurentian and with the Kalahari drift histories at the same time. Moreover, the age of 1122 Ma is important since it generally marks the timing of the amalgamation of the Rodinia supercontinent assembly after the breakup of Hudsonland (also known as Columbia or Nuna). Finally, with this new pole for Baltica we are able to study the proposed long-lasting unity of Baltica and Laurentia which formed the core of the Mesoproterozoic supercontinent. For example, Pesonen et al. (2003) have suggested that Laurentia and Baltica were connected for c. 600 Ma from 1.83–1.27 Ga with a separation which took place at c. 1.2 Ga. Several other authors have suggested a variety of configurations and lifecycles of this supercontinent (Karlström et al. 2001; Meert 2002; Zhao et al. 2004, and refs therein) which somewhat differs from the next supercontinent Rodinia that existed c. 1000–750 Ma ago. The new palaeomagnetic data from the Salla Diabase Dyke of Baltica allow us to study the timing of tectonic transition between supercontinents Hudsonland and Rodinia.

Geology and sampling

The Salla Dyke is located in the northeastern part of the Fennoscandian Shield (Fig. 1). There, Palaeoproterozoic supracrustal rocks cover the Archaean basement forming the Lapland Greenstone Belt (LGB), which extends from northern Norway through central Finnish Lapland to western Russia. The Finnish part of the LGB records a sedimentary and volcanic evolution spanning several hundreds of

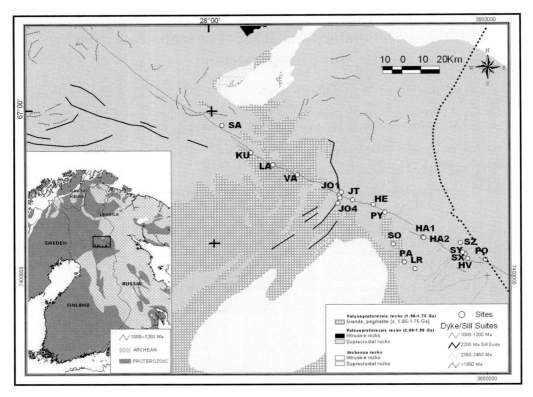

Fig. 1. Simplified geological map of the sampling area. The insert shows the study area (rectangle). Sampling sites are marked with letters. Dashed line marks the border between Finland and Russia.

millions of years (2.5–1.8 Ga) (Hanski & Huhma 2005). The WNW–ESE-striking Salla schist belt forms the southeastern extension of the Palaeoproterozoic LGB. This c. 100 km-long and 40 km-wide belt is divided by the Finnish–Russian border in a north–south direction (Manninen & Huhma 2001).

Metasedimentary rocks of the LGB in the Salla area are penetrated by a 2.2 Ga differentiated gabbro sill (Lauerma 1995; Hanski et al. 2001; Fig. 1). The upper part of the sill is composed of highly metamorphosed pyroxenite, which gradually changes to more gabbroic rocks that are the dominant rock type of the sill. At site Jokinenä (JO1-3, Fig. 1) the younger 1.12 Ga diabase dyke crosscuts the gabbro sill (Lauerma 1995). The contact of the younger dyke with the sill is visible at the northern side of the dyke.

The Salla Diabase Dyke is the youngest rock formation in the area, cutting the volcanic, metasedimentary and granitoid rocks (Lauerma 1995). The dyke forms an *en echelon* structure, and according to the aeromagnetic map it can be followed from the Russian side through Salla at least as far as central north Finland (Fig. 2), thus having a length of c. 150 km. The dyke is nearly vertical, it trends in a WNW–ESE direction, and its width varies typically from 60–100 m (locally up to 170 m). Some apophyses with widths of a few centimetres to decimetres have been found at site Jokinenä (JO1-3 in Fig. 1; Lauerma 1995). During the field campaign of 2006 we found a new c. 60 cm-wide apophysis from site Haltiavaara (HA2, Fig. 1).

The Salla Dyke has been described by Väänänen (1965), Lauerma (1995) and Manninen & Huhma (2001). The compositional structure of the dyke is symmetrical, and contacts with the host rocks (e.g. metavolcanic rocks and gabbro) are sharp. The fine-grained chilled margins have olivine basaltic composition that gradually changes to medium-grained layered pyroxene diabase. The main part of the dyke is glomeroporphyritic with phenocrysts of pyroxene and plagioclase, whereas the central part is reddish quartz diabase. Optical-microscopic analyses of the dyke indicate that flow differentiation is the main process causing the layered structure (Lauerma 1995).

Sampling of the Salla Dyke for palaeomagnetic studies began as early as in 1978. Since then samples have been taken during field campaigns

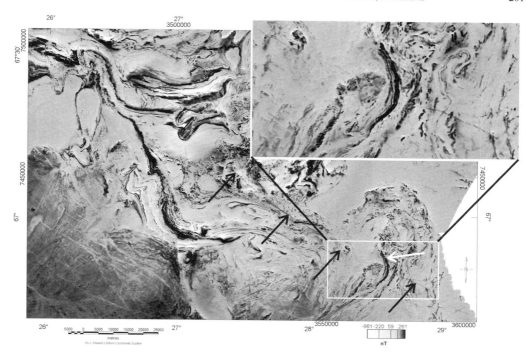

Fig. 2. Aeromagnetic total intensity map of Salla area, northern Finland. Black (white) arrows indicate the Salla Diabase Dyke (2.2 Ga gabbro sill).

in 1981, 1984, 1988, 1989, 2001, 2006 and 2007. Altogether 194 diabase cores and hand samples were collected from 15 sites. Orientation of the samples was made using either sun or magnetic compasses. Sampling was restricted due to the fact that there is poor exposure except on hills. Many of the outcrops are broken by frost action and turned to block fields. Sampling sites are shown in the geological map (Fig. 1).

In order to carry out baked-contact tests for the dyke, samples were taken also from the host rocks. Fifty-one metavolcanic samples (both baked and unbaked) from nine sites, twenty-eight gabbro samples (both baked and unbaked) from three sites, and six unbaked granite samples from one site were investigated. The 2.2 Ga gabbro sill at site Jokinenä 4 was sampled along a c. 200 m-long railway cutting (site JO4, Fig. 1), where the upper part of a 2.2 Ga layered sill can be seen, and at site Jokinenä 3 where the Salla Diabase Dyke cuts cross it (site JO1-3, Fig. 1).

Methods

Petrophysical and palaeomagnetic measurements have been carried out both at the Solid Earth Geophysics Laboratory of the University of Helsinki and at the Palaeomagnetic Laboratory of the Geological Survey of Finland.

Petrophysical properties, such as natural remanent magnetization (NRM), magnetic susceptibility and density, were measured for each specimen before palaeomagnetic and other rock magnetic studies. Remanence measurements were made using a 2G-Enterprise SQUID magnetometer and an Agico spinner magnetometer.

Palaeomagnetic studies included alternating-field (to peak fields of 160 or 100 mT) and thermal (up to temperatures of 680 °C) demagnetization of the specimens. Alternating field demagnetization was carried out using the 2G SQUID magnetometer static-coil system. Some samples were demagnetized using the Agico LDA-3 demagnetizer. Thermal cleanings were carried out with the single-axis Schonstedt TSD-1 furnace, the Magnetic Measurement thermal demagnetizer and the ASC thermal furnace of the palaeomagnetism laboratory of the Otago University in New Zealand. Measurement of magnetization was carried out using a 2G SQUID magnetometer, both in Finland and in New Zealand. During the thermal cleaning, the bulk magnetic susceptibility was monitored after each step in order to detect mineralogical changes due to heating. Specimens from 114 diabase samples and from 63 host rock samples (9 andesites, 21 gabbros, 4 granites, 4 amphibolites and 15 metavolcanic rocks) were thermally cleaned. Remanence components were identified using principal component analysis (Kirschvink 1980; Leino 1991). Statistical means were calculated according to Fisher (1953). The palaeopoles were plotted with GMAP program (Torsvik & Smethurst 1999).

In order to identify magnetic carriers, thermomagnetic analyses (susceptibility v. temperature) were carried out using an Agico CS2-KLY3 kappabridge. To avoid oxidation during heating, the specimens were heated in argon gas, from room temperature to 700 °C and back to room temperature. Curie points were determined from the maxima in the second derivatives of the thermomagnetic curves (Tauxe 2002; Petrovsky & Kapicka 2006). For determination of domain states of the magnetic carriers, hysteresis properties were measured using a Princeton Measurement Corporation MicroMag™3900 model Vibrating Sample Magnetometer (VSM). Scanning Electron Microscope (SEM; Jeol JSM-5900LV) analyses were performed in order to verify the nature of the magnetic carriers. Rock magnetic properties of the ferromagnetic minerals were studied using the Lowrie test (Lowrie 1990), which distinguishes the test utilizing the different coercivities and thermal unblocking temperature characteristics of the most common ferromagnetic minerals. Isothermal remanent magnetization was given at 1.2, 0.4 and 0.12 T fields along the z, y and x axis, respectively. Samples were then thermally demagnetized up to 680 °C. After each step, magnetic susceptibility and remanence were measured.

The anisotropy of magnetic susceptibility was measured for a selection of diabase specimens using the Agico KLY-3 kappabridge in order to define the possible deformation of the dyke (Hrouda 1982) and to characterize the magma flow pattern (Rochette et al. 1992).

Stability of magnetization was studied using the simplified version of the viscosity test, which defines the changes in magnetization due to viscous processes (Dunlop & Buchan 1977). It consists of measuring the magnetization of the specimens before (initial NRM, NRM_0) and after (stable NRM, NRM_{st}), storing them in a zero-field environment for three weeks. The viscosity coefficient S_v (%) can be determined using the formula: $S_v = [(NRM_0 - NRM_{st})/NRM_0] \times 100$.

Results

Petrophysical properties

The Salla Diabases generally yield stable palaeomagnetic results, but they are only moderately strongly magnetized. Mean results of density, magnetic susceptibility, remanence intensity (NRM) and Koenigsbergers ratios (Q-values) for the diabase and host rocks are shown in Table 1. Figure 3 shows the NRM v. susceptibility with isopleths of Q.

There is little variation of bulk petrophysical properties between different diabase sites. Q-values are around 1–1.5 (Fig. 3), suggesting that the remanence of the samples is carried by moderate to large multidomain (MD) grains. The remanence of the larger grains is less stable and can more likely acquire viscous remagnetization (Dunlop 1983). However, the intensity of the natural remanent magnetization of the samples is moderately high (average 695 mA/m; Fig. 3; Table 1) and indicates that the original magnetization may be preserved.

Petrophysical properties of the diabases and host rocks differ. Unbaked metavolcanic rock samples have two-orders-of-magnitude lower susceptibility and remanence intensity values than the diabase samples (Fig. 3). Q-values of such rocks are low, indicating that they do not contain stable magnetization, as was also observed during the palaeomagnetic study. Baked metavolcanic rock samples show lower density and two-orders-of-magnitude higher susceptibility and remanence values than unbaked metavolcanic samples, indicating that baking has clearly affected the mineralogy and composition of such rocks. Both unbaked and baked gabbros carry stable magnetizations. As with metavolcanics, baking has clearly affected the gabbros, since baked gabbros show higher susceptibility and remanence values than the unbaked gabbros (Fig. 3, Table 1). This is most probably due to the fact that the baked gabbros have been heated during the emplacement of the diabase dyke, when some alteration in mineralogy in high temperatures took place (oxidation, growth of magnetic minerals, etc.). The magnetic properties of the gabbros are discussed more briefly in the context of the baked-contact test. Granite samples show somewhat higher Q-values than metavolcanic rock samples, but palaeomagnetic study showed that their magnetization is not stable.

Rock magnetism – magnetic carriers of remanence

Figure 4 shows examples of results from rock magnetic studies.

Thermomagnetic analyses. Representative examples of the Curie point analyses are shown in Figure 4a. These analyses were carried out for one diabase sample from each site, except for sites Haltiavaara (HA1-HA2) and Jokinenä (JO1-JO3), where several samples were analysed. The samples showed variable irreversibilities of heating and cooling curves.

Table 1. *Petrophysical properties of diabase, gabbro and metavolcanic rocks of Salla area*

Rock type	B/N/n	D (kg/m^3)	K (uSI)	J (mA/m)	Q
Diabase	15*/197/780	2981 ± 22	17087 ± 3697	695 ± 340	1.0 ± 0.3
Baked gabbro	1/12*/61	2973 ± 66	39263 ± 19064	2238 ± 1602	1.6 ± 1.4
Unbaked gabbro	2/16*/65	3133 ± 22	9447 ± 7873	420 ± 443	0.8 ± 0.4
Baked metavolcanic rocks	3/23*/77	2792 ± 83	11449 ± 13449	447 ± 580	0.5 ± 0.4
Unbaked metavolcanic rocks	6/28*/130	2936 ± 41	656 ± 80	5 ± 3	0.2 ± 0.2
Granite	1/6*/20	2617 ± 2	262 ± 125	10 ± 3	1.3 ± 1.2

B/N/n is the number of sites/samples/specimens, where *denotes the level used in statistical calculations. Lat, latitude; Long, longitude of the sampling site; D, density; K, susceptibility; J, the intensity of the natural remanent magnetization; Q, the Koenigsberger's ratio calculated for a field of 50 μT. The standard deviations are calculated for each for the mean value.

For example, sample ST64 from HA2 (60 cm-wide diabase apophysis) shows strong irreversibility, indicating mineralogical alteration during heating. In contrast, samples JN1 (site JO3) and JS2 (site JT) are nearly reversible, indicating that these samples are suitable for thermal demagnetization. A few samples show only one ferromagnetic phase with Curie points of 550–580 °C (e.g. sample JS2; Fig. 4a). However, the majority of the samples show two ferromagnetic phases in their thermomagnetic curves: one with Curie temperatures at c. 300–340 °C and a second one with Curie temperatures at c. 550–580 °C. The lower temperature phase could be associated with pyrrhotite or titanomagnetite, whereas the latter one points to magnetite with varying Ti content. The prominent peak of a lower temperature phase in the sample ST64 indicates the presence of pyrrhotite. The concentration of pyrrhotite varies in samples, which is seen in thermomagnetic curves (Fig. 4a) as differing heights of the lower temperature peak with respect to general behaviour. When the thermomagnetic curves show pyrrhotite there is also the likelihood that the samples contain some pyrite. During heating, pyrite may oxidize to pyrrhotite and magnetite, which can be seen as the irreversible thermomagentic curves.

Scanning electron microscope studies. Scanning electron microscope studies verified the presence of magnetite, with variable Ti content, pyrrhotite and pyrite in the Salla Dyke samples. Examples of these are shown in Fig. 4b (lower image). Often within the same sample some magnetite grains look fairly fresh showing their primary shape (e.g. JS2 in Fig. 4b), whereas other magnetite grains

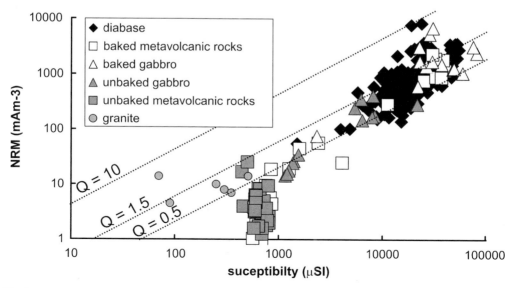

Fig. 3. Intensity of natural remanent magnetization v. magnetic susceptibility with equal Q lines of Salla diabase and host rock samples.

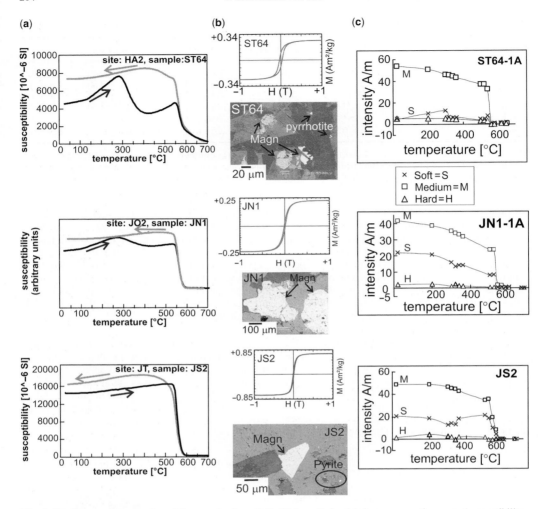

Fig. 4. Rock magnetic properties of the samples from Salla Diabase Dyke. (**a**) thermomagnetic curves (susceptibility v. temperature), (**b**) hysteresis loops (upper image) and Scanning Electron Microscope figures of magnetic minerals (lower image), (**c**) Lowrie test for the samples.

appear to be highly altered (e.g. ST64 and JN1 in Fig. 4b).

Hysteresis properties. Typically magnetite saturates at 0.3 T, titanomagnetites around 0.1–0.2 T and pyrrhotite at 0.5–1.0 T (O'Reilly 1984; Dekkers 1988). The hysteresis loops show that all the diabase specimens saturate in high fields up to 0.6 T (Fig. 4b, upper image), indicating the probable presence of soft and hard carriers of magnetization. Since there are both (titano)magnetite and pyrrhotite in the same samples, we would expect the presence of wasp-waisted hysteresis loops, but this is not observed. Thus we suppose that the amount of pyrrhotite is relatively small compared with the amount of (titano)magnetite. However, the presence of pyrrhotite makes the use of the hysteresis ratios (such as M_r/M_s and H_c/H_{cr}, where M_r is saturation remanence, M_s is saturation magnetization, H_c is coercive force and H_{cr} coercivity of remanence) and the Day plot (Day *et al.* 1977) less usable in terms of defining the magnetic domain sizes.

Lowrie test. In order to study the coercivities of ferromagnetic minerals, we performed the Lowrie test (Lowrie 1990) for the selected diabase samples. Isothermal remanent magnetization (IRM) was given in three orthogonal directions (see Methods above) and the intensity curves of thermal demagnetization of each component were plotted separately (Fig. 4c).

The nine studied specimens from six different sites show consistent results. In all samples the IRM of the components decay to zero below 600 °C, indicating that there is no sufficient amount of hematite in the samples. The intensities of the soft and hard components decay clearly around 300–330 and 520–580 °C, pointing to pyrrhotite and magnetite/titanomagnetite, respectively. The medium fraction dominates the samples. The intensity drops around 520–580 °C, and indicates that magnetite/titanomagnetite is the dominant magnetic mineral. Results thus indicate that medium- to soft-coercivity magnetite/titanomagnetite dominates and pyrrhotite exists to a lesser extent.

Anisotropy of susceptibility

For reliable palaeomagnetic studies, the obtained characteristic remanence direction should reflect accurately the Earth's ambient magnetic field direction at the time of acquisition. It has been suggested that the direction of remanence will not be significantly altered if the ratio between the maximum and minimum suceptibilities (degree of anisotropy) is lower than 5% (Hrouda 1982). Anisotropy of magnetic susceptibility (AMS) can record the flow direction of the magma, but later deformations may change it and cause additional deviations for the remanence directions (Dragoni et al. 1997; Elming & Mattsson 2001). AMS was measured on 109 diabase specimens. Measurements were conducted for both non-demagnetized and alternating-field (AF) demagnetized samples, since AF demagnetization does not cause significant changes in AMS (Puranen et al. 1992).

For Salla Diabase Dyke the degree of anisotropy is typically below 5%, indicating that the samples have no significant AMS, and the samples are thus suitable for palaeomagnetic study. Both magnetic lineation (L) and foliation (F) values are below 1.04, indicating that the dyke has not been affected by later deformations (Tarling & Hrouda 1993). The AMS data can be used to define the direction of magma flow (Dragoni et al. 1997; Elming & Mattsson 2001). When a basic magma flows in the crust, the magnetite grains are aligned with the long axes parallel to the direction of magma flow and the short axes perpendicular to it. In the Salla Dyke, the directions of maximum axis are highly variable, but the majority of the samples show shallow to moderate inclinations with no preferred orientation (e.g. strike of the dyke). The directions of the minimum axes vary greatly. Samples close to the contact show similar directions to the samples taken from the middle of the dyke. The problem with the Salla Diabase samples is that the concentration of pyrrhotite varies (discussed above). Large concentration of pyrrhotite may cause some biased AMS, which will not reflect the original magma flow. Wide scatter in directions of maximum and minimum axes may be due to two factors. First, pyrrhotite may affect the results, obscuring the original magma flow fabric; or second, the magma flow may have been originally turbulent (Hrouda 1982; Tarling & Hrouda 1993).

Palaeomagnetic results of diabase samples

The palaeomagnetic results of the diabase dyke and host rock sites are shown in Table 2. Examples of demagnetization behaviour of the diabase specimens are shown in Figure 5. Palaeomagnetic results of the other host rock sites are presented in the section 'Baked-contact test' below.

The diabase samples carry two remanence components. The first one is typically demagnetized during the first steps of alternating-field (AF) treatment (0–10 mT) and below 340 °C (Fig. 5a–c). It has variable directions and probably represents a viscous remanent magnetization (VRM). The majority of the samples (170 out of 194 samples) also show a harder, characteristic remanence (ChRM) with steep downward direction. The ChRM is isolated below 580 °C, suggesting that remanence resides in magnetite (Fig. 5d). Diabase samples responded similarly for both AF and thermal treatments (Fig. 5). In some specimens, however, AF cleaning was more effective in isolating the ChRM from the viscous component. AF treatment shows that the remanence of diabases is mainly fairly soft (Fig. 5a, c), yet the direction is stable. The finer-grained diabases show somewhat harder behaviour (Fig. 5b). The mean direction of the ChRM is D = 42.2°, I = 73.9° (k = 75.7°, α_{95} = 4.8°; 13 sites; Fig. 7). When using the reference location for central Baltica (Kajaani city, Lat = 64°N and Long = 28°E), the mean direction is D = 39.8° and I = 73.6°.

Results of the viscosity test (Dunlop & Buchan, 1977) on two specimens from site Haltiavaara 2 and Jokinenä 3 show that viscosity coefficients S_v of these specimens (sample ST44-3c: 3.0% and sample JM5-2b: 4.9%) are below 5%. In general, values below 5% imply that the remanent magnetization is stable and not due to viscous magnetization (Hrouda 1982; Tarling & Hrouda 1993).

Baked-contact test

Baked-contact tests can be used to verify that the observed remanence in the dyke is thermoremanence with a primary ambient field direction acquired during emplacement (Everitt & Clegg 1962). The test is based on the idea that an igneous intrusion heats the host rock at the contact above the unblocking temperatures of magnetic

Table 2. *Palaeomagnetic results for the Salla diabase dyke and for the gabbro sill*

Site	Site	B/N/n	Lat/Long	D (°)	I (°)	k	α$_{95}$ (°)	Plat (N°)	Plon (E°)	dp (°)	dm (°)	A$_{95}$ (°)	K	S (°)
					diabase									
Haltiavaara 1	HA1	66*/199	66.75/28.97	34	79	6	9	77	93	16	17	16		
Haltiavaara 2	HA2	7*/23	66.75/28.97	36	63	34	11	60	151	13	16	15		
Helppilä	HE	2*/3	66.82/28.75	48	87			71	43					
Hanhivaara 1	HV1	5*/14	66.71/29.17	52	63	5	37	55	131	46	59	52		
Hanhivaara 2	HV2	5*/8	66.7/29.2	25	69	41	12	71	157	17	21	19		
Jokinenä 1	JO1	5*/23	66.83/28.58	53	70	10	25	63	127	38	44	41		
Jokinenä 2	JO2	14*/40	66.83/28.58	40	80	20	9	75	86	17	18	17		
Jokinenä 3	JO3	16*/59	66.83/28.5	83	70	15	10	51	96	15	17	16		
Jormantie	JT	7*/20	66.82/28.62	32	76	56	8	77	111	14	15	15		
Kulvakkojänkä	KU	9*/16	66.92/28.18	36	69	26	10	68	139	15	18	16		
Pyhäjärvi	PY	11*/19	66.8/28.8	34	72	9	16	72	134	25	29	27		
Pahakuru	PA	18*/48	66.71/28.88	19	88	23	7	71	33	15	15	15		
Possolijärvi	PO	5*/12	66.7/29.2	38	67	37	13	64	142	17	21	19		
Mean		13*/170/484		42	74	76	5	71	113	8	9	8	27	15.6
				40 (D$_r$)	74 (I$_r$)									
					baked gabbro									
Jokinenä 1-3	JO1-3	1/14*/51	66.83/28.5	62	74	50	6	63	100	9	10	9	20	34
				60 (D$_r$)	74 (I$_r$)									
					unbaked gabbro									
Jokinenä 1-3 60 m to dyke	JO3	1/6*/16	66.83/28.5	338	57	73	8	59	243	8	12	9	53	12
				338 (D$_r$)	55 (I$_r$)									
Jokinenä 4 2 km to dyke	JO4	1/5*/9	66.82/28.58	338	61	38	13	63	245	15	19	16	24	17
				339 (D$_r$)	59 (I$_r$)									

Note: Mean pole (direction) has been calculated as a mean for poles (directions) of all the sites. D$_r$ (I$_r$) is the declination (inclination) calculated for Baltica's reference location (Lat: 64°N, Long: 28°E; Pesonen *et al.* 2003). Lat and Long are the latitude and longitude of the sampling site. B/N/n denotes number of sites/samples/specimens, where *denotes the statistical level used for mean calculations. D and I are the declination and inclination of the remanent magnetization component. k is Fisher's (1953) precision parameter for the direction. α$_{95}$ is radius of circle of 95% confidence of directions. Plat and Plon are the latitude and longitude for the virtual geomagnetic poles pole. *dp* and *dm* are the semi axes of the oval of 95% confidence of the pole. A95 is the radius of the circle of 95% confidence of the pole. K is Fisher's (1953) precision parameter for the pole. s is the estimated angular standard deviation.

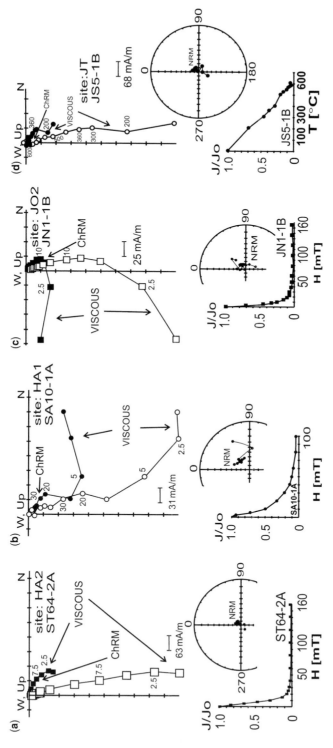

Fig. 5. Examples of alternating field (a–c) and thermal (d) demagnetization behaviour of specimens of the dyke. Upper row is showing orthogonal (Zijderveld) demagnetization diagrams. Open (closed) symbol denotes vertical (horizontal) plane. Middle row is showing stereoplots of observed direction of magnetization. Lower row is showing the relative intensity decay. Numbers in Figure a–c (d) denotes used cleaning af-field (temperature).

minerals. When cooling, the remanence carriers will acquire the same direction as the igneous intrusion (e.g. dyke). Farther away from the intrusion, the host rock retains its pre-intrusion remanence. In between, the samples can hold two remanences that are superimposed: the pre-intrusion remanence and the intrusion remanence (Everitt & Clegg 1962; McCelland-Brown 1981).

The gabbro and the metavolcanic samples close to the Salla Dyke were collected for a baked-contact test at site Jokinenä (JO1-4). For comparison, several Archaean unbaked metavolcanic and granite samples were collected far from the dyke. However, due to unstable behaviour during demagnetization, the metavolcanic samples failed to provide reliable data. Also, the host rocks collected for comparison show weak stability of the remanence and could therefore not be used to verify the primary nature of the dyke.

Only the *c.* 2.2 Ga-old gabbro samples provided magnetically stable results among sampled host rocks. They clearly give a positive baked-contact

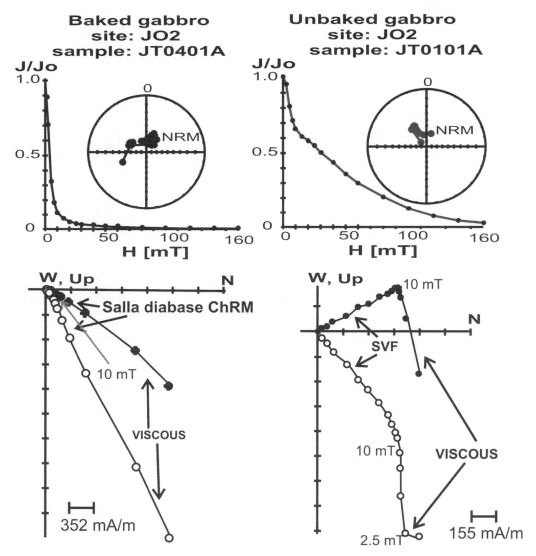

Fig. 6. Examples of positive baked-contact test. Left: baked gabbro (2.2 Ga) taken 30 m from diabase dyke at site Jokinenä. Right: unbaked gabbro taken 60 m from diabase dyke at site Jokinenä. Upper figures show the relative intensity decay and stereographic projections of directional data upon demagnetization. Lower figures show orthogonal (Zijderveld) demagnetization diagrams. Open (closed) symbol denotes vertical (horizontal) plane.

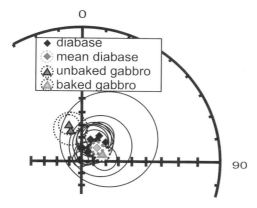

Fig. 7. Palaeomagnetic directions for diabase, baked gabbro and unbaked gabbro samples (Table 2), equal angle projection.

test (Figs 6 & 7). At site JO1-3, where the dyke width is 60–100 m, the remanence direction of the gabbro samples within 30 m of the contact is similar to that of the diabase dyke itself (D = 62°, I = 74°; Figs 6 & 7; Table 2). The baking effect is also seen in the increased remanence intensity and magnetic susceptibility values in the gabbro samples close to contact, whereas the unbaked gabbro samples show distinctly lower values (Fig. 3, Table 1). The unbaked gabbros, some 60 m away from the dyke (site JO3), show clearly different direction (D = 338°, I = 57°) from the baked gabbros, and their magnetization is harder (Fig. 6). Similar direction (D = 338°, I = 61°) was observed also from the gabbros 2 km away from the dyke (Fig. 7). The unbaked direction is similar to the typical Svecofennian c. 1.9–1.8 Ga remanence direction, which is found widely throughout Fennoscandia (Mertanen 1995). We interpret this direction to be an overprint caused by the Svecofennian Orogeny. However, the baked-contact test verifies that the Salla Dyke clearly carries a different natural remanent magnetization component than the host rocks.

Discussion

The characteristic, steep, northeastward pointing, remanence direction (D = 42.2°, I = 73.9°; Fig. 7) for Salla Diabase Dyke was separated from 170 samples (484 specimens). The ChRM yields a virtual geomagnetic pole (VGP) at Plat = 71°N and Plon = 113°E (A_{95} = 8.1°).

The 95%-error circle of this ChRM component (Table 2) encloses the direction of the Present Earth's field (D = 11°, I = 76°) on the sampling site. However, on the basis of the positive baked-contact test to the 2.2 Ga gabbro sill at site Jokinenä,

on the fact that the diabase carries clearly two components (viscous and stable ChRM; Fig. 5) and on the viscosity test that indicates stable remanent magnetization, we suggest that the origin of the ChRM component is thermal and represents the primary remanence of the diabase dyke acquired during the cooling of the dyke. Since it has been interpreted that the dyke was emplaced near the surface of the Earth (Väänänen 1965) where the cooling of the magma is relatively rapid, the U–Pb age (1122 ± 7 Ma; Lauerma 1995) of the dyke can be correlated with the timing of the remanence acquisition. Rock magnetic and microscopic studies (Curie point, hysteresis, Lowrie test and SEM studies, Fig. 4) show that magnetite in a wide range of coercivities is the carrier of the primary remanence.

Secular variation

In order to reconstruct the palaeogeographical position of Baltica during the magnetization of Salla Diabase Dyke, a virtual geomagnetic pole (VGP) was calculated for the characteristic remanence component (Table 2). Although the Salla Dyke is a fairly thick, vertical dyke intrusion (width 60–170 m and length 160 km) with apophysic dykelets, it is possible that in the case of only one dyke, that sampling does not average out secular variation (SV) (Buchan & Halls 1990). If the SV is not averaged out, it could cause additional error in the estimation of pole position. The difference between individual virtual geomagnetic poles (VGP) and the rotation axis/palaeomagnetic reference frame can be as much as 15°–20° (McElhinny & McFadden 2000).

In the case of one dyke, the only way to study if SV has been averaged out is to calculate the angular dispersion (between-site dispersion, s) of the site-mean virtual geomagnetic poles (Merrill & McElhinny 1983) at the site palaeolatitude. The s-value was calculated from K (dispersion of poles) and upper and lower limits of (s) were calculated using the methods by Cox (1969). The angular dispersion of Salla was calculated and compared with theoretical dispersion, for example with the model G by Merrill & McElhinny (1983) for the palaeolatitude of the sampling site (60°). We obtained a value of s = 15.6° +4.7°/−2.9° (Fig. 8) for the dispersion. Even though the upper limit falls on the model G curve, the mean (s) falls well below the SV-model curve of Merrill & McElhinny (1983), suggesting that the secular variation may not have been adequately averaged out. Therefore this might cause an additional error both to the location of the VGP of the Salla Dyke, and as a consequence, to the ancient location of Baltica based on this VGP.

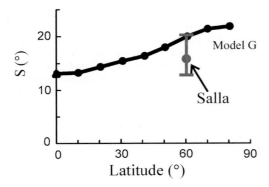

Fig. 8. Calculated between site dispersion of palaeomagnetic pole for Salla Diabase Dyke is plotted on the mean curve (model G) for secular variation, after Merrill & McElhinny (1983).

Implications for pre-Sveconorwegian apparent polar wander path of Baltica

The virtual geomagnetic pole (VGP; 71°N, 113°E) of the characteristic remanence component of the Salla Diabase Dyke is not close to any known Mesoproterozoic poles of Baltica (Fig. 9, Table 3). This VGP refines Baltica's apparent polar wander path (APWP) and creates a loop to the APWP between 1265 and 1042 Ma, just older than the Sveconorwegian Loop. Since we are using a VGP, which may differ from a real palaeopole by 15–20° (McElhinny & McFadden 2000) plus its own A_{95} (8.1°), the suggested loop may be anything between c. 50° and c. 100° (Salla Dyke VGP is used in Fig. 9 and drawn loop is 70°). After c. 1265 Ma the APWP makes a clockwise loop towards the Salla VPG until reaching again the equatorial 1100–1040 Ma poles (Bamble Intrusion and Laanila-Ristijärvi Dyke, Table 3). Before 1265 Ma, the path shows a small hinge.

The part before the Grenville Loop in Laurentia's APWP, called the Logan Loop (poles between 1267–1060 Ma; Robertson & Fahrig 1971; Pesonen 1978; Ernst & Buchan 1993), shows a similar shape to the pre-Sveconorwegian Loop of Baltica (Fig. 9). However, the older 1270–1110 Ma limb of the Logan Loop is poorly constrained (Ernst & Buchan 1993; Donadini 2007). The minimum length of the Logan Loop is 40° (Ernst & Buchan 1993; Donadini 2007), but there are only a few data between the older part of the loop, for example the 1235 Ma pole from Sudbury Dykes, and the younger part, for example the 1112 ± 2.4 Ma pole from Seagull Pluton (Table 3). There is one 1141 ± 2 Ma reliable pole from the Abitibi Dykes (Ernst & Buchan 1993). Similar to Baltica, the APWP of Laurentia also shows the small hinge before 1270 Ma (Fig. 9b). Kalahari-Grunehogna craton's APWP also shows a large loop with length of c. 60° during the time interval 1165–1000 Ma (Fig. 9c). Unfortunately there are large gaps in the APWP of Kalahari (Gose et al. 2004; Hanson et al. 2004). In particular, there are no palaeomagnetic data either between 1270 and 1165 Ma or between the combined Umkondo pole (c. 1109 Ma) and younger poles from the Namaqua-Natal metamorphic belt (Okiep copper district, Ntimbankula, Namaqua mean pole and Port Edward, Table 3). New palaeomagnetic data from well-dated rocks between 1270–1112 Ma are required from all three continents to test the similarity between Logan Loop of Laurentia, pre-Sveconorwegian Loop of Baltica and the 1.27–1.0 Ga segment of Kalahari-Grunehogna's APWP. All these show large APWP loops, which

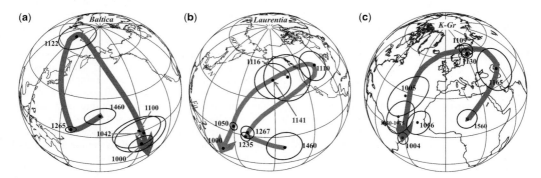

Fig. 9. (a) Selected Pre-Sveconorwegian poles for Baltica in present-day Baltican coordinates. Numbers are ages in Ma. (b) Logan Loop of Laurentia (Ernst & Buchan 1993; Donadini 2007) based on selected poles for Laurentia in present-day Laurentian coordinates, (c) Palaeomagnetic poles for the Kalahari-Grunehogna craton in the 1165–1000 Ma interval in present-day African coordinates. Grunehogna reconstructed to Africa by Euler parameter: Lat: −9.7°, Long: 328.8°, angle 56.3° (Powell & Li 1994). Used poles are listed in Table 3.

Table 3. *Selected Late Mesoproterozoic palaeomagnetic data from Baltica, Laurentia and Kalahari–Grunehogna cratons*

Pole	Age (Ma)	Plat	Plon	dp	dm	Reference
Baltica						
Mean Baltica	1512	39	171	16	16	Pesonen et al. 2003
Mean Baltica	1460	15	181	11	11	Salminen & Pesonen 2007
Mean Baltica	1265	4	158	4	4	Pesonen et al. 2003
Salla Diabase VGP	1122 ± 7	71	113	8	8	this work
Mean Baltica	1100	1	208	16	16	Pesonen et al. 2003
Laanila-Ristijarvi	1042 ± 50	−2.1	212	13	21	Mertanen et al. 1996
Protogine zone Dolerites	1000	−12	211	7	7	Pisarevsky & Bylund 1998
Laurentia						
Mean Laurentia	1460	−7	217	13	13	Salminen & Pesonen 2007
Mean MacKenzie Dykes	1267 +7/−3	4	190	5	5	Buchan & Halls 1990; Fahrig et al. 1981; Le-Cheminant & Heaman 1989
Sudbury Dykes	1235 +7/−3	−3	192	3	3	Schwarz & Buchan 1982; Dudas et al. 1994
Abitibi Dykes (N)	1141 ± 2	43	209	13	13	Ernst & Buchan 1993; Krogh et al. 1987
Seagull Pluton	1112 ± 1.4	42	233	2	2	Borradaile & Middleton 2006; Davis & Paces 1990
Arizona Dykes (R)	1110	46	257	12	12	Donadini 2007
Logan Sills	1109 +4/−2	49	220	4	4	Pesonen 1978; Davis & Green 1997
Nipigon Inspiration Sills	1109	45	218	2	2	Borradaile & Middleton 2006
Mean Laurentia	1116	46	227	13	13	this work
Mean Laurentia	1099	36	188	–	–	Pesonen et al. 2003
Nonesuch Shale + Freda sandstone	1050	6	179	3	3	Pesonen et al. 2003
Mean Laurentia	1000	−20	167			Pesonen et al. 2003
Kalahari-Grunehogna						
Van Dyk Mine Dike	1560–1610	12	14	6	8	Jones & McElhinny 1966
Premier Kimberlites	1165 ± 10	41	55	16	16	Powell et al. 2001
Ritscherflya Supergroup rotated to Kalahari	1130 ± 10	−8 −61 (rot)	232 29 (rot)	4	4	Powell & Li 1994; Powell et al. 2001
Borgmassivet Intrusion rotated to Kalahari	1130 ± 10	8 (−8) 62 (rot)	55 (235) 35 (rot)	3	3	Jones et al. 2003
Mean Grunehogna rotated to Kalahari	1130 ± 10	−8 61 (rot)	234 32 (rot)	4	4	Jones et al. 2003
Umkondo combined	1109 +2/−1	64	36	2	2	Hanson et al. 2004
Okiep copper district	1056 ± 20	8.8	339.5	7.6	7.6	Muller et al. 1978
Namaqua mean pole	1030–1075	9.4	324	13	13	Onstott et al. 1986
Ntimbankula pluton	1005	28	325	15	15	Mare & Thomas 1998
Port Edward pluton	1004 ± 5	−7.4	327.8	4.2	4.2	Gose et al. 2004

Plat and Plon are latitude and longitude of the palaeomagnetic pole. dp and dm are the semi-axes of the oval of 95% confidence of the pole. 'Rot' means that pole has been rotated.

appear to reach their apexes at different ages (Fig. 9). It is evident that Baltica, Laurentia and Kalahari experienced fast drift rates during these time intervals, but we need more coeval well-dated palaeomagnetic poles to test similarities between these APWPs.

1.12 Ga palaeoreconstruction of continents

Using the new 1.12 Ga virtual geomagnetic pole (VGP) for Baltica, we test the model of continuation of the proposed long-lasted unity of Baltica and Laurentia during the Mesoproterozoic (Pesonen et al. 2003; Salminen & Pesonen 2007 and refs therein). The 1122 Ma Salla VGP takes Baltica to a high palaeolatitude of c. 60° (Fig. 10a), but since we are using a VGP, Baltica may also occupy a palaeolatitude of 30° (see previous section). Laurentia's position is based on a 1116 Ma mean pole (Table 3), and it also occupies high palaeolatitudes. Palaeoreconstruction in Figure 10a is made using the individual c. 1.12 Ga poles to reconstruct each continent to its correct palaeolatitudinal position and evaluate the relationships between continents based on the 'closest approach' technique, which takes into account the maximum probable error limits of the poles. We use the 1.12 Ga Salla VGP for Baltica, where the error bar includes errors from both secular variation (15–20°; McElhinny & McFadden 2000) and A_{95} of Salla VGP (8.1°); the mean 1.12 Ga pole for Laurentia; the 1.14 Ga Mount Isa Lake View Dolerite pole for Australia; and the 1.13 Ga mean pole for Kalahari-Grunehogna (Table 4). Since the poles for Baltica and Laurentia do not have exactly the same age (Table 4) it is permissible that their ovals of confidence are just overlapping (Fig. 10a). However, if we calculate Baltica's Euler rotation in relation to Laurentia, we get Euler parameters of Lat: 53°, Long: 0.5° and angle +58°.

There are not exactly any coeval palaeomagnetic data for other continents. Closest in ages are Australia, Kalahari and Grunehogna (Table 4). Australia is reconstructed using the pole from Mount Isa Lake View Dolerite, which places Australia also at high palaeolatitudes, as in Pesonen et al. (2003). Other plates (Kalahari–Grunehogna) seem to occupy equatorial palaeolatitudes (both polarity options are shown) at 1.12 Ga.

Proposed long-lasting connection between Baltica and Laurentia during the Mesoproterozoic

Recent 1.46 Ga palaeomagnetic data of Salminen & Pesonen (2007) support the idea of Baltica–Laurentia unity between 1.76 and 1.27 Ga. We use the new VGP of Salla to test this proposed long-lasting unity between Baltica and Laurentia at 1.76–1.12 Ga. Moreover, this connection between Baltica and Laurentia also formed the core of the Mesoproterozoic supercontinent Hudsonland (also

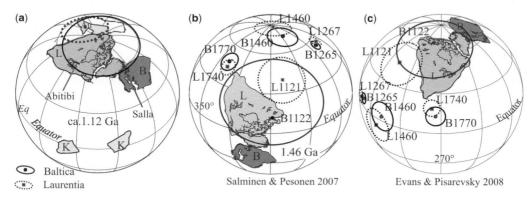

Fig. 10. Palaeoreconstructions of Mesoproterozoic connection between Baltica and Laurentia. Numbers are ages of used poles in million years. B (L) stands for Baltica (Laurentia). (a) Reconstruction at 1.12 Ga using the 'closest approach' technique where the continents are shown at their palaeolatitudes. Abitibi and Salla dykes are drawn in Laurentia and Baltica, respectively. The error bar of the Salla pole (B1122) includes the error caused by secular variation. (b) Reconstruction of Baltica and Laurentia using the 'closest approach' technique at 1.46 Ga, based on mean poles for Baltica and Laurentia (Euler pole for Baltica is Lat: 0°, Long: 102°, angle 79° and for Laurentia is Lat: 0, Long: 312, angle: −93.5). Also shown 1.76, 1.46, 1.27 and 1.12 Ga poles for Baltica and Laurentia by using this continental configuration (after Salminen & Pesonen 2007). (c) Reconstruction according to Evans & Pisarevsky (2008) where Baltica and its poles have been rotated to Laurentia's reference frame using the Euler parameters: Lat: 47.5°, Long: 1.5°, angle: +49°. Also shown 1.76, 1.46, 1.27 and 1.12 Ga poles for Baltica and Laurentia according to this continental configuration. Used data are listed in Table 4.

Table 4. *Palaeomagnetic poles used for reconstructions*

Craton	Formation	Plat (°)	Plong (°)	A_{95} (°)	Age (Ma)	Original references
			c. 1.12 Ga			
B	Salla Diabase Dyke VGP	71	113	8	1122 ± 7	this work
L	Mean Laurentia	46	227	17	1116	this work, see Table 3
Au	Mount Isa Lakeview Dolerite	−10	131	17	1140	Tanaka & Idnurm 1994; unpublished age cited in Wingate & Evans 2003
K–Gr	Mean Kalahari–Grunehogna	61	32	4	1130 ± 10	Jones *et al.* 2003
			c. 1.27 Ga			
B	Mean Baltica	4	158	4	1265	Pesonen *et al.* 2003
L	MacKenzie dykes	4	19	5	$1267 +7/-3$	Buchan & Halls 1990, LeCheminant & Heaman 1989
			c. 1.46 Ga			
B	Mean Baltica	15	181	11	c. 1460	Salminen & Pesonen 2007
L	Mean Laurentia	−7	217	13	1460	Salminen & Pesonen 2007
			c. 1.77 Ga			
B	Ropruchey–Shoksha	42	221	7	c. 1770	Pisarevsky & Sokolov 2001
L	Cleaver Dykes	19	263	6	$1740 +5/-4$	Irving *et al.* 2004

B, Baltica; L, Laurentia; Au, Australia; K, Kalahari; Gr, Grunehogna. Plat and Plong are latitude and longitude of the palaeomagnetic pole. A_{95} is the 95% confidence circle of the pole.

known as Columbia or Nuna; Meert 2002; Rogers & Santosh 2002; Pesonen *et al.* 2003; Zhao *et al.* 2003, 2004 and refs therein). Salminen & Pesonen (2007) propose a palaeoconfiguration at 1.46 Ga where northern Baltica is placed against northeastern Laurentia (Fig. 10b). This configuration is made using the 'closest approach' technique, where continents have been rotated to their correct palaeolatitudes. Calculated Euler parameters in relation to the continents' present positions are for Baltica, Lat: 0°, Long: 102°, angle 79°; and for Laurentia, Lat: 0°, Long: 312°, angle: −93.5°. Pole pairs of Baltica and Laurentia at *c.* 1.76 and 1.27 Ga allow the presented configuration (Fig. 10b, Table 4). Within error limits of the used poles, *c.* 1.76 Ga-old geologically similar Trans Scandinavian Igneous Belt (in Baltica) and Yavapai/Ketilidian (in Laurentia) belts become contiguous in this continental reconstruction. Moreover, there is geological and geophysical evidence, including the proposed common magma source for coeval dolerites in north and central Sweden and in Finland, and basalts and dolerites in NE Greenland at 1.27 Ga, supporting such a continental configuration (Elming & Mattsson 2001; Karlström *et al.* 2001).

The 1.12 Ga poles (VGP of Salla Dyke for Baltica and mean 1116 Ma palaeomagnetic pole for Laurentia, Table 3) are not overlapping, but their error ovals (for Salla Dyke VGP 23–28° and for Laurentia 13°; see previous section and Table 3) overlap in this continental reconstruction, which would imply that they allow the configuration. Thus there is new palaeomagnetic evidence for a long-lived (1.76–1.12 Ga) Baltica–Laurentia connection, although not conclusive since VGP is used for Baltica.

This particular configuration is also supported by other studies. Some researchers proposing a long-lived Baltica–Laurentia connection use somewhat similar configuration of Baltica and Laurentia during the Mesoproterozoic (Gower *et al.* 1990). For example, Evans & Pisarevsky (2008) rotate Baltica using Euler parameters: Lat: 47.5°, Long: 1.5°, angle: +49° in relation to Laurentia (Fig. 10c). This is a different approach than in Figure 10b, since Baltica and the poles of Baltica have been rotated to a Laurentian reference frame keeping Laurentia in its present position (based on a hypothetical configuration). Hence, this approach does not show the palaeolatitudes of the continents, although these could easily be computed and illustrated for the joined assembly. The connection between 1.76 and 1.12 Ga has been tested using this configuration (Fig. 10c). If similarly aged poles fall in identical locations following rotation, then the reconstruction is supported by the palaeomagnetic data. Since there is only slight difference between this and the configuration of Salminen & Pesonen (2007), the results are similar, allowing the Baltica–Laurentia connection between 1.76 and 1.12 Ga. However, we need to take into account the allowances concerning the Salla VGP.

The proposed configuration appears to be close to the standard Rodinia configuration (Pisarevsky et al. 2003 and refs therein). This raises a question of whether the Baltica–Laurentia entity lasted until Rodinia. Since the pole from the Salla Dyke is only a VGP, further study of this question requires new, accurately-dated and well-defined palaeomagnetic data.

Conclusions

The well-dated 1.12 Ga Salla Diabase Dyke yields a stable characteristic remanence component. The direction of remanence is close to the known direction of the present Earth's field at the sampling site, but a positive baked-contact test proves its primary nature. Rock magnetic studies indicate that magnetite is the remanence carrier. In spite of 13 different sites along the wide and long dyke for palaeomagnetic data, the results are obtained only from one dyke. Hence it is possible that the secular variation has not been sufficiently sampled out and the pole represents only a virtual geomagnetic pole (VGP).

This VGP is not close to any previously known Proterozoic poles of Baltica. Therefore it refines the pre-Sveconorwegian part of the apparent polar wander path (APWP) of Baltica. It adds a clockwise loop to the Baltica's APWP between 1265 and 1040 Ma, just before the Sveconorwegian Loop. Since the pole is a VGP, which may differ from a real palaeopole by $15-20°$ (McElhinny & McFadden 2000) plus its own A_{95} (8.1°), the suggested loop may be anything between $c.$ 50° and $c.$ 100°. A similarly aged Logan Loop (Robertson & Fahrig 1971; Ernst & Buchan 1993; Donadini 2007) of the Laurentian APWP, and the 1165–1000 Ma APWP of Kalahari–Grunehogna, also express long lengths, which may imply a common drift history for these continents. However, new palaeomagnetic data from well-dated rocks between 1270 and 1112 Ma are required from all of these continents to test the similarity between these loops of APWP. More evidence of the common history is the fact that the similarity between Baltica's and Laurentia's APWP already starts at 1.5 Ga. Both APWPs form a hinge after 1.5 Ga when heading to the 1.27 Ga part of the APWP.

The VGP for Salla Diabase Dyke takes Baltica to a high palaeolatitude of $c.$ 60° at 1.12 Ga, but since the VGP may differ from a real palaeomagnetic pole (McElhinny & McFadden 2000) the position of Baltica may also be as low as $c.$ 30°. At that time, Laurentia and Australia were located at high palaeolatitudes. Unfortunately, besides these examples at $c.$ 1.12 Ga, there are palaeomagnetic data only for Kalahari-Grunehogna, which is located in equatorial palaeolatitudes.

The extension of a proposed long-lived Mesoproterozoic connection between Baltica and Laurentia (Gower et al. 1990; Åhäll & Connelly 1998; Karlström et al. 2001; Pesonen et al. 2003 and refs therein) is permissible according to the new 1.12 Ga palaeomagentic data for Salla Diabase Dyke, and supports a long-lived (1.76–1.12 Ga) Baltica–Laurentia connection. However, when taking into account that the ages of the used poles for Baltica and Laurentia are not exactly coeval, and the data for Baltica come from only one dyke, where secular variation has not been adequately sampled, this interpretation needs further verification.

The authors thank S. Pisarevsky and D. Evans for their comments and suggestions, which improved our manuscript and made it clearer. G. Wilson is thanked for kindly letting us use his palaeomagnetism laboratory at University of Otago. S. Dayioglu is thanked for helping in sampling during the field trip in 2007. M. Lehtonen and B. Johansson are thanked for their help with SEM. M. Ehrnrooth's Foundation is warmly thanked for financial support towards Salminen's Ph.D. This paper contributes to the IGCP-509 project.

References

ÅHÄLL, K. I. & CONNELLY, J. 1998. Intermittent 1.53–1.13 Ga magmatism in western Baltica; age constraints and correlations within a postulated supercontinent. *Precambrian Research*, **92**, 1–20.

BORRADAILE, G. J. & MIDDLETON, R. S. 2006. Proterozoic palaeomagnetism in the Nipigon Embayment of Northern Ontario: Pillar Lake Lava, Waweig troctolite and Gunflint Formation tuffs. *Precambrian Research*, **157**, 526–536.

BUCHAN, K. L. & HALLS, H. C. 1990. Palaeomagnetism of Proterozoic mafic dyke swarms of the Canadian Shield. *In*: PARKER, A. J., RICKWOOD, P. C. & TUCKER, D. H. (eds) *Mafic Dykes and Emplacement Mechanism*. Balkema, Rotterdam, 209–230.

COX, A. 1969. Confidence limits for the precision parameter K. *Geophysical Journal of the Royal Astronomical Society*, **18**, 545–549.

DAVIS, D. W. & GREEN, J. C. 1997. Geochronology of the North American Midcontinent rift in western Lake Superior and implications for its geodynamic evolution. *Canadian Journal of Earth Sciences*, **34**, 476–488.

DAVIS, D. W. & PACES, J. B. 1990. Time resolution of geological events on the Keweenaw Peninsula and implications for the development of the Midcontinent rift system. *Earth and Planetary Science Letters*, **97**, 54–64.

DAY, R., FULLER, M. D. & SCHMIDT, V. A. 1977. Hysteresis properties of titanomagnetites: grain size and composition dependence. *Physics of the Earth and Planetary Interiors*, **13**, 260–267.

DEKKERS, M. J. 1988. Magnetic properties of natural pyrrhotite. Part I: behaviour of initial susceptibility and saturation-magnetization-related rock-magnetic parameters in a grain-size dependent framework. *Physics of the Earth and Planetary Interiors*, **52**, 376–393.

DONADINI, F. 2007. *Features of the Geomagnetic Fields During the Holocene and Proterozoic*. Ph.D. thesis, Univeristy of Helsinki, 188.

DRAGONI, M., LANZA, R. & TALLARICO, A. 1997. Magnetic anisotropy produced by magma flow: theoretical model and experimental data from Ferrar dolerite sills (Antarctica). *Geophysical Journal International*, **128**, 230–240.

DUDAS, F. O., DAVIDSON, A. & BETHUNE, K. M. 1994. Age of the Sudbury diabase dykes and their metamorphism in the Grenville Province, Ontario. *Radiogenic age and isotopic studies*. Report 8, Geological. Survey of Canada, Current Research 1994F, 97–106.

DUNLOP, D. J. 1983. Determination of domain structure in igneous rocks by alternating field and other methods. *Earth and Planetary Science Letters*, **63**, 353–367.

DUNLOP, D. J. & BUCHAN, K. L. 1977. Thermal remagnetization and the paleointensity record of metamorphic rocks. *Physics of the Earth and Planetary Interiors* **13**, 325–331.

ELMING, S.-Å. & MATTSSON, H. 2001. Post Jotnian basic intrusions in the Fennoscandian shield, and the breakup of Baltica from Laurentia: a palaeomagnetic and AMS study. *Precambrian Research*, **108**, 215–236.

ERNST, R. E. & BUCHAN, K. L. 1993. Palaeomagnetism of the Abitibi dyke swarm, southern Superior Province, and implications for the Logan Loop. *Canadian Journal of Earth Sciences*, **30**, 1886–1897.

EVANS, D. A. D. & PISAREVSKY, S. A. 2008. Plate tectonics on the early Earth? – Weighing the palaeomagnetic evidence. *In*: CONDIE, K. & PEASE, V. (eds) *When Did Plate Tectonics Begin?* Geological Society of America Special Paper, **440**, 249–263.

EVERITT, C. W. F. & GLEGG, J. A. 1962. A field test of palaeomagnetic stability. *Geophysical Journal of the Royal Astronomical Society*, **6**, 312–319.

FAHRIG, W. F., CHRISTIE, K. W. & JONES, D. L. 1981. Palaeomagnetism of the Bylot Basins: evidence for Mackenzie continental tensional tectonics. *In*: CAMPBELL, F. H. A. (ed.) *Proterozoic Basins of Canada*. Geological Survey of Canada, Paper 81–10, 303–312.

FISHER, R. 1953. Dispersion of sphere. *Proceedings of the Royal Society, London*, **A217**, 293–305.

GOSE, J. A., JOHNSTON, S. T. & THOMAS, R. J. 2004. Age of magnetization of Mesoproterozoic rocks from the Natal sector of the Namaqua-Natal belt, South Africa. *Journal of African Earth Sciences*, **40**, 137–145.

GOWER, C. F., RYAN, A. B. & RIVERS, T. 1990. Mid-Proterozoic Laurentia–Baltica: an overview of its geological evolution and a summary of the contributions made by this volume. *In*: GOWER, C. F., RIVERS, T. & RYAN, B. (eds) *Mid-Proterozoic Laurentia–Baltica*. Geological Association of Canada Special Paper, **38**, 1–20.

HANSON, R .E., CROWLEY, J. L. ET AL. 2004. Coeval large-scale magmatism in the Kalahari and Laurentian cratons during Rodinia assembly. *Science*, **304**, 1126–1129.

HANSKI, E. & HUHMA, H. 2005. Central Lapland Greenstone Belt. *In*: LEHTINEN, M., NURMI, P. & RÄMÖ, T. (eds) *Precambrian Bedrock of Finland – Key to the Evolution of the Fennoscandian Shield*. Elsevier, New York, 139–194.

HANSKI, E., WALKER, R. J., HUHMA, H. & SUOMINEN, I. 2001. The Os and Nd isotopic systematics of the c. 2.44 Ga Akanvaara and Koitelainen mafic layered intrusions in northern Finland. *Precambrian Research*, **109**, 73–102.

HROUDA, F. 1982. Magnetic anisotropy of rocks and its application in geology and geophysics. *Geophysical Survey*, **5**, 37–82.

IRVING, E., BAKER, J., HAMILTON, M. & WYNNE, P. J. 2004. Early Proterozoic geomagnetic field in western Laurentia: implications for paleolatitudes, local rotations and stratigraphy. *Precambrian Research*, **129**, 251–270.

JONES, D. L. & MCELHINNY, M. W. 1966. Palaeomagnetic correlation of basic intrusions in the Precambrian of Southern Africa. *Journal of Geophysical Research*, **71**, 543–552.

JONES, D. L., BATES, M. P., LI, Z. X., CORNER, B. & HODGKINSON, G. 2003. Palaeomagnetic results from the c. 1130 Ma Borgmassivet intrusions in the Ahlmannryggen region of Dronning Maud Land, Antarctica, and tectonic implications. *Tectonophysics*, **375**, 247–260.

KARLSTRÖM, K. E., ÅHÄLL, K.-I., HARLAN, S., WILLIAMS, M. L., MCLELLAND, J. & GEISSMAN, J. W. 2001. Long-lived (1.8–1.0 Ga) convergent orogen in southern Laurentia, its extensions to Australia and Baltica, and implications for refining Rodinia. *Precambrian Research*, **111**, 5–30.

KIRSCHVINK, J. L. 1980. The least-square line and plane and the analysis of palaeomagnetic data. *Geophysical Journal of the Royal Astronomical Society*, **62**, 699–718.

KROGH, T. E., CORFU, F. ET AL. 1987. Precise U–Pb isotopic ages of diabase dykes and gabbros using trace baddeleyite. *In*: HALLS, H. C. & FAHRIG, W. H. (eds) *Mafic Dyke Swarms*. Geological Association of Canada Special Paper, **33**, 147–152.

LAUERMA, R. 1995. *Kursun ja Sallan kartta-alueiden kalliopera. Summary: Pre-Quaternary rocks of the Kursu and Salla map-sheet areas*. Geological map of Finland 1:100 000, Explanation to the map of Pre-Quaternary Rocks, Sheets 3643, 4621 + 4623.

LECHEMINANT, A. N. & HEAMAN, L. M. 1989. Mackenzie igneous events, Canada: middle Proterozoic hotspot magmatism associated with ocean opening. *Earth and Planetary Science Letters*, **96**, 38–48.

LEINO, M. A. H. 1991. *Paleomagneettisten tulosten monikomponenttianalyysi pienimmän neliösumman menetelmällä*, Laboratory for Palaeomagnetism, Department of Geophysics, Geological Survey of Finland, Report Q29.1/91/2 (in Finnish).

LOWRIE, W. 1990. Identification of ferromagnetic minerals in a rock by coercivity and unblocking temperature properties. *Geophysical Research Letters*, **17**, 159–162.

MANNINEN, T. & HUHMA, H. 2001. A new U–Pb zircon constraint from the Salla schist belt, northern Finland. In: *Radiometric Age Determinations from Finnish Lapland and their Bearing on the Timing of Precambrian Volcano-sedimentary Sequences*. Geological Survey of Finland. Special Paper, **33**. Geological Survey of Finland, Espoo, 201–208.

MARÉ, L. P. & THOMAS, R. J. 1998. Palaeomagnetism and aeromagnetic modelling of the Mesoproterozoic Ntimbankulu Pluton, KwaZulu-Natal, South Africa: mushroom-shaped diapir? *Journal of African Earth Sciences*, **25**, 519–537.

MCCLELLAND-BROWN, E. 1981. Palaeomagnetic estimates of temperatures reached in contact metamorphism. *Geology*, **9**, 112–116.

MCELHINNY, M. W. & MCFADDEN, P. L. 2000. *Palaeomagnetism: Continents and Oceans*. Academic Press, San Diego.

MEERT, J. G. 2002. Palaeomagnetic evidence for a Palaeo-Mesoproterozoic supercontinent Columbia. *Gondwana Research*, **5**, 207–215.

MERRILL, R. T. & MCELHINNY, M. W. 1983. *The Earth's Magnetic Field: its History, Origin, and Planetary Perspective*. Academic Press, London.

MERTANEN, S. 1995. *Multicomponent remanent magnetizations reflecting the geological evolution of the Fennoscandian Shield – a palaeomagnetic study with emphasis of the Svecofennian orogeny*. Ph.D. thesis, University of Helsinki.

MERTANEN, S., PESONEN, L. J. & HUHMA, H. 1996. Palaeomagnetism and Sm–Nd ages of the Late Proterozoic diabase dykes in Laanila and Kautokeino, northern Fennoscandia. In: BREWER, T. S. (ed.) *Precambrian Crustal Evolution in the North Atlantic Region*. Geological Society, London, Special Publications, **112**, 331–358.

MULLER, J. A., MAHER, M. J. & SAAL, E. W. 1978. The significance of natural remanent magnetization in geophysical exploration in the Okiep copper district. In: VERWOERD, W. J. (ed.) *Mineralization in Metamorphic Terranes*. J. L. van Schaik Limited, Pretoria, 385–401.

ONSTOTT, T. C., HARGRAVES, R. B. & JOUBERT, P. 1986. Constraints on the tectonic evolution of the Namaqua province 11: reconnaissance palaeomagnetic and 40Ar/39Ar results from the Namaqua province and Kheis Belt. *Transactions of the Geological Society of South Africa*, **89**, 143–170.

O'REILLY, W. 1984. *Rock and Mineral Magentism*. Blackie, Glasgow.

PETROVSKY, E. & KAPICKA, A. 2006. On determination of the Curie point from thermomagnetic curves. *Journal of Geophysical Research*, **111**, B12S27, doi:10.1029/2006JB004507.

PESONEN, L. J. 1978. *Palaeomagnetic, Paleointensity and Paleosecular variation studies of Keweenawan igneous and baked contact rocks*. Ph.D. thesis, University of Toronto, 375.

PESONEN, L. J., ELMING, S.-Å. ET AL. 2003. Palaeomagnetic configuration of continents during the Proterozoic. *Tectonophysics*, **375**, 289–324.

PISAREVSKY, S. A. & BYLUND, G. 1998. Neoproterozoic palaeomagnetic directions in rocks from a key section of the Protogine Zone, southern Sweden. *Geophysical Journal International*, **133**: 185–200.

PISAREVSKY, S. A. & SOKOLOV, S. J. 2001. The magnetostratigraphy and a 1780 Ma palaeomagnetic pole from the red sandstones of Vazhinka River section, Karelia, Russia. *Geophysical Journal International*, **146**, 531–538.

PISAREVSKY, S. A., WINGATE, M. T. D., POWELL, C. MCA., JOHNSON, S. & EVANS, D. A. D. 2003. Models of Rodinia assembly and fragmentation. In: YOSHIDA, M., WINDLEY, B. F., DASGUPTA, S. & POWELL, C. MCA. (eds) *Proterozoic East Gondwana: Supercontinent Assembly and Breakup*. Geological Society, London, Special Publications, **206**, 35–55.

POWELL, C. MCA. & LI, Z. X. 1994. Reconstruction of the Panthalassan margin of Godwanaland. In: VEEVERS, J. J. & POWELL, C. MCA. (eds) *Permian-Triassic Pangean Basin and Foldbelts along the Pathalassan Margin of Gondwanaland*. Geological Society of America Memoir, **184**, 5–9.

POWELL, C. MCA., JONES, D. L., PISAREVSKY, S. & WINGATE, M. T. D. 2001. Palaeomagnetic constraints on the position of the Kalahari craton in Rodinia. *Precambrian Research*, **110**, 33–46.

PURANEN, R., PEKKARINEN, L. J. & PESONEN, L. J. 1992. Interpretation of magnetic fabrics in the Early Proterozoic dykes of Keuruu, central Finland. *Physics of the Earth and Planetary Interiors*, **72**, 68–82.

ROBERTSON, W. A. & FAHRIG, W. F. 1971. The great Logan palaeomagnetic loop – the polar wandering path from Canadian Shield rocks during Neohelikian era. *Canadian Journal of Earth Sciences*, **8**, 1355–1372.

ROCHETTE, P., JACKSON, M. & JACKSON, C. A. 1992. Rock magnetism and the interpretation of anisotropy of magnetic susceptibility. *Reviews of Geophysics*, **30**, 209–226.

ROGERS, J. J. W. & SANTOSH, M. 2002. Configuration of Columbia, a Mesoproterozoic supercontinent. *Gondwana Research*, **5**, 5–22.

SALMINEN, J & PESONEN, L. J. 2007. Palaeomagnetic and rock magnetic study of the Mesoproterozoic sill, Valaam Island, Russian Karelia. *Precambrian Research*, **159**, 212–230.

SCHWARZ, E. J. & BUCHAN, K. L., 1982. Uplift deduced from remanent magnetization; Sudbury area since 1250 Ma ago. *Earth and Planetary Science Letters*, **58**, 65–74.

TANAKA, H. & IDNURM, M. 1994. Palaeomagnetism of Proterozoic mafic intrusions and host rocks of the Mount Isa inlier, Australia: revisited. *Precambrian Research*, **69**, 241–258.

TARLING, D. H. & HROUDA, F. 1993. *The Magnetic Anisotropy of Rocks*. Chapman & Hall, London.

TAUXE, L. 2002. *The PMAG software package Online documentation for use with paleomagnetic Principles and Practice*. Kluwer Academic Publishers, Dordrecht, The Netherlands, 113.

TORSVIK, T. H. & SMETHURST, M. A. 1999. Plate tectonic modeling: virtual reality with GMAP. *Computer & Geosciences*, **25**, 395–402.

VÄÄNÄNEN, P. 1965. *Diabase Dyke at Salla*. M.Sc thesis, University of Helsinki (in finnish).

WINGATE, M. D. T. & EVANS, D. A. D. 2003. Palaeomagnetic constraints on the Proterozoic tectonic evolution of Australia. *Geological Society, London, Special Publications*, **206**, 77–91.

ZHAO, G., SUN, M., WILDE, S. & LI, S. 2003. Assembly, accretion and breakup of the Palaeo-Mesoproterozoic Columbia supercontinent: records in the North China craton. *Gondwana Research*, **6**, 417–434.

ZHAO, G., SUN, M., WILDE, S. A. & LI, S. 2004. A Palaeo-Mesoproterozoic supercontinent: assembly, growth and breakup. *Earth and Science Reviews*, **67**, 91–123.

Sm–Nd data for granitoids across the Namaqua sector of the Namaqua–Natal Province, South Africa

ÅSA PETTERSSON[1], DAVID H. CORNELL[1]*, MASAKI YUHARA[2] & YUKA HIRAHARA[3]

[1]*Department of Earth Science, Göteborg University, Box 460, SE-405 30 Göteborg, Sweden*

[2]*Department of Earth Science, Faculty of Science, Fukuoka University, Nanakuma 8-19-1, Fukuoka 814-0180, Jonan, Japan*

[3]*Japan Agency for Marine-Earth Science and Technology, 2-15, Natsushima, Yokosuka, 237-0061, Japan*

*Corresponding author (e-mail: cornell@gvc.gu.se)

Abstract: Sm–Nd data for rocks of granitic composition in an east–west traverse across the Namaqua Sector of the Mesoproterozoic c. 1.2 Ga Namaqua–Natal Province of southern Africa provide new evidence about the timing of crustal extraction from the mantle. Recent ion probe zircon dating has shown that, contrary to previous indications that pre-Namaqua basement had been preserved in parts of the Namaqua Sector, the majority of the magmatic rocks were emplaced during the 1.4–1.0 Ga Namaqua tectogenesis and very little U–Pb evidence of older precursors remains. Sm–Nd model ages show that Mesoproterozoic crustal extraction did occur, but was also strong in the Palaeoproterozoic, with local evidence for the existence of Archaean crust. The distribution of model ages generally corresponds with the established terranes, subdivided using lithostratigraphic and structural criteria. The northern part of the Bushmanland Terrane shows early Palaeoproterozoic to Archaean model ages and is clearly different from the southern part, which has Mesoproterozoic model ages. This supports the previously published results of a north–south traverse and justifies the separation of the Garies Terrane from the Bushmanland Terrane, though further subdivisions are not supported. The continuation of the Richtersveld Province east of Pofadder is supported by one sample, and the terrane boundary between the Bushmanland and Kakamas terranes is marked by an abrupt change from Palaeoproterozoic to Mesoproterozoic model ages. Model ages for the Kakamas and Areachap terranes do not distinguish them well. They suggest a Mesoproterozoic to late Palaeoproterozoic origin for both terranes, neither of which has a purely juvenile character. The influence of Palaeoproterozoic crust-forming events is clear in the Kaaien Terrane to the east, possibly reflecting reworking of the Kaapvaal Craton. The Namaqua Sector of the Province thus has a history of crustal extraction and evolution which reaches back locally to the Archaean, with major Palaeoproterozoic and Mesoproterozoic crust-forming events. This differs from the Natal Sector, which has a largely juvenile character related to a Mesoproterozoic Wilson cycle. Attempts to reconstruct the Mesoproterozoic Supercontinent Rodinia will have to take into account the extensive Nd-isotopic evidence for older crustal events in the Namaqua sector.

The Namaqua–Natal Province of southern Africa is composed of several terranes, coeval with the Mesoproterozoic tectonic events which led to the formation of Rodinia at around 1.1 Ga (Dalziel 1991; Hoffman 1991; Torsvik *et al.* 1996; Dalziel *et al.* 2000). It comprises granitoids and metamorphosed supracrustal rocks, and is wrapped around the western and southern parts of the Archaean Kaapvaal Craton (Fig. 1). The nomenclature and definition of the terranes used here is reviewed by Cornell *et al.* (2006), based on rock associations and available geochronology. Terrane boundaries largely follow late-tectonic shear zones, but the original suture zones are often obscured. The terrane definitions and boundaries are still uncertain in much of the Namaqua Sector, because of a lack of good geochronological data on the crustal evolution and pretectonic history. Traditionally, with respect to the Namaquan Orogeny, the granitoids have been grouped into pre-, syn- and post-tectonic, based on their structural complexity and degree of foliation, and many of the granitic gneisses and supracrustal units have been regarded as pretectonic relative to the first collision at 1.2 Ga, or to an event older than the Namaqua Wilson Cycle which started at about 1.4 Ga. However, recent ion probe zircon dating has shown that very few of the granitoids are older than 1.4 Ga (Robb *et al.* 1999; Raith *et al.* 2003; Clifford *et al.* 2004; Cornell & Pettersson 2007*a, b*; Bailie *et al.* 2007; Pettersson

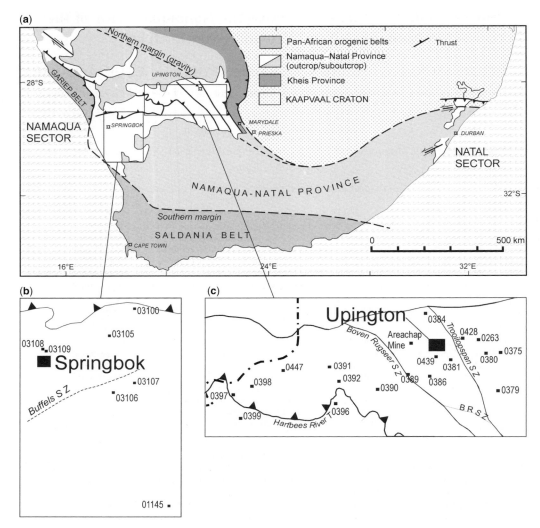

Fig. 1. (a) Precambrian framework of southern Africa, modified after Cornell *et al.* 2006. (b) Sample localities in the western field area, all samples have prefix 'DC'. (c) Sample localities for the eastern part of the field area with terrane boundaries. The samples shown have prefix 'DC' except for the seven from Areachap Mine, with prefix 'AP', which are not shown in detail.

et al. 2007). Many of them formed in two major magmatic periods, the early Namaquan tectonothermal event at 1.20–1.15 Ga and the late Namaquan, largely thermal event at 1.10–1.02 Ga, whereas a few may be related to breakup events or arc magmatism in the 1.4–1.2 Ga interval. Apart from the *c.* 2.0 Ga Richtersveld Subprovince, only the 1.82 Ga Gladkop Suite granitoids in the northwestern Bushmaland Terrane are still regarded as having been emplaced in pre-Namaqua tectonic cycles. This suggests that, if older crust exists, it has been almost completely reworked by crustal melting during the Namaqua tectogenesis.

Sm–Nd whole rock data is known to be largely unaffected by metamorphism and deformation (Hawkesworth & Van Calsteren 1984; Barovich & Patchett 1992). It has been used to trace episodes of crustal growth within continents, tectonic provinces and terranes and is an additional tool for plate reconstruction. T_{CHUR} ages are minimum ages for when material left the mantle as a melt with decreased Sm–Nd ratio relative to a chondritic mantle. T_{DM} ages represent approximate crustal residence ages, if derived from the depleted mantle of DePaolo *et al.* (1981). In the Namaqua–Natal Province, the mantle composition seems generally to have been between these two models (Cornell et al. 1986). This chapter presents Sm and Nd data from 32 granitoids, gneisses and supracrustal rocks from the Namaqua Sector of the Namaqua–Natal Province. This provides a window to see through the Namaquan magmatic and metamorphic events and sheds light on the origin and identity of the individual terranes.

Geological outline

The Namaqua Sector of the Namaqua–Natal Province has been subdivided differently by several authors (Hartnady et al. 1985; Stowe 1986; Eglington 2006). We follow Thomas et al. (1994) and Cornell et al. (2006) in this work, as shown in Figure 1.

The Richtersveld Subprovince comprises arc-related Orange River Group greenschist to amphibolite grade volcanic rocks, intruded by granitoids of the Vioolsdrif Suite, both of Palaeoproterozoic magmatic age (Reid 1979). The Bushmanland Terrane to the south contains a few Palaeoproterozoic granitoids in the Gladkop Suite south of the border with the Richtersveld Subprovince, but is generally underlain by Mesoproterozoic granitoids with minor metasediments in upper amphibolite to granulite grade (Robb et al. 1999; Raith et al. 2003; Clifford et al. 2004; Cornell & Pettersson 2007a). Crustal growth during the Palaeoproterozoic is reflected by Sm–Nd model ages from the Richtersveld subprovince (Reid et al. 1997) and the northwestern Bushmanland Terrane (Barton & Burger 1983; Clifford et al. 1995; Yuhara et al. 2001). The Sm–Nd data has been used to support the view that, despite their lithological and magmatic age differences, these two terranes might represent different crustal levels of a single Palaeoproterozoic continental block, intensively reworked by Namaqua tectonism and crustal remelting in the Bushmanland Terrane (Clifford et al. 1995).

The eastern part of the Namaqua Sector is composed of three terranes with NW-oriented boundaries. The Kakamas Terrane is composed of supracrustal rocks with only one published age of 1.2 Ga (Cornell & Pettersson 2007b) and many granitic intrusions with ages corresponding to the early and late Namaquan magmatic events. The Areachap Terrane is a narrow belt comprising amphibolite facies metavolcanic rocks with magmatic ages between 1.24 and 1.27 Ga and interpreted as an arc sequence, with indications of a juvenile character from Sm–Nd data at Prieska Copper Mines (Cornell et al. 1986). The definition, dating and extent of the Areachap and Kakamas Terranes is discussed by Cornell & Pettersson (2007b), but note that Eglington (2006) uses different boundaries. The Kaaien Terrane is composed of greenschist to amphibolite grade metaquartzitic rocks with detrital zircon ages of c. 1.77 Ga (Moen, unpublished data in Cornell et al. 2006), and contains two volcanic groups in elongate 'basins', the pre-collision 1.3 Ga Wilgenhoutsdrif Group (Moen, unpublished data in Cornell et al. 2006) and the post-collision (in reference to the Namaquan event) 1.17–1.09 Ga Koras Group (Gutzmer et al. 2000; Pettersson et al. 2007).

Methods

Samples were taken along an east–west profile, roughly perpendicular to the strike of most of today's inferred terrane boundaries, from Springbok in the west to east of Upington (Fig. 1).

The westernmost sample lies in the western part of Bushmanland, just north of Springbok (Fig. 1b). The easternmost sample is the Kalkwerf Gneiss, located in the Kaaien Terrane, c. 65 km SE of Upington (Fig. 1c). This profile complements the north–south profile in the Bushmanland Terrane investigated by Yuhara et al. (2001).

Samples of about 2 kg were chipped in the field and weathered material discarded. Rock chips were crushed in batches of 0.3 kg in a steel swing mill for about 15 seconds, then coned and halved down to 0.1 kg. Each rock powder was ground in an agate vessel for 3 minutes, then coned down to 20 g. The sample was then run in the agate mill for another 12–18 minutes to produce a fine powder.

Isotopic analyses of samples with prefix DC were done at Niigata University on MAT261 and MAT262 mass spectrometers, equipped with 5 and 9 Faraday cups, respectively. The $^{143}Nd/^{144}Nd$ ratios were normalized to $^{146}Nd/^{144}Nd$ 0.7219 and measured ratios corrected relative to the JB-1a reference sample with $^{143}Nd/^{144}Nd$ 0.512784. ^{149}Sm-^{150}Nd mixed spikes were used for isotope dilution analysis of Sm and Nd concentrations. All raw data is given at the 1σ-level and calculated ages at the 2σ level. Estimated errors are better than 0.1% for $^{147}Sm/^{144}Nd$ and 0.002% for $^{143}Nd/^{144}Nd$. For detailed method description, see Miyazaki & Shuto (1998), Hamamoto et al. (2000) and Yuhara et al. (2000).

The Areachap Mine samples (prefix AP) were analysed at the Open University, UK, as described by Cornell et al. (1986), except that for consistency $^{143}Nd/^{144}Nd$ ratios are reported relative to the JB-1a reference sample with $^{143}Nd/^{144}Nd$ 0.512784, equivalent to 0.512638 for BCR-1. Errors are better than 0.75% for $^{147}Sm/^{144}Nd$ and 0.004% for $^{143}Nd/^{144}Nd$. T_{CHUR} and T_{DM} ages were calculated using ISOPLOT 3.0 by Ludwig (2003) with $\lambda^{147}Sm$ 6.54×10^{-12}/year and the depleted mantle (DM) model of DePaolo et al. (1981).

Results and discussion

The results are presented in Table 1, together with sample rock types and coordinates. The data are plotted on isochron diagrams by terrane in Figure 2. Reference lines with 1.3 Ga slope through CHUR and DM are shown for comparison. Samples with $^{147}Sm/^{144}Nd$ ratios above 0.14 plot relatively close to CHUR and DM and the model ages from these are considered to be geologically uncertain.

Bushmanland Terrane

The Bushmanland samples are all plutonic granitoids, the Aroams Gneiss, Achab Gneiss, Banks Vlei Gneiss, Modderfontein Gneiss, two samples assigned to the more general Klein Namaqualand Suite, Concordia Granite, Landplaas Gneiss and Burtons Puts Gneiss. The T_{DM} ages of these rocks show a large spread and two groups can be distinguished in Figure 2a and geographically in Figure 3. The first group plots within the 1.3 Ga reference lines on Figure 2a, has Mesoproterozoic T_{DM} ages of 1422–1540 Ma and, with one exception, the samples lie in the southern part of the Bushmanland Terrane in Figure 3. The second group has older Palaeoproterozoic to Archaean (1923–2704 Ma) T_{DM} ages and lies in the northern part of the terrane. Sample DC03105 of the Klein Namaqualand Suite plots in the northern part but has a younger T_{DM} age of 1628 Ma, and is intermediate between the two groups.

This difference in model ages agrees with that documented following Hartnady *et al* (1985) and Yuhara *et al.* (2001), who identified high εNd and low εNd groups in their Nababeep Gneiss (equivalent to Klein Namaqualand Suite) samples, shown for comparison in Figure 2a. Their boundary agrees roughly with the amphibolite to granulite facies isograd (Robb *et al.* 1999), and corresponds to the Buffels Shear Zone, which is now recognized as a terrane boundary, shown in Figure 3, although still poorly defined to the east of Pofadder. We call the southern terrane the Garies Terrane following Yuhara *et al.* (2001) and Eglington (2006), corresponding to the Vaalputs Terrane of Andreoli *et al.* (2006). Sm–Nd model ages from boreholes south of the Namaqua sector (Eglington & Armstrong 2003) range between 1.2 and 2.4 Ga, showing that the young model ages from the Garies Terrane do not prevail as far south. We retain the name Bushmanland Terrane for the northern part which has a Palaeoproterozoic to Archaean signature from Springbok in the west (Clifford *et al.* 1995) to the boundary with the Kakamas Terrane along the Hartbees River Thrust in the east. The available Nd-isotope data does not support subdivisions within the northern Bushmanland Terrane, as proposed by Van Aswegen *et al.* (1987), followed by Eglington (2006) and Andreoli *et al.* (2006). Magnetic and structural maps show a general east–west fabric in Bushmanland that changes abruptly at the Bushmanland–Kakamas Terrane boundary to NW–SE, a trend which continues in the terranes to the east. The Bushmanland–Kakamas boundary also shows a distinct drop in Sm–Nd crustal residence ages.

Richtersveld Subprovince

Most previous dating in the Richtersveld Subprovince has been done in the western part, around Vioolsdrif (Fig. 3). The Richtersveld Subprovince is characterized by largely undeformed 2.0–1.9 Ga supracrustal volcanic rocks and intrusive granitoids, which have been described as a volcanic arc complex (Reid 1997, 1982; Kröner *et al.* 1983). The eastern part is more deformed and of amphibolite grade. Dewey (2006) proposed that the eastern and western parts represent crustal blocks with different levels. Little geochronology has been done east of Pofadder (Fig. 3). In this study, a sample of the Coboop Granite Gneiss from the easternmost part of the Subprovince, has an ion probe magmatic U–Pb zircon age of *c.* 1900 Ma. Its T_{DM} age of 2274 Ma agrees with the *c.* 2.2 Ga T_{DM} model ages of Reid (1997) from the western area. This result confirms the extension of the Richtersveld Subprovince east of Pofadder.

Kakamas Terrane

The Kakamas Terrane is characterized by amphibolite grade supracrustal rocks such as metapelite, mature quartzite and calcsilicate, which seem to represent a shelf sequence, together with metavolcanic quartzofeldpathic gneisses, both intruded by voluminous early- to late-Namaquan granitoids. A metasedimentary, probably volcanoclastic supracrustal rock of the Kenhardt Formation, ion probe dated by Cornell & Pettersson (2007b), yielded inherited zircon core domains of 1568, 1351 and 1296 Ma, a main domain of 1197 ± 5 Ma magmatic zircon and 1194 ± 23 Ma metamorphic rims. Other sampled rocks are two Friersdale Charnockites, two Riemvasmaak Gneiss samples, Vaalputs Granite, Polisiehoek Gneiss and Beenbreek Gneiss. The magmatic ages of these plutonic rocks, determined by ion probe zircon dating, vary from 1110–1210 Ma. Our samples have 1360–1618 Ma T_{DM} ages (Fig. 3), with one exception, the Polisiehoek Gneiss at 2244 Ma. However, its T_{CHUR} age is 1420 Ga and its true crustal residence age depends greatly on the mantle source because of its high Sm/Nd ratio, shown in Figure 2b. The samples all lie between the 1.3 Ga reference lines, thus all samples could have entered the crust at

Table 1. *Sample localities, rock types and Sm–Nd data.*

Stratigraphic name	Rock type	Sample no.	Latitude	Longitude	Terrane	Sm ppm	Nd ppm	$^{147}Sm/^{144}Nd$	Error %	$^{143}Nd/^{144}Nd$	Error %	T_{CHUR}	T_{DM}
Jannelsepan Fn	Amphibolite	AP4-2	29°03.750′	21°02.340′	Ar	2.70	7.88	0.2076	0.75	0.513009	0.004	5072	5337
Jannelsepan Fn	Phlogopite cordierite gneiss	AP4-20	29°03.750′	21°02.340′	Ar	3.04	13.51	0.1362	0.75	0.512322	0.004	798	1410
Jannelsepan Fn	Massive sulphide ore	AP4-23	29°03.750′	21°02.340′	Ar	0.273	0.973	0.1697	0.75	0.512476	0.004	918	1952
Jannelsepan Fn	Amphibolite	AP22-13	29°03.750′	21°02.340′	Ar	6.9	35.7	0.1168	0.75	0.511960	0.004	1294	1694
Jannelsepan Fn	Metapelite	AP4-7	29°03.750′	21°02.340′	Ar	11.9	43.8	0.1643	0.75	0.512608	0.004	142	1333
Jannelsepan Fn?	Augen gneiss	AP22-14	29°03.750′	21°02.340′	Ar	5.04	19.71	0.1546	0.75	0.512492	0.004	531	1411
Jannelsepan Fn?	Leucogneiss	AP22-7	29°03.750′	21°02.340′	Ar	5.23	26.27	0.1204	0.75	0.511834	0.004	1605	1966
Josling	Granite gneiss	DC0384	29°21.674′	21°08.194′	Ar	6.90	30.8	0.1359	0.1	0.512273	0.002	917	1499
Jannelsepan Fn?	Intermediate biotite–hornblende gneiss	DC0439	28°30.279′	21°12.378′	Ar	10.7	68.6	0.09433	0.1	0.511194	0.002	2144	2356
Leeuwdraai Fn, Koras Group	Felsic lava	DC0263	28°27.800′	21°40.800′	Kai	10.9	65.8	0.1000	0.1	0.511747	0.002	1404	1729
Wilgenhoutsdrif	Mylonitic felsic lava	DC0375	28°30.022′	21°43.864′	Kai	3.40	5.80	0.3522	0.1	0.513687	0.002	1027	701
Kalkwerf	Granite gneiss	DC0379	28°41.586′	21°48.730′	Kai	5.50	27.5	0.1214	0.1	0.511676	0.002	1944	2252
Swartkopsleegte Fn, Koras Group	Felsic lava	DC0380	28°24.878′	21°36.364′	Kai	15.2	77.0	0.1189	0.1	0.512061	0.002	1131	1570
Straussburg	Biotite granite gneiss	DC0381	28°27.180′	21°19.127′	Kai	11.6	58.0	0.1206	0.1	0.511967	0.002	1344	1752
Swanartz	Biotite granite	DC0428	28°24.794′	21°25.900′	Kai	7.45	47.0	0.09572	0.1	0.511505	0.002	1708	1978
Friersdale	Pyroxene–hornblende biotite granite	DC0386	28°37.819′	21°30.299′	Km	14.9	78.7	0.1143	0.1	0.512000	0.002	1181	1591
Vaalputs	Biotite granite	DC0389	28°41.542′	21°00.983′	Km	10.5	53.9	0.1174	0.1	0.512016	0.002	1196	1616
Friersdale	Pyroxene–hornblende biotite granite	DC0390	28°47.010′	20°47.717′	Km	14.7	77.4	0.1144	0.1	0.511983	0.002	1214	1618
Riemvasmaak	Biotite granite gneiss	DC0391	28°33.096′	20°17.185′	Km	15.2	70.9	0.1294	0.1	0.512181	0.002	1036	1549
Riemvasmaak	Hornblende granite gneiss	DC0392	28°38.406′	20°20.601′	Km	18.3	91.9	0.1201	0.1	0.512204	0.002	865	1360
Polisiehoek	Biotite granite gneiss	DC0447	28°50.135′	19°43.617′	Km	6.44	23.1	0.1683	0.1	0.512374	0.002	1420	2244
Beenbreek?	Biotite megacrystic gneiss	DC0398	28°50.133′	19°30.283′	Km	11.1	79.7	0.08440	0.1	0.511812	0.002	1122	1445
Coboop	Muscovite biotite augen granite gneiss	DC0397	28°52.223′	19°20.312′	Ri	3.30	24.6	0.08200	0.1	0.511072	0.002	2075	2274
Aroams	Biotite granite gneiss	DC0399	29°07.030′	19°24.258′	Bu	8.50	45.2	0.1140	0.1	0.511521	0.002	2054	2322

(*Continued*)

Table 1. *Continued*

Stratigraphic name	Rock type	Sample no.	Latitude	Longitude	Terrane	Sm ppm	Nd ppm	$^{147}Sm/^{144}Nd$	Error %	$^{143}Nd/^{144}Nd$	Error %	T_{CHUR}	T_{DM}
Achab	Biotite hornblende granite gneiss	DC03100	29°13.591'	18°54.769'	Bu	10.0	53.6	0.1127	0.1	0.511231	0.002	2543	2740
Klein Namaqualand Suite	Biotite hornblende granite gneiss	DC03105	29°28.133'	18°27.257'	Bu	7.20	43.5	0.09960	0.1	0.511818	0.002	1287	1628
Burtonsputs	Biotite granite gneiss	DC03106	30°04.666'	18°30.615'	Bu	6.20	31.9	0.1176	0.1	0.512067	0.002	1101	1540
Klein Namaqualand Suite	Biotite granite gneiss	DC03107	29°59.677'	18°47.033'	Bu	11.3	59.8	0.1138	0.1	0.512038	0.002	1104	1525
Concordia	Garnetiferous leucogranite	DC03108	29°35.939'	17°50.501'	Bu	10.4	53.1	0.1188	0.1	0.511693	0.002	1846	2161
Modderfontein	Biotite hornblende granite gneiss	DC03109	29°36.171'	17°50.666'	Bu	3.40	18.1	0.1136	0.1	0.511774	0.002	1583	1923
Landplaas Gneiss	Granite gneiss	DC01145	31°05.202'	19°14.003'	Bu	7.80	39.8	0.1188	0.1	0.512153	0.002	950	1422
Banks Vlei	Biotite granite gneiss	DC0396	28°57.460'	20°21.278'	Bu	5.93	36.6	0.09801	0.1	0.511427	0.002	1867	2121
CHUR used in this work								0.1966	0.1	0.512638	0.002		
DM present day								0.2136	0.1	0.513150	0.002		

Note: Tectono-stratigraphic abbreviations are: Ar, Areachap Terrane; Kai, Kaaien Terrane; Km, Kakamas Terane; Ri, Richtersveld Subprovince; Bu, Bushmanland Terrane.

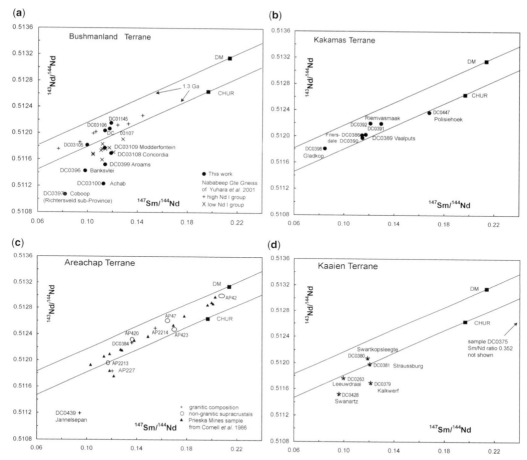

Fig. 2. Sm–Nd isochron diagrams showing the samples presented in this work, plotted by terrane, (**a**) Bushmanland Terrane; (**b**) Kakamas Terrane; (**c**) Areachap Terrane; (**d**) Kaaien Terrane. Reference lines with 1.3 Ga slope through CHUR and DM are shown for comparison.

about 1.3 Ga if derived from mantle sources varying from CHUR to DM. However, this seems improbable in view of the zircon dates up to 1568 Ma mentioned above. More likely, the Kakamas Terrane crust was extracted from a single mantle source like DM in the Mesoproterozoic between 1.6 and 1.2 Ga, with a possible minor Palaeoproterozoic component, indicated in the model age of the Polisiehoek Gneiss. These components were then melted to give rise to the 1110–1210 Ma magmatic granitoid rocks we sampled.

Areachap Terrane

Recent studies show that the magmatic ages of this arc sequence (Figs 1 & 3) range from 1241 ± 12 Ma (Pettersson *et al.* 2007) to 1275 ± 7 Ma (Cornell & Pettersson 2007*b*). Sm–Nd data from the Prieska Mines (Cornell *et al.* 1986), 50 km SW of Prieska, gave T_{DM} model ages around 1.3 Ga, suggesting that the Areachap Terrane is a juvenile oceanic arc terrane. Rocks analysed in this study are the Josling Granite, Jannelsepan migmatitic metadacite and seven samples from a borehole core of the Areachap Mine, north of Upington. With two exceptions, the samples plot between the 1.3 Ga reference lines through CHUR and DM in Figure 2c, although several of the non-granitoid samples have high Sm–Nd ratios and thus have model ages greatly dependant on the source chosen. The Jannelsepan Formation metadacite, with ion probe dates of 1241 ± 12 Ma for magmatism and 1165 ± 10 Ma for migmatization, has a Palaeoproterozoic T_{DM} age of 2366 Ma and a leucogneiss sample from Areachap Mine gives 1966 Ma. Together with the scatter of the samples which plot within the 1.3 Ga

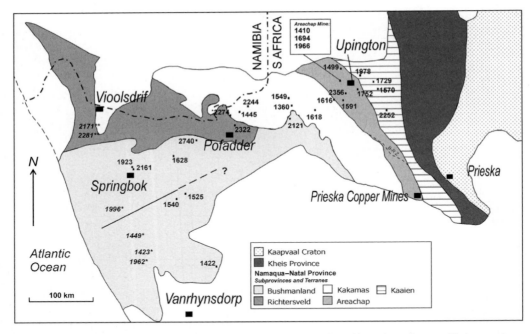

Fig. 3. T_{DM} model ages (Ma) of samples in the Namaqua Sector. Samples from this work are shown as filled rectangles. Selected recalculated ages from the literature are shown for comparison: *Yuhara et al. 2001, **Reid 1997. The line across the Bushmanland Terrane is the boundary between the Garies Terrane (to the south) and reduced Bushmanland Terrane.

reference lines, this indicates that the samples from this terrane were derived from more than one source, with one component being Palaeoproterozoic. This contradicts the idea, originated by Barton & Burger (1983) and extended by Cornell et al. (2006) and Cornell & Pettersson (2007b), that the Jannesepan Formation, Areachap Group and entire Areachap Terrane has a juvenile origin. The data from Prieska Mines is shown for comparison in Figure 2c and shows a remarkable correspondence with the new data, including two samples plotting below the reference line through DM 1.3 Ga with T_{DM} model ages around 2.0 Ga. These two samples are, however, from 25 km south of Prieska Mines and on the boundary between the Areachap and Kaaien terranes. Our data indicates that while many Areachap Terrane samples have Mesoproterozoic age and can be regarded as juvenile, the northern and possibly southern parts of the Areachap Terrane contain a Palaeoproterozoic component. Thus, an older crustal component was present in the magmatic arc or arcs which gave rise to the Areachap Group. The Sm–Nd model ages obtained for the Areachap Terrane are similar to the ages obtained for the Kakamas Terrane, which raises the question of the integrity of these two terranes. Whereas the Kakamas Terrane contains mature metasedimentary rocks suggesting a shelf environment, possibly followed by arc-related volcanism, the Areachap Terrane has only arc-related supracrustal formations without mature metasediments. In view of these different supracrustal assemblages, the two terranes are retained at this stage.

Kaaien Terrane

The Kaaien Terrane is dominated by c. 1.8 Ga metaquartzites with some younger intrusives and basins filled by bimodal volcanic suites. Our samples (Fig. 3) are from the 1370 Ma Swanartz Gneiss (Pettersson et al. 2007), c. 1100 Ma Straussburg Granite, the 1290 Ma Wilgenhoutsdrif Group (Moen, unpublished data in Cornell et al. 2006) and the Swartkopsleegte and Leeuwdraai Formations of the Koras Group, dated at 1170 and 1095 Ma, respectively by Pettersson et al. (2007) and confirming the 1170 Ma Swartkopsleegte age of Gutzmer et al. (2000). Only one sample plots between the 1.3 Ga reference lines in Figure 3d, reflecting an older crustal history for this terrane. The Koras Group and the Straussburg Granite with late-Namaqua magmatic origin, have T_{DM} between 1570 and 1752 Ma. Presence of older crustal material is confirmed by xenocrystic 2.1–1.7 Ga zircons in the Koras Group (Pettersson et al. 2007). The pre-Namaqua collision Swanartz Gneiss has a somewhat older model age of 1978 Ma. These ages suggest that a crust-forming event took place at the edge of the Kaapvaal Craton at some time during the Palaeoproterozoic, possibly during the enigmatic Kheis orogeny (c. 1.8 Ga). The pre-Namaquan Wilgenhoutsdrif Group sample gives an unrealistic model age at 701 Ma and our unpublished rare earth element data show that it is hydrothermally altered.

The Kalkwerf Gneiss, dated at 1290 Ma by Moen (unpublished data in Cornell et al. 2006) is synchronous with the supracrustal rift-related Wilgenhoutsdrif Group. The Kalkwerf Gneiss T_{DM} is 2252 Ma. Thus the 1370 Ma (Swanartz, granitic) and 1290 Ma (Wilgenhoutsdrif, bimodal) magmatism in the Kaaien Terrane involved melting of Palaeoproterozoic crust. The structural boundary between the adjoining Kheis Province and the Archaean Kaapvaal Craton is c. 60 km to the east, but geophysical evidence (Meixner & Peart 1984; James et al. 2001) indicates that the craton extends westwards beneath the Kheis Province and possibly the Kaaien Terrane. A component of Kaapvaal Craton crust in the Kalkwerf Gneiss might account for its old model age.

The Namaqua and Natal sectors

The Natal Sector comprises three terranes with a largely juvenile terrane composition (Eglington et al. 1989; Thomas et al. 1994), which are thought to have formed by accretion of several island arcs onto the Kaapvaal Craton between 1200 and 1135 Ma (Thomas et al. 1994; Jacobs et al. 1997). Only the Margate Terrane of the Natal Sector has yielded a few Palaeoproterozoic T_{DM} ages, whereas the Mzumbe Terrane yields ages between 1.2 and 1.4 Ga (Eglington 2006). By contrast, our study reconfirms earlier studies (Reid 1997; Yuhara et al. 2001; Eglington & Armstrong 2003) showing that much of the Namaqua Sector was derived from the mantle before the c. 1.4 Ga beginning of the Namaqua Wilson cycle, with juvenile contributions during the Namaqua cycle in the Areachap, Kakamas and possibly Garies terranes (Fig. 3). The heterogeneity in T_{DM} dates shown by most of the terranes suggest that they have a complex origin involving more than two crustal components prior to terrane assembly at 1.1–1.2 Ga. In Figure 4, our T_{DM} data for the Namaqua Sector indicate four main crust-forming events at 1.4, 1.6, 2.0 and 2.3 Ga, although these could be 100 Ma younger if a less depleted mantle is envisaged. These ages need to be further strengthened by U–Pb zircon dating. The former existence

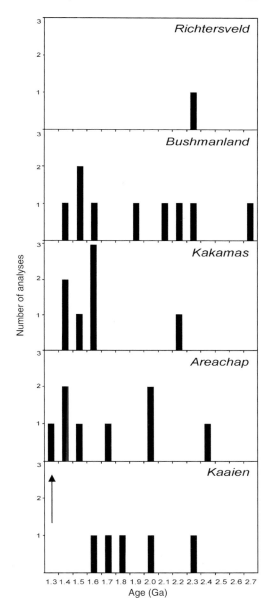

Fig. 4. A histogram of T_{DM} data of this work suggests that four main crust-forming events can be seen in the Namaqua sector of the Namaqua–Natal Province. The true crustal residence ages could be as much as 100 Ma younger than shown if a less depleted mantle than that of DePaolo et al. (1981) is envisaged.

of Archaean crust in the Bushmanland Terrane, although not found exposed at present, is documented by the 2740 Ma T_{DM} age of the Achab Gneiss and the discovery of a 2790 Ma zircon, dated by ion probe U–Pb, in a nearby metasediment by Pettersson et al. (2004). Whereas Dronning Maud Land in East Antarctica was probably a contiguous part of the Natal Sector (Groenewald et al. 1991; Jacobs et al. 1993; Frimmel 2004), the Namaqua Sector has no known counterpart.

Conclusions

1. With some exceptions, the pattern of Sm–Nd model ages in the Namaqua sector of the Namaqua–Natal province broadly corresponds to the established terrane boundaries.
2. The northern Bushmanland Terrane has a Palaeoproterozoic and partly Archaean crustal history. On the contrary, the southwestern part has a clearly younger Mesoproterozoic signature and the two parts probably had different geological histories before terrane assembly. This confirms that the Bushmanland Terrane should be subdivided into two terranes. The southern part should be called the Garies Terrane, following Hartnady et al. (1985) and Yuhara et al. (2001). The Buffels Shear Zone is most likely the northern boundary, as shown by previous workers, although its eastern continuation is not known. To the south, Phanerozoic sequences cover the Namaquan basement, although the extension must be limited, as shown from Sm–Nd borehole data (Eglington & Armstrong 2003). The northern part should continue as the Bushmanland Terrane without further subdivision.
3. T_{DM} ages decrease abruptly from Palaeoproterozoic to Mesoproterozoic across the eastern boundary between the Bushmanland and Kakamas terranes, confirming the Hartbees River Thrust as a distinct terrane boundary.
4. In the Kakamas and Areachap terranes, model ages are Mesoproterozoic to Palaeoproterozoic and the two terranes cannot be clearly distinguished from one another. However, the two terranes should be retained at this stage in view of their different supracrustal assemblages.
5. Sm–Nd model ages clearly indicate a heterogeneous source for the rocks in the Areachap Terrane in general and an old crustal component particularly in the north. It may consist of an assemblage of arc-related rocks that vary in character from oceanic to continental.
6. T_{DM} ages of Mesoproterozoic granotoids and felsic rocks in the Kaaien Terrane show that it is dominated by late Palaeoproterozoic crust. The data, combined with presence of xenocrystic zircons, indicate a crust-forming event at the edge of the Kaapvaal Craton at some time during the Palaeoproterozoic, possibly during the enigmatic c. 1.8 Ga Kheis orogeny. The 2252 Ma model age of the Kalkwerf Gneiss could reflect the influence of the Kaapvaal Craton, which may underlie the Kaaien Terrane.
7. The Namaqua Sector of the Namaqua–Natal Province carries the imprint of Palaeoproterozoic crust-forming events in most of its terranes, in contrast with the Natal Sector, where the crust originated mainly during the 1.4–1.0 Ga Namaqua tectogenesis.
8. This work and the accumulating body of Sm–Nd data will provide new constraints on matches of continental fragments for reconstructions of the Rodinia supercontinent.

We thank G. Moen for advice in the field, H. Kagami for technical support and C. Hawkesworth for providing Sm–Nd laboratory facilities to DHC and the Manager and P. Potgieter, Chief Geologist at Black Mountain Mine, for logistical support. We would also like to thank J. Jacobs and B. Eglington for improvements to the manuscript by their reviews. This work was financed by Swedish Research Council grant 2002-4274 to DHC.

References

ANDREOLI, M. G., HART, R. J., ASHWAL, L. D. & COETZEE, H. 2006. Correlations between U, Th content and metamorphic grade in the western Namaqualand Belt, South Africa with implications for radioactive heating of the crust. *Journal of Petrology*, **47**, 1095–1118.

BAILIE, R., ARMSTRONG, R. & REID, D. 2007. Composition and single zircon U–Pb emplacement ages of the Aggenys Granite Suite, Bushmanland, South Africa. *South African Journal of Geology*, **110**, 87–110.

BAROVICH, K. M. & PATCHETT, P. J. 1992. Behaviour of isotopic systematics during deformation and metamorphism: a Hf, Nd and Sr isotopic study of mylonitized granite. *Contribution to Mineralogy and Petrology*, **109**, 386–393.

BARTON, E. S. & BURGER, A. J. 1983. Reconnaissance isotopic investigations in the Namaqua mobile belt and implications for Proterozoic crustal evolution – Upington geotraverse. *In*: BOTHA, B. J. V. (ed.) *Geological Society of South Africa, Marshalltown. Special Publications*, **10**, 173–191.

CLIFFORD, T. N., BARTON, E. S., REX, D. C. & BURGER, A. J. 1995. A crustal progenitor for the intrusive anorthosite-charnockite kindred of the cupriferous Koperberg Suite, O'okiep District, Namaqualand, South Africa; new isotope data for the country rocks and the intrusives. *Journal of Petrology*, **16**, 154–188.

CLIFFORD, T. N., BARTON, E. S., STERN, R. A. & DUCHESNE, J. C. 2004. U–Pb zircon calendar Namaquan (Grenville) crustal events in the granulite-facies terrane of the O'okiep copper district of South Africa. *Journal of Petrology*, **45**, 669–691.

CORNELL, D. H. & PETTERSSON, Å. 2007a. Ion probe dating of the Achab Gneiss, a young basement to the Central Bushmanland Ore District? *Journal of African Earth Sciences*, **47**, 112–116.

CORNELL, D. H. & PETTERSSON, Å. 2007b. Ion probe zircon dating of metasediments from the Areachap and Kakamas Terranes, Namaqua–Natal Province and the stratigraphic integrity of the Areachap Group. *South African Journal of Geology*, **110**, 169–178.

CORNELL, D. H., HAWKESWORTH, C. J., VAN CALSTEREN, P. & SCOTT, W. D. 1986. Sm–Nd study of Precambrian crustal development in the Prieska-Copperton region, Cape Province. *Transactions of the Geological Society of South Africa*, **89**, 17–28.

CORNELL, D. H., THOMAS, R. J., GIBSON, R., MOEN, H. F. G., MOORE, J. M. & REID, D. L. 2006. Namaqua-Natal Province. *In*: JOHNSON, M. R., ANHAUESSER, C. R. & THOMAS, R. J. (eds) *Geology of South Africa*. Geological Society of South Africa and Council of Geoscience, Pretoria, 325–379.

DALZIEL, I. W. D. 1991. Pacific margins of Laurentia and East Antarctica-Australia as a conjugate rift pair: evidence and implications for an Eocambrian supercontinent. *Geology*, **19**, 598–601.

DALZIEL, I. W. D., MOSHER, S. & GAHAGAN, L. M. 2000. Laurentia-Kalahari collision and the assembly of Rodinia. *Journal of Geology*, **108**, 499–513.

DEPAOLO, D. J., LINN, A. MA. & SCHUBERT, G. 1981. Neodymium isotopes in the Colorado front range and crust-mantle evolution in the Proterozoic. *Nature*, **291**, 193–196.

DEWEY, J. F., ROBB, L. & VAN SCHALKWYK, L. 2006. Did Bushmanland extensionally unroof Namaqualand? *Precambrian Research*, **150**, 173–182.

EGLINTON, B. M. 2006. Evolution of the Namaqua–Natal Belt, southern Africa – A geochronological and isotope geochemical review. *Journal of African Earth Sciences*, **46**, 93–111.

EGLINTON, B. M. & ARMSTRONG, R. A. 2003. Geochronological and isotopic constraints on the Mesoproterozoic Namaqua–Natal Belt: evidence from deep borehole intersections in South Africa. *Precambrian Research*, **125**, 179–189.

EGLINTON, B. M., HARMER, R. E. & KERR, A. 1989. Isotope and geochemical constraints on Proterozoic crustal evolution in the south-eastern Africa. *Precambrian Research*, **45**, 159–174.

FRIMMEL, H. E. 2004. Formation of a late Mesoproterozoic supercontinent: The South Africa – East Antarctica connection. *In*: ERIKSSON, P. G., ALTERMANN, W., NELSON, D. R., MUELLER, W. U. & CATUNEANU, O. (eds) *The Precambrian Earth: Tempos and Events, Developments in Precambrian Geology*, **12**, Elsevier, Amsterdam, 240–255.

GROENEWALD, P. B., MOYES, A. B., GRANTHAM, G. H. & KRYNAUW, J. R. 1991. East Antarctic crustal evolution: geological constraints and modelling in western Dronning Maud Land. *Precambrian Research*, **75**, 231–251.

GUTZMER, J., BEUKES, N. J., PICKARD, A. & BARLEY, M. E. 2000. 1170 Ma SHRIMP age for Koras Group bimodal volcanism, Northern Cape Province. *South African Journal of Geology*, **103**, 32–37.

HAMAMOTO, T., YUHARA, M. *ET AL*. 2000. Rb, Sr, Sm and Nd separation from rocks, minerals and natural water using ion-exchange resin. *Science Reports, Niigata University, Series E (Geology)*, **15**, 49–58.

HARTNADY, C. J. H., JOUBERT, P. & STOWE, C. W. 1985. Proterozoic crustal evolution in southwestern Africa. *Episodes*, **8**, 236–244.

HAWKESWORTH, C. J. & VAN CALSTEREN, P. W. C. 1984. *In*: HENDERSON, P. (ed.) *Rare Earth Element Geochemistry*. Developments in Geochemistry 2, Elsevier, Amsterdam, 375–421.

HOFFMAN, P. F. 1991. Did the breakout of Laurentia turn Gondwanaland inside-out? *Science*, **252**, 1409–1412.

JACOBS, J., FALTER, M., THOMAS, R. J., KUNZ, J. & JEßBERGER, E. K. 1997. ^{40}Ar/^{39}Ar thermochronological constraints on the structural evolution of the Mesoproterozoic Natal Metamorphic Province, SE Africa. *Precambrian Research*, **86**, 71–92.

JACOBS, J., THOMAS, R. J. & WEBER, K. 1993. Accretion and indentation tectonics at the southern edge of the Kaapvaal Craton during the Kibaran (Grenville) Orogeny. *Geology*, **21**, 203–206.

JAMES, D. E., FOUCH, M. J., VANDECAR, J. C., VAN DER LEE, S. KAAPVAAL SEISMIC GROUP, 2001. Tectospheric structure beneath the southern Africa. *Geophysical Research Letters*, **28**, 2485–2488.

KRÖNER, A., BARTON, E. S., BURGER, A. J., ALLSOPP, H. L. & BERTRAND, J. M. 1983. The ages of the Goodhouse granite and grey gneisses from the marginial zone of the Richtersveld Province and their bearing on the timing of tectonic events in the Namaqua Mobile Belt. *Special Publication of the Geological Society of South Africa*, **10**, 123–129.

LUDWIG, K. R. 2003. Isoplot 3: a Geochronological Toolkit for Microsoft Excel. Berkeley *Geochronology Center Special Publication*, **4**, 1–71.

MEIXNER, H. M. & PEART, R. J. 1984. The Kalahari Drilling Project: a report on the geophysical and geological results of follow-up drilling to the aeromagnetic Survey of Botswana. *Geological Survey Department, Bullentin*, **27**.

MIYAZAKI, T. & SHUTO, K. 1998. Sr and Nd isotope ratios of twelve GSJ rock reference samples. *Geochemical Journal*, **32**, 345–350.

PETTERSSON, Å., CORNELL, D. H. & YUHARA, M. 2004. Evidence of an Archaean component in the Bushmanland terrane – Sm–Nd whole rock model ages and U–Pb dating. *Geoscience Africa 2004, Abstract Volume*. University of Witwatersrand, Johannesburg, South Africa, 515.

PETTERSSON, Å., CORNELL, D. H., MOEN, H. F. G., REDDY, S. & EVANS, D. 2007. Ion-probe dating of 1.2 Ga collision and crustal architecture in the Namaqua-Natal province of southern Africa. *Precambrian Research*, **158**, 79–92.

RAITH, J. G., CORNELL, D. H., FRIMMEL, H. E. & DE BEER, C. H. 2003. New insights into the geology of the Namaqua tectonic province, South Africa, from ion probe dating of detrital and metamorphic zircon. *Journal of Geology*, **111**, 347–366.

REID, D. L. 1979. Total rock Rb–Sr and U–Th–Pb isotopic study of Precambrian metavolcanic rocks in the lower Orange River region, southern Africa. *Earth and Planetary Science Letters*, **42**, 368–378.

REID, D. L. 1982. Age relationships within the Vioolsdrif batholith, lower Orange River region: II. A two-stage emplacement history, and the extent of Kibaran overprinting. *Transactions of the Geological Society of South Africa*, **85**, 105–110.

REID, D. L. 1997. Sm–Nd age and REE geochemistry of Proterozoic arc-related rocks in the Richtersveld Subprovince, Namaqua Mobile Belt, southern Africa. *Journal of African Earth Sciences*, **24**, 621–633.

ROBB, L. J., ARMSTRONG, R. A. & WATERS, D. J. 1999. The history of granulite-facies metamorphism and crustal growth from single zircon U–Pb geochronology; Namaqualand, South Africa. *Journal of Petrology*, **40**, 1747–1770.

STOWE, C. W. 1986. Synthesis and interpretation of structures along the north–eastern boundary of the Namaqua tectonic province, South Africa. *Transactions of the Geological Society of South Africa*, **89**, 185–198.

THOMAS, R. J., CORNELL, D. H., MOORE, J. M. & JACOBS, J. 1994. Crustal evolution of the Namaqua-Natal metamorphic province, Southern Africa. *South African Journal of Geology*, **97**, 8–14.

TORSVIK, T. H., SMETHURST, M. A. *ET AL*. 1996. Continental break-up and collision in the Neoproterozoic and Palaeozoic – A tale of Baltica and Laurentia. *Earth Science Reviews*, **40**, 229–258.

VAN ASWEGEN, G., STRYDOM, D. *ET AL*. 1987. The structural-stratigraphic development of part of the Namaqua metamorphic complex, South Africa – an example of Proterozoic major thrust tectonics. *In*: KRÖNER, A. (ed.) *Proterozoic Lithospheric Evolution. Geodynamics Series, American Geophysics Union*, **17**, 207–216.

YUHARA, M., HAMAMOTO, T., KONDO, H., IKAWA, T., KAGAMI, H. & SHUTO, K. 2000. Rb, Sr, Sm and Nd concentrations of GSJ,KIGAM and BCR-1 rock reference samples analyzed by isotope dilution method. *Science Reports, Niigata University, Series E (Geology)*, **15**, 23–34.

YUHARA, M., KAGAMI, H. & TSUCHIYA, N. 2001. Rb–Sr and Sm–Nd systematics of granite and metamorphic rocks in the Namaqualand Metamorphic Complex, South Africa: implications for evolution of marginal part of the Kaapvaal craton. *National Institute of Polar Research, Special Issue, Memoirs*, **55**, 127–144.

Geodynamic evolution of the 2.25–2.0 Ga Palaeoproterozoic magmatic rocks in the Man-Leo Shield of the West African Craton. A model of subsidence of an oceanic plateau

M. LOMPO*

Department of Geology, University of Ouagadougou O1 BP 1973 Ouagadougou O1, Burkina Faso

Corresponding author (e-mail: lompo@univ-ouaga.bf)

Abstract: The West African Craton is defined by the presence of Archaean and Palaeoproterozoic (Birimian) rocks. Since 1990, researchers have published several papers proposing various regional scale accretionary models for the evolution of the Birimian series and the Eburnean orogeny. The Birimian lithostratigraphic succession starts with a major sequence of tholeiitic pillow basalts with intercalations of sediments, overlain by detrital and carbonate sediments, associated with calc-alkaline volcanics and plutons. A global scheme for the geodynamical evolution can be reasonably constrained using the characteristics of early tholeiitic plutono-volcanic and granitoid rocks. Geochemical, geochronological and structural characteristics of Palaeoproterozoic magmatic rocks of the Man-Leo Shield distinguish three tholeiitic (PTH1, PTH2, PTH3) and two granitic (PAG, PBG) series and date three principal events that have characterized the 2250–2000 Ma period: (1) Event I (2250–2200 Ma) characterized by tholeiitic volcanism with a widespread first tholeiite generation (PTH1), interpreted as the eruption of a mantle plume in an oceanic basin floor; (2) Event II (2200–2150 Ma) marked by second (PTH2) and third (PTH3) tholeiite generations and a calc-alkaline magma setting; high granitization and amphibole-bearing (PAG) granitoid emplacement with greenstone belt deformation. This event has been favoured by a crustal subsidence in a broad synclinorium mega-structure followed by vertical tectonics; (3) Event III (2150–2000 Ma) characterized by main biotite $\pm$ muscovite (without amphibole) bearing (PBG) granitoid emplacement in a context of transcurrent deformation. There are many similarities between West African Palaeoproterozoic rocks and Archaean formations and their crustal evolution also may be similar.

The West African Craton (Fig. 1) is made up of two main areas of Archaean and Palaeoproterozoic rocks: the Reguibat Shield in the north and the Man-Leo Shield in the south. Some isolated Palaeoproterozoic rocks are found between the two shields within Neoproterozoic formations that are widespread in the east of the craton. In the Man-Leo Shield, the Palaeoproterozoic comprises Birimian terranes (2.2–2.0 Ga) (Abouchami *et al.* 1990; Boher *et al.* 1992; Taylor *et al.* 1992; Hirdes *et al.* 1996), which predominantly comprise granitic batholiths, plutono-volcanic and meta-sedimentary greenstone belts. Geodynamic evolutionary models have been proposed in reference to two concepts: (1) modern tectonics, for example horizontal plate tectonics, and (2) vertical movements related to differential gravity of magmas such as those described for some Archaean crustal evolution. For West African Palaeoproterozoic rocks, geochemical data interpreted with respect to Phanerozoic tectonic discriminantary diagrams indicate variable settings for the same data in the same area. For instance, Leube *et al.* (1990) proposed a continental rift for the Palaeoproterozoic rocks in Ghana, whereas arc or subduction related contexts have been often proposed in Mali (Liégois *et al.* 1991), Ivory Coast (Vidal & Alric 1994), Niger (Ama Salah *et al.* 1996; Soumaila *et al.* 2004), Burkina Faso (Béziat *et al.* 2000), Senegal (Ndiaye *et al.* 1997; Diallo 2001) and Guinea (Egal *et al.* 2002). For the entire craton, Abouchami *et al.* (1990) and Boher *et al.* (1992) have proposed mantle plume derived magmas on oceanic plateaus to explain the early stage of Palaeoproterozoic tholeiitic series emplacement, followed by island arcs on the top of the oceanic plateaus that then collided with the Man Archaean craton. For the Ivory Coast, Vidal *et al.* (1996) proposed a geodynamic scheme where juvenile tholeiitic magmas form in an oceanic context, followed by gravitational deformation and transcurrent shearing, with granitoid pluton emplacement and vertical isostatic readjustment; they argued this model in respect to Archaean crustal development. In Burkina Faso, Béziat *et al.* (2000) in a study of ultramafic-mafic and associated volcanic rocks, suggested that the Palaeoproterozoic crust of the West African Craton was heterogeneous and not the consequence of a single process of genesis;

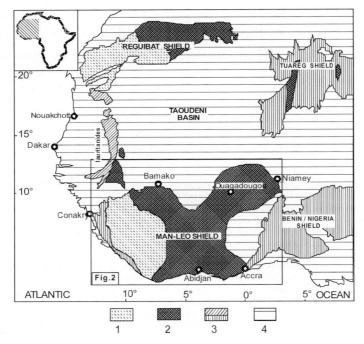

Fig. 1. West African Craton – Simplified geological map (adapted from fig. 1, in Boher *et al.* 1992 and fig. 2 in Ferré & Caby 2007) and location of Palaeoproterozoic rocks. 1, Archaean (3.5–2.6 Ga); 2, Palaeoproterozoic (2.25–2.0 Ga) greenstone and granitoids belts; 3, Hercynian and Pan African orogenic belts; 4, intracratonic Neoproterozoic basins and Phanerozoic cover.

they proposed both volcanic arc and oceanic plateau accretion. Nevertheless, isotopic data from all the West African Craton (Abouchami *et al.* 1990; Liégeois *et al.* 1991; Boher *et al.* 1992; Taylor *et al.* 1992; Cheilletz *et al.* 1994; Hirdes *et al.* 1996; Ama Salah *et al.* 1996; Ndiaye *et al.* 1997; Dia *et al.* 1997; Doumbia *et al.* 1998; Gasquet *et al.* 2003), indicate a juvenile source for the widespread Palaeoproterozoic volcanic and granitic rocks.

The present study uses geochemical analysis of mafic and granitic igneous rocks from Burkina Faso, considered in the light of published geochemical data from elsewhere in the Man-Leo Shield and their structural setting, to shed light on the apparently conflicting models for the evolution of the Birimian terranes and to propose a new model for their geodynamic evolution that critically considers development of uniformitarian plate-tectonic processes within the craton.

Geology

Palaeoproterozoic rocks of the Man-Leo Shield crop out in nine countries of West Africa (Fig. 2a): Burkina Faso, Ivory Coast, Ghana, Guinea, Liberia, Mali, Niger, Senegal and Togo. The general constitution shows plutono-volcanic and sedimentary rocks deformed and weakly metamorphosed in a regional greenschist facies event between widespread granitic batholiths sometimes displaying high-grade metamorphic contacts. In this study, these two principal units are termed Palaeoproterozoic greenstone belts and Palaeoproterozoic granitic rocks, which display a distinctive time evolution (Fig. 2b).

Palaeoproterozoic greenstone belts

Throughout the Man-Leo Shield, the Palaeoproterozoic (Birimian) greenstone belts are characterized by a lithostratigraphic succession comprising a lower major tholeiitic basaltic suite (pillowed basalts) with some intercalations of sedimentary rocks, overlain by predominant pelagic detrital and chemical sedimentary rocks (shales, sandstones, volcaniclastic and carbonate rocks) in association with calc-alkaline plutono-volcanic rocks (Bassot 1987; Leube *et al.* 1990; Abouchami *et al.* 1990; Sylvester & Attoh 1992; Feybesse & Milési 1994; Bossière *et al.* 1996; Hirdes *et al.* 1996; Pouclet *et al.* 1996). These formations are affected by greenschist facies metamorphism with amphibolite facies assemblages locally occurring in contact

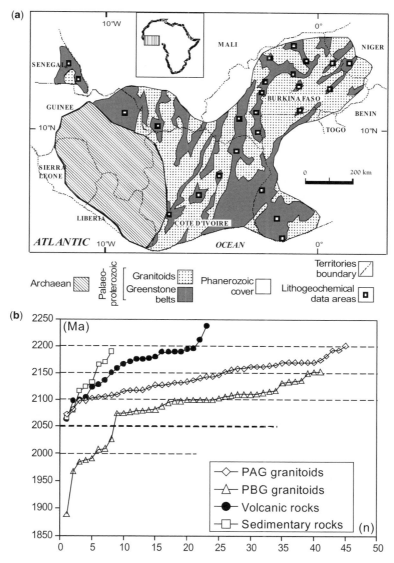

Fig. 2. (a) Simplified geological map of Man-Leo Shield and lithogeochemical data locations. (b) Time evolution of Palaeoproterozoic rocks. Data compiled from Dia (1988), Abouchami et al. (1990), Liégeois et al. (1991), Lompo (1991), Boher et al. (1992), Hirdes et al. (1992), Taylor et al. (1992), Diallo et al. (1993), Cheilletz et al. (1994), Davis et al. (1994), Ama Salah et al. (1996), Hirdes et al. (1996, 1998), Bossière et al. (1996), Ndiaye et al. (1997), Dia et al. (1997), Doumbia et al. (1998), Oberthur et al. (1998), Béziat et al. (2000), Egal et al. (2002), Hirdes & Davis (1998, 2002), Gasquet et al. (2003). PAG, Palaeoproterozoic amphibole-bearing granitoids types; PBG, Palaeoproterozoic biotite ± muscovite-bearing granitoids types; (n), number of samples from series.

with some granitoids. Available geochronological data (Fig. 2b) show that volcanic activities (lavas and pyroclastic rocks) and intrusive rocks (ultramafic, mafic, intermediary and acid rocks) were mainly produced between 2.25 and 2.10 Ga (data source: Abouchami et al. 1990; Liégeois et al. 1991; Lompo 1991; Taylor et al. 1992; Diallo et al. 1993; Ama Salah et al. 1996; Hirdes et al. 1996, 1998; Béziat et al. 2000; Hirdes & Davis 1998, 2002; Lahondière et al. 2002). The depositional age of sedimentary rocks ranges between 2.18 and 2.05 Ga (Vialette et al. 1990; Diallo et al. 1993; Davis et al. 1994; Bossière et al. 1996; Hirdes & Davis 2002).

Palaeoproterozoic granitoids and their emplacement mechanisms

Granitic rocks make up more than two-thirds of Palaeoproterozoic rocks in the shield and separate the linear greenstone belts. Several criteria have been used for the classification of these rocks: (i) structural and metamorphic considerations (Roques 1948; Arnould 1961; Ducellier 1963; Tagini 1971; Papon 1973; Hottin & Ouédraogo 1975; Bessoles 1977; Lemoine 1988; Vidal & Alric 1994); (ii) nature of the host rock, geochemistry and isotopic data considerations (Junner 1940; Leube et al. 1990; Hirdes et al. 1992, 1996; Doumbia et al. 1998; Gasquet et al. 2003).

All the Palaeoproterozoic granitic rocks of the Man-Leo Shield have a classic composition of tonalite–granodiorite–monzogranite, and a high content of magnetite and ilmenite (Lompo et al. 1995; Naba 1999; Naba et al. 2004). Using the mineral composition, these granitoids can be grouped into two main series, each with a distinctive time evolution (Fig. 2b), according to the presence or absence of amphibole (data source: Dia 1988; Lompo 1991; Boher et al. 1992; Hirdes et al. 1992; Cheilletz et al. 1994; Davis et al. 1994; Hirdes et al. 1996; Ndiaye et al. 1997; Dia et al. 1997; Doumbia et al. 1998; Oberthur et al. 1998; Egal et al. 2002; Hirdes & Davis 2002; Gasquet et al. 2003; Kagambega 2005): (1) at a macroscopic scale, amphibole bearing granitoids (PAG) are often layered or foliated (banded or gneissic texture). They are the oldest rocks (2.20–2.02 Ga) and constitute the widespread granitic batholiths of the shield; (2) the biotite $\pm$ muscovite-bearing granitoids without amphibole (PBG) series (2.15–1.95 Ga) rarely display visible anisotropies with elliptical or elongated shape and related dykes; they are intrusive in greenstone belts or in PAG granitoids.

The structure of the Palaeoproterozoic rocks in the Man-Leo Shield is dominated by a 'basin and dome' structure, overprinted by partitioned structures (Fig. 3a, b). The earliest structure is a regional NE–SW foliation with the global trends of greenstone belts and early granitic batholiths. An example of the 'basin and dome' map pattern is illustrated from the Boromo greenstone belt in Burkina Faso (Fig. 3b); field relationships are constrained by petrographic (Ducellier 1963; Napon 1988; Lompo 1991) and petro-structural analyses (Lompo 1991). Mylonitic deformation structures occur in north–south, NE–SW shear zones overprinting the regional foliation and ENE–WSW trending faults. Fieldwork on the shield show that granitic plutons have variable shapes, elliptical or circular, isolated or elongated, and are often embedded in larger batholiths (Milési et al. 1992; Pons et al. 1992, 1995; Lompo et al. 1995; Hirdes et al. 1996; Vidal et al. 1996; Doumbia et al. 1998; Caby et al. 2000; Egal et al. 2002; Gasquet et al. 2003; Castaing et al. 2003; Naba et al. 2004). Their setting is related to the different stages of the deformation.

Early deformation and related plutons. Early syn-metamorphic deformation is preserved in the greenstone belts by a sub-vertical schistosity developed across the entire shield: Senegal (Bertrand et al. 1989; Pons et al. 1992), Burkina Faso (Lompo et al. 1991; Lompo 1991), Niger (Pons et al. 1995), Ivory Coast (Vidal et al. 1996; Caby et al. 2000). This regional sub-vertical foliation is associated with a down-dip plunging lineation due to a general plane strain deformation. Structures in some areas of the shield indicate that shortening is >50%, synchronous with a steeply plunging to sub-horizontal stretching lineation (Lompo 1991; Pons et al. 1992, 1995; Caby et al. 2000). In the greenstone belts, the foliation is variably expressed because of the differences in material rheology (association of meta-volcanic and meta-sedimentary rocks). The regional foliation can display variable dips in some areas (Vidal 1987). The strike trajectory (Fig. 3a, b) is sinusoidal in accordance with the belt strike and pluton's shapes, or locally disturbed by the late deformation.

Emplacement mechanism studies of layered or banded early granitoids (PAG) have been carried out in some areas of the shield (Dupuis et al. 1991; Pons et al. 1992, 1995; Vidal et al. 1996; Caby et al. 2000). In the early granitoids (PAG), fabrics are S–L type, oriented in concordance with the schistosity in the greenstone host rocks, especially at the contact zones with high-grade metamorphic facies proving their syntectonic emplacement mechanism (Pons et al. 1995; Gasquet et al. 2003). According to these authors, sub-vertical stretching lineations are related to magma ascent during the granitoid emplacement and the gently plunging lineations are related to the main shear zones. Indeed, pressure and temperature $(P-T)$ conditions obtained in some widely-spaced areas (Fig. 3a) on the shield [south Ghana (John et al. 1999; Klemd et al. 2002), west Ivory Coast (Caby et al. 2000), north Burkina Faso (Debat et al. 2003) and east Burkina Faso (own observations)], give the same results at 5–6 kbar at 500–700 °C, similar to the conditions retrieved from the aureoles of the greenstone host rock as in the amphibole-bearing granitoids (PAG). These observations suggest that most of the PAG ganitoids are syntectonic intrusives in the greenstone terranes; the structural patterns of greenstone belts are indicative of a vertical tectonic regime resulting from the interference between the dynamics and thermal effects of pluton emplacement under the

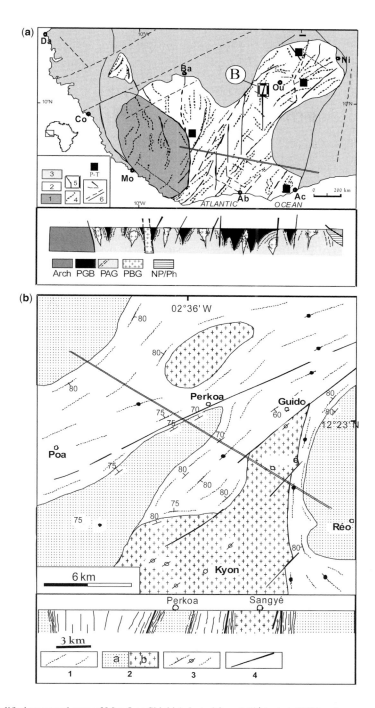

Fig. 3. (a) Simplified structural map of Man-Leo Shield (adapted from Milési *et al.* 1992) and cross-section location. 1, Archaean; 2, Palaeoproterozoic greenstone/granitoid belts; 3, Neoproterozoic and Phanerozoic cover; 4, regional foliation trajectories; 5, Mylonitic zones; 6, late and post–Palaeoproterozoic deformation; Ab, Abidjan; Ac, Accra; Ba, Bamako; Co, Conakry; Da, Dakar; Mo, Monrovia; Ni, Niger; Ou, Ouagadougou. Cross-section: Arch, Archaean; PGB, Palaeoproterozoic greenstone belts; PAG, Palaeoproterozoic amphibole-bearing granitoids types; PBG, Palaeoproterozoic biotite ± muscovite-bearing granitoids types; NP/Ph, Neoproterozoic/Phanerozoic cover; P–T, calculated pressure and temperature areas. (**b**) 'Dome and Basin'-like structures: example of field map in the Boromo greenstone belt [adapted from Ducellier (1963), Napon (1988) and Lompo (1991)]. 1, greenstone belt formation; 2, (a) PAG granitoids, (b) PBG granitoids; 3, main foliation trajectories with deep values; 4, mylonitic shear zones.

well-known NW–SE shortening that affected the entire shield (summarized in Gasquet et al. 2003).

Emplacement of the later granitoids. The well-described late granitoids PBG-type in Niger (Pons et al. 1995), Burkina Faso (Lompo et al. 1995; Naba et al. 2004; Vegas et al. 2008), Ivory Coast (Doumbia et al. 1998; Gasquet et al. 2003) and Senegal (Pons et al. 1992), are related to magma transport along fractures through a Palaeoproterozoic crust that deformed in a brittle nature. Some of the late granitoids have been emplaced along fractures: (i) parallel to the inferred regional NW–SE bulk shortening direction (Pons et al. 1995) or (ii) along NNE–SSW to NE–SW transpressional faults (Doumbia et al. 1998; Naba et al. 2004; Vegas et al. 2008). These works show that the PBG granitoids emplacement setting is a progressive transcurrent horizontal evolution from brittle-ductile to brittle conditions under the same strain pattern. The related structures (Fig. 3a) are sinistral north–south and dextral NE–SW shear zones.

Geochemistry

Methodology

Geochemical data are selected in respect to rock type and well-described lithologies (published and unpublished data) showing the least alteration. Palaeoproterozoic rocks in the greenstone belts are more or less affected by hydrothermal and/or meteoric water alteration. The selection of samples took into account the petrographic description, focusing on the less altered rocks and seeking a representative set of samples for each type of rock. Geochemical data of 215 samples with $SiO_2 = 35-65\%$ in the greenstone belts and 223 samples with $SiO_2 = 52-78\%$ in the granitoids, are from Burkina Faso, Ivory Coast, Ghana, Niger, Senegal and Guinea. In the tables, data are presented with averages but plots on diagrams are clusters of representative positions for Burkina Faso and average positions for the other areas in the Man-Leo Shield. Major (wt%) and trace elements used for rocks discrimination have been plotted but only the more significant ones are considered in this study. SiO_2, TiO_2, Al_2O_3, FeO ($Fe_2O_3^T$ or FeO^T), MgO, Ni, Ce and REE are used for the characterization of plutono-volcanic series in the greenstone belts and SiO_2, TiO_2, Al_2O_3, FeO ($Fe_2O_3^T$ or FeO^T), MgO, CaO, Na_2O, K_2O, Rb, Ba, Sr and REE for granitoids.

To discuss the geodynamical evolution of the Palaeoproterozoic magmatic rocks in the Man-Leo Shield, tholeiitic rocks in the greenstone belts and granitic rocks are taken in account because they are more represented and more studied. The tholeiitic series in the entire shield are the oldest magmatic rocks (Abouchami et al. 1990) and are more abundant in the greenstone belts (Bassot 1987; Zonou 1987; Leube et al. 1990; Abouchami et al. 1990; Sylvester & Attoh 1992; Pouclet et al. 1996; Ama Salah et al. 1996; Dia et al. 1997; Béziat et al. 2000; Diallo 2001; Soumaila et al. 2004). These series are the best records of the accretionary process during the critical period of post-Archaean transition. Calc-alkaline and ultramafic series are characterized using data from Burkina Faso (Béziat et al. 2000) and are not subject to a separate interpretation in this study because they can be included in the general process. The granitic series are grouped into two series according to amphibole presence, or not, and are considered as principal types of granitic rocks.

Greenstone belts magmatic series

Variation of the major elements (Table 1) allows the distinction of lavas and intrusive magmatic series (Fig. 4) in the Palaeoproterozoic greenstone belts of the Man-Leo Shield.

(1) Palaeoproterozoic tholeiitic series are essentially composed of massive basaltic lavas with pillow lavas and doleritic dykes (PTH) and gabbroic intrusive rocks (PTHI). They are characterized by a ratio $FeO/MgO = 1-5$, while $SiO_2 = 45-55\%$. The elements FeO (7–19%), MgO (3–10%) and Al_2O_3 (11–18%) have variable concentrations but REE concentrations clearly distinguish between three subgroups in the tholeiitic lavas (Fig. 5): (a) PTH1 lavas characterized by flat patterns (REE = 10–40 × chondrites); (b) PTH2 lavas with weakly depleted light elements (LREE = 4–10 × chondrites); and (c) PTH3 lavas with enriched light elements (LREE = 20–100 × chondrites). The PTHI intrusions (d) have more or less flat REE patterns (2–30 × chondrites) and often display Eu anomalies.

(2) Palaeoproterozoic calc-alkaline series (PCA) in the greenstone belts are represented by andesitic, dacitic and rhyolitic rocks (lavas, pyroclastic rocks and dykes). For $SiO_2 = 45-70\%$, $FeO/MgO = 0.3-3$. FeO (2–13%) varies inversely with SiO_2 but proportionally to MgO (0.1–15%). Al_2O_3 (6–18%) is inversely proportional to MgO. The REE variation (Fig. 5) shows that (e) calc-alkaline lavas (PCAL) are enriched in light REEs (LREE = 20–200 × chondrites), whereas (f) intrusive rocks (PCAI) display flat to weakly LREE-enriched patterns with Eu anomalies.

Table 1. Lithogeochemical data for representative Palaeoproterozoic plutono-volcanic rocks of Man-Leo Shield. Data compiled from Zonou (1987), Napon (1988), Leube et al.(1990), Abouchami et al. (1990), Lompo (1991), Ama Salah et al. (1991, 1996), Vidal et al. (1994), Béziat et al. (2000), Diallo (2001), Bumigeb (unpublished data). PTH1-PTH2-PTH3, Palaeoproterozoic tholeiitic lavas; PTHI, Palaeoproterozoic intrusive tholeiites; PCAL, Palaeoproterozoic calc-alkaline lavas; PCAI, Palaeoproterozoic calc-alkaline intrusive rocks; PUI, Palaeoproterozoic ultramafic intrusive rocks. (n), number of samples. West African Craton samples: Burkina Faso (50 PTH1, 25 PTH2, 18 PTH3); Ivory Coast (15 PTH1); Niger (7 PTH2, 7 PTH3); Ghana (33 PTH3); Senegal (8 PTH2). $FeO = Fe_2O_3^T$ or FeO^T.

Samples (n)	WEST AFRICAN CARTON								BURKINA FASO					
	PTH1		PTH2		PTH3		PTHI		PCAL		PCAI		PUI	
	65	(+/−)	40	(+/−)	58	(+/−)	21	(+/−)	11	(+/−)	13	(+/−)	7	(+/−)
SiO$_2$ (%)	49.95	1.82	49.25	1.62	52.07	1.82	49.34	1.08	56.53	4.28	52.92	3.07	41.46	3.90
TiO$_2$ (%)	1.14	0.36	0.99	0.33	1.08	0.28	0.86	0.38	0.71	0.21	0.80	0.75	0.19	0.12
Al$_2$O$_3$ (%)	13.89	0.94	14.08	0.96	14.59	1.37	15.05	1.65	14.17	0.88	13.15	3.74	6.48	3.44
Fe$_2$O$_3$ (%)	13.80	1.83	14.65	2.40	11.45	1.29	12.49	2.16	8.55	2.52	8.42	2.78	10.43	0.99
MgO (%)	5.85	1.06	6.76	1.41	5.72	1.42	6.96	1.25	5.47	2.58	8.84	3.67	26.42	7.49
Ni (ppm)	151.16	147.86	335.86	234.43	161.01	136.70	159.52	120.79	211.26	221.11	170.11	80.76	790.47	369.87
Ce (ppm)	11.24	3.97	4.68	1.09	26.11	12.07	8.48	4.96	29.93	8.24	23.49	20.64	4.84	4.66
La 139	4.85	1.80	1.86	0.45	12.32	5.81	3.65	2.16	16.12	10.36	10.80	10.90	2.34	2.10
Ce 140	11.24	3.97	4.68	1.09	26.11	12.07	8.82	5.04	28.10	12.00	23.57	21.72	4.84	4.66
Pr 141	1.75	0.62	0.85	0.43	3.54	1.43	1.30	0.73	3.99	2.02	3.08	2.62	0.67	0.61
Nd 145	8.87	2.82	4.87	1.18	15.79	6.12	6.49	3.47	16.57	7.38	13.37	10.15	3.04	2.87
Sm 147	2.71	0.84	2.03	0.67	3.62	1.05	1.83	0.84	3.51	1.69	3.21	2.30	0.72	0.64
Eu 151	0.98	0.25	0.78	0.16	1.10	0.28	0.78	0.28	0.95	0.31	0.88	0.36	0.25	0.21
Gd 157	3.56	0.97	2.96	0.77	3.75	0.90	2.42	1.02	3.47	1.94	3.55	2.60	0.75	0.53
Tb 159	0.64	0.17	0.58	0.15	0.62	0.15	0.41	0.17	0.52	0.25	0.61	0.48	0.13	0.08
Dy 161	4.18	1.07	4.05	1.16	3.74	0.87	2.81	1.05	3.21	1.63	3.85	3.13	0.88	0.55
Ho 165	0.93	0.22	0.89	0.28	0.78	0.18	0.61	0.22	0.68	0.34	0.85	0.71	0.19	0.10
Er 166	2.74	0.67	2.70	0.77	2.30	0.56	1.85	0.62	1.93	0.99	2.42	2.06	0.57	0.30
Tm 169	0.41	0.11	0.44	0.13	0.33	0.08	0.27	0.09	0.28	0.13	0.34	0.28	0.08	0.04
Yb 174	2.64	0.60	2.63	0.84	2.17	0.56	1.80	0.59	1.78	0.73	2.25	1.80	0.55	0.27
Lu 175	0.41	0.09	0.42	0.13	0.32	0.08	0.27	0.08	0.28	0.12	0.34	0.28	0.09	0.04

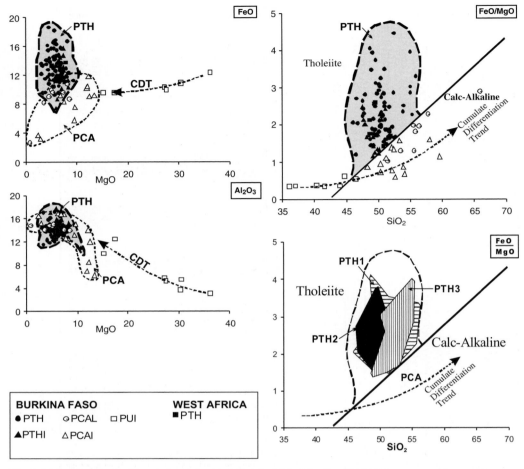

Fig. 4. Major elements (wt%) variation and magmatic series of the Man-Leo Shield Palaeoproterozoic greenstone rocks. FeO/MgO v. SiO$_2$ plot (Miyashiro 1974). PTH, Palaeoproterozoic tholeiitic series and characteristic fields of PTH1, PTH2 and PTH3; PTHI, Palaeoproterozoic intrusive tholeiites; PCA, Palaeoproterozoic calc-alkaline series with PCAL (lavas), PCAI (intrusive rocks); PUI, Palaeoproterozoic ultramafic intrusive rocks; CDT, cumulate differentiation trend.

(3) Palaeoproterozoic ultramafic intrusive rocks (PUI) are represented in this study by werhlite and dunite plutons associated with pyroxene cumulates (Béziat et al. 2000). They are characterized by high-grade MgO (15–38%) for FeO = 10–13% and SiO$_2$ = 35–45%. Al$_2$O$_3$ (3–13%) is inversely proportional to MgO. REE patterns (Fig. 5d) are weakly LREE-enriched.

Major elements concentrations suggest an evolution of a magmatic suite with tholeiitic, ultramafic and calc-alkaline rocks differentiation. In that magmatic suite, PTH1, PTH2 and PTH3 series have specific REE characterization and are more or less distinct according to their relative content in SiO$_2$, Al$_2$O$_3$, FeO and MgO. Each subgroup can be represented by a field on an FeO/MgO v. SiO$_2$ diagram (Fig. 4). For the same ratio FeO/MgO, the PTH3 subgroup has higher SiO$_2$ values (50% < SiO$_2$ > 55%) than PTH2 (45% < SiO$_2$ > 50%), whereas PTH1 is represented in the entire tholeiites (45% < SiO$_2$ > 55%). This distinction, through major elements, is in accordance with REE variation.

In respect to REE patterns, tholeiites intrusive rocks (PTHI) are close to PTH1 lavas. Calc-alkaline lavas (PCAL) and intrusive rocks (PCAI) and (PUI) are similar to PTH3, except that PUI are less enriched in REE than PTH3. In respect to

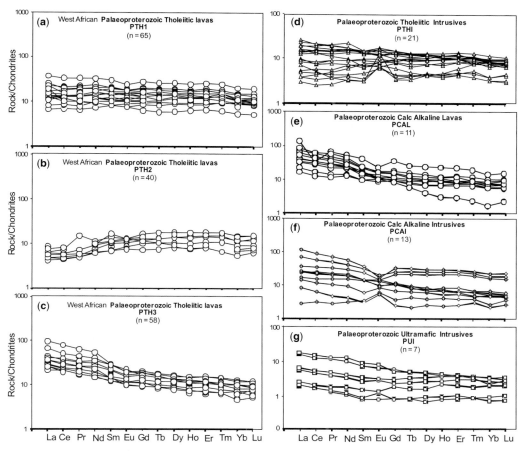

Fig. 5. Chondrite normalized rare earth element (REE) plots (Sun & McDonough 1989) for the Man-Leo Palaeoproterozoic greenstone rocks. REE patterns of plutono-volcanic series. (**a**) PTH1; (**b**) PTH2; (**c**) PTH3 Palaeoproterozoic tholeiitic lavas; (**d**) PTHI Palaeoproterozoic intrusive tholeiites; (**e**) PCAL Palaeoproterozoic calc-alkaline lavas; (**f**) PCAI Palaeoproterozoic calc-alkaline intrusive rocks; (**g**) PUI Palaeoproterozoic ultramafic intrusive rocks; (n), number of samples.

Miyashiro's (1974) discriminant boundary (Fig. 4), Béziat *et al.* (2000) showed that the ultramafic rocks (PUI) used in this work, according to their cumulative nature and the association with differentiated tholeiitic gabbros, appear linked to the calc-alkaline suite.

Geotectonic context of tholeiite emplacement. The tectonic context of the emplacement of PTH1, PTH2 and PTH3 is discussed in comparison between REE profiles and Al_2O_3, FeO contents (Fig. 6): (a) PTH1 with flat REE profiles are comparable to oceanic plateau basalts and Al_2O_3 content is similar to oceanic island basalts (OIB), although they are higher in FeO content; (b) PTH2, with weakly LREE-depleted profiles, are comparable to oceanic NMORB basalts but differ by having lower Al_2O_3 and higher FeO contents; (c) PTH3, with LREE-enriched profiles and higher Al_2O_3 content, are comparable to arc tholeiites but they have higher FeO contents; (d) in comparison to REE profiles of modern Pacific Ocean basalts (Storey *et al.* 1991) PTH1 and PTH2 are similar to rocks from the Nauru Basin (1 & 3), Ontong-Java (2) and Manihiki (4) oceanic plateau basalts; (e) in general, the Palaeoproterozoic tholeiites (PTH) have Al_2O_3 contents similar to OIB and some Archaean tholeiites from Canada and Australia; FeO content of PTH is more than OIB but less than Archaean tholeiites. In respect to Phanerozoic tectonic discriminant diagrams (Fig. 7) of Sun & McDonough (1989) and Jochum *et al.* (1990), Palaeoproterozoic tholeiites (PTH) have a similar evolution as Archaean tholeiites. According to

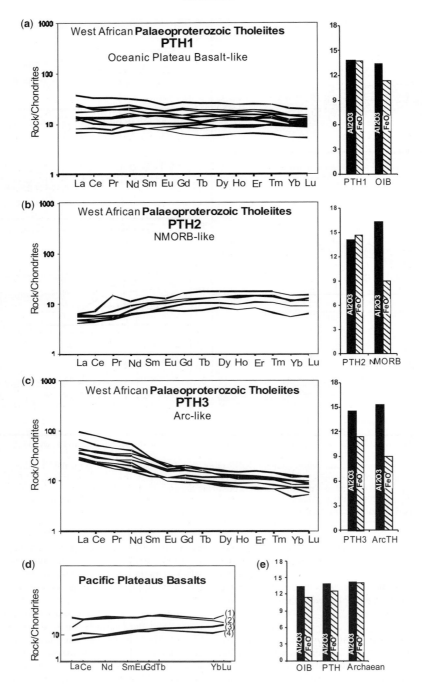

Fig. 6. Chondrite normalized rare earth elements (REE) patterns (Sun & McDonough 1989) and values of major elements (Al_2O_3, FeO) of the Palaeoproterozoic tholeiites (PTH) sub-groups. Comparison with Phanerozoic geotectonic settings. (**a**) PTH1 with flat REE profiles is like oceanic plateau basalts or OIB (oceanic island basalts); (**b**) PTH2 with weakly LREE-depleted REE profiles is like NMORB (normal mid-ocean ridge basalts); (**c**) PTH3 with weakly LREE-enriched REE profiles is like arc basalts; (**d**) REE profiles of the Pacific plateau basalts (Storey *et al.* 1991): (1) & (3) Nauru Basin; (2) Ontong-Java; (4) Manihiki Plateau; (**e**) Al_2O_3 and FeO values in comparison between PTH, Archaean and OIB tholeiites.

SiO$_2$ content, PTH are MORB-like but differ by having higher Ni and FeO content. Their lower contents of incompatible (Ce) and compatible (Ti) elements make PTH closer to island-arc volcanic rocks (IAV) but IAV have very low FeO and Ni content. Al$_2$O$_3$, FeO and Ni of PTH plots are common in the OIB field and suggest that the OIB geodynamic context is the best match for the PTH.

The three subgroups of tholeiites present three REE patterns, suggesting three different contexts of geodynamic setting. The principal characters of PTH with high FeO and lower Al$_2$O$_3$ content make them closer to modern oceanic island basalts (OIB), and REE profiles of the main tholeiites (PTH1 and PTH2) are similar to modern oceanic plateau basalts. Thus it is possible to imagine an oceanic basin setting for the primary PTH emplacement confirming the first stage of tholeiites emplacement proposed by Abouchami et al. (1990) and Boher et al. (1992). However, some of the tholeiites, especially PTH3, with lower FeO and REE patterns similar to modern arc tholeiites (Fig. 6c), may have a different tectonic context of emplacement and therefore should be related to the next stages of dynamic evolution discussed below.

Magmatic source of tholeiites. In the Al$_2$O$_3$ v. Mg# diagram (Fig. 7), Archaean rocks and modern OIB show two trends; according to Arndt (1998), the first trend of Al$_2$O$_3$ increasing with Mg# decreasing is characteristic of magnesium lavas, and the second trend of constant to decreasing Al$_2$O$_3$ with decreasing Mg# is characteristic of more highly developed magmas. PTH rocks are in accordance with the second trend and can be interpreted as magma fractionates of a plagioclase-bearing assemblage. According to the experimental work of Hirose & Kushiro (1993) on the melting rate of mantle peridotites, the depth of melting can be related to Al$_2$O$_3$ content. Mantle melting at low pressure produces Al$_2$O$_3$-rich magmas while melting at deeper levels results in Al$_2$O$_3$-poor melts. The experimental work of Langmuir & Hanson (1980) also shows that FeO content is directly related to depth of partial melting of the mantle. With the highest FeO content in the Palaeoproterozoic tholeiites (PTH1 and PTH2) with low Al$_2$O$_3$ content with respect to modern basalts, suggests that the main PTH are derived from partial melting from a deep source in the mantle similar to modern oceanic island basalts or Archaean tholeiites.

PTH3 displays higher content of Al$_2$O$_3$ and lower FeO content than other Palaeoproterozoic tholeiites; it is suggested that this magma derives from partial melting at a higher level in the mantle. On the other hand, their characteristics are close to island arc basalts (Fig. 6c) which suggests that the melt source could be hydrated and probably metasomatized. In a modern context, this can be explained by a subduction system where the top of the plate (in contact with seawater) participates in the melting. However, isotopic data from Abouchami et al. (1990), Taylor et al. (1992), Cheilletz et al. (1994), Hirdes et al. (1996), Ama Salah et al. (1996) and Dia et al. (1997), suggest a juvenile source for the Palaeoproterozoic tholeiites of the West African Craton. PTH3 could be derived from the partial melting of the top of the mantle which may be hydrated and metasomatized.

Béziat et al. (2000) undertook a petrogenetic study of mafic-ultramafic and calc-alkaline rocks, where it was possible to establish field relationships and mineral analysis, and concluded that the cumulate ultramafic rocks (PUI) and related gabbros used in this work, are derived from fractional crystallization of PTH-like basic magma.

Granitoids series and petrogenesis

Palaeoproterozoic granitoids of the Man-Leo Shield comprise two types distinguished by mineral composition, principally by the presence (PAG) or absence (PBG) of amphibole. This distinction is also outlined by the variation of chemical major and trace (REE) elements (Table 2). In the Q-A-P diagram (Fig. 8a), PAG (amphibole-bearing granitoids) with SiO$_2$ = 60-70% comprises diorites, tonalities and granodiorites. In the K$_2$O/Na$_2$O diagram (Fig. 8b), their composition is mainly tonalite/trondhjemite and granodiorite. PBG (biotite $\pm$ muscovite-bearing granitoids) with SiO$_2$ = 70-75% comprising granodiorite and monzogranite (Fig. 8a), or simply granite, granodiorite and adamellite (Fig. 8b). An oxide v. SiO$_2$ diagram (Fig. 9a) shows that PAG and PBG constitute two magmatic suites with constant Al$_2$O$_3$ (c. 15%), whilst the other elements (TiO$_2$ c. 0.5%; CaO c. 4%; Fe$_2$O$_3$ c. 4% and MgO c. 2%) in the PAG, and (TiO$_2$ < 0.3%; CaO c. 1.5%; FeO c. 2% and MgO c. 0.5%) in the PBG, are inversely proportional to SiO$_2$. REE (Fig. 10) are weakly fractionated (2 < La$_N$/Yb$_N$ > 40) in the PAG, whereas they are highly fractionated in the PBG (3 < La$_N$/Yb$_N$ > 165).

According to major chemical elements (Fig. 8a, b), Palaeoproterozoic granitoids (PAG and PBG) of the Man-Leo Shield constitute a medium potassic, calc-alkaline to calcic magmatic suite with PBG purely calcic. For the same Na$_2$O content, PBG ganitoids are more K$_2$O-rich. The two series are clearly distinguished by their isotopic age evolution (Fig. 2b) and the fractionation of REE.

According to the ratio La$_N$/Yb$_N$ v. Yb$_N$ (Fig. 11a), the Palaeoproterozoic granitoids (PAG and PBG) have similar characters as Archaean TTG (Tonalite-Trondhjemite-Granodiorite), with

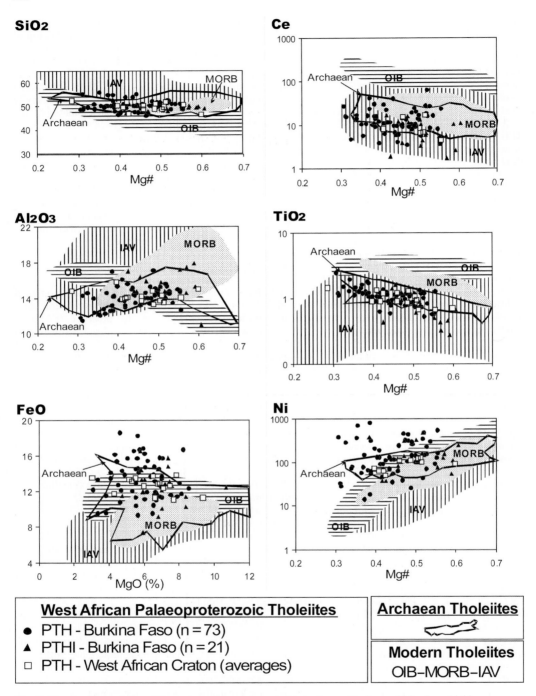

Fig. 7. Comparative variation of the major and trace elements for Archaean, Modern and Man-Leo shield Palaeoproterozoic tholeiites (PTH). Phanerozoic geotectonic fields of mid-ocean ridge basalts (MORB), oceanic island basalts (OIB) and intra-arc volcanic rocks (IAV) defined by Sun & McDonough (1989) and Jochum *et al.* (1990); Archaean tholeiites field is defined from Arndt (1998) compilation. Palaeoproterozoic tholeiitic lavas (PTH) and tholeiitic intrusive rocks (PTHI) from Burkina Faso and representative Palaeoproterozoic tholeiitic lavas (PTH) from different areas on the West African craton (Ivory Coast, Ghana, Niger, Senegal).

Table 2. Lithogeochemical data for representative Palaeoproterozoic granitic rocks of Man-Leo Shield. Data compiled from Lompo (1991), Ndiaye et al. (1997), Doumbia et al. (1998), John et al. (1999), Egal et al. (2002), Bumigeb (unpublished data). PAG, Palaeoproterozoic amphibole-bearing granitoids; averages data are from Burkina Faso (BF) (n = 84), Ghana (n = 2), Ivory Coast (n = 2), Senegal (n = 7) and Guinea (n = 10). PBG, biotite ± muscovite-bearing (without amphibole) granitoids; averages data are from Burkina Faso (BF) (n = 103), Ivory Coast (n = 2), Senegal (n = 5) and Guinea (n = 2). $FeO = Fe_2O_3^T$ or FeO^T.

Area	PAG										PBG							
	BF		GHANA		I.C.		SENEGAL		GUINEE		BF		I.C.		SENEGAL		GUINEE	
Samples (n)	84	(+/−)	2	(+/−)	4	(+/−)	7	(+/−)	10	(+/−)	103	(+/−)	2	(+/−)	5	(+/−)	2	(+/−)
SiO_2 (%)	66.93	5.70	68.44	0.30	66.01	3.93	63.71	3.69	63.30		72.60	3.54	71.58	3.62	71.80	1.16	71.85	3.04
TiO_2 (%)	0.46	0.22	0.34	0.06	0.50	0.24	0.64	0.14	0.66		0.27	0.22	0.24	0.18	0.28	0.14	0.25	0.12
Al_2O_3 (%)	14.70	1.49	14.65	0.54	15.32	1.74	15.43	0.98	15.80		14.09	1.41	15.21	1.07	14.70	0.35	14.45	1.20
Fe_2O_3 (%)	4.34	2.04	3.53	0.13	3.39	1.59	5.00	1.29	5.41		2.00	1.16	1.66	1.15	1.63	0.66	2.20	0.77
MgO (%)	2.08	1.55	1.81	0.40	1.52	0.90	2.72	0.91	2.20		0.55	0.50	0.83	0.91	0.40	0.21	0.55	0.21
CaO (%)	3.85	1.90	3.62	0.04	3.57	1.17	3.54	1.15	4.20		1.74	0.90	1.44	0.91	1.04	0.53	1.30	0.57
Na_2O (%)	4.02	0.84	4.28	0.42	4.15	0.41	3.91	0.47	4.50		3.96	0.74	4.12	0.68	3.62	0.30	3.95	0.78
K_2O (%)	2.28	1.12	2.19	0.42	2.22	0.88	3.12	0.81	2.66		3.75	1.38	3.24	1.84	5.14	0.50	4.93	0.62
Rb (ppm)	73.32	39.03	61.00	15.56	83.80	43.11	222.71	105.91	92.00		140.47	67.06	151.25	117.26	262.20	65.82	190.50	40.31
Ba (ppm)	815.10	442.13	710.00	62.23	934.75	268.02	620.29	108.42	1205.00		854.40	530.68	522.33	34.88	601.80	264.23	1213.00	528.92
Sr (ppm)	656.86	391.86	602.50	12.02	791.75	216.20	463.43	82.92	703.00		390.39	301.99	315.17	260.92	185.20	113.59	290.00	210.72
La 139	27.85	19.31	18.50	10.61	28.74	13.05	37.58	11.29	46.70		35.69	25.93	12.45	3.56	40.44	17.46	65.15	9.69
Yb 174	1.60	1.36	0.80	0.00	1.72	1.25	1.30	0.20	1.20		1.03	1.16	0.49	0.23	0.39	0.09	0.85	0.07

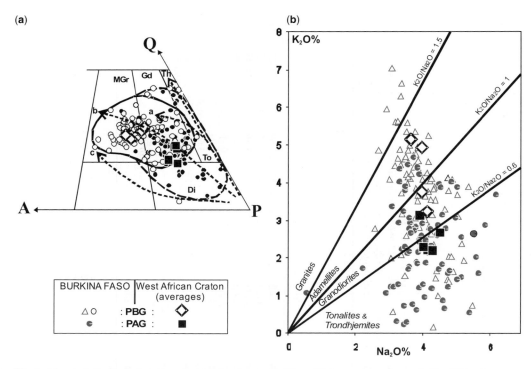

Fig. 8. Chemical discrimination diagrams showing the composition of Palaeoproterozoic granitoids of Man-Leo Shield. (**a**) Q–A–P diagram with trends of the evolution of classic magmatic series (Lameyre & Bowden 1982). Th, tholeiitic series; a, K$_2$O-poor calc-alkaline series; b, medium K$_2$O calc-alkaline series and c, K$_2$O-rich calc-alkaline series. Di, diorite; To, tonalite; Gd, granodiorite; MGr: monzogranite. (**b**) K$_2$O v. Na$_2$O diagram (Harpum 1963). PAG: Palaeoproterozoic amphibole-bearing granitoids; PBG: Palaeoproterozoic biotite ± muscovite-bearing granitoids; averages data plots for the West African Craton are from Burkina Faso ($n = 84$), Ghana ($n = 2$), Ivory Coast ($n = 4$), Senegal ($n = 7$) and Guinea ($n = 10$). PBG: biotite ± muscovite-bearing (without amphibole) granitoids; averages data plots for the West African Craton are from Burkina Faso ($n = 103$), Ivory Coast ($n = 2$), Senegal ($n = 5$) and Guinea ($n = 2$).

a high fractionation of REE (La$_N$/Yb$_N$ = 3–160) indicating an enrichment of LREE and depletion of HREE. For Jahn (1998), these characteristics suggest fractionation of plagioclase ± hornblende ± pyroxene during fractional crystallization or partial melting at low pressure (≤ 5 kbar). Some geochemical experiments indicate that Archaean TTG are produced from partial melting of garnet-bearing amphibolite or quartz eclogite with a residue dominated by amphibole and/or garnet-bearing (Arth & Hanson 1975; Barker & Arth 1976; Glikson 1979; Jahn et al. 1981; Martin et al. 1983; Jahn & Zhang 1984; Condie 1986; Shirey & Hanson 1986; Drummond & Defant 1990). In the Sun & McDonough (1989) diagram (Fig. 11a), REE plots of PAG and PBG are in accordance with curves of liquid composition from eclogite or 25% garnet-bearing melt. Variation of Rb, Sr and Ba is compatible with partial melting followed by fractionated crystallization for PAG and PBG; in the Rb/Sr v. Rb and Ba v. Sr diagram of Martin (1985) (Fig. 11b, c), PAG are preferentially in the field of partial melting (PM) of eclogite or mantle, whereas PBG are mainly in the field of fractionated crystallization (FC).

According to experimental data, Palaeoproterozoic granitoids of the Man-Leo Shield derive from partial melting of eclogite or mantle and produce amphibole ± magnetite ± ilmenite-bearing residues (PAG) and crystallization of PBG without amphibole. Geochronology ages indicate that PBG granitoids are younger than PAG′ and studies on PBG-type (Lompo et al. 1995; Hirdes et al. 1996; Doumbia et al. 1998; Naba et al. 2004) show that late PBG emplaced in the brittle crust, are derived from partial melting of PAG. Experimental data (above) showed also that partial melting did not exceed 20%,<900°C and <8 kbar conditions for these types of rock. In respect to Drummond & Defant's (1990) REE discrimination diagram

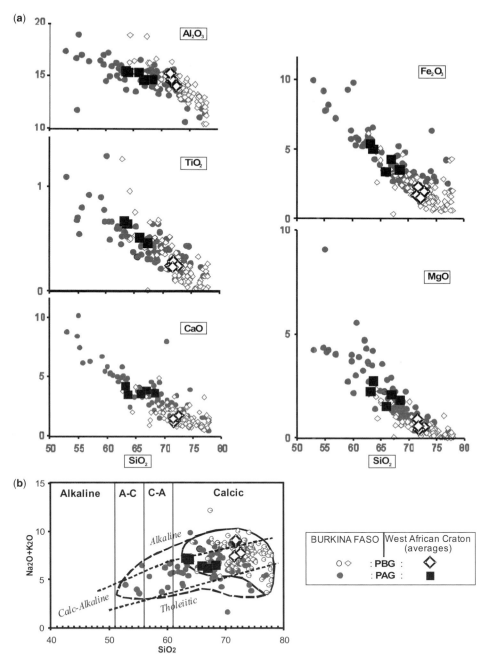

Fig. 9. Chemical discrimination diagrams of the magmatic suites of the Palaeoproterozoic granitoids in the Man-Leo Shield. (a) Oxides v. SiO_2 diagrams showing a single trend for PAG (Palaeoproterozoic amphibole-bearing granitoids) and PBG (Palaeoproterozoic biotite $\pm$ muscovite-bearing granitoids) series. (b) $Na_2O + K_2O$ v. SiO_2 diagram with chemical character and magmatic suites distinction (MacDonald & Katsura 1964). PAG averages data plots for the West African Craton are from Burkina Faso ($n = 84$), Ghana ($n = 2$), Ivory Coast ($n = 4$), Senegal ($n = 7$) and Guinea ($n = 10$). PBG averages data plot for the West African Craton are from Burkina Faso ($n = 103$), Ivory Coast ($n = 2$), Senegal ($n = 5$) and Guinea ($n = 2$).

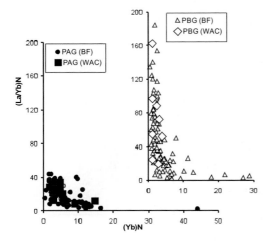

Fig. 10. La$_N$/Yb$_N$ v. Yb$_N$ showing REE fractionation and distinction of the two granitic series (PAG, PBG) of Palaeoproterozoic granitoids of the Man-Leo shield. PAG, Palaeoproterozoic amphibole-bearing granitoids; averages data plots for the West African Craton (WAC) are from Burkina Faso (BF) ($n = 84$), Ghana ($n = 2$), Ivory Coast ($n = 4$), Senegal ($n = 7$) and Guinea ($n = 10$). PBG, biotite $\pm$ muscovite-bearing (without amphibole) granitoids; averages data plots for the WAC are from Burkina Faso (BF) ($n = 103$), Ivory Coast ($n = 2$), Senegal ($n = 5$) and Guinea ($n = 2$).

(Fig. 11a), PAG and PBG magmas are different from classic island arc magmas. Isotopic data from Liégeois *et al.* (1991), Boher *et al.* (1992), Taylor *et al.* (1992), Cheilletz *et al.* (1994), Hirdes *et al.* (1996), Ndiaye *et al.* (1997), Dia *et al.* (1997), Doumbia *et al.* (1998) and Gasquet *et al.* (2003), allow us to conclude that the earliest of these granitoids are mantle source derived.

A model of geodynamic evolution

The widespread Palaeoproterozoic magmatic rocks that are common throughout the entire Man-Leo Shield are characterized by a predominance of tholeiitic magmas and abundance of layered or foliated granitoids. Greenstone belts are straightened and have a schistose fabric imposed on them during early granitoid emplacement. Tholeiites frequently display flat REE patterns or have enriched LREEs, low Al$_2$O$_3$ and high FeO content in comparison with actual tholeiites. They also have high SiO$_2$ and low Ce and Ti, making them similar to island arc volcanic rocks. Granitoids have highly fractionated REEs (La/Yb)N with low values of (Yb)N. εNd is positive (c. +2) (Dupré *et al.* 1984; Shirey & Hanson 1986; Machado *et al.* 1986; Abouchami *et al.* 1990; Boher *et al.* 1992) and suggests evolution of a juvenile crust. Some Archaean granite-greenstone belts comprise petrographical and geochemical similarities with Palaeoproterozoic rocks of the Man-Leo Shield and are thought to have been formed by mantle plume processes (Dostal & Mueller 1997; Hollings & Wyman 1999; Hollings *et al.* 1999; Wyman 1999; Tomlinson & Condie 2001; Wyman *et al.* 2002; Sandeman *et al.* 2006). In respect to these authors, tholeiites with flat REE patterns and ultramafic rocks (Archaean komatiites), are commonly related to mantle plumes; even associated basalts with oceanic island and oceanic arc basaltic characteristics, are termed arc plume associated greenstone belts. This study shows that the Palaeo-proterozoic (Birimian) rocks of the Man-Leo Shield show similarities with Archaean terranes and lack the features that typify Phanerozoic subduction settings. These Palaeoproterozoic rocks may therefore be formed by processes similar to those which generated the Archaean rocks.

Subduction or arc models proposed for the Palaeoproterozoic rocks of the Man-Leo Shield are not well constrained for many reasons: (i) according to chemical data, there is no contribution of older rocks in early Palaeoproterozoic tholeiitic magmas (Abouchami *et al.* 1990; Boher *et al.* 1992); (ii) according to field relationships, there is no expressed metamorphic zonation across the belts, and there are no lithospheric-scale thrusts described anywhere in the entire shield. Instead, the Palaeoproterozoic belts of the Man-Leo Shield attest to homogeneous metamorphic conditions along horizontal profiles through the shield and trace out 'dome and basin' geometries with extensive domains of flat-lying fabrics with stretching lineations and vertical shear zones; these structural and metamorphic patterns have been observed in many Archaean and Palaeoproterozoic belts (Cagnard *et al.* 2007), suggesting the particularity of deformation modes of weak lithospheres under compression. The spatial extent of these rocks (>500 km in a NW–SE cross-section on the shield), after considerable tectonic flattening, argues against the subduction of a unique plate, producing magmas with the same characters across the entire shield, or the juxtaposition of many small subducted plates similar to Phanerozoic tectonics.

The first stage of accretion can be considered as mantle plume evolution in an oceanic basin context, in accordance with Aboucahmi *et al.* (1990) and Boher *et al.* (1992). On the other hand, the next stages producing late tholeiites (PTH3) and calc-alkaline (PCA) magmas with arc-related signatures, appear contrary to the rheological and structural framework described above. I suggest that similar to the Archaean evolution, the lithosphere

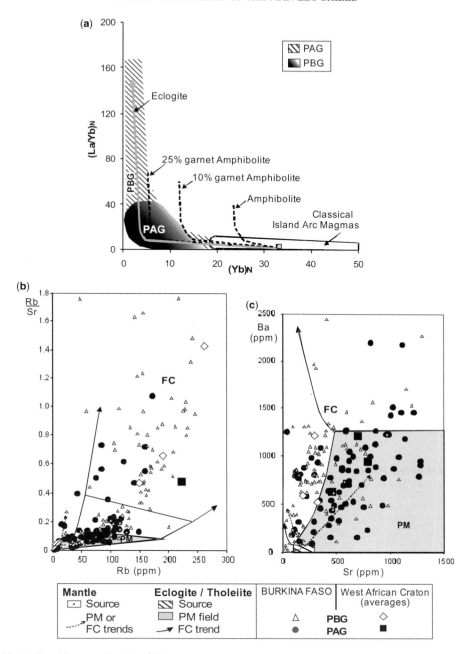

Fig. 11. (a) Chondrite normalized La_N/Yb_N v. Yb_N (Sun & McDonough 1989) plot for Palaeoproterozoic granitoids in the Man-Leo Shield. Equilibrium melting curves with amphibolitic, garnet amphibolitic and eclogitic mineral residues. Classical island arc magmas evolved via fractional crystallization (from Drummond & Defant 1990). (b) Rb/Sr v. Rb and (c) Ba v. Sr diagrams (Martin 1985) showing the position of Palaeoproterozoic granitoids of the Man-Leo Shield in relation to partial melting (PM) and fractional crystallization (FC) trends. Source is from mantle or eclogite/tholeiitic basalts composition. PAG, Palaeoproterozoic amphibole-bearing granitoids; averages data plots for the West African craton (WAC) are from Burkina Faso (BF) ($n = 84$), Ghana ($n = 2$), Ivory Coast ($n = 4$), Senegal ($n = 7$) and Guinea ($n = 10$). PBG, Palaeoproterozoic biotite $\pm$ muscovite-bearing (without amphibole) granitoids; averages data plots for the West African Craton (WAC) are from Burkina Faso (BF) ($n = 103$), Ivory Coast ($n = 2$), Senegal ($n = 5$) and Guinea ($n = 2$).

during the Palaeoproterozoic behaved in a highly ductile manner in the Man-Leo Shield under regional NW–SE shortening providing widespread PAG granitoids (Pons et al. 1995; Vidal et al. 1996; Gasquet et al. 2003). This resulted in widespread folding rather than faulting. So subduction models cannot easily explain vertical tectonics and foundering of oceanic crust, nor in the first stage of accretion (Ama Salah et al. 1996), nor in the next stages with crustal thickening (Boher et al. 1992).

The geodynamic model proposed in this chapter emphasizes the latter stages of magmatism that followed the emplacement of tholeiites. It is based on the ductile behaviour of the lithosphere at this time, which as it was shortened, foundered by subsidence processes instead of breaking and sinking into the asthenosphere as a rigid plate by subduction processes.

Geodynamic evolution of the Palaeoproterozoic magmatic rocks

The geochemical, geochronological and structural characteristics of the Palaeoproterozoic magmatic rocks, allow the timing of the principal magmatic events to be worked out. The subdivision is not strict, but outlines the periods of intensive activity; according to these dates, there are both syn- and post-volcanic granitoid emplacement. Three principal magmatic events (Fig. 12) have characterized the 2250–2000 Ma period: (a) Event I (2250–2200 Ma) is characterized by tholeiitic volcanism with a widespread PTH setting by predominant mantle plume evolution; (b) Event II (2200–2150 Ma) is marked by PAG granitoid emplacement with greenstone belt deformation in a calc-alkaline magma setting; vertical tectonics characterize this event; (c) Event III (2150–2000 Ma) is characterized by PBG granitoid emplacement in a context of transcurrent movements.

These three events are interpreted as follows (Fig. 13):

(a) 2300–2250 Ma. Palaeoproterozoic oceanic basin floor – 'Oceanic lithosphere?'. Studies on the Palaeoproterozoic tholeiitic pillowed basalts and pelagic sedimentary rocks (pelites and carbonatites) on the Man-Leo Shield (Abouchami et al. 1990; Boher et al. 1992; Bossière et al. 1996; Pouclet et al. 1996) suggest that emplacement occurred in a marine basin. The presence of a banded iron formation in the eastern margin of the Man Archaean nucleus confirms an oceanic setting (Fig. 13a) for Palaeoproterozoic rocks in this area.

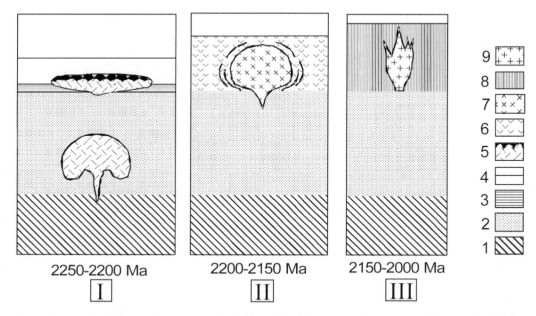

Fig. 12. Magmatic events: I, predominant mantle plume activities and PTH tholeiites deposition (adapted scheme from Boher et al. 1992); II, diapiric emplacement of PAG Palaeoproterozoic amphibole-bearing granitoids; III, PBG Palaeoproterozoic biotite ± muscovite-bearing (without amphibole) granitoids emplacement. 1, asthenosphere; 2, lithosphere; 3, ocean floor; 4, seawater level; 5, tholeiitic magma and pillowed basalts; 6, plutono-volcanic and sedimentary rocks; 7, PAG granitoids; 8, greenstone and PAG formations; 9, PBG granitoids.

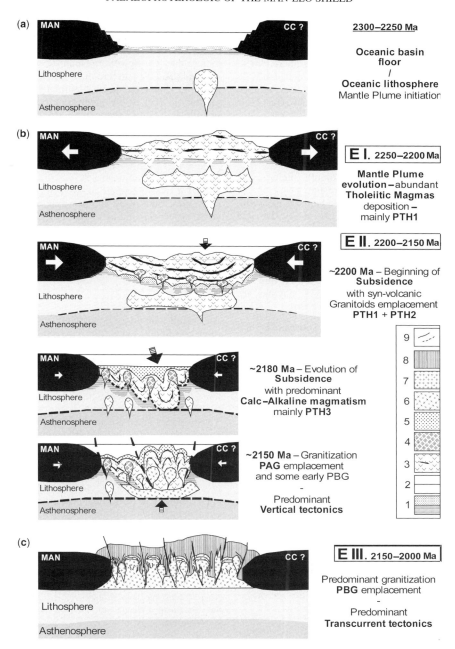

Fig. 13. Geodynamic evolution scheme of the Palaeoproterozoic magmatic rocks in the Man-Leo Shield. Scales are not respected. (a) Early context: oceanic basin floor with Man-Archaean nucleus (MAN) and probably Congo Craton (CC) merges. (b) Magmatic event I (E I) mantle plumes with early Palaeoproterozoic tholeiites (PTH1) deposition and next stages of event II (E II) subsidence with Palaeoproterozoic tholeiites (PTH2 and PTH3) deposition and amphibole-bearing granitoids (PAG) emplacement in predominant vertical tectonics. (c) Magmatic event III (E III) with main biotite ± muscovite-bearing granitoids (PBG) emplacement in transcurrent tectonics. 1, hydrated and metasomatized lithosphere; 2, seawater level; 3, volcanic and sedimentary intercalations rocks; 4, plutonic intrusion rocks; 5, pelagic sedimentary rocks; 6, PAG-type granitoids; 7, PBG-type granitoids; 8, Palaeoproterozoic greenstone belts unit; 9, shearing deformation zones.

Without data on the mechanism of formation of this ocean between two Archaean nuclei, we can consider a 'proto-oceanic crust' with a near mantle composition, similar to oceanic lithosphere, where the top is in contact with seawater and could be hydrated and metasomatized. This top surface could present some rare sedimentary fragments derived from continents.

(b) 2250–2000 Ma. Main Palaeoproterozoic magmatic rock formation.

Event I *(2250–2200 Ma).* Mantle plume evolution and tholeiitic magma intrusion.

Recent isotopic data (Abouchami et al. 1990; Liégeois et al. 1991; Boher et al. 1992; Ama Salah et al. 1996; Doumbia et al. 1998) indicate a juvenile origin of all the early volcanic and granitic rocks. Hence, PTH1 and PTH2 could be derived from a deep mantle source by partial melting (cf. Magnetic source of tholeiites, above). At this first stage (Fig. 13b-E.I), mainly PTH1 magmas are deposited on the floor of an ocean basin.

Event II *(2200–2150 Ma).* Subsidence and vertical movements.

c. 2200 Ma – Beginning of the subsidence. PTH3 magmas derived from partial melting of the top of the mantle (cf. Magnetic source of tholeiites, above). Without structural weaknesses such as suture zones to explain the plunging of the oceanic plate, we can reasonably imagine another mechanism such as subsidence of the basin floor (Fig. 13b-E.II) due to tholeiitic (PTH1 and some PTH2) volcanism. A depression by subsidence (Gorman et al. 1978) could thus allow the upper part of the mantle to attain the deeper mantle temperatures. Favourable arguments are: (1) the presence of underlying plumes with material escape; (2) the quantity of plutono-volcanic rocks (widespread thick tholeiitic rocks known across the entire shield) formed during the period of accretion; the overload of this material could create a balance of density favourable to depression with basin margin convergence. These arguments show that the megastructure has subsided.

c. 2180–2150 Ma. Evolution of the subsidence. The depression of the mega-structure would favour the partial melting of the base of the pile and also refold the edges of the synclinorium (Fig. 13b-E.II). This melting could produce PTH3-type and calc-alkaline magmas widespread across the entire shield and considered by all workers as posterior to PTH1 and PTH2 types. The mantle signature of early granitoids (PAG) and the other characters (cf. Granitoids series and petrogenesis, above) show that PAG derived from partial melting of the top of mantle and probably a part of underlying basaltic pile. At this stage, the main deformation is dominated by vertical movements due to upwelling granitic magma. The global NE–SW preferential orientation of the major structures (cf. Early deformation and related plutons, above) in the greenstone belts and early granitic batholiths (shapes and magmatic fabrics) suggests that the whole system evolved during NW–SE shortening.

(c) Event III. (2150–2000 Ma). Late granite intrusion and horizontal tectonics.

Event III is characterized by PBG emplacement (Fig. 13c). This study and previous works (Hirdes et al. 1996; Doumbia et al. 1998; Naba et al. 2004) suggest that most of these granitoids derived from partial melting of the underlying amphibole bearing granitoids (PAG). The setting is progressively evolved towards transcurrent tectonics.

Conclusion

An overall scheme for geodynamic evolution of the Man-Leo Shield during the Palaeoproterozoic can be reasonably constrained using the characteristics of early tholeiitic plutono-volcanic and granitoid rocks. Geochemical, geochronological and structural characteristics of these rocks allow us to date three principal events which have characterized the 2250–2000 Ma period in the whole shield. The proposed schematic model of geodynamic evolution is a simplified presentation and scales are not respected. Nevertheless, chemical and structural constraints have been taken into account.

The characteristics of the Archaean and Palaeoproterozoic rocks and tectonic context present very close affinities. It seems that there is no fundamental change in crustal evolution from the Archaean to the Palaeoproterozoic period in this region.

This synthesis has been prepared with respect to the research topics of the IGCP 509 project. I gratefully acknowledge Project Leaders for this initiative. Thanks to David Evans and particular thanks to Mark Jessell and Alan Collins for help to improve the presentation of the document. Thanks to referees and anonymous contributors.

References

ABOUCHAMI, W., BOHER, M., MICHARD, A. & ALBARÈDE, F. 1990. A major 2.1 Ga event of mafic magmatism in West Africa: an early stage of crustal accretion. *Journal of Geophysical Research*, **95**, 17 605–17 629.

AMA SALAH, I. 1991. *Pétrogaphie et relations structurales des formations métavolcaniques et sédimentaires du Birimien du Niger occidental. Problème de l'accrétion crustale au Protérozoique inférieur*. Thèse de l'Université d'Orléans.

AMA SALAH, I., LIÉGEOIS, J. P. & POUCLET, A. 1996. Evolution d'un arc insulaire océanique birimien précoce au Liptako nigérien (Sirba): géologie, géochronologie et géochimie. *Journal of African Earth Sciences*, **22**, 235–254.

ARNDT, N. 1998. Magmatisme ultrabasique et basique au Précambrien. *In*: HAGEMANN, R. & TREUIL, M. (eds) *Introduction à la géochimie et ses applications*, CEA **1**, 287–317.

ARNOULD, A. 1961. *Etude géologique des migmatites et des granites précambriens du Nord-Est de la Côte d'Ivoire et de la Haute Volta méridionale*. Rapport DGPM, Abidjan, BRGM, Mémoire 3.

ARTH, J. G. & HANSON, G. N. 1975. Geochemistry and origin of the early Precambrian crust of Northeastern Minnesota. *Geochimica et Cosmochimica Acta*, **39**, 325–362.

BARKER, F. & ARTH, J. G. 1976. Generation of trondhjemite-tonalitic liquids and Archean bimodal trondhjemite-basalt suites. *Geology*, **4**, 596–600.

BASSOT, J. P. 1987. Le complexe volcano-plutonique calco-alcalin de la rivière Daléma (Est Sénégal): discussion de sa signification géodynamique dans le cadre de l'orogénie éburnéenne (Protérozoïque inférieur). *Journal of African Earth Sciences*, **6**, 505–519.

BERTRAND, J. M., DIA, A., DIOH, E. & BASSOT, J. P. 1989. Réflexions sur la structure interne du craton Ouest-Africain au Sénégal oriental et confins guinéomaliens. *Comptes Rendus de l'Académie des Sciences*, Paris, T.309, **II**, 751–756.

BESSOLES, B. 1977. *Géologie de l'Afrique. Le craton ouest-africain*. Mémoires BRGM, Paris.

BÉZIAT, D., BOURGES, F., DEBAT, P., LOMPO, M., MARTIN, F. & TOLLON, F. 2000. A Palaeoproterozoic ultramafic-mafic assemblage and associated volcanic rocks of the Boromo greenstone belt: Burkina Faso: fractionates originating from island-arc volcanic activity in the West African Craton. *Precambrian Research*, **101**, 25–47.

BOHER, M., ABOUCHAMI, W., MICHARD, A., ALBARÈDE, F. & ARNDT, N. 1992. Crustal growth in West Africa at 2.1 Ga. *Journal of Geophysical Research*, **97**, 345–369.

BOSSIÈRE, G., BONKOUNGOU, I., PEUCAT, J. J. & PUPIN, J. P. 1996. Origin and age of Paleoproterozoic conglomerates and sandstones of the Tarkwaian group in Burkina Faso, West Africa. *Precambrian Research*, **80**, 153–172.

CABY, R., DELORS, C. & AGOH, O. 2000. Lithologie, structures et métamorphisme des formations birimiennes dans la région d'Odienné (Côte d'Ivoire): rôle majeur du diapirisme des plutons et des décrochements en bordure du craton de Man. *Journal of African Earth Sciences*, **30**, 351–374.

CAGNARD, F., GAPAIS, D. & BARBEY, P. 2007. Collision tectonics involving juvenile crust: the example of the southern Finnish Svecofennides. *Precambrian Research*, **154**, 125–141.

CASTAING, C. & LE METOUR, J. *ET AL*. 2003. *Carte géologique et minière du Burkina Faso à 1/1000000*. Bureau de recherches géologiques et minières, Orléans.

CHEILLETZ, A., BARBEY, P., LAMA, C., PONS, J., ZIMMERMANN, J. L. & DAUTEL, D. 1994. Age de refroidissement de la croûte juvénile Birimienne d'Afrique de l'Ouest, données U–Pb, Rb–Sr et K–Ar sur les formations à 2.1 Ga au SW-Niger. *Comptes Rendus de l'Académie des Sciences Paris*, **319**(II), 435–442.

CONDIE, K. C. 1986. Origin and early growth rate of continents. *Precambrian Research*, **32**, 261–278.

DAVIS, D. W., HIRDES, W., SCHALTEGGER, U. & NUNOO, E. A. 1994. U–Pb constraints on deposition and provenance of Birimian and gold-bearing Tarkwaian sediments in Ghana, West Africa. *Precambrian Research*, **67**, 89–107.

DEBAT, P., NIKIÉMA, S. *ET AL*. 2003. A new metamorphic constraint for the Eburnean orogeny from Palaeoproterozoic formations of the Man shield (Aribinda and Tampelga countries, Burkina Faso). *Precambrian Research*, **123**, 47–65.

DIA, A. 1988. *Caractères et significations des complexes magmatiques et métamorphiques du secteur de Sandikounda–Laminia (Nord de la boutonnière de Kédougou, Est Sénégal). Un modèle géodynamique du Birimien de l'Afrique de l'Ouest*. Thèse de l'Université de Dakar.

DIA, A., VAN SCHMUS, W. R. & KRÖNER, A. 1997. Isotopic constraints on the age and formation of a Palaeoproterozoic volcanic arc complex in the Kedougou Inlier, eastern Senegal, West Africa. *Journal of African Earth Sciences*, **24**, 197–213.

DIALLO, D. P. 2001. Le paleovolcanisme de la bordure occidentale de la boutonnière de Kédougou, Paléoprotérozoïque du Sénégal Oriental: Incidences géotectoniques. *Journal of African Earth Sciences*, **32**(4), 919–940.

DIALLO, D. P., DEBAT, P., ROCCI, G., DIA, A., NGOM, P. M. & SYLLA, M. 1993. Pétrographie et géochimie des roches méta-volcanodétritiques et méta-sédimentaires du Protérozoïque Inférieur du Sénégal Oriental dans le supergroupe de Mako (Sénégal, Afrique de l'Ouest): incidences géotectoniques. Publications occasionnelles CIFEG, 1993/23, 11–15.

DOSTAL, J. & MUELLER, W. U. 1997. Komatiite flooding of a rifted Archean rhyolitic arc complex: geochemical signature and tectonic significance of the Stoughton-Roquemaure Group, Abitibi greenstone belt, Canada. *Journal of Geology*, **105**, 545–563.

DOUMBIA, S., POUCLET, A., KOUAMELAN, A., PEUCAT, J. J., VIDAL, M. & DELOR, C. 1998. Petrogenesis of juvenil-type Birimian (Palaeoproterozoic) granitoids in central Côte d'Ivoire, West Africa: geochemistry and geochronology. *Precambrian Research*, **87**, 33–63.

DRUMMOND, M. S. & DEFANT, M. J. 1990. A model of trondhjemite-tonalite-dacite genesis and crustal growth via slab melting: Archean to modern comparisons. *Journal of Geophysical Research*, **95**, 21 503–21 521.

DUCELLIER, J. 1963. *Contribution à l'étude des roches cristallines et métamorphiques du Centre et du Nord de la Haute-Volta*. Mémoire BRGM 10.

DUPRÉ, B., CHAUVEL, C. & ARNDT, N. T. 1984. Pb and Nd isotopic study of two Archean komatiite flows from Alexo, Ontario. *Geochimica et Cosmochimica Acta*, **48**, 1965–1972.

DUPUIS, D., PONS, J. & PROST, A. E. 1991. Mise en place de plutons et caractérisation de la déformation birimienne au Niger occidental. *Comptes Rendus de l'Académie des Sciences, Paris*, **321**, 769–776.

EGAL, E., THIÉBLÉMONT, D. *ET AL.* 2002. Late Eburnean granitization and tectonics along the western and northwestern margin of the Archean Kénéma-Man domain (Guinea, West African Craton). *Precambrian Research*, **117**, 57–84.

FERRÉ, E. C. & CABY, R. 2007. Granulite facies metamorphism and charnockite plutonism: examples from the Neoproterozoic belt of northern Nigeria. *Proceedings of Geologists' Association*, **18**, 1–8.

FEYBESSE, J. L. & MILESI, J. P. 1994. The Archaean/Proterozoic contact zone in West Africa: a mountain belt of décollement thrusting and folding on a continental margin related to 2.1 Ga convergence of Archaean cratons? *Precambrian Research*, **69**, 199–227.

GASQUET, D., BARBEY, P., ADU, M. & PAQUETTE, J. L. 2003. Structure, Sr–Nd isotope geochemistry and zircon U–Pb geochronology of the granitoids of the Dabakala area (Côte d'Ivoire): evidence for a 2.3 Ga crustal growth event in the Palaeoproterozoic of West Africa? *Precambrian Research*, **127**, 329–354.

GLIKSON, A. Y. 1979. Early Precambrian tonalite-trondhjemite sialic nuclei. *Earth Science Reviews*, **15**, 1–73.

GORMAN, B. E., PEARCE, T. H. & BIRKETT, T. C. 1978. On the structure of Archaean greenstone belts. *Precambrian Research*, **6**, 23–41.

HARPUM, J. R. 1963. Petrographie classification of granitic rocks by partial chemical analysis. *Tanganyika Geological Survey*, **10**, 80–88.

HIRDES, W. & DAVIS, D. W. 1998. First U–Pb zircon age of extrusive volcanism in the Birimian Supergroup of Ghana/West Africa. *Journal of African Earth Sciences*, **27**, 291–294.

HIRDES, W. & DAVIS, D. W. 2002. U–Pb geochronology of Paleoproterozoic rocks in the southern part of the Kedougou-Kéniéba Inlier, Senegal, West Africa: evidence for diachronous accretionary development of the Eburnean province. *Precambrian Research*, **118**, 83–99.

HIRDES, W., DAVIS, D. W. & EISENLOHR, B. N. 1992. Reassessment of Proterozoic ages in Ghana on the basis of U–Pb zircon and monazite dating. *Precambrian Research*, **56**, 89–96.

HIRDES, D., DAVIES, D. W., LÜDTKE, G. & KONAN, G. 1996. Two generations of Birimian (Palaeoproterozoic) volcanic belts in northeastern Côte d'Ivoire (West Africa): consequences for the Birimian controversy. *Precambrian Research*, **80**, 173–191.

HIROSE, K. & KUSHIRO, I. 1993. Partial melting of dry peridotites at high pressures: determination of compositions of melts segregated from peridotite using aggregates of diamond. *Earth Planetary Science Letters*, **114**, 477–489.

HOLLINGS, P. & WYMAN, D. 1999. Trace element and Sm–Nd systematics of volcanic and intrusive rocks from the 3 Ga Lumby Lake Greenstone belt, Superior Province: Evidence for Archean plume-arc interaction. *Lithos*, **46**, 189–213.

HOLLINGS, P., WYMAN, D. & KERRICH, R. 1999. Komatiite-basalt-rhyolite volcanic associations in Northern Superior Province greenstone belt: significance of plume-arc interaction in the generation of the proto-continental Superior Province. *Lithos*, **46**, 137–161.

HOTTIN, G. & OUÉDRAOGO, O. F. 1975. *Notice explicative de la carte géologique à 1/1000 000 du Burkina Faso*. Direction de la Geologie et des Mines, Ouagadougou. Bureau de recherches géologiques et minières, Orléans.

JAHN, B. M. 1998. Géochimie des granitoïdes archéens et de la croûte primitive. *In*: HAGEMANN, R. & TREUIL, M. (eds) *Introduction à la géochimie et ses applications*, CEA **1**, 359–391.

JAHN, B. M. & ZHANG, Z. Q. 1984. Archean granulite gneisses from Eastern Hebei Province, China: rare earth geochemistry and tectonic implications. *Contribution to Mineralogy and Petrology*, **85**, 224–243.

JAHN, B. M., GLIKSON, A. Y., PEUCAT, J. J. & HICKMAN, A. H. 1981. REE geochemistry and isotopic data of Archean silicic volcanics and granitoids from the Pilbara Block, Western Australia: implication for the early crustal evolution. *Geochimica et Cosmochimica Acta*, **45**, 1633–1652.

JOCHUM, K. P., ARNDT, N. T. & HOFMANN, A. W. 1990. Nb–Th–La in komatiites and basalts: constraints on komatiite petrogenesis and mantle evolution. *Earth Planetary Science Letters*, **107**, 272–289.

JOHN, T., KLEMD, R., HIRDES, W. & LOH, G. 1999. The metamorphic evolution of the Palaeoproterozoic (Birimian) volcanic Ashanti belts (Ghana, West Africa). *Precambrian Research*, **98**, 11–30.

JUNNER, N. R. 1940. Geology of the Gold Coast and Western Togoland. *Gold Coast Geological Survey*, **11**, 1–40.

KAGAMBÉGA, N. 2005. *Typologie des granitoïdes Paleoproterozoïques (Birimien) du Burkina Faso, Afrique de l'Ouest. Approche pétrologique dans la région de Pô*. Thèse de 3è cycle de l'Université Cheik Anta Diop de Dakar.

KLEMD, R., HÜNKEN, U. & OLESCH, M. 2002. Metamorphism of the country rocks hosting gold-sulfide-bearing quartz veins in the Palaeoproterozoic southern Kibi-Winneba belt (SE-Ghana). *Journal of African Earth Sciences*, **35**, 199–211.

LAHONDÈRE, D., TIÉBLEMONT, D., TEGYEY, M., GUERROT, C. & DIABATÉ, B. 2002. First evidence of early Birimian (2.21 Ga) volcanic activity in Upper Guinea: the volcanics and associated rocks of the Niani suite. *Journal of African Earth Sciences*, **35**, 417–432.

LAMEYRE, J. & BOWDEN, P. 1982. Plutonic rocks types series: discrimination of various granitoid series and related rocks. *Journal of Volcanology and Geothermal Research*, **14**, 169–186.

LANGMUIR, C. H. & HANSON, G. N. 1980. An evaluation of major element heterogeneity in the mantle source of basalts. *Philosophical Transactions of the Royal Society of London, A*, **297**, 383–407.

LEMOINE, S. 1988. *Évolution géologique de la région de Dabakala (NE de la Côte d'Ivoire) au Protérozoïque Inférieur. Possibilité d'extension au reste de la Côte* d'Ivoire et au Burkina Faso: similitudes et différences: les linéaments grenville-Ferkessédougou et Grand Cess-Niakaramandougou. Thèse d'État, Université de Clermont-Ferrand, France.

LEUBE, A., HIRDES, W., MAUER, R. & KESSE, G. 1990. The early Proterozoic birimian supergroup of Ghana and some aspect of its associated gold mineralisation. Precambrian Research, 46, 139–165.

LIÉGEOIS, J. P., CLAESSENS, W., CAMARA, D. & KLERKX, J. 1991. Short-lived Eburnian orogeny in southern Mali. Geology, tectonics, U–Pb and Rb–Sr geochronology. Precambrian Research, 50, 111–136.

LOMPO, M. 1991. Etude structurale et géologique des séries birimiennes de la région de Kwademen, Burkina Faso, Afrique de l'Ouest. Evolution et contrôle structural des minéralisations sulfurées et aurifères pendant l'Eburnéen. Thèse Université de Clermont Fd, II, France.

LOMPO, M., CABY, R. & ROBINEAU, B. 1991. Evolution structurale du Birimien au Burkina Faso – exemple de la ceinture de Boromo-Goren dans le secteur de Kwademen (Afrique de l'Ouest). Comptes Rendus Académie des Sciences, Paris, 313, 945–950.

LOMPO, M., BOURGES, F., DEBAT, P., LESPINASSE, P. & BOUCHEZ, J. L. 1995. Mise en place d'un pluton granitique dans la croûte birimienne fragile: fabrique magnétique du pluon de Tenkodogo (burkina Faso). Comptes Rendus de l'Académie des Sciences, Paris, 320(II), 1211–1218.

MACDONALD, G. A. & KATSURA, T. 1964. Chemical composition of Hawaiian lavas. Journal of Petrology, 5, 82–113.

MACHADO, N., BROOKS, C. & HART, S. R. 1986. Determination of initial $^{87}Sr/^{86}Sr$ and $^{143}Nd/^{144}Nd$ in primary minerals from mafic and ultramafic rocks: experimental procedure and implications for the isotopic characteristics of the Archaean mantle under the Abitibi greenstone belt. Geochimica et Cosmochimica Acta, 50, 2335–2348.

MARTIN, H. 1985. Nature, origine et évolution d'un segment de croûte continentale archéenne: contraintes chimiques et isotopiques. Exemple de la Finlande orientale. Mémoire CAESS, 1, Rennes.

MARTIN, H., CHAUVEL, C. & JAHN, B. M. 1983. Major and trace element geochemistry and crustal evolution of Archaean granodioritic rocks from eastern Finland. Precambrian Research, 21, 159–180.

MILÉSI, J. P., LEDRU, P., FEYBESSE, J. L., DOMMANGET, A. & MARCOUX, E. 1992. Early Proterozoic ore deposits and tectonics of the Birimian oroegnic belt, West Africa. Precambrian Research, 58, 305–344.

MIYASHIRO, A. 1974. Volcanic rock series in island arcs and active continental margins. American Journal of Science, 274, 321–355.

NABA, S. 1999. Structure et mode de mise en place de plutons granitiques enboités: exemple de l'alignement plutonique paléoproterozoïque de Tenkodogo-Yamba dans l'Est du Burkina Faso. Thèse de l'Université C.A.D. de Dakar.

NABA, S., LOMPO, M., DEBAT, P., BOUCHEZ, J. L. & BÉZIAT, D. 2004. Structure and emplacement model for late-orogenic Palaeoproterozoic granitoids: the Tenkodogo-Yamba elongate pluton (Eastern Bourkina Faso). Journal of African Earth Sciences, 38, 41–57.

NAPON, S. 1988. Le gisement d'amas sulfuré (Zn-Ag) de Perkoa dans la province du Sangyé (Burkina Faso – Afrique de l'Ouest) Cartographie. Étude pétrographique, géochimique et métallogénique. Thèse de Doctorat, Université de Franche-Comté, France.

NDIAYE, P. M., DIA, A. ET AL. 1997. Donnée pétrogaphiques, géochimiques et géodynamiques nouvelles sur les granitoïdes du Paléoproérozoïque du Supergroupe de Dialé-Daléma (Sénégal Oriental): implications pétrogénétiques et géodynamiques. Journal of African Earth Sciences, 25, 193–208.

OBERTHÜR, T., VETTER, U., DAVIS, D. W. & AMANOR, J. A. 1998. Age constraints on gold mineralization and Paleoproterozoic crustal evolution in the Ashanti belt of southern Ghana. Precambrian Research, 89, 129–143.

PAPON, A. 1973. Géologie et minéralogie du Sud-Ouest de la Côte d'Ivoire. Mémoire BRGM, 80.

PONS, J., OUDIN, C. & VALERO, J. 1992. Kinematics of large syn-orogenic intrusions: example of the Lower Proterozoic Saraya batholith (Eastern Senegal). Geologische Rundschau, 81, 473–486.

PONS, J., BARBEY, P., DUPUIS, D. & LEGER, J. M. 1995. Rheological behaviour of a 2.1 Ga juvenile crust recorded by the mechanism of pluton emplacement. The Birimian of southern Niger. Precambrian Research, 70, 281–301.

POUCLET, A., VIDAL, M., DELOR, C., SIMEON, Y. & ALRIC, G. 1996. Le volcanisme birimien du nord-est de la Côte-d'Ivoire, mise en évidence de deux phases volcano-tectoniques distinctes dans l'évolution géodynamique du Paléoprotérozoïque. Bulletin de la Société Géologique de France, 167, 529–541.

ROQUES, M. 1948. Le Précambrien de l'Afrique Occidentale Française. Bulletin de la Société Géologique de France, 18, 589–628.

SANDEMAN, H. A., HANMER, S., TELLA, S., ARMITAGE, A. A., DAVIS, W. J. & RYAN, J. J. 2006. Petrogenesis of Neoarchaean volcanic rocks of the MacQuoid supracrustal belt: a back-arc setting for the northwestern Hearne subdomain, western Churchill Province, Canada. Precambrian Research, 144, 140–165.

SHIREY, S. B. & HANSON, G. N. 1986. Mantle heterogeneity and crustal recycling in Archean granite-greenstone belts: evidence from Nd isotopes and trace elements in the Rainy Lake area, Superior Province, Ontario, Canada. Geochimica et Cosmochimica Acta, 50, 2631–2651.

SOUMAILA, A., HENRY, P. & ROSSY, M. 2004. Contexte de mise en place des roches basiques de la ceinture de roches vertes birimiennes de Diagorou-Darbani (Liptako, Niger, Afrique de l'Ouest): plateau océanique ou environnement d'arc/bassin arrière-arc océanique. Comptes Rendus Géosciences, 336, 1137–1147.

STOREY, M., MAHONEY, J. J., KROENKE, L. W. & SAUNDERS, A. D. 1991. Are oceanic plateaus sites of komatiite formation? Geology, 19, 376–379.

SUN, S. S. & MCDONOUGH, W. F. 1989. Chemical and isotopic systematics of oceanic basalt: implication for mantle composition and processes. *In*: SAUNDERS, A. D. & NORRY, M. J. (eds) *Magmatism in the Ocean Basins*. Geological Society, London, Special Publications, **42**, 313–345.

SYLVESTER, P. J. & ATTOH, K. 1992. Lithostratigraphy and composition of 2.1 Ga greenstone belts of the West African Craton and their bearing on crustal évolution and the Archean-Proterozoic boundary. *Journal of Geology*, **100**, 377–393.

TAGINI, B. 1971. *Esquisse structurale de la Côte d'Ivoire. Essai de géotectonique régionale*. Thèse de l'Université de Lausanne, Marseille.

TAYLOR, P. N., MOORBATH, S., LEUBE, A. & HIRDES, W. 1992. Early Proterozoic crustal evolution in the Birimian of Ghana: constraints from geochronology and isotope geochemistry. *Precambrian Research*, **56**, 97–111.

TOMLINSON, K. Y. & CONDIE, K. C. 2001. Archaean mantle plumes: evidence from greenstone belt geochemistry. *In*: ERNST, R. E. & BUCHAN, K. L. (eds) *Mantle Plumes: Their Identification Through Time*. Geological Society of America Bulletin Special Paper, **352**, 341–357.

VEGAS, N., NABA, S., BOUCHEZ, J. L & JESSELL, M. 2008. Structure and emplacement of granite plutons in the Paleoproterozoic crust of Eastern Burkina Faso: rheological implications. *International Journal of Earth Sciences (Geologisches Rundschau)*, **97**, 1165–1180.

VIALETTE, Y., BASSOT, J. P. & MERGOIL-DANIEL, J. 1990. Mesure de l'âge des niveaux calcaro-dolomitiques du Protérozoïque inférieur Sénégalais. 15è Colloque de Géologie Africaine, CIFEG, Paris, Publication occasionnelle.

VIDAL, M. 1987. Les déformations éburnéennes de l'unité birrimienne de la Comoé (Côte d'Ivoire). *Journal of African Earth Sciences*, **6**, 141–152. CIFEG, Paris, Publication occasionnelle.

VIDAL, M. & ALRIC, G. 1994. The Palaeoproterozoic (Birimian) of Haute-Comoé in the West African craton, Ivory Coast: a transtensional back-arc basin. *Precambrian Research*, **65**, 207–229.

VIDAL, M., DELOR, C., POUCLET, A., SIMÉON, Y. & ALRIC, G. 1996. Evolution géodynamique de l'Afrique de l'Ouest entre 2.2 Ga et 2 Ga: le style 'archéen' des ceintures vertes et des ensembles sédimentaires birimiens du nord-est de la Côte d'Ivoire. *Bulletin de la Société Géologique de France*, **167**, 307–319.

WYMAN, D. A. 1999. A 2.7 Ga depleted tholeiite suite: evidence of plume-arc interaction in the Abitibi greenstone belt, Canada. *Precambrian Research*, **97**, 27–42.

WYMAN, D. A., KERRICH, R. & POLAT, A. 2002. Assembly of Archean cratonic mantle lithosphere and crust: plume-arc interaction in the Abitibi-Wawa subduction-accretion complex. *Precambrian Research*, **115**, 37–62.

ZONOU, S. 1987. *Les formations leptyno-amphibolitiques et le complexe volcanique et volcano-sédimentaire du Protérozoïque inférieur de Bouroum-nord (Burkina Faso – Afrique de l'Ouest). Etude pétrographique, géochimique, approche pétrogénétique et évolution géodynamique*. Thèse de l'Université de Nancy.

U–Pb and Sm–Nd constraints on the nature of the Campinorte sequence and related Palaeoproterozoic juvenile orthogneisses, Tocantins Province, central Brazil

MARIA E. S. D. GIUSTINA*, CLAUDINEI G. DE OLIVEIRA, MÁRCIO M. PIMENTEL, LUCIANA V. DE MELO, REINHARDT A. FUCK, ELTON L. DANTAS & BERNHARD BUHN

Instituto de Geociências, Campus Universitário Darcy Ribeiro, Universidade de Brasília (UnB), CEP 70910-900, Brasília, DF, Brazil

Corresponding author (e-mail: maria_emilia@unb.br)

Abstract: The Palaeoproterozoic era was the most important stage of crustal accretion in the South American Platform, being responsible for the development of several magmatic arcs, which represent approximately 35% of the present-day continental crust.

The recently mapped Campinorte volcano-sedimentary sequence and associated plutonic rocks represent this Palaeoproterozoic history in the northern Brasília Belt, central Brazil. The sequence consists of metapsammites and metapelites, with interbedded lenses of gondites and metacherts, as well as rhyolite and pyroclastic deposits. Tonalite, granodiorite and granite crystallized between c. 2.18 and 2.16 Ga, as indicated by U–Pb zircon analyses. Sm–Nd T_{DM} model ages range between c. 2.1 and 2.7 Ga, with ε_{Nd} values ranging from −2.14 to +3.36, indicating the dominantly juvenile nature of the original magmas. A LA-ICPMS provenance study of zircon grains from a quartzite sample reveals a single sediment source with Palaeoproterozoic age. The data presented here provide new information on the Palaeoproterozoic juvenile crust of central Brazil and suggest correlation with other Palaeoproterozoic provinces, especially the Birimian Belt in West African Craton and those of the Guiana Shield, thus contributing to reconstruction of the Columbia supercontinent.

The Palaeoproterozoic era (2.5–1.6 Ga) represents the main episode of crustal growth recorded on present-day continents. In the South-American Platform, two distinct accretionary phases are observed, one between c. 2.3 and 2.1 Ga and another at 2.1–1.8 Ga (Sato & Siga 2000, 2002).

In Brazil, Palaeoproterozoic units constitute the basement of most Neoproterozoic belts and are also exposed in large areas within the São Francisco and Amazonian Craton, which has led to global correlation with the West African and Congo cratons (Abouchami et al. 1990; Ledru et al. 1994; Feybesse et al. 1998; Teixeira et al. 2000; Brito Neves et al. 2000, 2001; Zhao et al. 2002; Barbosa & Sabaté 2004; Duarte et al. 2004; Klein et al. 2005; Lerouge et al. 2006; Noce et al 2007).

In central Brazil, Palaeoproterozoic rock units form the sialic basement of the supracrustal sequences of the Tocantins Province, which is a large Neoproterozoic orogenic area including the Brasília, Araguaia and Paraguay fold belts (Fig. 1).

The Campinorte volcano-sedimentary sequence and associated plutonic rocks (Kuyumjian et al. 2004; Oliveira et al. 2006) has been recently mapped in the northern portion of the Brasília Belt and represents this Palaeoproterozoic episode of crustal accretion. Its lithological association is similar to other Palaeoproterozoic units described in the Tocantins Province. However, its geotectonic significance has remained uncertain due to the absence of geochronological and isotopic data.

Therefore, this study aims to determine the U–Pb and Sm–Nd isotopic composition of the Campinorte volcano-sedimentary sequence and related plutonic rocks, in order to constrain the igneous crystallization ages, as well as to verify the sedimentary provenance of the supracrustal lithologies. The data presented here provide new constraints that will help in our understanding of the geological evolution and tectonic significance of Palaeoproterozoic rock associations in central Brazil.

Geological setting

The area investigated in the present study is located within the Brasília Belt (Fig. 1), which is an orogen developed as a result of the closure of the Brasilides Ocean during the Neoproterozoic era. This orogenic belt is divided into four main

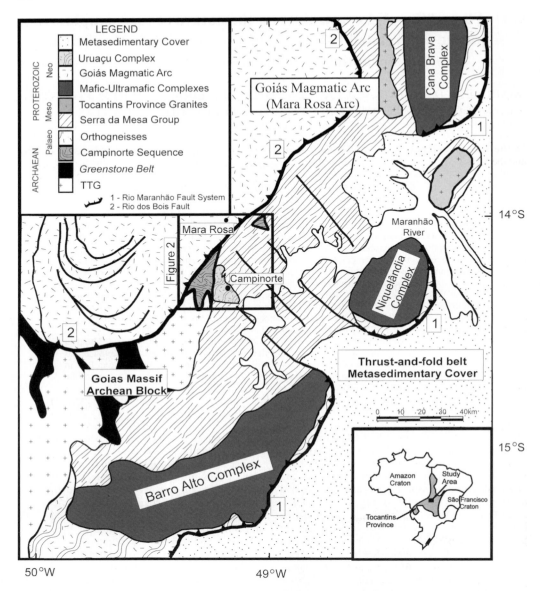

Fig. 1. Geological sketch map of the northwestern part of the Brasília Belt (after Fuck 1994).

sectors: (i) a thrust-and-fold belt, consisting of Neoproterozoic metasedimentary sequences; (ii) the metamorphic core, comprising high-grade rocks and ultra-high temperature granulites; (iii) the Goiás Magmatic Arc, which represents a juvenile terrain exposed in both the northern (Mara Rosa Arc) and southern Brasilia Belt (Arenópolis Arc); and (iv) the Goiás Massif, interpreted as a microcontinent accreted to the continental margin at the end of the Neoproterozoic (Brito Neves & Cordani 1991; Fuck et al. 1994; Pimentel et al. 2000; Fuck et al. 2005).

The Campinorte sequence is part of the Goiás Massif, which consists of an Archaean nucleus surrounded by Palaeoproterozoic orthogneisses, covered by Palaeo- to Mesoproterozoic platformal metasedimentary rocks of the Serra da Mesa Group (Fig. 1). The eastern margin of the Goiás Massif is marked by a seismic and gravimetric discontinuity, interpreted as a suture produced by closure of the

Brasilides Ocean at the end of the Neoproterozoic (Marangoni et al. 1995; Soares et al. 2006). At the surface, this suture zone is represented by the Rio Maranhão fault system (Soares et al. 2006).

The Archaean core contains typical tonalite-trondhjemite-granodiorite (TTG) terrains and greenstone belt sequences. The narrow supracrustal sequences are confined between TTG orthogneisses forming dome-and-keel structures (Fig. 1). The U–Pb analyses reveal ages between 2.8 and 2.7 Ga for the igneous crystallization of the TTG protoliths, and the Sm–Nd isotopic composition points to dominantly juvenile magmatism, with limited crustal contamination (Jost et al. 2006).

Exposures of the Campinorte sequence are limited by the Hidrolina Dome in the south and by the Rio dos Bois Fault in the west, which separates it from the Neoproterozoic Mara Rosa Magmatic Arc (Kuyumjian et al. 2004; Oliveira et al. 2006) (Fig. 2). It consists of supracrustal (volcano-sedimentary) and plutonic rocks with T_{DM} Sm–Nd model ages between 2.5 and 2.3 Ga, with positive initial ε_{Nd} values (Oliveira & Pimentel 1998; Kuyumjian et al. 2004). U–Pb analyses obtained in zircon grains from a metagranite associated with the Campinorte sequence yielded a discordia upper intercept age of $2176 +12/-9$ Ma (Pimentel et al. 1997), which led these authors to suggest that

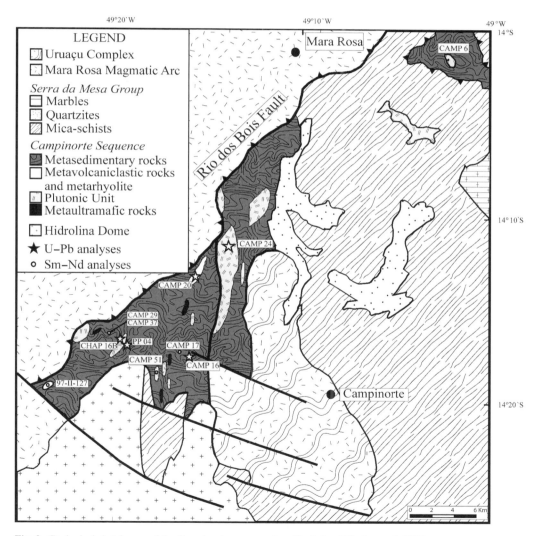

Fig. 2. Geological sketch map of the Campinorte sequence (modified after Oliveira et al. 2006).

this terrane may be part of the westernmost margin of the São Francisco craton.

The Uruaçu complex is separated from the adjacent units by regional-scale NW–SE and north–south shear zones and comprises a variety of para- and orthogneisses with amphibolite to granulite mineral assemblages. U–Pb crystallization ages for zircon grains from felsic and mafic orthogneisses range from c. 690–650 Ma, and zircon overgrowths indicate metamorphic ages around 650 Ma, suggesting that the metamorphism was partly concomitant or only slightly younger than the intrusion of granitoids and mafic bodies (Giustina et al. 2009). The Uruaçu complex is interpreted as the metamorphic nucleus of the Brasiliano orogen and might be related to delamination, uplift and extension following ocean closure and continental collision between the Amazonian and São Francisco cratons (Giustina et al. 2009).

The Campinorte sequence

The Campinorte sequence (Fig. 2) consists of a volcano-sedimentary sequence and a number of associated granitic bodies. The Campinorte volcano-sedimentary sequence is dominantly detrital, and original structures and features such as primary bedding are observed in places. The original stratigraphic sequence, however, cannot be recognized, since the geological units have been intensely reworked during Palaeoproterozoic and Brasiliano-Pan-African deformations.

Quartz micaschist, with variable amounts of carbonaceous matter, is the main rock type of this unit. Psammitic rocks include micaceous, thinly-laminated quartzite, as well as orthoquartzite that forms massive beds, centimetres to a few metres thick. Chemical deposits such as gondites, with spessartite and cummingtonite, and metacherts have been identified, and constitute m-scale lenses within the metasedimentary rocks. Felsic volcaniclastic rocks exist as small, elongate bodies interlayered within the metasedimentary sequence. Pyroclastic deposits are the most representative lithologies, and are composed of rhyolitic crystal metatuffs to metalapilli tuffs. Rhyolites are rare and contain idiomorphic K-feldspar and anhedral blue quartz phenocrysts, immersed in a fine-grained quartz-feldspar groundmass.

This supracrustal unit has been subjected to at least two deformational events (Oliveira et al. 2006) and has been metamorphosed under greenschist facies conditions.

Meta-ultramafic rocks form tectonic slices emplaced within the Campinorte sequence. These schists are intensely weathered, and contain variable amounts of talc, actinolite-tremolite and chlorite.

The plutonic bodies associated with the Campinorte sequence include metatonalites, metagranodiorites and metagranites. In general, they are NNE- trending elongate plutons showing varied degrees of deformation. Contacts with the supracrustal sequence are tectonic, and intrusive features such as enclaves or metamorphic aureoles have not been observed. Like the supracrustal unit, the plutonic unit does not include mafic or intermediate igneous rocks. The main granite bodies are represented by the Mundinho and Viúva intrusions, which host gold deposits related to Neoproterozoic transcurrent shear zones (Oliveira et al. 2004).

Analytical procedures

U–Pb geochronology

U–Pb isotopic analyses were performed in the Geochronology Laboratory of the University of Brasília and followed the analytical procedure described by Buhn et al. (2009). Zircon concentrates were extracted from c. 10 kg rock samples, using a Frantz magnetic separator. Mineral concentrates were hand-picked under a binocular microscope to obtain fractions of similar size, shape and colour.

For the conventional U–Pb analyses, fractions were dissolved in concentrated HF and HNO_3 ($HF:HNO_3 = 4:1$) using microcapsules in Parr-type bombs. A mixed $^{205}Pb-^{235}U$ spike was used. Chemical extraction followed standard anion exchange technique, using Teflon microcolumns, following modified procedures from Krogh (1973). Pb and U were loaded together on single Re filaments with H_3PO_4 and Si gel, and isotopic analyses were carried out on a Finnigan MAT-262 multi-collector mass spectrometer equipped with secondary electron multiplier-ion counting, at the Geochronology Laboratory of the University of Brasília. Procedure blanks for Pb, at the time of analyses, were better than 20 pg. Data reduction and age calculations were performed using the PBDAT (Ludwig 1993) and ISOPLOT-Ex (Ludwig 2001a, b) software. Errors for isotopic ratios are quoted at 2σ.

For LA-ICPMS analyses, hand-picked zircon grains were mounted in epoxy blocks and polished to obtain a smooth surface. Backscattered electron images were obtained using a scanning electron microscope, in order to investigate the internal structures of the zircon crystals prior to analysis.

Before LA-ICPMS analyses, mounts were cleaned by carefully rinsing with dilute (c. 2%) HNO_3. Once fully dry, the samples were mounted together with a GJ1 standard zircon (Jackson et al. 2004) in a specially adapted laser cell and loaded into a New Wave UP213 Nd:YAG laser ($\lambda = 213$ nm), linked to a Thermo Finnigan Neptune Multi-collector ICP-MS. Helium was used as the carrier gas and mixed with argon before entering the ICP. The laser was run at a frequency of 10 Hz and energy of 34%. In order to avoid downhole fractionation during ablation, we opted for a raster scan of about 70 μm in total diameter with a spot size of 30 μm.

The Pb isotopes 204, 206 and 207 were collected with ion counters; ^{238}U was analysed on a Faraday cup. The content of ^{202}Hg was monitored on an ion counter for correction of the isobaric interference between ^{204}Hg and ^{204}Pb. The signals during ablation were taken in 40 cycles of 1 sec each. For data evaluation, only coherent intervals of signal response were considered. Data reduction was performed with an Excel spreadsheet developed by one of the authors, which considers blank values, zircon standards composition and error propagation. The ^{204}Pb signal intensity was calculated using a natural ^{202}Hg/^{204}Hg ratio of 4.346. A common-Pb correction was applied for zircon grains with ^{206}Pb/^{204}Pb lower than 1000, applying a common-lead composition following the Stacey & Kramers (1975) model. Plotting of U–Pb data was performed by ISOPLOT v.3 (Ludwig 2003). Errors for isotopic ratios are presented at the 2σ level.

Sm–Nd isotopic analyses

Sm–Nd isotopic analyses followed the method described by Gioia & Pimentel (2000) and were carried out at the Geochronology Laboratory of the University of Brasília. Whole rock powders (c. 50 mg) were mixed with ^{149}Sm–^{150}Nd spike solution and dissolved in Savillex capsules. Sm and Nd extraction of whole-rock samples followed conventional cation exchange techniques, using Teflon columns containing LN-Spec resin (HDEHP – diethylhexil phosphoric acid supported on PTFE powder). Sm and Nd samples were loaded on Re evaporation filaments of double filament assemblies and the isotopic measurements were carried out on a multi-collector Finnigan MAT 262 mass spectrometer in static mode. Uncertainties for Sm/Nd and ^{143}Nd/^{144}Nd ratios are better than $\pm 0.5\%$ (2σ) and $\pm 0.005\%$ (2σ), respectively, based on repeated analyses of international rock standards BHVO-1 and BCR-1. The ^{143}Nd/^{144}Nd ratios were normalized to ^{146}Nd/^{144}Nd of 0.7219 and the decay constant used was 6.54×10^{-12} a^{-1}. The T_{DM} values were calculated using the model of De Paolo (1981).

Samples and results

U–Pb and Sm–Nd analyses were performed on samples from the Campinorte sequence and associated plutonic rocks. Analytical results are presented in Tables 1 to 4, and Figure 2 shows the sample locations.

U–Pb results

Volcano-sedimentary sequence. A micaceous quartzite (sample CAMP 16), interlayered with quartz-mica schist, was investigated by LA-ICPMS to determine the provenance pattern of the Campinorte metasedimentary rocks. Zircon grains are yellowish to light-brown in colour, and present well preserved prismatic habit and faces, indicating limited transport. Forty spots, with twenty-nine concordant analyses yielded only Palaeoproterozoic ages, around 2.2 Ga (Fig. 3, Table 1), which may represent a good estimate for the average age of the source rocks of the original sandstone. This very homogeneous zircon population demonstrates that the sediments of the Campinorte sequence were derived entirely from local arc sources and settled in a proximal depositional setting, with restricted transport. It also establishes the maximum age for the deposition of the protoliths.

Sample CAMP20 is a rhyolitic metatuff that contains three distinct zircon populations: (i) the dominant fraction is composed by light-brown to yellowish prismatic crystals, with smooth and well-defined faces, which are interpreted as being magmatic; (ii) pink to brown grains with corroded surface typical of detrital grains, which normally show preserved prismatic shapes; and (iii) metamict zircon grains. Three zircon fractions from the igneous population were investigated by conventional ID-TIMS and yielded an upper intercept age of 2179 ± 4 Ma (MSWD = 0.62; Fig. 4, Table 2), interpreted as the age of crystallization of the original magma. This is compatible with the maximum depositional age of c. 2.2 Ga for the supracrustal unit discussed above.

Plutonic unit. The Mundinho mylonitic granite is characterized by K-feldspar phenocrysts set in a quartz, biotite and plagioclase groundmass. Its elongated shape and subvertical foliation are controlled by a north–south trending shear zone. Sample CAMP24 contains pink elongated prismatic zircon grains, which present well-defined faces. Five from six zircon fractions revealed variably discordant compositions indicating an upper-intercept age of 2173 ± 10 Ma (MSWD = 0.54; Fig. 5, Table 2), interpreted as being representative of the igneous crystallization age.

Zircon grains from metatonalite CHAP16B consist of pink, clear and elongated prisms with well-formed faces. Five concordant zircon grains out of nine fractions analysed by ID-TIMS resulted in an upper intercept age of 2163 ± 1 Ma

Table 1. Summary of LA-ICPMS data for sample CAMP 16

Spot	$^{206}Pb/^{204}Pb$	Isotopic ratios								Ages							
		$^{207}Pb/^{206}Pb$	2σ	$^{206}Pb/^{238}U$	2σ	$^{207}Pb/^{235}U$	2σ	$^{207}Pb/^{206}Pb$	2σ	$^{206}Pb/^{238}U$	2σ	$^{207}Pb/^{235}U$	2σ	rho
2	23920	0.1372	0.0125	0.3912	0.0333	7.3994	0.6302	2192	157.8	2129	152.6	2161	73.5	0.93
2	1594	0.1133	0.0029	0.3390	0.0077	6.1325	0.1386	1854	45.1	1882	36.9	1995	19.5	0.84
3	793	0.1257	0.0177	0.3745	0.0531	7.0151	0.9759	2359	240.6	2090	244.3	2229	116.7	0.85
3	–	0.1365	0.0039	0.4168	0.0119	7.8473	0.2234	2184	49.3	2246	53.8	2214	25.3	0.62
6	1527	0.1382	0.0109	0.4053	0.0321	7.5660	0.5921	2204	137.4	2193	145.5	2181	67.9	0.89
7	–	0.1386	0.0046	0.4006	0.0134	7.6541	0.2561	2209	57.5	2172	61.4	2191	29.6	0.93
8	326731	0.1366	0.0015	0.3825	0.0040	7.2029	0.0751	2184	18.6	2088	18.6	2137	9.3	0.96
9	4005	0.1426	0.0043	0.3887	0.0121	7.6414	0.2371	2259	52.4	2117	55.7	2190	27.5	0.93
9	19417	0.1367	0.0016	0.4042	0.0056	7.6165	0.1057	2185	20.1	2188	25.7	2187	12.4	0.93
11	1901	0.1082	0.0033	0.3231	0.0092	5.8220	0.1647	1769	52.6	1805	44.5	1950	24.2	0.75
12	7284	0.1370	0.0048	0.4133	0.0147	7.8066	0.2771	2189	60.7	2230	66.6	2209	31.5	0.87
12	22866	0.1364	0.0028	0.4190	0.0089	7.8825	0.1666	2182	35.6	2256	40.1	2218	18.9	0.94
13	352	0.0997	0.0041	0.2893	0.0128	5.3393	0.2346	2353	71.1	1701	63.6	2015	36.9	0.91
13	12791	0.1367	0.0037	0.3868	0.0113	7.2893	0.2135	2185	47.7	2108	52.5	2147	25.8	0.94
13	2933	0.1354	0.0223	0.3315	0.0548	6.1906	1.0234	2170	287.1	1846	260.0	2003	135.1	0.91
14	14586	0.1370	0.0155	0.4087	0.0467	7.7204	0.8822	2189	197.2	2209	210.3	2199	97.9	0.90
16	10268	0.1356	0.0026	0.4033	0.0078	7.5429	0.1454	2172	33.5	2184	35.6	2178	17.1	0.90
17	770	0.1118	0.0025	0.3332	0.0069	6.2323	0.1294	2262	39.1	1886	33.5	2072	18.0	0.70
17	–	0.1353	0.0028	0.4064	0.0083	7.5812	0.1553	2168	36.4	2199	38.1	2183	18.2	0.90
18	9313	0.1388	0.0023	0.3884	0.0064	7.4324	0.1225	2212	28.3	2116	29.7	2165	14.6	0.85
20	3669	0.1389	0.0162	0.3898	0.0446	7.4647	0.8538	2213	201.7	2122	203.6	2169	97.6	0.72
21	6007	0.1364	0.0033	0.4084	0.0100	7.6815	0.1884	2182	41.8	2208	45.7	2194	21.8	0.91
21	3467	0.1380	0.0079	0.3558	0.0231	6.7689	0.4395	2202	99.3	1962	108.9	2082	55.9	0.95
22	4095	0.1384	0.0071	0.3946	0.0204	7.5319	0.3890	2208	88.8	2144	93.5	2177	45.3	0.90
22	21193	0.1355	0.0051	0.4012	0.0145	7.4972	0.2704	2171	66.2	2175	66.2	2173	31.8	0.91
25	1107	0.1087	0.0054	0.3234	0.0159	5.7436	0.2811	2217	85.8	1832	77.1	2109	41.5	0.75
26	8930	0.1357	0.0021	0.4136	0.0060	7.7424	0.1119	2174	26.6	2232	27.2	2202	12.9	0.90
26	6077	0.1368	0.0044	0.3966	0.0132	7.4793	0.2492	2187	55.9	2153	60.7	2171	29.4	0.80
27	42438	0.1366	0.0030	0.4004	0.0088	7.5440	0.1662	2185	37.8	2171	40.5	2178	19.6	0.93
31	454	0.1065	0.0026	0.3124	0.0090	5.6953	0.1643	2267	41.6	1804	44.3	2029	24.6	0.92
31	–	0.1349	0.0046	0.4069	0.0135	7.5702	0.2514	2163	59.5	2201	61.6	2181	29.4	0.92
33	29159	0.1386	0.0022	0.3975	0.0060	7.5978	0.1139	2210	27.3	2157	27.4	2185	13.4	0.95
36	15547	0.1373	0.0036	0.4006	0.0106	7.5843	0.2004	2194	46.1	2172	48.5	2183	23.4	0.78
41	7576	0.1359	0.0023	0.4005	0.0066	7.5045	0.1246	2175	29.1	2171	30.5	2174	14.8	0.93

Table 2. Summary of ID-TIMS data for the Campinorte sequence and associated plutonic unit

Fractions	Size (mg)	U (ppm)	Pb (ppm)	Th (ppm)	$^{206}Pb/^{204}Pb$	Radiogenic ratios						Ages			
						$^{207}Pb/^{235}U$	(pct)	$^{206}Pb/^{238}U$	(pct)	rho	$^{207}Pb/^{206}Pb$	$^{206}Pb/^{238}U$	$^{207}Pb/^{235}U$	$^{207}Pb/^{206}Pb$	Ma

Supracrustal Unit
CAMP 20
CAMP20-1	0.012	169.6	73.5	90.7	309.78	6.531	1.12	0.349	1.11	0.990	0.1358	1929	2050	2174	2.7
CAMP20-2	0.025	146.0	58.0	43.5	616.03	6.406	1.00	0.343	0.97	0.974	0.1356	1899	2033	2172	4.0
CAMP20-3	0.014	152.3	51.2	77.7	282.98	5.010	1.26	0.270	1.24	0.986	0.1348	1539	1821	2161	3.7

Plutonic Unit
CAMP2 24
CAMP24-X	0.009	134.2	56.2	–	353.56	6.419	1.72	0.351	1.72	0.994	0.1327	1938	2035	2134	3
CAMP24-Y	0.01	147.1	73.5	–	201.55	6.670	1.36	0.363	1.31	0.963	0.1334	1995	2069	2143	6
CAMP24-I	0.037	52.3	23.4	–	674.28	7.298	1.55	0.391	1.48	0.952	0.1354	2127	2149	2169	8
CAMP24-J	0.4	58.9	24.1	–	414.62	6.177	1.36	0.339	1.32	0.969	0.1321	1882	2001	2127	6
CAMP24-G	0.03	74.2	28.6	–	1881.71	6.267	0.63	0.343	0.58	0.927	0.1325	1901	2014	2131	4
CAMP24-H	0.048	82.2	31.3	–	1861.47	6.208	0.47	0.341	0.44	0.953	0.1322	1889	2006	2128	3

CHAP 16B
CHAP 16B D4	0.055	214.1	70.1	19.8	2105.54	5.264	0.37	0.300	0.37	0.991	0.1271	1693	1863	2059	0.9
CHAP 16B D15	0.126	127.1	51.5	8.6	3797.05	6.692	0.33	0.365	0.32	0.974	0.1328	2001	2072	2136	1.3
CHAP 16B D4b	0.077	127.8	51.9	14.1	4221.21	6.564	0.24	0.360	0.24	0.978	0.1324	1980	2055	2130	0.9
CHAP 16B V	0.093	210.4	89.8	11.7	7122.31	7.018	0.40	0.380	0.35	0.996	0.1338	2078	2114	2148	0.6
CHAP 16B Y	0.053	112.6	44.4	20.5	4536.21	6.549	0.92	0.359	0.92	0.997	0.1324	1977	2053	2130	1.1

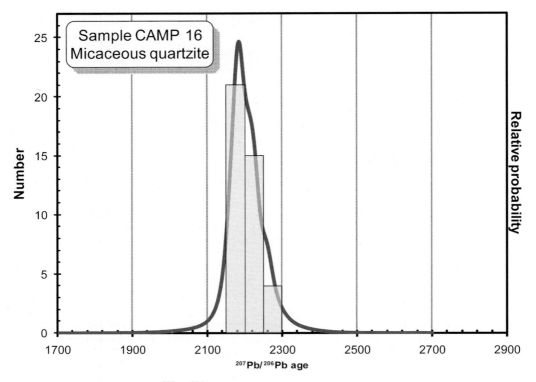

Fig. 3. Probability density plot of $^{207}Pb/^{206}Pb$ ages obtained from detrital zircon grains within sample CAMP 16.

(MSWD = 0.76; Fig. 6, Table 2), which is also interpreted as the age of igneous crystallization.

Sample PP04 derives from a mylonitic tonalite. It contains pink, elongate prismatic zircon grains with well-defined faces. Inclusions or fractures are rare. Under BSE imaging, grains are homogeneous and do not show a well-defined zoning pattern (Fig. 7). Five spot analyses carried out by LA-ICPMS yielded a concordia age of 2158 ± 8 Ma (MSWD = 0.5; Fig. 8, Table 3), which is taken as the best estimate for the crystallization age of the igneous protolith.

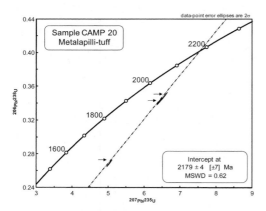

Fig. 4. Concordia diagram for ID-TIMS zircon analyses of metapyroclastic rock from the Campinorte sequence. Data-point error ellipses are 2σ.

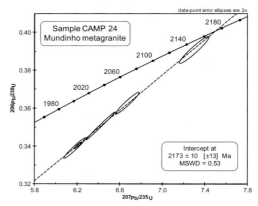

Fig. 5. Concordia diagram for ID-TIMS zircon analyses of Mundinho metagranite. Data-point error ellipses are 2σ.

Table 3. *Summary of LA-ICPMS data for sample PP04*

Spot	$^{206}Pb/^{204}Pb$	Isotopic ratios							Ages					
		$^{207}Pb/^{206}Pb$	2σ	$^{206}Pb/^{238}U$	2σ	$^{207}Pb/^{235}U$	2σ	$^{207}Pb/^{206}Pb$	2σ	$^{206}Pb/^{238}U$	2σ	$^{207}Pb/^{235}U$	2σ	rho
6	−19251	0.1346	0.0045	0.3987	0.0136	7.3992	0.2532	2158	58.9	2163	62.6	2161	30.2	0.922
7	19275	0.1353	0.0036	0.3926	0.0106	7.3263	0.1980	2168	46.2	2135	48.9	2152	23.9	0.861
18	13035	0.1339	0.0036	0.3994	0.0104	7.3755	0.1921	2150	46.7	2166	47.7	2158	23.0	0.863
19	15240	0.1346	0.0022	0.3976	0.0065	7.3799	0.1200	2159	28.5	2158	29.7	2159	14.4	0.841
20	16956	0.1348	0.0040	0.3886	0.0093	7.2228	0.1729	2162	51.6	2116	43.0	2139	21.1	0.838

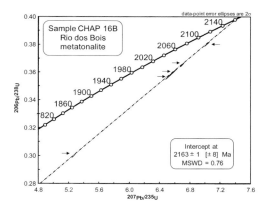

Fig. 6. Concordia diagram for ID-TIMS zircon analyses of a metatonalite (sample CHAP 16B). Data-point error ellipses are 2σ.

Nd isotopic data

Volcano-sedimentary sequence. Samples selected for Sm–Nd analyses represent all different metasedimentary rock types of the Campinorte sequence (Table 4). Since the lithological association and provenance pattern discussed above point to a proximal depositional site in an arc setting, it is expected that the metasedimentary samples will display Nd-isotopic compositions which are similar to the surrounding Palaeoproterozoic igneous arc rocks.

$^{147}Sm/^{144}Nd$ values for the metasedimentary rocks vary from 0.10–0.14, and T_{DM} model ages are between 2073 and 2686 Ma. The ε_{Nd} values range from −2.1 to +3.4. The data suggest derivation from juvenile Palaeoproterozoic crust, with restricted Archaean contribution (Fig. 9). Sample CAMP-37 has a $^{147}Sm/^{144}Nd$ ratio which is significantly higher when compared to the other samples, and therefore the calculated T_{DM} model age might be overestimated. In addition, one gondite sample (CAMP 29) has positive ε_{Nd} of +2.5 and most probably represents the metamorphic product of chemical sediment formed by hydrothermal activity with a larger mantle input.

Plutonic unit. Sm–Nd data were obtained for 4 samples of metagranitic rocks (Table 4). They present $^{147}Sm/^{144}Nd$ ratios between 0.09 and 0.13, and T_{DM} model ages vary mostly between 2250 and 2360 Ma. The ε_{Nd} values at T = 2.17 Ga are positive (0.7–2.1), indicating that this igneous event was juvenile (Fig. 9).

Discussion

The 2.2–2.0 Ga crustal accretion episode is known from many Palaeoproterozoic provinces worldwide,

Table 4. *Sm–Nd isotopic data for the Campinorte sequence and associated plutonic unit*

Sample	Rock type	Sm (ppm)	Nd (ppm)	^{147}Sm/^{144}Nd	^{143}Nd/^{144}Nd	$\varepsilon_{(0)}$	$\varepsilon_{(2,\ 17)}$	T_{DM} (Ma)
Supracrustal Unit								
CAMP 37	Quartzite	6.409	27.897	0.1389	0.511726	−17.78	−1.68	2686
CAMP 29	Gondite	7.332	38.417	0.1154	0.511606	−18.74	2.55	2216
CAMP 17	Metapelite	5.914	34.892	0.1025	0.511521	−21.79	3.36	2073
97-II-127	Metarhyolite	38.584	222.404	0.1049	0.511217	−27.72	−2.14	2552
Plutonic Unit								
CAMP 51	Metagranite	3.045	20.625	0.0892	0.511174	−28.56	1.42	2278
CAMP 24	Metagranite	12.001	55.854	0.1299	0.511849	−15.39	1.84	2292
CAMP 6	Metagranite	4.005	24.087	0.1005	0.511283	−26.43	0.39	2360
CHAP16B	Metagranite	4.857	28.990	0.1013	0.511337	−25.38	1.23	2303

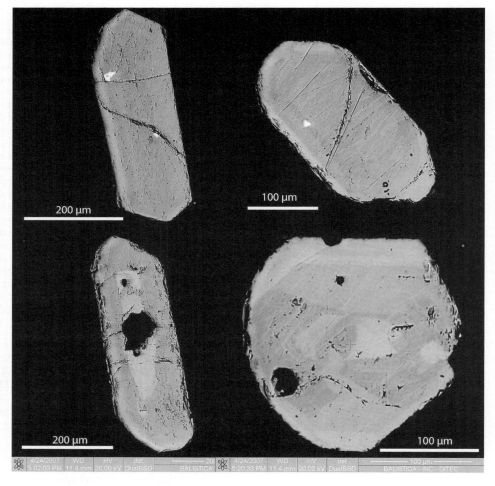

Fig. 7. BSE images of sample PP04 zircon grains.

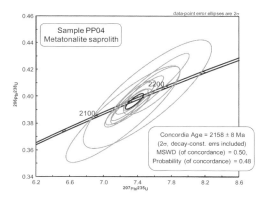

Fig. 8. Concordia diagram for LA-ICP-MS analyses of zircon grains from a tonalitic saprolite. Data-point error ellipses are 2σ.

and geological similarities among them allow global correlations, attempts of continental reconstruction and testing of the Mesoproterozoic Columbia (also known as Nuna) supercontinent hypothesis (Rogers & Santosh 2002; Zhao et al. 2002; 2004). One important feature of this reconstruction is the relationship between South America and Africa, which has been previously suggested by many authors during the last decades (Fig. 10; Abouchami et al. 1990; Feybesse et al. 1998; Teixeira et al. 2000; Brito Neves et al. 2000, 2001; Barbosa & Sabaté 2004; Duarte et al. 2004; Klein et al. 2005; Lerouge et al. 2006; Noce et al. 2007). Palaeoproterozoic continental crust is abundant in central Brazil, but is not continuous since it is mostly covered by younger sediments of Brasiliano orogenic belts.

Geological and isotopic data presented in this study suggest the correlation between the Campinorte sequence and similar provinces described in other cratons, in particular the Birimian Belt in the West African Craton and the Transamazonian Belt in the Guiana Shield. Nevertheless, since the Goiás Massif is an allochthonous block and there is no certainty of its original palaeogeographic position, such a correlation will only be accurate when palaeomagnetic data become available.

Greenstone belts from the Birimian Province present a lithostratigraphic association comparable to that described herein, except for the absence of mafic and intermediate volcanism in the Campinorte sequence (Leube et al. 1990; Davis et al. 1994; Hein

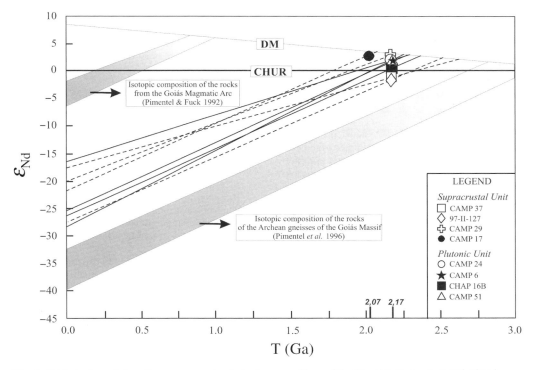

Fig. 9. Nd-isotopic evolution diagram comparing isotopic compositions of the Campinorte sequence and related intrusive rocks. Compositional fields of the Mara Rosa Magmatic Arc (Junges et al. 2002) and Goiás Archaean gneisses (Pimentel et al. 1996) are also shown.

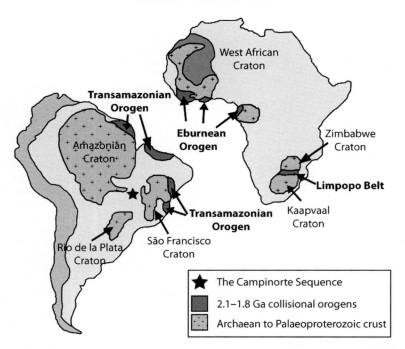

Fig. 10. Location of Palaeoproterozoic orogens including the Campinorte sequence and South American and African cratons (modified after Ledru *et al.* 1994; Zhao *et al.* 2002, 2004).

et al. 2004). The Birimian supracrustal sequences were intruded by several juvenile tonalite to granite plutons, with ages ranging from 2.19–2.12 Ga (Hirdes *et al.* 1992, 1996; Davis *et al.* 1994; Doumbia *et al.* 1998; Hirdes & Davis 2002), comparable to those small plutonic bodies recorded in the Campinorte sequence.

Moreover, in the Guiana Shield such Palaeoproterozoic greenstone belts show supracrustal associations which are very similar to the Campinorte supracrustal unit, with associated plutonic activity ranging from 2.18–2.13 Ga (Delor *et al.* 2003; McReath & Faraco 2006). The metamorphic event recorded at *c.* 2.1 Ga in the Transamazonian Province, which resulted in ultra-high temperature assemblages (Roever *et al.* 2003), may be represented in the Palaeoproterozoic context of the Campinorte sequence. A sillimanite-cordierite-garnet paragranulite that appears in a few outcrops near the Campinorte sequence yields U–Pb ages within the late Rhyacian interval (*c.* 2100 Ma; Giustina pers. comm.), reinforcing a possible match with the Transamazonian crust of the Guiana Shield.

Conclusion

The new U–Pb and Sm–Nd isotopic data presented in this study allow some relevant conclusions regarding the nature and evolution of the Palaeoproterozoic continental crust in central Brazil:

- The lithological association of detrital, chemical and volcaniclastic rocks suggests a marine depositional environment, next to an emergent Palaeoproterozoic volcanic arc. These rocks were subjected to greenschist facies metamorphism.
- U–Pb provenance studies performed on a quartzite sample from the supracrustal unit identifies a single juvenile source, and sets the maximum depositional age at *c.* 2190 Ma. The metasedimentary rocks from the Campinorte sequence were generated, therefore, from the erosion of the surrounding Palaeoproterozoic arc rocks.
- Zircon grains from the investigated granitic rocks yield ages ranging from 2.18–2.16 Ga. No inheritance was observed, suggesting that no Archaean crust has been involved in the genesis of the granitic rocks. The data suggest that the plutonic bodies related to the Campinorte sequence are relict fragments of the Palaeoproterozoic episode of crustal growth exposed in the Goiás Massif.
- The Sm–Nd isotopic data for the supracrustal rocks of the Campinorte sequence and related plutonic bodies, with positive ε_{Nd} values and

T_{DM} ranging from 2216–2552 Ma, demonstrate the juvenile character of these rocks. The data suggest that deposition of the sedimentary succession took place concomitantly with the crustal accretion processes in a volcanic arc system, with limited or no contribution from Archaean sources.

- The isotopic signature of the intrusions related to the Campinorte sequence contrasts with that from the São Francisco Craton, as initially proposed by Pimentel *et al.* (1997), in which the Palaeoproterozoic crust is dominantly a product of reworking processes rather than juvenile additions. Thus, the Sm–Nd signature of the plutonic rocks associated with the Campinorte sequence confirms the allochthonous nature of this block, which accreted to the western margin of the Brasília Belt during the Neoproterozoic Brasiliano–Pan-African orogenesis.
- The lithological associations combined with the Nd and U–Pb isotopic/geochronologic characteristics of the Campinorte volcano-sedimentary sequence and associated plutonic rocks are similar to those observed in other Palaeoproterozoic provinces described in the world, especially those of the Birimian Belt of the West African Craton and of the Transamazonian Belt in the Guiana Shield.

Support from CNPq (470183/04-7) and FAPDF/CNPq (Pronex 193.000.106/2004) research grants is thankfully acknowledged. C. Oliveira, M. Pimental, R. Fuck, and E. Dantas are CNPq research fellows. Miustina and Lelo thank CNPq fellowships. The authors thanks S. Lenharo (DITEC-Brazilian Federal Police) for the zircon BSE images and S. Junges, S. Araújo and B. Lima for providing laboratory assistance. Authors also thank C. Valeriano and D. Evans for their constructive reviews of the manuscript.

References

ABOUCHAMI, W., BOHER, M., MICHARD, A. & ALBAREDE, F. 1990. A major 2.1 Ga event of mafic magmatism in West Africa: an early stage of crustal accretion. *Journal of Geophysical Research*, **95**, 17 605–17 629.

BARBOSA, J. S. F. & SABATÉ, P. 2004. Archean and Palaeproterozoic crust of the São Francisco Craton, Bahia, Brazil: geodynamic features. *Precambrian Research*, **133**, 1–27.

BRITO NEVES, B. B. & CORDANI, U. G. 1991. Tectonic evolution of South America during the late Proterozoic. *Precambrian Research*, **53**, 23–40.

BRITO NEVES, B. B., SANTOS, E. J. & VAN SHMUS, W. R. 2000. Tectonic history of the Borborema Province, Northeastern Brazil. *In*: CORDANI, U. G., MILANI, E. J., THOMAZ FILHO, A. & CAMPOS, D. A. (eds) *Tectonic Evolution of South America*. 31st. International Geological Congress, Rio de Janeiro–RJ, 151–182.

BRITO NEVES, B. B., VAN SCHMUS, W. R. & FETTER, A. H. 2001. Noroeste da África-Nordeste do Brasil (Província Borborema): ensaio comparativo e problemas de correlação. *Geologia USP-Série Científica*, **1**(1), 59–78.

BUHN, B., PIMENTEL, M. M., MATTEINI, M. & DANTAS, E. L. 2009. High spatial resolution analysis of Pb and U isotopes for geochronology by laser ablation multi-collector inductively coupled plasma mass spectrometry (LA-MC-ICP-MS). *Anais da Academia Brasileira de Ciências*, **81**, 1–16.

DAVIS, D. W., HIRDES, W., SCHALTEGGER, U. & NUNOO, E. A. 1994. U–Pb constraints on deposition and provenance of Birimian and gold-bearing Tarkwaian sediments in Ghana, West Africa. *Precambrian Research*, **67**, 89–107.

DELOR, C., LAHONDÈRE, D. *ET AL*. 2003. Transamazonian crustal growth and reworking as revealed by the 1:500 000-scale geological map of French Guiana (2nd edn) *Géologie de la France*, **2-3-4**, 5–57.

DE PAOLO, D. J. 1981. A neodymium and strontium isotopic study of the Mesozoic calc-alkaline granitic batholiths of the Sierra Nevada and Peninsular Ranges, California. *Journal of Geophysical Research*, **86**, 10 470–10 488.

DOUMBIA, S., POUCLET, A., KOUAMELAN, A., PEUCAT, J. J., VIDAL, M. & DELOR, C. 1998. Petrogenesis of juvenile-type Birimian (Palaeoproterozoic) granitoids in Central Côte-d'Ivoire, West Africa: geochemistry and geochronology. *Precambrian Research*, **87**, 33–63.

DUARTE, B. P., VALENTE, S. C., HEILBRON, M. & CAMPOS NETO, M. C. 2004. Petrogenesis of the orthogneisses of the Mantiqueira Complex, Central Ribeira Belt, SE Brazil: an Archaean to Palaeoproterozoic basement unit reworked during the Pan-African Orogeny. *Gondwana Research*, **7**(2), 437–450.

FEYBESSE, J. L., JOHAN, V. *ET AL*. 1998. The West Central African belt: a model of 2.5–2.0 Ga accretion and two-phase orogenic evolution. *Precambrian Research*, **87**, 151–216.

FUCK, R. A. 1994. A Faixa Brasília e a compartimentação tectônica na Província Tocantins. *In*: SBG, *Simpósio de Geologia do Centro-Oeste, 4*, Anais, 184–187.

FUCK, R. A., PIMENTEL, M. M., SOARES, J. E. P. & DANTAS, E. L. 2005. Compartimentação de Faixa Brasília. *In*: SBG, *Simpósio de Geologia do Centro-Oeste*, Anais, **9**, 26–27.

GIOIA, S. M. C. L. & PIMENTEL, M. M. 2000. The Sm–Nd isotopic method in the Geochronology Laboratory of the University of Brasília. *Anais da Academia Brasileira de Ciências*, **72**, 219–245.

GIUSTINA, M. E. S. D., OLIVEIRA, C. G., PIMENTEL, M. M. & BUHN, B. 2009. Neoproterozoic magmatism and high-grade metamorphism in the Goiás Massif: new LA-MC-ICPMS U–Pb and Sm–Nd data and implications for collisional history of the Brasília Belt. *Precambrian Research*, **172**, 67–79.

HEIN, K. A. A., MOREL, V., KAGONÉ, O., KIEMDE, F. & MAYES, K. 2004. Birimian lithological succession and structural evolution in the Gorem segment of the Boromo-Gorem Greenstone Belt, Burkina Faso. *Journal of African Earth Sciences*, **39**, 1–23.

HIRDES, W. & DAVIS, D. W. 2002. U–Pb Geochronology of Palaeoproterozoic rocks in the southern part of the Kedougou-Kéniéba Inlier, Senegal, West Africa: evidence for diachronous accretionary development of the Eburnean Province. *Precambrian Research*, **118**, 83–89.

HIRDES, W., DAVIS, D. W. & EISENLOHR, B. N. 1992. Reassessment of Proterozoic granitoid ages in Ghana on the basis of U/Pb zircon and monazite dating. *Precambrian Research* **56**, 89–96.

HIRDES, W., DAVIS, D. W., LÜDTKE, G. & KONAN, G. 1996. Two generations of Birimian (Palaeoproterozoic) volcanic belts in northeastern Côte d'Ivoire (West Africa), as demonstrated by precise U–Pb mineral dating: consequences for the 'Birimian controversy'. *Precambrian Research*, **80**, 173–191.

JACKSON, S. E., PEARSON, N. J., GRIFFIN, W. L. & BELOUSOVA, E. A. 2004. The application of laser ablation-inductively coupled plasma-mass spectrometry to in situ U–Pb zircon geochronology. *Chemical Geology*, **211**, 47–69.

JOST, H., FUCK, R. A. ET AL. 2006. Geologia e Geocronologia do Complexo Uvá, Bloco Arqueano de Goiás. *Revista Brasiliera de Geociências*, **35**(4): 559–572.

JUNGES, S. L., PIMENTEL, M. M. & MORAES, R. 2002. Nd isotopic study of the Neoproterozoic Mara Rosa Arc, central Brazil: implications for the evolution of the Brasília Belt. *Precambrian Research*, **117**, 101–118.

KLEIN, E. L., MOURA, C. A. V. & PINHEIRO, B. L. S. 2005. Palaeoproterozoic crustal evolution of the São Luís Craton, Brazil: evidence from zircon geochronology and Sm–Nd isotopes. *Gondwana Research*, **8**, 177–186.

KROGH, T. E. 1973. A low-contamination method for hydrothermal decomposition of zircon and extraction of U and Pb for isotopic age determinations. *Geochimica et Cosmochimica Acta*, **37**, 485–494.

KUYUMJIAN, R. M., OLIVEIRA, C. G., CAMPOS, J. E. G. & QUEIROZ, C. L. 2004. Geologia do limite entre os terrenos arqueanos e o arco magmático de Goiás na região de Chapada-Campinorte, Goiás. *Revista Brasiliera de Geociências*, **34**(3), 329–334.

LEDRU, P., JOHAN, V., MILÉSI, J. P. & TEGYEY, M. 1994. Markers of the last stages of the Palaeoproterozoic collision: evidence for a 2 Ga continent involving circum-South Atlantic provinces. *Precambrian Research*, **69**, 169–191.

LEROUGE, C., COCHERIE, A. ET AL. 2006. SHRIMP U–Pb zircon age evidence for Palaeoproterozoic sedimentation and 2.05 Ga. Transamazonian belt of the NE Brazil and Central Africa. *Journal of African Earth Sciences*, **44**(4–5), 413–427.

LEUBE, A., HIRDES, W., MAUER, R. & KESSE, G. O. 1990. The early Proterozoic Birimian Supergroup of Ghana and some aspects of its associated gold mineralization. *Precambrian Research*, **46**, 139–165.

LUDWIG, K. R. 1993. PBDAT. *A computer program for processing Pb–U–Th isotope data*. USGS Open File Report, **34**, 88–542.

LUDWIG, K. R. 2001a. *Squid 1.02. A User's Manual*. BGC Special Publication 2, Berkeley, 19.

LUDWIG, K. R. 2001b. *User's Manual for Isoplot/Ex v. 2.47. A Geochronological Toolkit for Microsoft Excel*. BGC Special Publication 1a, Berkeley, 55.

LUDWIG, K. R. 2003. *User's Manual for Isoplot/Ex v. 3.00. A Geochronological Toolkit for Microsoft Excel*. BGC Special Publication 4, Berkeley, 71.

MARANGONI, Y., ASSUMPÇÃO, M. & FERNANDES, E. P. 1995. Gravimetria em Goiás, Brasil. *Revista Brasiliera de Geofísica*, **13**, 205–219.

MCREATH, I. & FARACO, M. T. L. 2006. Palaeoproterozoic greenstone-granite belts in Northern Brazil and former Guyana Shield–West African Craton Province. *Geologia USP Série Científica*, **5**(2), 49–63.

NOCE, C. M., PEDROSA-SOARES, A. C., SILVA, L. C., ARMSTRONG, R. & PIUZANA, D. 2007. Evolution of polycyclic basement complexes in the Araçuaí Orogen, based on U–Pb SHRIMP data: implications for Brazil–Africa links in Paleoproterozoic time. *Precambrian Research*, **159**(1–2), 60–78.

OLIVEIRA, C. G. & PIMENTEL, M. M. 1998. *Geology of the northern part of the Hidrolina dome and Sm–Nd isotopic data for metavolcanic and granitoid rocks in the Campinorte region, Goiás, central Brazil*. 14th International Conference on Basement Tectonics, Ouro Preto-MG, Anais, 55–56.

OLIVEIRA, C. G., PIMENTEL, M. M., MELO, L. V. & FUCK, R. A. 2004. The cooper-gold and gold deposits of the Neoproterozoic Mara Rosa Magmatic Arc, central Brazil. *Ore Geology Reviews*, **25**, 285–299.

OLIVEIRA, C. G., OLIVEIRA, F. B., DANTAS, E. L. & FUCK, R. A. 2006. *Programa Geologia do Brasil – Folha Campinorte*. FUB/CPRM, Brasília, 124.

PIMENTEL, M. M., FUCK, R. A. & SILVA, J. L. H. 1996. Dados Rb-Sr e Sm–Nd da região de Jussara-Goiás-Mossâmedes (GO), e o limite entre terrenos antigos do Maciço de Goiás e o Arco Magmático de Goiás. *Revista Brasiliera de Geociências*, **26**, 61–70.

PIMENTEL, M. M., WHITEHOUSE, M. J., VIANA, M. G., FUCK, R. A. & MACHADO, N. 1997. The Mara Rosa Arc in the Tocantins Province: further evidence for Neoproterozoic crustal accretion in Central Brazil. *Precambrian Research*, **81**, 299–310.

PIMENTEL, M. M., FUCK, R. A., JOST, H., FERREIRA FILHO, C. F. & ARAÚJO, S. M. 2000. The basement of the Brasília Belt and the Goiás Magmatic Arc. *In*: CORDANI, U. G., MILANI, E. J., THOMAZ FILHO, A. & CAMPOS, D. A. (eds) *Tectonic Evolution of South America*. 31st International Geological Congress, Rio de Janeiro, 195–229.

ROEVER, E. W. F., LAFON, J. M., DELOR, C., COCHERIE, A., ROSSI, P., GUERROT, C. & POTREL, A. 2003. The Bakhuis ultrahigh-temperature granulite belt (Suriname): 1. Petrological and geochronological evidence for a counterclockwise P–T path at 2.07–2.05 Ga. *Géologie de la France*, **2-3-4**, 175–205.

ROGERS, J. J. W. & SANTOSH, M. 2002. Configuration of Columbia, a Mesoproterozoic supercontinent. *Gondwana Research*, **5**, 5–22.

SATO, K. & SIGA, O. 2000. Evidence of the superproduction of the continental crust during Palaeoproterozoic in South American Plataform. Implications regarding the interpretative value of the Sm–Nd model ages. *Revista Brasiliera de Geociências*, **30**, 126–129.

SATO, K. & SIGA, O. 2002. Rapid growth of continental crust between 2.2 to 1.8 Ga in the South American Plataform: integrated Australian, European, North American and SW USA crustal evolution study. *Gondwana Research*, **5**, 165–173.

SOARES, J. E. P., BERROCAL, J., FUCK, R. A., MOONEY, W. D. & VENTURA, D. B. R. 2006. Seismic characteristics of central Brazil crust and upper mantle: a deep seismic refraction study. *Journal of Geophysical Research*, **111**, 1029–1060.

STACEY, J. S. & KRAMERS, J. D. 1975. Approximation of terrestrial lead isotope evolution by a two-stage model. *Earth and Planetary Science Letters*, **26**, 207–221.

TEIXEIRA, W., SABATÉ, P., BARBOSA, J., NOCE, C. M. & CARNEIRO, M. A. 2000. Archean and Palaeoproterozoic evolution of the São Francisco craton, Brazil. *In*: CORDANI, U. G., MILANI, E. J., THOMAZ FILHO, A. & CAMPOS, D. A. (eds) *Tectonic Evolution of South America*. 101–137.

ZHAO, G., CAWOOD, P. A., WILDE, S. A. & SUN, M. 2002. Review of global 2.1–1.8 Ga orogens: implications for a pre-Rodinia supercontinent. *Earth Science Reviews*, **59**, 125–162.

ZHAO, G., SUN, M., WILDE, S. A. & LI, S. 2004. A Palaeo-Mesoproterozoic supercontinent: assembly, growth and breakup. *Earth Science Reviews*, **67**, 91–123.

Evidence for 2.35 to 2.30 Ga juvenile crustal growth in the northwest Borborema Province, NE Brazil

TICIANO J. S. DOS SANTOS[1]*, ALLEN H. FETTER[2,3], W. RANDALL VAN SCHMUS[2] & PETER C. HACKSPACHER[4]

[1]*Instituto de Geociências, Universidade Estadual de Campinas (UNICAMP), PO Box 6152, CEP 13081-970, Campinas, SP, Brazil*

[2]*Department of Geology, University of Kansas, Lawrence, Kansas 66045-7613, USA*

[3]*US Nuclear Regulatory Commission 11555 Rockville Pike, Rockville, MD 20852, USA*

[4]*Instituto de Geociências e Ciências Exatas, Universidade Estadual Paulista (UNESP), PO Box 13506-900, Rio Claro, SP, Brazil*

Corresponding author (e-mail: ticiano@ige.unicamp.br)

Abstract: The *c.* 600 Ma Brasiliano Borborema Province of NE Brazil comprises a complex collage of Precambrian crustal blocks cut by a series of continental-scale shear zones. The predominant basement rocks in the province are 2.1–2.0 Ga Transamazonian gneisses of both juvenile and reworked nature. U–Pb zircon and Sm–Nd whole-rock studies of tonalite-trondhjemite-granodiorite basement gneisses in the NW Ceará or Médio Coreaú domain in the northwestern part of the Borborema Province indicate that this represents a continental fragment formed by 2.35–2.30 Ga juvenile crust. This block has no apparent genetic affinity with any other basement gneisses in the Borborema Province, and it does not represent the tectonized margin of the *c.* 2.1–2.0 Ga São Luis Craton to the NW. The petrological and geochemical characteristics, as well as the Nd-isotopic signatures of these gneisses, are consistent with their genesis in an island arc setting. This finding documents a period of crustal growth during a period of the Earth's history which is known for its tectonic quiescence and paucity of crust formation.

The study area is situated in the NW part of the Borborema Province (Almeida *et al.* 1981), NE Brazil, which formed as a result of Neoproterozoic convergence of the Amazonian, West African, São Luis and São Francisco-Congo cratons during the assembly of West Gondwana. The northern portion of this province comprises three major tectonic domains: Rio Grande do Norte domain (RGND), Ceará Central domain (CCD) and Médio Coreaú domian (MCD, Fig. 1). These domains comprise a collage of Archaean crustal fragments or nuclei, large tracts of Palaeoproterozoic basement or massifs, Palaeoproterozoic to Neoproterozoic supracrustal fold belts and cover sequences, and Neoproterozoic syn- to late-kinematic Brasiliano/Panafrican granitoid plutons (Van Schmus *et al.* 1995). To the NW of the Transbrasiliano lineament, a series of gneissic units constitutes the Granja-Senador Sá massif of the Médio Coreaú domain (Fig. 2). The shear zones that pervade the NW Borborema Province (Transbrasiliano lineament and Senador Pompeu lineament) range from localized accommodation zones within individual crustal domains to major tectonic lineaments that transect the entire province and can be traced into Africa (Kandi fault and Ile Ife fault; Caby 1989; Trompette 1994; Santos *et al.* 2008).

These major tectonic lineaments commonly mark the boundaries between different crustal blocks or domains, but in some cases they simply cut across them (Granja, Senador Sá and Água Branca shear zones; Fig. 2). U–Pb and Sm–Nd geochronological data from the Central Ceará domain (Martins *et al.* 1998; Fetter 1999; Fetter *et al.* 2000; Castro *et al.* 2004) and from the São Luis–West African Craton (Abouchami *et al.* 1990; Boher *et al.* 1992; Vanderhaeghe *et al.* 1998; Doumbia *et al.* 1998; Klein and Moura 2001; Klein *et al.* 2005) show an important link between these two terranes; crustal growth in these two regions occurred mainly during the middle Palaeoproterozoic, between 2.1 and 2.2 Ga. The Médio Coreaú domain exists between the São Luis–West African Craton and the CCD (Fig. 1); it comprises migmatitic tonalitic to granodioritic gneisses that constitute the greater portion of the complex, with subordinate charnockites, kinzigites and enderbites (Nogueira *et al.* 1989; Gaudette *et al.* 1993). With the exception of the localized kinzigites, Palaeoproterozoic supracrustal rocks have

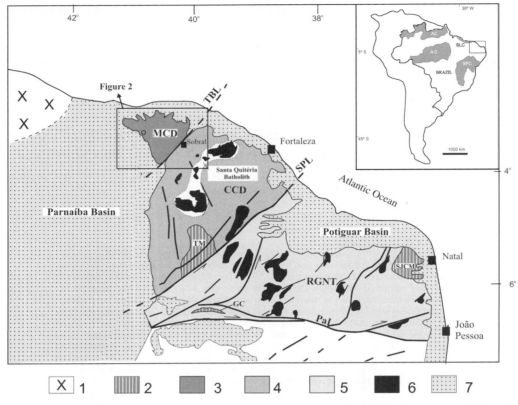

Fig. 1. Geological map of the northern part of the Borborema Province showing the main lithotectonic units. Box in NW Ceara shows Figure 2. 1, São Luis craton; 2, Archaean terrains; 3, Médio Coreaú domain; 4, Ceará Central domain; 5, Rio Grande do Norte terrain; 6, Brasiliano granites; 7, Phanerozoic sediments. Am, Amazonian Craton; SFC, São Francisco Craton; SLC, São Luis Craton; PaL, Patos lineament; SPL, Senador Pompeu lineament; TBL, Transbrasilian lineament, TM, Tróia Massif; GC, Granjeiro Complex; SJCM, São José de Campestre Massif.

not been identified within this massif. Initial attempts by Hackspacher et al. (1991) and Gaudette et al. (1993) to date the orthogneissic basement in the MCD yielded a range of ages spanning more than a billion years. By plotting a series of Rb–Sr analyses from various orthogneisses in the region, Hackspacher et al. (1991) obtained a Rb–Sr whole-rock isochron age of 3.134 ± 134 Ma for the Granja Massif. Subsequent zircon studies of orthogneisses from the same massif yielded $^{207}Pb/^{206}Pb$ evaporation ages ranging from 2.25–2.03 Ga (Gaudette et al. 1993). Based on these ages, Gaudette et al. (1993) concluded that the rocks of this domain formed during Transamazonian orogenesis.

The aim of this article is to present new U–Pb and Sm–Nd isotopic data from the basement rocks of the Médio Coreaú domain and to use them to define a tectonic configuration that involved the continental crust of the Ceará Central domain and the São Luis–West African Craton during the assembly of a middle Palaeoproterozoic supercontinent.

Geological setting

Archaean crust constitutes only a small component of the crustal framework of the Borborema Province. It is found in the three different Archaean blocks: the Tróia Massif (TM), the Granjeiro Complex (GC) and the São Jose do Campestre Massif (SJCM; Fig. 1). Geochronological data suggest that the three crustal fragments are unrelated, and presumably have different origins. The Granjeiro Complex is the youngest of the Archaean domains, c. 2540 Ma (Silva et al. 2002) and is characterized by juvenile crustal signatures. The Tróia Massif comprises juvenile and reworked Archaean crust, which ranges in age from c. 2675 Ma to around 2857 Ma (Fetter 1999).

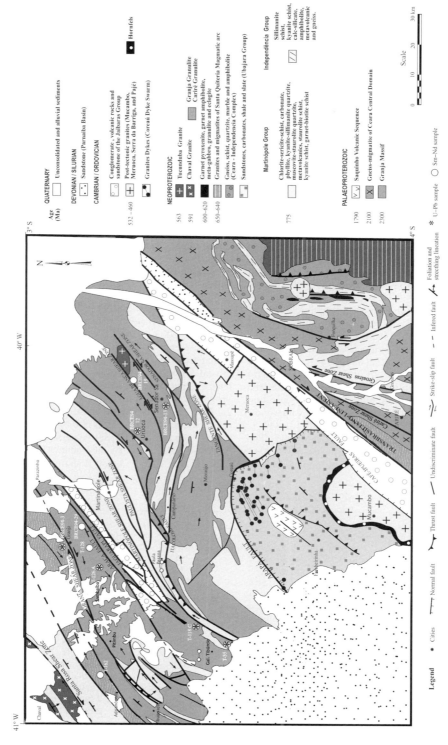

Fig. 2. Geological map of NW Ceará (Médio Coreaú domain) showing the main geological units and locations of zircon and Sm–Nd samples from basement gneisses.

Rocks of the São José do Campestre Massif in the eastern part of the Rio Grande do Norte terrane (Van Schmus et al. 1995; Silva et al. 2002; Dantas et al. 2004) are also Archaean, but most of the rocks there are considerably older than those of the Tróia Massif, ranging in age from 3.45–2.70 Ga (Dantas et al. 2004). Both juvenile and reworked crust are present in the São José do Campestre Massif with depleted mantle Nd model ages (TDM) values ranging from 3.73–3.18 Ga. With the exception of the Granjeiro Complex in the Rio Grande do Norte terrane, all of the Archaean crustal fragments in the Borborema, Nigerian and Hoggar provinces appear to have been derived from the São Francisco and Congo cratons.

Basement of the Médio Coreaú domain comprises: (a) the Granja Massif, a high-grade complex composed mainly of orthogneisses with tonalite-trondhjemite-granodiorite (TTG) affinities; (b) amphibolite gneisses, leucogranites, pyrigarnite and garnet amphibolite interlayered with kinzigites, mafic granulites, enderbites and leptinites; and (c) migmatites (Caby & Arthaud 1986; Santos et al. 2002). The TTG orthogneisses are mainly composed of biotite gneiss, amphibole gneiss (hornblende gneiss and ferrohastingsite gneiss) and leucogneiss with modal predominance of plagioclase (andesine-oligoclase). Locally this unit shows intense deformation along transcurrent shear zones. This unit crops out in the General Tibúrcio and Pitimbu localities. The granulitic ortho- and paragneisses consist of mafic granulite, enderbite and kinzigite gneiss. Thermobarometric values indicate granulitization temperature around 750–840 °C and pressure of 7–8 kbar (Nogueira Neto 1996). The kinzigite gneisses are aluminous rocks composed of kyanite, sillimanite and almandine garnet, and the enderbite corresponds to an orthopyroxene gneiss of tonalitic composition. Gama (1992) defined a high-temperature mylonitization with ortho- and clinopyroxene crystal-plastic deformation. Textural features common in the mafic granulite and enderbite gneiss are symplectites (clinopyroxene + plagioclase ± amphibole intergrowths) and reaction coronas around garnet that indicate isothermal decompression and isobaric cooling, respectively (Nogueira Neto 2000). The migmatites are banded, stromatic and crosscut by shear bands with incipient migmatization. The granular granitic leucosomes suggest pervasive percolation of high-temperature fluids during shearing. Santos (1993) defines this unit as the youngest of the basement, with two recorded phases of migmatization. The first one is marked by a progressive development of stromatic to nebulitic structures, and the second phase is characterized by late granitic injection. K/Ar and Rb/Sr data yield Neoproterozoic (Novaes et al. 1979; Prado et al. 1981; Nascimento et al. 1981), Mesoproterozoic (Cavalcante 1993), Palaeoproterozoic (Hackspacher et al. 1991; Gaudette et al. 1993; Santos 1993; Abreu et al. 1993) and Archaean ages (Hackspacher et al. 1991). Pb–Pb evaporation zircon data obtained from the Granja gneiss yield an age of 2028 ± 38 Ma, whereas the Riacho da Gangorra gneiss yields an age of 2253 ± 27 Ma (Gaudette et al. 1993). These previous data show crustal reworking during the Archaean, Palaeoproterozoic and Neoproterozoic, but the main crustal growth occurred during the Transamazonian (2.2–2.1 Ga), and the youngest thermal event occurred during the Brasiliano (around 600 Ma). Table 1 summarizes the most significant geochronological data from the Médio Coreaú basement. The basement rocks of the Médio Coreaú domain were affected

Table 1. *Geochronological data from Médio Coreaú basement*

Rock	Locality	Method	Age (Ma)	Reference
Migmatite	Granja	K/Ar horn	566 ± 17	Almeida et al. (1968)
Gneiss	South Granja	K/Ar	545 ± 16	Brito Neves et al. (1974)
Migmatite	16 km N of Coreaú	K/Ar	566 ± 17	Brito Neves et al. (1974)
Gneiss	South Granja	Rb/Sr wr	1150 ± 176	Brito Neves (1975)
Gneiss	Granja	Rb/Sr wr	1700	Novaes et al. (1979)
Migmatite	Granja	Rb/Sr wr	1609 ± 26	Nascimento et al. (1981)
Gneiss	Gal. Tibúrcio	Rb/Sr wr	3134 ± 94	Hackspacher et al. (1991)
Granulite	North Granja	Rb/Sr wr	1915 ± 19	Abreu & Lafon (1991)
Granulite	North Granja	Rb/Sr wr	1929 ± 60	Abreu & Lafon (1991)
Migmatite	Riacho da Gangorra	Rb/Sr wr	2400	Gama Jr (1992)
Gneiss	North Granja	Pb/Pb wr	2028 ± 38	Gaudette et al. (1993)
Gneiss	Riacho da Gangorra	Pb/Pb	2253 ± 27	Gaudette et al. (1993)
Gneiss	Riacho da Gangorra	Rb/Sr wr	2320 ± 277	Gaudette et al. (1993)
Granulite	Granja	Ar/Ar amph	575.1 ± 4.8	Monié et al. (1997)
Granulite	Granja	Ar/Ar biot	558.7 ± 2.5	Monié et al. (1997)

amph, amphibole; biot, biotite; horn, hornblende; wr, whole rock.

by two phases of intracratonic extension at 1785 Ma and 775 Ma, resulting in the deposition of widespread volcano-sedimentary supracrustal rocks, represented by the Saquinho volcanic sequence (Santos et al. 2002) and Martinópole and Ubajara groups (Fetter et al. 2003), respectively. Syn-, late- and post-tectonic (late Neoproterozoic to Cambrian) granitoids generated during the Brasiliano orogeny intrude both basement and supracrustal units.

Analytical methods

U–Pb and Sm–Nd data shown in Tables 2 and 3 were obtained at the Isotope Geochemistry Laboratory (IGL), Department of Geology and Kansas University Center for Research, University of Kansas, Lawrence. Full details of the sample preparation and analytical methods are given in Fetter (1999); essential points are summarized here.

Mineral fractions were prepared, dissolved and Pb and U were purified using procedures modified after Krogh (1973, 1982) and Parrish (1987) using a ^{205}Pb–^{235}U tracer solution. Isotopic ratios were measured with a VG sector mass spectrometer using Faraday/Daly multicollector mode for samples with strong signals (>200 millivolts for ^{206}Pb) and single-collector mode using the Daly detector for weaker signals. Both Pb and U isotopic compositions were analysed on single Re filaments using silica gel and phosphoric acid and corrected for average mass discrimination of $0.12 \pm 0.05\%$ per mass unit for Faraday collectors and $0.18 \pm 0.05\%$ per mass unit for the Daly collector mode (based on replicate analyses of common Pb standard SRM 981). Uranium fractionation was monitored by replicate analyses of SRM U-500. Uncertainties in U/Pb ratios due to uncertainties in fractionation and mass spectrometry were $\pm 0.5\%$; in some instances weak signals caused uncertainties up to 2%. Radiogenic Pb isotopes were calculated by correcting for modern blank Pb and for original nonradiogenic original Pb corresponding to the Stacey & Kramers (1975) model Pb for the approximate age of the sample. Decay constants and isotopic ratios used in the age calculations are those listed by Steiger & Jäger (1977). Total procedure blanks over the course of analyses ranged from 2–25 pg for lead and 0.5–4 pg for uranium. Zircon data (Table 2) were regressed using ISOPLOT (Ludwig 2001). Uncertainties in concordia intercept ages are given at the 2σ level.

For Sm–Nd analysis, rock powders and mineral concentrates were loaded in 23 ml PTFE Teflon Parr bombs and spiked with a mixed ^{149}Sm/^{150}Nd tracer. Dissolution and Sm–Nd separation followed methods modified after Patchett & Ruiz (1987). Samarium samples were loaded onto single tantalum filaments with 0.25 M phosphoric acid and analysed as Sm^+ in static multicollector or single collector mode. Nd samples were loaded onto single rhenium filaments with AG50W-X8 cation resin and 0.25 M phosphoric acid and analysed as Nd^+ in a dynamic multicollector mode to collect between 50 and 100 ratios at 1V of ^{144}Nd signal intensity. All analyses are adjusted for instrumental bias assuming a ^{143}Nd/^{144}Nd ratio of 0.511860 for the La Jolla Nd standard; Nd-isotopic compositions are normalized to a ^{146}Nd/^{144}Nd ratio of 0.7219. During the course of these analyses, Nd blanks ranged from 500 to <100 pg, with corresponding Sm blanks of 100 to <50 pg. Depleted mantle model ages (TDM) were calculated from the equation $\varepsilon Nd\ (T) = 0.25T^2 - 3T + 8.5$ (DePaolo 1981), where T is the model age. $\varepsilon Nd\ (t)$ values were calculated using the U–Pb ages listed in Table 3, where t is the measured or inferred crystallization age of the rock.

Results

Although ^{207}Pb/^{206}Pb evaporation-method ages from zircons are considered to be more reliable than Rb–Sr ages, there are still limitations to the technique. With the ^{207}Pb/^{206}Pb evaporation method no uranium data are obtained, thus it is neither possible to observe zircon Pb loss trajectories nor to plot discordia through any data. In addition, it is very difficult to evaluate the effects of xenocrystic components in a zircon population. These complications may preclude obtaining accurate protolith crystallization ages. Due to the uncertainties from the previous studies, some of the localities sampled by Hackspacher et al. (1991) and Gaudette et al. (1993) were sampled again for U–Pb and Sm–Nd analyses (Fig. 2). Sm–Nd whole-rock data of these basement gneisses were the primary tools used to investigate the crustal residence of this domain, with U–Pb zircon analyses used to determine protolith crystallization ages. Five samples were selected for analysis based on their geographic distribution and zircon yields. Four orthogneiss samples, ranging from granodiorite to tonalite in composition, were selected to determine the crystallization ages of the igneous protoliths, and a fifth sample, a granulite-grade metasedimentary rock (kinzigite), was chosen to evaluate zircon provenance and to determine the age of metamorphism. The U–Pb results from these samples indicate that this domain formed prior to the c. 2.1 Ga Transamazonian orogeny, with crustal growth spanning about 70 Ma.

Samples BRCE94-12 and T-118C, a tonalitic gneiss and granodioritic gneiss, respectively, yield the oldest U–Pb zircon ages in this domain.

Table 2. U–Pb zircon data for rocks from the Granja Complex

Fraction[a]	Size (mg)	U ppm	Pb ppm	$^{206}Pb/^{204}Pb$ (obs.)	$^{207}Pb*/^{235}U$	$2\sigma\%$	$^{206}Pb*/^{238}U$	$2\sigma\%$	(rho)	$^{207}Pb*/^{206}Pb$	$2\sigma\%$	$^{206}Pb*/^{238}U$ Age (Ma)	2σ	$^{207}Pb*/^{235}U$ Age (Ma)	2σ	$^{207}Pb*/^{206}Pb$ Age (Ma)	2σ
T-35A																	
M(−2)	0.063	119	38	6345	5.3428	0.58	0.29212	0.58	0.990	0.13265	0.06	1.652	10	1.876	11	2.133	1.1
M(−1)	0.078	112	30	3563	4.0571	0.90	0.23666	0.89	0.996	0.12434	0.08	1.369	12	1.646	15	2.019	1.5
T-118C																	
NM(−1)	0.021	278	56	2554	3.1796	0.71	0.19707	0.64	0.910	0.11702	0.30	1.160	7	1.452	10	1.911	5.3
M(−1)	0.027	280	58	4371	3.3558	0.60	0.20402	0.57	0.949	0.11930	0.19	1.197	7	1.494	9	1.945	3.4
M(−1)	0.033	248	63	1231	4.2309	0.71	0.23973	0.69	0.975	0.12800	0.16	1.385	10	1.680	12	2.071	2.8
M(0)	0.009	500	105	2306	3.3716	0.79	0.20479	0.74	0.946	0.11940	0.26	1.201	9	1.498	12	1.947	4.6
T-126[a]																	
NM(−2)	0.054	217	79	4695	6.0857	0.53	0.32090	0.51	0.975	0.13754	0.12	1.794	9	1.988	11	2.197	2.0
M(−2)	0.059	169	58	3517	5.6163	0.55	0.30090	0.53	0.979	0.13537	0.11	1.794	10	1.919	11	2.169	1.9
M(−1)	0.052	216	66	4435	5.0491	0.54	0.27622	0.52	0.976	0.13258	0.12	1.794	9	1.828	10	2.132	2.0
M(0)	0.028	194	71	1381	6.0613	0.66	0.32034	0.64	0.975	0.13723	0.15	1.794	11	1.985	13	2.192	2.6
M(1)	0.011	309	82	886	4.1732	1.02	0.23842	0.99	0.977	0.12695	0.22	1.794	18	1.669	17	2.056	3.8
BRCE94 − 3																	
NM(−2)L	0.004	255	91	2537	6.1970	0.68	0.33829	0.67	0.989	0.13286	0.10	1.879	13	2.004	14	2.136	1.8
M(−1)L	0.091	667	209	41540	5.4807	0.47	0.31126	0.46	0.994	0.12771	0.05	1.747	8	1.898	9	2.066	0.9
M(0)L	0.085	690	226	97354	5.8669	0.46	0.32720	0.46	0.994	0.13005	0.05	1.824	8	1.956	9	2.098	0.9
M(1)L	0.082	747	245	51953	5.8430	1.40	0.32730	1.40	0.999	0.12946	0.07	1.825	26	1.953	27	2.091	1.2
M(−2)R	0.002	555	209	6918	6.175	0.58	0.34338	0.57	0.998	0.13024	0.09	1.905	11	2.001	12	2.101	1.6
M(−1)R	0.005	780	297	15912	6.2469	0.64	0.34681	0.63	0.991	0.13064	0.09	1.919	12	2.011	13	2.107	1.5
M(0)R	0.001	1007	338	1110	5.7093	0.85	0.32892	0.84	0.993	0.32892	0.10	1.833	15	1.933	16	2.041	1.8
BRCE94 − 12																	
NM(−2)	0.036	76	36	1060	8.2688	0.75	0.40234	0.70	0.949	0.14905	0.23	2.180	15	2.260	17	2.335	4.0
M(−2)	0.012	108	52	1922	8.6495	0.99	0.41867	0.97	0.985	0.14984	0.17	2.254	22	2.302	23	2.344	2.9
M(−1)	0.010	86	42	1310	8.7797	1.38	0.42420	1.38	0.995	0.15011	0.14	2.280	31	2.315	32	2.347	2.5
M(0)	0.004	127	62	1138	8.6463	0.87	0.41836	0.86	0.993	0.14989	0.11	2.253	19	2.301	20	2.345	1.8

Total U and Pb concentrations corrected for analytical blank.
$^{206}Pb/^{204}Pb$ not corrected for blank or non-radiogenic Pb.
Radiogenic Pb corrected for blank and initial Pb; U corrected for blank.
Ages given in Ma using decay constants recommended by Steiger & Jäger (1977).
T-126 intercept ages (sphene not included): Upper-intercept age = 2286 ± 6 Ma; Lower-intercept age = 549 ± 17 Ma; MSWD = 0.7.
T-126 intercept ages (all mineral fractions regressed): Upper-intercept age = 2288 ± 2 Ma; Lower-intercept age = 554 ± 4 Ma; MSWD = 0.8.
[a] Magnetic fraction.
*Radiogenic Pb corrected for blank and initial Pb; U corrected for blank.

Table 3. *Sm–Nd whole-rock data from Granja Complex, NE Brazil*

Sample number	Nd ppm	Sm ppm	$^{147}Sm/^{144}Nd$	$^{143}Sm/^{144}Nd$	±	2σ	$\varepsilon Nd\,(0Ma)^1$	$\varepsilon Nd\,(600Ma)^1$	$(t)^3$ Ga	$\varepsilon Nd\,(t)^4$	$T\,(DM)^5$ Ga
BRCE94-3	26.96	4.93	0.11052	0.511334	±	10	−25.4	−18.9	2.33	0.4	2.54
BRCE94-12	59.72	8.32	0.08423	0.510960	±	9	−32.7	−24.1	2.35	0.5	2.46
BRCE94-13	17.01	2.65	0.09411	0.511012	±	8	−31.7	−23.9			2.61
BRCE94-19	31.85	3.60	0.06825	0.510717	±	8	−37.5	−27.7			2.45
BRCE94-29	9.27	1.55	0.10076	0.511216	±	8	−27.7	−20.4			2.48
T-35A	13.58	1.88	0.08371	0.511003	±	8	−31.9	−23.3	2.3	1.5	2.40
T-35B	14.83	1.94	0.07907	0.510949	±	9	−32.9	−24.0	2.35	1.9	2.38
T-118C	13.79	2.01	0.08825	0.511034	±	8	−31.3	−23.0	2.35	1.5	2.45
T-126	58.39	9.95	0.10304	0.511256	±	10	−27.0	−19.8	2.35	0.7	2.47
T-128	41.87	7.09	0.10236	0.511233	±	8	−27.4	−20.2			2.49
T-162	39.15	5.87	0.09069	0.511098	±	8	−30.0	−21.9			2.42

Notes: (1) Calculated assuming $^{143}Nd/^{144}Nd$ today = 0.512638 with data normalized to $^{146}Nd/^{144}Nd$ = 0.72190. Epsilon Nd (today) = $[(^{143}Nd/^{144}Nd\,[\text{sample now}])/0.512638) - 1] \times 10^4$. (2) Epsilon Nd (600Ma) = $((^{143}Nd/^{144}Nd\,[\text{sample 600Ma}]/^{143}Nd/^{144}Nd\,[\text{CHUR, 600Ma}]) - 1) \times 10^4$. CHUR = Chondritic Uniform Reservoir (DePaolo 1981). (3) Calculated following model of DePaolo (1981).

Analyses of four fractions of colourless, euhedral zircons from sample BRCE94-12 yielded an upper-intercept age of 2356 ± 7 Ma (Fig. 3, Table 2). Because the data from these zircons are collinear and plot close to concordia, the upper-intercept age is considered to reflect the primary crystallization age of this orthogneiss. Four fractions of pink, turbid euhedral zircons from sample T-118C were also analysed and yielded strongly discordant analyses indicating an upper-intercept age of 2358 ± 19 Ma (Fig. 4, Table 2). The higher degree of Pb loss in these zircons may be due to radiation damage caused by their higher uranium content (Table 2). In any case, the upper-intercept age obtained for this sample is consistent with other zircon ages of the gneisses in this domain and is interpreted to be a good estimate of the protolith crystallization age.

Two other samples of tonalitic gneiss from the NW Ceará domain have crystallization ages that are younger. Five zircon fractions and one titanite fraction of sample T-126A plot in a co-linear array, intersecting concordia at 2288 ± 2 Ma and 554 ± 4 Ma (Fig. 5, Table 2). The older age is interpreted as the protolith crystallization age and the younger age is considered to reflect a younger partial resetting age. Two zircon fractions from sample T-35A were analysed and by themselves yield an upper-intercept age of 2271 ± 11 Ma (Table 2). Although two zircon fractions are considered inadequate for a rigorous age determination, these data indicate that sample T-35A is roughly the same age as the other gneiss samples. This is supported by the plot of these two analyses in Figure 6.

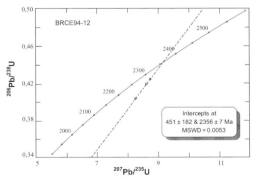

Fig. 3. Concordia plot for four nearly concordant zircon fractions from sample CE94-12. A Model 1 (Ludwig 2001) solution for all four fractions yields an upper intercept age of 2356.1 ± 6.6 Ma and a lower intercept age of 458 ± 180 Ma, with MSWD = 0.081 and probability of fit = 0.92. We interpret the upper-intercept age as the primary magmatic age of the rock.

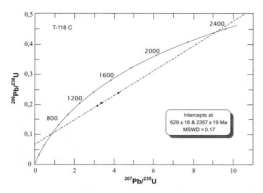

Fig. 4. Concordia plot for four discordant zircon fractions from sample T-118C. These fractions are discordant with respect to their primary Palaeoproterozoic age, reflecting strong inluence from the Brasiliano orogeny. A Model 1 (Ludwig 2001) solution for all four fractions yielding an upper-intercept age of 2358 ± 20 Ma and a lower-intercept age of 630 ± 18 Ma, with MSWD = 0.18 and probability of fit = 0.83. We interpret the upper-intercept age as the primary magmatic age of the rock.

Figure 6 shows a composite plot for the four samples discussed above. The general co-linearity of the data reinforces the conclusion that these samples represent c. 2.36–2.27 Ga magmatism.

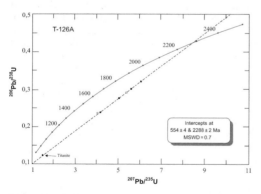

Fig. 5. Concordia plot for five moderately discordant zircon fractions from sample T-126A (closed squares). Two moderately discordant zircon fractions from sample T-35A are represented by open squares and are not included in regression. A Model 1 solution for all five fractions yields an upper-intercept age of 2286.4 ± 5.6 Ma and a lower-intercept age of 549 ± 17 Ma, with MSWD = 0.83 and probability of fit = 0.48. We interpret the upper-intercept age as the primary magmatic age of the rock. The titanite analysis was excluded from regression, but it is compatible with the zircon age, and including it in the regression does not significantly change the age.

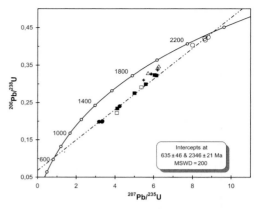

Fig. 6. Composite plot for all zircon and titanite data showing general concordance for samples CE94-12 (open circles), T-126A (closed squares), T-35A (open squares) and T-118C (closed circles) (Figs 3–5). The regression for these data should only be considered as a reference chord, not as a definitive age determiantion. Both types of zircon for sample CE94-03 (Table 2) plot distinctly above the composite concordia (open trangles and crosses), suggesting that this sample may be slightly younger overall, but nonetheless related to early Palaeoproterozoic crust.

Since the more discordant data yielded the younger ages, it is possible that they also crystallized at approximately 2.36 Ga and the chord has been rotated slightly by cryptic Pb loss. The regression for these data should only be considered as a reference chord, not as a definitive age determination.

U–Pb zircon studies from the kinzigite, sample BRCE94-3, are also shown in Figure 6. Two distinct zircon populations were identified in this kinzigite: spherical, clear, pale-yellow grains and elongate (3:1 aspect ratios), subhedral, clear to slightly turbid brown grains. Because these two zircon populations were significantly different with respect to both colour and morphology, it was considered that they might have had different origins, that is detrital igneous grains v. metamorphic grains. Spherical, multifaceted ('soccer ball') metamorphic zircons have been documented in some granulite-grade rocks (Toteu et al. 1994), thus the spherical, clear grains were considered as possible metamorphic zircon of either Transamazonian or Brasiliano age. The elongate zircons were not considered to be metamorphic candidates because they were somewhat turbid and have igneous zoning. Analyses of three fractions of the spherical grains (S) yield an upper intercept age of 2341 ± 76 Ma, and four fractions of elongate zircons (L) yield an imprecise upper intercept age of 2277 ± 290 Ma (Table 2). Since these ages fall within the age range of the

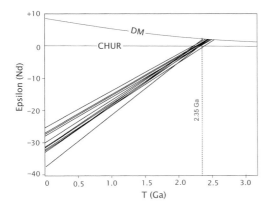

Fig. 7. Nd evolution diagram showing data from the NW Ceará domain. Dashed line perpendicular to the *x*-axis is the age reference line for this domain. These data are also presented in Table 3.

orthogneisses in this domain, it is suggested that the original sediments were derived from them, even though they do not plot on the same composite discordia defined by the four magmatic samples (Fig. 6). Clearly, more detailed studies are needed to resolve this question.

Sm–Nd data obtained from 12 basement gneiss samples widely distributed within the domain yield TDM ages between 2.61 and 2.38 Ga (Table 3, Fig. 7; Fetter *et al.* 2000). The TDM ages from the suite of rocks analysed by Hackspacher *et al.* (1991) range between 2.49 and 2.38 Ga. Even without any zircon data it is clear that these rocks are not Mesoarchaean, as originally proposed by those workers. Only the oldest TDM age of 2.61 Ga is suggestive of Archaean material, but zircon data do not support this. The εNd (t = crystallization age) values calculated for the gneisses of this domain are all positive, and the TDM ages of the gneisses also correspond closely with their U–Pb zircon ages (Table 3). These data indicate that this domain is composed of mostly juvenile early Palaeoproterozoic crust. In contrast, the adjoining orthogneiss basement in the Ceará Central domain is distinctly younger (middle Palaeoproterozoic) and comprises a composite collage of reworked and juvenile crust. Both the antiquity and juvenile nature of the gneisses of the Médio Coreaú domain suggest that it evolved as an isolated entity, that is an island arc complex, prior to the main Transamazonian orogeny.

Conclusions

Based on geochronological data from the northern portion of NE Brazil and northwestern Africa (Van Schmus *et al.* 1995; Potrel *et al.* 1998; Thiéblemont *et al.* 2001; Silva *et al.* 2002; Dantas *et al.* 2004), it is possible to suggest that Archaean crustal fragments in the Borborema, Nigerian and Hoggar provinces were derived from the São Francisco and Congo cratons during a rifting episode that preceded the *c.* 2.1 Ga Transamazonian–Eburnian orogeny. This hypothesis finds support when it is recognized that large portions of the Palaeoproterozoic terrains, mainly from the São Luis Craton and Rio Grande do Norte domain, incorporated significant older crustal material, probably via subduction and melting of Archaean detritus from the rifted cratons. Although many of the Palaeoproterozoic terranes in the Borborema Province were enriched by Archaean crust, large tracts of Palaeoproterozoic gneisses grew and evolved free from its influence, exhibiting juvenile signatures. Rifting of the Archaean crust probably also involved the generation and subsequent closure of Palaeoproterozoic oceanic lithosphere. The continuous convergence was responsible for the development of the 2.1 Ga Transamazonian–Eburnian orogeny, an accretionary-collisional event that culminated with the fusion of diverse juvenile and reworked Palaeoproterozoic terranes, along with smaller Archaean fragments. In this context, the Granja Complex constitutes a singular terrane in West Gondwana. This complex is essentially composed of juvenile rocks and corresponds to an early stage of continental crust generation in an active subduction zone prior to the dominant crustal growth in the São Luis–West African Craton and Ceará Central domain. The crustal growth event around 2.4–2.3 Ga probably occurred in many areas, since similar ages have been identified in the Rio Grande do Norte Domain of the Borborema Province (Dantas *et al.* 2008), the Dabakala area of the West African Craton (Gasquet *et al.* 2003), southwestern portions of the Churchill Craton, Canada (Hartlaub *et al.* 2007) and the southeastern part of the Amazonian Craton (Vasquez *et al.* 2006).

References

ABREU, F. A. M. & LAFON, J. M. 1991. Granulitos Transamazônicos no Cinturão de Cisalhamento Noroeste do Ceará. Atas do XIV Simpósio de Geologia do Nordeste, **29**, 229–231.

ABREU, F. A. M., HASUI, Y. & GORAYEB, P. S. S. 1993. Grábens eopaleozóicos do oeste cearense – considerações sobre as seqüências lito-estratigráficas. *In*: *Simpósio Geologia do Nordeste*. 15. Atas, Boletim Núcleo Nordeste da SBG, **13**, 29–31.

ABOUCHAMI, W., BOHER, M., MICHARD, A. & ALBAREDE, F. 1990. A major 2.1 Ga old event of mafic magmatism in West Africa: an early stage of crustal accretion. *Journal of Geophysical Research*, **95**, 17 605–17 629.

ALMEIDA, F. F. M., MELCHEER, G. C., CORDANI, U. G., KAWASHITA, K. & VANDOROS, P. 1968. Radiometric age determinations from northern Brazil. *Boletim da Sociedade Brasileira de Geologia*, **17**, 3–14.

ALMEIDA, F. F. M., HASUI, Y., BRITO NEVES, B. B. & FUCK, R. A. 1981, Brazilian structural provinces: an introduction. *Earth Science Reviews*, **17**, 1–29.

BOHER, M., ABOUCHAMI, W., MICHARD, A., ALBARPDE, F. & ARNDT, N. T. 1992. Crustal growth in West Africa at 2.1 Ga. *Journal of Geophysical Research*, **97**, 345–369.

BRITO NEVES, B. B. 1975. Regionalização tectônica do precambriano nordestino. Thesis, Universidade de São Paulo (unpublished).

BRITO NEVES, B. B., VANDOROS, P., PESSOA, D. A. R. & CORDANI, U. 1974. Reavaliação dos dados geocronológicos do Pré-Cambriano do Nordeste brasileiro. XXVIII Congresso Brasileiro de Geologia. Porto Alegre, **6**, 261–271.

CABY, R. 1989. Precambrian terranes of Benin-Nigeria and northeast Brazil and the Late Proterozoic south Atlantic fit. *In:* DALLMEYER, R. D. (ed.) *Terranes in the Circum-Atlantic Palaeozoic Orogens*. Geological Society of America, Special Paper, **230**, 145–158.

CABY, R. & ARTHAUD, M. 1986. Major Precambrian nappes of the Brazilian belt, Ceará, northeast Brazil. *Geology*, **14**, 871–874.

CASTRO, N. A., BASEI, M. A. S. & CAMPOS NETO, M. C. 2004. Geocronologia e evolução tectônica Proterozóica do Domínio Ceará Central (Região entre Madalena e Taperuaba, Província Borborema, NE do Brasil). *Simp. 40 Anos de Geocronologia no Brasil*, USP-São Paulo, 74.

CAVALCANTE, J. DE C. 1993. Estratigrafia Precambriana do Estado do Ceará – Uma síntese. *In: Simpósio Geologia do Nordeste*. 15. Atas. Natal, Boletim Núcleo Nordeste da SBG. **13**, 313–316.

DANTAS, E. L., VAN SCHMUS, W. R. *ET AL.* 2004. The 3.4–3.5 Ga São José do Campestre Massif, NE Brazil: remnants of the oldest crust in South America. *Precambrian Research*, **130**, 113–137.

DEPAOLO, D. J. 1981. A neodymium and strontium isotopic study of the Mesozoic calc-alkaline granitic batholiths of the Sierra Nevada and Peninsular Ranges, California. *Journal of Geophysical Research*, **86**, 10 470–10 488.

DOUMBIA, S., POUCLET, A., VIDAL, M., KOUAMELAN, A., PEUCAT, J. J., VIDAL, M. & DELOR, C. 1998. Petrogenesis of juvenile-type Birimian (Palaeoproterozoic) granitoids in central Côte d'Ivoire, West Africa: geochemistry and geochronology. *Precambrian Research*, **87**, 33–63.

FETTER, A. H. 1999. *U–Pb and Sm–Nd constraints on the crustal framework and geologic history of Ceará State, NW Borborema Province, NE Brazil: implications for the assembly of Gondwana*. Ph.D. thesis, University of Kansas, USA.

FETTER, A. H., VAN SCHMUS, W. R., DOS SANTOS, T. J. S., ARTHAUD, M. & NOGUEIRA NETO, J. A. 2000. U–Pb and Sm–Nd geochronological constraints on the crustal evolution and basement architecture of Ceará State, NW Borborema Province, NE Brazil: implications for the existence of the Palaeoproterozoic supercontinent 'Atlantica'. *Revista Brasileira de Geociências*, **30**, 102–106.

FETTER, A. H., SARAIVA DOS SANTOS, T. J. *ET AL.* 2003. Evidence for Neoproterozoic continental arc magmatism in the Santa Quitéria batholith of Ceará State, NW Bornorema Province, NE Brazil: implications for the assembly of West Gondwana. *Gondwana Research*, **6**, 265–273.

GAMA, T. 1992. *Geologia do setor nordeste da zona de cisalhamento de Granja – noroeste do Ceará*. Curso de Pós-graduação em Geociências, Universidade Federal do Pará. Ph.D. thesis.

GASQUET, D., BARBEY, P., ADOU, M. & PAQUETTE, J. L. 2003. Structure, Sr–Nd isotope geochemistry and zircon U–Pb geochronology of the granitoids of the Dabakala area (Côte d'Ivoire): evidence for a 2.3 Ga crustal growth event in the Palaeoproterozoic of West Africa? *Precambrian Research*, **127**, 329–354.

GAUDETTE, H. E., ABREU, F. A. M., LAFON, J. M. & GORAYEB, P. S. S. 1993. Evolução Transamazônica do cinturão de cisalhamento noroeste do Ceará: nova evidências geocronológicas. XV Simpósio de Geologia do Nordeste, Resumos, 316–319, Natal.

HACKSPACHER, P. C., SANTOS, T. J. S. & LAFON, J. M. 1991. Evolução geocronológica do complexo gnaissico-migmatítico do NW Ceará. XIV Simpósio de Geologia do Nordeste, Recife, 260–271.

HARTLAUB, R. P., HEAMAN, L. M., CHACKO, T. & ASHTON, K. E. 2007. Circa 2.3 Ga magmatism of the Arrowsmith Orogeny, Uranium City Region, Western Churchill Craton, Canada. *The Journal of Geology*, **115**, 181–195.

KLEIN, E. L. & MOURA, C. A. V. 2001. Age constraints on granitoids and metavolcanic rocks of the São Luis Craton and Gurupi Belt, northern Brazil: implications for lithostratigraphy and geological evolution. *International Geology Review*, **43**, 237–253.

KLEIN, E. L., MOURA, C. A. V. & PINHEIRO, L. S. 2005. Palaeoproterozoic crustal evolution of the São Luis Craton, Brazil: evidence from zircon geochronology and Sm–Nd isotopes. *Gondwana Research*, **8**, 177–186.

KROGH, T. E. 1973. A low contamination method for hydrothermal decomposition of zircon and extraction of U and Pb for isotopic age determinations. *Geochimica et Cosmochimica Acta*, **37**, 485–494.

KROGH, T. E. 1982. Improved accuracy of U–Pb zircon ages by the creation of more concordant systems using an air abrasion technique. *Geochimica et Cosmochimica Acta*, **46**, 637–649.

LUDWIG, K. R. 2001. *Isoplot/Ex (rev. 2.49), a geochronological toolkit for Microsoft Excel*. Berkeley Geochronology Center Special Publication No. 1a, University of California, Berkeley.

MARTINS, G., OLIVEIRA, E. P., SOUZA FILHO, C. R. & LAFON, J. M. 1998. Geochemistry and geochronology of the Algodões sequence, Ceará, NE Brazil: a Palaeoproterozoic magmatic arc in the Central Ceará domain of the Borborema Province? *XL Congresso Brasileiro de Geologia*, Belo Horizonte, Anais, 28.

MONIÉ, P., CABY, R. & ARTHAUD, M. H. 1997. The Neoproterozoic Brasiliano orogeny in northeast

Brazil: $^{40}Ar/^{39}Ar$ ages and petrostructural data from Ceará. *Precambrian Research*, **81**, 241–264.

NASCIMENTO, D. A., GAVA, A., PIRES, J. L. & TEIXEIRA, W. 1981. Geologia da folha sa 24–Fortaleza. *In*: *Projeto Radam Brasil*. DNPM, Rio de Janeiro. **21**, 23–212.

NOGUEIRA NETO, J. A. 1996. Evolução Metamórfica das Faixas Granulíticas de Granja e Cariré (NW do Ceará). *In*: Seminário de Pós-Graduação em Geociências, Rio Claro. IGCE/UNESP. Res. Exp. 328–333.

NOGUEIRA NETO, J. A. 2000. Evolução Geodinâmica das Faixas Granulíticas de Granja e Cariré, Extremo Noroeste da Província Borborema. Ph.D. thesis, UNESP–Rio Claro.

NOGUEIRA NETO, J. A., MARQUESM, F., JR., NERI, T. F. O. & PEDREIRA, L. H. S. T. 1989. Contribuição à geologia do município de Chaval (CE). *In*: *Atas do XIII Simpósio de Geologia do NE*. Fortaleza, SBG, Bol. **11**, 27–28.

NOVAES, F. R. G., BRITO NEVES, B. B. & KAWASHITA, K. 1979. Reconhecimento cronoestratigráfico da região nordeste do Estado do Ceará. *In*: *Atas do XI Simpósio de Geologia do NE*. Natal, 93–110.

PARRISH, R. R. 1987. An improved micro-capsule for zircon dissolution in U–Pb geochronology. *Isotope Geoscience*, **66**, 99–102.

PATCHETT, P. J. & RUIZ, J. 1987. Nd isotopic ages of crust formation and metamorphism in the Precambrian of eastern and southern Mexico. *Contributions to Mineralogy and Petrology*, **96**, 523–528.

POTREL, A., PEUCAT, J. J. & FANNING, C. M. 1998. Archean crustal evolution of the West African Craton: example of the Amsaga Area (Reguibat Rise). U–Pb and Sm–Nd evidence for crustal growth and recycling. *Precambrian Research*, **90**, 107–117.

PRADO, F. S., MENDONA, J. C. G., MORAES, J. B. A., NEDEIROS, M. F., ANDRADE, P. R. S. & MEDEIROS, R. P. 1981. *ProjetoMartinópole*. DNPM/CPRM, Relatório Final, v. 5.

SANTOS, T. J. S. 1993. *Aspectos geológicos de uma área a sudoeste de Granja, região noroeste do Ceará*. UNESP – Rio Claro. Dissertação de Mestrado. 159.

SANTOS, T. J. S., SOUZA, G. M., QUEIROZ, H. B., NOGUEIRA NETO, J. A. & PARENTE, C. V. 2002. Tafrogênese estateriana no embasamento palaeoproterozóico do NW da Província Borborema: Uma aboradagem petrográfica, geoquímica e geocronológica. *In*: XLI Congresso Brasileiro de Geologia, João Pessoa. 337. *Actas V Symposium on South American Isotope Geology*, Argentina, 1–4.

SANTOS, T. J. S., FETTER, A. H. & NETO, J. A. N. 2008. South American and African Pan-Gondwana correlations: the link between the northwestern part of the Borborema Province (NE Brazil) and southwestern Pharusian (Dahomey) belt (SW central Africa). *In*: PANKHURST, R. J., TROUW, R. A. J., BRITO NEVES, B. B. & DE WIT, M. J. (eds) *West Gondwana: pre-Cenozoic correlations across the South Atlantic region*. Geological Society, London, Special Publications, **294**, 101–119.

SILVA, DA L. C., ARMSTRONG, R. *ET AL*. 2002. Reavaliação da evolucão geológica em terrenos pré-cambrianos brasileiros com base em novos dados U–Pb SHRIMP, Parte III: províncias Borborema, Mantiqueira Meridional e Rio Negro-Juruena. *Revista Brasileira de Geociências*, **32**, 529–544.

STACEY, J. S. & KRAMERS, J. D. 1975. Approximation of terrestrial lead isotope evolution by a two stage model. *Earth and Planetary Science Letters*, **26**, 207–251.

STEIGER, R. H. & JÄGER, E. 1977. Subcommission on Geochronology: convention on the use of decay constants in geo- and cosmochronology. *Earth and Planetary Science Letters*, **28**, 359–362.

THIÉBLEMONT, D., DELOR, C. M. *ET AL*. 2001. A 3.5 Ga granite–gneiss basement in Guinea: further evidence for early archean accretion within the West African Craton. *Precambrian Research*, **108**, 179–194.

TOTEU, S. F., VAN SCHMUS, W. R., PENAYE, J. & NYOBÉ, J. B. 1994. U–Pb and Sm–Nd evidence for Eburnian and Pan-African high-grade metamorphism in cratonic rocks of southern Cameroon. *Precambrian Research*, **67**, 321–347.

TROMPETTE, R. 1994. *Geology of Western Gondwana, Pan-African – Brasiliano Aggregation of South America and Afric*. A. A. Balkema, Rotterdam, Brookfield, 350.

VANDERHAEGHE, O., LEDRU, P., THIÉBLEMONT, D., EGAL, E., COCHERIE, A., TEGYEY, M. & MILESI, J. P. 1998. Contrasting mechanism of crustal growth: geodynamic evolution of the Palaeoproterozoic granite-greenstone belts of French Guiana. *Precambrian Research*, **92**, 165–193.

VAN SCHMUS, W. R., BRITO NEVES, B. B., HACKSPACHER, P. & BABINSKI, M. 1995. U/Pb and Sm/Nd geochronolgic studies of eastern Borborema Province, northeastern Brazil: initial conclusions. *Journal of South American Earth Sciences*, **8**, 267–288.

VASQUEZ, M. L., MACAMBIRA, M. J. B. & ARMSTRONG, R. A. 2006. SHRIMP ages for granitoids from Bacajá domain, southeastern Amazonian craton, Brazil; new evidence of Siderian accretion. *Actas V Symposium on South American Isotope Geology*, Argentina, 250–253.

SHRIMP U–Pb c. 1860 Ma anorogenic magmatic signatures from the NW Himalaya: implications for Palaeoproterozoic assembly of the Columbia Supercontinent

SANDEEP SINGH[1]*, A. K. JAIN[1] & MARK E. BARLEY[2]

[1]*Department of Earth Sciences, Indian Institute of Technology Roorkee, Roorkee – 247 667, India*

[2]*The School of Earth and Geographical Sciences, University of Western Australia, 35 Stirling Highway, Crawley 6009, Western Australia, Australia*

Corresponding author (e-mail: sandpfes@iitr.ernet.in)

Abstract: The basal parts of the Higher Himalayan Crystallines (HHC), Lesser Himalayan sedimentary sequences and mylonite zone at the base of Main Central Thrust (MCT) within the NW Himalaya clearly demonstrate anorogenic magmatic signatures at around 1860 Ma, as indicated by SHRIMP U–Pb zircon ages from Bandal granitoids, Kulu–Bajura mylonite and Wangtu granitoids along the Sutlej Valley, Himachal Pradesh. Some of the zircon crystals contain older cores mostly extending back to 2600 Ma. We report for the first time a 3000 Ma old zircon core from Wangtu granitoids, which indicates reworking of ensialic Archaean crust during the assembly of the Columbia Supercontinent between 2.1 and 1.8 Ga. During the Himalayan collisional tectonics, the reworked Archaean and Palaeoproterozoic crust was imbricated and placed adjacent to each other in the Higher Himalayan Crystallines, the Inner Lesser Himalayan window zone and the Kulu–Bajura Nappe.

The Himalaya is the youngest evolving active mountain belt in the world, displaying various geodynamic processes and the development of distinct tectonic units (Fig. 1; Gansser 1964; Le Fort 1975; Honegger et al. 1982; Coward et al. 1982; Valdiya 1989; Searle et al. 1993; Thakur 1993; Hodges 2000; Jain et al. 2002; Yin 2006). This highest mountain chain in the world extends laterally for about 2500 km from Nanga Parbat (8126 m, 33°15′N:74°36′E) in the west to Namche Barwa (7756 m, 29°37′N:95°15′E) in the east and has a width of 250–300 km. The geology of this orogenic belt has been summarized in recent syntheses (Hodges 2000; Jain et al. 2002; Steck 2003; Yin 2006), following an initial attempt by Gansser (1964). These syntheses mostly deal with the Cenozoic collisional tectonics with very little emphasis on the pre-Himalayan history.

The pre-Himalayan history has been mainly constrained by the presence of various granitoids, whose ages were assigned from the Proterozoic to Late Cenozoic, based on field relationships, nature of xenoliths, degree of metamorphism, petrographical similarities and structural trends, prior to the application of isotopic dating methods (McMahon 1884; Greisbach 1893; Auden 1935; Wadia 1928, 1957; and others). In the Himalaya, the first radiometric age was reported by Jager et al. (1971) with a Rb–Sr whole-rock isochron for the Mandi Granite, followed by Rb–Sr age determinations by Bhanot et al. (1974, 1975, 1976), Frank et al. (1977), Mehta (1977) and others from the NW Himalaya. These ages clearly indicate the presence of pre-Himalayan granitoids, now metamorphosed to gneiss, and ranging in age from 2.0–0.5 Ga (Singh 2001, 2005; Singh & Jain 2003). These rock bodies are mostly confined to the Higher Himalayan Crystallines (HHC) and the North Himalayan Gneissic Domes.

The HHC contain three main types of granitoids ranging in age from 2600 Ma to as young as 2 Ma along distinct zones and can be grouped into a Proterozoic belt, Pan-African belt and the collision-related Cenozoic Higher Himalayan Leucogranite (HHL) belt, respectively (see Singh & Jain 2003 for details). The rocks from the basal zone HHC, between the Main Central Thrust in the south and the Vaikrita Thrust in the north, are generally fine- to coarse-grained megacrystic gneiss, associated with metasediments and amphibolites, and are the oldest in the Himalaya (Table 1). These are variously called the Iskere gneiss, Kotla and Shang gneiss (in Pakistan), Rameshwar granitoid, Bandal granitoid, Kulu–Bajura mylonitic gneiss, Wangtu Granitic Complex, Naitwar, Hanuman Chatti, Bhatwari, Namik, Gwalda, Chailli, Ghuttu,

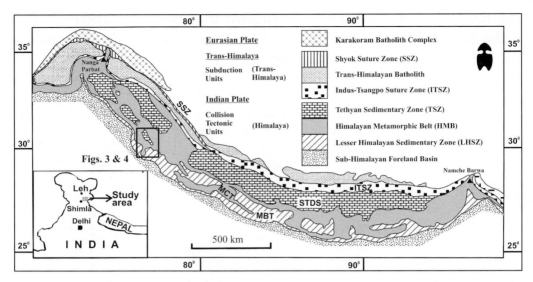

Fig. 1. Simplified regional geological map of the Himalaya in plate tectonic framework. SSZ, Shyok Suture Zone; ITSZ, Indus Tsangpo Suture Zone; STDS, South Tibetan Detachment System; MCT, Main Central Thrust; MBT, Main Boundary Thrust. Compiled from published data.

Chirpatiya, Rihee-Ganga, Ramgarh, Almora–Askot–Dhramgarh gneiss (all in India), to the Lingtse, Darjeeling-Sikkim, Bomdila and Kalaktang granite gneiss (in NE Himalaya). These bodies are meta-aluminous to peraluminous in composition, with S-type characteristics. They have SiO_2 concentrations between 60 and 76% and plot in the Within Plate Granite (WPG) field in the Rb v. Y + Nb discrimination diagram of Pearce et al. (1984), indicating their anorogenic character (Fig. 2).

These data play an important role in the reconstruction of the pre-Himalayan configuration, which has not been attempted prior to Pan-African times (Le Fort et al. 1987). Recently, much attention has been paid to reconstruction of the Meso-Neoproterozoic Rodinian supercontinent, but the position of the Himalaya (Greater India) has not been discussed in detail (Dalziel 1991, 1992; Powell et al. 1993; Li et al. 1996; Dalziel et al. 2000). The concept of 'Greater India' postulates the northern extension of the Indian sub-continent in the Himalaya (Argand 1924; Veevers et al. 1971, 1975). This configuration was widely accepted in the Gondwana fit of the Indian Plate with the Australian and Antarctica plates, where Greater India was placed next to Western Australia (Powell et al. 1988). A recent synthesis by Ali & Aitchison (2005) details the 'Greater India' concept. The Greater India acquires places/gaps north of the Indian Plate containing rocks of present-day Lesser Himalayan Sedimentary Zone and Higher Himalayan Crystallines. The existence of Greater India can be traced back to the pre-Rodinian supercontinent of Columbia in the Palaeo- to Mesoproterozoic (Rogers & Santosh 2002; Zhao et al. 2002, 2003; Santosh et al. 2003; Zhao et al. 2004). Between 2.1 and 1.8 Ga, various collisional orogenies took place, as listed by Zhao et al. (2003). Within India, the southern and northern Indian blocks sutured along the c. 1.8 Ga Central Indian Tectonic Zone (Jain et al. 1991; Mazumder et al. 2000). According to the configuration proposed by Zhao et al. (2003), the 'Greater India' zone was placed within the plate set-up to the north of the Central Indian Tectonic Zone. Timing also characterizes subduction-related magmatism along the continental margin of the Columbia Supercontinent (Karlstrom et al. 2001; Rogers & Santosh 2002; Condie 2002; Zhao et al. 2002, 2003). Further more, during Mesoproterozoic times (1.6–1.2 Ga), anorogenic anorthosite-magnesite-charnockite granite and rapakivi granites were generated, probably due to an extensive underplating, preceding the breakup of the Columbian Supercontinent (Rogers & Santosh 2002; Zhao et al. 2003) with contemporaneous plate-scale kimberlite emplacement due to plume activity. That resulted in the dispersion of the cratonic blocks of Columbia (Zhao et al. 2002, 2003).

The present work reports Palaeoproterozoic U–Pb SHRIMP ages of granite gneiss and mylonite from parts of the Greater Indian crust in the NW Himalaya and assesses its significance for the assembly of the Columbia Supercontinent.

Table 1. *Previous geochronological data for Proterozoic granitoid bodies of the High Himalayan Crystalline Belt*

Body	Age	Initial ($^{86}Sr/^{87}Sr)_i$	References
Iskere gneiss	1852 ± 14 Ma (U–Pb Zr)		Zeitler et al. 1989
	2500 Ma (Nd model age)		Whittington et al. 1999
Kotla Orthogneiss	1839 ± 9 Ma (U–Pb Zr)		DiPietro & Isachsen 1997
	1836 ± 1 Ma (U–Pb Zr)		DiPietro & Isachsen 2001
Shang Granodiorite	1864 ± 4 Ma (U–Pb Zr)		DiPietro & Isachsen 2001
Rameshwar granite	1820 ± 130 (6 Pt.)	0.7114 ± 0.0118	Trivedi et al. 1984
Bandal	1220 ± 100 (3 Pt.)	0.748 ± 0.100	Bhanot et al. 1976, 1979
	1840 ± 0.0027 (4 Pt.)	0.7083	Frank et al. 1977
Nirath-Baragaon	1430 ± 150 (6 Pt.)	0.746	Bhanot et al. 1978
Wangtu–Jeori	2075 ± 86 (6 Pt.)	0.7074 ± 0.01	Kwatra et al. 1986
Wangtu granite	1866 ± 10 Ma (U–Pb Zr)		Singh 1993; Singh et al. 1994
Magladgad granite stock	2068 ± 5 Ma (U–Pb Zr)		Singh 1993; Singh et al. 1994
Wangtu granite gneiss (CPG)	1895 ± 64 Ma (6 pt)	0.7044 ± 0.0072	Rao et al. 1995
Wangtu granite gneiss (FGG)	1895 ± 64 Ma (6 pt)	0.7044 ± 0.0072	Rao et al. 1995
Naitwar, Tons Valley	1811 ± 133 (4 Pt.)	0.707 ± 0.017	Singh et al. 1986
Hanuman chatti, Yamuna Valley	1972 ± 102 (4 Pt.)	0.703 ± 0.010	Singh et al. 1986
Bhatwari, Bhagirathi Valley	2047 ± 119 (4 Pt.)	0.706 ± 0.007	Singh et al. 1986
Munsiari	1830 ± 200 (1894*) (2 Pt)	0.725	Bhanot et al. 1977
	1890 ± 155 (1956*) (3 Pt.)	0.725 ± 0.010	Bhanot et al. 1977
Almora–Askot	1620 ± 90 (4 Pt.)	0.749 ± 0.007	McPowell et al. 1979
Askot Dharamgarh	1795 ± 30 (9 Pt.)	0.7090 ± 0.0015	Pandey et al. 1981
Namik	1910 ± 88 (5 Pt.)	0.724 ± 0.013	Singh et al. 1985
Tawaghat	1906 ± 220 (9 Pt.)	0.724 ± 0.012	Singh et al. 1985
Ghuttu, Bhilganga Valley	1763 ± 116 (5 Pt.)	0.727 ± 0.029	Singh et al. 1985
Chirpatiya khal, Bhilganga Valley	1708 ± 131 (4 Pt.)	0.732 ± 0.147	Singh et al. 1985
Rihee–Gangi, Bhilganga Valley	1841 ± 86 (4 Pt.)	0.710 ± 0.010	Singh et al. 1986
Chailli, Bhilganga Valley	2121 ± 60 (4 Pt.)	0.710 ± 0.20	Raju et al. 1982
Ramgarh granite	1765 ± 60 (11 Pt.)	0.7235 ± 0.0046	Trivedi et al. 1984
Gwaldam granite	1300 ± 80 (5 Pt.)	0.793	Pandey et al. 1981
	1700 ± 70 (6 Pt.)	0.7375 ± 0.0127	Trivedi et al. 1984
Lingtse granite, NE Himalaya	1075 ± 28 Ma	0.7001	Paul et al. 1982
	1678 (Rb–Sr whole rock)		Paul et al. 1996
Darjeling–Sikkim granite gneiss	1792 (Pb–Pb age)		Paul et al. 1996
Bomdila gneiss	1874 ± 24 Ma (U–Pb Zr)		Rao 1998
	1827 ± 95 Ma (U–Pb Zr)		Rao 1998
Kalaktang granite	1706 ± 80 Ma (errorchron)	0.7055 ± 0.0066	Rao 1998

*Equivalent age using new decay constant of Steiger & Jager (1977).

Geological framework

The Himalaya has evolved due to collision of the Indian Plate with the Eurasian Plate (Fig. 1; Thakur 1993; Hodges 2000; Jain et al. 2002; Yin 2006). The northern margin of the Indian Plate, Greater India, provides ample evidence of the evolution of remobilized Proterozoic basement and sedimentary cover in the continuous Himalayan Metamorphic Belt (HMB) (Fig. 1) during the Himalayan collision following closure of the Neo-Tethys Ocean, largely due to subduction of the Indian Plate (Searle 1996). The HMB is overlain by the Neoproterozoic–Eocene platform sequence of the Tethyan Sedimentary Zone (TSZ). The basement to the TSZ has been regionally deformed, metamorphosed and thrust southward along the Main Central Thrust (MCT) and its numerous splays. The central part of the dismembered Indian Plate exposes parts of the Early Proterozoic to Early Palaeozoic sedimentary basins in the Lesser Himalaya, which have experienced marine transgressions during the Permian and the Eocene (Valdiya 1995). The whole succession is thrust further southwestwards

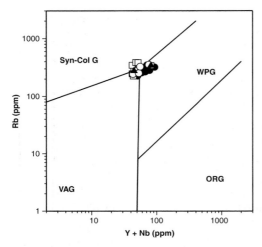

Fig. 2. Rb v. Y + Nb discrimination diagram of Pearce *et al.* (1984) with data from Wangtu granitoids (filled circle – Rao *et al.* 1995; open circle – our own data, Singh *et al.* 1995), Kulu–Bajura Nappe mylonite (filled triangle – Miller *et al.* 2000) and Bandal granitoids (open square – coarse-grained gneiss of Sharma & Rashid 2001).

over the Cenozoic foredeep sedimentary deposits along the Main Boundary Thrust (MBT) which, in turn, overrides the Holocene Indo-Gangetic alluvium along the Main Frontal Thrust (MFT). The southern undeformed margin of the Indian Plate comprises the Proterozoic Aravalli–Bundelkhand–Shillong basement and the Vindhyan platform that is largely covered by the recent alluvium (Jain *et al.* 2002 and references therein).

The area under investigation lies along the Sutlej Valley in Himachal Pradesh, between Baragaon, Nirath, Jeori and Wangtu (Fig. 3). The lowermost tectono-stratigraphic unit of the Lesser Himalayan Proterozoic sedimentary zone is exposed in the main Kulu–Rampur Window and other smaller windows. It is overthrust by the Kulu–Bajura Nappe of the mylonitized augen gneiss and the Higher Himalayan Crystallines (HHC) belt along the Main Central Thrust (MCT, Singh 1993 and references therein).

The Kulu–Rampur Window

The Lesser Himalaya comprises a Proterozoic and Early Palaeozoic volcano-sedimentary sucession, deposited over a long time span through a number of sedimentary cycles. These platform-type sequences are exposed in the outer and inner Lesser Himalayan belts. The outer belt is exposed as a continuous belt in the frontal parts, whereas the inner belt constitutes several tectonic windows due to erosion of the overlying allochthonous metamorphic nappes. Along the Sutlej Valley, these constitute part of the inner Lesser Himalayan window, which is exposed in the main Kulu–Rampur Window and smaller windows at Charoeta to the west of the Kulu–Rampur Window, the Luhri-I and the Luhri-II windows further downstream along the Sutlej river and the Shali, making a long linear window in the westernmost part of the area (Fig. 3, Singh 1993).

The main Kulu–Rampur Window is approximately 50 km long and forms a NW–SE trending linear belt of Early to Middle Proterozoic sedimentary sequence along with the Rampur Volcanics (Sm–Nd whole-rock age of 2509 ± 94 Ma, Bhat & Le Fort 1992; and Pb–Pb single zircon evaporation ages for the Rampur metabasalts of 1800 ± 13 Ma, Miller *et al.* 2000) and the overlying Manikaran Quartzite. The sequence is overlain by carbonaceous phyllite and dolomitic limestone, containing thin bands of greyish chert. A pluton having concordant relationship with the country rock; the Bandal granitoid (Rb–Sr whole-rock isochron age of mixed rocks is 1840 ± 70 Ma, Frank *et al.* 1977), intrudes the sedimentary succession in the core of the Kulu–Rampur Window (Bhanot *et al.* 1976; Frank *et al.* 1977; Sharma 1977). It is a porphyroclastic granite gneiss with K-feldspar, quartz and plagioclase augen, with biotite $\pm$ muscovite and accessory zircon, apatite and iron oxides. The pluton is concordant with the country rocks with several tongs and apophyses into the Manikaran Quartzite. The granitoid occurs as an elongated dome with its hinge almost parallel to the trends of the Kulu–Rampur Window. The granitoids are strongly foliated along their margin, although they are locally massive. The granularity varies from fine- to coarse-grained. At the contact, rocks show reduction of grain size and recrystallization. Megacrysts and xenoliths have asymmetric tails and in combination with S-C-shear fabrics, indicate a top-to-SW sense of ductile shearing. The feldspars are locally tabular in shape, with graphic and perthitic intergrowths.

The Kulu–Bajura Nappe

Overlying the Kulu–Rampur Window zone, a dismembered and cliff-forming basal thrust sheet to the HHC has a typical lithological association of highly mylonitized augen gneiss of about 800 m thick. The Kulu–Bajura Nappe, a sub-nappe to the Higher Himalayan Crystalline–Jutogh Nappe is bounded by the Kulu Thrust at its base and the MCT–Jutogh Thrust at its top. Considering the unique occurrence of highly mylonitized augen gneiss with a distinct L-S fabric, and its tectonized sharp lower and upper contacts, only this lithology

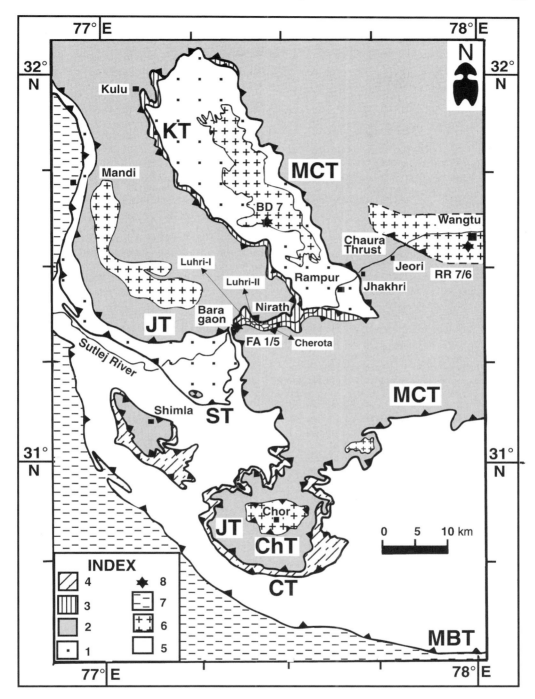

Fig. 3. Regional geological map of Himachal Pradesh. 1, Kulu-Rampur-Shali Window; 2, Higher Himalayan Crystallines (HHC); 3, Kulu–Bajura Nappe; 4, Chail Nappe; 5, Simla–Deoban–Garhwal Group; 6, granitoid/granite gneiss; 7, Sub-Himalaya Tertiary Belt; MCT, Main Central Thrust; JT, Jutogh Thrust; KT, Kulu Thrust; ChT, Chor Thrust; CT, Chail Thrust; MBT, Main Boundary Thrust; ST, Suketi Thrust. Compiled from published data and our own observations.

has been incorporated in the Kulu–Bajura Nappe. The underlying carbonaceous phyllite/slate and the Manikaran quartzite have their gradational contacts and have been included as parts of the Kulu–Rampur Window zone. On the basis of a six-point Rb–Sr whole-rock isochron, the augen gneiss has been assigned an age of 1430 ± 150 Ma, with an initial high $^{87}Sr/^{86}Sr$ ratio of 0.746, indicating derivation from remobilized crustal material (Bhanot et al. 1978).

The mylonitized augen gneiss contains large blue quartz and feldspar megacrysts. It is a grey to dark-grey, hard, compact, highly sheared mylonite with elongated and deformed alkali feldspar, quartz, plagioclase, muscovite and biotite megacrysts which are embedded in an extremely fine-grained, greenish, greasy-looking and foliated groundmass of the same minerals. Due to intense shearing along the margins, megacrysts have undergone grain-size reduction, with recrystallization and development of blastomylonite and ultramylonite. The megacrysts show flattened mantles and elongated tails of either dynamically recrystallized or reaction-softened minerals. Locally, extremely thin tails of intensely strained megacrysts are folded, where they taper off away from the megacryst. In such cases, ribbon structures parallel the preferred mica orientation. The mylonitized augen gneiss contains a well-developed mylonitic foliation due to preferred orientation and alternating quartz-feldspar-rich bands and muscovite-biotite- chlorite bands. On all scales, chlorite and mica are seen anastomosing between quartz and feldspar megacrysts along S- and C-planes, wherever shearing is prominent. The general trend of the mylonitic foliation is nearly ENE–WSW with low to moderate dips towards NNE or SSW. The most prominent character of augen mylonite is the presence of a very strong stretching/mineral lineation on the main mylonitic foliation, giving it a L-S tectonite character. This lineation is marked by strong preferred orientation of mica and streaked alkali feldspar, quartz and plagioclase augen. Both lineation and foliation are folded due to later deformations, which have produced doubly plunging antiformal windows along the Sutlej Valley, indicating that large-scale open folding has been superposed upon the mylonitization.

The Higher Himalayan Crystallines (HHC)

The Higher Himalaya Crystallines (HHC) of the NW Himalaya are exposed along the Great Himalayan Range and constitutes the 'Precambrian Basement' to the Phanerozoic Tethyan sedimentary pile (Bhargava et al. 1991). The HHC predominantly incorporates pelitic and psammitic metamorphosed sequences and deformed granitoids and gneiss with ages obtained by Rb–Sr whole-rock and U–Pb systematics of 400–500, 1800–2000 and 2500 Ma; which were all subsequently reworked during the Himalayan orogeny (Zeitler et al. 1989; Treloar et al. 1989; Pognante et al. 1990; Searle et al. 1992; Miller et al. 2000, 2001; Singh & Jain 2003; Singh 2005). The upper contact of the HHC with Tethyan sediments is marked by the Zanskar Shear Zone (ZSZ)/Trans-Himadri Thrust/South Tibetan Detachment System and other associated faults (Searle 1986; Herren 1987; Valdiya 1989; Burchfiel et al. 1992; Patel et al. 1993; Vannay & Grassemann 1998; Jain et al. 2002; Yin 2006).

The Jakhri–Wangtu–Karcham–Akpa region along the Sutlej Valley provides a good cross-section of the HHC belt. A thick stack of about 15–20 km of the HHC is exposed along the NE flank of the Kulu–Rampur Window beyond Jakhri. The HHC comprises the basal Jeori Formation and the overlying Wangtu Granitic Complex (WGC), overlain by the Vaikrita Group along the Sutlej Valley. The Jeori Formation, which lies between the MCT and Chaura Thrust, was originally delineated on the basis of deformation and strain patterns by Singh & Jain (1993) and confirmed by fission track ages (Jain et al. 2000; Thiede et al. 2004). Garnetiferous mica-schist/gneiss is the most abundant lithology in the basal parts of the HHC in the immediate vicinity of the MCT between Jakhri and Jeori. Fine- to medium-grained, dark-coloured and highly foliated amphibolite bodies are occasionally exposed with the Jeori Formation. These are mostly 1–10 m thick, except for a mappable band of about 300 m thickness near Jeori. Along a stream near Jeori, known as Maglad Khad, discordant fine-grained grey granite, now occurring as gneiss along with late-stage aplite and pegmetite, intrudes the metapelite and fine-grained banded biotite gneiss as concordant to discordant igneous bodies. The grey granite has a conventional U–Pb zircon age of 2068 ± 5 Ma, which occurred along the metamorphic banding (Singh et al. 2006).

The Wangtu granite gneiss is exposed in the northeastern parts of the Jeori Formation. It is a monotonous medium- to coarse-grained, porphyroclastic granitic gneiss having a tectonized contact along the Chaura Thrust. It imperceptibly grades into undeformed granitoid with a Rb–Sr whole-rock and conventional U–Pb zircon age of about 1.8 Ga near Wangtu (Singh 1993; Singh et al. 1994; Rao et al. 1995). These gneisses contain numerous xenoliths of highly foliated metamorphic rocks. The Wangtu granite gneiss is overthrust by garnetiferous mica-schist, staurolite-kyanite schist/gneiss, sillimanite schist/gneiss, calc-silicate, augen gneiss, migmatite and leucogranite of the Vaikrita Group (Sharma

1977; Vannay & Grassemann 1998) along the Vaikrita Thrust at Karcham (Fig. 4). The Wangtu granitoids are mainly porphyroclastic with light to dark-grey gneissic components due to variation in the amount of biotite. The megacrysts are made up of quartz, K-feldspar and plagioclase, which are wrapped around by biotite-muscovite flakes. Feldspar megacrysts are buff, elongated and slightly weathered, whereas quartz is greyish, vitreous and fresh in character. S-C shear fabrics and augen with asymmetric tails indicate a top-to-SW overthrust sense of movement. The body predominantly contains plagioclase with lesser amounts of K-feldspar, quartz and biotite. The accessory minerals include apatite, zircon and opaques. Locally, plagioclase develops myrmekitic growth possibly due to partial melting and recrystallization. Some of the plagioclase is zoned indicating disequilibrium conditions during crystallization. Locally, epidote develops at the expense of plagioclase as a result of fluid activity during subsequent deformation and metamorphism. A few feldspar crystals show sericitization, indicating incipient retrogression or fluid activity.

Analytical procedure

Unweathered samples of the Bandal granitoid (BD-7) from the Lesser Himalayan sequence, mylonitized augen gneiss from the Kulu–Bajura Nappe (FA 1/5) and the Wangtu granite gneiss (RR 7/6) of the HHC belt were prepared using standard mineral separation procedures, by crushing, sieving and heavy liquid separation. All samples yielded a large number of zircons that were hand-picked to obtain grains for the SHRIMP analyses. More than 250 zircon grains from each sample were randomly selected and mounted in epoxy resin with the CZ3 Sri Lankan zircon standard ($^{206}Pb/^{238}U = 0.0914$, corresponding to an age of 564 Ma; Pidgeon et al. 1994) on 25 mm diameter mounts. The polished mounts were examined utilizing back-scattered electron (BSE) and cathodoluminescence (CL) imaging, using a JEOL 6400 SEM at the Centre for Microscopy and Microanalysis (CMM) at the University of Western Australia. Sites for SHRIMP analysis were then selected on the basis of these images. Analyses were performed on the SHRIMP-II based at Curtin University, following the techniques described by Smith et al. (1998) and also Singh (2003).

Results

Bandal granitoid. The sample from the main Bandal granitoid is an almost undeformed to

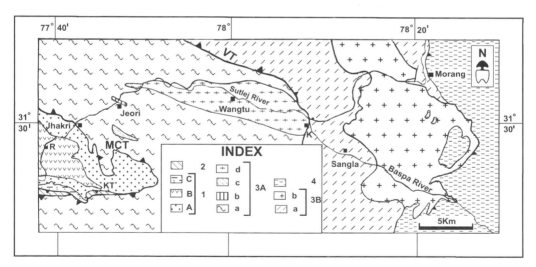

Fig. 4. Geological map of the Higher Himalayan Crystallines along the Sutlej–Baspa Valleys, Himachal Pradesh. 1, Lesser Himalayan (LH) Rampur Group – (A) Manikaran Quartzite, (B) Rampur Volcanics, (C) Carbonaceous phyllite. 2, Kulu – Bajura Nappe – augen mylonite. 3, Higher Himalayan Crystallines (HHC) – (A) Jeori Group, staurolite/garnetiferous schist, banded biotite gneiss, augen gneiss (a), amphibolite (b), quartz mica schist (c), and Wangtu granite gneiss/granite (d), (B) Karcham Group – garnet/ staurolite/kyanite/sillimanite schist/gneiss, calc-silicate, augen gneiss and migmatite (a) and Akpa leucogranite (b). 4, Tethyan Sedimentary Zone (TSZ). Abbreviations: KT, Kulu Thrust; MCT, Main Central Thrust; VT, Vaikrita Thrust; R, Rampur; K, Karchham. Compiled from published data and our own observations.

weakly deformed granitoid. The zircons have well-defined euhedral crystal shapes with sharp edges and pointed pyramidal terminations. The average length to width ratio is about 3:2. The crystals are mostly transparent without any cores; however, a few show clouded grains and fluid inclusions. CL images of the zircon show oscillatory growth and sector zoning (Fig. 5a, b, c). Th/U values are >0.1 and indicate primary magmatic zircons (Williams & Claesson 1987). Nineteen spots on fifteen zircon grains were analysed and nine near-concordant analyses from both rims and cores yield a near concordant ^{207}Pb/^{206}Pb weighted mean age of 1866 ± 4 Ma (Table 2, Fig. 5d). The zircons from the Bandal granitoid do not exhibit any older component.

Kulu–Bajura mylonite. The Kulu–Bajura mylonite sample is from the main Sutlej Valley section between Nirath and Baragaon. The sample is strongly mylonitized followed by the development of biotite megacrysts. The zircons are euhedral crystals with sharp pointed pyramidal terminations. The average length to width ratio is about 3:1.5. The crystals are mostly transparent with a few indicating the presence of cores; a few zircons are clouded and some contain fluid inclusions. CL images of the zircon show sector zoning (Fig. 6a) with Th/U values >0.1 and indicate primary magmatic zircons (Williams & Claesson 1987). Thirty-nine spots on thirty-one zircon grains were analysed and fifteen near-concordant analyses yield a ^{207}Pb/^{206}Pb weighted mean age of 1856 ± 3 Ma (Table 3, Fig. 6b). Older cores with ages of 2048, 2330, 2515 and 2612 Ma have also been recorded from the sample (Table 3, Fig. 6b).

Wangtu granioid. The Wangtu granitoid is a massive body, with a tectonized contact with the country rock and contains numerous xenoliths of biotite schist and banded gneiss. The central portion of the body is almost undeformed; it is monotonous and fine- to coarse-grained. The sample is from Wangtu Bridge along National Highway 22A. The majority of zircons are euhedral prisms, commonly with sharp edges and pointed pyramidal terminations. The average length to width ratio is about 3:1. The crystals are mostly transparent with a few showing the presence of cores. There are also a few twinned crystals, clouded grains and crystals containing fluid inclusions. CL images of the zircon show oscillatory and sector zoning (Fig. 7a, b, c). The Th/U ratio is >0.1 and typical of primary magmatic origin

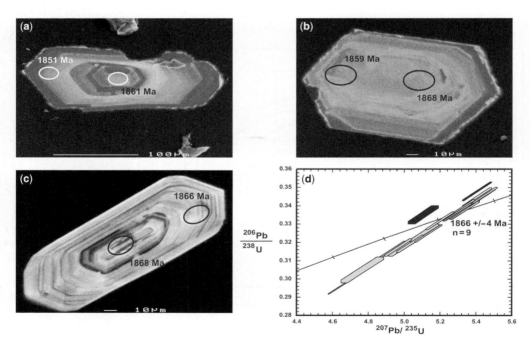

Fig. 5. (a) CL image of zircon (grain 7) from the Bandal granitoids showing oscillatory growth zoning. (b) CL image of zircon (grain 9) from the Bandal granitoids. (c) CL image of zircon (grain 15) from the Bandal granitoids showing oscillatory growth zoning. (d) Concordia diagram for Bandal granitoids showing weighted mean ^{207}Pb/^{206}Pb age from 9 analyses.

(a)
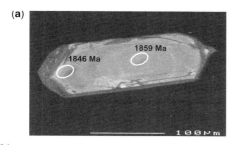

(b)
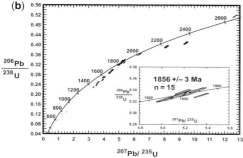

Fig. 6. (a) CL image of zircon (grain 25) from the Kulu–Bajura mylonite, also indicating location of spots of SHRIMP analyses along with ages. (b) Concordia diagram for Kulu–Bajura mylonite showing weighted mean $^{207}Pb/^{206}Pb$ age from 15 analyses. The lower intercept around 500 indicates widespread thermal activity around Pan-African magmatism within the Himalaya.

(Williams & Claesson 1987). Thirty-two spots were analysed on twenty-six zircon grains and out of these eighteen near-concordant analyses of cores as well as rims (Fig. 7a, b) yield a weighted mean $^{207}Pb/^{206}Pb$ age of 1865 ± 4 Ma (Table 4, Fig. 7d) with a few discordant rim or grain ages. Older cores of 2180 and 3000 Ma (Table 4, Fig. 7d) have also been observed and indicate the involvement of Archaean crust. The 3000 Ma age is the first report of such an old age from the Himalayan domain.

Discussions and conclusion

The evolution of the Himalaya has been explained as the product of collision tectonics between the Indian and Eurasian plates during the Eocene (Leech et al. 2005). However, the timing of collision is poorly constrained, ranging between c. 65 and 40 Ma. The most cited age is between 55 and 50 Ma. The discrepancy in age has resulted from different approaches to defining the timing of continent–continent collision. On the basis of terrestrial faunas, the age is 65 Ma (Jaeger et al. 1989; Rage et al. 1995). A similar age has also been suggested on the basis of lithospheric plate reorganization in the Indian Ocean (Courtillot et al. 1986) and on the basis of possible tectonic mélange emplacement along the Indian Plate margin due to the initial collision (Searle 1983, 1986; Beck et al. 1996). The palaeomagnetic data from the Indian ocean floor also give an age of 65 Ma, recording a change in the direction and rate of plate motion (Klootwijk et al. 1992). However, according to Dewey et al. (1988) and Le Pichon et al. (1992), the initial collision started at about 45 Ma, which is incompatible with both the early foreland-basin stratigraphical record from Pakistan to Nepal, that is dominated by early Eocene detrital sediments from a volcanic arc and ophiolites (Critelli & Garzanti 1994; DeCelles et al. 1998; Najman & Garzanti 2000). On the basis of sedimentological constraints, Treloar & Coward (1991) and Rowley (1996), suggested that the first collision took place in the Western syntaxis at around 55–50 Ma and it became younger in age from the central to eastern part of the range between 50 and 45 Ma. Guillot et al. (2003) estimated the initiation age of collision to be c. 55 Ma, based on the stratigraphy, palaeomagnetism, geochronology and tectono-physics in the NW Himalaya. However, on the basis of age of the UHP metamorphism and the possible geometry of the subducting plate, Leech et al. (2005, 2006, 2007) suggested that initial collision took place no later than 57 Ma.

The large-scale loading, buckling and stripping-off of masses, along with continuous subduction of the Indian Plate under the Eurasian Plate, has caused about 2000 km of crustal shortening during the last c. 40 Ma (Windley 1988). However, the estimated values of this crustal shortening by different workers vary over a wide range, possibly due to different times of collision, neglecting the northern extension of the Indian Plate below Tibet and the different rate of convergence throughout the collision period. The palaeomagnetic data indicate convergence of the Indian Plate against the Eurasian Plate at about 2150–2300 km (Dewey et al. 1988; Le Pichon et al. 1992) or 3000 km (Molnar & Tapponnier 1975; Replumaz & Tapponnier 2003). However, various data on Himalayan shortening based on different data range from 1250 ± 250 km (Achache et al. 1984; Besse et al. 1984; Powell et al. 1988; Dewey et al. 1988; Patzelt et al. 1996; Matte et al. 1997) to c. 1100 km (Gulliot et al. 2003), c. 900 km (Le Pichon et al. 1992) between c. 700 km and 1500 km (Patzelt et al. 1996) and c. 650 km (DeCelles et al. 2002). Shortening estimates within the various sectors of the Himalaya have also been constrained based on the balance cross-sectioning which varies from west to east. In Pakistan within the Hazara Syntaxis, it is about 470 km (Coward & Butler 1985), within

Table 2. SHRIMP U–Pb data for the Bandal granitoids

Spot	U/ppm	Th/U	f^{206}	$^{207}Pb/^{206}Pb$	+/-	$^{208}Pb/^{206}Pb$	+/-	$^{206}Pb/^{238}U$	+/-	$^{207}Pb/^{235}U$	+/-	%Conc	Age $^{207}Pb/^{206}Pb$	+/-
1-rim	1958	0.321	<0.1	0.1147	0.0001	0.0921	0.0002	0.3374	0.0050	5.3340	0.0799	100	1875	2
3-core	2139	0.298	<0.1	0.1105	0.0006	0.1216	0.0015	0.3350	0.0050	5.1032	0.0845	103	1807	11
4-rim	226	0.499	<0.1	0.1133	0.0007	0.1426	0.0015	0.3050	0.0071	4.7649	0.1183	93	1853	11
7-rim	2455	0.049	<0.1	0.1132	0.0002	0.0134	0.0002	0.3136	0.0071	4.8924	0.1124	95	1851	2
7-core	4551	0.072	<0.1	0.1138	0.0001	0.0214	0.0001	0.2984	0.0068	4.6828	0.1072	90	1861	2
8-rim	923	0.227	<0.1	0.1141	0.0004	0.0662	0.0007	0.3198	0.0073	5.0313	0.1180	96	1866	6
9-rim	1118	0.263	<0.1	0.1137	0.0002	0.0763	0.0004	0.3446	0.0051	5.4014	0.0820	103	1859	3
9-core	1451	0.299	<0.1	0.1143	0.0002	0.0868	0.0003	0.3441	0.0051	5.4209	0.0816	102	1868	3
10-core	736	0.227	<0.1	0.1139	0.0004	0.0704	0.0008	0.3252	0.0048	5.1085	0.0806	97	1863	7
11-rim	842	0.237	<0.1	0.1135	0.0003	0.0688	0.0005	0.3172	0.0047	4.9645	0.0762	96	1856	4
12-core	1840	0.204	<0.1	0.1138	0.0002	0.0590	0.0002	0.3383	0.0050	5.3084	0.0797	101	1861	3
13-core	375	0.291	<0.1	0.1145	0.0004	0.0812	0.0007	0.3377	0.0051	5.3295	0.0846	100	1871	6
13-rim	1078	0.309	<0.1	0.1144	0.0003	0.0901	0.0005	0.3192	0.0047	5.0330	0.0771	96	1870	4
14-core	2106	0.591	<0.1	0.1126	0.0001	0.1663	0.0003	0.3478	0.0051	5.4017	0.0809	104	1843	2
15-core	1692	0.286	<0.1	0.1142	0.0002	0.0819	0.0003	0.3449	0.0051	5.4325	0.0818	102	1868	3
15-rim	1243	0.272	<0.1	0.1141	0.0002	0.0800	0.0003	0.3368	0.0050	5.2987	0.0802	100	1866	3
16-core	1647	0.401	<0.1	0.1146	0.0002	0.1193	0.0004	0.3249	0.0048	5.1316	0.0774	97	1873	3
17-core	1186	0.386	<0.1	0.1136	0.0002	0.1122	0.0004	0.3317	0.0049	5.1967	0.0789	99	1858	3
18-core	1479	0.378	<0.1	0.1141	0.0002	0.1112	0.0003	0.3364	0.0050	5.2924	0.0797	100	1866	3

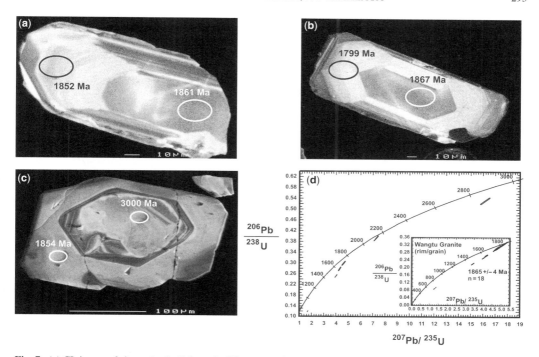

Fig. 7. (a) CL image of zircon (grain 1) from the Wangtu granitoids showing faint oscillatory growth zoning. (b) CL image of zircon (grain 3) from the Wangtu granitoids showing faint oscillatory growth zoning. (c) CL image of zircon (grain 44) from the Wangtu granitoids with an older core of 3000 Ma. (d) Concordia diagram for Wangtu granitoid showing weighted mean ^{207}Pb/^{206}Pb age from 18 analyses. The lower intercept indicate thermal activity at later stage.

NW India it is between 480 and 591 km (Searle 1986; Srivastava & Mitra 1994; Searle et al. 1997), in western Nepal it is between 485 and 669 km (DeCelles et al. 2001; Murphy & Yin 2003) and in eastern Nepal it is between 318 and 419 km (Schelling & Arita 1991; Schelling 1992; Ratschbacher et al. 1994).

Such a large convergence and shortening led workers to propose the 'Greater India' concept. Greater India is placed north of the Indian Plate, containing rocks of the present-day Lesser Himalayan Sedimentary Zone and Higher Himalayan Crystallines. The Proterozoic basement and its sedimentary cover present within the Lesser and Higher Himalayan zone appear to have been involved in the Cenozoic Himalayan remobilization due to collision. The orthoquartzite-volcanic association in the Lesser Himalayan window zone from Kistwar, Kulu–Rampur and the inner Lesser Himalayan sedimentary belt of Garhwal–Kumaon in NW Himalaya is of great interest in tracing the Proterozoic evolutionary trends and mutual relationships with the granitoids (Sharma 1977; Bhat & Le Fort 1992). These Proterozoic interstratified, transitional tholeiitic to alkali basalts, are relatively enriched in incompatible trace elements, like basalt erupted in a 'plume' setting (Bhat 1987). Although Sm–Nd whole-rock ages of the Mandi–Darla–Rampur volcanics (Bhat & Le Fort 1992, 1993) and of the Garhwal Volcanics (Bhat et al. 1998) suggest intense rift-related volcanism around 2.5 Ga, zircon separated from the Rampur metabasalts indicated zircon evaporation ages of 1.8 Ga (Miller et al. 2000) within the Himalayan orogenic belt. These orthoquartzite-volcanic platform sequences also contain contemporary 'Within Plate' granitoids with ages around 1.85 Ga (this work, whole-rock Rb–Sr isochron of 1861 ± 32 Ma from Kistwar Window and 1840 ± 70 Ma (Bandal bodies; Frank et al. 1977; Miller & Frank 1992).

In Himachal Pradesh, interstratified volcanics, associated with orthoquartzite-limestone of the Lesser Himalaya, probably represent the oldest phase of Proterozoic magmatism around 2.0–1.85 Ga. Within the Himalayan metamorphic nappes, the earliest magmatic phase intrudes the still older pelitic and banded biotite gneiss. It is likely that the thinly banded biotite gneiss and pelitic sequence of the HHC may represent the original geosynclinal facies of shale-sandstone alternation, probably of Archaean age, as indicated by 3.0 Ga older cores in this study. This setting

Table 3. *SHRIMP U–Pb data for the Kulu–Bajura mylonite*

Spot	U/ppm	Th/U	f^{206}	^{207}Pb/^{206}Pb	+/−	^{208}Pb/^{206}Pb	+/−	^{206}Pb/^{238}U	+/−	^{207}Pb/^{235}U	+/−	%Conc	Age ^{207}Pb/^{206}Pb	+/−
1-core	482	0.230	<0.1	0.1094	0.0006	0.0866	0.0010	0.2593	0.0031	3.9112	0.0529	83	1790	9
1-core	582	0.088	<0.1	0.1099	0.0004	0.0218	0.0006	0.2916	0.0035	4.4169	0.0572	92	1797	7
1-core	2081	0.122	<0.1	0.0780	0.0003	0.0426	0.0007	0.1172	0.0014	1.2614	0.0163	62	1148	9
1-rim	994	0.202	<0.1	0.1053	0.0005	0.0685	0.0007	0.2484	0.0053	3.6086	0.0811	83	1720	8
2-core	332	0.349	<0.1	0.1138	0.0004	0.1023	0.0007	0.3318	0.0040	5.2071	0.0690	99	1861	7
3-core	456	0.265	<0.1	0.1134	0.0004	0.0739	0.0007	0.3255	0.0039	5.0892	0.0664	98	1854	7
4-core	489	0.322	<0.1	0.1136	0.0004	0.0930	0.0007	0.3340	0.0040	5.2336	0.0678	100	1858	7
5-core	471	0.134	<0.1	0.1131	0.0004	0.0379	0.0005	0.3364	0.0040	5.2434	0.0680	101	1849	7
6-core	597	0.259	<0.1	0.1263	0.0004	0.0772	0.0006	0.3668	0.0044	6.3886	0.0813	98	2048	5
6-rim	2231	0.045	<0.1	0.0822	0.0003	0.0520	0.0005	0.0954	0.0011	1.0811	0.0135	47	1251	7
7-core	394	0.652	<0.1	0.1136	0.0005	0.1882	0.0010	0.3359	0.0041	5.2616	0.0694	100	1858	7
8-core	348	0.271	<0.1	0.1136	0.0004	0.0786	0.0005	0.3405	0.0041	5.3314	0.0701	102	1857	7
9-core	655	0.289	<0.1	0.1137	0.0003	0.0834	0.0004	0.3393	0.0040	5.3167	0.0663	101	1859	5
9-rim	415	0.217	<0.1	0.1119	0.0005	0.0615	0.0007	0.3364	0.0041	5.1900	0.0684	102	1830	7
10-rim	2707	0.071	<0.1	0.0850	0.0004	0.1183	0.0008	0.1100	0.0013	1.2902	0.0166	51	1317	9
11-core	773	0.264	<0.1	0.1126	0.0003	0.0757	0.0005	0.3236	0.0038	5.0246	0.0626	98	1842	5
12-core	106	0.567	<0.1	0.1658	0.0012	0.1650	0.0022	0.4110	0.0056	9.3925	0.1512	88	2515	12
13-rim	600	0.254	<0.1	0.1137	0.0003	0.0753	0.0004	0.3391	0.0040	5.3137	0.0666	101	1859	5
13-core	112	0.263	<0.1	0.1486	0.0012	0.0897	0.0022	0.3964	0.0054	8.1214	0.1356	92	2330	14
14-core	381	0.532	<0.1	0.1127	0.0004	0.1571	0.0008	0.3320	0.0040	5.1588	0.0679	100	1843	7
15-core	768	0.215	<0.1	0.1756	0.0004	0.0591	0.0004	0.5212	0.0061	12.6200	0.1539	104	2612	3
16-rim	717	0.367	<0.1	0.1174	0.0003	0.1028	0.0004	0.3308	0.0039	5.3534	0.0663	96	1917	5
17-rim	828	0.138	<0.1	0.1082	0.0003	0.0425	0.0005	0.2718	0.0032	4.0569	0.0509	88	1770	6
18-rim	597	0.287	<0.1	0.1108	0.0003	0.0878	0.0007	0.2895	0.0034	4.4224	0.0576	90	1812	7
19-rim	563	0.285	<0.1	0.1045	0.0004	0.1068	0.0006	0.2290	0.0027	3.2984	0.0426	78	1705	7
20-rim	598	0.359	<0.1	0.1109	0.0004	0.1069	0.0006	0.2988	0.0036	4.5704	0.0580	93	1815	6
21-rim	914	0.114	<0.1	0.1060	0.0003	0.0375	0.0005	0.2704	0.0032	3.9516	0.0495	89	1732	6
22-core	750	0.163	<0.1	0.1135	0.0003	0.0471	0.0004	0.3290	0.0039	5.1483	0.0642	99	1856	5
23-rim	779	0.435	<0.1	0.1052	0.0004	0.1070	0.0006	0.2503	0.0029	3.6285	0.0459	84	1717	6
24-core	2079	0.098	<0.1	0.0872	0.0003	0.0559	0.0005	0.1280	0.0015	1.5391	0.0192	57	1364	7
24-rim	479	0.323	<0.1	0.1132	0.0004	0.0927	0.0007	0.3331	0.0040	5.1974	0.0675	100	1851	7
25-core	481	0.155	<0.1	0.1137	0.0004	0.0431	0.0006	0.3343	0.0040	5.2404	0.0678	100	1859	7
25-rim	692	0.225	<0.1	0.1129	0.0003	0.0652	0.0004	0.3315	0.0039	5.1583	0.0642	100	1846	5
26-rim	1018	0.101	<0.1	0.1072	0.0004	0.0431	0.0005	0.2381	0.0028	3.5202	0.0445	79	1753	6
27-rim	503	0.334	<0.1	0.1094	0.0004	0.1474	0.0009	0.3000	0.0036	4.5235	0.0590	95	1789	7
28-rim	739	0.444	<0.1	0.1091	0.0003	0.1275	0.0006	0.2894	0.0034	4.3528	0.0545	92	1784	5
29-core	431	0.240	<0.1	0.1137	0.0004	0.0694	0.0007	0.3350	0.0040	5.2540	0.0689	100	1860	7
30-core	458	0.137	<0.1	0.1142	0.0004	0.0377	0.0006	0.3348	0.0040	5.2727	0.0688	100	1868	7
31-core	556	0.377	<0.1	0.1127	0.0004	0.1076	0.0007	0.3228	0.0038	5.0174	0.0646	98	1844	6

Table 4. SHRIMP U–Pb data for the Wangtu granite gneiss

Spot	U/ppm	Th/U	f^{206}	$^{207}Pb/^{206}Pb$	+/−	$^{208}Pb/^{206}Pb$	+/−	$^{206}Pb/^{238}U$	+/−	$^{207}Pb/^{235}U$	+/−	%Conc	Age $^{207}Pb/^{206}Pb$	+/−
1-core	288	0.383	<0.1	0.1138	0.0005	0.1095	0.0010	0.3210	0.0074	5.0364	0.1214	96	1861	8
1-rim	584	0.320	<0.1	0.1132	0.0004	0.0922	0.0007	0.2934	0.0067	4.5808	0.1078	90	1852	6
3-core	360	0.444	<0.1	0.1138	0.0005	0.1365	0.0011	0.2733	0.0063	4.2870	0.1030	84	1861	8
3-rim	1497	0.088	<0.1	0.1100	0.0005	0.0274	0.0009	0.1282	0.0029	1.9437	0.0464	43	1799	8
5-core	212	0.499	<0.1	0.1142	0.0006	0.1441	0.0012	0.3332	0.0051	5.2447	0.0878	99	1867	9
6-core	751	0.182	<0.1	0.1149	0.0003	0.0606	0.0004	0.2512	0.0037	3.9796	0.0613	77	1878	4
8-rim	354	0.257	<0.1	0.1136	0.0006	0.0820	0.0011	0.2648	0.0040	4.1486	0.0686	81	1858	9
10-core	533	0.097	<0.1	0.1172	0.0004	0.0275	0.0003	0.2469	0.0057	3.9902	0.0940	74	1914	6
12-core	382	0.452	<0.1	0.1362	0.0004	0.1261	0.0007	0.3845	0.0089	7.2207	0.1701	96	2180	5
14-core	521	0.099	<0.1	0.1141	0.0003	0.0318	0.0003	0.2940	0.0067	4.6249	0.1085	89	1866	5
14-rim	258	0.292	<0.1	0.1142	0.0005	0.0843	0.0009	0.3124	0.0072	4.9189	0.1192	94	1867	8
18-core	295	0.501	<0.1	0.1141	0.0004	0.1467	0.0008	0.3219	0.0049	5.0664	0.0809	96	1866	6
19-core	529	0.253	<0.1	0.1149	0.0004	0.0728	0.0006	0.3003	0.0069	4.7553	0.1119	90	1878	6
29-core	382	0.157	<0.1	0.1147	0.0004	0.0458	0.0007	0.3047	0.0046	4.8196	0.0771	91	1875	7
31-rim	241	0.536	<0.1	0.1135	0.0006	0.1574	0.0013	0.3265	0.0050	5.1091	0.0852	98	1856	9
33-core	1853	0.297	<0.1	0.1050	0.0003	0.1021	0.0007	0.1182	0.0027	1.7114	0.0401	42	1714	6
33-rim	2519	0.535	<0.1	0.1053	0.0008	0.0872	0.0018	0.0872	0.0020	1.2671	0.0317	31	1720	14
44-rim	171	0.278	<0.1	0.1133	0.0009	0.0788	0.0018	0.3054	0.0072	4.7726	0.1224	93	1854	14
44-core	135	0.468	<0.1	0.2226	0.0008	0.1287	0.0012	0.5211	0.0122	15.9945	0.3878	90	3000	6
45-rim	789	0.240	<0.1	0.1121	0.0004	0.0838	0.0006	0.2201	0.0050	3.4000	0.0799	70	1833	6
48-core	362	0.440	<0.1	0.1144	0.0005	0.1329	0.0010	0.3038	0.0046	4.7937	0.0773	91	1871	7
49-core	719	0.359	<0.1	0.1147	0.0004	0.1060	0.0008	0.2609	0.0060	4.1264	0.0978	80	1875	7
58-core	2143	0.185	<0.1	0.1039	0.0003	0.0684	0.0006	0.0951	0.0022	1.3619	0.0319	35	1695	6
60-core	358	0.292	<0.1	0.1141	0.0004	0.0832	0.0008	0.3100	0.0071	4.8756	0.1162	93	1865	7
60-rim	344	0.228	<0.1	0.1134	0.0005	0.0619	0.0009	0.2997	0.0069	4.6858	0.1125	91	1854	8
61-core	375	0.436	<0.1	0.1140	0.0005	0.1275	0.0010	0.3132	0.0072	4.9241	0.1177	94	1865	8
62-core	288	0.527	<0.1	0.1147	0.0005	0.1519	0.0011	0.3204	0.0074	5.0665	0.1221	96	1875	8
64-rim	442	0.426	<0.1	0.1142	0.0004	0.1317	0.0009	0.2933	0.0067	4.6181	0.1098	89	1867	7
65-core	506	0.381	<0.1	0.1135	0.0004	0.1121	0.0007	0.3116	0.0072	4.8756	0.1152	94	1856	6
66-rim	615	0.064	<0.1	0.1139	0.0003	0.0185	0.0004	0.2992	0.0069	4.6992	0.1103	91	1863	5
67-core	566	0.279	<0.1	0.1136	0.0003	0.0797	0.0004	0.3124	0.0072	4.8940	0.1144	94	1858	5
68-core	152	0.358	<0.1	0.1144	0.0005	0.1030	0.0009	0.3160	0.0050	4.9822	0.0843	95	1870	9

appears to be similar to the rift-controlled basinal setting of the Aravalli, Dharwar and other early to middle Proterozoic basins of Peninsular India.

This leads us to believe that an Archaean/Palaeoproterozoic ensialic rift basin developed within the Columbia Supercontinent, lying within the present Himalayan orogenic belt. The Columbia Supercontinent began to accrete between 1.9 and 1.8 Ga and contained all major former continents, with the east coast of India possibly attached to Western North America, and southern Australia against Western Canada (Rogers & Santosh 2002). At that time, most of South America rotated so that the Western edge of modern-day Brazil lined up with Eastern North America, forming a continental margin that extended into the southern edge of Scandinavia (Rogers & Santosh 2002; Condie 2002; Zhao et al. 2002). The orthoquartzite-volcanic platform sediments of the inner Lesser Himalayan sedimentary sequence now exposed in the window zone lay in the southern parts within the Lesser Himalayan terrane and geosynclinal pelites of the Higher Himalaya further to the north. Here, mafic volcanism and the 'Within Plate' felsic magmatism were more likely to be associated with the adiabatic rise of a mantle plume through an extremely attenuated lithosphere, rather than ocean floor spreading, as proposed by Bhat (1987). Geochemical signatures from these anorogenically emplaced 1.8–2.0 Ga 'Within Plate granites' suggest partial melting of the lower and middle Archaean continental crust to generate the granitoids. All these bodies contain metamorphosed xenoliths and are characterized by ubiquitously high $^{87}Sr/^{86}Sr$ initial ratios, possibly due to the rise of Palaeoproterozoic small mantle plumes or hot spots within the extended crust, both in the Lesser and Higher Himalayan terranes. At 1.8 Ga, such processes were widespread and generated the Wangtu granitoid, Bandal granitoid and Kulu–Bajura mylonitized augen gneiss; all these were emplaced within extended Archaean to Palaeoproterozoic ensialic basins, which were subsequently deformed and imbricated during the Cenozoic Himalayan collisional tectonics.

The data clearly indicate that Palaeoproterozoic (c. 1860 Ma) felsic magmatic rocks are not restricted to a single tectonic zone; rather they are widespread in the Lesser Himalayan Sedimentary Zone as well as within the basal part of the Higher Himalayan Crystallines (HHC). In addition, granite gneiss also forms a continuous zone of mylonites at the base of the Main Central Thrust (MCT), all along the Himalayan range. The geochemical data clearly indicate the anorogenic character of this body, with remobilization of older crustal material. The presence of a 3000 Ma zircon core, reported for the first time from the Himalaya, clearly indicates reworking/remobilization of the Archaean sialic crust within the Himalaya. These bodies have been further imbricated and placed in adjacent nappes due to collisional tectonics during the Himalayan orogeny.

Fieldwork was funded by the Department of Science and Technology (DST) of India under various projects. Ion-Probe zircon analyses were carried out on SHRIMP-II operated by a consortium consisting of Curtin University of Technology, the Geological Survey of Western Australia and the University of Western Australia, with the support of the Australian Research Council. We thank S. Reddy and R. Mazumder for their patience with us during manuscript preparation, and R. Sorkhabi and S. Wilde for critically going through the manuscript.

References

ACHACHE, J., COURTILLOT, V. & ZHOU, Y. X. 1984. Palaeaogeographic and tectonic evolution of Southern Tibet since Middle Cretaceous times: new palaeomagnetic data and synthesis. *Journal of Geophysical Research*, **89**, 10 311–10 339.

ALI, J. R. & AITCHISON, J. C. 2005. Greater India. *Earth Science Review*, **72**, 169–188.

ARGAND, E. 1924. La tectonique de l'Asie. *Proceedings of the 13th International Geological Congress*, **7**, 171–372.

AUDEN, J. B. 1935. Traverses in the Himalaya. *Record Geological Survey India*, **69**, 123–167.

BECK, R. A., BURBANK, D. W., SERCOMBE, W. J., KHAN, A. M. & LAWRENCE, R. D. 1996. Late Cretaceous ophiolite obduction and Palaeocene India-Asia collision in the westernmost Himalaya. *Geodynamica Acta*, **9**, 114–144.

BESSE, J., COURTILLOT, V., POZZI, J. P., WESTPHAL, M. & ZHOU, Y. X. 1984. Palaeomagnetic estimates of crustal shortening in the Himalayan thrusts and Zangbo suture. *Nature*, **311**, 621–626.

BHANOT, V. B., GILL, J. S., ARORA, R. P. & BHALLA, J. K. 1974. Radiometric dating of the Dalhousie granite. *Current Science*, **43**, 208.

BHANOT, V. B., GOEL, A. K., SINGH, V. P. & KWATRA, S. K. 1975. Rb–Sr radiometric studies in the Dalhousie and Rohtang area. *Current Science*, **44**, 219.

BHANOT, V. B., BHANDARI, A. K., SINGH, V. P. & KANSAL, A. K. 1976. Precambrian, 1220 my. Rb–Sr whole rock isochron age for Bandal granite, Kulu Himalaya, Himachal Pradesh. *Abstract Himalayan Geology Seminar*, Delhi, 197.

BHANOT, V. B., SINGH, V. P., KANSAL, A. K. & THAKUR, V. C. 1977. Early Proterozoic Rb–Sr whole rock age for Central Crystalline gneiss of Higher Himalaya, Kumaun. *Geological Society of India*, **18**, 90–91.

BHANOT, V. B., KWATRA, S. K., KANSAL, A. K. & PANDEY, B. K. 1978. Rb–Sr whole rock age for Chail Series of Northwestern Himalaya. *Journal of Geological Society of India*, **19**, 224–227.

BHANOT, V. B., BHANDARI, A. K., SINGH, V. P. & KANSAL, A. K. 1979. Geochronological and geological studies of granites of Higher Himalaya, Northeast of Manikaran, H.P. *Journal of Geololgical Society of India*, **20**, 90–94.

BHARGAVA, O. N., BASSI, U. K & SHARMA, R. K. 1991. The Crystalline thrust sheets, age of metamorphism Himachal Himalaya. *Indian Minerals*, **45**, 1–18.

BHAT, M. I. 1987. Spasmodic rift reactivation and its role in the pre-orogenic evolution of the Himalaya Region. *In*: GUPTA, H. K. (ed.) *Deep Seated Processes in Collision Zones, Tectonophysics*, **134**, 103–127.

BHAT, M. I. & LE FORT, P. 1992. Sm–Nd age and petrogenesis of Rampur metavolcanic rocks, NW-Himalaya: late Archaean relics in the Himalayan belt. *Precambrian Research*, **56**, 191–210.

BHAT, M. I. & LE FORT, P. 1993. Nd-isotopic study of the Late Archaean Continental Tholeiites, NW Lesser Himalaya: a case of Ocean Island Basalt source for Continental Tholeiites. *Journal of Himalayan Geology*, **4**, 1–13.

BHAT, M. I., CLAESSON, S., DUBEY, A. K. & PANDE, K. 1998. Sm–Nd age of the Garhwal-Bhowali volcanics, western Himalayas: vestiges of the Late Archaean Rampur flood basalt province of the northern Indian craton. *Precambrian Research*, **87**, 217–231.

BURCHFIEL, B. C., ZHILIANG, C., HODGES, K. V., YUPING, L., ROYDEN, L. H., DENG, C. & XUE, J. 1992. The South Tibetan detachment system, Himalayan orogen: extension contemporaneous with and parallel to shortening in a collisional mountain belt. *Geological Society of America Special Paper* **269/41**.

CONDIE, K. C. 2002. Breakup of a Palaeoproterozoic supercontinent. *Gondwana Research*, **5**, 41–43.

COURTILLOT, V., BESSE, J., VANDAMME, D., MONTIGNY, R., JAEGER, J. & CAPETTA, H. 1986. Deccan flood basalts at the Cretaceaous/Tertiary boundary? *Earth and Planetary Science Letters*, **80**, 361–374.

COWARD, M. P. & BUTLER, R. H. W. 1985. Thrust tectonics and the deep structure of the Pakistan Himalaya. *Geology* **13**, 417–420.

COWARD, M. P., JAN, M. Q., REX, D., TARNEY, J., THIRLWALL, M. & WINDLEY, B. F. 1982. Geotectonic framework of the Himalaya of North Pakistan. *Journal of Geological Society of London*, **139**, 299–308.

CRITELLI, S. & GARZANTI, E. 1994. Provenance of the Lower Tertiary Murree redbeds (Hazara-Kashmir syntaxis, Pakistan) and initial rising of the Himalayas. *Sedimentary Geology*, **89**, 265–284.

DALZIEL, I. W. D. 1991. Pacific margins of Laurentia and East Antarctic-Australia as a conjugate rift pair: evidence and implications for an Eocambrian supercontinent. *Geology*, **19**, 598–601.

DALZIEL, I. W. D. 1992. Antarctic: a tale of two supercontinents? *Annual Review of Earth and Planetary Science*, **20**, 501–526.

DALZIEL, I. W. D., MOSHER, S. & GAHAGAN, L. M. 2000. Laurentia-Kalahari collision and the assembly of Rodinia. *Journal of Geology*, **108**, 499–513.

DECELLES, P. G., GEHRELS, G. E., QUADE, J. & OJHA, T. P. 1998. Eocene-early Miocene foreland basin development and the history of Himalayan thrusting, western and central Nepal. *Tectonics*, **17**, 741–765.

DECELLES, P. G., ROBINSON, D. M., QUADE, J., OJHA, T. P., GARZIONE, C. N., COPELAND, P. & UPRETI, B. N. 2001. Stratigraphy, structure, and tectonic evolution of the Himalayan fold-thrust belt in western Nepal. *Tectonics*, **20**, 487–509.

DECELLES, P. G., ROBINSON, D. M. & ZANDT, G. 2002. Implications of shortening in the Himalayan fold-thrust belt for uplift of the Tibetan Plateau. *Tectonics*, **21**, 1062, doi:10.1029/2001TC001322.

DEWEY, J. F., SHAKELTON, R. M., CHANG, C. & SUN, Y. 1988. The tectonic evolution of the Tibetan Plateau. *Philosophical Transaction of Royal Society of London*, **327**(A), 379–413.

DIPIETRO, J. A. & ISACHSEN, C. E. 1997. An early Proterozoic age for Precambrian rock units in the Indus Syntaxis, NW Himalaya, Pakistan. *In*: Abstracts volume. 12th Himalayan-Karakoram-Tibet workshop, Roma, Italy, 137–138.

DIPIETRO, J. A. & ISACHSEN, C. E. 2001. U–Pb zircon ages from the Indian plate in northwest Pakistan and their significance to Himalayan and pre-Himalayan geologic history. *Tectonics*, **20**, 510–525.

FRANK, W., THONI, M. & PURTSCHELLER, F. 1977. Geology and petrography of Kulu-South Lahul area. *Colloquium International Center for Natural Research Sciences*, **33**, 147–172.

GANSSER, A. 1964. Geology of the Himalayas. Wiley, New York.

GRIESBACH, C. L. 1893. Notes on the Central Himalaya. *Records of Geological Survey of India*, **26**, 19–25.

GUILLOT, S., GARZANTI, E., BARATOUX, D., BARATOUX, D., MARQUER, D., MAHEO, G. & DE SIGOYER, J. 2003. Reconstructing the total shortening history of the NW Himalaya. *Geochemical Geophysical and Geosystem G3*, **4**, 1064, doi:10.1029/2002GC000484.

HERREN, E. 1987. Zanskar shear zone: northeast–southwest extension within the Higher Himalaya. *Geology*, **15**, 409–413.

HODGES, K. V. 2000. Tectonics of the Himalaya and southern Tibet from two decades perspectives. *Geological Society of American Bulletin*, **112**, 324–350.

HONEGGER, K., DIETRICH, V., FRANK, W., GANSSER, A., THONI, M. & TROMMSDORF, V. 1982. Magmatism and metamorphism in the Ladakh Himalayas (the Indus-Tsangpo suture zone). *Earth and Planetary Science Letters*, **60**, 253–292.

JAGER, E., BHANDARI, A. K. & BHANOT, V. B. 1971. Rb–Sr age determinations on biotites and whole rock samples from the Mandi and Chor granites, Himachal Pradesh, India. *Eclogae Geologicae Helvetiae*, **64**, 521–527.

JAEGER, J. J., COURTILLOT, V. & TAPPONNIER, P. 1989. Palaeontological view of the ages of the Deccan Traps, the Cretaceous/tertiary boundary, and the India/Asia collision. *Geology*, **17**, 316–319.

JAIN, A. K., KUMAR, D., SINGH, S., KUMAR, A. & LAL, N. 2000. Timing, quantification and tectonic modelling of Pliocene-Quaternary movements in the NW Himalaya: evidence from fission track dating. *Earth and Planetary Science Letters*, **179**, 437–451.

JAIN, A. K., SINGH, S. & MANICKAVASAGAM, R. M. 2002. Himalayan collision tectonics. *Gondwana Research Group Memoir* 7/114.

JAIN, S. C., YEDEKAR, D. B. & NAIR, K. K. K. 1991. Central India shear zone: a major Precambrian crustal boundary. *Journal of Geological Society of India*, **37**, 521–531.

KARLSTROM, K. E., AHALL, K. I., HARLAN, S. S., WILLIAMS, M. L., MCLELLAND, J. & GEISSMAN, J. W. 2001. Long-lived (1.8–1.0 Ga) convergent orogen in southern Laurentia, its extensions to Australia and Baltica, and implications for refining Rodinia. *Precambrian Research*, **111**, 5–30.

KLOOTWIJK, C. T., GEE, J. S., PEIRCE, J. W., SMITH, G. M. & MCFADDEN, P. 1992. An early India-Asia contact: palaeomagnetic constraints from Ninetyeast Ridge, ODP Leg 121. *Geology*, **20**, 395–398.

KWATRA, S. K., BHANOT, V. B., KAKAR, R. K. & KANSAL, A. K. 1986. Rb–Sr radiometric ages of the Wangtu Gneissic Complex, Kinnaur district, Higher Himachal Himalaya. *Bulletin of Indian Geologists' Association*, **19**, 127–130.

LEECH, M. L., SINGH, S., JAIN, A. K., KLEMPERER, S. L. & MANICKAVASAGAM, R. M. 2005. The onset of India-Asia continental collision: early, steep subduction required by the timing of UHP metamorphism in the western Himalaya. *Earth and Planetary Science Letters*, **234**, 83–97.

LEECH, M. L., SINGH, S., JAIN, A. K., KLEMPERER, S. L. & MANICKAVASAGAM, R. M. 2006. Reply to comment by P. J. O'Brien on: 'The onset of India-Asia continental collision: early, steep subduction required by the timing of UHP metamorphism in the western Himalaya' by MARY L. LEECH, S. SINGH, A. K. JAIN, SIMON L. KLEMPERER & R. M. MANICKAVASAGAM. Earth and Planetary Science Letters 234 (2005) 83–97. *Earth and Planetary Science Letters*, **245**, 817–820.

LEECH, M. L., SINGH, S. & JAIN, A. K. 2007. Zircon reveals complex history in the UHP Tso Morari Complex, western Himalaya. *International Geological Review*, **49**, 313–328.

LE FORT, P. 1975. Himalaya, the collided range: present knowledge of the continental arc. *American Journal of Science*, **275**(A), 1–44.

LE FORT, P., CUNEY, M., DENIEL, C., LANORDS, C. F., SHEPPARD, N. F., UPRETI, B. N. & VIDAL, P. 1987. Crustal generation of the Himalayan leucogranite. *Tectonophysics*, **134**, 39–57.

LE PICHON, X., FOURNIER, M. & JOLIVET, L. 1992. Kinematics, topography and extrusion in the India-Eurasia collision. *Tectonics*, **11**, 1085–1098.

LI, Z. X., ZHANG, L. & POWELL, C. MCA. 1996. Positions of the East Asian cratons in the Neoproterozoic Supercontinent Rodinia. *Australian Journal of Earth Sciences*, **43**, 593–604.

MATTE, P., MATTAUER, M., JOLIVET, J. M. & GRIOT, D. A. 1997. Continental subductions beneath Tibet and the Himalayan orogeny: a review. *Terra Nova*, **9**, 264–270.

MAZUMDER, R., BOSE, P. K. & SARKAR, S. 2000. A commentary on the tectono-sedimentary record of the pre-2.0 Ga continental growth of India vis-a-vis a possible pre-Gondwana Afro Indian supercontinent. *Journal of Asian Earth Sciences*, **30**, 201–217.

MCMAHON, C. A. 1884. Microscopic structures of some Himalayan granites and gneissose granites. *Records Geological Survey of India*, **17**, 53–73.

MEHTA, P. K. 1977. Rb–Sr geochronology of the Kulu-Mandi belt: its implications for the Himalayan tectonogenesis. *Geologische Rundschau*, **66**, 156–175.

MILLER, C. & FRANK, W. 1992. Geochemistry and isotope geology of Proterozoic and Early Palaeozoic granitoids in the NW Himalaya. *Abstract 7th Himalayan-Karakoram-Tibet Workshop*, Oxford, 58.

MILLER, C., KLOTZLI, U., FRANK, W., THONI, M. & GRASEMANN, B. 2000. Proterozoic crustal evolution in the NW Himalaya (India) as recorded by circa 1.80 Ga mafic and 1.84 Ga granitic magmatism. *Precambrian Research*, **103**, 191–206.

MILLER, C., THONI, M., FRANK, W., GRASEMANN, B., KLOTZLI, U., GUNTLI, P. & DRAGANITS, E. 2001. The early Palaeozoic magmatic event in the northwest Himalaya, India: source, tectonic setting and age of emplacement. *Geological Magazine*, **138**, 237–251.

MOLNAR, P. & TAPPONNIER, P. 1975. Cenozoic tectonics of Asia: effects of a continental collision. *Science*, **189**, 419–426.

MURPHY, M. A. & YIN, A. 2003. Structural evolution and sequence of thrusting in the Tethyan fold-thrust belt and Indus-Yalu suture zone, southwest Tibet. *Geological Society of American Bulletin*, **115**, 21–34.

NAJMAN, Y. & GARZANTI, E. 2000. An integrated approach to provenance studies: reconstructing early Himalayan palaeogeography and tectonic evolution from Tertiary foredeep sediments, Northern India. *Geological Society of American Bulletin*, **112**, 435–449.

PANDEY, B. K., SINGH, V. P., BHANOT, V. B. & MEHTA, P. K. 1981. Rb–Sr geochronological studies of the gneissic rocks of the Ranikhet and Masi area of Almora Crystallines, Lesser Himalaya, Kumaun, U.P. *Abstract Himalayan Geology 12th Seminar*.

PATEL, R. C., SINGH, S., ASOKAN, A., MANICKAVASAGAM, R. M. & JAIN, A. K. 1993. Extensional tectonics in the Himalayan orogen Zanskar, NW India. *In*: TRELOAR, P. J. & SEARLE, M. P. (eds.) *Himalayan Tectonics*. Geological Society, London Special Publications, **74**, 445–459.

PATZELT, A., LI, H., WANG, J. & APPEL, E. 1996. Palaeomagnetism of Cretaceous to Tertiary sediments from southern Tibet: evidence for the extent of the northern margin of India prior to the collision with Eurasia. *Tectonophysics*, **259**, 259–284.

PAUL, D. K., CHANDY, K. C., BHALLA, J. K., PRASAD, N. & SENGUPTA, N. R. 1982. Geochronology and geochemistry of Lingtse gneiss, Darjeeling-Sikkim Himalayas. *Indian Journal of Earth Sciences*, **9**, 11–17.

PAUL, D. K., MCNAUGHTON, N. J., CHATOPADYA, S. & RAY, K. K. 1996. Geochronology and Geochemistry of the Lingtse Gneiss, Darjeeling-Sikkim Himalaya: revisited. *Journal of Geological Society of India*, **48**, 497–506.

PEARCE, J. A., HARRIS, N. B. W. & TINDLE, A. G. 1984. Trace element discrimination diagram for the tectonic interpretation of granitic rocks. *Journal of Petrology*, **25**, 956–983.

PIDGEON, R. T., FURFARO, D., KENNEDY, A., VAN BRONSWJK, W. & TODT, W. 1994. Calibration of the CZ3 zircon standard for the Curtin SHRIMP II. *US Geological Survey Circular*, **1107/251**.

POGNANTE, U., CASTELLI, D., BENNA, P., GENOVESE, G., OBRELI, F., MEIER, M. & TONARINI, S. 1990. The crystalline units of the High Himalayas in the Lahul-Zanskar region (northwest India): metamorphic-tectonic history and geochronology of the collided and imbricated Indian plate. *Geological Magazine*, **127**, 101–116.

POWELL, C., MCA., CRAWFORD, A. R., ARMSTONG, R. L., PRAKASH, R. & WYNNE-EDWARDS, H. R. 1979. Reconnaissance Rb–Sr dates for the Himalayan Central Gneiss, Northwest India. *Indian Journal of Earth Sciences*, **6**, 139–151.

POWELL, C. MCA., ROOTS, S. R. & VEEVERS, J. J. 1988. Pre-breakup continental extension in East Gondwanaland and the early opening of the eastern Indian Ocean. *Tectonophysics*, **155**, 261–283.

POWELL, C. MCA., LI, Z. X., MCELHINNY, M. W., MEERT, J. G. & PARK, J. K. 1993. Palaeomagnetic constraints on the Neoproterozoic breakup of Rodinia and the mid-Cambrian formation of Gondwanaland. *Geology*, **21**, 889–892.

RAGE, J. C., CAPPETTA, H. ET AL. 1995. Collision age. *Nature*, **375**, 286.

RAO, P. S. 1998. Kameng Orogeny, 1.8–1.9 Ga. from the isotopic evidence of the Bomdila orthogneisses, Kameng Sector, NEFA, India. *Geological Bulletin, University of Peshawar*, **31**, 159–162.

RAO, D. R., SHARMA, K. K. & GOPALAN, K. 1995. Granitoid rock of Wangtu Gnessic Complex, Himachal Pradesh: an example of in situ fractional crystallization and volatile action. *Journal of Geological Society of India*, **46**, 5–14.

RAJU, B. N. V., CHABRIA, T., PRASAD, R. N., MAHADEVAN, T. M. & BHALLA, N. S. 1982. Early Proterozoic Rb–Sr isochron age for Central Crystalline, Bhilangana valley, Garhwal Himalaya. *Himalayan Geology*, **12**, 196–205.

RATSCHBACHER, L., FRISCH, W., LIU, G. & CHEN, C. 1994. Distributed deformation in southern and western Tibet during and after the India–Asia collision. *Journal of Geophysical Research*, **99**, 19 817–19 945.

REPLUMAZ, A. & TAPPONNIER, P. 2003. Reconstruction of the deformed collision zone Between India and Asia by backward motion of lithospheric blocks. *Journal of Geophysical Research*, **108**(B6), 2285, doi:10.1029/2001JB000661.

ROGERS, J. J. W. & SANTOSH, M. 2002. Configuration of Columbia, a Mesoproterozoic Supercontinent. *Gondwana Research*, **5**, 5–22.

ROWLEY, D. B. 1996. Age of initiation of collision between India and Asia: a review of Stratigraphic data. *Earth and Planetary Science Letters*, **145**, 1–13.

SANTOSH, M., YOKOYAMA, K., BIJU-SEKHAR, S. & ROGERS, J. J. W. 2003. Multiple tectonothermal events in the granulite block of Southern India revealed from EPMA dating: implications on the history of supercontinent. *Gondwana Research*, **6**, 27–61.

SCHELLING, D. 1992. The tectonostratigraphy and structure of the eastern Nepal Himalaya. *Tectonics*, **11**, 925–943.

SCHELLING, D. & ARITA, K. 1991. Thrust tectonics, crustal shortening, and the structure of the far-eastern Nepal, Himalaya. *Tectonics*, **10**, 851–862.

SEARLE, M. P. 1983. Stratigraphy, structure and evolution of the Tibetan-Tethys zone in Zanskar and the Indus suture zone in the Ladakh Himalaya. *Transactions of Royal Society of Edinburgh Earth Sciences*, **73**, 205–219.

SEARLE, M. P. 1986. Structural evolution and sequence of thrusting in the High Himalayan, Tibetan Tethyan and Indus suture zones of Zanskar and Ladakh, western Himalaya. *Journal of Structural Geology*, **8**, 923–936.

SEARLE, M. P. 1996. Cooling history, Erosion, Exhumation and Kinematics of the Himalaya–Karakoram–Tibet orogenic belt. *In*: YIN, A. & HARRISON, T. M. (eds) *Tectonics of Asia Rubey Symposium volume*. Los Angeles, 110–137.

SEARLE, M. P., WATERS, D. J., REX, D. C. & WILSON, R. N. 1992. Pressure, temperature and time constraints on Himalayan metamorphism from eastern Kashmir and western Zanskar. *Journal of the Geological Society, London*, **149**, 753–773.

SEARLE, M. P., METCALFE, R. P., REX, A. J. & NORRY, M. J. 1993. Field relations, petrogenesis and emplacement of the Bhagirathi leucogranite, Garhwal Himalaya. *In*: TRELOAR, P. J. & SEARLE, M. P. (eds) *Himalayan Tectonics*. Geological Society, London, Special Publications, **74**, 429–444.

SEARLE, M. P., PARRISH, R. R., HODGES, K. V., HURFORD, A., AYRES, M. W. & WHITEHOUSE, M. J. 1997. Shishma Pangma Leucogranite, South Tibetan Himalaya: field relations, geochemistry; age, origin and emplacement. *Journal of Geology*, **105**, 295–317.

SHARMA, K. K. & RASHID, S. A. 2001. Geochemical evolution of peraluminous palaeoproterozoic bandal orthogneiss NW, Himalaya, Himachal Pradesh, India: implications for the ancient crustal growth in the Himalaya. *Journal of Asian Earth Sciences*, **19**, 413–428.

SHARMA, V. P. 1977. Geology of the Kulu–Rampur belt, Himachal Pradesh. *Geological Survey of India Memoir* **106**, 235–407.

SINGH, S. 1993. Collision Tectonics: metamorphics and geochronological constraints from parts of Himachal Pradesh, NW-Himalaya. Ph.D. thesis, University of Roorkee, Roorkee.

SINGH, S. 2001. Status of geochronological studies in Himalaya: a review. *Journal of Indian Geophysical Union*, **5**, 57–72.

SINGH, S. 2003. Conventional and SHRIMP U–Pb zircon dating of the Chor Granitoid, Himachal Himalaya. *Journal of Geological Society of India*, **62**, 614–626.

SINGH, S. 2005. A review of U–Pb ages from Himalayan Collisonal Belt. *Journal of Himalayan Geology*, **26**, 61–76.

SINGH, S. & JAIN, A. K. 1993. Deformation and strain pattern in parts of the Jutogh Nappe along the Sutlej valley, in Jeori-Wangtu region, Himachal Pradesh, India. *Journal of Himalayan Geology*, **4**, 41–55.

SINGH, S. & JAIN, A. K. 2003. Himalayan granitoids. *In*: SINGH, S. (ed.) *Granitoids of the Himalayan Collisional Belt*. Journal of the Virtual Explorer, **11**, 1–20.

SINGH, S., CLAESSON, S., JAIN, A. K., SJOBERG, H., GEE, D. G., MANICKAVASAGAM, R. M. &

ANDREASSON, P. G. 1994. Geochemistry of the Proterozoic peraluminous granitoids from the Higher Himalayan Crystalline, HHC and Jutogh Nappe, Himachal Pradesh, India. *Journal of Geological Society of Nepal*, **10**, 125.

SINGH, S., JAIN, A. K. & MANICKAVASAGAM, R. M. 1995. Comment on the paper 'Granitoid rock of Wangtu Gnessic Complex, Himachal Pradesh: an example of in situ fractional crystallisation and volatile action' by D. RAMESHWAR RAO, KEWAL K. SHARMA & K. GOPALAN, *Journ. Geol. Soc. India*, **46**(1), 5–14. *Journal of Geological Society of India*, **46**, 682–685.

SINGH, S., CLAESSON, S., JAIN, A. K., GEE, D. G., ANDREASSON, P. G. & MANICKAVASAGAM, R. M. 2006. 2.0 Ga granite of the lower package of the Higher Himalayan Crystallines (HHC), MagladKhad, Sutlej Valley, Himachal Pradesh, India. *Journal of Geological Society of India*, **67**, 295–300.

SINGH, V. P., BHANOT, V. B. & SINGH, R. P. 1985. Geochronology of the granitic and gneissic rocks from Munsiari, Namik and Tawaghat areas of the Central Crystalline Zone, Kumaun Himalaya, U.P. *3rd National Symposium on Mass Spectrometry*, Hyderabad, September 22–24.

SINGH, V. P., SINGH, R. P. & BHANOT, V. B. 1986. Rb–Sr isotopic studies for the granitic and gneissic rocks of Almora area of Almora Crystallines, Kumaun Himalaya, U.P. *4th National Symposium on Mass Spectrometry*, Bangalore, EPS-1, 1–4.

SMITH, J. B., BARLEY, M. E., GROOVES, D. I., KRAPEZ, B., MCNAUGHTON, N. J., BICKLE, M. J. & CHAPMAN, H. J. 1998. The Sholl Shear Zone, west Pilbara: evidence for a terrane boundary from integrated tectonic analysis, SHRIMP U—Pb age-dating and granitoid geochemistry. *Precambrian Research*, **88**, 143–171.

SRIVASTAVA, P. & MITRA, G. 1994. Thrust geometries and deep structure of the outer and Lesser Himalaya, Jumoan and Garhwal (India): implications for evolution of the Himalayan fold and thrust belt. *Tectonics*, **13**, 89–109.

STECK, A. 2003. Geology of the NW Indian Himalaya. *Eclogae Geologicae Helvetiae*, 147–213.

STEIGER, R. H. & JAGER, E. 1977. Convention on the use of decay constants in geo- and cosmochronology. *Earth and Planetary Science Letters*, **36**, 359–362.

THAKUR, V. C. 1993. Geology of the Western Himalaya. *Physics and Chemistry of Earth*, **19**, 113–116.

THIEDE, R. C., BOOKHAGEN, B., ARROWSMITH, J. R., SOBEL, E. R. & STRECKER, M. R. 2004. Climatic control on rapid exhumation along the Southern Himalayan Front. *Earth and Planetary Science Letters*, **222**, 791–806.

TRELOAR, P. J. & COWARD, M. P. 1991. Indian plate motion and shape: constraints on the geometry of the Himalaya orogen. *Tectonophysics*, **191**, 189–198.

TRELOAR, P. J., BROUGHTEN, R. D., WILLIAMS, M. P., COWARD, M. P. & WINDLEY, B. F. 1989. Deformation, metamorphism, and imbrication of Indian plate, south of the Main mantle thrust, north Pakistan. *Journal of Metamorphic Geology*, **7**, 111–125.

TRIVEDI, J. R., GOPALAN, K. & VALDIYA, K. S. 1984. Rb–Sr ages of granitic rocks within the Lesser Himalayan nappes, Kumaun, India. *Journal of Geological Society of India*, **25**, 641–653.

VALDIYA, K. S. 1989. Trans-Himadri intracrustal fault and basement upwarps south of Indus-Tsangpo Suture Zone. *In*: MALINCONICO, L. L. & LILLIE, R. J. (eds) *Tectonics of the western Himalaya*. Geological Society of America Special Paper, **232**, 153–168.

VALDIYA, K. S. 1995. Proterozoic sedimentation and Pan-African geodynamic development in the Himalaya. *Precambrian Research*, **74**, 35–55.

VANNAY, J. C. & GRASEMANN, B. 1998. Inverted metamorphism in the High Himalaya of Himachal Pradesh (NW India): phase equilibria versus thermobarometry. *Schweitzerische Mineralogische und Petro Graphische Mitteilurgen*, **78**, 107–132.

VEEVERS, J. J., JONES, J. G. & TALENT, J. A. 1971. Indo-Australia stratigraphy and the configuration and dispersal of Gondwana. *Nature*, **229**, 383–388.

VEEVERS, J. J., POWELL, C. M. & JOHNSON, B. D. 1975. Greater India's place in Gondwanaland and in Asia. *Earth and Planetary Science Letters*, **27**, 383–387.

WADIA, D. N. 1928. The geology of the Poonch State, Kashmir and adjacent parts of the Panjab. *Memoir Geological Survey of India*, **51**.

WADIA, D. N. 1957. *Geology of the India*. McMillan & Co., London.

WHITTINGTON, A., FOSTER, G., HARRIS, N., VANCE, D. & AYERS, M. 1999. Lithostratigraphic correlation in the Western Himalya – an isotopic approach. *Geology*, **27**, 585–588.

WILLIAMS, L. S. & CLAESSON, S. 1987. Isotopic evidence for the Precambrian provenance and Caledonian metamorphism of the high-grade paragneisses from the Seve Nappes, Scandinavian Caledonides, II. Ion microprobe zircon U–Pb–Th. *Contributions to Mineralogy and Petrology*, **97**, 205–217.

WINDLEY, B. F. 1988. Tectonic framework of the Himalaya, Karakoram and Tibet, and problems of their evolution. *Philosophical Transaction of Royal Society of London*, Series A, Mathematical and Physical Sciences, **326**, 3–16.

YIN, A. N. 2006. Cenozoic tectonic evolution of the Himalayan orogen as constrained by along-strike Variation of structural geometry, exhumation history, and foreland sedimentation. *Earth Science Review*, **76**, 1–131.

ZIETLER, P. K., SUTTER, J. F., WILLIAMS, I. S., ZARTMAN, R & TAHIRKHELI, R. A. K. 1989. Geochronology and temperature history of the Nanga Parbat Haromosh massif, Pakistan. *Geological Society of America Special Paper* **232**, 1–22.

ZHAO, G. C., CAWOOD, P. A., WILDE, S. A. & SUN, M. 2002. Review of global 2.1–1.8 Ga orogens: implications for a pre-Rodinia supercontinent. *Earth Science Review*, **59**, 125–162.

ZHAO, G. C., SUN, M. & WILDE, S. A. 2003. Correlations between the Eastern Block of the North China Craton and the South Indian Block of the Indian Shield: an Archaean to Palaeoproterozoic link. *Precambrian Research*, **122**, 201–233.

ZHAO, G. C., SUN, M., WILDE, S. A. & LI, S. 2004. A Palaeo-Mesoproterozoic supercontinent: assembly, growth and breakup. *Earth Science Review*, **67**, 91–123.

Palaeoproterozoic seismites (fine-grained facies of the Chaibasa Formation, east India) and their soft-sediment deformation structures

R. MAZUMDER[1]*, J. P. RODRÍGUEZ-LÓPEZ[2], M. ARIMA[3] & A. J. VAN LOON[4]

[1]*Geological Studies Unit and Fluvial Mechanics Laboratory, Indian Statistical Institute, 203 B.T. Road, Kolkata 700108, India*

[2]*Grupo de Análisis de Cuencas Sedimentarias UCM-CAM, Departamento de Estratigrafía, Instituto de Geología Económica UCM-CSIC, Facultad de Ciencias Geológicas, Universidad Complutense, Ciudad Universitaria, 28040 Madrid, Spain*

[3]*Department of Environment and Natural Sciences, Graduate School of Environment and Information Sciences, Yokohama National University, 79-7, Tokiwadai, Hodogaya Yokohama 240-8501, Japan*

[4]*Institute of Geology, Adam Mickiewicz University, Maków Polnych 16, 61-606 Poznan, Poland*

Corresponding author (e-mail: mrajat@isical.ac.in)

Abstract: Metamorphosed shales, heterolithic deposits and sandstones build up the Palaeoproterozoic Chaibasa Formation in east India. The shales (referred here to as the fine-grained facies) comprising mudstone (clay and silt size) with some minor amounts of very fine to fine sandstone were deposited below storm wave base in a deep marine basin that simultaneously underwent tectonic activity. This fine-grained facies contains strongly deformed layers, intercalated between undeformed layers. Sedimentological analysis of the deformations indicates that they formed while still in an unconsolidated or slightly consolidated state, partly during and after sedimentation, but before being covered by younger sediments. The types of deformation structures indicate an earthquake-induced origin. Thus, they should be considered as seismites. The soft-sediment deformation structures in the seismites show a wide variety of shapes and other characteristics that appear to depend on their relative position to the epicentre of the earthquake.

The entirely siliciclastic Chaibasa Formation in east India is 6–8 km thick. It rests partly on an Archaean granitic basement, partly on the terrestrial Dhanjori Formation and is overlain by the Dhalbhum Formation (Bose *et al.* 1997; Mazumder & Sarkar 2004; Mazumder 2005) (Fig. 1). The age of the Chaibasa Formation has not been established, but the underlying Dhanjori mafic volcanics are 2100 Ma (Roy *et al.* 2002*a*), whereas the minimum age of the Dalma Lavas that conformably overlie the Singhbhum Group (Bhattacharya & Bhattacharya 1970), to which the Chaibasa Formation belongs, are 1600 Ma old (Roy *et al.* 2002*b*). Thus, the Chaibasa Formation is of late Palaeoproterozoic age (Mazumder 2003, 2005).

The rocks underwent several post-depositional deformation phases and greenschist to amphibolite facies metamorphism that turned the sandstones into quartzites, and the shales, sometimes with alternating intercalations of very fine sandstones and shales that are interpreted as turbidites (Bose *et al.* 1997), into mica schists (Naha 1965; Saha 1994).

Characteristics of the Chaibasa Formation

The Chaibasa Formation is built of repeated alternations of quartzites (metamorphosed sandstones), heterolithic units (very fine quartzite/schist intercalations) and mica schists (metamorphosed mudstones) (Bhattacharya 1991; Bose *et al.* 1997; Bhattacharya & Bandyopadhaya 1998; Mazumder 2005; Fig. 2).

The sandstone facies consists of very fine to fine sandstones that locally may be muddy. The individual sandstone units range from 5–45 m in thickness, but show considerable changes in thickness laterally (Fig. 2). The individual layers in these units are up to 2.5 m thick, but also show lateral thickness variations. Many sandstones show cross-lamination. The units show tidal rhythms (Bose *et al.* 1997; Mazumder 2004) and the sandstone units are interpreted as shallow-marine deposits reflecting periods during which the tectonically active basin was either uplifted or had been filled by muds to such a thickness that the previously deep-marine character had changed to shallow-marine. Thus,

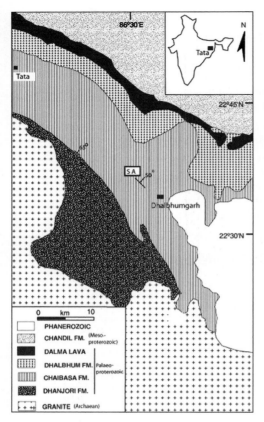

Fig. 1. Simplified geological map showing the disposition of the Chaibasa Formation and its adjacent formations (modified after Saha 1994; Mazumder et al. 2006); the Chaibasa and Dhalbhum formations together constitute the Singhbhum Group.

the deposition of the formation probably took several hundreds of millions of years to become several kilometres thick. It is therefore only logical that, under conditions of tectonic activity, shallow-marine and deep-marine nature intervals alternated (see Bose et al. (1997) and Mazumder (2005) for a discussion on the conditions under which the Chaibasa sequence was deposited).

The heterolithic facies consists of alternating layers of siltstone to very fine sandstone and shales. The relatively coarse layers of this unit range in thickness from 5–40 cm, most of them being 12–15 cm-thick. They show cross-bedding (including hummocky cross-stratification) or horizontal lamination and scour structures (25–30 cm deep, 1–1.5 m wide). The shale layers have variable thicknesses (2–60 cm) (Bose et al. 1997). We interpret this facies as transitional between the sandstone and the shale facies, being relatively distal but at a depth where the bottom was frequently affected by waves.

The fine-grained facies consists of metamorphosed shales (clay and silt fraction) with some minor amounts of quartzite (very fine sand fraction), often in the form of thin laminae but there are also some layers of quartzite. The individual units can reach thicknesses of over 100 m (Bose et al. 1997). No structures that might be ascribed to wave action have been found; in combination with the small grain size, this strongly suggests deposition of the sediments below storm wave base. Details of this facies, from which the soft-sediment deformation structures will be described and analysed, are provided in the next section.

Particularly in the upper part of the fine-grained facies, layers with soft-sediment deformation structures are abundant (Fig. 3) and show a wide variety of forms and complexity, including small clastic dykes, which Montenat et al. (2007) justifiably prefer to call 'sedimentary dykes', cone structures, load casts, pseudonodules, etc. (Fig. 4). The structures are interbedded with undeformed intervals, and have the same lithological composition as the undeformed layers. The lateral extent of the deformed layers is unknown because they cannot be traced beyond about a kilometre (which is, in fact, a remarkable distance for the occurrence of sedimentary deformations within one layer), because of limited exposure. It seems nevertheless that they have an even wider extent, because similarly deformed layers can be found far apart in more or less comparable stratigraphic positions, but lack of sufficient marker horizons makes precise correlation tentative.

Characteristics of the fine-grained facies

The fine-grained facies of the Chaibasa Formation occasionally contains, apart from the predominant shales, some thin, very fine-grained sandstones. The shales mostly consist of units about 75 m thick without any differentiation in granulometry that can be observed with the naked eye, so that it is difficult to discern individual layers. The very fine sandstones rarely exceed 4 cm in thickness, and are more commonly measurable in millimetres (laminae). A few may reach a thickness of up to 50 cm. These relatively thick layers of very fine sandstone generally have an exposure-wide extent which gives them a sheet-like appearance. Probably they are turbidites (Bose et al. 1997) and show horizontal lamination and/or ripple cross-lamination (Fig. 5).

The shale without sandstone intercalations in this fine-grained facies is, in general, massive, mica-rich and dark in colour (Fig. 5). Although no true sandstone intercalations occur in this type of deposit, horizontal light-coloured sub-laminae of

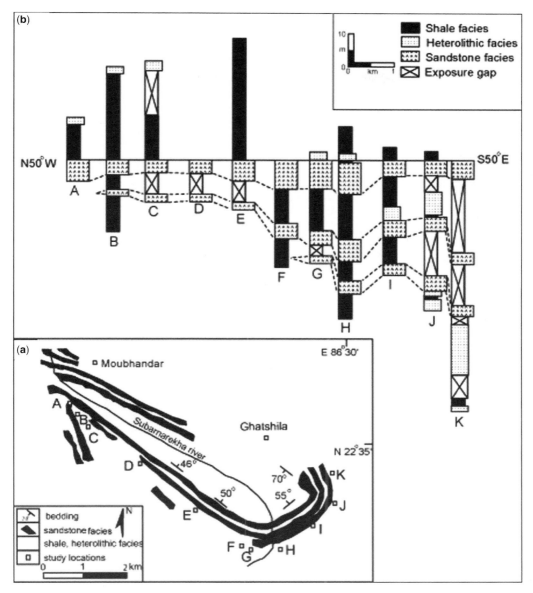

Fig. 2. The Chaibasa Formation. (**a**) Geological map showing lithological units of the Chaibasa Formation in and around Ghatshila (modified after Naha 1965). (**b**) Lateral and vertical lithofacies transitions of the Chaibasa sediments. Study locations are marked in Figure 2a (modified after Mazumder *et al.* 2006).

fine sandstone may be present locally. They show normal grading and light are in colour. These fine-grained units with sand/mud sub-laminae are separated from each other by sand-free, sheet-like mudstone beds with an average thickness of 3 cm, showing gradational bases and sharp tops. The ripple cross-lamination that is occasionally present (amplitude = 0.7–1.6 cm, wavelength = 2–6 cm) is distinctly asymmetric in profile with a tendency to climb (Fig. 6) and occurs in isolated ripple trains vertically separated by a set of horizontal laminae with an average thickness of 1.5 cm. Draping laminae on the lee faces of ripples are a common feature. These accretionary laminae tend to become horizontal, thus filling the ripple troughs completely and passing upwards gradually

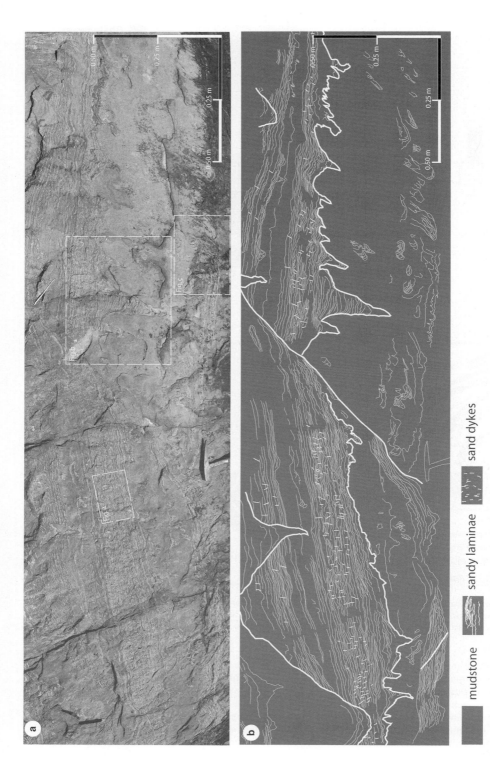

Fig. 3. Photo-mosaic (a) and derived drawing (b) of the soft-sediment deformation structures in part of an exposure of the Chaibasa fine-grained facies. The sequence of events resulting in the complex deformations is explained in Figure 16. The exposure (more or less perpendicular to the stratification) shows, among other deformations; a synsedimentary fault (Fig. 10); cone-shaped structures (Fig. 8); contorted mud balls (Fig. 11); pillow structures (Figs 7 & 14); load casts and pseudonodules (Figs 12 & 13). Details of the deformation structures in the areas marked by the rectangular boxes are presented in Figure 4.

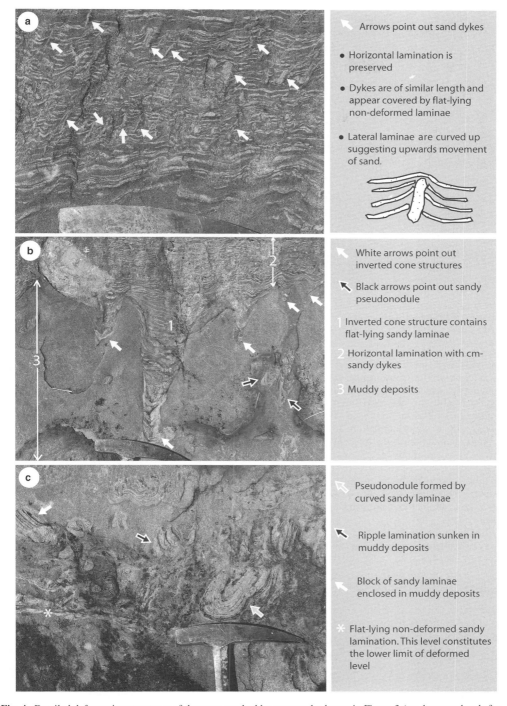

Fig. 4. Detailed deformation structures of the areas marked by rectangular boxes in Figure 3 (see hammer heads for scale). (a) Irregular lamination, contorted by numerous small-scale dykes of fluidized very fine sand; (b) Inverted cone structures with several other small-scale deformation structures; (c) Chaotic bed with sagged ripples and pseudonodule.

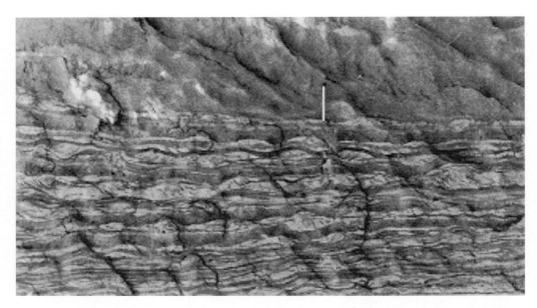

Fig. 5. Chaibasa fine-grained facies with intercalated thin laminae of very fine sand (light-coloured); the sandstone laminae show either horizontal or ripple cross-lamination. At places, the ripple trains are vertically stacked. Matchstick 4.2 cm.

into horizontal laminae. The individual layers, that in cross-section show ripple cross-lamination throughout, all have sharp, often fluted bases, and pass gradually into the horizontal laminae just mentioned (fig. 12d in Bose et al. 1997). The sets showing ripple cross-lamination at the base and horizontal lamination at the top range in thickness from 5 to occasionally 45 cm and are typically 15 cm thick. These thicker units are vertically separated from one another by sheet-like shale layers with an average thickness of 3 cm. These shale layers without coarser material are gradational but their tops are sharp.

The fine-grained facies is completely devoid of any wave-generated structure and is therefore likely to be a relatively deep-marine sediment formed below the storm wave base (Bose et al. 1997; Mazumder 2002, 2005). The relatively thick units that are separated by shales without thin sandy intercalations are probably products of episodic waning flows, possibly contour currents (an interpretation based on the various contributions to Rebesco & Camerlenghi 2008). Asymmetric ripples (Figs 5 & 6) imply traction. The draping laminae on the lee side of the ripples indicate weakening of the currents to the extent that ripple migration was no longer possible, but that sweeping of the suspension fall-out to the current shadow zone on the lee side of the ripples was still possible (Bose et al. 1997). The gradual upward transition from

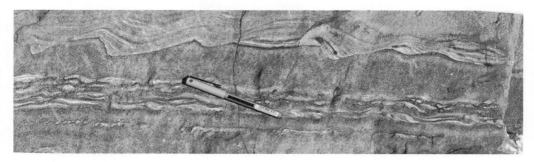

Fig. 6. Strongly asymmetric ripples within the Chaibasa fine-grained facies; note vertically stacked and sagged ripple trains within alternating shale and fine-sandstone laminae (pen length 12 cm).

draping laminae to horizontal laminae indicates a further slow decline in tractive force, eventually allowing settling of the suspension load. Grading of the fine sandstone laminae corroborates this contention. With a deep-sea distal current under a persistent suspension cloud, the depositional system appears to be similar to that of the hemiturbidites described by Stow & Wetzel (1990). The sand-free shales in between the thicker units show normal grading and are considered as indigenous pelagic deposits settled from suspension between events that whirled up sediment from the sea floor, or that supplied suspended load by contour currents, low-density mass flows or any other type of hyperpycnal flow.

The sediment is very fine-grained, without any coarse material, ruling out supply by a nearby river. Possibly the sediments had their transient source in the distal shelf. Abundant slump scars, bevelled bed-edges and slump folds in the Chaibasa heterolithic facies, detailed in Bose et al. (1997) and Mazumder (2002), support this interpretation. The persistent alternations of finer- and coarser-grained laminae may be due to basal shear that prevented mud floccules to settle unless their diameter exceeded a threshold value (Stow & Bowen 1978, 1980).

Objectives and approach

The objectives of the present study were to provide an inventory of the various types of soft-sediment deformation structures in the Chaibasa shales (fine-grained sediments deposited in a deep-marine facies contain much less, and much less variegated, soft-sediment deformation structures than the Chaibasa shales) and to analyse their genesis. This analysis was carried out following the approach suggested by Owen (1987), who pointed out that such a genetic analysis should be based on the reconstruction of: (1) the timing of deformation, and (2) the deformational mechanism(s).

Timing of the deformational processes

It has been well-known for several decades that deformations cannot only take place when a sediment has been covered by younger layers (post-depositional deformations), but also during the depositional process (syndepositional deformations) or after deposition but before the sediment is covered by a younger layer meta-depositional deformations (Nagtegaal 1963; Allen 1982; Owen 1995).

Whether deformations are syndepositional or meta-depositional, is of great importance for their genetic interpretation. In the first case, seismic influence, may easily affect the subaqueous sediments that are still being formed and that have not started to consolidate. On the other hand, meta-depositional deformations, which develop most commonly within about half a metre from the sedimentary surface, but which may occasionally (depending on the deformational mechanism and the forces involved) also affect deeper sediments, are related to surficial sediments that had already undergone some degree of consolidation. Relatively few mechanisms are capable of meta-depositional deformation, but earthquake-induced shocks can generate recurrent horizontal shear strain triggering a thixotropic effect in unconsolidated deposits, resulting in an instantaneous segregation of the liquid from the solid sedimentary phase (Montenat et al. 2007).

Analysis of the deformational mechanisms

Because the Chaibasa shales have been deposited in an active tectonic setting (Mazumder 2005), the analysis of the deformations (Mazumder et al. 2006) was directed first at the distinction between tectonic and non-tectonic (sedimentary) deformations. These may be interrelated, as tectonic shocks can lead to pressure gradients that induce the deformation of surficial, non-consolidated sediments. It is well-known that particularly the topmost sedimentary layer, if consisting of material that is susceptible to shock (earthquake)-induced deformations, can become strongly disorganized. Such layers, which tend to show more or less similar deformation structures over large distances, are known as 'seismites'.

The term 'seismite' has been introduced by Seilacher (1969) for a layer (or a set of layers) with earthquake-generated soft-sediment deformation structures (Neuendorf et al. 2005). Seismites are probably the best proof of earthquakes in the history of Earth, although their recognition is difficult, partly due to the fact that unconsolidated sediments that are easily deformed by earthquake-induced shocks, are commonly also susceptible to disturbances not related to seismicity. Yet the detection of unmistakable fingerprints of seismic events in ancient successions is, however difficult, of paramount importance for insight into basin dynamics (Seth et al. 1990).

Sedimentological evidence of earthquakes other than in the form of seismites is scarce in the sedimentary record and the interpretation of such earthquake-induced phenomena is often even more questionable than the recognition of seismites (Seilacher 1984). For instance, faulting at a basin's margin may result from earthquakes that also induce mass failure, but subsequent mass flows facilitated by the fault-created steep slope need in no way be directly related to seismic activity. Therefore it is difficult to separate the two genetic groups of mass-flow deposits within a single lithological unit.

Recognition of seismites is hampered by the fact that earthquakes are just one of the trigger mechanisms that induce soft-sediment deformation. Other mechanisms are: (1) pressure changes resulting from, for instance, the passage of storm currents, breaking waves or flood surges; (2) oversteepening of slopes; (3) gravitational density flows; and (4) rapid loading of sediments (Van Loon & Wiggers 1976b; Allen 1982; Owen 1987, 1995, 1996; Van Loon & Brodzikowski 1987; Maltman 1994; Rossetti 1999; Rossetti & Goes 2000; Jones & Omoto 2000; and refs therein). In addition to these processes, differential compaction of sediments can also create soft-sediment deformation structures, although the process may be slow.

The following combination of characteristic of layers with abundant soft-sediment deformation structures is commonly considered diagnostic for an earthquake origin (Sims 1973, 1975; Scott & Price 1988; Moretti & Tropeano 1996; Obermeier 1996; Alfaro et al. 1997; Rossetti 1999; Jones & Omoto 2000; Bose et al. 2001): (1) the restriction of deformational structures to discrete stratigraphic levels; (2) lateral continuity of the deformed character over large distances; (3) recurrence of such deformed layers through time; (4) a consistent deflection of palaeocurrent trends from their usual pattern across the deformed levels; (5) confinement between undeformed strata (or strata with a distinctly different origin of the deformations); and (6) a preferred association with wedges of intraclastic breccias, conglomerates and massive sandstones.

Our investigation was directed at finding out whether specific deformed layers in the Chaibasa shales should be considered as seismites. We analysed all types of soft-sediment deformation that were encountered in the Chaibasa shales, and interpreted their syn- or meta-depositionally nature. Finally, we reconstructed their genesis to find a possible seismic origin.

In the following sections, the soft-sediment deformation structures in the fine-grained facies of the Chaibasa Formation will be described. We group these structures according to their syndepositional or meta-depositional origin. As the difference between syndepositional and metadepositional deformations, and the way in which they can be distinguished from one another, has been described extensively in the literature (Nagtegaal 1963; Allen 1982; Owen 1995), we will not detail the precise arguments for our interpretation of the timing here.

Syndepositional deformation structures

Deformation structures in the Chaibasa shales that must have formed syndepositionally show a wide variety of shapes and other characteristics. On the basis of their forms they can be grouped into the following types.

Pillows

Several levels in the fine sandy intercalations that are present within the fine-grained facies are characterized by pillow-shaped deformations that can be traced laterally over tens to hundreds of metres, and that occur within these levels without significant lateral interruptions. They are more common in the thicker sandy layers. The pillows have obtained a U-shape because they have sunk deep into the underlying mud, commonly having parallel vertical contacts, whereas their bottoms are rounded (Fig. 7). Their internal lamination tends to follow the outer shape. The size of the pillows (max. height 35 cm; max. diameter 12 cm) changes gradually laterally within a single level. Within individual structures, the degree of deformation decreases upward, and younger lamina sets may truncate the underlying older sets. This implies repeated sagging during ongoing sedimentation, which proves the syndepositional character.

Pillow structures like these have commonly been interpreted as due to loading of denser sediment in a two-layer system (Allen 1982 & references therein). The loading process, which should be considered as a response to unstable density contrasts or lateral variations in load (Owen 2003), may also be triggered by mechanisms such as differential

Fig. 7. Pillow beds consisting of predominantly clay- and silt-sized sediments, with thin laminae of very fine sand (light-coloured) that make the deformations better visible. Loading has resulted in plastic deformation, and very local fluidization has made the original lamination disappear in places.

compaction due to rapid sediment accumulation, wave activity inducing stresses and ice growth in glacigenic sediments (Ricci Lucchi 1995). A seismic origin for such pillow structures has also been suggested (Montenat et al. 1987; Cojan & Thiry 1992; Guiraud & Plaziat 1993; Rodríguez-Pascua et al. 2000).

The Chaibasa pillow structures developed in sediments that do not show sufficient density contrasts that could cause loading without a forceful trigger mechanism, because the density of water-saturated fine sands is barely higher than that of water-saturated mud (Anketell et al. 1970). The layers in which the pillows developed were below storm wave base (Bose et al. 1997), so that wave-induced stresses could not have been the cause. Ice growth in these deep-sea sediments can also be excluded, so a syndepositional seismically-induced origin is the most likely. The formation of the pillows was probably initiated by shock-induced liquefaction of some specific layers within sets of sandy layers, followed by an upward movement (into the direction of the lowest pressure) of the liquefied silty fine sand.

Cone-shaped contorted laminae sets

A completely different type of soft-rock deformations consist of inverted cone-shaped structures (Fig. 8). The cones are filled with concave upward laminae of shale and silty sandstone in multiple sets. The younger sets truncate the older ones and the degree of concavity of the laminae decreases in an upward direction from one set to the next. These alternations of fine sandy and muddy laminae are generally absent outside the cones (Fig. 8).

The configuration and the internal structure of the cones suggest that cracks formed on the subaqueous mud surface and that a current laden with mud and fine sand filled these cracks (Cozzi 2000). This can occur when a hydraulic jump is created in the underflow on top of the crack so that the particles can settle within the cracks.

The erosion at the base of each lamina set and the upward decreasing deformation rate suggest repeated enlargement of the cracks. Consequently, the crack-filling laminae sank plastically down into the cracks, thus creating new accomodation space. Because this process took place repeatedly, the oldest (lowermost) set of laminae was exposed to the largest number of deformation phases and thus became most deformed, whereas the youngest (uppermost) set suffered no deformation at all. The origin of the cracks is not clear. Cozzi (2000) mentions some cracks in a shallow-marine environment but does not provide a good explanation. In addition, the cracks described by Cozzi are somewhat different from those in the Chaibasa Formation. Considering the repeated enlargement of the cracks and the successive infillings, a series of successive events must have taken place within a relatively short time. The interpretation is that earthquake-induced shocks have been the trigger mechanism; the main reason for this interpretation is that no other mechanisms are known that might result in such cracks. Opening of cracks, filling up with sediment, reopening, etc. under subaquous conditions has been mentioned from other recently seismically active areas; the process has been detailed by Vachard et al. (1987) and Montenat et al. (2001).

An alternative explanation might be that the sea floor was deformed by synaeresis cracks (Plummer & Gostin 1981), but these structures are not well understood (Hounslow 2003) and apparently involve changes in volume of clay minerals induced by salinity changes (Collinson & Thompson 1982), expulsion of fluids from colloidal suspensions, or removal of water from mud layers that are in contact with brines produced by evaporation (McLane 1995). The Chaibasa shales do not show any indications of such conditions: on the contrary, the conditions seem to have been very constant over long periods (the whole succession most probably was deposited during a time span of several hundreds of millions of years), apart from some short-lasting events such as earthquakes. Tanner (2003) mentions that intrastratal synaeresis can be caused by earthquake-induced ground

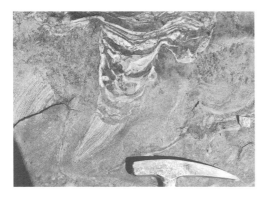

Fig. 8. Inverted cone-shaped contorted laminae sets, as seen in a section almost perpendicular to the bedding plane; note that the cones are filled with concave upward laminae of shale and silty sandstone in multiple sets. The morphological resemblance with ice-wedge casts (Montenat et al. 2007) makes it likely that cracks were formed in the seafloor (possibly by extension tectonics) that became filled with fine-grained sediment supplied by traction; this process must have taken place during widening of the cracks.

motion. This is consistent with Pratt (1998), who states that cracks in subaqueous argillaceous sediments may result from earthquake-induced dewatering. Another alternative is that the cracks are diastasis cracks (Cowan & James 1992), which have been described also from bathyal conditions (Shiki & Yamazaki 1996).

Infilled depressions

One shale layer contains several depressions that all show at least one steep wall in cross-section (fig. 4 in Mazumder et al. 2006) The exposed depressions all have more or less identical sizes, on average 48 cm deep and 36 cm wide (the length cannot be established, as no 3-D exposures are present). The depressions represent erosional forms, as the laminae of the adjacent shales (and some very thin fine sandstones) are truncated by the margins of the depressions, which are themselves filled with comparable alternating laminae of shale and very fine sandstone (Fig. 9) that partly, particularly in the deepest part of the depressions, end abruptly against the depressions' walls and partly, and ever more clearly, follow the slope of the walls as they have a higher position in the infilling. None of the depressions has a symmetrical shape as the lower parts of the walls are steeper at one side than at the other side.

The laminae immediately underlying the structures do not exhibit any significant degree of deformation, so is must be concluded that the depressions are erosional features that later were gradually filled up; first with laminae that could not be deposited on the steep lower walls but (when the ongoing sedimentation had partly filled the depression until the level that the walls became more gently inclined) later following the then sedimentary surface within the depression (following the inclined walls and laterally continuing outside the depression). The individual sedimentary laminae/layers in the depressions are relatively thick in the centre of the depression and thin towards the margin, indicating that the depressions acted as a sediment trap.

The geometry of the structures resembles that of the meso-scale scours described by Sarkar et al. (1991) from a near-shore region, and described by Cheel & Middleton (1993) from an offshore environment. The formation of such erosional forms in a cohesive mud substratum is also known from turbidity currents (Mutti & Normak 1987). It is unlikely, however, that turbidity currents were involved in this case, as all turbidites found in this part of the Chaibasa Formation are thin and fine-grained, whereas the erosional structures have a depth of various decimetres.

Although the infilled depressions have many characteristics in common with 'normal' channels (the occurrence of one steep and one less inclined side seem to point at meandering channels), it is remarkable that they must be ascribed to relatively high-energy events, whereas such events left no other traces in the sediment. The infilling of the structures also lacks any sign of a lag deposit, which makes these depressions even more enigmatic. It was interpreted earlier that at least some of the depressions could have formed due to the lifting of sediment blocks by the successive phases of under- and overpressure during the passage of tsunami waves (Mazumder et al. 2006), but renewed analysis of the depressions showed features (e.g. the succession on the less steep side of these depressions) that contradict this view.

Since the sedimentary unit in which the depressions occur was not significantly deformed (only some minor deformations were observed but these could not unambiguously be ascribed to a seismic shock), this unit should, however, not be called a seismite.

Fig. 9. Cross-section through infilled depression within the Chaibasa shale (hammer length 42 cm). The infilling points at a synsedimentary filling of the depression.

Sagged ripple trains

Vertically stacked and sagged ripple trains occur within a succession of alternating shale and fine-sandstone laminae (Figs 5 & 6). The sagging is generally more pronounced in the lower part of the stack. The laminae of very fine sandstone in the shale underneath the sagged ripple trains roughly follow the undulating base of the sagged ripples.

Ripple trains can repeatedly form and sag (Dzulynski & Kotlarczyk 1962; Dzulynski & Slaczka 1965; Dzulynski 1996). The rippled sediment under investigation has evidently been deposited in a number of phases, but before the underlying mud could release the excess pore water (Dzulynski & Slaczka 1965; Reineck & Singh 1980).

The sagging of the ripple trains must be due to partial liquefaction of the underlying mud. The liquefaction may have been caused by an earthquake-induced shock, but may, alternatively, be due to simple pressure resulting from differential loading (Lowe 1975; Jones & Omoto 2000).

Metadepositional deformations

The typical metadepositional deformations in the Chaibasa shale show a wide variety of types. They include: (1) penecontemporaneous faults; (2) contorted mud balls; (3) pseudonodules; (4) chaotic layers; and (5) a collapse structure.

Penecontemporaneous faults

Fault planes, commonly upthrusts, up to some 55 cm long and dipping 5–12°, occur in the upper part of the Chaibasa shales, without affecting the under- and overlying layers (Fig. 10). They show the following characteristics: (1) progressive squeezing of the laminae towards the fault planes; (2) a spreading in orientaton of the faults within each individual layer; (3) evidence of liquefaction, mostly at the top of the fault planes, and a decreasing degree of liquefaction downwards; and (4) planar or sigmoidal fault planes that consist of masses of shale, as commonly found along shear planes.

The position of the faulted layers intercalated with unfaulted layers is evidence of a metadepositional origin, and the faults themselves indicate that some compaction had taken place before the faulting, but differential compaction apparently did not play a role, as no indications for this process are present (the lateral changes in thickness of laminae in the direct neighbourhood of the fault are a result of shearing). The upthrusting of the slightly compacted sediments indicates that some lateral stress must have been exerted. A shock, presumably due to an earthquake, seems the most logical cause of such a stress, and thus a seismic shock must be considered as the most likely explanation for the faulting.

Contorted mud balls in massive sandstone

Some of the rare massive layers of very fine sandstone that are thicker (c. 30 cm) than most other sand layers in this fine-grained facies of the Chaibasa Formation, occasionally contain 'floating' balls of shale (Fig. 11) that internally show contorted laminae of silty to fine sandy material with upturned margins. The structures obviously consist of sediment that sank into the underlying layer and eventually became detached from the overlying parent mud layers (pseudonodules).

The loading of mud balls into sand is difficult to explain unless the sand also had a relatively low density. This may have been the result of a strong hydrostatic stress that developed within the sandy layers under an impermeable mud cover (Schwab & Lee 1988). The configuration indicates late-stage breakup of the overlying mud layers and also that liquefaction of the fine sand probably destroyed any primary sedimentary structures in the fine sands, reflected now by the massive fabric of the sandstone layers.

The structures do not provide a clue for the trigger that resulted in the liquefaction, but a suddenly increased hydrostatic stress seems the most likely. The quiet depositional environment, reflected by the fine-grained character of the sediments, does not provide any clue for such a sudden increase in

Fig. 11. Contorted balls of shale within layers of very fine sandstone of the fine-grained facies of the Chaibasa Formation. (coin diameter 2.5 cm). The balls of shale still show internal lamination indicating that the, then, slightly consolidated mud was deformed plastically, whereas the sandstone does not show any internal structure, suggesting that the sand was liquefied during the deformational process.

Fig. 10. Penecontemporaneous thrust within the Chaibasa shale (after Mazumder 2005). The fault plane shows in cross-section a characteristic, slightly sigmoidal shape.

hydrostatic stress, particularly since the floating mud balls are the only deformation structures in this part of the section. An 'outside' trigger therefore seems the most probable; this may, considering the synsedimentary tectonics in the basin, well have been an earthquake-induced shock.

Levels with small-scale pseudonodules

The original bedding in the shales is locally indicated by a high concentration of small crescent-shaped pseudonodules of very fine sandstone. The frequency of these pseudonodules diminishes downwards. Their lateral distribution is irregular: they form clusters with an average thickness of 8 cm (Fig. 12). The form of the pseudonodule clusters suggests lateral variability in the resistance of the depositional substratum. Lateral variation in thickness of the original sandy layer is highly unlikely, as such lateral variations are observed nowhere in this facies.

The pseudonodule levels can be explained satisfactorily only by assuming that a layer of very fine sand was deposited on top of a water-saturated mud unit of lower density. The sand started to sink down in the mud, forming load casts and eventually pseudonodules. The vertical distribution of the pseudonodules can be explained by increasing compaction and resistance to deformation of the muddy substratum with depth. This resistance is also reflected in the upturned edges of many pseudonodules. As the mud showed apparently some resistance, it may have been slightly compacted (which is also suggested by the shape of the pseudonodule clusters). Compaction, however little developed, would under quiet conditions have led to gradual loading of the sand layer, resulting (after continued loading) in a series of pseudonodules beside each other (Van Loon & Wiggers 1976b), but not in clusters of pseudonodules that contain two or even more pseudonodules on top of each other. It seems therefore likely that the loading process was not a quiet one but rather the result of a sudden, uncommon disturbance. Such a disturbance needs a trigger. This was most likely an earthquake-induced shock, as experiments (Kuenen 1965) have shown that vertical 'accumulations' of saucer-like pseudonodules can form as a consequence of vibrations affecting the sand layer and the overlying sediment (Reineck & Singh 1980).

Chaotic layers

The internal structure of some shale layers, which are up to 70 cm thick, is chaotic. These layers look tabular in form in outcrop. They show, between their sharp lower and upper contacts with undeformed layers with the same granulometry (and thus porosity and, particularly, permeability) deformation structures (well visible because of the presence of laminae of very fine sandstone) such as convolutions, pillows (both mainly in the upper parts of these layers) together with load casts and pseudonodules (mainly in the lower part)

Fig. 12. High concentration of small saucer-shaped pseudonodules of very fine sandstone within the Chaibasa shale; the distribution of the pseudonodules is irregular and they form elongated clusters. Pseudonodule clusters of this type have been produced experimentally, exclusively by vibration (Kuenen 1965).

Fig. 13. Chaotic shale layer (section oblique to the bedding plane) with loadcasts and a pseudonodule cluster.

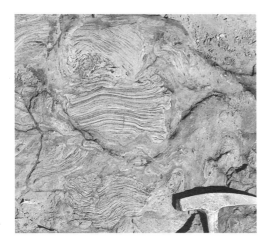

Fig. 14. Chaotic Chaibasa shale layer with exceptionally well-developed convolutions and pillows.

Fig. 15. Collapse structure (shown by arrows) within the fine-grained facies. Note the difference in deformation between the lower (units L1 and L2) and upper (U) beds. In both units, the intensity of deformation decreases upwards.

(Figs 13 & 14); these load structures have rotated in such a way that they are now upside down. Apart from these plastically deformed structures, liquefaction took place locally. Comparable deformations have been described frequently from seismites (see fig. 17a in Merriam 2005).

The individual layers with the chaotic deformations must have been affected by a stress that had the same intensity everywhere, as there are no significant lateral changes in the intensity of the deformations. The undisturbed layers below the deformed layers suggest that the deformation affected only the surficial sediments; the undisturbed layers on top of the disturbed layer indicate that the deformation must have taken place before the affected layer was covered with younger sediments. This suggests that the deformation took place because entrapped pore water was forced to find a way out (Williams 1966, 1969, 1970; Owen 1995). The mechanism that triggered this process cannot be reconstructed on the basis of the structures themselves but it cannot have been the weight of an overburden, as the overlying sediments are undeformed. It must therefore be deduced that, like in the case of the mud balls described above, an 'outside' mechanism, such as an earthquake-induced shock, has been responsible.

Collapse structure

Two layers with a locally strange contact (Fig. 15) are present in places where the shale contains some fine sandstone (Mazumder *et al.* 2006). The lowermost layers are most intensely deformed, and their internal layering follows exactly the bending of the base of the overlying unit; this proves that the structure cannot be ascribed to erosion (scouring). The lower layer shows chaotic deformation, whereas the overlying layer shows no internal deformation, except for some vertical fractures/cracks (the origin of these fractures is still not clear, but the fractures are comparable to those described above about the penecontemporaneous faults and the block-shaped depressions). A slab of the upper layer penetrates the chaotic mass of the lower layer.

The chaotic character of the lower layer must be due to a stress field that forced water-saturated sediment to move laterally, thus creating some kind of underpressure that caused collapse of the overlying layer. As the intensity of the deformations in the lower layer diminishes in an upward direction, and as there is no internal deformation within the upper layer, is must be deduced that the force causing the deformation came from beneath. The force was apparently not strong enough to cause deformation upwards of the contact plane between the two layers, which seems to have served as a shear plane. This suggests that the chaotic deformation took place in a confined state, underneath the sedimentary surface.

Discussion of the possible trigger mechanisms

Almost all structures described above can be generated under a wide variety of conditions. However, the clusters of pseudonodules, with several saucer-shaped specimens above each other, have only been produced experimentally as a result of vibrations (Kuenen 1965). Such vibrations occur under natural conditions almost exclusively as a result of earthquakes. These pseudonodule clusters are therefore strong evidence for earthquakes as the responsible mechanism. This interpretation is supported by the fact that the area at the time formed part of a tectonically active basin. The occurrence of synsedimentary upthrusts can be explained only by lateral pressure, for which – under the environmental conditions that seem to have lasted during sedimentation of the entire fine-grained facies without any noticeable interruption (apart from temporary non-sedimentation) – only seismic activity can be considered as a logical cause.

The other syndepositional and metadepositional soft-sediment deformation structures can all be explained as resulting from other deformational processes, but some of the possible genetic interpretations are mutually inconsistent, and other possible interpretations do not fit the environmental conditions (i.e. very quiet marine environment below storm wave base). In addition, most of the abundant soft-sediment deformation structures in the fine-grained facies cannot be explained by mass-flow transport (fine-grained turbidites occur, but high-density flows with sediment in a plastic state are absent). Moreover, sediments were deposited below storm wave base, which also excludes wave-related deformational mechanisms.

The only mechanism that can explain all types of soft-sediment deformation structures is a series of shocks (sometimes apparently at relatively short intervals, sometimes with longer intervals), sometimes in combination with repeated phases of hydrostatic under- and overpressure; in this context the two main earthquake-induced mechanisms for soft-sediment deformation are: (1) liquefaction of sediment generated by superficial shear waves; and (2) violent expulsion of pore water and liquefied sediment, generated by compression waves (Montenat et al. 2007). Considering the commonly very quiet sedimentary environment where only fine-grained sediment accumulated (the coarsest material being very fine sand), the shocks can be ascribed only to the tectonic activity of the basin.

The above does not imply that *all* soft-sediment deformations were triggered by an earthquake-induced shock. In almost all environments, even in tectonically quiet ones, some deformations may occur, particularly if the content of silt is high (as this grain size is very susceptible to deformations) and also if sediments with contrasting granulometry are present (Montenat et al. 2007). Because the fine-grained facies of the Chaibasa Formation, consists primarily of metamorphosed shales (derived from a combination of clay and silt), and because some sand was also present (although almost exclusively very fine sand), the sediment characteristics were favourable for soft-sediment deformation. The fact that apparently some consolidation of the muddy basin floor took place at some intervals (as a result of compaction and/or some interruption in sediment accumulation), thus providing a vertical change in physical characteristics after renewed sedimentation, made the topmost sediments even more prone to deformation. Although the above combination of characteristics favoured soft-sediment deformation, and although most soft-sediment deformation structures can be explained by different processes, it seems that a considerable part of the deformation was triggered by seismic shocks. This is even more likely because of the stratigraphic distribution of the soft-sediment deformation structures: they are concentrated in relatively few layers that are interbedded between undeformed layers, sometimes with several deformed layers within a short vertical succession. Although the deformed horizons cannot be traced without interruption over lateral distances of several kilometres (because of lack of exposures), the exposure-wide continuation of the deformed levels and the spatial distribution of the various types of deformation strongly favour their interpretation as layers with deformations that were triggered by earthquakes (Rodríguez-López et al. 2007), sometimes apparently with aftershocks (Seilacher 1984; Seth et al. 1990; Ricci Lucchi 1995; Bose et al. 1997, 2001; Jones & Omoto 2000). The deformed layers should therefore be considered as seismites.

Another strong argument for an earthquake-induced shock is the collapse structure. It must be deduced that a shear plane developed between the layer with a chaotic fabric and the overlying internally undeformed layer that collapsed. Earthquake waves, if not sufficiently strong, cannot propagate across shear planes (Schwab & Lee 1988) and this explains the contrasting degrees of deformation in the two layers.

Pillow structures have also often been attributed to seismicity (Montenat et al. 1987; Cojan & Thiry 1992; Roep & Everts 1992; Guiraud & Plaziat 1993; Rodríguez-Pascua et al. 2000). Experiments applying repeated jerks also resulted in similar pillows (Allen 1982; Owen 1996). An earthquake origin of these pillows in the Chaibasa shales is also suggested: (1) because the affected deposits formed below the storm wave base; (2) by their complex and multi-stage deformations revealing repeated seiches; and (3) by their lateral continuity and gradual lateral change in size and complexity. In addition, they are confined to certain stratigraphic levels, although the granulometric characteristics are the same throughout the entire fine-grained facies, which is prone to deformation.

Conclusions

Several of the soft-sediment deformation structures show characteristics that indicate earthquakes as a

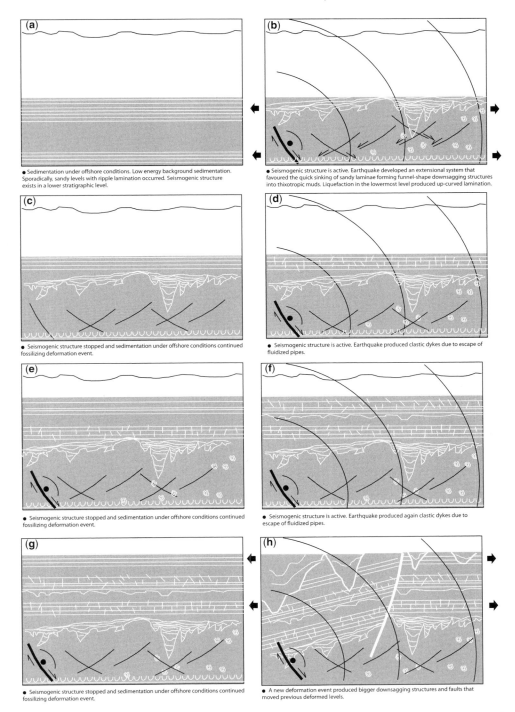

Fig. 16. Reconstruction of the sequence of events that can explain the complex soft-sediment deformation structures in the Chaibasa fine-grained facies as a result of seismic activity. Several phases of tectonic quiescence and of earthquake activity must have alternated.

trigger. The most logical reconstruction of the successive phases of seismic deformation as recorded in the Chaibasa fine-grained facies is as follows (Fig. 16). After an interval of tectonic quiescence (Fig. 16a), an earthquake develops faults and results in an extensional system that gives rise to down-sagging of sandy laminae into the underlying muds (Fig. 16b) that have become thixotropic due to the shock-induced stress. Then a new interval of tectonic rest (Fig. 16c) is followed by a new shock, resulting in the escape of pore water from the surficial sandy and muddy laminae. After the water has escaped, fluidized sand fills the escape pathways, thus forming sandy clastic dykes (Fig. 16d). During a third phase of tectonic quiescence the deformed layers become buried under a new cover of horizontal muddy and sandy laminae (Fig. 16e). A new earthquake then results in a new phase of clastic-dyke formation (Fig. 16f), comparable to the situation described before. Ongoing sedimentation of sandy and muddy layers during a next quiet interval (Fig. 16g) 'fossilizes' the buried deformation structures. Finally, a new, relatively heavy or nearby shock triggers large down-sagging structures (Fig. 16h, left-hand side) and results in faulting of the previously deformed levels. This sequence of events explains almost all soft-sediment deformation; no other sequence of events can do so satisfactorily.

In addition to the deformations that can be ascribed only to seismic activity, numerous structures can also be explained (if not better) due to earthquakes than as a result of other triggers. This all strongly supports the interpretation of most of the soft-sediment deformation structures in specific levels of the Chaibasa fine-grained facies as palaeoseismic features (although, obviously, not all deformations need necessarily be due to seismic shocks). Considering that the depositional setting of this formation, particularly during the later phase of sedimentation was in a tectonically active area, the seismite character of these layers becomes even more logical. The occurrence of turbidites in this succession further strengthens this hypothesis.

The authors are grateful to A. George and G. Owen for their critical, constructive and painstaking reviews, and to S. Reddy for useful suggestions in the field. RM is grateful to S. Pal, Director, Indian Statistical Institute (ISI), for providing funding for a fieldtrip in the Singhbhum crustal province in March 2007 and to J. K. Ghosh (ISI), B. S. Mazumder (ISI), P. K. Bose, S. Sarkar (Jadavpur University) and W. Altermann (Munich University) for inspiring discussions and motivations. Fieldwork by AJvL for this research project was supported by the Foundation Dr Schürmannfonds, grant no. 34/2006. The Japan Society for the Promotion of Science (JSPS) also supported the project. This is an IGCP-509 contribution.

References

ALFARO, P., MORETTI, M. & SORIA, J. M. 1997. Soft-sediment deformation structures induced by earthquakes (seismites) in Pliocene lacustrine deposits (Guadix-Baza basin, central Betic Cordillera). *Eclogae Geologicae Helvetiae*, **90**, 531–540.

ALLEN, J. R. L. 1982. Sedimentary structures – their character and physical basis. *Developments in Sedimentology*, **30A**, 593; **30B**, 663.

ANKETELL, J. M., CEGLA, J. & DZULYNSKI, S. 1970. On the deformational structures with reversed density gradients. *Annales Societatis Geologorum Poloniae*, **40**, 3–29.

BHATTACHARYA, D. S. & BHATTACHARYA, T. K. 1970, Geological and geophysical investigations of a basaltic layer in Archaean terrain of eastern India. *Geological Society of America Bulletin*, **81**, 3073–3078.

BHATTACHARYA, H. N. 1991. A reappraisal of the depositional environment of the Precambrian metasediments around Ghatshila-Galudih, eastern Singhbhum. *Journal of the Geological Society of India*, **37**, 47–54.

BHATTACHARYA, H. N. & BANDYOPADHAYA, S. 1998. Seismites in a Proterozoic tidal succession, Singhbhum, Bihar, India. *Sedimentary Geology*, **119**, 239–252.

BOSE, P. K., MAZUMDER, R. & SARKAR, S. 1997. Tidal sandwaves and related storm deposits in the transgressive Protoproterozoic Chaibasa Formation, India. *Precambrian Research*, **84**, 63–81.

BOSE, P. K., SARKAR, S., CHAKRABORTY, S. & BANERJEE, S. 2001. Overview of the Meso- to Neoproterozoic evolution of the Vindhyan basin, central India. *Sedimentary Geology*, **141**, 395–419.

CHEEL, R. J. & MIDDLETON, G. V. 1993. Directional scours on transgressive surface: examples from southern Ontario, Canada. *Journal of Sedimentary Petrology*, **63**, 393–397.

COJAN, I. & THIRY, M. 1992. Seismically induced deformation structures in Oligocene shallow-marine and eolian coastal sands (Paris Basin). *Tectonophysics*, **206**, 79–89.

COLLINSON, J. D. & THOMPSON, T. B. 1982. *Sedimentary structures*. George Allen & Unwin, London.

COWAN, C. A. & JAMES, N. P. 1992. Diastasis cracks: mechanically generated synaeresis-like cracks in upper Cambrian shallow water oolite and ribbon carbonates. *Sedimentology*, **39**, 1101–1118.

COZZI, A. 2000. Synsedimentary tensional features in Upper Triassic shallow-water platform carbonates of the Carnian Prealps (northern Italy) and their importance as palaeostress indicators. *Basin Research*, **12**, 133–146.

DZULYNSKI, S. 1996. Erosional and deformational structures in single sedimentary beds: a genetic approach. *Rocznik Polskiego Towarzystwa Geologicznego*, **64**, 101–189.

DZULYNSKI, S. & KOTLARCZYK, J. 1962. On load-casted ripples. *Rocznik Polskiego Towarzystwa Geologicznego*, **30**, 214–241.

DZULYNSKI, S. & SLACZKA, A. 1965. On ripple-load convolution. *Bulletin de l'Académie Polonaise des Science, Série des Sciences de la Terre*, **13**, 135–139.

GUIRAUD, M. & PLAZIAT, J. C. 1993. Seismite in the fluviatile Bima sandstones: identification of palaeoseisms and discussion of their magnitudes in a Cretaceous synsedimentary strike-slip basin (Upper Benue, Nigeria). *Tectonophysics*, **225**, 493–522.

HOUNSLOW, M. W. 2003. Septarian concretions. *In*: MIDDLETON, G. V. (ed.) *Encyclopedia of Sediments and Sedimentary Rocks*. Kluwer Academic Publishers, Dordrecht, 657–659.

JONES, A. P. & OMOTO, K. 2000. Towards establishing criteria for identifying trigger mechanisms for soft sediment deformation: a case study of Late Pleistocene lacustrine sands and clays, Onikobe and Nakayamadaira Basins, northeastern Japan. *Sedimentology*, **47**, 1211–1226.

KUENEN, PH. H. 1965. Value of experiments in geology. *Geologie en Mijnbouw*, **44**, 22–36.

LOWE, D. R. 1975. Water escape structures in coarse-grained sediments. *Sedimentology*, **22**, 157–204.

MALTMAN, A. J. 1994. *The Geological Deformation of Sediments*. Chapman & Hall, London.

MAZUMDER, R. 2002. *Sedimentation history of the Dhanjori and Chaibasa formations, eastern India and its implications*. Unpublished Ph.D. Dissertation, Jadavpur University, Kolkata.

MAZUMDER, R. 2003. Correlations between the Eastern Block of the North China Craton and the South Indian Block of the Indian Shield: an archaean to palaeoproterozoic link – Comment. *Precambrian Research*, **127**, 379–380.

MAZUMDER, R. 2004. Implications of lunar orbital periodicities from Chaibasa tidal rhythmite of late Palaeoproterozoic age. *Geology* **32**, 841–844.

MAZUMDER, R. 2005. Proterozoic sedimentation and volcanism in the Singhbhum crustal province, India and their implications. *Sedimentary Geology*, **176**, 167–193.

MAZUMDER, R. & SARKAR, S. 2004. Sedimentation history of the Palaeoproterozoic Dhanjori Formation, Singhbhum, eastern India. *Precambrian Research*, **130**, 269–289.

MAZUMDER, R., VAN LOON, A. J. & ARIMA, M. 2006. Soft-sediment deformation structures in the Earth's oldest seismites. *Sedimentary Geology*, **186**, 19–26.

MCLANE, M. 1995. *Sedimentology*. Oxford University Press, New York.

MERRIAM, D. F. 2005. North America – Continental interior. *In*: SELLEY, R. C., COCKS, L. R. M. & PLIMER, I. R. (eds) *Encyclopedia of Geology*, **4**, 21–36.

MONTENAT, C., BARRIER, P. & OTT D'ESTEVOU, P. 2001. Some aspects of the recent tectonics in the Strait of Messina, Italy. *Tectonophysics*, **194**, 203–215.

MONTENAT, C., BARRIER, P., OTT D'ESTEVOU, P. & HIBSCH, C. 2007. Seismites: an attempt at critical analysis and classification. *Sedimentary Geology*, **196**, 5–30.

MONTENAT, C., OTT D'ESTEVOU, P. & MASSE, P. 1987. Tectonic-sedimentary characters of the Betic Neogene basins evolving in a crustal transcurrent shear zone (SE Spain). *Bulletin – Centre de Recherches Exploration Production Elf Aquitaine*, **11**, 1–22.

MORETTI, M. & TORPEANO, M. 1996. Structure sedimentary deformative (sismiti) nei depositi Tirrenimani de Bari. *Memorie della Società Geologica Italiana*, **51**, 485–500.

MUTTI, E. & NORMAK, W. R. 1987. Comparing examples of modern and ancient turbidite systems: problems and concepts. *In*: LEGGET, J. K & ZUFFA, G. G. (eds) *Marine Clastic Sedimentology: Concepts and Case Studies*. Graham and Trotman, London, 1–38.

NAGTEGAAL, P. J. C. 1963. Convolute lamination, metadepositional ruptures and slumping in an exposure near Pobla de Segur (Spain). *Geologie en Mijnbouw*, **42**, 363–374.

NAHA, K. 1965. Metamorphism in relation to stratigraphy, structure and movements in parts of east Singhbhum, Eastern India. *Quarterly Journal of the Geological, Mining and Metallurgical Society of India*, **37**, 41–88.

NEUENDORF, K. E., MEHL, J. P. & JACKSON, J. A. 2005. *Glossary of Geology*. American Geological Institute, Alexandria.

OBERMEIER, S. F. 1996. Use of liquefaction-induced features for paleoseismic analysis. *Engineering Geology*, **44**, 1–76.

OWEN, G. 1987. Deformation processes in unconsolidated sands. *In*: JONES, M. E. & PRESTON, R. M. F. (eds) *Deformation of Sediments and Sedimentary Rocks*. Geological Society, London, Special Publications, **29**, 11–24.

OWEN, G. 1995. Soft sediment deformation in upper Proterozoic Torridonian sandstones (Applecross Formation) at Torridon, Northwest Scotland. *Journal of Sedimentary Research*, **A65**, 495–504.

OWEN, G. 1996. Experimental soft-sediment deformation: structures formed by liquefaction of unconsolidated sands and some ancient examples. *Sedimentology*, **43**, 279–293.

OWEN, G. 2003. Load structures: gravity-driven sediment mobilization in the shallow subsurface. *In*: VAN RENSBERGEN, P., HILLIS, R. R., MALTMAN, A. J. & MORLEY, C. K. (eds) *Subsurface Sediment Mobilization*. Geological Society, London, Special Publications, **216**, 21–34.

PLUMMER, P. S. & COSTIN, V. A. 1981. Shrinkage cracks: desiccation or synaeresis? *Journal of Sedimentary Research*, **51**, 1147–1156.

PRATT, B. R. 1998. Syneresis cracks: subaqueous shrinkage in argillaceous sediments caused by earthquake induced dewatering. *Sedimentary Geology*, **117**, 1–10.

REBESCO, M. & CAMERLENGHI, A. (eds) 2008. *Contourites*. Developments in Sedimentology **60**. Elsevier, Amsterdam.

REINECK, H. E. & SINGH, I. B. 1980. *Depositional Sedimentary Environments*. Springer-Verlag, Berlin.

RICCI LUCCHI, F. 1995. Sedimentological indicators of palaeoseismicity. *In*: SERVA, L. & SLEMMONS, D. B. (eds) *Perspective in Palaeoseismology*. Association of Engineering Geologists, Special Publication, **6**, 7–17.

ROEP, TH. B. & EVERTS, A. J. 1992. Pillow-beds: a new type of seismites? An example from an Oligocene turbidite fan complex, Alicante, Spain. *Sedimentology*, **39**, 711–724.

RODRÍGUEZ-LÓPEZ, J. P., MELÉNDEZ, N., SORIA, A. R., LIESA, C. L. & VAN LOON, A. J. 2007. Lateral variability of ancient seismites related to differences in sedimentary facies (the synrift Escuha Formation, mid-Cretaceous, eastern Spain. *Sedimentary Geology*, **201**, 461–484.

RODRÍGUEZ-PASCUA, M. A., CALVO, J. P., VICENTE, G. D. & GOMEZ-GRAS, D. 2000. Soft-sediment deformation structures interpreted as seismites in lacustrine sediments of the Prebetic Zone, SE Spain, and their potential use as indicators of earthquake magnitudes during the Late Miocene. *Sedimentary Geology*, **135**, 117–135.

ROSSETTI, D. F. 1999. Soft sediment deformation structures in late Albian to Cenomanian deposits, Sao Luis Basin, northern Brazil: evidence for palaeoseismicity. *Sedimentology*, **46**, 1065–1081.

ROSSETTI, D. F. & GOES, A. M. 2000. Deciphering the sedimentological imprint of palaeoseismic events: an example from the Aptian Codo Formation, northern Brazil. *Sedimentary Geology*, **135**, 137–156.

ROY, A., SARKAR, A., JEYAKUMAR, S., AGGRAWAL, S. K. & EBIHARA, M. 2002*a*. Sm–Nd age and mantle characteristics of the Dhanjori volcanic rocks, Eastern India. *Geochemical Journal*, **36**, 503–518.

ROY, A., SARKAR, A., JEYAKUMAR, S., AGGRAWAL, S. K. & EBIHARA, M. 2002*b*. Mid-Proterozoic plume-related thermal event in Eastern Indian craton: evidence from trace elements, REE geochemistry and Sr–Nd isotope systematics of basic-ultrabasic intrusives from Dalma Volcanic Belt. *Gondwana Research*, **5**, 133–146.

SAHA, A. K. 1994. Crustal evolution of Singhbhum-North Orissa, eastern India. *Geological Society of India Memoir*, **27**.

SARKAR, S., BOSE, P. K. & BANDYOPADHAYA, S. 1991. Intertidal occurrence of mesoscale scours in the Bay of Bengal, India and their implications. *Sedimentary Geology*, **75**, 29–37.

SCHWAB, W. C. & LEE, H. J. 1988. Causes of two slope-failure types in continental-slope sediment, northeastern Gulf of Alaska. *Journal of Sedimentary Petrology*, **58**, 1–11.

SCOTT, B. & PRICE, S. 1988. Earthquake-induced structures in young sediments. *Tectonophysics*, **147**, 167–170.

SEILACHER, A. 1969. Fault-graded beds interpreted as seismites. *Sedimentology*, **13**, 155–159.

SEILACHER, A. 1984. Sedimentary structures tentatively attributed to seismic events. *Marine Geology*, **55**, 1–12.

SETH, A., SARKAR, S. & BOSE, P. K. 1990. Synsedimentary seismic activity in an immature passive margin basin (Lower Member of the Katrol Formation, Upper Jurassic, Kutch, India). *Sedimentary Geology*, **68**, 279–291.

SHIKI, T. & YAMAZAKI, T. 1996. Tsunami-induced conglomerates in Miocene upper bathyal deposits, Chita Peninsula, central Japan. *Sedimentary Geology*, **104**, 175–188.

SIMS, J. D. 1973. Earthquake induced structures in sediments of Van Norman Lake, San Fernando, California. *Science*, **182**, 161–163.

SIMS, J. D. 1975. Determining earthquake recurrence intervals from deformational structures in young lacustrine sediments. *Tectonophysics*, **29**, 144–152.

STOW, D. A. V. & BOWEN, A. J. 1978. Origin of lamination in deep sea fine-grained sediments. *Nature*, **274**, 324–328.

STOW, D. A. V. & BOWEN, A. J. 1980. A physical model for the transport and sorting of fine grained sediment by turbidity currents. *Sedimentology*, **27**, 31–46.

STOW, D. A. V. & WETZEL, A. 1990. Hemiturbidite: a new type of deep water sediment. *Proceedings of the Deep Sea Drilling Projects*, **116**, 25–34.

TANNER, P. W. G. 2003. Syneresis. *In*: MIDDLETON, G. V. (ed.) *Encyclopedia of sediments and sedimentary rocks*. Kluwer Academic Publishers, Dordrecht, 718–720.

VACHARD, D., BARRIER, P., MONTENAT, C. & OTT D'ESTEVOU, P. 1987. Dykes neptuniens, brèches internes et éboulis cimentés des escarpements de faille du Détroit de Messine au Plio-Quaternaire. *Documents et Traveaux de l'IGAL (Paris)*, **11**, 27–41.

VAN LOON, A. J. & BRODZIKOWSKI, K. 1987. Problems and progress in the research on soft_sediment deformations. *Sedimentary Geology*, **50**, 167–193.

VAN LOON, A. J. & WIGGERS, A. J. 1976*b*. Metasedimentary 'graben' and associated structures in the lagoonal Almere Member (Groningen Formation, The Netherlands). *Sedimentary Geology*, **16**, 237–254.

WILLIAMS, G. E. 1966. Palaeogeography of the Torridonian Applecross Group. *Nature*, **209**, 1303–1306.

WILLIAMS, G. E. 1969. Characteristics and origin of a Precambrian Pediment. *Journal of Geology*, **77**, 183–207.

WILLIAMS, G. E. 1970. Origin of disturbed bedding in Torridon Group sandstones. *Scottish Journal of Geology*, **6**, 409–411.

Correlations and reconstruction models for the 2500–1500 Ma evolution of the Mawson Continent

JUSTIN L. PAYNE[1]*, MARTIN HAND[1], KARIN M. BAROVICH[1], ANTHONY REID[2] & DAVID A. D. EVANS[3]

[1]*Continental Evolution Research Group, School of Earth and Environmental Sciences, University of Adelaide, SA 5005, Australia*

[2]*Geological Survey Branch, Primary Industries and Resources South Australia, GPO Box 1671, Adelaide, SA 5001, Australia*

[3]*Department of Geology and Geophysics, Yale University, 210 Whitney Avenue, New Haven, CT 06520-8109, USA*

*Corresponding author (e-mail: justin.payne@adelaide.edu.au)

Abstract: Continental lithosphere formed and reworked during the Palaeoproterozoic era is a major component of pre-1070 Ma Australia and the East Antarctic Shield. Within this lithosphere, the Mawson Continent encompasses the Gawler–Adélie Craton in southern Australia and Antarctica, and crust of the Miller Range, Transantarctic Mountains, which are interpreted to have assembled during *c.* 1730–1690 Ma tectonism of the Kimban–Nimrod–Strangways orogenies. Recent geochronology has strengthened correlations between the Mawson Continent and Shackleton Range (Antarctica), but the potential for Meso- to Neo-proterozoic rifting and/or accretion events prevent any confident extension of the Mawson Continent to include the Shackleton Range. Proposed later addition (*c.* 1600–1550 Ma) of the Coompana Block and its Antarctic extension provides the final component of the Mawson Continent. A new model proposed for the late Archaean to early Mesoproterozoic evolution of the Mawson Continent highlights important timelines in the tectonic evolution of the Australian lithosphere. The Gawler–Adélie Craton and adjacent Curnamona Province are interpreted to share correlatable timelines with the North Australian Craton at *c.* 2500–2430 Ma, *c.* 2000 Ma, 1865–1850 Ma, 1730–1690 Ma and 1600–1550 Ma. These common timelines are used to suggest the Gawler–Adélie Craton and North Australian Craton formed a contiguous continental terrain during the entirety of the Palaeoproterozoic. Revised palaeomagnetic constraints for global correlation of proto-Australia highlight an apparently static relationship with northwestern Laurentia during the *c.* 1730–1590 Ma time period. These data have important implications for many previously proposed reconstruction models and are used as a primary constraint in the configuration of the reconstruction model proposed herein. This palaeomagnetic link strengthens previous correlations between the Wernecke region of northwestern Laurentia and terrains in the eastern margin of proto-Australia.

This chapter outlines the Palaeoproterozoic to early Mesoproterozoic tectono-thermal evolution of the Mawson Continent of Australia and Antarctica (Fig. 1). The Mawson Continent comprises the Gawler Craton, South Australia, and the correlative coastal outcrops (e.g. Cape Hunter and Cape Denison) of Terre Adélie and George V Land in Antarctica and various other terrains of East Antarctica (Fig. 1, Oliver & Fanning 1997; Goodge *et al.* 2001; Fitzsimons 2003). Perhaps the most notable feature of the Mawson Continent is its lack of exposure. Excluding the flat-lying *c.* 1590 Ma Gawler Range Volcanics, the Gawler Craton portion is estimated to contain <5% basement exposure in an area approximately 530 800 km^2 (slightly smaller than France). The Antarctic component of the Mawson Continent contains even less exposure. Despite the impediment of limited basement exposure, numerous tectonic reconstruction models have been proposed to account for the evolution of the Mawson Continent and its interaction with other Proterozoic terrains, particularly other portions of the current Australian continent (Borg & DePaolo 1994; Daly *et al.* 1998; Betts *et al.* 2002; Dawson *et al.* 2002; Fitzsimons 2003; Giles *et al.* 2004; Betts & Giles 2006; Wade *et al.* 2006).

From: REDDY, S. M., MAZUMDER, R., EVANS, D. A. D. & COLLINS, A. S. (eds) *Palaeoproterozoic Supercontinents and Global Evolution.* Geological Society, London, Special Publications, **323**, 319–355.
DOI: 10.1144/SP323.16   0305-8719/09/$15.00 © Geological Society of London 2009.

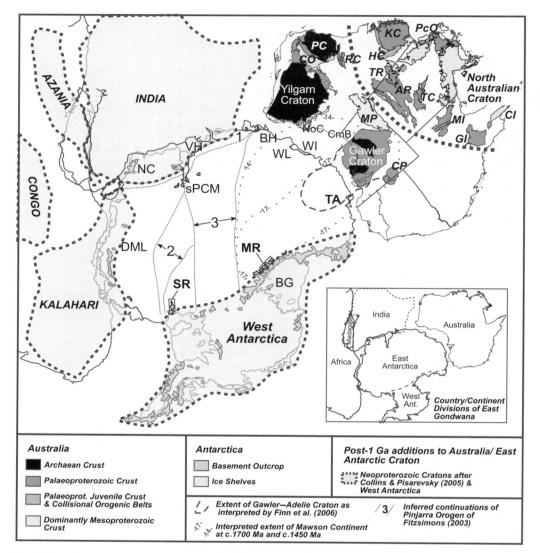

Fig. 1. Map of East Gondwana (modified from Collins & Pisarevsky 2005) displaying pre-Gondwana terrain locations in Antarctica (after Boger *et al.* 2006) and pre-1 Ga crustal provinces of Australia (after Betts *et al.* 2002; Payne *et al.* 2008). East Antarctica terrains are: BG, Beardmore Glacier; BH, Bunger Hills, DML, Dronning Maud Land; NC, Napier Complex, sPCM, southern Prince Charles Mountains; VH, Vestfold Hills, WI, Windmill Islands; WL, Wilkes Land. Bold abbreviations are MR, Miller Range; SR, Shackleton Range; and TA, Terre Adélie Craton, which have all experienced *c.* 1700 Ma tectonism (see text). Australian pre-1070 Ma terrains are: AR, Arunta Region, CI, Coen Inlier; CmB, Coompana Block; CO, Capricorn Orogen; CP, Curnamona Province; GI, Georgetown Inlier; HC, Halls Creek Orogen; KC, Kimberley Craton; MI, Mt Isa Inlier; MP, Musgrave Province; NoC, Nornalup Complex; PC, Pilbara Craton; PcO, Pine Creek Orogen; RC, Rudall Complex; TC, Tennant Creek Region; and TR, Tanami Region.

In attempting to reconstruct the evolution of the Mawson Continent, a particularly intriguing aspect of the geology of the Mawson Continent, and the Australian Proterozoic in general, is the comparative lack of evidence for subduction-related magmatism. The few Australia-wide examples of documented late Palaeoproterozoic and early Mesoproterozoic subduction-related magmatism can be summarized as follows: *c.* 1850 Ma magmatism associated with the accretion of the Kimberley Craton (Sheppard *et al.* 1999); volumetrically minor granites of the 1770–1750 Ma

Calcalkaline–Trondhjemite (CAT) Suite in the Arunta Region (Foden et al. 1988; Zhao & McCulloch 1995); the bimodal 1620–1600 Ma St Peter Suite of the Gawler Craton (Swain et al. 2008) and 1600–1550 Ma magmatism in the Musgrave Province (Wade et al. 2006). Reconnaissance geochemical data indicates c. 1690 Ma magmatism in the Warumpi Province on the southern margin of the Arunta Region may also represent subduction-related magmatism (Scrimgeour et al. 2005). In contrast, Palaeoproterozoic orogenic belts preserved in Laurentia and Baltica are commonly associated with identifiable subduction-related magmatism (e.g. Gandhi et al. 2001; Theriault et al. 2001; Ketchum et al. 2002; Mueller et al. 2002; Ansdell 2005; Whitmeyer & Karlstrom 2007; Åhall & Connelly 2008 and references therein). In addition, these orogenic belts are commonly quasi-linear belts often with associated inverted back-arc basins, accreted micro-continents and island arcs (e.g. Ketchum et al. 2002; St-Onge et al. 2006; Åhall & Connelly 2008 and references therein). This is in stark contrast to many orogenic events within Palaeo- to Mesoproterozoic Australia (e.g. summary of Betts & Giles 2006), which lack these elements and are commonly diffuse craton-wide events.

An example of the complex tectonic systems preserved in the Mawson Continent and Australia is the 1730–1690 Ma Strangways and Kimban orogenies in the Arunta Region and Gawler Craton, respectively. The Strangways Orogeny is preceded by the interpreted subduction-related CAT Suite magmatism, and forms a cornerstone of the argument for a long-lived accretionary system on the southern margin of the North Australian Craton (e.g. Betts & Giles 2006). However, the Strangways Orogeny has a very limited east–west extent and, as discussed later, is not easily reconcilable with an east–west trending accretionary margin. The temporally equivalent Kimban Orogeny does not preserve evidence for subduction-related magmatism (Hand et al. 2007) and has a craton-wide distribution, and the (current) aggregate geometry of Kimban and Strangways deformation is not readily reconcilable with an east–west trending linear plate margin setting.

The style of many tectonic events within the Palaeo- to Mesoproterozoic of Australia (McLaren et al. 2005) suggest that long-lived, pseudo-linear continental margins such as Phanerozoic Andean or Caledonian systems are not readily reconcilable with the geological record of the Australian Proterozoic. This appears to be a fundamental difference with continents such as Laurentia. While these differences are obviously generalized, reconstruction models for the Mawson Continent and Australia must take into account the nature of tectonic events within Australia, and be driven by available geological constraints rather than a priori postulated plate-tectonic models. McLaren et al. (2005) goes some way towards explaining many of the phenomena of the Australian Proterozoic by attributing high geothermal gradients and the predominance of high temperature–medium- to low-pressure metamorphism to the high heat producing nature of the North Australian Palaeoproterozoic crust. Although it seems likely that high heat production played a role in shaping the character of the Mawson Continent and Australia, it does not resolve many of the issues surrounding existing reconstruction models for the evolution of the Mawson Continent and Australia. This review provides a revised model for the 2500–1500 Ma evolution of the Mawson Continent for the purpose of outlining event correlations in associated terranes, thus enabling the revision of continent reconstruction models for the 2500–1500 Ma period.

The Mawson Continent

The name 'Mawson Continent' was first used to describe the Archaean–Mesoproterozoic southern Australian Gawler Craton and correlated terrains in Antarctica (Fanning et al. 1996). Alternative nomenclature has since included 'Mawson Block' (Oliver & Fanning 1996; Wingate et al. 2002a; Finn et al. 2006; Mikhalsky et al. 2006) and 'Mawson Craton' (Condie & Myers 1999; Fitzsimons 2003; Bodorkos & Clark 2004b). As 'continent' has first precedence and is a non-genetic descriptor, we favour the use of 'Mawson Continent' over alternative names.

The extent of the Mawson Continent is uncertain due to the extensive Neoproterozoic to Phanerozoic cover in Australia, and ice and snow cover in Antarctica. The Gawler Craton and the directly correlated coastal outcrops of Terre Adélie and George V Land in Antarctica (Oliver & Fanning 1997), form the nucleus of the Mawson Continent. In addition to these regions, the unexposed Coompana Block in South and Western Australia (Fig. 1) is often considered part of the Mawson Continent (Condie & Myers 1999; Bodorkos & Clark 2004a). In Antarctica, the Mawson Continent is commonly extended to include Palaeoproterozoic crust in the Miller and Shackleton ranges of the Trans-Antarctic Mountains (Fanning et al. 1999; Goodge et al. 2001). A recent compilation of airborne and satellite magnetic geophysical data (Finn et al. 2006) has suggested that fundamental differences in crustal petrophysical properties exist between the Gawler and Adélie cratons on the one hand, and the Miller Range and remainder of the East Antarctic Shield on the other. This is supported by differing geological evolutions of the various terranes with the presence of c. 1700 Ma tectonism considered as evidence for a single continent in the late Palaeoproterozoic period (Fanning et al. 1999; Goodge et al. 2001). In this review we adopt the terminology 'Mawson Continent' for the region encompassing the Gawler Craton, Terre Adélie Craton, Miller Range and Coompana Block. The former three of these domains are presumed to have acted as a coherent crustal fragment during the Proterozoic and early to mid-Phanerozoic after initial amalgamation at c. 1700 Ma, with proposed later addition of the Coompana Block at c. 1600–1550 Ma. The Mawson Continent was subsequently divided during the breakup of Gondwana Land.

The following section presents the tectonic histories for the proposed components of the Mawson Continent, which forms the basis for ensuing discussion on its c. 2500–1500 Ma amalgamation and evolution.

The Gawler Craton

The Gawler Craton (Fig. 2) is composed of late Archaean–early Palaeoproterozoic supracrustal and magmatic lithologies which are surrounded, overlain and intruded by Palaeoproterozoic (2000–1610 Ma) and Mesoproterozoic (1590–1490 Ma) units (Daly et al. 1998; Ferris et al. 2002; Swain et al. 2005b; Fanning et al. 2007; Hand et al. 2007). Tectonic domains have been delineated for the Gawler Craton based upon the interpretation of Total Magnetic Intensity (TMI) and gravity datasets combined with available geological evidence (Fig. 3, Ferris et al. 2002). These domains largely represent variations in structural trends and extent of crustal re-working as opposed to fundamental terrane boundaries (Hand et al. 2007).

Late Archaean–Early Palaeoproterozoic. The late Archaean stratigraphy in the central Gawler Craton consists of metasedimentary, volcanic and granite-greenstone lithologies (c. 2560–2500 Ma, Daly & Fanning 1993; Swain et al. 2005b) that were deformed during the Sleafordian Orogeny (2460–2430 Ma, Daly et al. 1998; McFarlane 2006). The c. 2560–2500 Ma Devil's Playground Volcanics and Dutton Suite are interpreted to have formed in a magmatic arc setting, which terminated shortly before or during the Sleafordian Orogeny (Swain et al. 2005b). Sleafordian Orogeny magmatism includes the Kiana Granite suite (c. 2460 Ma, Fanning et al. 2007) and leucogranites of the Whidbey Granite (c. 2445 Ma, Jagodzinski et al. 2006). Metamorphic grade of the Sleafordian Orogeny ranges from sub-greenschist to granulite facies (Daly & Fanning 1993). Peak metamorphism is recorded by $P-T$ estimates of 800–850 °C and c. 7.5 kbar (Tomkins & Mavrogenes 2002) and 750–800 °C and 4.5–5.5 kbar (Teasdale 1997) for localities within the Mulgathing Complex (Fig. 2). Sleafordian Orogeny-aged structures in the central Gawler Craton (Mulgathing Complex) consist of shallowly NNE–NE plunging folds which have been subjected to some degree of block rotation by later shear zone movement (Teasdale 1997; Direen et al. 2005).

Circa 2000 Ma Miltalie Event. The Miltalie Event represents the first recognized tectonic activity after approximately 400 Ma of tectonic quiescence following the Sleafordian Orogeny (Webb et al. 1986; Daly et al. 1998). The Miltalie Gneiss has protolith magmatic ages of 2002 ± 15 Ma and 1999 ± 13 Ma (Fanning et al. 2007). Our field observations indicate that the Miltalie Gneiss map unit (Parker 1983) incorporates apparently metasedimentary lithologies; however, the age of this sequence is yet to be determined, and the tectonic setting of the Miltalie Gneiss and its protoliths has not yet been constrained.

2000–1860 Ma sediment deposition and the c. 1850 Ma Cornian Orogeny. The Miltalie Gneiss is overlain by sequences of the Hutchison Group which are interpreted to have been deposited on a passive margin in the time interval 2000–1860 Ma (Parker 1993; Schwarz et al. 2002). Final sedimentation prior to the onset of the Cornian Orogeny is relatively tightly constrained by the Bosanquet Formation volcanics at 1866 ± 10 Ma (Fanning et al. 2007). Apparently time equivalent sediment deposition is also evident in the Corny Point region of the Yorke Peninsula (Howard et al. 2007). The Cornian Orogeny (1850–1840 Ma, Reid et al. 2008) has associated voluminous felsic magmatism of the Donington Suite (Hoek & Schaefer 1998; Reid et al. 2008). This orogenic event produced ESE striking structural fabrics overprinted by east–west striking folds and late south-side down extensional ductile shearing. Metamorphism associated with the Cornian Orogeny is represented by a clockwise $P-T$ path with peak metamorphic conditions of c. 750 °C and c. 6 kbar (Reid et al. 2008). The Donington Suite intrusions and Cornian Orogeny appear to be restricted to east of the Kalinjala Shear Zone in the southern Gawler Craton (Fig. 2). Temporally equivalent magmatic lithologies also exist as basement in the Olympic Dam region in the central eastern Gawler Craton (Fig. 2, Jagodzinski 2005), indicating the Cornian Orogeny system affected much of what is now the eastern Gawler Craton.

1800–1740 Ma magmatism and sedimentation. The 60 Myr period from approximately 1800–1740 Ma marks an interval of extensive sediment deposition

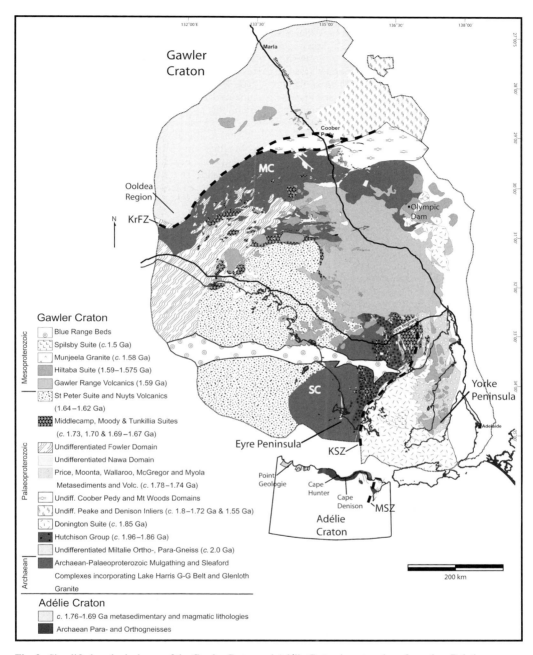

Fig. 2. Simplified geological map of the Gawler Craton and Adélie Craton in restored configuration. Relative positions of the two cratons after Oliver & Fanning (1997). Geology of the Gawler Craton after Fairclough et al. (2003). Geology of the Adélie Craton after Pelletier et al. (2002) and Ménot et al. (2005). Abbreviations are: KSZ, Kalinjala Shear Zone; KrFZ, Karari Fault Zone; MC, Mulgathing Complex; and SC, Sleafordian Complex.

across much of the Gawler Craton. In the southern Gawler Craton this includes the Myola Volcanics (1791 ± 4 Ma, Fanning et al. 1988), McGregor Volcanics and Moonabie Formation (c. 1740 Ma, Fanning et al. 1988), Wallaroo Group (Parker 1993; Cowley et al. 2003), Price Metasediments (c. 1770 Ma, Oliver & Fanning 1997) and successions previously incorporated into the Hutchison Group (Szpunar et al. 2006). Volcanic sequences in the Wallaroo Group constrain the age of deposition to 1772 ± 14 Ma (Wardang Volcanics, Fanning et al. 2007), 1753 ± 8 Ma (Moonta Porphyry rhyolite, Fanning et al. 2007) and 1740 ± 6 Ma (Mona Volcanics, Fanning et al. 2007). In the northern Gawler Craton, metasedimentary lithologies intersected in drillholes in the Nawa Domain are interpreted to have been deposited in the interval c. 1750–1730 Ma (Payne et al. 2006) and may correlate with the Peake Metamorphics within the Peake and Denison Inliers in the northeastern Gawler Craton (Hopper 2001). The depositional ages for the protoliths of the Peake Metamorphics are constrained by ages of 1789 ± 10 and 1740 ± 6 Ma for the Tidnamurkuna and Spring Hill volcanics, respectively (Fanning et al. 2007). The Wirriecurrie Granite in the Peake and Denison Inlier is constrained to 1787 ± 8 Ma (Parker 1993; Fanning et al. 2007) and is interpreted to have formed in an intracontinental setting that sampled a c. 2500 Ma subduction-modified mantle source (Hopper 2001).

1730–1690 Ma Kimban Orogeny. The Kimban Orogeny is interpreted as the most pervasive orogenic event in the Gawler Craton (Daly et al. 1998; Fanning et al. 2007; Hand et al. 2007; Payne et al. 2008). Geochronology has confirmed it to be a craton-wide event with metamorphic and syntectonic magmatic ages in the range 1730–1690 Ma reported from the southern (Vassallo 2001; Fanning et al. 2007), western (Fowler Domain, Teasdale 1997) and northern Gawler Craton (Hopper 2001; Betts et al. 2003; Payne et al. 2008).

Kimban Orogeny structures in the southern Gawler Craton formed during dextral transpression (Parker 1993; Vassallo & Wilson 2001, 2002) with NE trending-structures curving to north trends further north. In the northern Gawler Craton, within the Peake and Denison Inlier, early north–south trending structures are correlated with the Kimban Orogeny (Hopper 2001). Payne et al. (2008) suggest the Kimban Orogeny is expressed by the prominent NE trends in the regional aeromagnetic (TMI) data for the northwestern Gawler Craton.

Craton-wide metamorphic conditions of the Kimban Orogeny are poorly constrained, but are represented by 625–650 °C/5.5–6.5 kbar to 700–750 °C/8–9 kbar in the western Gawler Craton (Teasdale 1997), and 600–675 °C/5–7 kbar to 800–850 °C/7–9 kbar in the southern Gawler Craton (Parker 1993; Tong et al. 2004). However, considerable variation in metamorphic grade is observed within the Kimban Orogeny as evidenced by the regions in the southern Gawler Craton that preserve greenschist-facies metamorphism (Price Metasediments, Oliver & Fanning 1997) adjacent to granulite regions.

Magmatism associated with the Kimban Orogeny is represented by the Middlecamp, Moody and Tunkillia suites. The Middlecamp Suite in the eastern Gawler Craton is a pre- to early Kimban Orogeny granite suite with ages in the range 1737 ± 7–1726 ± 7 Ma (Fanning et al. 2007). The Moody Suite is later in the Kimban Orogeny (1720 ± 9–1701 ± 12 Ma) and contains intrusives ranging from hornblende-bearing granitoids to muscovite-bearing leucogranites (Fanning et al. 2007). The Tunkillia Suite is constrained to 1690–1670 Ma (Ferris & Schwarz 2004) and interpreted to be a post-tectonic magmatic suite based upon Nd-isotope and trace element geochemistry (Payne 2008).

Localized syn-Kimban Orogeny basin formation is also preserved in the central Gawler Craton. Here the c. 1715 Ma Labyrinth Formation is typified by upward coarsening sequences (Cowley & Martin 1991). The depositional environment is interpreted to be within a fault-bounded basin with sediments derived from local sources (Daly et al. 1998). A rhyolite within the Labyrinth Formation constrains the timing of deposition to c. 1715 ± 9 Ma (Fanning et al. 2007).

1660 Ma Ooldean Event. The Ooldean Event as defined by Hand et al. (2007) is currently constrained to 1659 ± 6 Ma (Fanning et al. 2007) and is represented by UHT metamorphic conditions of c. 950 °C and 10 kbar (Teasdale 1997). The mineral assemblage associated with these $P-T$ conditions defines a fine-grained mylonitic fabric which overprints a high-grade metamorphic assemblage (Teasdale 1997). The earlier assemblage is interpreted to represent Kimban Orogeny metamorphism (c. 1690 Ma, Teasdale 1997; Payne et al. 2008). The tectonic setting of the Ooldean Event is unconstrained and evidence for the event is currently confined to the Ooldea region (Fig. 2). Elsewhere, at c. 1660 Ma, the Gawler Craton appears to have undergone extension as evidenced by the deposition of the Tarcoola Formation (Cowley & Martin 1991; Daly et al. 1998), and also potentially sedimentary packages in the Mt Woods region (Fig. 2, Betts et al. 2003; Skirrow et al. 2006). Further petrographically constrained geochronology is required to determine if the c. 1660 Ma age obtained by

Fanning et al. (2007) does represent the age of UHT metamorphism, as this age has not been found in subsequent geochronology studies of the Ooldea 2 lithologies (Payne et al. 2008).

Palaeo-Mesoproterozoic transition events. The Gawler Craton preserves evidence of a complex sequence of events in the period from 1630–1540 Ma. The co-magmatic mafic and felsic intrustions of the St Peter Suite (1620–1608 Ma, Fig. 2, Flint et al. 1990) have been interpreted to have a subduction-related petrogenesis (Swain et al. 2008). St Peter Suite magmatism was followed by the voluminous and metallogenically significant Gawler Range Volcanics (GRV, c. 1592 Ma) and 1595–1575 Ma Hiltaba Suite intrusives (Flint 1993; Daly et al. 1998; Budd 2006). The GRV and Hiltaba Suite have previously been interpreted as an anorogenic magmatic event and linked to a plume (Flint 1993; Creaser 1995). However, new evidence of contemporaneous high-grade metamorphism and deformation in the Mt Woods and Coober Pedy Ridge domains (Skirrow et al. 2006; Fanning et al. 2007) has led to a suggestion of a syntectonic setting for the GRV–Hiltaba event. However, this syntectonic setting does not negate the potential role of a mantle plume in magma generation or necessarily require a direct causal link between the tectonism and magmatism (Betts et al. 2007). Deformation and low-grade metamorphism at this time is also reported (Direen & Lyons 2007; Hand et al. 2007) from the Eyre Peninsula region (Foster & Ehlers 1998) and in the Wallaroo Group (Conor 1995). Deformation and metamorphism associated with the GRV–Hiltaba event occurred shortly before, and within uncertainty of, the c. 1565–1540 Ma Kararan Orogeny (Hand et al. 2007).

The Kararan Orogeny, as defined by Hand et al. (2007), represents the final episode of high-grade metamorphism and deformation within the Gawler Craton (Teasdale 1997; Fraser & Lyons 2006; Fanning et al. 2007; Hand et al. 2007; Payne et al. 2008) before a final period of shear-zone activity and subsequent cratonization at c. 1450 Ma (Webb et al. 1986; Fraser & Lyons 2006). Evidence for the Kararan Orogeny is largely restricted to the northern and western Gawler Craton, with peak metamorphic conditions of 800 °C and 10 kbar recorded in the Fowler Domain (Teasdale 1997) and granulite-grade metamorphism in the Coober Pedy and Mabel Creek Ridge regions (Fanning et al. 2007; Payne et al. 2008). East–west to NE-trending structures evident in regional aeromagnetic data of the northern Gawler Craton and in outcrop in the Peake and Denison Inliers, are interpreted to have formed during the Kararan Orogeny (Hopper 2001; Payne et al. 2008). The extent of shear zone activity at c. 1450 Ma (Fraser & Lyons 2006) and its influence of Gawler Craton geometry is yet to be constrained.

Archaean–Palaeoproterozoic Antarctica

Three regions of outcropping Palaeoproterozoic geology are commonly assigned to the Mawson Continent in Antarctica: Terre Adélie Craton, the Miller Range and the Shackleton Range (Fig. 1, Fitzsimons 2003; Finn et al. 2006). The Terre Adélie Craton (represented by outcrop in George V Land and Terre Adélie Land) represents the conjugate rifted margin of the Gawler Craton (Stagg et al. 2005). This correlation is supported by reconstruction of the rifted margin and correlation of Archaean to Palaeoproterozoic lithologies of the Eyre Peninsula, Gawler Craton, and George V and Terre Adélie Land (Oliver & Fanning 1997).

Terre Adélie Craton. The Terre Adélie Craton is known from c. 400 km of discontinuous outcrops along the coast of George V and Terre Adélie Land (Fig. 2, Peucat et al. 2002). The outcrop consists of late Archaean–Early Palaeoproterozoic gneisses with overlying Palaeoproterozoic metasedimentary lithologies that directly correlate to lithologies and tectono-thermal events of the Gawler Craton (Oliver & Fanning 1997). Granodiorites at Cape Denison yield a crystallization age of c. 2520 Ma (correlating with the Dutton Suite of the Gawler Craton) and, along with metasedimentary lithologies, were deformed during 2440–2420 Ma orogenesis (Monnier 1995). Garnet-cordierite granites, outcropping from Point Martin to Cape Denison, were generated during c. 2440 Ma orogenesis (Monnier 1995) and are similar to garnet-cordierite granites of the Whidbey Granite association (c. 2443 Ma, Daly & Fanning 1993; Jagodzinski et al. 2006). Deformation associated with c. 2440 Ma orogenesis displays a NW–SE trend associated with NE–SW shortening (Ménot et al. 2005).

Migmatitic Palaeoproterozoic metasedimentary lithologies at Pointe Geologie are intruded syntectonically by dolerites and gabbros, which show mingling relationships with anatectic melts (Peucat et al. 1999). The Cape Hunter Phyllite is suggested to be a stratigraphic equivalent which has only experienced greenschist facies metamorphism (Oliver & Fanning 1997). Structures within the Cape Hunter Phyllite are sub-vertical with an approximate north–south orientation (Oliver & Fanning 1997). The Point Geologie HT–LP event is constrained by monazite ages of 1694 ± 2 and 1693 ± 2 Ma (U–Pb TIMS multi-grain dissolution and evaporation, respectively, Peucat et al. 1999). U–Pb SHRIMP zircon ages constrain metamorphism to 1696 ± 11 Ma with a reported

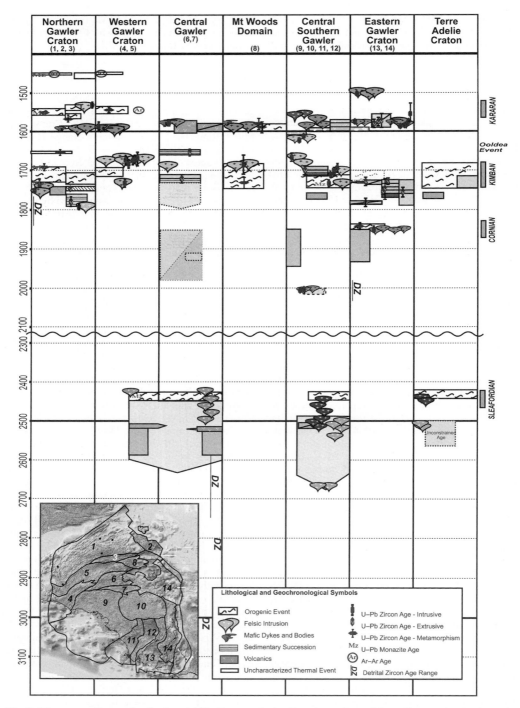

Fig. 3. Time-space diagram for Gawler–Adélie Craton and other East Antarctica and Australian terrains. Regions of the Gawler Craton correspond to domains as numbered and represented in inset map. Domains are: 1, Nawa Domain; 2, Peake and Denison Inliers; 3, Coober Pedy Ridge Domain; 4, Fowler Domain; 5, Christie Domain; 6, Wilgena Domain; 7, Lake Harris Greenstone Domain: 8, Mount Woods Domain; 9, Nuyts Domain; 10, Gawler Range Domain; 11, Coulta Domain; 12, Cleve Domain; 13, Spencer Domain; and 14, Olympic Domain. T-S plot for Gawler Craton

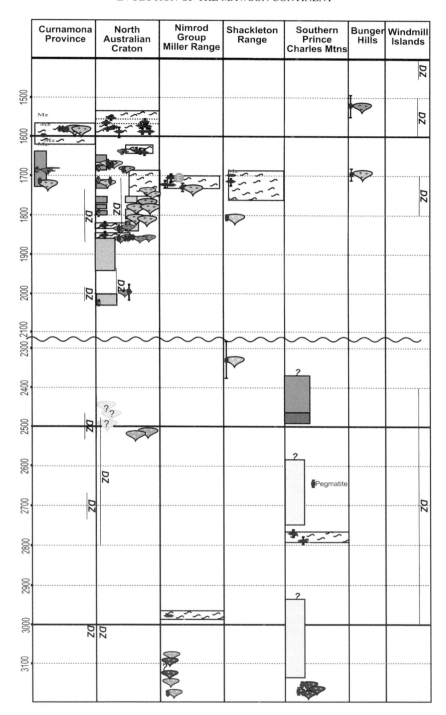

Fig. 3. (*Continued*) modified after Ferris *et al.* (2002) with additional data from Peucat *et al.* (1999, 2002), Holm (2004), Jagodzinski *et al.* (2006), Fraser & Lyons (2006), Payne *et al.* (2006), Howard *et al.* (2007) and Hand *et al.* (2007). Data sources for remainder of T-S plots are as discussed in text. North Australian Craton timeline is a simplified representation with greater detail provided in text and Figures 8 and 9. E, eclogite facies metamorphism.

maximum deposition age of 1740–1720 Ma from zircon cores (Peucat et al. 1999). This suggests the metasediments of the Point Geologie region were deposited in a basin setting associated with the development of the early Kimban Orogeny (after c. 1720 Ma, potentially equivalent to the Labyrinth Formation) and were metamorphosed under HT–LP conditions during the late Kimban Orogeny (c. 1690 Ma).

Miller Range. The Nimrod Group is a relatively localized group of Archaean–Palaeoproterozoic exposures in the Miller Range within the Trans-Antarctic Mountains (Fig. 1). The Nimrod Group includes quartzofeldspathic, mafic and calc-silicate gneiss, pelitic schist, amphibolite, orthogneiss, along with relict eclogite and ultramafic pods (Goodge et al. 2001).

Protolith U–Pb zircon ages from the gneisses are in the age range 3290–3060 Ma with metamorphism at c. 2975 Ma (Bennett & Fanning 1993; Goodge & Fanning 1999). These U–Pb data agree with Nd-isotope evidence (Borg & DePaolo 1994) for initial crustal growth in the mid- to late-Archaean (Goodge et al. 2001). Late Palaeoproterozoic metamorphism is constrained by U–Pb SHRIMP zircon ages of 1723 ± 14 and 1720 Ma for biotite-hornblende gneisses and 1723 ± 29 Ma for relict eclogites (Goodge et al. 2001). Syntectonic orthogneisses yield a U–Pb SHRIMP zircon age of 1730 ± 10 Ma. This orogenic event was termed the Nimrod Orogeny by Goodge et al. (2001). The presence of eclogitic material indicates a likely collisional-setting with the potential for a proximal terrane boundary (Kurz & Froitzheim 2002 and referencers therein). Structural information for the Nimrod Orogeny is largely overprinted and/or re-oriented by reworking during the Phanerozoic Ross Orogeny.

Shackleton Range. The lithologies of the Shackleton Range (Fig. 1) preserve a complex history of early Palaeoproterozoic to Cambrian tectonism. Links with the Nimrod Group and Gawler Craton are based upon the correlation of age-equivalent Palaeoproterozoic tectonism (Fanning et al. 1999; Goodge et al. 2001; Zeh et al. 2004). The southern Shackleton Range consists of upper amphibolite-granulite ortho- and paragneisses that were metamorphosed at 1763 ± 32 Ma as constrained by a Rb–Sr whole-rock isochron age (Schubert & Will 1994; Talarico & Kroner 1999). The northern Shackleton Range records extensive overprinting and tectonic juxtaposition by the c. 500 Ma (Pan-African) Ross Orogeny. Orthogneisses in the Haskard Highland and La Grange Nunataks preserve U–Pb zircon protolith ages of 2328 ± 47 and 1810 ± 2 Ma, respectively (Brommer et al. 1999; Zeh et al. 1999). Upper amphibolite to granulite facies metamorphism of these orthogneisses and associated paragneisses is constrained by a 1715 ± 6 Ma U–Pb zircon age from a syntectonic leucosome (Brommer et al. 1999) and a U–Pb monazite age of 1737 ± 3 Ma (Zeh et al. 1999). $P-T$ data from these locations and the central-northern Shackleton Ranges indicate Barrovian-style metamorphism with peak conditions of 630–750 °C and 7–11 kbar (Schubert & Will 1994; Brommer et al. 1999; Zeh et al. 1999). Monazite and zircon from Meade Nunatak in the NE Shackleton Range record metamorphic ages of c. 1700 and 1686 ± 2 Ma, respectively (Zeh et al. 2004), suggesting it too was part of a regional c. 1730–1690 Ma tectono-thermal event. Interpreted cooling ages of 1650–1550 Ma, provided by Sm–Nd, Rb–Sr and K–Ar data, are reported from both the northern and southern Shackleton Ranges (Zeh et al. 2004).

Assembling the Mawson Continent

The regions considered as key components of the Mawson Continent (Gawler and Terre Adélie cratons, Miller Range and Shackleton Range) share few similar tectono-thermal events, with the basis for comparison being solely c. 1700 Ma tectonism (Fig. 3). The exceptions to this generality are the Gawler and Terre Adélie cratons which have coincident late Archaean to early Proterozoic histories. The numerous tectono-thermal and relatively precise geographical correlations between these two (Oliver & Fanning 1997) means they can be considered to have formed a continuous crustal block from the Archaean until Cretaceous rifting. Hence we propose the name of Gawler–Adélie Craton to refer to this terrain. In this section we discuss the validity of assigning Antarctic terrains to the Mawson Continent and the various proposals for the Palaeo- to early Mesoproterozoic extent of the Mawson Continent.

The Gawler–Adélie–Miller Range–Shackleton Range 1700 Ma connection

The near identical timing of Nimrod Group metamorphism at 1730–1723 Ma with early Kimban Orogeny metamorphism (1730–1720 Ma) is suggestive of a related and possibly contiguous tectono-thermal event involving both regions. Given the presence of relict eclogite (c. 1730 Ma, Goodge et al. 2001) and the lack of evidence for later terrane accretion, we suggest that the 1730–1690 Ma Kimban–Nimrod Orogeny records the accretion of the Miller Range terrain to the Gawler–Adélie Craton. The suture zone that accommodated this amalgamation is potentially at or near the location of the Nimrod Group, as suggested by the presence of eclogite-facies metamorphic lithologies within this sequence.

The Shackleton Range is located approximately 3500 km from outcrops of the Gawler–Adélie Craton and approximately 1800 km from the Nimrod Group in the Miller Range (Fig. 1). Despite these distances, the geochronology of Zeh et al. (2004) highlights a temporal correlation between c. 1730–1690 Ma tectonism in all three terrains. Nd-isotope data of Borg & DePaolo (1994) indicate a Palaeoproterozoic model age (T_{DM} = 2.2–1.6 Ga) for granite source regions along the Transantarctic Mountains from Victoria Land to the Beardmore Glacier region, in contrast to Mesoproterozoic model ages beyond the Beardmore Glacier (Fig. 1). This change in crustal evolution is utilized by Fitzsimons (2003) to suggest that one of three possible paths for the c. 550–500 Ma Pinjarra Orogen (Path 3, Fig. 1) bisects the Transantarctic Mountains between the Miller and Shackleton ranges, with the latter considered a Neoproterozoic or Cambrian addition to the proto-East Antarctic Shield. The non-unique nature of bulk-rock Nd-isotope data implies that Mesoproterozoic model ages may result from a variety of processes, and the Shackleton Range may still have formed part of the Mawson Continent. However, given the high degree of uncertainty, for the purposes of this review the Shackleton Range is excluded from the Mawson Continent.

The western extent of the Mawson Continent

The western extent of the Mawson Continent (eastern extent in Antarctica) is unclear. The Coompana Block and Nornalup Complex in Australia (Fig. 1, Fitzsimons 2003; Bodorkos & Clark 2004a) and Bunger Hills and Windmill Islands in Antarctica (Fig. 1, Fitzsimons 2000, 2003) have typically been assigned to the Mawson Continent. The vast majority of this region, marked on Figure 1, is unexposed, with the current level of geophysical characterization insufficient to adequately constrain potential crustal-scale terrane boundaries.

Of the above regions, the Bunger Hills is the only location to preserve pre-1500 Ma crust, with two magmatic protolith conventional U–Pb zircon ages of 1699 ± 15 and 1521 ± 29 Ma (Fig. 3, Sheraton et al. 1992). The Windmill Islands preserve metasedimentary units similar in age and provenance to lithologies of the Nornalup Complex (c. 1400–1340 Ma, see discussion of Fitzsimons 2003 and references therein). As summarized by Fitzsimons (2003), the Nornalup Complex of the Albany–Fraser Belt is predominantly composed of syn-orogenic granites (1330–1290 Ma) with preserved c. 1440 Ma zircon xenocrysts and paragneisses with an interpreted depositional range of c. 1550–1400 Ma (Nelson et al. 1995). The non-outcropping Coompana Block has a single chronological constraint of 1505 ± 7 Ma for a juvenile, anorogenic orthogneiss intersected by drillhole (Wade et al. 2007).

The vast majority of the Coompana–Albany–Fraser–Wilkes region appears to be composed of distinctly different crust to the Gawler–Adélie Craton. The aeromagnetic signature of the Coompana Block appears different to that of the Gawler–Adélie Craton, with numerous large, approximately circular magnetic lows interpreted to represent undeformed plutons (Cowley 2006) that are not present on the Gawler–Adélie Craton. The apparently younger magmatism and basement (c. 1500–1400 Ma) and slightly more juvenile nature of the magmatism is also not consistent with a continuous Archaean-floored Palaeoproterozoic continent. However, metasedimentary lithologies from this region preserve old Nd-isotope model ages (e.g. 3.2–2.4 Ga from Windmill Islands, Post 2001) and pre-1500 Ma detrital zircon ages that appear to be consistent with derivation from a Gawler–Adélie Craton source (Fig. 3, Post 2001). This may indicate some genetic link with the Gawler–Adélie Craton at or after c. 1500 Ma. The model adopted herein suggests accretion of the Coompana Block and Antarctic equivalents to the Mawson Continent at c. 1600–1550 Ma (Betts & Giles 2006).

The extent of the Mawson Continent south of the Wilkes Province in East Antarctica (Fig. 1) has little constraint. As noted by Boger et al. (2006) the Archaean lithologies of the southern Prince Charles Mountains share no common timelines with the Archaean crust of the Mawson Continent. It was previously thought that the southern Prince Charles Mountains represented a distinct terrane to the adjoining Napier Complex and Vestfold Hills until Palaeozoic amalgamation (Fitzsimons 2000; Boger et al. 2001). However, recent detrital zircon geochronology suggests these terranes may have been amalgamated as early as the late Archaean–early Palaeoproterozoic (Phillips et al. 2006). This further highlights the differing evolution of these terranes compared to the Gawler–Adélie Craton and Mawson Continent. It would appear that the southern Prince Charles Mountains and associated terranes were not part of the Palaeoproterozoic Mawson Continent, with probable amalgamation with the East Antarctic Shield occurring during a later event such as the Pinjarra Orogen associated with East Gondwana-Land assembly (Fitzsimons 2003).

Palaeomagnetic constraints

As a general principle, palaeomagnetic data are well-suited to disproving proposed cratonic connections, but they cannot definitively substantiate any particular reconstruction. Nonetheless, long-lived connections between two or more cratons can be supported by a well-populated palaeomagnetic database as follows. One craton, along with its palaeomagnetic apparent polar wander (APW) path, can be rotated by the same Euler parameters into the reference frame of another craton. If this results in both the direct spatial juxtaposition of those cratons, and also superposition of the two APW paths with precise age matches, then a long-lived, direct connection is allowable for the given interval of time represented by the palaeomagnetic poles. The more poles that rotate into alignment, the more powerful the connection is supported; and if there are tectono-stratigraphic similarities that are brought together in the reconstruction, even more compelling does that model become. The palaeomagnetic technique requires assumptions of a constant-radius Earth and geocentric-axial-dipole (GAD) hypothesis for the Earth's magnetic field, the latter verified to first order by Evans (2006).

In the case of the Palaeo-Mesoproterozoic interval, a sparse palaeomagnetic dataset exists from each of the three blocks considered here: Mawson Continent, North Australian Craton and Laurentia. Data from Australia are summarized by Idnurm (2000) and Wingate & Evans (2003) and those from Laurentia are reviewed by Irving et al. (1972, 2004) and supplemented by a new isotopic age of c. 1590 Ma (Hamilton & Buchan 2007) for the Western Channel Diabase (Irving et al. 1972). The Mawson Continent data, namely the pole from the Gawler Range Volcanics, are first restored to North Australia by an Euler pole with parameters (18°S, 134°E, 51°CCW), as described above with minor modifications from Giles et al. (2004). Thereafter, this pole and the 1725–1640 Ma APW path from the McArthur Basin and Lawn Hill Platform (Idnurm 2000) are rotated to the Laurentian reference frame by an Euler pole with parameters (31.5°N, 98°E, 102.5°CCW). As shown in Figure 4, these rotations bring the Australian cratons in direct juxtaposition with northwestern Laurentia, in a reconstruction that is reminiscent of the SWEAT model for Rodinia (Moores 1991). Although the Rodinian SWEAT hypothesis has been shown to be palaeomagnetically untenable for the ages 755 Ma (Wingate & Giddings 2000), 1070 Ma (Wingate et al. 2002b) and c. 1200 Ma (Pisarevsky et al. 2003), our figured proto-SWEAT reconstruction appears attractive for the pre-Rodinian interval of 1740–1590 Ma.

A long-lived connection between reconstructed Australian cratons and Laurentia through to the end of the Palaeoproterozoic begs the question of when it could have initially formed, and when it ultimately fragmented. In principle, successively older palaeomagnetic poles can be compared in the same rotated reference frame until discordance of a precisely coeval pole pair is identified, and this provides a maximum age estimate for the formation of the cratonic juxtaposition. Laurentia assembled c. 1810 Ma (St-Onge et al. 2006), thus prior to that age we must compare data only from its more proximal components to the proto-SWEAT juxtaposition, that is, the Slave Craton and conjoined portions of the Churchill Province. A preliminary APW path for those regions is developing (Buchan et al. 2007; Evans & Raub 2007), but comparative data from ages immediately older than 1800 Ma in the North Australian Craton are lacking. It thus remains unclear when our proposed connections between Australian cratons and NW Laurentia initiated (Fig. 5). Presumably, there were collisions associated with the Barramundi Orogeny, as discussed above, but the current palaeomagnetic database is inadequate to test whether small-scale Wilson cycles of separation and reunification occurred between Australian and Laurentian blocks (as queried in Fig. 6b, Option 2).

A long-lived connection as proposed here also begs the question of when it fragmented. Successively younger palaeomagnetic poles, of precisely the same ages across all of the reconstructed cratons, will ultimately result in a discordance, which then provides the minimum age of breakup. For our proposed reconstruction, the Laurentian database following 1590 Ma is well constrained by high-quality poles (Evans & Pisarevsky 2008), but the Australian database lacks high-reliability results until c. 1200 Ma from the Albany–Fraser belt and southern Yilgarn Craton (Pisarevsky et al. 2003). Those poles are broadly compatible with our reconstruction, which permits the intriguing possibility that the North (combined with West) Australian craton remained fixed to NW Laurentia until the late Mesoproterozoic, while the Mawson Continent rifted from Laurentia and rotated into Albany–Fraser–Musgrave orogenesis at c. 1300 Ma. Such kinematics would be similar to the Neogene rotation of Arabia away from Africa and towards collision with Eurasia, those larger continents being relatively stationary. If so, then mid-ocean ridge propagation into the originally unified Laurentia–Mawson plate could be manifested by mafic magmatism of either 1470 Ma (Sears et al. 1998) or 1370 Ma (Doughty & Chamberlain 1996) in western North America. High-quality palaeomagnetic data from the Gawler Craton in the interval following 1590 Ma are needed to test this hypothesis.

Existing tectonic reconstruction models

In recent years a significant number of tectonic reconstruction models that address the evolution of

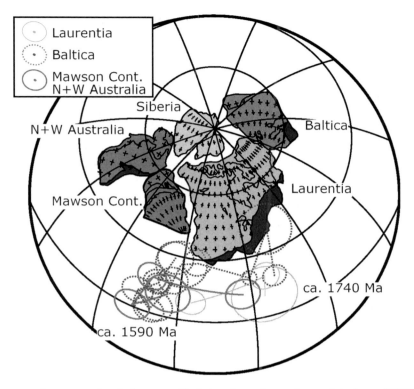

Fig. 4. Palaeomagnetic reconstruction, in the present North American reference frame, of the 'proto-SWEAT' connection between Australia and Laurentia, along with possibly adjacent cratons Baltica and Siberia, for the interval c. 1740–1590 Ma, and likely immediately earlier and later times. For geographic reference, late Mesoproterozoic ('Grenvillian') orogenic belts are shown in dark grey. As discussed in the text, the Mawson Continent (Gawler, Terre Adélie and proposed adjacent regions of Antarctica) is reconstructed to the united North and West Australian cratons by the Euler parameters (18°S, 134°E, +51°CCW), largely following Giles et al. (2004) but with minor modifications. Thereafter, the reconstructed Australian craton in North Australian reference frame is rotated to Laurentia (31.5°N, 098°E, +102.5°). Following Smethurst et al. (1998), again with minor modifications, the northwestern portion of Siberia is restored to the Aldan Shield by closing the Devonian Vilyuy rift (60°N, 115°E, −25°). Thereafter, the reconstructed Siberian Craton in Aldan reference frame is rotated to Laurentia (77.1°N, 113.2°E, +138.7) with minor modifications from the Rodinia models of Rainbird et al. (1998) and Li et al. (2008). Baltica is restored to Laurentia in the NENA connection of Gower et al. (1990) as quantified (47.5°N, 001.5°E, +49°) by Evans & Pisarevsky (2008). Palaeomagnetic poles are rotated by the same parameters as their host cratons, sharing the same colour codes. Australian paleomagnetic poles are selected from Idnurm (2000), inclusion here requiring satisfaction of a field-stability test on the age of magnetization. Laurentian and Baltic poles are illustrated in Evans & Pisarevsky (2008).

the Mawson Continent, or components thereof, have been proposed (Daly et al. 1998; Karlstrom et al. 2001; Betts et al. 2002; Dawson et al. 2002; Giles et al. 2002, 2004; Fitzsimons 2003; Direen et al. 2005; Betts & Giles 2006; Wade et al. 2006). Discussions regarding many of these models can be found in Betts & Giles (2006), Payne et al. (2006, 2008) and Hand et al. (2007). For the purposes of this review we focus on the most recent tectonic reconstruction model for the Mawson Continent, Betts & Giles (2006), and earlier versions of this model.

Betts & Giles (2006) present a model for the 1800–1000 Ma tectonic evolution of Proterozoic Australia that builds upon concepts first published in Betts et al. (2002) and subsequently revised in Giles et al. (2002, 2004). A primary characteristic of each of these models is the presence of a long-lived accretionary margin on the southern margin of the North Australian Craton (Fig. 5). In each of the Betts & Giles models (listed above), the Kimban Orogeny is interpreted to align/connect with the Strangways Orogeny in the Arunta Region to form a roughly east–west trending

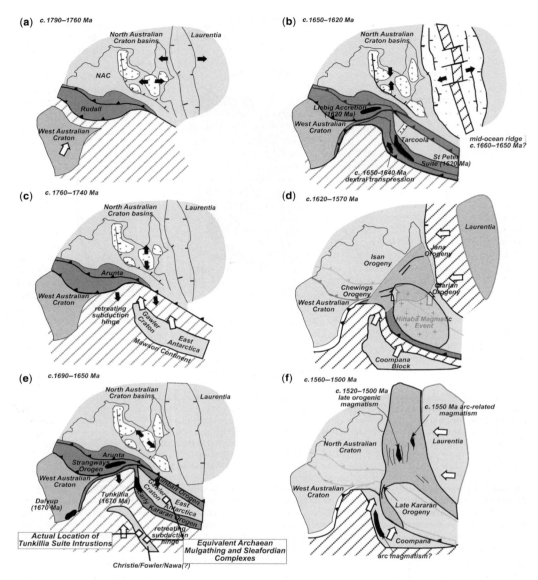

Fig. 5. Reconstruction model of Betts & Giles (2006) showing hypothesized multiple accretion events in the Gawler Craton. Specific tectono-thermal events highlighted are those referred to within the text.

collisional belt (Fig. 5). This rotation is formalized by utilizing palaeomagnetic data at c. 1590 Ma to infer a counter-clockwise rotation of 52° about an Euler pole of 136°E and 25°S (modern-day coordinates, Giles et al. 2004) with minor modification here to accommodate the broader 1740–1590 Ma dataset (Gawler to North Australia: −18°, 134°, 51° CCW). By assigning an active margin to the southern North Australian Craton, Giles et al. (2002, 2004) place the intracratonic McArthur and Mt Isa basins in northern Australia into a far-field extensional back-arc setting.

In the models of Betts et al. (2002) and Giles et al. (2002, 2004), the Gawler Craton was thought to have accreted to the North Australian Craton during the Kimban–Strangways Orogeny. This was revised in Betts & Giles (2006) such that crust east of the Kalinjala Shear Zone in the Gawler Craton was originally part of the North Australian Craton and the proto-Gawler Craton (namely c. 2560–2420 Ma

Sleafordian Complex and 200–1780 Ma cover sequences) was the colliding terrain during the Kimban–Strangways Orogeny (Fig. 5b). The remainder of the Gawler Craton is then interpreted to have accreted at c. 1690–1650 Ma (Fig. 5c, d). Orogenesis at 1565–1500 Ma in the North and South Australian Cratons is interpreted to be related to collision with Laurentia along the eastern margin of the North Australian Craton.

The model of Betts & Giles (2006) honours many geological constraints and consequently a number of aspects of this model are adopted in the reconstruction model proposed herein. However, there are a number of apparent inconsistencies between the model of Betts & Giles (2006) and known geological constraints from the North Australian Craton and Mawson Continent. The long-lived southern accretionary margin of Betts & Giles (2006) is not readily reconcilable with the apparent lack of evidence for subduction and collisional orogenesis in the 1760–1690 Ma time period outside of the easternmost Arunta Region (Claoue-Long et al. 2008). The lack of evidence for subduction and/or collision along the strike of the proposed southern margin cannot be attributed to preservation because the remainder of the Arunta Region is one of the best preserved and exposed regions of the Australian Proterozoic. Second, within the Gawler Craton the proposed division into three terranes at c. 1690 Ma is largely unsupported and in some cases irreconcilable with geological constraints. The distinction of the eastern Gawler Craton (east of the Kalinjala Shear Zone) from the proto-Gawler Craton Archaean is largely based on the lack of 1850 Ma Donington Suite granitoids and associated Cornian Orogeny to the west of the Kalinjala Shear Zone (Fig. 2). Recently collected detrital zircon Hf-isotope data potentially supports the distinction as it suggests pre-1850 Ma sedimentary rocks deposited east of the Kalinjala Shear Zone were not sourced from the currently outcropping Gawler Craton (Howard et al. 2007). In contrast, the presence of temporally equivalent sediment deposition on either side of the proposed suture at both pre-1850 Ma (Fanning et al. 2007; Howard et al. 2007) and 1780–1740 Ma time periods (Daly et al. 1998; Cowley et al. 2003; Fanning et al. 2007) suggests some common tectonic context for the two regions prior to the Kimban Orogeny. Metasedimentary rocks west of the Kalinjala Shear Zone, deposited after 1850 Ma, have a high proportion of c. 1860–1850 Ma detrital zircons (Jagodzinski 2005). This suggests that the proto-Gawler Craton was already associated with a significant volume of 1850 Ma magmatic lithologies prior to the 1730–1690 Ma Kimban Orogeny. In addition, there is little or no evidence for deformation of the proposed overriding plate during the Kimban Orogeny, even immediately adjacent to the proposed suture: major deformation in the Yorke Peninsula region is synchronous with c. 1590 Ma Hiltaba Suite intrusion (Cowley et al. 2003).

The second proposed accretionary event of Betts & Giles (2006, Fig. 5c) is difficult to reconcile with current geological constraints. The Tunkillia Suite (1690–1670 Ma) in the central and western Gawler Craton has previously been suggested to represent subduction-related magmatism based upon trace-element tectonic discrimination diagrams (Teasdale 1997; Betts & Giles 2006). This classification has since been demonstrated to be questionable (post-tectonic petrogenesis, Payne 2008) and hence there is no evidence for subduction beneath the Gawler Craton at this time. Regardless of the tectonic setting of the Tunkillia Suite, the bulk of its magmatism occurs on the proposed underthrust Nawa–Christie–Fowler plate of Betts & Giles (2006), meaning the proposed model is internally inconsistent. The proposed allochthonous Nawa–Christie–Fowler collider also separates the late-Archaean Mulgathing (on Christie Plate, Fig. 5c) and Sleafordian Complexes, despite their identical Archaean–early Palaeoproterozoic tectonic history (Swain et al. 2005b). Furthermore, the metamorphic and magmatic expressions of the Kimban Orogeny effectively stitch the Gawler Craton together at 1730–1690 Ma (Payne et al. 2008), which argues against any younger accretionary events.

Towards a unified model

Internal architecture of the Gawler Craton

Due to the large degrees of freedom in reconstruction models for the Palaeoproterozoic, an exact geometry of the continental blocks is commonly not required and typically not possible. However, the potential for large intra-cratonic architectural rearrangements must be assessed to ascertain the validity of utilizing palaeomagnetic and structural geology constraints. At the craton-scale, the Gawler–Adélie Craton has an architecture in which the Archaean to Palaeoproterozoic lithologies appear to have been wrapped around a younger Palaeo- to Mesoproterozoic core (namely St Peter Suite magmatic lithologies) in the regional interpreted geology (Fig. 2). This architecture has led to suggestions incorporating a hypothesized 'bending' of a previously more linear Gawler–Adélie Craton through oroclinal folding (Swain et al. 2005a), or large degrees of lateral movement through late left-lateral shear zone movement (Direen et al. 2005). Proponents of a bending or transposition of the Gawler Craton cite the north–south to NNE–SSW structural trends in the

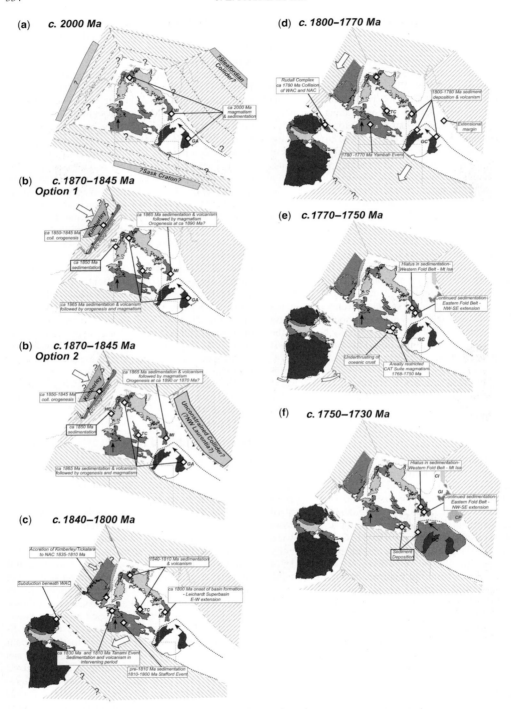

Fig. 6. Proposed reconstruction model for the development and tectonic evolution of the Mawson Continent. Small black arrow in Gawler–Adélie Craton and Arunta Region represents current north. Large arrows represent potential plate movement directions. Figure 6b provides two alternative scenarios for the *c.* 1850 Ma timeline. Going forward from this timeline, Option 1 is adopted but the geometry can be readily exchanged such that rifting in the

(g) *c. 1730–1690 Ma*

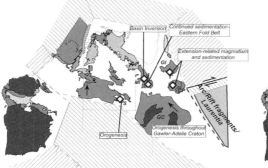

(j) *c. 1600–1580 Ma*

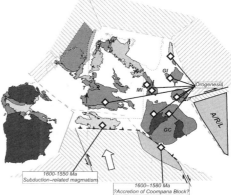

(h) *c. 1690–1650 Ma*

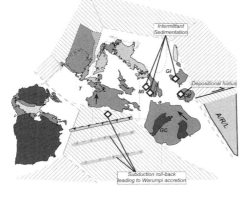

(k) *c. 1580–1540 Ma*

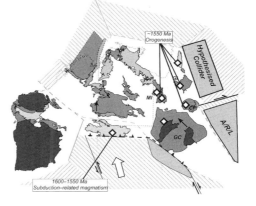

(i) *c. 1640–1600 Ma*

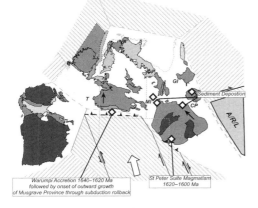

Fig. 6. (*Continued*) northeastern North Australian Craton relates to the rifting away of the collided terrain. Terrain abbreviations are: A, Arunta Region; CI, Coen Inlier; CP, Curnamona Province; GC, Gawler-Adelie Craton; GI, Georgetown Inlier; HC, Halls Creek; MI, Mount Isa Inlier; PI, Pine Creek; R, Rudall Complex; S, Strangways Complex; T, Tanami Region; TC, Tennant Creek/Davenport Region; and W, Warumpi Province. Striped grey regions represent oceanic lithosphere, dashed black lines represent active plate boundaries, grey dashed lines represent inactive boundaries, dotted line represents extent of the Gawler–Adelie crust. Greyscale shading of Australian terrains as per shading in Figure 1.

southeastern Gawler Craton and Adélie Craton and NE–SW to east–west structural trends in the northern and northwestern Gawler Craton as supporting evidence. If this hypothesis is correct, it implies that oroclinal bending or transposition of the northern parts of the craton occurred between c. 1608 and 1592 Ma, as the upper Gawler Range Volcanics (in the core of the apparently-arcuate shaped cratonic domains) are relatively flat-lying and undeformed (Daly et al. 1998). Because convincing evidence demonstrating the bending of the Gawler Craton has not been recorded, we adopt the simplest model for the Palaeoproterozoic geometry of the Gawler–Adélie Craton, restoring only the Mesozoic rifting between Australia and Antarctica (Figs 1 & 2).

Evolution of the Gawler–Adélie Craton and Mawson Continent in a global setting

This section outlines a new model for the evolution of the Gawler–Adélie Craton and Mawson Continent (Fig. 6). The model focuses on identifying correlatable timelines within other Archaean–Mesoproterozoic terrains and summarizing all correlations in a geological constraint-driven model.

Late Archaean–early Palaeoproterozoic. The first correlatable tectonic cycle on the Gawler–Adélie Craton is late Archaean magmatism and sedimentation, including subduction-related magmatism of the Dutton suite and Devil's Playground Volcanics (c. 2560–2520 Ma) and the Sleafordian Orogeny at c. 2460–2430 Ma (Swain et al. 2005b). Convergent tectonic settings of this age are rare among the world's cratons (Fig. 7) and the Sleafordian Orogeny in particular represents a relatively uncommon timeline. Three terrains record evidence for similar-aged metamorphism/orogenesis with peak metamorphism in the earliest Palaeoproterozoic: the North Australian Craton, the Sask Craton and the very poorly-known North Korean peninsula.

The c. 2520 Ma magmatism in the Gawler Craton has temporal equivalents in the North Australian Craton in the Pine Creek Inlier and Tanami Region (Lally 2002; Cross et al. 2005; Crispe et al. 2007). The nature and timing of metamorphism of the North Australian Archaean lithologies is yet to be reliably constrained. Recent reconnaissance geochronology has reported an age of 2473 ± 12 Ma from three analyses of zircon rims in a c. 2633 Ma orthogneiss (Worden et al. 2006). Within basement inliers in the Mt Isa region, McDonald et al. (1997) report 2500–2420 Ma magmatic ages for the Black Angel Gneiss and identified subduction-related arc-geochemical signatures. The classification of the Black Angel Gneiss as a magmatic Archaean-age lithology is disputed, with alternative c. 1850 Ma interpreted ages suggested (Page & Sun 1998). Furthermore, detailed evaluation of geochemical data used to identify a subduction-related petrogenesis is not provided in McDonald et al. (1997). In addition to Archaean basement lithologies of the North Australian Craton, ages of 2500–2450 Ma are obtained from detrital zircon geochronology of metasedimentary lithologies within the eastern Arunta Region (Wade et al. 2008). These provenance data provide further evidence for similar-aged Archaean to early Palaeoproterozoic lithologies within the North Australian and Gawler–Adelie cratons. The similarity in age, and potentially tectonic setting, of tectono-thermal events in the late Archaean and early Palaeoproterozoic lithologies of the North Australian and Gawler–Adélie Cratons appears to continue throughout the Palaeoproterozoic (see below). Based upon the outlined temporal correlations and apparent longevity of interaction between the two cratons, we suggest they formed a single continental domain in the late Archaean–early Palaeoproterozoic.

The crustally evolved Nd-isotope composition of Archaean lithologies within the Gawler Craton (Nd-depleted mantle model ages – c. 3.4–2.8 Ga, Swain et al. 2005b) and presence of detrital zircon ages (2720–2600 Ma with minor inheritance at c. 3000–2800 Ma) that are not consistent with derivation from the currently exposed Gawler–Adélie Craton (Swain et al. 2005b), suggest that the Gawler–Adélie Craton was built upon pre-existing Mesoarchaean crust. The Sask Craton, Trans-Hudson Orogen, Laurentia, yields evidence of Mesoarchaean crust that is consistent with potential Gawler–Adelie Craton protoliths (Chiarenzelli et al. 1998) and records orogenesis similar in age to the Sleafordian Orogeny (Chiarenzelli et al. 1998; Rayner et al. 2005). Magmatic lithologies and inherited zircons within the Sask Craton (Chiarenzelli et al. 1998; Ashton et al. 1999; Rayner et al. 2005) correspond to all major detrital zircon age populations in the late Archaean metasedimentary lithologies of the Gawler–Adélie Craton (Fig. 7, Swain et al. 2005b). Magmatism at c. 2520–2450 Ma with metamorphic reworking of the crust at c. 2450 Ma (Rayner et al. 2005) correlates well with the late Archaean–early Palaeoproterozoic lithologies of the Gawler–Adélie Craton. The combined evidence suggests the Sask Craton could have been contiguous with the Gawler–Adélie and North Australian cratons, and also provides potential equivalent lithologies for the unexposed basement to the Gawler–Adélie Craton lithologies.

In addition to the North Australian Craton and Sask Craton, Sleafordian-age metamorphism is also recorded in the North Korean Peninsula (Zhao et al. 2006a). The North Korean Peninsula records

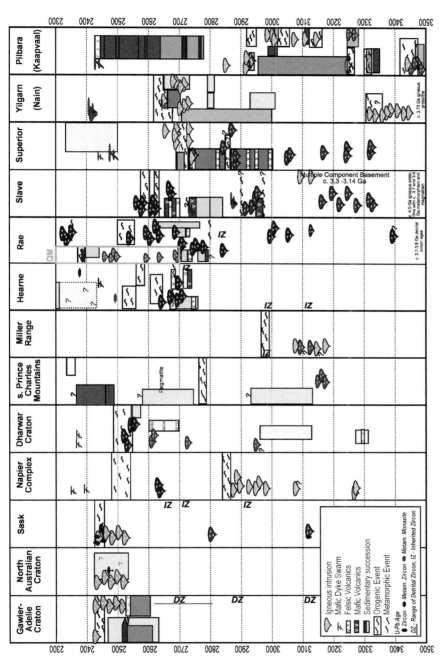

Fig. 7. Geological event plot displaying tectono-thermal evolutions of the Archaean components of the Mawson Continent and various other Archaean cratons/terranes. For simplicity, cratons which have previously been highlighted as similar to those portrayed in the figure (and generally not suitable correlatives of the Gawler–Adélie Craton) are noted in brackets in the column heading. Gray rectangle within Rae Craton column represents the Queen Maud Block (QM). Age-constrained metamorphic mineral growth that is not demonstrated to represent orogenesis is represented as a 'Metamorphic Event'. Data sources not mentioned in text are: *Yilgarn*: Cassidy *et al.* (2006) and references therein; *Pilbara*: Hickman (2004), Pawley *et al.* (2004), Trendall *et al.* (2004) and references therein; *Napier Complex*: Kelly & Harley (2005) and references therein; *Sask*: Chiarenzelli *et al.* (1998), Ashton *et al.* (1999), Rayner *et al.* (2005); *Superior*: Stott (1997), Bleeker (2003), Tomlinson *et al.* (2004), Lin *et al.* (2006), Melynk *et al.* (2006), Parks *et al.* (2006), Percival *et al.* (2006); *Slave*: Bleeker *et al.* (1999), Bowring & Williams (1999), Davis & Bleeker (1999), Bleeker (2003), Ketchum *et al.* (2004), Goodwin *et al.* (2006); *Rae*: Bethune & Scammell (2003), Carson *et al.* (2004), Berman *et al.* (2005), Hartlaub *et al.* (2004, 2005, 2007) and references therein; *Hearne*: Davis *et al.* (2004, 2006), Hanmer *et al.* (2004), MacLachlan *et al.* (2005), van Breemen *et al.* (2007) and references therein; *Queen Maud*: Schultz *et al.* (2007); *Dharwar*: Nutman *et al.* (1992, 1996), Peucat *et al.* (1993, 1995), Chadwick *et al.* (2001), Vasudev *et al.* (2000), Jayananda *et al.* (2006), Halls *et al.* (2007).

magmatism at c. 2640 and 2540 Ma with c. 2460–2430 Ma metamorphism (Zhao et al. 2006a). Further work is required to better characterize this terrain and its relationship with the eastern block of the North China Craton (Zhao et al. 2006a), but these preliminary data suggests possible links with the Gawler–Adélie and North Australian cratons.

Immediately prior to the time of the Sleafordian Orogeny, the western block of the North China Craton (Zhao et al. 2006b and references therein), Dharwar Craton (India, Friend & Nutman 1991), Napier Complex and Vestfold Hills (Antarctica, Kelly & Harley 2005; Zulbati & Harley 2007) and Rae and Hearne cratons (Laurentia, see Fig. 7 caption for references) underwent orogenesis over the interval 2550–2470 Ma with peak metamorphism prior to 2500 Ma. The Rae Craton and Queen Maud Block (Fig. 7) also record a later episode(s) of orogenesis in the 2390–2320 Ma period (Arrowsmith Orogeny, Schultz et al. 2007). The relationship of the Sleafordian Orogeny with deformation in these terrains is unclear; however, given the close temporal relationship of orogenesis in relation to the apparent extensional or cratonized state of most other Archaean cratons (Bleeker 2003), the potential for palaeogeographic proximity is worthy of further consideration.

Circa 2000–1850 Ma rifting and sedimentation. Both the North Australian Craton and Gawler–Adélie Craton share a period of inactivity (c. 2440–2050 Ma) prior to the onset of sedimentation and felsic magmatism (Daly et al. 1998; Fanning et al. 2007; Worden et al. 2008), possibly representing continental rifting and breakup. In order to honour the late Archaean link proposed with the Sask Craton, the Sask Craton must have rifted away from the Gawler–Adélie and North Australian cratons sometime prior to c. 1830 Ma collision and incorporation into the Trans-Hudson Orogen (Ansdell 2005; Rayner et al. 2005). The c. 2000 Ma timeframe appears to be the most suitable time for the required rifting of the Sask Craton away from the proposed Gawler–Adelie and North Australian craton lithosphere. That sedimentation and magmatism developed within previously stabilized continental domains and is followed by an extended period of sediment deposition is also circumstantial evidence for a rift or extensional setting at c. 2000 Ma (as proposed by Daly et al. 1998). The central Pine Creek Inlier metasedimentary and volcanic lithologies suggest deposition in east-deepening tilted-block basins with coarsening of fluvial fan material indicating topographic relief to the west and approximately east–west directed extension at the time of deposition (c. 2020–2000 Ma, Worden et al. 2008 and references therein). McDonald et al. (1997) outline a subduction-related magmatic event in the Mt Isa inlier at c. 2000 Ma. As the age of this magmatism is disputed (Page & Sun 1998), we have not included it in the model proposed here.

Globally, the 2000 Ma timeline is late in the time period typically assigned to the final breakup of late Archaean supercontinent/s before the onset of extensive continent amalgamation starting at c. 1900 Ma (Trans-Hudson and North China Orogenies, Condie 2002, 2004; St-Onge et al. 2006; Zhao et al. 2006b and references therein).

Circa 1890–1810 Ma orogenesis and magma generation. The period of the Palaeoproterozoic from c. 1950–1800 Ma is commonly cited as representing the final amalgamation of the proposed supercontinent Nuna or Columbia (Zhao et al. 2002; Rogers & Santosh 2003; Zhao et al. 2004; Kusky et al. 2007). In the case of continents such as Laurentia and Baltica, this period is relatively well-characterized and has been demonstrated to represent the amalgamation of multiple cratons and continent stabilization prior to a period of terrain accretion. By contrast, the events of this period in the Gawler–Adélie and North Australian Cratons are yet to be fully understood.

The c. 1850 Ma Cornian Orogeny in the eastern Gawler Craton is effectively delineated by the areal extent of the syntectonic Donington Suite (Daly et al. 1998; Reid et al. 2008). This areal distribution has led to the aforementioned proposal of a collisional suture along the Kalinjala Shear Zone during the Kimban Orogeny (Betts & Giles 2006). The model proposed herein differs from the Betts & Giles (2006) model, instead interpreting the Kalinjala Shear Zone as an intra-cratonic shear zone. This interpretation is consistent with the information outlined earlier in this review, which does not support a c. 1730–1690 Ma continental suture along the Kalinjala Shear Zone (Betts & Giles 2006) and retains the Gawler–Adélie and North Australian Cratons as a single entity at this time.

The North Australian Craton records metamorphism and deformation similar in age to the Cornian Orogeny in a number of terrains. The stepwise accretion of the Kimberley Craton to the western margin of the North Australian Craton is the best understood of these events. The Kimberley Craton is interpreted to have collided with the Tickalara arc at c. 1850–1845 Ma (Sheppard et al. 1999, 2001; Griffin et al. 2000), before both terrains were accreted to the North Australian Craton at c. 1835–1810 Ma (Fig. 6b, c, Sheppard et al. 2001). Within the interior of the North Australian Craton, events in the Tennant Creek–Davenport and Pine Creek regions of the North Australian Craton (Fig. 8) correlate temporally with events in the Gawler–Adélie Craton. Within the Pine Creek

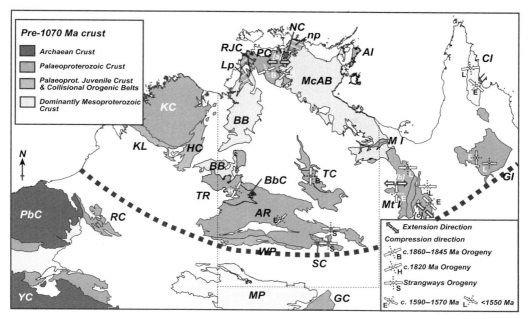

Fig. 8. Geology of the North Australian Craton with orientation of principle stress for tectonic events in the Palaeo- to early Mesoproterozoic. Thick dashed line represents widely used southern extent of the North Australian Craton. Region abbreviations are: AI, Arnhem Inlier; AR, Arunta Region; BB, Victoria-Birindudu Basin; BbC, Billabong Complex (Archaean); CI, Coen Inlier; GC, Gawler Craton; GI, Georgetown Inlier; HC, Halls Creek Orogen; KC, Kimberley Craton; KL, King Leopold Orogen; McAB, McArthur Basin; MI, Murphy Inlier; Mt I, Mt Isa Inlier; MP, Musgrave Province; NC, Nanumbu Complex (Archaean); PC, Pine Creek Orogen with white hatched regions representing; Lp, Litchfield Province and np, Nimbuwah Province; PbC, Pilbara Craton; RC, Rudall Complex; RJC, Rum Jungle Complex (Archaean); SC, Strangways Complex (stippled region); TC, Tennant Region; TR, Tanami Reigon; WP, Warumpi Province; and YC, Yilgarn Craton. Shortening directions for 1860–1845 Ma (Pine Creek, Tennant Region), c. 1820 Ma (Halls Creek Orogeny, Tanami Event) tectonic events, Strangways Orogeny, c. 1600–1570 Ma (Early Isan and equivalent and Chewings Orogeny, Blewett & Black 1998; Boger & Hansen 2004) and 1550–1520 Ma (Late Isan and equivalent, Black et al. 1998; Boger & Hansen 2004) tectonic events. Dominant extension directions at time of basin formation represent: (**a**) c. 2000 Ma in Pine Creek Orogen; (**b**) c. 1800 Ma in Mt Isa Inlier; (**c**) 1780–1750 Ma in Mt Isa Inlier. Refer to text for references.

region basin development, including volcanics at 1864 ± 3–1861 ± 4 Ma, is terminated by metamorphism and deformation at 1853 ± 4 Ma with deformation complete by 1847 ± 1 Ma (Carson et al. 2008; Worden et al. 2008 and references therein). This deformation and metamorphism is contemporaneous with the deposition of the upper Halls Creek Group on the western margin of the North Australian Craton (Olympio Formation, Blake et al. 1998). The setting of the Halls Creek Group has been equated to passive margin sedimentary sequences (Sheppard et al. 1999), and importantly, these do not show evidence for c. 1850 Ma orogenesis. South of the Halls Creek Orogen, recent research has identified a c. 1865 Ma volcanic sequence in the western Tanami Region (Bagas et al. 2008). This sequence is the first of this age identified in the region and is suggested to have undergone tectonism as early as 1850 Ma (Bagas et al. 2008). If this proves to be the case, it further highlights the extent of c. 1850 Ma tectonism in the North Australian Craton.

In the eastern North Australian Craton, the Mt Isa Inlier records a similar sequence of sedimentary and igneous events to the Pine Creek and Tennant regions and Gawler Craton, with volcanics at c. 1865 Ma (Page & Williams 1988), and voluminous felsic magmatism of the Kalkadoon-Ewen Batholith. However, metamorphism and deformation (and magmatism) within the Mt Isa Inlier is considered to have occurred significantly prior to that of the Pine Creek region and Gawler Craton. Metamorphism (the basis for the Barramundi Orogeny nomenclature) is constrained by a single SHRIMP I U–Pb zircon age of 1890 ± 8 Ma (Page & Williams 1988). However, Bierlein et al. (2008) suggest this age is older than the age of metamorphism and assign an orogenic

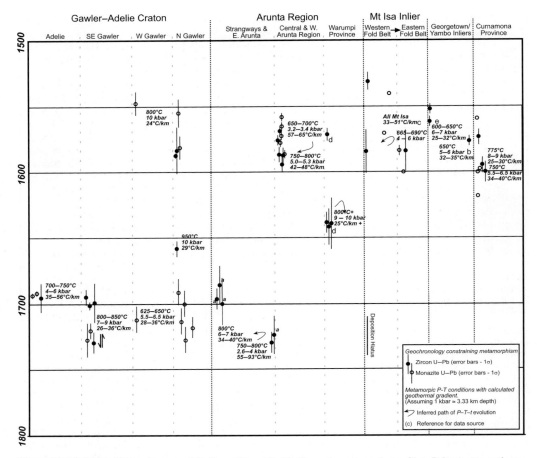

Fig. 9. Detailed Time-Space diagram of $P-T$ conditions, $P-T$ paths and geochronology of late Palaeoproterozoic to early Mesoproterozoic tectono-thermal events in northern and southern Australia. Source data is referred to in text. Information not referred to in text is: (**a**) Claoue-Long & Hoatson (2005); (**b**) Blewett & Black (1998), Blewett et al. (1998) and references therein; (**c**) Mark et al. (1998) with regional geothermal gradients from Foster & Rubenach (2006) and references therein; (**d**) Scrimgeour et al. (2005); (**e**) Black et al. (1998), Boger & Hanson (2004).

age of c. 1870 Ma based upon U–Pb zircon ages of granitic intrusions. Given the poorly constrained nature of metamorphism, further work is required to resolve the relationship of Mt Isa tectonism with c. 1850 Ma orogenesis elsewhere in North Australian and Gawler–Adélie Cratons.

Two alternative models are envisaged to accommodate the widely distributed metamorphism and tectonism within the 1870–1845 Ma time period. The first (Fig. 6b, Option 1) considers the c. 1850 Ma tectonism to be a far-field effect of the Kimberley accretion. Close timing and event duration correlations between accretion of the Kimberley Craton to the Tickalara Arc and events throughout the North Australian Craton supports such a model. However, the lack of deformation within the geographically closer, passive margin Olympio Formation, seemingly contradicts this model. The second scenario (Fig. 6b, Option 2) considers there to have been multiple active margins during this period, potentially with collisional orogenesis occurring on both the eastern and western margins of the North Australian Craton (similar to suggestion of Betts et al. 2002). In such a model, tectonism in the Gawler–Adélie, Pine Creek, Mt Isa and Tennant Creek-Davenport regions would be related to a separate collisional event on the eastern or northeastern margin of Australia. Support for such a scenario is provided by the suggested arc-affinities of the geochemical signature for basement granitoids within the Mt Isa Inlier (McDonald et al. 1997; Bierlein & Betts 2004). The apparent earlier timing of tectonism within the Mt Isa Inlier, compared to the remainder of the North Australian Craton, may be consistent with a convergent margin in this region.

At c. 1830 and 1810 Ma, the Tanami Region underwent metamorphism and deformation associated with the Tanami Event (Crispe et al. 2007). The latter part of this event is temporally equivalent to the Stafford Event in the Arunta Region (Scrimgeour 2003). These events are magmatically dominated systems with metamorphism driven by magmatic heat advection and, particularly in the case of the Arunta Region, show limited deformation (Scrimgeour 2003; White et al. 2003). The Tanami and Stafford Events (and Murchison Event in Tennant Region) are coeval with the final accretion of the combined Kimberley/Tickalara terrains to the North Australian Craton. As no other plate margin activity is evident within the North Australian Craton at this time, we agree with recent suggestions (Crispe et al. 2007; Worden et al. 2008) that c. 1830–1810 Ma magmatism and deformation within the Tanami and Arunta regions is related to the accretion of the Kimberley/Tickalara crust to the North Australian Craton (Fig. 6c).

Circa 1800–1770 Ma. The Rudall Complex, Western Australia, is interpreted to record the collision of the North Australian and West Australian cratons during the period c. 1795–1765 Ma (Fig. 6d). This timing is constrained by the age of granitoid intrusions which are interpreted to pre- and post-date high-pressure metamorphism (c. 800 °C, 12 kbar) interpreted to record the collision event (Smithies & Bagas 1997; Bagas 2004). The geometry of subduction leading to this interpreted collision is effectively unconstrained. Smithies & Bagas (1997) propose a NE-dipping subduction geometry that is also adopted by Betts & Giles (2006). We also adopt a NE-dipping subduction geometry but recognize that subduction is equally as likely to have occurred with a SW-dipping geometry and further research is required to resolve this issue.

Apparently synchronous with collision in the Rudall Complex is the 1780–1770 Ma Yambah Event (previously termed Early Strangways) in the Arunta Region (Hand & Buick 2001; Scrimgeour 2003). The Yambah Event is a dominantly magmatic event that does not appear to represent major crustal thickening (Scrimgeour 2003). In the central Arunta Region, Hand & Buick (2001) suggest the Yambah Event was associated with NE–SW directed shortening. A similar-style event occurred in the Tanami Region at c. 1800–1790 Ma (Crispe et al. 2007), dominated by magmatism with WSW–ENE to east–west shortening represented by thrust faulting (Wygralak et al. 2005; Crispe et al. 2007). In our proposed model the compressional deformation associated with these two events is linked to the ongoing collision with the West Australian Craton to the SW and west of the Tanami and Arunta Regions. Within the Mt Isa Inlier, basin formation and sedimentation was initiated at c. 1800 Ma and continued until basin inversion at c. 1740 Ma, as represented by the Leichhardt Superbasin. Initial extension was east–west-directed (O'Dea et al. 1997) up to and including the deposition of the Eastern Creek Volcanics. Neumann et al. (2006) revise the Mt Isa stratigraphy such that the Eastern Creek Volcanics were emplaced at the initiation of the Myally Supersequence (c. 1780–1765 Ma, Neumann et al. 2006). The Myally Supersequence was previously interpreted to have been deposited in a north–south extensional regime (O'Dea et al. 1997) but this interpretation may no longer be accurate due to the recently revised stratigraphy (see Foster & Austin 2008 for discussion). Initial c. 1800 Ma basin extension is interpreted to represent the onset of intracontinental rifting, either within a trailing edge of the North Australian Craton, or between the North Australian Craton and another protocontinent, potentially Laurentia, as hypothesized in Betts & Giles (2006).

Circa 1770–1740 Ma. Within the Arunta Region, c. 1770–1750 Ma CAT suite magmatism has been identified as having a subduction-related petrogenesis (Foden et al. 1988; Zhao & McCulloch 1995) and has led to the proposal of long-lived north-dipping subduction under the southern margin of the North Australian Craton (Scott et al. 2000; Giles et al. 2002; Betts & Giles 2006). However, the CAT suite is localized in the easternmost Arunta Region and has a much smaller volume than the predominant 1780–1770 Ma granites, which do not have a subduction-related petrogenesis and are found throughout the southern part of the North Australian Craton (Zhao & McCulloch 1995). The model proposed herein relates the generation of the CAT Suite to rotation of the West Australian Craton comparative to the North Australian Craton following the initial c. 1780 Ma interpreted collision between the two cratons. This is interpreted to result in the rupture and forced under-thrusting of the intervening oceanic crust under the eastern Arunta region (Fig. 6e), generating the CAT Suite magmatism. In the western Mt Isa Inlier, the c. 1765–1750 Ma period represents a break in sedimentation between the Myally and Quilalar Supersequences (Neumann et al. 2006). Conversely, within the Eastern Fold Belt of the Mt Isa Inlier, this period correlated to NW–SE directed extension during the deposition of the Malbon Group (Potma & Betts 2006).

CAT Suite magmatism in the Arunta Region ceased approximately 20–30 Ma prior to the onset of the Strangways Orogeny. The intervening period appears to have been dominated by short-lived basin formation, c. 1750–1740 Ma (Fig. 6e), in the eastern Arunta Region and northern Gawler–Adélie Craton (Payne et al. 2006; Wade et al. 2008). We hypothesize that this may have represented the increased influence of subduction/oceanic crust consumption to the east of the continent, prior to the proposed accretion with Laurentia, resulting in an extensional regime in the Australian plate. Sedimentation also resumed within the Western Fold Belt of the Mt Isa at this time in a sag basin setting (Neumann et al. 2006 and references therein).

Circa 1730–1690 Ma orogeny and sediment deposition.

The c. 1730–1690 Ma Kimban Orogeny timeline has been nominated as a primary correlation event for Proterozoic continental reconstruction models (Goodge et al. 2001; Karlstrom et al. 2001; Giles et al. 2004; Betts & Giles 2006). Equivalent timelines exist in Antarctica (outlined Miller and Shackleton range events) and the Arunta region of the North Australian Craton (Strangways Orogeny). The Yavapai Orogeny in southern Laurentia also records similar timing (Duebendorfer et al. 2001; Jessup et al. 2006 and references therein).

The Strangways Orogeny in the eastern Arunta Region is currently constrained to c. 1730–1690 Ma (Möller et al. 2003; Claoue-Long & Hoatson 2005; Maidment et al. 2005; Clarke et al. 2007), consistent with the timing of the Kimban Orogeny. Within the Strangways Complex (Figs 6, 8 & 9), peak $P-T$ conditions of 800 °C and up to 8 kbar are recorded (Ballevre et al. 1997; Möller et al. 2003). High-grade gneissosity is deformed by upright folds with a near vertical, north–south trending foliation defined by sillimanite and biotite in garnet-cordierite-quartz metapelites (Hand et al. 1999). In the eastern part of the Strangways Metamorphic Complex, Strangways-age metamorphism is consistent with near isobaric heating-cooling paths that reach peak metamorphic conditions of 2.6–4.0 kbar and 750–800 °C (Deep Bore Metamorphics) with metamorphic zircon rims recording a SHRIMP U–Pb age of 1730 ± 7 Ma (Scrimgeour et al. 2001; Scrimgeour & Raith 2002). A north–south trending upright sillimanite-bearing fabric (Scrimgeour et al. 2001) post-dates melt crystallization, and either represents a second event after minor isobaric cooling or a pressure increase after melt crystallization in an anticlockwise $P-T$ path (Scrimgeour et al. 2001). Metamorphic grade decreases to greenschist facies to the NW (Shaw et al. 1975; Warren & Hensen 1989; Scrimgeour et al. 2001; Scrimgeour & Raith 2002).

Effects of the Strangways Orogeny extend with decreasing intensity into the Tanami in the form of a NW-trending belt of magmatism (Scrimgeour 2003). Within the Tennant Creek region there is Strangways-age magmatism and low-grade metamorphism (Compston & McDougall 1994; Compston 1995). The above summary of the evidence for the Strangways Orogeny highlights the approximate north–south trend of the compressive phase of the orogeny, supported by north–south trending structures and east–west decrease in metamorphic grade, and the limited east–west extent of c. 1730–1690 Ma deformation in the southern North Australian Craton (Scrimgeour 2003). Hence the Strangways Orogeny does not appear to provide strong evidence for a previously proposed active southern margin of the North Australian Craton at this time (Giles et al. 2002; Betts & Giles 2006).

Given the metamorphic and structural characteristics of the Kimban and Strangways orogenies we consider the reconstruction model in Figure 6g the most appropriate. This interpretation considers the current-day eastern margin of Proterozoic Australia to have undergone active rifting initiating at c. 1800 Ma. The conjugate rifted fragment may be Laurentian, if earlier collision at c. 1850 Ma (Fig. 6b, Option 2) brought the North Australian Craton and Laurentia together. If so, this rifting is similar to that proposed by Betts et al. (2002). Following the collision of the West and North Australian Craton, and possibly related to this event, active rifting ceased and the consumption of oceanic crust commenced in a subduction zone to the west of current-day Laurentia. This resulted in the accretion of proto-Australia to the western margin of the Laurentian plate at c. 1730–1720 Ma in the configuration allowed by palaeomagnetic constraints (Fig. 4). We suggest this collision was centred around the margin of the Gawler–Adélie Craton crust, with margin geometry resulting in an initial dextral margin. In the proposed scenario we interpret the Miller Range to represent an orphaned fragment of Laurentia, possibly a fragment of the Slave Craton. Although the Archaean history of the Miller Range is extremely poorly-known it is not inconsistent with that of the Slave Craton, as both terrains contain evidence for pre-3100 Ma magmatic lithologies and c. 2980 Ma metamorphism (Fig. 7). An interpreted period of intra-orogenic crustal relaxation and/or extension is marked by sedimentation and volcanism within the Gawler–Adélie Craton (c. 1715–1710 Ma, Labyrinth Formation) prior to a second period of high-grade metamorphism at c. 1700–1690 Ma (Fanning et al. 2007; Payne et al. 2008). This is interpreted to represent a second episode of compressional deformation, either within a single collisional event or perhaps representing the consumption of a back-arc basin and accretion of an arc and proto-Australia to the main Laurentian continent.

Sedimentary basins in the Mt Isa Inlier in northern Australia record differing histories during the 1730–1690 Ma period. Within the western Mt Isa region a basin inversion event at c. 1740–1710 Ma represents east–west shortening (Betts 1999) and is followed by a short period of sedimentation (including basal conglomerates) and volcanism (Bigie Formation and Fiery Creek Volcanics, c. 1710 Ma). Subsequent deposition of the Prize Supersequence does not commence until c. 1688 Ma (Neumann et al. 2006). These event timings correlate well with periods of compression and the intervening sedimentation/volcanism recorded in the eastern Arunta Region and Gawler–Adélie Craton. The Eastern Fold Belt of the Mt Isa Inlier does not record evidence of basin inversion in the 1740–1710 Ma period but, in the revised stratigraphy of Foster and Austin (2008), also did not undergo rifting and sedimentation in the period 1730–1680 Ma. The mechanisms for this difference in tectonic history are uncertain. We provide a speculative hypothesis that the Kalkadoon/Leichardt belt basement inlier may have provided some form of structural continuity with deforming crust to the south, resulting in weak compression of crust to the west of the inlier, i.e. the Western Fold Belt.

Evidence to support the adopted model linking Australia and Laurentia is found in the Yukon region of northwestern Laurentia. There the Wernecke Supergroup is deposited some time during the c. 1840–1710 Ma period, intruded by the c. 1710 Ma Bonnet Plume River intrusions, deformed during the Racklan Orogeny and unconformably overlain by the Slab Volcanics (Thorkelson et al. 2001, 2005; Laughton et al. 2005). The Racklan Orogeny produced tight north-trending, east-verging folds that are related to the main schistosity development with peak metamorphic temperatures of 450–550°C (Laughton et al. 2005; Thorkelson et al. 2005 and references therein). These structures were overprinted by open to tight, south-verging folds. The timing of the Racklan Orogeny is uncertain but is pre-1600 Ma as constrained by the cross-cutting Wernecke Breccia (Laughton et al. 2005). Thorkelson et al. (2005) note that 'few if any of the (Bonnet Plume) intrusions' are foliated, suggesting the possibility of a post-kinematic petrogenesis; however, given the relatively low grade of deformation and metamorphism the apparent lack of foliation does not preclude a pre-kinematic petrogenesis. The model proposed here suggests that the initial east–west compression relates to the accretion of proto-Australia to the Laurentian plate. The Racklan Orogeny may be correlated to the intracratonic Forward Orogeny further to the east (Cook & MacLean 1995; Thorkelson et al. 2005). The Forward Orogeny is represented by NW–SE directed compression with south- and north-directed vergent folds and thrust faults followed by later wrench faulting along more northerly directed faults (Cook & MacLean 1995). The Forward Orogeny is constrained by syntectonic sedimentation and volcanism at 1663 ± 8 Ma (Bowring & Ross 1985). The apparent difference in direction of compression between the two orogenic systems highlights the uncertainty regarding the correlation between the Racklan and Forward orogenies, and the need for direct geochronological constraints on Racklan Orogeny metamorphism and deformation.

Palaeomagnetic constraints for proto-Australia with respect to Laurentia result in an open passive margin east of the Georgetown/Curnamona region in the rotated model (Fig. 4). Although potentially a fortuitous coincidence, this palaeogeometry allows for the extension and basin development in the Georgetown/Curnamona region as recorded by the Etheridge and Willyama sequences and associated magmatism (Black et al. 2005; Stevens et al. 2008 and references therein). Recorded extension commenced at c. 1700 Ma in the Georgetown region and c. 1720–1715 Ma in the Curnamona Province. Basin formation within the Curnamona Province is approximately synchronous with the period of short-lived volcanism and sedimentation recorded in the Gawler–Adélie Craton (Labyrinth Formation and Point Geologie migmatite protolith) and Western Fold Belt of the Mt Isa Inlier. Unlike the western Mt Isa Inlier, sediment deposition continued in the Curnamona Province until c. 1690–1680 Ma before a hiatus in deposition until c. 1650 Ma (Conor 2004; Page et al. 2005; Stevens et al. 2008 and references therein). The Georgetown Inlier appears to have recorded continuous deposition until Mesoproterozoic orogenesis. The continuation of rifting and sedimentation after c. 1720 Ma is interpreted to represent the continued extension of the northeastern North Australian Craton crust due to the continued consumption of oceanic crust within a subduction zone further to the east or northeast of proto-Australia.

A potential problem with the model proposed here, is the possibility for Siberia to occupy a position to the NW of Laurentia during the Mesoproterozoic (Fig. 4), as would be the case for the hypothesized extension of Rodinia reconstruction models backwards in time to the early Mesoproterozoic (Frost et al. 1998; Rainbird et al. 1998; Pisarevsky & Natapov 2003; Pisarevsky et al. 2008). This would remove the possibility of an ocean-facing margin to the east of northeastern Australia and hence require alternative mechanisms to explain the continued extension (1700–1600 Ma) and subsequent c. 1550 Ma orogenesis in the Mt Isa, Georgetown and Coen Inlier regions.

Circa 1690–1620 Ma accretion, UHT metamorphism and sedimentation. The interpreted accretion of the Warumpi Province to the southern margin of the Arunta Region at c. 1640–1620 Ma (Scrimgeour et al. 2005) provides the only direct evidence for plate margin processes during the c. 1690–1620 Ma time period. The proposed model considers this to have occurred via south-dipping subduction beneath the Warumpi Province initiating shortly after the Kimban–Strangways Orogeny (Close et al. 2005; Scrimgeour et al. 2005). This model provides a petrogenetic framework for c. 1690 Ma interpreted subduction-related magmatism in the Warumpi Province (Close et al. 2005; Scrimgeour et al. 2005). It also satisfies geophysical constraints that image a south-dipping boundary in the lithospheric mantle extending beneath the Warumpi Province (Selway 2007). Given the potential antiquity of hypothesized oceanic lithosphere between the Gawler–Adélie Craton and West Australian Craton subduction initiation may have occurred through oceanic plate foundering with subsequent subduction-zone roll-back resulting in the accretion of the Warumpi Province to the Arunta Region. Alternatively, in the proposed model the advancement of the Warumpi plate is interpreted to relate to the docking of the proto-Australian plate with Laurentia, re-establishing the previous regime of differing relative movement of proto-Australia and oceanic lithosphere to its south. The transform margin interpreted along the southwestern margin of the Gawler–Adélie Craton provides the mechanism for advancement of the Warumpi Province and is also interpreted to relate to the formation and exhumation of the Ooldea ultra-high temperature granulite lithologies at c. 1660 Ma.

Convergent tectonism at c. 1650–1620 Ma is largely absent from the remainder of the Mawson Continent and West and North Australian cratons. Minor basin formation is evidenced at c. 1650 Ma within the Gawler–Adélie Craton, leading to the deposition of the Tarcoola Formation. The McArthur Basin, Mt Isa Inlier, Curnamona Province and Georgetown Inlier all underwent active sedimentation during the c. 1660–1600 Ma period, and do not appear to record basin inversion associated with the accretionary Leibig Orogeny (Black et al. 2005; Betts et al. 2006; Neumann et al. 2006; Foster & Austin 2008; Stevens et al. 2008 and references therein). The continued basin development is interpreted to reflect the continued extensional nature of the northeastern margin of proto-Australia. Palaeomagnetic constraints allow static relative positions of proto-Australia and Laurentia until at least c. 1590 Ma (Fig. 4), suggesting the two continents did not separate prior to then. We interpret this to indicate the continued extension within northeastern proto-Australia may have been related to continued consumption of oceanic crust north of the Laurentia plate until final c. 1550 Ma collisional orogenesis.

Circa 1620–1500 Ma arc-magmatism and orogenesis. Subduction-related magmatism was initiated in the southern Gawler Craton at c. 1620 Ma as represented by the St Peter Suite (Flint et al. 1990; Fanning et al. 2007; Swain et al. 2008). This is interpreted to represent the onset of subduction outboard of the Warumpi accretion, progressing to northward-dipping subduction under the trailing edge of the Warumpi Province at c. 1600 Ma. This results in the formation of the interpreted arc-related magmatism in the Musgrave Province from 1.6–1.55 Ga (Wade et al. 2006). A north-dipping subduction zone may also provide mechanisms for c. 1590–1550 Ma magmatism in the Rudall Complex (Maidment & Kositcin 2007). Convergence along the southern margin of proto-Australia is synchronous with the early stages of the intracratonic Olarian, Isan and Chewings orogenies (Hand 2006). Shortening directions for these events are represented in Figure 8. For orogenesis in the North Australian Craton, the initial phase of orogenesis represents north–south or NW–SE compression constrained to c. 1590–1570 Ma (Hand & Buick 2001; Betts et al. 2006). Early Olarian Orogeny structures are typically north–south to NE–SW trending and are associated with peak metamorphism (Wilson & Powell 2001; Conor 2004 and references therein). Rotation of the Gawler–Adélie Craton and Curnamona Province into the alignment utilized in the proposed model results in the early Olarian Orogeny structures implying an approximately north–south to NE–SW compression direction. Although a simplistic assessment of the palaeo-tectonic stress orientations, this broadly correlates to the transport direction recorded in the inliers of the North Australian Craton (Hand & Buick 2001). Circa 1600–1580 Ma orogenesis within Australia is consistent with compression caused by the proposed north-dipping subduction beneath the Musgrave Province. Alternatively or additionally, as hypothesized by Betts & Giles (2006), intracratonic orogenesis at c. 1600–1580 Ma may be linked to a collision on the palaeo-southern margin of the Gawler–Adélie Craton in which crust now forming the Coompana Block was accreted. This is potentially consistent with the cessation of St Peter Suite magmatism at c. 1600 Ma. Recent evidence for crustal thickening and anatexis reported from the southern

Gawler–Adélie Craton (Payne 2008) may also support such a scenario. As discussed by Betts & Giles (2006), the petrogenesis of the voluminous GRV and Hiltaba Suite can be considered to represent a mantle plume coincident with tectonothermal activities occurring in the over-riding plate and may not be genetically linked to these events.

The late Isan Orogeny and related events record a distinct second phase of deformation. In the Mt Isa, Georgetown and Coen inliers, the principal direction of compression is approximately east–west at 1560–1550 Ma (Blewett & Black 1998; Boger & Hansen 2004). The westerly directed transport direction is consistent with a collision along the northeastern margin of proto-Australia (Betts & Giles 2006), who suggest this event represents collision of Australia with Laurentia. This proposed scenario is not supported by our interpretation of the existing palaeomagnetic constraints as Australia and Laurentia would already have been amalgamated at this stage. Currently the colliding terrain is unidentified. As outlined in Betts & Giles (2006), the potential exists for subduction to have occurred beneath the Georgetown Inlier leading up to 1550 Ma orogenesis based upon granite geochemistry (Champion 1991).

Subsequent to c. 1550 Ma tectonism, proto-Australia does not appear to have undergone orogenesis until the c. 1300–1100 Ma Musgrave and Albany-Fraser (Camacho & Fanning 1995; Condie & Myers 1999; Bodorkos & Clark 2004b). As summarized in Betts & Giles (2006), the proto-Australian continent underwent a period of extension, basin formation and magmatism in the c. 1500–1400 Ma period. In the Mawson Continent this includes c. 1500 Ma anorogenic magmatism in the Coompana Block (Wade et al. 2007) and the formation of the Cariewerloo Basin in the Gawler–Adélie Craton (Cowley 1993). Palaeomagnetic constraints for c. 1070 Ma (Wingate et al. 2002b) indicate proto-Australia and Laurentia did not stay amalgamated in their c. 1590 Ma configuration for the duration of the Mesoproterozoic, suggesting the two separated at some point in the 1590–1070 Ma period. As suggested by Betts & Giles (2006), this may have occurred during the 1500–1400 Ma time period. However, as outlined above, palaeomagnetic data permit the continued association of northern Australia and Laurentia until as late as c. 1200 Ma.

Conclusions

In constructing the proposed model we have attempted to draw together geological and palaeomagnetic constraints to provide an internally consistent reconstruction of the evolution of the Mawson Continent and associated proto-Australia terrains. The model highlights the presence of comparable timelines in the North Australian Craton and Gawler–Adélie Craton for the duration of the Palaeoproterozoic. This is interpreted such that the two 'cratons' formed a single entity in the late Archaean to middle Proterozoic. As with other recently proposed models (Betts et al. 2002; Giles et al. 2004; Betts & Giles 2006), this model highlights the complexity and longevity of interaction between the so-called North and South Australian cratons.

The identification of reliable palaeomagnetic data for the c. 1750–1730 Ma and c. 1595 Ma timelines provides new constraints on the palaeogeometry of proto-Australia and Laurentica. These data indicate Australia and Laurentia may have been contiguous from c. 1730–1595 Ma. The proposed geometry lends additional weight to previous suggestions (Thorkelson et al. 2005 and references therein) for correlations between basement geology of the Wernecke Mountains and the eastern Australian Proterozoic.

Although the proposed model will probably not provide the ultimate answer to the evolution of the Mawson Continent and Australia, it is hoped that it will provide some insight into poorly understood events within Australia, and will stimulate further investigations on continental reconstruction models for the Palaeoproterozoic.

This work was supported by ARC Linkage Grant LP 0454301 and contributes to UNESCO-IGCP Project 509. A. Collins, I. Scrimgeour, D. Giles, N. Rayner, L. Bagas and colleagues at the University of Adelaide, Monash University and PIRSA are thanked for comments and numerous discussions. P. Betts and K. Ansdell are thanked for thorough and constructive reviews which greatly improved the manuscript and proposed reconstruction model.

Cited references published by Mines and Energy South Australia or Primary Industries and Resources South Australia (PIRSA) are available through the web-based information server 'SARIG' (http://www.pir.sa.gov.au/minerals/sarig).

References

ÅHALL, K. I. & CONNELLY, J. N. 2008. Long-term convergence along SW Fennoscandia: 330 m.y. of Proterozoic crustal growth. *Precambrian Research*, **161**, 452–474.

ANSDELL, K. M. 2005. Tectonic evolution of the Manitoba-Saskatchewan segment of the Palaeoproterozoic Trans-Hudson Orogen, Canada. *Canadian Journal of Earth Sciences*, **42**, 741–759.

ASHTON, K. E., HEAMAN, L. M., LEWRY, J. F., HARTLAUB, R. P. & SHI, R. 1999. Age and origin of the Jan Lake Complex: a glimpse at the buried Archaean craton of the Trans-Hudson Orogen. *Canadian Journal of Earth Sciences*, **36**, 185–208.

BAGAS, L. 2004. Proterozoic evolution and tectonic setting of the northwest Paterson Orogen, Western Australia. *Precambrian Research*, **128**, 475–496.

BAGAS, L., BIERLEIN, F. P., ENGLISH, L., ANDERSON, J., MAIDMENT, D. & HUSTON, D. L. 2008. An example of a Palaeoproterozoic back-arc basin: petrology and geochemistry of the *c*. 1864 Ma Stubbins Formation as an aid towards an improved understanding of the Granites–Tanami Orogen, Western Australia. *Precambrian Research*, **166**, 168–184.

BALLEVRE, M., HENSEN, B. J. & REYNARD, B. 1997. Orthopyroxene-andalusite symplectites replacing cordierite in granulites from the Strangways Range (Arunta block, central Australia): a new twist to the pressure-temperature history. *Geology*, **25**, 215–218.

BENNETT, V. C. & FANNING, C. M. 1993. A glimpse of the cryptic Gondwana shield: Archaean and Proterozoic ages from the central Transantarctic Mountains. *Geological Society of America Abstracts with Programs*.

BERMAN, R. G., SANBORN-BARRIE, M., STERN, R. A. & CARSON, C. J. 2005. Tectonometamorphism at *c*. 2.35 and 1.85 Ga in the Rae Domain, western Churchill Province, Nunavut, Canada: insights from structural, metamorphic and *in situ* geochronological analysis of the southwestern Committee Bay Belt. *Canadian Mineralogist*, **43**, 409–442.

BETHUNE, K. M. & SCAMMELL, R. J. 2003. Geology, geochronology, and geochemistry of Archaean rocks in the Eqe Bay area, north-central Baffin Island, Canada: constraints on the depositional and tectonic history of the Mary River Group of northeastern Rae Province. *Canadian Journal of Earth Sciences*, **40**, 1137–1167.

BETTS, P. G. 1999. Palaeoproterozoic mid-basin inversion in the northern Mt Isa terrane, Queensland. *Australian Journal of Earth Sciences*, **46**, 735–748.

BETTS, P. G. & GILES, D. 2006. The 1800–1100 Ma tectonic evolution of Australia. *Precambrian Research*, **144**, 92–125.

BETTS, P. G., GILES, D., LISTER, G. S. & FRICK, L. R. 2002. Evolution of the Australian lithosphere. *Australian Journal of Earth Sciences*, **49**, 661–695.

BETTS, P. G., VALENTA, R. K. & FINLAY, J. 2003. Evolution of the Mount Woods Inlier, northern Gawler Craton, Southern Australia: an integrated structural and aeromagnetic analysis. *Tectonophysics*, **366**, 83–111.

BETTS, P. G., GILES, D., MARK, G., LISTER, G. S., GOLEBY, B. R. & AILLERES, L. 2006. Synthesis of the proterozoic evolution of the Mt. Isa Inlier. *Australian Journal of Earth Sciences*, **53**, 187–211.

BETTS, P. G., GILES, D., SCHAEFER, B. F. & MARK, G. 2007. 1600–1500 Ma hotspot track in eastern Australia: implications for Mesoproterozoic continental reconstructions. *Terra Nova*, **19**, 496–501.

BIERLEIN, F. P. & BETTS, P. G. 2004. The Proterozoic Mount Isa Fault Zone, northeastern Australia: is it really a *c*. 1.9 Ga terrane-bounding suture? *Earth and Planetary Science Letters*, **225**, 279–294.

BIERLEIN, F. P., BLACK, L. P., HERGT, J. & MARK, G. 2008. Evolution of pre-1.8 Ga basement rocks in the western Mt. Isa Inlier, northeastern Australia – Insights from SHRIMP U-Pb dating and *in-situ* Lu–Hf analysis of zircons. *Precambrian Research*, **163**, 159–173.

BLACK, L. P., GREGORY, P., WITHNALL, I. W. & BAIN, J. H. C. 1998. U–Pb zircon age for the Etheridge Group, Georgetown region, North Queensland; implications for relationship with the Broken Hill and Mt Isa sequences. *Australian Journal of Earth Sciences*, **45**, 925–935.

BLACK, L. P., WITHNALL, I. W., GREGORY, P., OVERSBY, B. S. & BAIN, J. H. C. 2005. U–Pb zircon ages from leucogneiss in the Etheridge Group and their significance for the early history of the Georgetown region, north Queensland. *Australian Journal of Earth Sciences*, **52**, 385–401.

BLAKE, D. H., TYLER, I. M., GRIFFIN, T. J., SHEPPARD, S., THORNE, A. M. & WARREN, R. G. 1998. Geology of the Halls Creek 1:100 000 Sheet area (4461), Western Australia. Australian Geological Survey Organisation, Canberra.

BLEEKER, W. 2003. The late Archaean record: a puzzle in *c*. 35 pieces. *Lithos*, **71**, 99–134.

BLEEKER, W., KETCHUM, J., JACKSON, V. & VILLENEUVE, M. 1999. The Central Slave Basement Complex, Part I: its structural topology and autochthonous cover. *Canadian Journal of Earth Sciences*, **36**, 1083–1109.

BLEWETT, R. S. & BLACK, L. P. 1998. Structural and temporal framework of the Coen Region, north Queensland: implications for major tectonothermal events in east and north Australia. *Australian Journal of Earth Sciences*, **45**, 597–609.

BLEWETT, R. S., BLACK, L. P., SUN, S. S., KNUTSON, J., HUTTON, L. J. & BAIN, J. H. C. 1998. U–Pb zircon and Sm–Nd geochronology of the Mesoproterozoic of North Queensland: implications for a Rodinian connection with the Belt supergroup of North America. *Precambrian Research*, **89**, 101–127.

BODORKOS, S. & CLARK, D. J. 2004a. Evolution of a crustal-scale transpressive shear zone in the Albany–Fraser Orogen, SW Australia: 2. Tectonic history of the Coramup Gneiss and a kinematic framework for Mesoproterozoic collision of the West Australian and Mawson cratons. *Journal of Metamorphic Geology*, **22**, 713–731.

BODORKOS, S. & CLARK, D. J. 2004b. Evolution of a crustal-scale transpressive shear zone in the Albany-Fraser Orogen, SW Australia: 2. Tectonic history of the Coramup Gneiss and a kinematic framework for Mesoproterozoic collision of the West Australian and Mawson cratons. *Journal of Metamorphic Geology*, **22**, 713–731.

BOGER, S. D. & HANSEN, D. 2004. Metamorphic evolution of the Georgetown Inlier, northeast Queensland, Australia; evidence for an accreted Palaeoproterozoic terrane? *Journal of Metamorphic Geology*, **22**, 511–527.

BOGER, S. D., WILSON, C. J. L. & FANNING, C. M. 2001. Early Palaeozoic tectonism within the East Antarctic craton: the final suture between east and west Gondwana? *Geology*, **29**, 463–466.

BOGER, S. D., WILSON, C. J. L. & FANNING, C. M. 2006. An Archaean province in the southern Prince Charles Mountains, East Antarctica: U–Pb evidence for *c*. 3170 Ma plutonism and *c*. 2780 Ma partial melting and orogenesis. *Precambrian Research*, **145**, 207–228.

BORG, S. G. & DEPAOLO, D. J. 1994. Laurentia, Australia and Antarctica as a Late Proterozoic supercontinent: constraints from isotopic mapping. *Geology*, **22**, 307–310.

BOWRING, S. A. & ROSS, G. M. 1985. Geochronology of the Narakay Volcanic Complex – Implications for the age of the Coppermine Homocline and Mackenzie Igneous Events. *Canadian Journal of Earth Sciences*, **22**, 774–781.

BOWRING, S. A. & WILLIAMS, I. S. 1999. Priscoan (4.00–4.03 Ga) orthogneisses from northwestern Canada. *Contributions to Mineralogy and Petrology*, **134**, 3–16.

BROMMER, A., MILLAR, I. L. & ZEH, A. 1999. Geochronology, structural geology and petrology of the northwestern Lagrange Nunataks, Shackleton Range, Antarctica. *Terra Antartica*, **6**, 269–278.

BUCHAN, K. L., VAN BREEMEN, O. & LE CHEMINANT, A. N. 2007. Towards a Palaeoproterozoic Apparent Polar Wander Path (APWP) for the Slave Province: palaeomagnetism of precisely-dated mafic dyke swarms. *Geological Association of Canada Abstracts*.

BUDD, A. 2006. The Tarcoola Goldfield of the Central Gawler Gold Province, and the Hiltaba association Granites, Gawler Craton, South Australia. Ph.D. thesis, Australian National University.

CAMACHO, A. & FANNING, C. M. 1995. Some isotopic constraints on the evolution of the granulite and upper amphibolite facies terranes in the Eastern Musgrave Block, Central Australia. *Precambrian Research*, **71**, 155–181.

CARSON, C. J., BERMAN, R. G., STERN, R. A., SANBORN-BARRIE, M., SKULSKI, T. & SANDEMAN, H. A. I. 2004. Age constraints on the Palaeoproterozoic tectonometamorphic history of the Committee Bay region, western Churchill Province, Canada: evidence from zircon and in situ monazite SHRIMP geochronology. *Canadian Journal of Earth Sciences*, **41**, 1049–1076.

CARSON, C. J., WORDEN, K. E., SCRIMGEOUR, I. & STERN, R. A. 2008. The Palaeoproterozoic evolution of the Litchfield Province, western Pine Creek Orogen, northern Australia: insights from SHRIMP U–Pb zircon and in-situ monazite geochronology. *Precambrian Research*, **166**, 145–167.

CASSIDY, K. F., CHAMPION, D. C. *ET AL.* 2006. A revised geochronological framework for the Yilgarn Craton, Western Australia, Geological Survey of Western Australia, Record 2006/8.

CHADWICK, B., HEGDE, G. V., NUTMAN, A. P. & VASUDEV, V. N. 2001. Syenite emplacement during accretion of the Late Archaean Dharwar batholith, South India: SHRIMP U/Pb age and structure of the Koppal pluton, Karnataka. *Journal of the Geological Society of India*, **58**, 381–390.

CHAMPION, D. C. 1991. The felsic granites of far north Queensland. Unpublished Ph.D. thesis, Australian National University.

CHIARENZELLI, J., ASPLER, L., VILLENEUVE, M. & LEWRY, J. 1998. Early Proterozoic evolution of the Saskatchewan Craton and its Allocthonous Cover, Trans-Hudson Orogen. *Journal of Geology*, **106**, 247–267.

CLAOUE-LONG, J. C. & HOATSON, D. M. 2005. Proterozoic mafic-ultramafic intrusions in the Arunta Region, central Australia. Part 2: event chronology and regional correlations. *Precambrian Research*, **142**, 134–158.

CLAOUE-LONG, J. C., MAIDMENT, D., HUSSEY, K. & HUSTON, D. 2008. The duration of the Strangways Event in central Australia: evidence for prolonged deep crust processes. *Precambrian Research*, **166**, 246–262.

CLARKE, G. L., WHITE, R. W., LUI, S., FITZHERBERT, J. A. & PEARSON, N. J. 2007. Contrasting behaviour of rare earth and major elements during partial melting in granulite facies migmatites, Wuluma Hills, Arunta Block, central Australia. *Journal of Metamorphic Geology*, **25**, 1–18.

CLOSE, D. F., SCRIMGEOUR, I., EDGOOSE, C. J., WINGATE, M. D. & SELWAY, K. 2005. Late Palaeoproterozoic oblique accretion of a 1690–1660 Ma magmatic arc onto the North Australian Craton. *In*: WINGATE, M. D. & PISAREVSKY, S. A. (eds) *Supercontinents and Earth Evolution Symposium*. Geological Society of Australia, Abstracts No. 81, Perth.

COLLINS, A. S. & PISAREVSKY, S. A. 2005. Amalgamating eastern Gondwana: the evolution of the Circum-Indian Orogens. *Earth Science Reviews*, **71**, 229–270.

COMPSTON, D. M. 1995. Time constraints on the evolution of the Tennant Creek Block, Northern Australia. *Precambrian Research*, **71**, 107–129.

COMPSTON, D. M. & MCDOUGALL, I. 1994. Ar-40–Ar-39 and K–Ar Age constraints on the Early Proterozoic Tennant Creek Block, Northern Australia, and the age of its gold deposits. *Australian Journal of Earth Sciences*, **41**, 609–616.

CONDIE, K. C. 2002. The supercontinent cycle: are there two patterns of cyclicity? *Journal of African Earth Sciences*, **35**, 179–183.

CONDIE, K. C. 2004. Supercontinents and superplume events: distinguishing signals in the geologic record. *Physics of the Earth and Planetary Interiors*, **146**, 319–332.

CONDIE, K. C. & MYERS, J. S. 1999. Mesoproterozoic Fraser Complex: geochemical evidence for multiple subduction-related sources of lower crustal rocks in the Albany–Fraser Orogen, Western Australia. *Australian Journal of Earth Sciences*, **46**, 875–882.

CONOR, C. H. H. 1995. *An interpretation of the geology of the Maitland and Wallaroo 1:100 000 sheet areas*. Env 8886, Department of Primary Industries and Resources South Australia, Adelaide.

CONOR, C. H. H. 2004. *Geology of the Olary Domain, Curnamona Province, South Australia*. Primary Industries and Resources South Australia, RB 2004/8, Adelaide.

COOK, D. G. & MACLEAN, B. C. 1995. The intracratonic Palaeoproterozoic Forward orogeny, and implications for regional correlations, Northwest Territories, Canada. *Canadian Journal of Earth Sciences*, **32**, 1991.

COWLEY, W. M. 1993. Cariewerloo Basin. *In*: DREXEL, J. F., PREISS, W. V. & PARKER, A. J. (eds) *The Geology of South Australia*. Geological Survey of South Australia, 139–142.

COWLEY, W. M. 2006. Solid geology of South Australia: peeling away the cover. *MESA Journal*, **43**, 4–15.

COWLEY, W. M. & MARTIN, A. R. 1991. *Kingoonya, 1:250 000 Geological Map Explanatory Notes*. Department of Mines and Energy, South Australia, Adelaide.

COWLEY, W. M., CONOR, C. H. H. & ZANG, W. 2003. New and revised Proterozoic stratigraphic units on northern Yorke Peninsula. MESA Journal, **29**, 46–58.

CREASER, R. A. 1995. Neodymium isotopic constraints for the origin of Mesoproterozoic Felsic Magmatism, Gawler-Craton, South Australia. *Canadian Journal of Earth Sciences*, **32**, 460–471.

CRISPE, A. J., VANDENBERG, L. C. & SCRIMGEOUR, I. R. 2007. Geological framework of the Archaean and Palaeoproterozoic Tanami Region, Northern Territory. *Mineralium Deposita*, **42**, 3–26.

CROSS, A., CLAOUE-LONG, J. C., SCRIMGEOUR, I., AHMAD, M. & KRUSE, P. D. 2005. Summary of results. Joint NTGS-GA geochronology project: Rum Jungle, basement to southern Georgina Basin and eastern Arunta Region 2001–2003. Northern Territory Geological Survey Record 2005–2006.

DALY, S. J. & FANNING, C. M. 1993. Archaean. *In*: DREXEL, J. F., PREISS, W. V. & PARKER, A. J. (eds) *The Geology of South Australia – the Precambrian*, Bulletin 54. Mines and Energy, South Australia, Adelaide.

DALY, S. J., FANNING, C. M. & FAIRCLOUGH, M. C. 1998. Tectonic evolution and exploration potential of the Gawler Craton, South Australia. *In*: HODGSON, I. & HINCE, B. (eds) *Geology and Mineral Potential of Major Australian Mineral Provinces*. AGSO Journal of Australian Geology and Geophysics, 145–168.

DAVIS, W. & BLEEKER, W. 1999. Timing of plutonism, deformation, and metamorphism in the Yellowknife Domain, Slave Province, Canada. *Canadian Journal of Earth Sciences*, **36**, 1169–1187.

DAVIS, W. J., HANMER, S. & SANDEMAN, H. A. 2004. Temporal evolution of the Neoarchean Central Hearne supracrustal belt: rapid generation of juvenile crust in a suprasubduction zone setting. *Precambrian Research*, **134**, 85–112.

DAVIS, W. J., HANMER, S., TELLA, S., SANDEMAN, H. A. & RYAN, J. J. 2006. U–Pb geochronology of the MacQuoid supracrustal belt and Cross Bay plutonic complex: key components of the northwestern Hearne subdomain, western Churchill Province, Nunavut, Canada. *Precambrian Research*, **145**, 53–80.

DAWSON, G. C., KRAPEZ, B., FLETCHER, I. R., MCNAUGHTON, N. J. & RASMUSSEN, B. 2002. Did late Palaeoproterozoic assembly of proto-Australia involve collision between the Pilbara, Yilgarn and Gawler cratons? Geochronological evidence from the Mount Barren Group in the Albany–Fraser Orogen of Western Australia. *Precambrian Research*, **118**, 195–220.

DIREEN, N. G. & LYONS, P. 2007. Crustal setting of iron-oxide Cu–Au mineral systems of the Olympic dam region, South Australia: insights from potential fields data. *Economic Geology*, **102**.

DIREEN, N. G., CADD, A. G., LYONS, P. & TEASDALE, J. 2005. Architecture of Proterozoic shear zones in the Christie Domain, western Gawler Craton, Australia: geophysical appraisal of a poorly exposed orogenic terrane. *Precambrian Research*, **142**, 28–44.

DOUGHTY, P. T. & CHAMBERLAIN, K. R. 1996. Salmon River Arch revisited: new evidence for 1370 Ma rifting near the end of deposition in the Middle Proterozoic Belt basin. *Canadian Journal of Earth Sciences*, **33**, 1037–1052.

DUEBENDORFER, E. M., CHAMBERLAIN, K. R. & JONES, C. S. 2001. Palaeoproterozoic tectonic history of the Cerbat Mountains, northwestern Arizona: Implications for crustal assembly in the southwestern United States. *Geological Society of America Bulletin*, **113**, 575–590.

EVANS, D. A. D. 2006. Proterozoic low orbital obliquity and axial-dipolar geomagnetic field from evaporite palaeolatitudes. *Nature*, **444**, 51–55.

EVANS, D. A. D. & RAUB, T. M. D. 2007. Growth of Laurentia from a palaeomagnetic perspective: critical review and update. *Geological Association of Canada Abstracts*.

EVANS, D. A. D. & PISAREVSKY, S. A. 2008. Plate tectonics on early Earth? Weighing the palaeomagnetic evidence. *In*: CONDIE, K. C. & PEASE, V. (eds) *When Did Plate Tectonics Begin?* Geological Society of Americ Special Paper, 249–263.

FAIRCLOUGH, M. C., SCHWARZ, M. & FERRIS, G. J. 2003. *Interpreted Crystalline Basement Geology of the Gawler Craton, South Australia*. Primary Industries and Resources South Australia.

FANNING, C. M., FLINT, R. B., PARKER, A. J., LUDWIG, K. R. & BLISSETT, A. H. 1988. Refined Proterozoic Evolution of the Gawler Craton, South Australia, through U–Pb Zircon Geochronology. *Precambrian Research*, **40–1**, 363–386.

FANNING, C. M., MOORE, D. H., BENNETT, V. C. & DALY, S. J. 1996. The 'Mawson Continent'; Archaean to Proterozoic crust in the East Antarctic Shield and Gawler Craton, Australia; a cornerstone in Rodinia and Gondwanaland. *In*: KENNARD, J. M. (ed.) *Geoscience for the Community; 13th Australian Geological Convention*. Geological Society of Australia, Abstracts 41, 135.

FANNING, C. M., MOORE, D. H., BENNETT, V. C., DALY, S. J., MENOT, R. P., PEUCAT, J. J. & OLIVER, R. L. 1999. The 'Mawson Continent'; the East Antarctic Shield and Gawler Craton, Australia. *In*: SKINNER, D. N. B. (ed.) *8th International Symposium on Antarctic Earth Sciences*. Royal Society of New Zealand, Wellington, New Zealand.

FANNING, C. M., REID, A. J. & TEALE, G. S. 2007. *A Geochronological Framework for the Gawler Craton, South Australia*. Bulletin 55, Geological Survey, Primary Industries and Resources South Australia.

FERRIS, G. J. & SCHWARZ, M. 2004. Definition of the Tunkillia Suite, western Gawler Craton. *MESA Journal*, **34**, 32–41.

FERRIS, G., SCHWARZ, M. & HEITHERSAY, P. 2002. The geological framework, distribution and controls of Fe-oxide and related alteration, and Cu–Au mineralisation in the Gawler Craton, South Australia: Part 1: geological and tectonic framework. *In*: PORTER, T. (ed.) *Hydrothermal Iron Oxide Copper-Gold & Related Deposits: A Global Perspective*. PGC Publishing, Adelaide.

FINN, C. A., GOODGE, J. W., DAMASKE, D. & FANNING, C. M. 2006. Scouting craton's edge in palaeo-Pacific Gondwana. *In*: FUTTERER, D. K., DAMASKE, D., KLEINSCHMIDT, G., MILLER, H. & TESSENSOHN, F. (eds) *Antarctica: Contributions to Global Earth Sciences*. Springer, Heidelberg.

FITZSIMONS, I. C. W. 2000. A review of tectonic events in the East Antarctic Shield and their implications for Gondwana and earlier supercontinents. *Journal of African Earth Sciences*, **31**, 3–23.

FITZSIMONS, I. C. W. 2003. Proterozoic basement provinces of southern and southwestern Australia, and their correlation with Antarctica. *In*: YOSHIDA, M., WINDLEY, B. F. & DASGUPTA, S. (eds) *Proterozoic East Gondwana: Supercontinent Assembly and Breakup*. Geological Society, London, Special Publications, **206**, 93–130.

FLINT, R. B. 1993. Mesoproterozoic. *In*: DREXEL, J. F., PREISS, W. V. & PARKER, A. J. (eds) *The Geology of South Australia – The Precambrian*, Bulletin 54. Mines and Energy, South Australia, Adelaide.

FLINT, R. B., RANKIN, L. R. & FANNING, C. M. 1990. Definition: the Palaeoproterozoic St. Peter Suite of the western Gawler Craton. Quarterly Geological Notes, Geological Survey of South Australia.

FODEN, J. D., BUICK, I. S. & MORTIMER, G. E. 1988. The petrology and geochemistry of granitic gneisses from the East Arunta Inlier, central Australia; implications for Proterozoic crustal development. *Precambrian Research*, **40–41**, 233–259.

FOSTER, D. A. & EHLERS, K. 1998. Ar-40–Ar-39 thermochronology of the southern Gawler craton, Australia: implications for Mesoproterozoic and Neoproterozoic tectonics of east Gondwana and Rodinia. *Journal of Geophysical Research Solid Earth*, **103**, 10 177–10 193.

FOSTER, D. R. W. & RUBENACH, M. J. 2006. Isograd pattern and regional low-pressure, high-temperature metamorphism of pelitic, mafic and calc-silicate rocks along an east–west section through the Mt Isa Inlier. *Australian Journal of Earth Sciences*, **53**, 167–186.

FOSTER, D. R. W. & AUSTIN, J. R. 2008. The 1800–1610 Ma stratigraphic and magmatic history of the Eastern Succession, Mount Isa Inlier, and correlations with adjacent Palaeoproterozoic terranes. *Precambrian Research*, **163**, 7–30.

FRASER, G. L. & LYONS, P. 2006. Timing of Mesoproterozoic tectonic activity in the northwestern Gawler Craton constrained by $^{40}Ar/^{39}Ar$ geochronology. *Precambrian Research*, **151**, 160–184.

FRIEND, C. R. L. & NUTMAN, A. P. 1991. Shrimp U–Pb Geochronology of the Closepet Granite and Peninsular Gneiss, Karnataka, South-India. *Journal of the Geological Society of India*, **38**, 357–368.

FROST, B. R., AVCHENKO, O. V., CHAMBERLAIN, K. R. & FROST, C. D. 1998. Evidence for extensive Proterozoic remobilization of the Aldan shield and implications for Proterozoic plate tectonic reconstructions of Siberia and Laurentia. *Precambrian Research*, **89**, 1–23.

GANDHI, S. S., MORTENSEN, J. K., PRASAD, N. & VAN BREEMEN, O. 2001. Magmatic evolution of the southern Great Bear continental arc, northwestern Canadian Shield: geochronological constraints. *Canadian Journal of Earth Sciences*, **38**, 767–785.

GILES, D., BETTS, P. & LISTER, G. 2002. Far-field continental backarc setting for the 1.80–1.67 Ga basins of northeastern Australia. *Geology*, **30**, 823–826.

GILES, D., BETTS, P. G. & LISTER, G. S. 2004. 1.8–1.5 Ga links between the North and South Australian Cratons and the Early–Middle Proterozoic configuration of Australia. *Tectonophysics*, **380**, 27–41.

GOODGE, J. W. & FANNING, C. M. 1999. 2.5 billion years of punctuated Earth history as recorded in a single rock. *Geology*, **27**, 1007–1010.

GOODGE, J. W., FANNING, C. M. & BENNETT, V. C. 2001. U–Pb evidence of c. 1.7 Ga crustal tectonism during the Nimrod Orogeny in the Transantarctic Mountains, Antarctica: implications for Proterozoic plate reconstructions. *Precambrian Research*, **112**, 261–288.

GOODWIN, A. M., LAMBERT, M. B. & UJIKE, O. 2006. Geochemical and metallogenic relations in volcanic rocks of the southern Slave Province: implications for late Neoarchean tectonics. *Canadian Journal of Earth Sciences*, **43**, 1835–1857.

GOWER, C. F., RYAN, A. B. & RIVERS, T. 1990. Mid-Proterozoic Laurentia-Baltica: an overview of its geological evolution and a summary of the contributions made by this volume. *In*: GOWER, C. F., RIVERS, T. & RYAN, A. B. (eds) *Mid-Proterozoic Laurentia-Baltica*. Geological Association of Canada.

GRIFFIN, T. J., PAGE, R. W., SHEPPARD, S. & TYLER, I. M. 2000. Tectonic implications of Palaeoproterozoic post-collisional, high-K felsic igneous rocks from the Kimberley region of northwestern Australia. *Precambrian Research*, **101**, 1–23.

HALLS, H. C., KUMAR, A., SRINIVASAN, R. & HAMILTON, M. A. 2007. Palaeomagnetism and U–Pb geochronology of easterly trending dykes in the Dharwar craton, India: feldspar clouding, radiating dyke swarms and the position of India at 2.37 Ga. *Precambrian Research*, **155**, 47–68.

HAMILTON, M. A. & BUCHAN, K. L. 2007. U–Pb geochronology of the Western Channel Diabase, Wopmay Orogen: implications for the APWP for Laurentia in the earliest Mesoproterozoic. *Geological Association of Canada Abstracts*, 35–36.

HAND, M. 2006. Intracratonic Orogeny in Mesoproterozoic Australia. *In*: LYONS, P. & HUSTON, D. L. (eds) *Evolution and Metallogenesis of the North Australian Craton*. Geoscience Australia, Record 2006/16, Alice Springs.

HAND, M. & BUICK, I. S. 2001. Polymetamorphism and reworking of the Reynolds and Anmatjira Ranges, central Australia. *In*: MILLER, J. A., HOLDSWORTH, R. E., BUICK, I. S. & HAND, M. (eds) *Continental Reactivation and Reworking*. Geological Society, London, Special Publications, **184**, 237–260.

HAND, M., MAWBY, J., MILLER, J. M., BALLEVRE, M., HENSEN, B. J., MOLLER, A. & BUICK, I. S. 1999. *Tectonothermal Evolution of the Harts and Strangways Range Region, eastern Arunta Inlier, Central Australia*. Geological Society of Australia, SGGMP Field Guide No. 4.

Hand, M., Reid, A. & Jagodzinski, E. A. 2007. Tectonic framework and evolution of the Gawler Craton, Southern Australia. *Economic Geology*, **102**, 1377–1395.

Hanmer, S., Sandeman, H. A. *et al*. 2004. Geology and Neoarchaean tectonic setting of the Central Hearne supracrustal belt, Western Churchill Province, Nunavut, Canada. *Precambrian Research*, **134**, 63–83.

Hartlaub, R. P., Heaman, L. M., Ashton, K. E. & Chacko, T. 2004. The Archean Murmac Bay Group: evidence for a giant archean rift in the Rae Province, Canada. *Precambrian Research*, **131**, 345–372.

Hartlaub, R. R., Chacko, T., Heaman, L. M., Creaser, R. A., Ashton, K. E. & Simonetti, A. 2005. Ancient (Meso- to Palaeoarchaean) crust in the Rae Province, Canada: evidence from Sm–Nd and U–Pb constraints. *Precambrian Research*, **141**, 137–153.

Hartlaub, R. P., Heaman, L. M., Chacko, T. & Ashton, K. E. 2007. Circa 2.3 Ga magmatism of the Arrowsmith Orogeny, Uranium City region, western Churchill craton, Canada. *Journal of Geology*, **115**, 181–195.

Hickman, A. H. 2004. Two contrasting granite-greenstone terranes in the Pilbara Craton, Australia: evidence for vertical and horizontal tectonic regimes prior to 2900 Ma. *Precambrian Research*, **131**, 153–172.

Hoek, J. D. & Schaefer, B. F. 1998. Palaeoproterozoic Kimban mobile belt, Eyre Peninsula: timing and significance of felsic and mafic magmatism and deformation. *Australian Journal of Earth Sciences*, **45**, 305–313.

Holm, O. 2004. *New geochronology of the Mount Woods Inlier and Central Gawler Gold Province, Gawler Craton – State of Play Conference, abstract volume*. RB 2004/18, Primary Industries and Resources South Australia, Adelaide.

Hopper, D. J. 2001. Crustal Evolution of Paleo- to Mesoproterozoic rocks in the Peake and Denison Ranges, South Australia. Ph.D. thesis, University of Queensland, Brisbane.

Howard, K. E., Reid, A. J., Hand, M., Barovich, K. M. & Belousova, E. A. 2007. Does the Kalinjala Shear Zone represent a palaeo-suture zone? Implications for distribution of styles of Mesoproterozoic mineralisation in the Gawler Craton. *Minerals and Energy South Australia Journal*, **43**, 6–11.

Idnurm, M. 2000. Towards a high resolution Late Palaeoproterozoic-earliest Mesoproterozoic apparent polar wander path for northern Australia. *Australian Journal of Earth Sciences*, **47**, 405–429.

Irving, E., Donaldson, J. A. & Park, J. K. 1972. Palaeomagnetism of the Western Channel Diabase and associated rocks, Northwest Territories. *Canadian Journal of Earth Sciences*, **9**, 960–971.

Irving, E., Baker, J., Hamilton, M. A. & Wynne, P. J. 2004. Early Proterozoic geomagnetic field in western Laurentia: implications for paleolatitudes, local rotations stratigraphy. *Precambrian Research*, **129**, 251–270.

Jagodzinski, E. A. 2005. *Compilation of SHRIMP U–Pb geochronological data, Olympic Domain, Gawler Craton, 2001–2003*. Geoscience Australia, Record 2005/20.

Jagodzinski, E. A., Black, L. P. *et al*. 2006. *Compilation of SHRIMP U–Pb Geochronological Data, for the Gawler Craton, South Australia 2005*. Geochronology Series Report Book 2005–1. Primary Industries and Resources South Australia, Report Book 2006/20.

Jayananda, M., Chardon, D., Peucat, J. J. & Capdevila, R. 2006. 2.61 Ga potassic granites and crustal reworking in the western Dharwar craton, southern India: tectonic, geochronologic and geochemical constraints. *Precambrian Research*, **150**, 1–26.

Jessup, M. J., Jones, J. V., Karlstrom, K. E., Williams, M. L., Connelly, J. N. & Heizler, M. T. 2006. Three Proterozoic orogenic episodes and an intervening exhumation event in the Black Canyon of the Gunnison region, Colorado. *Journal of Geology*, **114**, 555–576.

Karlstrom, K. E., Åhall, K.-I., Harlan, S. S., Williams, M. L., McLelland, J. & Geissman, J. W. 2001. Long-lived (1.8–1.0 Ga) convergent orogen in southern Laurentia, its extensions to Australia and Baltica, and implications for refining Rodinia. *Precambrian Research*, **111**, 5–30.

Kelly, N. M. & Harley, S. L. 2005. An integrated microtextural and chemical approach to zircon geochronology: refining the Archaean history of the Napier Complex, east Antarctica. *Contributions to Mineralogy and Petrology*, **149**, 57–84.

Ketchum, J. W. F., Culshaw, N. G. & Barr, S. M. 2002. Anatomy and orogenic history of a Palaeoproterozoic accretionary belt: the Makkovik Province, Labrador, Canada. *Canadian Journal of Earth Sciences*, **39**, 711–730.

Ketchum, J. W. F., Bleeker, W. & Stern, R. A. 2004. Evolution of an Archaean basement complex and its autochthonous cover, southern Slave Province, Canada. *Precambrian Research*, **135**, 149–176.

Kurz, W. & Froitzheim, N. 2002. The exhumation of eclogite-facies metamorphic rocks – a review of models confronted with examples from the Alps. *International Geology Review*, **44**, 702–743.

Kusky, T., Li, J. & Santosh, M. 2007. The Palaeoproterozoic North Hebei Orogen: North China Craton's collisional suture with the Columbia supercontinent. *Gondwana Research*, **12**, 4–28.

Lally, J. H. 2002. *Stratigraphy, Structure, and Mineralisation, Rum Jungle Mineral Field, Northern Territory*, Northern Territory Geological Survey Record 2002–2005.

Laughton, J. R., Thorkelson, D. J., Brideau, M. A., Hunt, J. A. & Marshall, D. D. 2005. Early Proterozoic orogeny and exhumation of Wernecke Supergroup revealed by vent facies of Wernecke Breccia, Yukon, Canada. *Canadian Journal of Earth Sciences*, **42**, 1033–1044.

Li, Z. X., Bogdanova, S. V. *et al*. 2008. Assembly, configuration, and break-up history of Rodinia: a synthesis. *Precambrian Research*, **160**, 179–210.

Lin, S., Davis, D. W., Rotenberg, E., Corkery, M. T. & Bailes, A. H. 2006. Geological evolution of the northwestern superior province: clues from geology, kinematics, and geochronology in the Gods Lake Narrows area, Oxford–Stull terrane, Manitoba. *Canadian Journal of Earth Sciences*, **43**, 749–765.

MACLACHLAN, K., DAVIS, W. J. & RELF, C. 2005. U/Pb geochronological constraints on neoarchean tectonism: multiple compressional events in the north-western Hearne Domain, Western Churchill Province, Canada. *Canadian Journal of Earth Sciences*, **42**, 85–109.

MAIDMENT, D. W. & KOSITCIN, N. 2007. Time-Space evolution of the Rudall Complex and eastern Pilbara Craton. *In*: NEUMANN, N. L. & FRASER, G. L. (eds) *Geochronological Synthesis and Time-Space plots for Proterozoic Australia*. Record 2007/06 – ISBN 978 1 921236 29 7, Geoscience Australia, Canberra, 140–149.

MAIDMENT, D. W., HAND, M. & WILLIAMS, I. S. 2005. Tectonic cycles in the Strangways Metamorphic Complex, Arunta Inlier, central Australia: geochronological evidence for exhumation and basin formation between two high-grade metamorphic events. *Australian Journal of Earth Sciences*, **52**, 205–215.

MARK, G., PHILLIPS, G. N. & POLLARD, P. J. 1998. Highly selective partial melting of pelitic gneiss in Cannington, Cloncurry Disctrict, Queensland. *Australian Journal of Earth Sciences*, **45**, 169–176.

MCDONALD, G. D., COLLERSON, K. D. & KINNY, P. D. 1997. Late Archaean and Early Proterozoic crustal evolution of the Mount Isa block, northwest Queensland, Australia. *Geology*, **25**, 1095–1098.

MCFARLANE, C. R. M. 2006. Palaeoproterozoic evolution of the Challenger Au Deposit, South Australia, from monazite geochronology. *Journal of Metamorphic Geology*, **24**, 75–87.

MCLAREN, S., SANDIFORD, M. & POWELL, R. 2005. Contrasting styles of proterozoic crustal evolution: a hot-plate tectonic model for Australian terranes. *Geology*, **33**, 673–676.

MELNYK, M., DAVIS, D. W., CRUDEN, A. R. & STERN, R. A. 2006. U–Pb ages constraining structural development of an Archaean terrane boundary in the Lake of the Woods area, western Superior Province, Canada. *Canadian Journal of Earth Sciences*, **43**, 967–993.

MÉNOT, R. P., PÊCHER, A., ROLLAND, Y., PEUCAT, J. J., PELLETIER, A., DUCLAUX, G. & GUILLOT, S. 2005. Structural setting of the Neoarchean Terrains in the Commonwealth Bay Area (143–145 °E), Terre Adélie Craton, East Antarctica. *Gondwana Research*, **8**, 1–9.

MIKHALSKY, E. V., BELIATSKY, B. V., SHERATON, J. W. & ROLAND, N. W. 2006. Two distinct precambrian terranes in the Southern Prince Charles Mountains, East Antarctica: SHRIMP dating and geochemical constraints. *Gondwana Research*, **9**, 291–309.

MÖLLER, A., HENSEN, B. J., ARMSTRONG, R. A., MEZGER, K. & BALLEVRE, M. 2003. U–Pb zircon and monazite age constraints on granulite-facies metamorphism and deformation in the Strangways Metamorphic Complex (central Australia). *Contributions to Mineralogy and Petrology*, **145**, 406–423.

MONNIER, O. 1995. Le socle protérozoïque de Terre Adélie (Antarctique Est): son evolution tectono-métamorphique et sa place dans les reconstitutions du Proto-Gondwana. Ph.D. thesis, Universite Saint Etienne, France.

MOORES, E. M. 1991. Southwest U.S.-East Antarctica (SWEAT) connection: a hypothesis. *Geology*, **19**, 425–428.

MUELLER, P. A., HEATHERINGTON, A. L., KELLY, D. M., WOODEN, J. L. & MOGK, D. W. 2002. Palaeoproterozoic crust within the Great Falls tectonic zone: implications for the assembly of southern Laurentia. *Geology*, **30**, 127–130.

NELSON, D. R., MYERS, J. S. & NUTMAN, A. P. 1995. Chronology and evolution of the Middle Proterozoic Albany-Fraser Orogen, Western Australia. *Australian Journal of Earth Sciences*, **42**, 481–495.

NEUMANN, N. L., SOUTHGATE, P. N., GIBSON, G. M. & MCINTYRE, A. 2006. New SHRIMP geochronology for the Western Fold Belt of the Mt Isa Inlier: developing a 1800–1650 Ma event framework. *Australian Journal of Earth Sciences*, **53**, 1023–1039.

NUTMAN, A. P., CHADWICK, B., RAMAKRISHNAN, M. & VISWANATHA, M. N. 1992. Shrimp U–Pb Ages of Detrital Zircon in Sargur Supracrustal Rocks in Western Karnataka, Southern India. *Journal of the Geological Society of India*, **39**, 367–374.

NUTMAN, A. P., CHADWICK, B., RAO, B. K. & VASUDEV, V. N. 1996. SHRIMP U/Pb zircon ages of acid volcanic rocks in the Chitradurga and Sandur groups, and granites adjacent to the Sahdur schist belt, Karnataka. *Journal of the Geological Society of India*, **47**, 153–164.

O'DEA, M. G., LISTER, G. S., BETTS, P. G. & POUND, K. S. 1997. A shortened intraplate rift system in the Proterozoic Mount Isa terrain, NW Queensland, Australia. *Tectonics*, **16**, 425–441.

OLIVER, R. L. & FANNING, C. M. 1996. 'Mawson Block' metamorphism; a preliminary synthesis. *In*: BUICK, I. S. & CARTWRIGHT, I. (eds) *Evolution of Metamorphic Belts*. Geological Society of Australia, Abstracts 42.

OLIVER, R. L. & FANNING, C. M. 1997. Australia and Antarctica; precise correlation of Palaeoproterozoic terrains. *In*: RICCI, C. A. (ed.) *The Antarctic Region; Geological Evolution and Processes; Proceedings of the VII International Symposium on Antarctic Earth Sciences*. International Symposium on Antarctic Earth Sciences. Terra Antarctica Publication, Siena, Italy, 163–172.

PAGE, R. W. & WILLIAMS, I. S. 1988. Age of the Barramundi Orogeny in Northern Australia by means of ion microprobe and conventional U–Pb zircon studies. *Precambrian Research*, **40–1**, 21–36.

PAGE, R. W. & SUN, S. S. 1998. Aspects of geochronology and crustal evolution in the Eastern Fold Belt, Mt Isa Inlier. *Australian Journal of Earth Sciences*, **45**, 343–361.

PAGE, R. W., CONOR, C. H. H., STEVENS, B. P. J., GIBSON, G. M., PREISS, W. V. & SOUTHGATE, P. N. 2005. Correlation of Olary and Broken Hill Domains, Curnamona Province: possible relationship to Mount Isa and other North Australian Pb–Zn–Ag-bearing successions. *Economic Geology*, **100**, 663–676.

PARKER, A. J. 1983. *Cowell–1:50 000 geology map sheet*. Department of Mines and Energy, South Australia, Adelaide.

PARKER, A. J. 1993. Palaeoproterozoic. *In*: DREXEL, J. F., PREISS, W. V. & PARKER, A. J. (eds) *The Geology of*

South Australia – the Precambrian, Bulletin 54. Mines and Energy, South Australia, Adelaide.

PARKS, J., LIN, S. F., DAVIS, D. & CORKERY, T. 2006. New high-precision U–Pb ages for the Island Lake greenstone belt, northwestern Superior Province: implications for regional stratigraphy and the extent of the North Caribou terrane. *Canadian Journal of Earth Sciences*, **43**, 789–803.

PAWLEY, M. J., VAN KRANENDONK, M. J. & COLLINS, W. J. 2004. Interplay between deformation and magmatism during doming of the Archaean Shaw Granitoid Complex, Pilbara Craton, Western Australia. *Precambrian Research*, **131**, 213–230.

PAYNE, J. L., BAROVICH, K. & HAND, M. 2006. Provenance of metasedimentary rocks in the northern Gawler Craton, Australia: implications for palaeoproterozoic reconstructions. *Precambrian Research*, **148**, 275–291.

PAYNE, J. L. 2008. Palaeo- to Mesoproterozoic evolution of the Gawler Craton: isotopic, geochronological and geochemical constraints. Ph.D. thesis, University of Adelaide.

PAYNE, J. L., HAND, M., BAROVICH, K. & WADE, B. P. 2008. Temporal constraints on the timing of high-grade metamorphism in the northern Gawler Craton: implications for construction of the Australian Proterozoic. *Australian Journal of Earth Sciences*, **55**, 623–640.

PELLETIER, A., GAPAIS, D., MENOT, R. P. & PEUCAT, J. J. 2002. Tectonique transpressive en terre Adélie au Palaéoprotérozoïque (Est Antarctique). *Comptes Rendus Geoscience*, **334**, 505–511.

PERCIVAL, J. A., MCNICOLL, V. & BAILES, A. H. 2006. Strike-slip juxtaposition of c. 2.72 Ga juvenile arc and >2.98 Ga continent margin sequences and its implications for Archaean terrane accretion, western Superior Province, Canada. *Canadian Journal of Earth Sciences*, **43**, 895–927.

PEUCAT, J. J., MAHABALESWAR, B. & JAYANANDA, M. 1993. Age of Younger Tonalitic Magmatism and Granulitic Metamorphism in the South Indian Transition Zone (Krishnagiri Area) – Comparison with older peninsular gneisses from the Gorur–Hassan Area. *Journal of Metamorphic Geology*, **11**, 879–888.

PEUCAT, J. J., BOUHALLIER, H., FANNING, C. M. & JAYANANDA, M. 1995. Age of the Holenarsipur Greenstone-Belt: relationships with the surrounding gneisses (Karnataka, South-India). *Journal of Geology*, **103**, 701–710.

PEUCAT, J. J., MÉNOT, R. P., MONNIER, O. & FANNING, C. M. 1999. The Terre Adélie basement in the East-Antarctica Shield: geological and isotopic evidence for a major 1.7 Ga thermal event; comparison with the Gawler Craton in South Australia. *Precambrian Research*, **94**, 205–224.

PEUCAT, J. J., CAPDEVILA, R., FANNING, C. M., MÉNOT, R. P., PECORA, L. & TESTUT, L. 2002. 1.6 Ga felsic volcanic blocks in the moraines of the Terre Adélie craton, Antarctica: comparisons with the Gawler Range volcanics, South Australia. *Australian Journal of Earth Sciences*, **49**, 831–845.

PHILLIPS, G., WILSON, C. J. L., CAMPBELL, I. H. & ALLEN, C. M. 2006. U–Th–Pb detrital zircon geochronology from the southern Prince Charles Mountains, East Antarctica – defining the Archaean to Neoproterozoic Ruker Province. *Precambrian Research*, **148**, 292–306.

PISAREVSKY, S. A. & NATAPOV, L. M. 2003. Siberia and Rodinia. *Tectonophysics*, **375**, 221–245.

PISAREVSKY, S. A., WINGATE, M. T. D. & HARRIS, L. B. 2003. Late Mesoproterozoic (c. 1.2 Ga) palaeomagnetism of the Albany–Fraser orogen: no pre-Rodinia Australia–Laurentia connection. *Geophysical Journal International*, **155**, F6–F11.

PISAREVSKY, S. A., NATAPOV, L. M., DONSKAYA, T. V., GLADKOCHUB, D. P. & VERNIKOVSKY, V. A. 2008. Proterozoic Siberia: a promontory of Rodinia. *Precambrian Research*, **160**, 66–76.

POST, N. J. 2001. Unravelling Gondwana fragments: an integrated structural, isotopic and petrographic investigation of the Windmill Islands, Antarctica. Ph.D. thesis.

POTMA, W. A. & BETTS, P. G. 2006. Extension-related structures in the Mitakoodi Culmination: impications for the nature and timing of extension, and effect on later shortening in the eastern Mt Isa Inlier. *Australian Journal of Earth Sciences*, **53**, 55–67.

RAINBIRD, R. H., STERN, R. A., KHUDOLEY, A. K., KROPACHEV, A. P., HEAMAN, L. M. & SUKHORUKOV, V. I. 1998. U–Pb geochronology of Riphean sandstone and gabbro from southeast Siberia and its bearing on the Laurentia-Siberia connection. *Earth and Planetary Science Letters*, **164**, 409–420.

RAYNER, N. M., STERN, R. A. & BICKFORD, M. E. 2005. Tectonic implications of new SHRIMP and TIMS U–Pb geochronology of rocks from the Sask Craton, Peter Lake Domain, and Hearne margin, Trans-Hudson Orogen, Saskatchewan. *Canadian Journal of Earth Sciences*, **42**, 635–657.

REID, A., HAND, M., JAGODZINSKI, E. A., KELSEY, D. & PEARSON, N. J. 2008. Palaeoproterozoic orogenesis within the southeastern Gawler Craton, South Australia. *Australian Journal of Earth Sciences*, **55**, 449–471.

ROGERS, J. J. W. & SANTOSH, M. 2003. Supercontinents in earth history. *Gondwana Research*, **6**, 357–368.

SCHUBERT, W. & WILL, T. 1994. Granulite-facies rocks of the Shackleton Range, Antarctica – Conditions of formation and preliminary petrogenetic implications. *Chemie Der Erde-Geochemistry*, **54**, 355–371.

SCHULTZ, M. F. J., CHACKO, T., HEAMAN, L. M., SANDEMAN, H. A., SIMONETTI, A. & CREASER, R. A. 2007. Queen Maud block: a newly recognized palaeoproterozoic (2.4–2.5 Ga) terrane in northwest Laurentia. *Geology*, **35**, 707–710.

SCHWARZ, M., BAROVICH, K. M. & HAND, M. 2002. A plate margin setting for the evolution of the southern Gawler Craton; evidence from detrital zircon and Sm–Nd isotopic data of the Hutchison Group. *In*: PREISS, W. V. (ed.) *Geoscience 2002: Expanding Horizons*. Abstracts vol. 67, Geological Society of Australia, Adelaide.

SCOTT, D. L., RAWLINGS, D. J., PAGE, R. W., TARLOWSKI, C. Z., IDNURM, M., JACKSON, M. J. & SOUTHGATE, P. N. 2000. Basement framework and geodynamic evolution of the Palaeoproterozoic superbasins of north-central Australia: an integrated review of geochemical, geochronological and geophysical data. *Australian Journal of Earth Sciences*, **47**, 341–380.

SCRIMGEOUR, I. 2003. *Developing a Revised Framework for the Arunta Region*, Annual Geoscience Exploration Seminar (AGES) 2003. Northern Territory Geological Survey, 2003–2001, Alice Springs, Australia.

SCRIMGEOUR, I. & RAITH, J. G. 2002. A sapphirine-phlogopite-cordierite paragenesis in a low-P amphibolite facies terrain, Arunta Inlier, Australia. *Mineralogy and Petrology*, **75**, 123–130.

SCRIMGEOUR, I., SMITH, J. B. & RAITH, J. G. 2001. Palaeoproterozoic high-T, low-P metamorphism and dehydration melting in metapelites from the Mopunga Range, Arunta Inlier, central Australia. *Journal of Metamorphic Geology*, **19**, 739–757.

SCRIMGEOUR, I. R., KINNY, P. D., CLOSE, D. F. & EDGOOSE, C. J. 2005. High-T granulites and polymetamorphism in the southern Arunta Region, central Australia: evidence for a 1.64 Ga accretional event. *Precambrian Research*, **142**, 1–27.

SEARS, J. W., CHAMBERLAIN, K. R. & BUCKLEY, S. N. 1998. Structural and U–Pb geochronologic evidence for 1.47 Ga rifting in the Belt basin, western Montana. *Canadian Journal of Earth Sciences*, **35**, 467–475.

SELWAY, K. 2007. *Magnetotelluric Experiments in Central and Southern Australia and their Implications for Tectonic Evolution*. Ph.D. thesis, University of Adelaide.

SHAW, R. D., WARREN, R. G., SENIOR, B. R. & YEATES, A. N. 1975. *Geology of the Alcoota 1:250 000 sheet area, N. T.*, Bureau of Mineral Resources, Geology and Geophysics, Canberra, Record 1975/100.

SHEPPARD, S., TYLER, I. M., GRIFFIN, T. J. & TAYLOR, W. R. 1999. Palaeoproterozoic subduction-related and passive margin baslts in the Halls Creek Orogen, northwest Australia. *Australian Journal of Earth Sciences*, **46**, 679–690.

SHEPPARD, S., GRIFFIN, T. J., TYLER, I. M. & PAGE, R. W. 2001. High- and low-K granites and adakites at a Palaeoproterozoic plate boundary in northwestern Australia. *Journal of the Geological Society, London*, **158**, 547–560.

SHERATON, J. W., BLACK, L. P. & TINDLE, A. G. 1992. Petrogenesis of plutonic rocks in a Proterozoic granulite-facies terrane – the Bunger Hills, East Antarctica. *Chemical Geology*, **97**, 163–198.

SKIRROW, R., FAIRCLOUGH, M. ET AL. 2006. *Iron Oxide Cu–Au (-U) Potential Map of the Gawler Craton, South Australia*. (1st edn) 1:500 000 scale. Geoscience Australia, Canberra, ISBN: 1 920871 76 4.

SMETHURST, M. A., KHRAMOV, A. N. & TORSVIK, T. H. 1998. The Neoproterozoic and Palaeozoic palaeomagnetic data for the Siberian platform: from Rodinia to Pangea. *Earth Science Reviews*, **43**, 1–24.

SMITHIES, R. H. & BAGAS, L. 1997. High pressure amphibolite-granulite facies metamorphism in the Paleoproterozoic Rudall Complex, central Western Australia. *Precambrian Research*, **83**, 243–265.

ST-ONGE, M. R., SEARLE, M. P. & WODICKA, N. 2006. Trans-Hudson Orogen of North America and Himalaya–Karakoram–Tibetan Orogen of Asia: structural and thermal characteristics of the lower and upper plates. *Tectonics*, **25**, TC4006, doi: 10.1029/2005TC001907.

STAGG, H. M. J., COLWELL, J. B. ET AL. 2005. *Geological Framework of the Continental Margin in the Region of the Australian Antarctic Territory*. Geoscience Australia Record, 2004/25.

STEVENS, B. P. J., PAGE, R. W. & CROOKS, A. 2008. Geochronology of Willyama Supergroup metavolcanics, metasediments and contemporaneous intrusions, Broken Hill, Australia. *Australian Journal of Earth Sciences*, **55**, 301–330.

STOTT, G. M. 1997. The Superior Province, Canada. *In*: DE WIT, M. J. & ASHWAL, L. D. (eds) *Greenstone Belts*. Oxford University Press, Oxford, 480–507.

SWAIN, G., BAROVICH, K., HAND, M. & FERRIS, G. J. 2005a. Proterozoic magmatic arcs and oroclines: St Peter Suite, Gawler Craton, SA. *In*: WINGATE, M. D. & PISAREVSKY, S. A. (eds) *Supercontinents and Earth Evolution Symposium*. Geological Society of Australia Inc. Abstracts 81.

SWAIN, G., WOODHOUSE, A., HAND, M., BAROVICH, K., SCHWARZ, M. & FANNING, C. M. 2005b. Provenance and tectonic development of the late Archaean Gawler Craton, Australia: U–Pb zircon, geochemical and Sm-Nd isotopic implications. *Precambrian Research*, **141**, 106–136.

SWAIN, G., BAROVICH, K., HAND, M., FERRIS, G. J. & SCHWARZ, M. 2008. Petrogenesis of the St Peter Suite, southern Australia: Arc magmatism and Proterozoic crustal growth of the South Australian Craton. *Precambrian Research*, **166**, 283–296.

SZPUNAR, M., HAND, M. & BAROVICH, K. M. 2006. Insights into age and provenance of the Palaeoproterozoic Hutchison Group, southern Gawler Craton, South Australia. *In*: REID, A. (ed.) *Primary Industries and Resources SA, University of Adelaide and Monash University, ARC Linkage Program*: 2006 Reporting day. Primary Industries and Resources South Australia, RB 2007/01.

TALARICO, F. & KRONER, U. 1999. Geology and tectonometamorphic evolution of the Read Group, Shackleton Range: a part of the Antarctic Craton. *Terra Antarctica*, **6**, 183–202.

TEASDALE, J. 1997. *Methods for Understanding Poorly Exposed Terranes: the Interpretive Geology and Tectonothermal Evolution of the Western Gawler Craton*. Ph.D. thesis, University of Adelaide, Adelaide.

THERIAULT, R. J., ST-ONGE, M. R. & SCOTT, D. J. 2001. Nd isotopic and geochemical signature of the paleoproterozoic Trans-Hudson Orogen, southern Baffin Island, Canada: implications for the evolution of eastern Laurentia. *Precambrian Research*, **108**, 113–138.

THORKELSON, D. J., MORTENSEN, J. K., CREASER, R. A., DAVIDSON, G. J. & ABBOTT, J. G. 2001. Early Proterozoic magmatism in Yukon, Canada: constraints on the evolution of northwestern Laurentia. *Canadian Journal of Earth Sciences*, **38**, 1479–1494.

THORKELSON, D. J., ABBOTT, J. A., MORTENSEN, J. K., CREASER, R. A., VILLENEUVE, M. E., MCNICOLL, V. J. & LAYER, P. W. 2005. Early and Middle Proterozoic evolution of Yukon, Canada. *Canadian Journal of Earth Sciences*, **42**, 1045–1071.

TOMKINS, A. G. & MAVROGENES, J. A. 2002. Mobilization of gold as a polymetallic melt during pelite anatexis at the Challenger deposit, South Australia: a metamorphosed Archean gold deposit. *Economic Geology*, **97**, 1249–1271.

TOMLINSON, K. Y., STOTT, G. M., PERCIVAL, J. A. & STONE, D. 2004. Basement terrane correlations and crustal recycling in the western Superior Province: Nd isotopic character of granitoid and felsic volcanic rocks in the Wabigon subprovince, N. Ontario, Canada. *Precambrian Research*, **132**, 245–274.

TONG, L., WILSON, C. J. L. & VASSALLO, J. J. 2004. Metamorphic evolution and reworking of the Sleaford Complex metapelites in the southern Eyre Peninsula, South Australia. *Australian Journal of Earth Sciences*, **51**, 571–589.

TRENDALL, A. F., COMPSTON, W., NELSON, D. R., DE LAETER, J. R. & BENNETT, V. C. 2004. SHRIMP zircon ages constraining the depositional chronology of the Hamersley Group, Western Australia. *Australian Journal of Earth Sciences*, **51**, 621–644.

VAN BREEMEN, O., HARPER, C. T., BERMAN, R. G. & WODICKA, N. 2007. Crustal evolution and Neoarchaean assembly of the central-southern Hearne domains: evidence from U–Pb geochronology and Sm–Nd isotopes of the Phelps Lake area, northeastern Saskatchewan. *Precambrian Research*, **159**, 33–59.

VASSALLO, J. J. 2001. *Tectonic Investigation in the Gawler Craton*. Ph.D. thesis, University of Melbourne, Melbourne.

VASSALLO, J. J. & WILSON, C. J. L. 2001. Structural repetition of the Hutchison Group metasediments, Eyre Peninsula, South Australia. *Australian Journal of Earth Sciences*, **48**, 331–345.

VASSALLO, J. J. & WILSON, C. J. L. 2002. Palaeoproterozoic regional-scale non-coaxial deformation; an example from eastern Eyre Peninsula, South Australia. *Journal of Structural Geology*, **24**, 1–24.

VASUDEV, V. N., CHADWICK, B., NUTMAN, A. P. & HEGDE, G. V. 2000. Rapid development of the Late Archaean Hutti schist belt, northern Karnataka: implications of new field data and SHRIMP U/Pb zircon ages. *Journal of the Geological Society of India*, **55**, 529–540.

WADE, B. P., BAROVICH, K. M., HAND, M., SCRIMGEOUR, I. R. & CLOSE, D. F. 2006. Evidence for early Mesoproterozoic arc magmatism in the Musgrave Block, central Australia: implications for Proterozoic crustal growth and tectonic reconstructions of Australia. *Journal of Geology*, **114**, 43–63.

WADE, B. P., PAYNE, J. L., BAROVICH, K. M., HAND, M. & REID, A. 2007. Petrogenesis of the 1.5 Ga magmatism in the Coompana Block–Filling the magmatic gap in Proterozoic Australia. *Australian Journal of Earth Sciences*, **54**, 1089–1102.

WADE, B. P., HAND, M., MAIDMENT, D. W., CLOSE, D. F. & SCRIMGEOUR, I. R. 2008. Origin of metasedimentary and igneous rocks from the Entia Dome, eastern Arunta Region, central Australia: a U–Pb LA-ICPMS, SHRIMP, and Sm–Nd isotope study. *Australian Journal of Earth Sciences*, **166**, 370–386.

WARREN, R. G. & HENSEN, B. J. 1989. The P–T evolution of the Proterozoic Arunta Block, central Australia, and implications for tectonic evolution. *In*: DALY, J. S., CLIFF, R. A. & YARDLEY, B. W. D. (eds) *Evolution of Metamorphic Belts*. Geological Society, London, Special Publications, **43**, 349–355.

WEBB, A. W., THOMSON, B. P., BLISSETT, A. H., DALY, S. J., FLINT, R. B. & PARKER, A. J. 1986. Geochronology of the Gawler Craton, South-Australia. *Australian Journal of Earth Sciences*, **33**, 119–143.

WHITE, R. W., POWELL, R. & CLARKE, G. L. 2003. Prograde metamorphic assemblage evolution during partial melting of metasedimentary rocks at low pressures: migmatites from Mt Stafford, central Australia. *Journal of Petrology*, **44**, 1937–1960.

WHITMEYER, S. J. & KARLSTROM, K. E. 2007. Tectonic model for the Proterozoic growth of North America. *Geosphere*, **3**, 220–259.

WILSON, C. J. L. & POWELL, R. 2001. Strain localisation and high-grade metamorphism at Broken Hill, Australia: a view from the Southern Cross area. *Tectonophysics*, **335**, 193–210.

WINGATE, M. D. & EVANS, D. A. D. 2003. Palaeomagnetic constraints on the Proterozoic tectonic evolution of Australia. *In*: YOSHIDA, M. & WINDLEY, B. F. (eds) *Proterozoic East Gondwana: Supercontinent Assembly and Breakup*. Geological Society, London, Special Publications, **206**, 77–91.

WINGATE, M. D. & GIDDINGS, J. W. 2000. Age and palaeomagnetism of the Mundine Well dyke swarm, Western Australia: implications for an Australia-Laurentia connection at 755 Ma. *Precambrian Research*, **100**, 335–357.

WINGATE, M. D., PISAREVSKY, S. A. & EVANS, D. A. D. 2002a. New palaeomagnetic constraints on Rodinia connections between Australia and Laurentia, Geological Society of America, 2002 Annual Meeting. Abstracts with programs – Geological Society of America, 34, 6, Boulder.

WINGATE, M. T. D., PISAREVSKY, S. A. & EVANS, D. A. D. 2002b. Rodinia connection between Australia and Laurentia; no SWEAT, no AUSWUS? *Terra Nova*, **14**, 121–128.

WORDEN, K. E., CLAOUE-LONG, J. C., SCRIMGEOUR, I. & DOYLE, N. J. 2006. Summary of Results. Joint NTGS-GA Geochronology Project: Pine Creek Orogen and Arunta Region, January–June 2004. Northern Territory Geological Survey, Record 2006-2005.

WORDEN, K. E., CARSON, C. J., SCRIMGEOUR, I. R., LALLY, J. H. & DOYLE, N. J. 2008. A revised Palaeoproterozoic chronostratigraphy for the Pine Creek Orogen, northern Australia: evidence from SHRIMP U–Pb zircon geochronology. *Precambrian Research*, **166**, 122–144.

WYGRALAK, A. S., MERNAGH, T. P., HUSTON, D. L., AHMAD, M. & GOLD, R. 2005. *Gold Mineral System of the Tanami Region*. Northern Territory Geological Survey, Report 18, Darwin.

ZEH, A., MILLAR, I. L., KRONER, U. & GORZ, I. 1999. The structural and metamorphic evolution of the Northern Haskard Highlands, Shackleton Range, Antarctica. *Terra Antarctica*, **6**, 249–268.

ZEH, A., MILLAR, I. L. & HORSTWOOD, M. S. A. 2004. Polymetamorphism in the NE Shackleton Range, Antarctica: constraints from petrology and U–Pb, Sm–Nd, Rb–Sr TIMS and *in situ* U–Pb LA-PIMMS dating. *Journal of Petrology*, **45**, 949–973.

Zhao, G. C., Cawood, P. A., Wilde, S. A. & Sun, M. 2002. Review of global 2.1–1.8 Ga orogens: implications for a pre-Rodinia supercontinent. *Earth Science Reviews*, **59**, 125–162.

Zhao, G. C., Sun, M., Wilde, S. A. & Li, S. Z. 2004. A Palaeo-Mesoproterozoic supercontinent: assembly, growth and breakup. *Earth Science Reviews*, **67**, 91–123.

Zhao, G. C., Cao, L., Wilde, S. A., Sun, M., Choe, W. J. & Li, S. Z. 2006a. Implications based on the first SHRIMP U–Pb zircon dating on Precambrian granitoid rocks in North Korea. *Earth and Planetary Science Letters*, **251**, 365–379.

Zhao, G. C., Sun, M., Wilde, S. A., Li, S. Z., Liu, S. W. & Zhang, H. 2006b. Composite nature of the North China granulite-facies belt: tectonothermal and geochronological constraints. *Gondwana Research*, **9**, 337–348.

Zhao, J. X. & McCulloch, M. T. 1995. Geochemical and Nd isotopic systematics of granites from the Arunta Inlier, central Australia; implications for Proterozoic crustal evolution. *Precambrian Research*, **71**, 265–299.

Zulbati, F. & Harley, S. L. 2007. Late Archaean granulite facies metamorphism in the Vestfold Hills, East Antarctica. *Lithos*, **93**, 39–67.

Index

Page numbers in *italic* denote figures. Page numbers in **bold** denote tables.

Achab Gneiss 222, **224**, 228
Adélie Craton *see* Terre Adélie Craton; Gawler-Adélie Craton
Africa, southern, IGCP 509 database system
    time-slice maps 44, *43*
    time–space correlation chart 37–44, *40*, *41*, *42*
Akitkan Formation 130
    Nd isotope data **134–135**, 139–140
Akitkan Group
    geochronology 148, 149, 153, 155–156
    geology 146, *147*, 148
    palaeomagnetism and U–Pb dates 145–161
Akitkan orogenic belt *128*, 129, 130, 146, *147*
Aldan superterrane 66, 127, *128*, 146, *147*
Almora–Askot–Dhramgarh gneiss 284, **285**
Amazonian craton 66
Anabar superterrane 127, *128*, 129, 146, *147*
Angara orogenic belt *128*, 129
anisotropy, magnetic susceptibility, Salla Diabase Dyke 205
Antarctica, Mawson Continent 321, 325–328
Areachap Terrane 221, *226*
    granitoids, Sm–Nd data **223**, 225–226
Arenópolis Arc 256
Aroams Gneiss 222, **224**
Arunta Region *320*, 321, 333, *334–335*, 341, 342
atmosphere, evolution 9–11

baddeleyite, U–Pb dating, Baltic Shield intrusions 172, 173–189
Baikal Group 146
Baikal terrane *128*, 129, 130
    crustal growth 140
    Nd isotope data **133–135**, *136*, *137*, 139–140
baked-contact test, Salla Diabase Dyke 205, 208–209
Baltic Shield 199, *200*
    geology *166*, 167
    ore-bearing intrusions 165–196
Baltica
    apparent polar wander path 210, 214

palaeomagnetic data 210, **211**
virtual geomagnetic pole 209, 210
Baltica–Laurentia reconstruction 199, 209–210, 212–214, *331*
Bandal granitoid 283, **285**, 286, 296
    SHRIMP U–Pb dating 289–290, **292**
banded iron formation 4, 8, 9–10
    Algoma-type 10
    Superior-type 8, 10
Banks Vlei Gneiss 222, **224**
Bao'an, metamorphism 57
Barramundi Orogeny 330
Beenbreek Gneiss 222, **223**
Bhatwari Gneiss 283, **285**
Birimian Belt, correlation
    Campinorte sequence 265–266
    North Hebei Orogen 64–65
Birusa terrane *128*, 129
    crustal growth 140
    Nd isotope data 131, **132**, *136–137*
Black Angel Gneiss 336
Black Reef Formation 37
bolides 14
Bomdila Granite Gneiss 284, **285**
Borborema Province, northwest
    crustal growth 271–279
    geology 272–275
    Sm–Nd isotope analysis 275, **277**, 279
    U–Pb dating 275–279
Boromo greenstone belt 234, *235*
Brasília Belt, geology 255–256, *256*
Bunger Hills *320*, *327*, 329
Burtons Puts Gneiss 222, **224**
Bushmanland Terrane 221, *226*
    granitoids, Sm–Nd data 222, **223**, *225*, 228
Bushveld igneous event 40

calcium carbonate, sea-floor 10, 11
Campinorte sequence 255–267, *256*
    correlation, Birimian and Transamazonian belts 265–266
    geology 255–258, *256*, *257*
    metasedimentary rocks 256, *257*, 259, 263
    plutonic unit 256–257, *257*, 259, 262, 263
    Sm–Nd isotope analysis 259, 263, **264**, *265*, 266–267
    U–Pb dating 258–262, **263**, *265*, 266–267
'Canfield Ocean' model 10
cap carbonate 13
Cape Hunter Phyllite 325

Capricornia 6
    see also Columbia Supercontinent
carbon isotope anomaly 10–11
CAT Suite magmatism 321, 341
Ceará Central domain 271, *272*, 279
Central Orogenic Belt, North China Craton **52**, 53, 57
Chaibasa Formation 301–316, *302*
    deformation mechanisms 313–316
Chail Nappe 287
Chail Thrust 287
Chailli gneiss 283, **285**
Chandil Formation *302*
Chang Cheng Series 59, *60*, 61
Changzhougou Formation 59
chaotic layers, Chaibasa Formation *305*, 312–313
Chaura Thrust 288
Chaya palaeopole 159
Chaya suite 146, *147*, 148
    geochronology and palaeomagnetism 149, **150**, 151–153, **154**, *155*
Chieress dykes 146
Chirpatiya gneiss 284, **285**
Chor Thrust 287
chromite 179, 191
Chuanlinggou Formation 59, 61
Chunatundra intrusion *180*, 183, **187**, *192*
Chuya granite complex *128*
    crustal growth 140
    Nd isotope data **135**, *137*, 139–140
Chuya Group 130
climate, evolution 11–12
CO_2 fluid inclusions, North China Craton 58–59
Coboop Granite Gneiss 222, **223**
Columbia Supercontinent 6, 7, 49, **50**, 51, 73
    correlation, North China Craton **50**, 51, 63–66
    Greater India 284, 296
    reconstruction *50*
Concordia Granite 222, **224**
cones, contorted laminae, Chaibasa Formation *304*, *305*, 309–310
Congo Craton 274, 279
continents, emergence of 6
convection, mantle 3–4
Coompana Block 322, 329, 344
core, evolution 2–3
Cornian Orogeny 322, 338

INDEX

crust, continental
  growth 5–6
    northwest Borborema
      Province 271–279
    Siberian craton, Sm–Nd isotope
      analysis 130–140

Dahongyu Formation 61
Dalma Lava 301, *302*
Daqingshan, khondalite belt
  geological background 75–76
  tectono-thermal events 73–94
Daqingshan-Wulashan terrane 54
Darjeeling–Sikkim Granite Gneiss
  284, **285**
DateView database 28, 33–35, 43
deformation
  Lüliang Massif 103, 105–109,
    122
  Man-Leo Shield 234
  Mawson Continent 339, 341, 345
  North China Craton 51, 53, 54, 57
  soft sediment, Chaibasa
    Formation 302, 303, *304*,
    305, 306–316
    metadepositional structures
      311–313
    syndepositional structures
      308–311
depressions, infilled, Chaibasa
  Formation 310
Dhalbhum Formation 301, *302*
Dhanjori Formation 301, *302*
Donington Suite 338
Dunite Block 179, *180*, **187**, **194**
dykes, mafic
  Baikal terrane 129, 130, 140
  North China Craton 56, 57, 59,
    64, 66
  Salla Diabase Dyke 199–214
  Superia 8

Earth, cooling 3–4
earthquakes, deformation 307–308,
  313–316
East Scandinavian Large Igneous
  Province, magma 191, 196
Eastern Block, North China Craton
  52, 53
Eburnian Belt, correlation with
  Northern Hebei orogen
  64–65
Elim Group 37
enderbite, Médio Coreaú domain 274
Epupa Complex 65
eukaryotes, fossil record 12
evaporites 11

faults
  penecontemporaneous, Chaibasa
    Formation *304*, 311
  thrust, Lüliang Massif 105
Fedorov Block 188, *189*
  mineralization 191

Fedorovo-Pansky Layered Complex
  165, *166*
  geology and petrology 167–170
  magmatism 186, 188–189
  mineralization **194**
  Sm–Nd analysis 186, **187**, 188,
    193, *194*
  U–Pb dating *174*, 175, 177,
    178, 179, **187**
Fenno-Karelian Belt 165
  geochronology **187**
  magmatism 189, 195–196
  mineralization 191
Fennoscandian Shield *see* Baltic
  Shield
folding, Lüliang Massif 105, 109, *110*
foliation
  Lüliang Massif 105, *106*
  Man-Leo Shield 234
Forward Orogeny 343
Friersdale Charnockites 222, **223**
Fuping Block 99, *104*, 120, 122
Fuping Massif, correlation with
  Lüliang Massif 120–122

Gaoyuzhuang Formation 61
Garies terrane 222, *226*, 227
Gawler Craton 319, 321, 322–325,
  323, *326*, 344
  accretion to North Australian
    Craton 332–333
  internal architecture 333, 336
  Mawson Continent assemblage
    328–329
  proto-Gawler Craton 332–333
  see also Gawler–Adélie Craton
Gawler Range Volcanics 325, 336
Gawler–Adélie Craton, evolution
  329, 333, *334–335*, 336, *337*,
  342, 343, 344–345
geodynamo 2–3
geomagnetic field, palaeointensity 2
geomagnetic poles, virtual 209, 210,
  212, **213**
Georgetown Inlier *320*, 343–344
  deformation 345
Ghuttu Gneiss 283, **285**
glaciation and oxygenation 13–14
Goiás Magmatic Arc 256
Goiás Massif 256–257
gold
  hydrothermal 12
  orogenic 9
Goloustnaya terrane *128*, 129, 130
  crustal growth 140
  Nd isotope data **133**, *137*,
    138–139
Gondwana, Mawson continent
  320, 322
granite
  Lüliang Group 103
  Northern Hebei orogen 57
    correlation 65–66
  Siberian craton 129–130, 146

granitoids
  Man-Leo Shield 234, *235*, 236,
    241, **243**, 244, *245*,
    246, *247*
  Namaqua–Natal Province
    219–228
  pre-Himalayan 283
Granja Complex **276**, **277**, 279
Granja Massif 271, 274
Granjeiro Complex 272
granulite, North China Craton
  55–58
Great Oxidation Event 9, 10
Greater India 284, 285, 293
greenstone belts, Man-Leo Shield
  232–233, *235*, 236–241
Grenvilleland 6
Gromia 12
Grypania spiralis 12
Guanghua Group 61–62
Guiana Shield, correlation,
  Campinorte sequence
  265–266
Gwalda gneiss 283, **285**

Hanuman Chatti Gneiss 283, **285**
Hengshan Massif 54, 57
  correlation with Lüliang Massif
    120–122
Hidrolina Dome 257
Higher Himalayan Crystalline
  Belt 283, **285**, *287*,
  288–289, 296
Higher Himalayan Leucogranite
  Belt 283
Hiltaba Suite 325
Himalaya, pre-Himalayan history
  283, 284
Himalaya, northwest 283–296
  geology *284*, 285–286
  SHRIMP U–Pb dating
    289–296
Himalayan Metamorphic Belt
  284, 285
Himchal Pradesh, anorogenic
  magmatic signatures
  286–296
Horodyskia 12
Hudsonland 6
Huijiazhuang granite 103
Hutuo Group, correlation with
  Yejishan Group 120, 122
hysteresis, Salla Diabase Dyke 204

ice ages 9, 13
IGCP 509 database system 27–45
  functionality 32–33
  future development 44, 45
  outputs 33–37
  time-slice maps, southern Africa
    44, *43*
  time–space correlation chart,
    southern Africa 37–44,
    40, *41*, *42*

INDEX

Imandra Lopolith *166*
　geology and petrology 171–172
　mineralization 189
　Sm–Nd analysis **187**, **193**, *194*
　U–Pb dating *174*, 183–184, **185**, 186, **187**
Indian Plate 285–286
　collision with Eurasian Plate 291
Inner Mongolia-North Hebei Orogenic Belt *see* North Hebei Orogenic Belt
Irkut-Kitoy domain *128*, 129
　Nd isotope data **133**, *136*, *137*, 138
iron *see* banded iron formation
Isan Orogeny 345
Iskere Gneiss 283, **285**

Jannelsepan Formation **223**, 225
Jatuli event *see* Lomagundi-Jatuli event
Jeori Formation 288
Ji'an Group 61–62
Jiao-Liao-Ji Belt *52*, 53, 54, 61–63
　metamorphism 51, 62
　tectonic models 62–63
Jiehekou Group 101
Jining khondalite terrane 54, 55–56, **92–93**
Jokinenä, Salla Diabase Dyke 200, 201
Josling Granite **223**, 225
Jourma ophiolite 65
Jutogh Nappe 286
Jutogh Thrust 286, *287*

Kaaien Terrane 221, *226*
　granitoids, Sm–Nd data **223**, *225*, 226–227
Kaapvaal Craton 37, 39, 44, *42*, *220*, 227
Kakamas Terrane 221, *226*
　granitoids, Sm–Nd data 222, **223**, 225
Kalahari–Grunehogna, palaeomagnetic data 210, **211**, 212, **213**
Kalaktang granite gneiss 284, **285**
Kalinjala Shear Zone 333, 338
Kalkwerf Gneiss **223**, 227
Kaltygey Cape granulites 148
Kaoko Belt 65
Kararan Orogeny 325
Kenorland 6, 7, 8
Keurusselkä bolide impact crater 14
Kheis orogeny 227
Khibelen suite 146, *147*, 148
　geochronology and palaeomagnetism **150**, 153, 155–158
khondalite belt, North China Craton 53, 54–57
　Daqingshan area 73–94, *74*
Kimban Orogeny 321, 324, 328, 331, 332–333, 342
Kimberley Craton *320*, 338, 340

Kimberley Domain 37
kinzigite, Médio Coreaú domain 274, 275, 278
Kitoy granite 129
Klein Namaqualand Suite 222, **224**
Kocherikovo granite 146, 148
Kola Belt 165
　layered intrusions, geology and petrology 167–173
　magmatism 186, 189, 195
　mineralization 171, 191, **194**
　Sm–Nd analysis 186, **187**, **188**, *194*
　U–Pb isotope analysis 173–186
Kola Peninsula, ore deposits 189, 191
Kolvitsa instrusion 165, *166*
Koras Group **223**, 226–227
Kotla Gneiss 283, **285**
Kulu Thrust 286, *287*
Kulu–Bajura mylonite 283, 288, 296
　SHRIMP U–Pb dating 290, **294**
Kulu–Bajura Nappe 286, *287*, 288
Kulu–Rampur Window 286, *287*
Kuonamka dykes 146

lamination, Chaibasa Formation 302–303, *304*, *305*, 306, *307*, 309–310
Landplaas Gneiss 222, **224**
Lapland Greenstone Belt 199–200
large igneous provinces 4, 6, 15
　East Scandinavia 191
Laurentia 7
　link with Australia 342, 343, 345
　Logan Loop 210
　palaeomagnetic data 210, **211**, 330
Laurentia–Baltica reconstruction 199, 209–210, 212–214, *331*
Laurentia–Siberia reconstruction 146, 159–160
Leeuwdraai Formation **223**, 226
Lesser Himalayan Sedimentary Zone *284*, 285, 286, 296
Liaohe Group 61–62
life, fossil 12–13
Limpopo Belt 40–44, *42*
lineation
　Lüliang Massif 105, *107*, 109, *111*
　Man-Leo Shield 234
Lingtse Granite Gneiss 284, **285**
Logan Loop, Laurentia 210
Lomagundi sediment 41
Lomagundi-Jatuli event 10–11, 13, 14
Longquanguan Thrust 120
Low Grade Mafic Unit, Wutaishan Massif 120
Lowrie test, Salla Diabase Dyke 204–205
Luchaogou porphyritic granite 103
Lucknow Unit 37
Lüliang Complex 56

Lüliang Group 101, *102*, 103
Lüliang klippe *100*, 103, *104*, 105, 109, 120
Lüliang Massif 99–122, *100*
　correlation with Hengshan–Wutaishan–Fuping massifs 120–122
　deformation 105–109, 122
　foliation 105, *106*
　geochronology 109–119
　lithotectonic units 101–103
　metamorphism 109
　shear zones 105, 108
　structural analysis 103–109
　tectonic evolution 122
　U–Pb zircon dating 116–119
　U–Th/Pb EPMA monazite dating 111–119
Lüliang Nappe 103, 105, 108
Lüliang Ocean 120, 122
Lüliangshan 99, *100*, 103, *104*
Luyashan porphyritic charnockite 103

'magma ocean' 3
magmatism
　arc-related
　　Lüliang Massif 122
　　Mawson Continent 322, 344
　Baltic Shield 186–189
　Daqingshan 86–87
　Himalaya 293, 296
　Man-Leo Shield 231–250
　Mawson Continent 336, 338, 339, 341, 342, 344
　subduction-related
　　Australia 320–321
　　Mawson Continent 325, 344
magnetization, remanent, Salla Diabase Dyke 202–209
Magondi Belt 41–42, *42*
Main Boundary Thrust *284*, 286, *287*
Main Central Thrust 283, *284*, 285, *287*, 296
Main Frontal Thrust 286
Main Ridge intrusion 165, *166*
　geology and petrology 171
　mineralization **194**
　U–Pb isotope analysis **187**
Makganyene Formation 13
Malaya Kosa palaeopole 159
Malaya Kosa suite 146, *147*, 148
　palaeomagnetism 158
Man-Leo Shield 231–250, *232*, *235*
　deformation 234
　geochemistry 236–246
　geodynamic evolution model 246, 248–250
　geology 232–236
　granitoids 234, *235*, 236, 241, *243*, 244, *245*, 246, *247*
　greenstone belts 232–233, 236–241

Manikaran Quartzite 286, 288
mantle
  depletion events 5
  evolution 3–4
  'magma ocean' 3
mantle plumes 4
  Baltic Shield 165–166, 191, 196
  Himalaya 296
  Man-Leo Shield 246, *248*
  Mawson Continent 345
  North China Craton 63
Mapedi Unit 37
Mara Rosa Magmatic Arc 256, 257
Margate Terrane 227
Mawson Continent 7
  Antarctica 325–328
  assemblage 328–329
  evolution 319–345
  Gawler Craton 322–325
  palaeomagnetic data 329–330
  reconstruction models 330–345
  western extent 329
Médio Coreaú domain 271, *272, 273*, 274
Meidaizhao Group 75, 76
  zircon, SHRIMP dating *80, 82, 83*, **90–91**
Mesoproterozoic, 'thermal catastrophe' 3, 4
metallogenesis, Baltic Shield 189, 191
metamorphism
  Jiao-Liao-Ji Belt 61–62, 63
  khondalite belt 55–57, 78, 86–89
  Limpopo and Magondi belts 40–44
  Lüliang Massif 109
  Mawson Continent 322, 324–325, 328, 338, 339–341, 342, 343, 344
  North China Craton 51
  Siberian craton 129, 130
  ultrahigh temperature
    Daqingshan 88
    Northern Hebei orogen 50, 53, 54, 57–59
    correlation 64–66
migmatite
  Lüliang Massif 103, 120
  Médio Coreaú domain 274
Miller Range *320*, 322, *327*, 328
  Mawson Continent assemblage 328–329, *337*, 342
Miltalie Gneiss 322
mineralization, Baltic Shield 171, 172, 186, 187, 188, 189, 191
minerals 8–9
Modderfontein Gneiss 222, **224**
monazite, U–Th/Pb EPMA dating 111–119

Monchegorsk Layered Complex 165, *166*
  geology and petrology 170–171
  Sm–Nd analysis **192**
  U–Pb dating *174*, 179, *180*, **181–182**, 183
Monchepluton
  geology and petrology 170–171
  mineralization 186, 188, 191, **194**
  REE data *190*, **190**
  Sm–Nd analysis **187, 192, 193**, *194*
  U–Pb dating *174*, 179, **187**
Monchetundra intrusion 171, *174, 180*, 183, **187**, 188–189, **192, 193**, *194*
Mt Bol'shaya Varaka, U–Pb dating 183, *184*, **185**
Mt Generalskaya instrusion 165, *166*
  geology and petrology 167
  Sm–Nd analysis **187, 193**, *194*
  U–Pb dating 173, *174*, 175, **176, 187**
Mt Isa Inlier *320*, 338, 339–340, 341
  deformation 345
  sedimentation 343, 344
Mt Nyud 179, *180*, **187**
Mt Travyanaka 179, *180*, **187**
mud balls, contorted, Chaibasa Formation *304*, 311–312
Mundinho mylonitic granite 259, 262
Myally Supersequence 341
mylonite
  Kulu–Bajura Nappe 283, 288, 290, **294**, 296
  Mundinho mylonitic granite 259, 262
Mzumbe Terrane 227

Naitwar gneiss 283, **285**
Namaqua Sector
  geology 221
  granitoids, Sm–Nd data 220–228
Namaqua–Natal Province, Namaqua granitoids 219–228
Namik gneiss 283, **285**
Napier Complex *320*, 329, *337*
nappes, Lüliang Massif 103, 105, 108, 122
Natal Sector 227–228
Nd isotope data
  Siberian craton 131–140
  see also Sm–Nd isotope analysis
NENA (Northern Europe–North America) 7
Nimrod Group *327*, 328–329
Nimrod Orogeny 328
Nornalup Complex *320*, 329
North Australian Craton *327, 339*
  accretionary margin 331–333
  evolution 336, *337*, 338–341, 342, 343, 344
  palaeomagnetic data 330

North China Craton 49–66, *52*, 73–94
  correlation with Columbia Supercontinent *50*, 51, 63–66
  geological evolution 51–59
  Lüliang Massif 99–122
  rifting and breakup 63
  tectonism 53–59, *60*
North Himalayan Gneissic Domes 283
North Korean Peninsula 336, 338
North-Baikal volcano-plutonic belt 146, *147*, 148
Northern Belt *see* Kola Belt
Northern Hebei orogen 49, *50*, 51, *52*, 53, 54, 74
  correlation 64–66
  UHT metamorphism 50, 53, 54, 57–59
Nuna 2, 6, 7, 8 *see also* Columbia Supercontinent

ocean, evolution 9–11
Okun Group 146, *147*
Olanga complex 165
Olarian Orogeny 344
Olenek province 127, *128*
Onot-Erma domain *128*, 129
  Nd isotope data *136*, 137–138
Ooldean Event 324–325
Ordos Terrane 73, *74*
orogenesis, Mawson Continent 344–345, 388
Orthogneiss and Volcanite Unit *100, 102*, 103, 105, 109, 120
Ostrovsky intrusion 179, *180*, **187, 192**
oxidation, atmospheric 9, 10–11
oxygenation 9
  and glaciation 13–14

palaeointensity, geomagnetic field 2–3
palaeomagnetism
  Akitkan Group 145–161
  Mawson Continent 329–330
  Salla Diabase Dyke 205, **206**
Palaeopangaea 8
palaeopoles, Siberia 146, 159–160
Pechenga rift 167
Perovskite–post-perovskite phase transition 4
pillow structures, Chaibasa Formation *304*, 308–309, *313*, 314
Pine Creek Inlier *320*, 338
Pinjarra Orogen *320*, 329
Piriwiri sediment 41, 42
plate tectonics
  initiation 8
  modelling 4

INDEX

Pointe Geologie metasediments 325, 328
Polisiehoek Gneiss 222, **223**
Pretoria Group 37
Prikhibin'je, U–Pb dating 183, *184*, **185**
Prince Charles Mountains, southern 320, 327, 329, *337*
prokaryotes, fossil record 12
proto-Australia 7
    link with Laurentia 342–343, 345
Protopangaea 8
pseudonodules, Chaibasa Formation *304*, *305*, 312, 313
Pyrshin instrusion 165, *166*, 171

quartzite
    Chaibasa Formation 301
    Manikaran Quartzite 286, 288

Racklan Orogeny 343
Rameshwar granitoid 283, **285**
Ramgarh gneiss 284, **285**
Rampur Volcanics 286
Rb–Sr analysis, Baltic Shield intrusions 173, 186, **193**
Reguibat Shield 231, *232*
remanence *see* magnetization, remanent
rhythmites, tidal 14
Richtersveld Subprovince 221, *226*
    granitoids, Sm–Nd data 222, **223**
Riemvasmaak Gneiss 222, **223**
Rihee-Ganga gneiss 284, **285**
Rio dos Bois Fault 257
Rio Grande do Norte domain 271, *272*, 274, 279
Rio Maranhao fault system 257
Rio Negro-Juruena Belt, correlation with Northern Hebei orogen 64–65
ripple trains, sagged, Chaibasa Formation *305*, 310–311
Rodinia 6, 7
    reconstruction 284, 330
Rondonia Belt, correlation with Northern Hebei orogen 64–65
Ross Orogeny 328
Rudall Complex *320*, 341

St Peter Suite magmatism 321, 325, 333, *335*, 344
Salla Diabase Dyke
    geology 199–201
    palaeomagnetism 202–209, 212, **213**
    petrophysics 202, **203**
    secular variation 209
sandstone, Chaibasa Formation 301–307
Sanggan Group 75–76
    zircon 85, **90–91**

Hf isotope composition 83–84
SHRIMP U–Pb dating 76, 77, 78, *79*, *80*
São Francisco Craton 274, 279
São Jose do Campestre Massif 272, 274
São Luis Craton–West African Craton correlation 271, 279
sapphirine, Northern Hebei Orogen 57–58
Sarma Group 130, 146
Sask Craton 336, *337*, 338
Sayan granite 129
schist, Chaibasa Formation 301
Sclavia 7, 8
secular variation, Salla Diabase Dyke 209
sedimentation 6
    Jiao-Liao-Ji Belt 61–62
    Limpopo Belt 39–44
    Mawson Continent 342–344
Seertenshan Group 75
Segwagwa Group 37
seismites, Chaibasa Formation 307–308, 313–316
Senador Pompeu lineament 271, *272*
Shackleton Range *320*, *327*, 328
    Mawson Continent assemblage 328–329
shale, Chaibasa Formation 302–307
    chaotic *305*, 312–313
Shang gneiss 283, **285**
Sharizhalgai terrane *128*, 129–130
    crustal growth 140
    Nd isotope data *132*, *136*, 137–138
Shoemaker bolide impact crater 14
Shumikhin granites 129, 146
Shumikhin palaeopole 159
Siberia
    palaeomagnetism 7, 146
    see also Laurentia–Siberia reconstruction
Siberian Craton
    Akitkan Group 145–161
    crustal growth 127–140
    geology *128*, 129–130, 146, *147*, 148
    metamorphism 130
    Sm–Nd isotope analysis 130–140
Silverton Shale 37
Simla–Deoban–Garhwal Group 287
Slave Craton 330, *337*, 342
Sleafordian Orogeny 322, 336, 338
Sm–Nd isotope analysis
    Baltic Shield intrusions 172–173, 175, 186, **187**, **188**, **192**, **193**
    Campinorte sequence 259, 263, **264**, *265*, 266–267
    Namaqua granitoids 220–228

northwest Borborema Province 275, **277**, 279
Siberian craton 130–140
'Snowball Earth' 13
Sopcheozero chromite deposit 179
Southern Belt *see* Fenno-Karelian Belt
Soutpansberg Group 44
Stafford Event 341
Stanovoy province 127, *128*
sterane biomarkers 12–13
Strangways Orogeny 321, 331, 332–333, 342
StratDB database 28, *29*, 30–31, 33–35, 37, 44
Straussburg Granite **223**, 226
stromatolites 12
strontium isotopes, seawater 11
Sub-Himalayan Tertiary Belt 287
subduction
    Gawler Craton 344
    Rudall Complex 341
    Warumpi Province 344
Sudbury bolide impact crater 14
Suketi Thrust 287
sulphide 9
sulphur isotopes, non-mass-dependent 9, 13
Sun, 'faint young' 13, 14
supercontinents 6–8
Superia 7, 8
Superior–Siberia reconstruction 159, *160*
Superior–Wyoming–Karelia reconstruction *195*
Sutlej Valley, anorogenic magmatic signatures 286–296
Svecofennian Belt, correlation with Northern Hebei orogen 64, 65–66
Swanartz Gneiss **223**, 226, 227
Swartkopsleegte Formation **223**, 226

Tanami Event 341
Tangershang granite 103, 120
Tarim Block *50*, 51
    correlation 64
tectonism
    Mawson Continent 338–340
    North China Craton 51, 53–59, *60*, 62–63
Terre Adélie Craton *320*, 321, *323*, 325–328, *326*
    Mawson Continent assemblage 328–329
    see also Gawler–Adélie Craton
Tethyan Sedimentary Zone 285
'thermal catastrophe', Mesoproterozoic 3, 4
thermomagnetic analysis, Salla Diabase Dyke 202–203, *204*
tholeiite, Man-Leo Shield 236–241, *242*
Tickalara Arc 338, 340

time-slice maps, southern Africa, IGCP 509 database system 44, *43*
time-space correlation chart, southern Africa, IGCP 509 database system 37–44, *40*, *41*, *42*
Timeball Hill Formation 13
Tocantins Province 255–267, *256*
Tornio intrusion 165
Tornio-Näränkävaara Belt 165
tourmaline, Jiao-Liao-Ji Belt 62
Trans-Hudson Orogen 336, 338
Trans-North China Belt 73
  Lüliang Massif 99–122
  tectonic models 99
Trans-North China Suture *100*, 108, 120
Transamazonian Belt, correlation
  Campinorte sequence 265–266
  Northern Hebei orogen 64–65
Transantarctic Mountains 329
Transbrasiliano lineament 271
Transvaal Supergroup 37
Tróia Massif 272
TTG
  Campinorte Sequence 257
  Granja Massif 274
  Lüliang Massif 101, 122
    migmatite basement 103
  Siberian Craton 129
Tuanshanzi Formation 61
Tuguiwula, UHT meteamorphism 57–58
Tungus province 127, *128*, 129
Tunkillia Suite 333
turbidites, Lüliang Massif *100*, 101, *102*

U–Pb dating
  Akitkan Group 148–151, 155–156
  Baltic Shield intrusions 172–189
  Campinorte sequence 258–262, **263**, *265*, 266–267
  Lüliang Massif 116–119
  northwest Borborema Province 275–279
  northwest Himalaya 289–296

U–Th/Pb EPMA dating, monazite, Lüliang Massif 111–119
Umbarechka Block, U–Pb dating 183, 184
Upper Lüliang Thrust *100*, 105, 108, 120
Upper Wutai Thrust 120
Urayualagua shield 66
Urik-Iya terrane *128*, 129, 130
  crustal growth 140
  Nd isotope data **132**, *137*, 138
Uruaçu complex *257*, 258

Vaalbara 7–8
Vaalputs Granite 222, **223**
Vaikrita Group 288–289
Vaikrita Thrust 283, 289
Vestfold Hills *320*, 329
Vredefort bolide impact crater 14
Vryburg Formation 37
Vurechuaivench Foothills, ore deposits 191

Wangtu Granitic Complex 283, **285**, 288–289, 296
  SHRIMP U–Pb dating 290–291, *293*, **295**
Warumpi Province, subduction 321, 344
Waterberg Group 37, 44
Wernecke Supergroup 343
West African Craton 231, *232*
  correlation
    Campinorte sequence 265–266, 274, 279
  São Luis Craton 271, 279
West Australian Craton 341
Western Block, North China Craton 51, *52*, 99
Western Pansky Block, ore deposits 188, 191
Wilgenhoutsdrif Group **223**, 226–227
Windmill Islands *320*, *327*, 329
Wulashan Group 75–76
  zircon 85–86, **90–91**
  Hf isotope composition 84–85, 89

SHRIMP dating 77, 78, *79*, *80*, 81, *82*, 83
Wutai Complex 56
Wutaishan Massif, correlation with Lüliang Massif 120–122

Yambah Event 341
Yarrabubba bolide impact crater 14
Yarva-Varaka intrusion 179, *180*, **192**
Yejishan Group 103, *110*
  correlation with Hutuo Group 120, 122
Yinshan Terrane 73, *74*
Yunzhongshan 99, *100*, 103, *104*, 120

Zhaertai Group 75
Zhujiafang shear zone 57
Zimbabwe Craton 42, 44
Zimvaalbara 7
zircon
  Akitkan Group, U–Pb dating 148–151, 155–156
  Baltic Shield intrusions, U–Pb dating 172, 173–189
  Campinorte sequence, U–Pb dating 258–262, **263**, *265*, 266–267
  khondalite belt 55–56
  Lüliang Massif, U–Pb dating 116–119
  Meidaizhao Group 86, **90**
    SHRIMP dating 77, *80*, *82*, 83
  North China Craton, SHRIMP dating 56
  northwest Borborema Province, U–Pb dating 275–279
  northwest Himalaya, SHRIMP dating 289–296
  Sanggan Group 85, 89, **90**
    Hf isotope composition 83–84, 89
    SHRIMP dating 76, *77*, 78, *79*, *80*
  Wulashan Group 85–86, **90**
    Hf isotope composition 84–85, 89
    SHRIMP dating 77, 78, *79*, *80*, 81, *82*, 83